T0325598

## COMBUSTION THERMODYNAMICS AND DYNAMICS

*Combustion Thermodynamics and Dynamics* builds on a foundation of thermal science, chemistry, and applied mathematics that will be familiar to most undergraduate aerospace, mechanical, and chemical engineers to give a first-year graduate level exposition of the thermodynamics, physical chemistry, and dynamics of advection-reaction-diffusion. Special effort is made to link notions of time-independent classical thermodynamics with time-dependent reactive fluid dynamics. In particular, concepts of classical thermochemical equilibrium and stability are discussed in the context of modern nonlinear dynamical systems theory. The first half emphasizes time-dependent spatially homogeneous reaction, while the second half considers effects of spatially inhomogeneous advection and diffusion on the reaction dynamics. Attention is focused on systems with realistic detailed chemical kinetics as well as simplified kinetics. Many mathematical details are presented, and several quantitative examples given. Topics include foundations of thermochemistry, reduced kinetics, reactive Navier-Stokes equations, reaction-diffusion systems, laminar flame, oscillatory combustion, and detonation.

Joseph M. Powers is a professor in the Department of Aerospace and Mechanical Engineering at the University of Notre Dame. His research uses computational science to consider the dynamics of high-speed reactive fluids, especially as it applies to verification and validation of complex multiscale systems. He has held positions at the NASA Lewis Research Center, the Air Force Research Laboratory, the Los Alamos National Laboratory, and the Chinese Academy of Sciences. He is editor-in-chief of the AIAA's *Journal of Propulsion and Power*, an Associate Fellow of AIAA, and a member of APS, ASME, the Combustion Institute, and SIAM. He is the recipient of numerous teaching awards.

# Combustion Thermodynamics and Dynamics

**Joseph M. Powers**
University of Notre Dame

Shaftesbury Road, Cambridge CB2 8EA, United Kingdom

One Liberty Plaza, 20th Floor, New York, NY 10006, USA

477 Williamstown Road, Port Melbourne, VIC 3207, Australia

314–321, 3rd Floor, Plot 3, Splendor Forum, Jasola District Centre, New Delhi – 110025, India

103 Penang Road, #05–06/07, Visioncrest Commercial, Singapore 238467

Cambridge University Press is part of Cambridge University Press & Assessment,
a department of the University of Cambridge.

We share the University's mission to contribute to society through the pursuit of
education, learning and research at the highest international levels of excellence.

www.cambridge.org
Information on this title: www.cambridge.org/9781107067455



First published 2016

*A catalogue record for this publication is available from the British Library*

*Library of Congress Cataloging-in-Publication data*
Names: Powers, Joseph, 1961– author.
Title: Combustion thermodynamics and dynamics / Joseph M. Powers.
Description: New York, NY : Cambridge University Press, [2016] | Includes
bibliographical references and index.
Identifiers: LCCN 2015047025 | ISBN 9781107067455 (hardback : alk. paper)
Subjects: LCSH: Combustion – Mathematical models. | Thermodynamics –
Mathematics.
Classification: LCC QD516 .P667 2016 | DDC 541/.361 – dc23
LC record available at http://lccn.loc.gov/2015047025

ISBN 978-1-107-06745-5 Hardback

# Contents

# Preface

This book considers mathematical modeling of combustion, in particular, its time-independent thermodynamics and its relation to time-dependent dynamics. A major goal is to more fully incorporate the methods and language of nonlinear dynamical systems analysis (e.g., equilibria, phase space, sources, sinks, saddles, and limit cycles) into the pedagogy of traditional combustion theory. A second major goal is to consider problems that show how the mechanisms of advection, reaction, and diffusion influence the multiscale features of combustion systems' evolution in space and time. The largest fraction of the book is an exposition of some standard material of combustion science. This is accompanied by original work of the author that has been a part of his graduate course lectures and some of the specialized work of the author, his students, and colleagues on relevant topics, especially model reduction, thermodynamics of irreversible processes, identification of length and time scales of one-dimensional unsteady systems, multiscale dynamics, and detonation theory, which has been adapted from studies that have appeared in the archival combustion literature.

The focus is on deterministic continuum models of gas phase combustion, solution methods, detailed development of analytical results, and physical interpretations. As computational methods and hardware expand in their capability, it is useful to take stock of what deterministic modeling can offer, and some of our examples are to this end. Indeed, practical combustion problems abound that do not yield to deterministic continuum methods. Nevertheless, the rapid insights for causality they afford will long render such models as playing a leading role in combustion science.

This book arose from lecture notes for AME 60636, Fundamentals of Combustion, a graduate course taught since 1994 in the Department of Aerospace and Mechanical Engineering of the University of Notre Dame. Many undergraduates with standard preparation in thermodynamics, fluid dynamics, linear algebra, and differential equations have successfully completed the course. The book can guide a semester-long course, although some topics may need to be omitted.

Part I is devoted to time-independent thermodynamics of reactive mixtures and time-dependent systems that are restricted to spatially homogeneous reaction, thus avoiding the significant complications that come with advection and diffusion. Chapter 1 gives a discussion of the reaction dynamics of some simple but realistic time-dependent gas phase chemistry. These examples bring to the fore many of the important topics of the book: posing of combustion problems as nonlinear dynamical systems, identification of equilibria, time stability of equilibria via local linear analysis, phase space analysis, and full nonlinear dynamics. Next, in the spirit of physical chemistry, the thermodynamics of reacting gas mixtures is presented. Chapter 2 considers Dalton's mixture theory. Chapter 3 presents the thermodynamics of reacting mixtures, including equilibrium thermochemistry. Chapter 4 considers the time dynamics of a single reaction, followed by its multistep equivalent in Chapter 5. Special attention is given to the topic of irreversible entropy production and its interplay with combustion dynamics. A small discussion of the large topic of model reduction is given in Chapter 6, focusing mainly on dynamical systems aspects; a brief consideration of the significantly complicating effects of diffusion closes the chapter and serves as a bridge to Part II.

Part II considers various combinations of advection and diffusion within reactive systems. Chapter 7 presents the reactive Navier-Stokes equations with detailed kinetics and multicomponent diffusion. Chapter 8 presents an idealized linear model of advection-reaction-diffusion with a simplicity that allows many features of multiscale dynamics to be exposed. Chapter 9 returns to nonlinear dynamics of systems with reaction and diffusion. Chapter 10 considers the well-studied field of premixed one-dimensional laminar flames in the context of a simple advection-reaction-diffusion model that admits a compact presentation as a dynamical system. Chapter 11 briefly considers systems that do not relax to a stationary equilibrium but rather to a long time limit cycle. We close in Chapter 12 with an extended discussion of one-dimensional detonation theory as it is connected with nonlinear dynamics. Each chapter is concluded with a few exercises appropriate for homework. The problems are either self-contained or may need standard information from a thermodynamics text. Instructors with access to software for detailed chemical kinetics can and should be able to develop problems harnessing these tools that enable consideration of a broader range of mixtures important for engineering applications. Because these software tools rapidly change and rely on specific computing systems, we have included no detailed descriptions.

Quantitative predictions are presented in detail to enable the reader to reproduce most results. Often more significant digits than are justified by experiment are presented to this end. Typically, ideal gases are considered with either modestly sized models of detailed kinetics (e.g., $H_2$-air, oxygen dissociation, or nitrous oxide formation) or one-step kinetics. The book is more general than a monograph and more focused than a comprehensive text. Largely absent are important topics in combustion well covered in the cited literature: turbulence and spray modeling, multidimensional effects, radiation, experimental validation, and so on. Most of these topics are of sufficient complexity that they do not readily lend themselves to compact analysis as a dynamical system.

Many of the chapters reflect the significant interaction of the author with his students, colleagues, and teachers, with support from the U.S. National Science Foundation, NASA, and the U.S. Department of Energy. The author is grateful to the long years of dedicated, patient scholarship shown by his PhD students in combustion over the years: M. J. Grismer, K. A. Gonthier, S. Singh, A. K. Henrick, A. N. Al-Khateeb, J. D. Mengers, and C. M. Romick. Their work infuses this book, as does that of colleagues S. Paolucci and T. D. Aslam. And it is hoped that the guidance and wisdom of advisers H. Krier and D. S. Stewart, teacher J. D. Buckmaster, colleague P. B. Butler, and dozens of friends in the broader combustion community are reflected in the text.

# PART I

## Reactive Systems

# 1 Introduction to Chemical Kinetics

*Poca favilla gran fiamma seconda.*[1] (A great flame follows a little spark.)
—Dante Alighieri (1265–1321), *Paradiso*, Canto I, l. 34

In physics, *kinetics* is the study of motion and its causes; it is more commonly known as *dynamics*. Typically it connotes an evolution of the state of a mechanical system as time advances. In chemistry, this notion is slightly modified; for reacting systems, *chemical kinetics* connotes the transformation over time of a system from one chemical state to another. It is less commonly known as *chemical dynamics*. In contrast, the classical definition of *thermodynamics* is the science that deals with heat and work and those properties of matter that relate to heat and work. In most introductory thermodynamics texts, efforts are made to remove most references to time and to consider transformations from one equilibrium state to another. Although a useful construct, this removal of time places limits on theories based solely on equilibrium thermodynamics. For example, thermodynamics is obviously relevant to internal combustion engines; such engines rely on a transformation of chemical energy to thermal energy to mechanical energy. And it is essential that the combustion be completed in a well-defined time for the engine to operate effectively. It thus should be obvious that any prediction of the behavior of an internal combustion engine needs to draw on both equilibrium thermodynamics and time-dependent chemical kinetics. It is this interplay of thermodynamics and dynamics that is the subject of this book.

Our discussion in this and the next five chapters will be confined to systems that are *spatially homogeneous*. That is to say, whereas the systems exist in a finite geometric space, the variables describing the system do not vary with location. They will, however, often vary with time. Spatially homogeneous systems cannot account for the important physical mechanisms of advection and diffusion. These mechanisms are important in a wide variety of combustion systems and will be considered in Chapters 7–12. To summarize, then, our strictly thermodynamic systems will be considered independent of space and time and will typically be described by algebraic equations. Our chemical kinetic systems will draw on thermodynamics and consider evolution in time as described by ordinary differential equations. These ordinary

[1]This verse from *La Divina Commedia* was selected in a spirit of optimism that the poet's better known entreaty from that opus's more topical *Inferno, "Lasciate ogne speranza, voi ch'entrate".* (All hope abandon, ye who enter in), Canto III, l. 1, will prove to be irrelevant.

differential equations will have an inhomogeneous forcing function related to chemistry, but there is no forcing due to spatial inhomogeneity. When we introduce advection and diffusion, our system will be modeled by partial differential equations with variation in time and space. For such systems, chemical and spatial inhomogeneities both drive the time evolution.

We begin in this chapter by considering some illustrative examples from gas phase chemical kinetics, the driving physical mechanism in the dynamics and thermodynamics of gas phase combustion. The examples are drawn from realistic physical systems but focus on highly simplified limits that have pedagogical value. In this chapter, our approach is exploratory and motivational. Systematic treatment of the underlying theory is given in later chapters. Because such development requires lengthy exposition, it is useful to have a flavor of the dynamics of thermochemistry for simple problems before embarking in Chapter 2 on a more rigorous path.

In this chapter, we first briefly introduce a gas phase kinetic model for a multicomponent continuum mixture and present example mechanisms based on (1) *irreversible* hydrogen-oxygen reactions and then more complicated (2) *reversible* hydrogen-oxygen reactions. Hydrogen-oxygen reactions are posed because (1) they are one of the simplest practical gas phase combustion systems, (2) they have sufficient complexity to illustrate some of the challenges of gas phase combustion modeling, and (3) the model has been well validated against a variety of experiments. From this, we consider a set of problems addressing some of the simplest chemistry possible: *dissociation* of diatomic oxygen into monatomic oxygen and its reverse *recombination* reaction, followed by a slightly more complicated problem of nitric oxide formation. The dynamics of both systems are considered in an isothermal, isochoric environment. We then turn to an even simpler kinetics model that will be a paradigm throughout the book for studying combustion. This simple model asserts transformation of a generic species $A$ into a generic species $B$ accompanied by heat release. Dynamics of this model are studied in an adiabatic, isochoric environment, thus introducing what is known as *thermal explosion theory*. Such a theory is useful for predicting the phenomenon of a slowly progressing exothermic reaction that suddenly accelerates to a fast reaction with large heat release. We close the chapter by simple exposition of such a thermal explosion predicted for a realistic hydrogen-air system.

## 1.1 A Gas Phase Kinetic Model

Let us consider the reaction of $N$ molecular chemical species composed of $L$ elements via $J$ chemical reactions. Let us assume that the gas is an ideal mixture of ideal gases that satisfies Dalton's[2] law of partial pressures (see Sec. 2.3). Full discussion of the entire model encompasses most of Chapters 2–5. The reaction will be considered to be driven by *molecular collisions*. We will not model individual collisions but instead attempt to capture their collective effect.

An example of a model of such a reaction is listed in Table 1.1. There we find an $N = 9$ species, $J = 37$ step irreversible reaction mechanism for an $L = 3$ hydrogen-oxygen mixture from Maas and Warnatz (1988), with corrected $f_{H_2}$ from Maas and Pope (1992). The symbol M represents an arbitrary *third body* and is an inert participant in the reaction, such as argon, a noble gas that cannot react chemically. Third bodies M are not restricted to noble gases. In some reactions, certain molecules

[2]John Dalton (1766–1844), English chemist.

Table 1.1. *Nine-Species, Thirty-Seven-Step Irreversible Reaction Mechanism for a Hydrogen-Oxygen Mixture*

| $j$ | Reaction | $a_j$ | $\beta_j$ | $\overline{\mathcal{E}}_j$ |
|---|---|---|---|---|
| 1 | $O_2 + H \rightarrow OH + O$ | $2.00 \times 10^{14}$ | 0.00 | 70.30 |
| 2 | $OH + O \rightarrow O_2 + H$ | $1.46 \times 10^{13}$ | 0.00 | 2.08 |
| 3 | $H_2 + O \rightarrow OH + H$ | $5.06 \times 10^{4}$ | 2.67 | 26.30 |
| 4 | $OH + H \rightarrow H_2 + O$ | $2.24 \times 10^{4}$ | 2.67 | 18.40 |
| 5 | $H_2 + OH \rightarrow H_2O + H$ | $1.00 \times 10^{8}$ | 1.60 | 13.80 |
| 6 | $H_2O + H \rightarrow H_2 + OH$ | $4.45 \times 10^{8}$ | 1.60 | 77.13 |
| 7 | $OH + OH \rightarrow H_2O + O$ | $1.50 \times 10^{9}$ | 1.14 | 0.42 |
| 8 | $H_2O + O \rightarrow OH + OH$ | $1.51 \times 10^{10}$ | 1.14 | 71.64 |
| 9 | $H + H + M \rightarrow H_2 + M$ | $1.80 \times 10^{18}$ | −1.00 | 0.00 |
| 10 | $H_2 + M \rightarrow H + H + M$ | $6.99 \times 10^{18}$ | −1.00 | 436.08 |
| 11 | $H + OH + M \rightarrow H_2O + M$ | $2.20 \times 10^{22}$ | −2.00 | 0.00 |
| 12 | $H_2O + M \rightarrow H + OH + M$ | $3.80 \times 10^{23}$ | −2.00 | 499.41 |
| 13 | $O + O + M \rightarrow O_2 + M$ | $2.90 \times 10^{17}$ | −1.00 | 0.00 |
| 14 | $O_2 + M \rightarrow O + O + M$ | $6.81 \times 10^{18}$ | −1.00 | 496.41 |
| 15 | $H + O_2 + M \rightarrow HO_2 + M$ | $2.30 \times 10^{18}$ | −0.80 | 0.00 |
| 16 | $HO_2 + M \rightarrow H + O_2 + M$ | $3.26 \times 10^{18}$ | −0.80 | 195.88 |
| 17 | $HO_2 + H \rightarrow OH + OH$ | $1.50 \times 10^{14}$ | 0.00 | 4.20 |
| 18 | $OH + OH \rightarrow HO_2 + H$ | $1.33 \times 10^{13}$ | 0.00 | 168.30 |
| 19 | $HO_2 + H \rightarrow H_2 + O_2$ | $2.50 \times 10^{13}$ | 0.00 | 2.90 |
| 20 | $H_2 + O_2 \rightarrow HO_2 + H$ | $6.84 \times 10^{13}$ | 0.00 | 243.10 |
| 21 | $HO_2 + H \rightarrow H_2O + O$ | $3.00 \times 10^{13}$ | 0.00 | 7.20 |
| 22 | $H_2O + O \rightarrow HO_2 + H$ | $2.67 \times 10^{13}$ | 0.00 | 242.52 |
| 23 | $HO_2 + O \rightarrow OH + O_2$ | $1.80 \times 10^{13}$ | 0.00 | −1.70 |
| 24 | $OH + O_2 \rightarrow HO_2 + O$ | $2.18 \times 10^{13}$ | 0.00 | 230.61 |
| 25 | $HO_2 + OH \rightarrow H_2O + O_2$ | $6.00 \times 10^{13}$ | 0.00 | 0.00 |
| 26 | $H_2O + O_2 \rightarrow HO_2 + OH$ | $7.31 \times 10^{14}$ | 0.00 | 303.53 |
| 27 | $HO_2 + HO_2 \rightarrow H_2O_2 + O_2$ | $2.50 \times 10^{11}$ | 0.00 | −5.20 |
| 28 | $OH + OH + M \rightarrow H_2O_2 + M$ | $3.25 \times 10^{22}$ | −2.00 | 0.00 |
| 29 | $H_2O_2 + M \rightarrow OH + OH + M$ | $2.10 \times 10^{24}$ | −2.00 | 206.80 |
| 30 | $H_2O_2 + H \rightarrow H_2 + HO_2$ | $1.70 \times 10^{12}$ | 0.00 | 15.70 |
| 31 | $H_2 + HO_2 \rightarrow H_2O_2 + H$ | $1.15 \times 10^{12}$ | 0.00 | 80.88 |
| 32 | $H_2O_2 + H \rightarrow H_2O + OH$ | $1.00 \times 10^{13}$ | 0.00 | 15.00 |
| 33 | $H_2O + OH \rightarrow H_2O_2 + H$ | $2.67 \times 10^{12}$ | 0.00 | 307.51 |
| 34 | $H_2O_2 + O \rightarrow OH + HO_2$ | $2.80 \times 10^{13}$ | 0.00 | 26.80 |
| 35 | $OH + HO_2 \rightarrow H_2O_2 + O$ | $8.40 \times 10^{12}$ | 0.00 | 84.09 |
| 36 | $H_2O_2 + OH \rightarrow H_2O + HO_2$ | $5.40 \times 10^{12}$ | 0.00 | 4.20 |
| 37 | $H_2O + HO_2 \rightarrow H_2O_2 + OH$ | $1.63 \times 10^{13}$ | 0.00 | 132.71 |

*Note:* Units of $a_j$ are $(\text{mol/cm}^3)^{\left(1-\nu'_{M,j}-\sum_{i=1}^{N}\nu'_{ij}\right)}/\text{s/K}^{\beta_j}$. The parameter $\beta_j$ is dimensionless. Units of $\overline{\mathcal{E}}_j$ are kJ/mol. Third-body collision efficiencies with M are $f_{H_2} = 1.00$, $f_{O_2} = 0.35$, and $f_{H_2O} = 6.5$.

that could react do not; their presence. However, is necessary for the collision-based reaction to occur. The adjective "third" is traditional but not always accurate. For instance, in Reaction 13 of Table 1.1, $O + O + M \rightarrow O_2 + M$, there are three bodies colliding, O, O, and M. But in Reaction 14 of Table 1.1, $O_2 + M \rightarrow O + O + M$, there are two bodies colliding, $O_2$ and M. So for Reaction 14 of Table 1.1, M is better called a second body; however, such usage is uncommon.

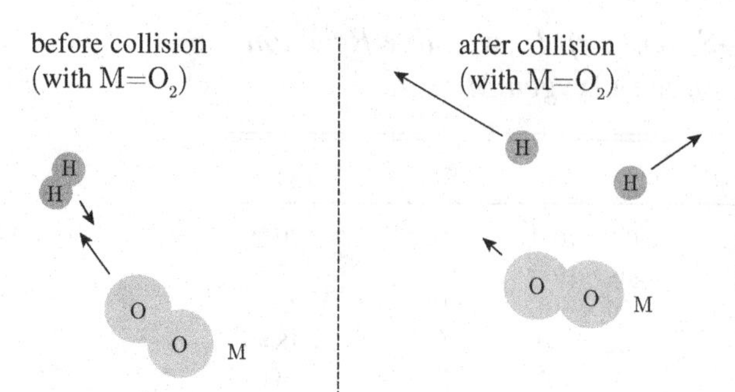


Figure 1.1. Sketch of a chain-initiation reaction following molecular collision. Here, M is $O_2$.

We need not worry yet about $f_{H_2}$, $f_{O_2}$, or $f_{H_2O}$. These are known as *collision efficiency coefficients*. They account for the fact that some collisions are different than others and allow for fine-tuning of the model to better match experimental data. Other mechanisms, not shown here, additionally include adjustments in the rate coefficients due to pressure. Details are given by Kee et al. (2003, Chapter 9), and many others. This model of so-called *detailed chemical kinetics* can accurately describe the dynamics of the formation of the dominant product, $H_2O$, from dominant reactants, $H_2$ and $O_2$. The study of detailed chemical kinetics began in the early twentieth century; contributions of Bodenstein,[3] Semenov,[4] and Hinshelwood[5] were seminal (e.g., Semenoff, 1935).

Although the overall reaction may be described by a so-called *global kinetics* model such as $2H_2 + O_2 \rightarrow 2H_2O$, the global kinetics mechanism does not accurately reflect the actual intermediate steps of reaction. The global mechanism suggests that two diatomic hydrogen molecules need to collide with a diatomic oxygen molecule to form two water molecules. This in fact does not happen. Instead, reaction of a mixture of $H_2$ and $O_2$ may commence with a *chain-initiation reaction*, such as Reaction 10 in Table 1.1:

$$H_2 + M \rightarrow H + H + M, \tag{1.1}$$

in which the fuel $H_2$ collides with *any* molecule to form the two *free radicals* H and H. We sketch a single collision for Reaction 10 in Fig. 1.1. Relative to the somewhat stable species $H_2$, the free radical species H is highly reactive. Chain-initiation reactions are often endothermic, as it takes some energy to break the bonds of either $H_2$ or $O_2$. This and other chain-initiation reactions provide radicals for various *chain-branching reactions*, such as Reaction 1 in Table 1.1:

$$O_2 + H \rightarrow OH + O. \tag{1.2}$$

Here, there is one free radical H among the reactants and two free radicals, OH and O, among the products. Also possible are *chain-propagation reactions*, such as Reaction 5 in Table 1.1:

$$H_2 + OH \rightarrow H_2O + H. \tag{1.3}$$

[3]Max Ernst August Bodenstein (1871–1942), German physical chemist.

[4]Nikolai Nikolaevich Semenov (1896–1986), Soviet physicist, chemist; 1956 Nobel laureate in Chemistry; often spelled "Semenoff."

[5]Sir Cyril Norman Hinshelwood (1897–1967), English physical chemist; 1956 Nobel laureate in Chemistry.

Such reactions, relative to chain-initiation reactions, are often nearly thermally neutral. Moreover, there is no net change in the number of free radicals. Here, one free radical OH is among the reactants and one free radical H is among the products. When enough free radicals accumulate, exothermic *chain-termination reactions* begin to occur to form the dominant final product. For example, collision of two free radical molecules of OH occurs in Reaction 7 of Table 1.1 to form the product $H_2O$:

$$OH + OH \rightarrow H_2O + O. \tag{1.4}$$

It is the chain-termination steps in which significant heat release usually occurs, temperature rises are often observed, and the number of free radicals is reduced.

The one-sided arrows indicate that each individual reaction is considered to be irreversible. For nearly each reaction, a separate reverse reaction is listed; thus, pairs of irreversible reactions can in some sense be considered to model reversible reactions. For example, Reaction 1 in Table 1.1 is the reverse of Reaction 2:

$$O_2 + H \rightarrow OH + O, \tag{1.5}$$

$$OH + O \rightarrow O_2 + H. \tag{1.6}$$

Each of the elementary reactions is described by a set of parameters. For the $j$th reaction, we have the *collision frequency coefficient* $a_j$, the *temperature-dependency exponent* $\beta_j$, and the *activation energy* $\overline{\mathcal{E}}_j$. These are explained in short order. Each individual reaction is not in fact modeling a single collision. Instead, it is modeling the cumulative effect of a large number of collisions of the reactants.

Table 1.2. *Nine-Species, Nineteen-Step Reversible Reaction Mechanism for a Hydrogen-Oxygen Mixture*

| $j$ | Reaction | $a_j$ | $\beta_j$ | $\overline{\mathcal{E}}_j$ |
|---|---|---|---|---|
| 1 | $H_2 + O_2 \rightleftharpoons OH + OH$ | $1.70 \times 10^{13}$ | 0.00 | 47780 |
| 2 | $OH + H_2 \rightleftharpoons H_2O + H$ | $1.17 \times 10^{9}$ | 1.30 | 3626 |
| 3 | $H + O_2 \rightleftharpoons OH + O$ | $5.13 \times 10^{16}$ | −0.82 | 16507 |
| 4 | $O + H_2 \rightleftharpoons OH + H$ | $1.80 \times 10^{10}$ | 1.00 | 8826 |
| 5 | $H + O_2 + M \rightleftharpoons HO_2 + M$ | $2.10 \times 10^{18}$ | −1.00 | 0 |
| 6 | $H + O_2 + O_2 \rightleftharpoons HO_2 + O_2$ | $6.70 \times 10^{19}$ | −1.42 | 0 |
| 7 | $H + O_2 + N_2 \rightleftharpoons HO_2 + N_2$ | $6.70 \times 10^{19}$ | −1.42 | 0 |
| 8 | $OH + HO_2 \rightleftharpoons H_2O + O_2$ | $5.00 \times 10^{13}$ | 0.00 | 1000 |
| 9 | $H + HO_2 \rightleftharpoons OH + OH$ | $2.50 \times 10^{14}$ | 0.00 | 1900 |
| 10 | $O + HO_2 \rightleftharpoons O_2 + OH$ | $4.80 \times 10^{13}$ | 0.00 | 1000 |
| 11 | $OH + OH \rightleftharpoons O + H_2O$ | $6.00 \times 10^{8}$ | 1.30 | 0 |
| 12 | $H_2 + M \rightleftharpoons H + H + M$ | $2.23 \times 10^{12}$ | 0.50 | 92600 |
| 13 | $O_2 + M \rightleftharpoons O + O + M$ | $1.85 \times 10^{11}$ | 0.50 | 95560 |
| 14 | $H + OH + M \rightleftharpoons H_2O + M$ | $7.50 \times 10^{23}$ | −2.60 | 0 |
| 15 | $H + HO_2 \rightleftharpoons H_2 + O_2$ | $2.50 \times 10^{13}$ | 0.00 | 700 |
| 16 | $HO_2 + HO_2 \rightleftharpoons H_2O_2 + O_2$ | $2.00 \times 10^{12}$ | 0.00 | 0 |
| 17 | $H_2O_2 + M \rightleftharpoons OH + OH + M$ | $1.30 \times 10^{17}$ | 0.00 | 45500 |
| 18 | $H_2O_2 + H \rightleftharpoons HO_2 + H_2$ | $1.60 \times 10^{12}$ | 0.00 | 3800 |
| 19 | $H_2O_2 + OH \rightleftharpoons H_2O + HO_2$ | $1.00 \times 10^{13}$ | 0.00 | 1800 |

*Note:* Units of $a_j$ are $(\text{mol/cm}^3)^{\left(1-\nu'_{M,j}-\sum_{i=1}^{N}\nu'_{ij}\right)}/\text{s}/\text{K}^{\beta_j}$. The parameter $\beta_j$ is dimensionless. Units of $\overline{\mathcal{E}}_j$ are cal/mol. Third-body collision efficiencies with M are $f_5(H_2O) = 21$, $f_5(H_2) = 3.3$, $f_{12}(H_2O) = 6$, $f_{12}(H) = 2$, $f_{12}(H_2) = 3$, and $f_{14}(H_2O) = 20$.

It is perhaps more common to describe chemical kinetics systems as being composed of reversible reactions. Such reactions are usually denoted by two-sided arrows. One such system for hydrogen-oxygen combustion is reported by Powers and Paolucci (2005), and is listed in Table 1.2. While more compact than the model of Table 1.1, reversible reactions pose more mathematical complexity. As such we delay their study until we have given examples of irreversible kinetic systems.

## 1.2 Isothermal, Isochoric Kinetics

For simplicity, we first focus attention on cases in which the temperature $T$ and volume $V$ are both constant. Such conditions are known as *isothermal* and *isochoric*, respectively. The volume contains a mixture of $N$ reacting species each with a number $n_i$ of moles. We restrict attention here to *closed* systems. Such systems do not exchange mass with their environment and thus have constant total mass $m$. The system is allowed to exchange energy with its surroundings. For typically exothermic combustion reactions, energy exchange is necessary to maintain a constant temperature. A sketch of this configuration is given in Fig. 1.2. We define the *molar concentration* $\overline{\rho}_i$ of species $i$ as the number of moles, $n_i$, per volume:

$$\overline{\rho}_i = n_i/V, \qquad i = 1, \ldots, N. \tag{1.7}$$

It has units of mol/cm$^3$. We might also call this the *molar density*. Our notation for molar concentration, $\overline{\rho}_i$, is nonstandard. In general, we will take the "bar" notation to indicate a property on molar basis; equivalent properties without the bar will indicate the property to be on a mass basis. For example, the mass density of species $i$ will be denoted $\rho_i$ and have units of g/cm$^3$. More commonly, most properties with a bar will be per unit mol; their equivalent without a bar will be per unit mass. For example, the *molar specific internal energy* of species $i, \overline{e}_i$, may have units of erg/mol; the *mass specific internal energy* $e_i$ would have units erg/g. Quantities on a mol basis are related to those on a mass basis by appropriately scaling by the *molecular mass*, $M_i$, the mass in g of a mol of species $i$. To convert from molar density to mass density for species $i$, we have

$$\rho_i = \overline{\rho}_i M_i. \tag{1.8}$$

To convert specific internal energy from a molar to mass basis, we have

$$e_i = \overline{e}_i / M_i. \tag{1.9}$$

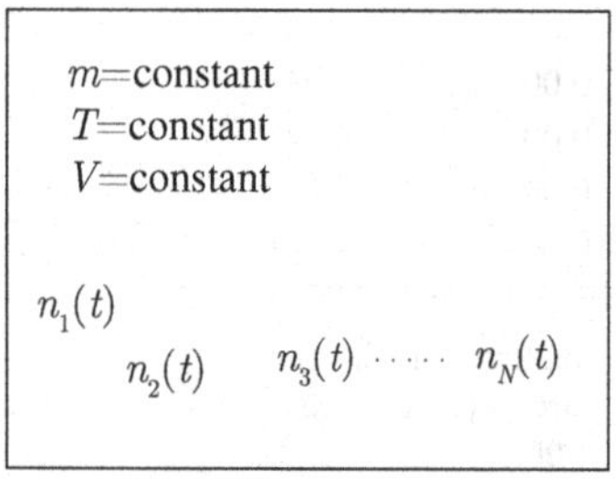

$m$=constant
$T$=constant
$V$=constant

$n_1(t)$
$n_2(t)$ $n_3(t)$ ..... $n_N(t)$

Figure 1.2. Configuration for closed, isothermal, isochoric, spatially homogeneous reaction of $N$ species.

A more common notation for molar concentration, which we generally avoid, is given by square brackets, for example, $\overline{\rho}_{O_2} = [O_2]$. Our notation makes obvious the symmetries between molar and mass concentrations in a variety of relations to be discussed in Chapter 2.

### 1.2.1 O-$O_2$ Dissociation

One of the simplest physical examples is provided by the dissociation of $O_2$ into its atomic component O, accompanied by a recombination of O into $O_2$. A good fundamental treatment of elementary reactions of this type is given by Vincenti and Kruger (1965), in their detailed monograph.

#### Pair of Irreversible Reactions

We begin by assuming that the dissociation and recombination process is described by a pair of irreversible reactions from Table 1.1. To begin, let us focus for now only on reactions $j = 13$ and $j = 14$ from Table 1.1 in the limiting case in which $T$ and $V$ are constant.

**MATHEMATICAL MODEL.** The two reactions are

$$\text{Recombination 13} \quad O + O + M \rightarrow O_2 + M, \tag{1.10}$$

$$\text{Dissociation 14} \quad O_2 + M \rightarrow O + O + M, \tag{1.11}$$

with

$$a_{13} = 2.90 \times 10^{17} \left(\frac{\text{mol}}{\text{cm}^3}\right)^{-2} \frac{\text{K}}{\text{s}}, \quad \beta_{13} = -1.00, \; \overline{\mathcal{E}}_{13} = 0 \; \frac{\text{kJ}}{\text{mol}}, \tag{1.12}$$

$$a_{14} = 6.81 \times 10^{18} \left(\frac{\text{mol}}{\text{cm}^3}\right)^{-1} \frac{\text{K}}{\text{s}}, \quad \beta_{14} = -1.00, \; \overline{\mathcal{E}}_{14} = 496.41 \; \frac{\text{kJ}}{\text{mol}}. \tag{1.13}$$

Also, we have the collision efficiency for $O_2$ as $f_{O_2} = 0.35$. For O, we take by default that $f_O = 1$. We envision Reaction 13 as one in which two O atoms and a third body (either O or $O_2$) collide simultaneously. Two of the O atoms combine to form a single $O_2$ molecule, and the third body remains as it was. Similarly, in Reaction 14, an $O_2$ molecule collides with a "third" body (in reality, a second body), either $O_2$ or O. The collision is of sufficient strength to induce the $O_2$ to dissociate into two O atoms. Again, the "third" body remains as it was. We sketch a single collision with $M = O_2$ for Reaction 14 in Fig. 1.3.

The irreversible nature of the reaction is indicated by the one-sided arrow. Although they participate in the overall hydrogen oxidation problem, these two reactions are in fact self-contained and can thus be considered in isolation. So, let us just consider that we have only oxygen in our volume with $N = 2$ species, O and $O_2$, $J = 2$ reactions (those being 13 and 14), and $L = 1$ element, that being O. We will take $i = 1$ to correspond to O and $i = 2$ to correspond to $O_2$.

We recast the mechanism as

$$\nu'_{1,13}O + \nu'_{2,13}O_2 + \nu'_{M,13}M \rightarrow \nu''_{1,13}O + \nu''_{2,13}O_2 + \nu''_{M,13}M, \tag{1.14}$$

$$\nu'_{1,14}O + \nu'_{2,14}O_2 + \nu'_{M,14}M \rightarrow \nu''_{1,14}O + \nu''_{2,14}O_2 + \nu''_{M,14}M, \tag{1.15}$$

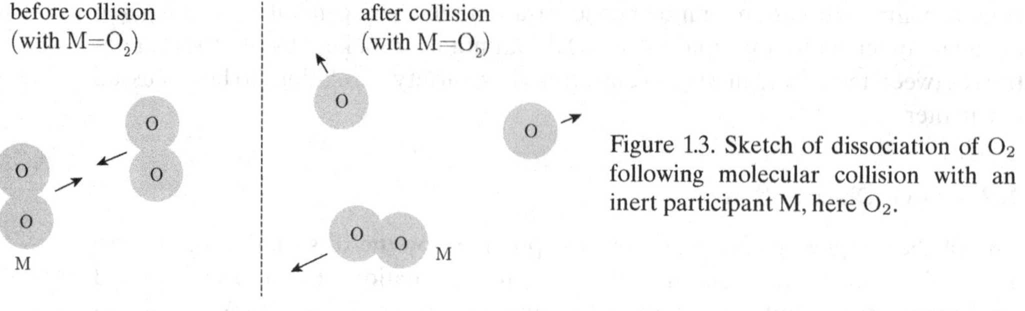


Figure 1.3. Sketch of dissociation of $O_2$ following molecular collision with an inert participant M, here $O_2$.

with the *forward stoichiometric coefficients*, $\nu'_{ij}$, and the *reverse stoichiometric coefficients*, $\nu''_{ij}$, taking the values

$$\nu'_{1,13} = 2,\ \nu'_{2,13} = 0,\ \nu'_{M,13} = 1,\ \nu''_{1,13} = 0,\ \nu''_{2,13} = 1,\ \nu''_{M,13} = 1, \tag{1.16}$$

$$\nu'_{1,14} = 0,\ \nu'_{2,14} = 1,\ \nu'_{M,14} = 1,\ \nu''_{1,14} = 2,\ \nu''_{2,14} = 0,\ \nu''_{M,14} = 1, \tag{1.17}$$

so that the mechanism is

$$(2)\mathrm{O} + (0)\mathrm{O}_2 + (1)\mathrm{M} \rightarrow (0)\mathrm{O} + (1)\mathrm{O}_2 + (1)\mathrm{M}, \tag{1.18}$$

$$(0)\mathrm{O} + (1)\mathrm{O}_2 + (1)\mathrm{M} \rightarrow (2)\mathrm{O} + (0)\mathrm{O}_2 + (1)\mathrm{M}. \tag{1.19}$$

As can be seen from Table 1.1, the units of $a_j$ are unusual. There is no settled set of units within the combustion community. For the most part, we make the common choice of cgs-based units (centimeter-gram-second), most often used in chemistry; occasionally, we use MKS-based units (meter-kilogram-second), most often used in physics and engineering. For reaction $j = 13$, we find the exponent for the $(\mathrm{mol/cm^3})$ portion of the units for $a_{13}$ to be

$$1 - \nu'_{M,j} - \sum_{i=1}^{N} \nu'_{ij} = 1 - \nu'_{M,13} - \left(\nu'_{1,13} + \nu'_{2,13}\right) = 1 - 1 - (2 + 0) = -2. \tag{1.20}$$

Thus, the units of $a_{13}$ are $(\mathrm{mol/cm^3})^{-2}/\mathrm{s}/\mathrm{K}^{-1}$. Similarly, for reaction $j = 14$, we find the exponent for the $(\mathrm{mol/cm^3})$ portion of the units for $a_{14}$ to be

$$1 - \nu'_{M,j} - \sum_{i=1}^{N} \nu'_{ij} = 1 - \nu'_{M,14} - \left(\nu'_{1,14} + \nu'_{2,14}\right) = 1 - 1 - (0 + 1) = -1. \tag{1.21}$$

Recall that in the cgs system, $1\ \mathrm{erg} = 1\ \mathrm{dyne\ cm} = 10^{-7}\ \mathrm{J} = 10^{-10}\ \mathrm{kJ}$. Recall also that the cgs unit of force is the dyne and that $1\ \mathrm{dyne} = 1\ \mathrm{g\ cm/s^2} = 10^{-5}\ \mathrm{N}$. So, for cgs, we have

$$\overline{\mathcal{E}}_{13} = 0\ \frac{\mathrm{erg}}{\mathrm{mol}}, \quad \overline{\mathcal{E}}_{14} = 496.41\ \frac{\mathrm{kJ}}{\mathrm{mol}}\left(10^{10}\ \frac{\mathrm{erg}}{\mathrm{kJ}}\right) = 4.9641 \times 10^{12}\ \frac{\mathrm{erg}}{\mathrm{mol}}. \tag{1.22}$$

We next introduce a common model for chemical dynamics of spatially homogeneous systems. We thus consider dependency only on time $t$, and not on space, because of neglect of modeling advection or diffusion. The standard model for spatially homogeneous chemical kinetics, which is generalized and discussed in more detail in Chapters 4 and 5, induces the following two ordinary differential equations for the temporal evolution of O and $O_2$ molar concentrations:

$$\frac{d\overline{\rho}_O}{dt} = \underbrace{-2\,\underbrace{a_{13}T^{\beta_{13}}\exp\left(\frac{-\overline{\mathcal{E}}_{13}}{\overline{R}T}\right)}_{k_{13}(T)}\overline{\rho}_O\overline{\rho}_O\overline{\rho}_M}_{r_{13}} \underbrace{+2\,\underbrace{a_{14}T^{\beta_{14}}\exp\left(\frac{-\overline{\mathcal{E}}_{14}}{\overline{R}T}\right)}_{k_{14}(T)}\overline{\rho}_{O_2}\overline{\rho}_M}_{r_{14}}, \tag{1.23}$$

$$\frac{d\overline{\rho}_{O_2}}{dt} = \underbrace{\underbrace{a_{13}T^{\beta_{13}}\exp\left(\frac{-\overline{\mathcal{E}}_{13}}{\overline{R}T}\right)}_{k_{13}(T)}\overline{\rho}_O\overline{\rho}_O\overline{\rho}_M}_{r_{13}} \underbrace{-\underbrace{a_{14}T^{\beta_{14}}\exp\left(\frac{-\overline{\mathcal{E}}_{14}}{\overline{R}T}\right)}_{k_{14}(T)}\overline{\rho}_{O_2}\overline{\rho}_M}_{r_{14}}. \tag{1.24}$$

Detailed exposition is given in Sec. 5.1. Equation (1.24) has its analog in the forward reaction of Eq. (5.13). The symbol $\overline{R}$ is the universal gas constant, for which

$$\overline{R} = 8.314\,\frac{\text{J}}{\text{mol K}}\left(\frac{10^7\text{ erg}}{\text{J}}\right) = 8.314\times 10^7\,\frac{\text{erg}}{\text{mol K}}. \tag{1.25}$$

Because the universal gas constant is on a per unit mol basis, we give it a "bar." We also use the common notation of a temperature-dependent portion of the dynamics of reaction $j$, $k_j(T)$, as we write the equivalent of the upcoming Eq. (5.17):

$$k_j(T) = a_j T^{\beta_j}\exp\left(\frac{-\overline{\mathcal{E}}_j}{\overline{R}T}\right), \qquad j = 1,\ldots,J. \tag{1.26}$$

Sometimes $k_j(T)$ is known as the *reaction rate coefficient* for reaction $j$. The actual *reaction rates* for Reactions 13 and 14, $r_{13}$ and $r_{14}$, are

$$r_{13} = k_{13}\overline{\rho}_O\overline{\rho}_O\overline{\rho}_M, \qquad r_{14} = k_{14}\overline{\rho}_{O_2}\overline{\rho}_M. \tag{1.27}$$

See Eq. (5.14) for a more general form. The reaction rates are proportional to the product of the molar concentrations and thus can be inferred to be related to the probability of molecular collisions. For example, if there is no O, there is no chance of a collision to induce Reaction 13, so the rate is zero.

The system of equations (1.23)–(1.24) can be written as

$$\frac{d\overline{\rho}_O}{dt} = -2r_{13} + 2r_{14}, \qquad \frac{d\overline{\rho}_{O_2}}{dt} = r_{13} - r_{14}. \tag{1.28}$$

More simply, using Gibbs's[6] boldfaced notation for vectors and matrices,[7] Eqs. (1.28) can be written in the form of the upcoming Eq. (5.344):

$$\frac{d\overline{\boldsymbol{\rho}}}{dt} = \boldsymbol{\nu}\cdot\mathbf{r}. \tag{1.29}$$

Here, we have taken

$$\overline{\boldsymbol{\rho}} = \begin{pmatrix}\overline{\rho}_O\\ \overline{\rho}_{O_2}\end{pmatrix}, \quad \boldsymbol{\nu} = \begin{pmatrix}-2 & 2\\ 1 & -1\end{pmatrix}, \quad \mathbf{r} = \begin{pmatrix}r_{13}\\ r_{14}\end{pmatrix}. \tag{1.30}$$

[6] Josiah Willard Gibbs (1839–1903), American mechanical engineer.

[7] One must recognize that these vectors and matrices are not associated with any ordinary physical geometry; they can be thought of as residing within a geometry whose coordinates are defined by thermochemical variables. This unusual space is often of dimension greater than three.

Here, $\boldsymbol{\nu}$ is the matrix of *net stoichiometric coefficients* given for this case by

$$\boldsymbol{\nu} = \boldsymbol{\nu}'' - \boldsymbol{\nu}' = \begin{pmatrix} \nu''_{1,13} - \nu'_{1,13} & \nu''_{1,14} - \nu'_{1,14} \\ \nu''_{2,13} - \nu'_{2,13} & \nu''_{2,14} - \nu'_{2,14} \end{pmatrix} = \begin{pmatrix} 0-2 & 2-0 \\ 1-0 & 0-1 \end{pmatrix}, \tag{1.31}$$

$$= \begin{pmatrix} -2 & 2 \\ 1 & -1 \end{pmatrix}. \tag{1.32}$$

See discussion surrounding Eq. (5.3) for more details. In general, we will have $\overline{\boldsymbol{\rho}}$ be a column vector of dimension $N \times 1$, $\boldsymbol{\nu}$ will be a rectangular matrix of dimension $N \times J$ of rank $R$, and $\mathbf{r}$ will be a column vector of length $J \times 1$. So, Eqs. (1.28) take the form

$$\frac{d}{dt}\begin{pmatrix} \overline{\rho}_O \\ \overline{\rho}_{O_2} \end{pmatrix} = \begin{pmatrix} -2 & 2 \\ 1 & -1 \end{pmatrix}\begin{pmatrix} r_{13} \\ r_{14} \end{pmatrix}. \tag{1.33}$$

The rank $R$ of $\boldsymbol{\nu}$ is $R = L = 1$, because $\det \boldsymbol{\nu} = 0$, and $\boldsymbol{\nu}$ has at least one nonzero element. Let us also define an *element-species matrix* $\boldsymbol{\phi}$ of dimension $L \times N$. The component of $\boldsymbol{\phi}$, $\phi_{li}$, represents the number of element $l$ in species $i$. Most commonly, $\boldsymbol{\phi}$ will be full rank. Because we can have $L < N$, $L = N$, or $L > N$, we may expect the rank of $\boldsymbol{\phi}$ to be either $L$ or $N$. Here, we have $L < N$, and $\boldsymbol{\phi}$ is of dimension $1 \times 2$. The matrix $\boldsymbol{\phi}$ is in fact

$$\boldsymbol{\phi} = (\phi_{11} \quad \phi_{12}) = (1 \quad 2). \tag{1.34}$$

This represents the fact that we have $L = 1$ elements, that being O. So for species 1, O, we have $\phi_{11} = 1$ atom of O in O. For species 2, $O_2$, we have $\phi_{12} = 2$ atoms of O in $O_2$. Element conservation is guaranteed by insisting that $\boldsymbol{\nu}$ be constructed such that it takes the form of the upcoming Eq. (5.4):

$$\boldsymbol{\phi} \cdot \boldsymbol{\nu} = \mathbf{0}. \tag{1.35}$$

Element conservation is the essence of *stoichiometric balance*, a feature of all chemical reactions, in contrast to nuclear reactions, for which elements may not be conserved. In the language of linear algebra, we can say that each of the column vectors of $\boldsymbol{\nu}$ lies in the right null space of $\boldsymbol{\phi}$. For our example, we see that Eq. (1.35) holds:

$$\boldsymbol{\phi} \cdot \boldsymbol{\nu} = (1 \quad 2)\begin{pmatrix} -2 & 2 \\ 1 & -1 \end{pmatrix} = (0 \quad 0). \tag{1.36}$$

The first equation

$$\phi_{11}\nu_{1,13} + \phi_{12}\nu_{2,13} = 0, \qquad (1)(-2) + (2)(1) = 0, \tag{1.37}$$

speaks to the conservation of O in Reaction 13. The second equation

$$\phi_{11}\nu_{1,14} + \phi_{12}\nu_{2,14} = 0, \qquad (1)(2) + (2)(-1) = 0, \tag{1.38}$$

speaks to the conservation of O in Reaction 14.

Let us take as initial conditions

$$\overline{\rho}_O(t=0) = \widehat{\overline{\rho}}_O, \qquad \overline{\rho}_{O_2}(t=0) = \widehat{\overline{\rho}}_{O_2}. \tag{1.39}$$

Now, M represents an arbitrary third body with which the reactant or reactants can collide. If all the collision efficiency coefficients are unity, we have a simple formula for the concentration of the third body:

$$\overline{\rho}_{\mathrm{M}} = \sum_{i=1}^{N} \overline{\rho}_i. \tag{1.40}$$

Some kinetic models, such as in Table 1.1, give a collision efficiency coefficient $f_i$ for some species. For such models, we infer that

$$\overline{\rho}_{\mathrm{M}} = \sum_{i=1}^{N} f_i \overline{\rho}_i. \tag{1.41}$$

Here, $f_i = 1$, unless otherwise stated. Other kinetic models, such as that in Table 1.2, give customized values of the collision efficiency coefficients for each species in each reaction. In such cases, we can generally represent the collision efficiency coefficients as a $J \times N$ matrix $F_{ji}$. Unless otherwise specified, all of the $F_{ji}$ terms are unity. The molar concentration of the third body for reaction $j$ is then

$$\overline{\rho}_{\mathrm{M}_j} = \sum_{i=1}^{N} F_{ji} \overline{\rho}_i. \tag{1.42}$$

We only have two types of potential third bodies, and our collision efficiency coefficient is appropriate for Eq. (1.41) with $f_{\mathrm{O}_2} = 0.35$. By default, $f_{\mathrm{O}} = 1$, so here we have

$$\overline{\rho}_{\mathrm{M}} = 0.35\overline{\rho}_{\mathrm{O}_2} + \overline{\rho}_{\mathrm{O}}. \tag{1.43}$$

Thus, the reaction dynamics, Eqs. (1.23, 1.24), reduce to

$$\begin{aligned}\frac{d\overline{\rho}_{\mathrm{O}}}{dt} &= -2a_{13}T^{\beta_{13}} \exp\left(\frac{-\overline{\mathcal{E}}_{13}}{\overline{R}T}\right) \overline{\rho}_{\mathrm{O}}\overline{\rho}_{\mathrm{O}} \left(0.35\overline{\rho}_{\mathrm{O}_2} + \overline{\rho}_{\mathrm{O}}\right) \\ &\quad + 2a_{14}T^{\beta_{14}} \exp\left(\frac{-\overline{\mathcal{E}}_{14}}{\overline{R}T}\right) \overline{\rho}_{\mathrm{O}_2} \left(0.35\overline{\rho}_{\mathrm{O}_2} + \overline{\rho}_{\mathrm{O}}\right), \end{aligned} \tag{1.44}$$

$$\begin{aligned}\frac{d\overline{\rho}_{\mathrm{O}_2}}{dt} &= a_{13}T^{\beta_{13}} \exp\left(\frac{-\overline{\mathcal{E}}_{13}}{\overline{R}T}\right) \overline{\rho}_{\mathrm{O}}\overline{\rho}_{\mathrm{O}} \left(0.35\overline{\rho}_{\mathrm{O}_2} + \overline{\rho}_{\mathrm{O}}\right) \\ &\quad - a_{14}T^{\beta_{14}} \exp\left(\frac{-\overline{\mathcal{E}}_{14}}{\overline{R}T}\right) \overline{\rho}_{\mathrm{O}_2} \left(0.35\overline{\rho}_{\mathrm{O}_2} + \overline{\rho}_{\mathrm{O}}\right). \end{aligned} \tag{1.45}$$

Equations (1.44)–(1.45) with Eqs. (1.39) represent two nonlinear ordinary differential equations in two unknowns $\overline{\rho}_{\mathrm{O}}$ and $\overline{\rho}_{\mathrm{O}_2}$. We seek the behavior of these two species concentrations as a function of time.

Systems of nonlinear equations are generally difficult to integrate analytically and usually require numerical solution. Before embarking on a numerical solution, we simplify as much as we can. Equations (1.44) and (1.45) can be linearly combined to form

$$\frac{d\overline{\rho}_{\mathrm{O}}}{dt} + 2\frac{d\overline{\rho}_{\mathrm{O}_2}}{dt} = \frac{d}{dt}\left(\overline{\rho}_{\mathrm{O}} + 2\overline{\rho}_{\mathrm{O}_2}\right) = 0. \tag{1.46}$$

We can integrate and apply the initial conditions (1.39) to get

$$\overline{\rho}_{\mathrm{O}} + 2\overline{\rho}_{\mathrm{O}_2} = \widehat{\overline{\rho}}_{\mathrm{O}} + 2\widehat{\overline{\rho}}_{\mathrm{O}_2} = \text{constant}. \tag{1.47}$$

The fact that this algebraic constraint exists for all time is a consequence of the conservation of mass of each O element. It can also be thought of as the conservation of number of O atoms.

Standard linear algebra provides a robust way to find the constraint of Eq. (1.47). We can use elementary row operations to cast Eq. (1.33) into a so-called row-echelon form. Here, our goal is to get a linear combination that on the right side has an upper triangular form. To achieve this, add twice the second equation from Eq. (1.33) with the first to form a new equation to replace the second equation. This gives

$$\frac{d}{dt}\begin{pmatrix}\overline{\rho}_O \\ \overline{\rho}_O + 2\overline{\rho}_{O_2}\end{pmatrix} = \begin{pmatrix}-2 & 2 \\ 0 & 0\end{pmatrix}\begin{pmatrix}r_{13} \\ r_{14}\end{pmatrix}. \tag{1.48}$$

Obviously the second equation is one we obtained earlier, $d/dt(\overline{\rho}_O + 2\overline{\rho}_{O_2}) = 0$, and this induces our algebraic constraint. Equation (1.48) can be recast as

$$\underbrace{\begin{pmatrix}1 & 0 \\ 1 & 2\end{pmatrix}}_{\mathbf{L}^{-1}}\frac{d}{dt}\begin{pmatrix}\overline{\rho}_O \\ \overline{\rho}_{O_2}\end{pmatrix} = \underbrace{\begin{pmatrix}-2 & 2 \\ 0 & 0\end{pmatrix}}_{\mathbf{U}}\begin{pmatrix}r_{13} \\ r_{14}\end{pmatrix}. \tag{1.49}$$

This is of the matrix form

$$\mathbf{L}^{-1}\cdot\mathbf{P}\cdot\frac{d\overline{\boldsymbol{\rho}}}{dt} = \mathbf{U}\cdot\mathbf{r}. \tag{1.50}$$

Here, $\mathbf{L}$ and $\mathbf{L}^{-1}$ are $N \times N$ lower triangular matrices of full rank $N$ and thus invertible. The matrix $\mathbf{U}$ is upper triangular of dimension $N \times J$ and with the same rank as $\boldsymbol{\nu}$, $R \geq L$. The matrix $\mathbf{P}$ is a permutation matrix of dimension $N \times N$. It is never singular and thus always invertable. It is used to effect possible row exchanges to achieve the desired form; often row exchanges are not necessary, in which case $\mathbf{P} = \mathbf{I}$, the $N \times N$ identity matrix. Equation (1.50) can be manipulated to form the original Eq. (1.33) via

$$\frac{d\overline{\boldsymbol{\rho}}}{dt} = \underbrace{\mathbf{P}^{-1}\cdot\mathbf{L}\cdot\mathbf{U}}_{\boldsymbol{\nu}}\cdot\mathbf{r}. \tag{1.51}$$

What we have done is the standard linear algebra decomposition of $\boldsymbol{\nu} = \mathbf{P}^{-1}\cdot\mathbf{L}\cdot\mathbf{U}$. This is sometimes known as an *LU decomposition*.

We can also decompose the algebraic constraint, Eq. (1.47), in a nonobvious way that is more useful for larger systems. We can first rearrange it to say

$$\overline{\rho}_{O_2} = \widehat{\overline{\rho}}_{O_2} - \frac{1}{2}\left(\overline{\rho}_O - \widehat{\overline{\rho}}_O\right). \tag{1.52}$$

Defining now a new state variable associated with the element O to be $\xi_O = \overline{\rho}_O - \widehat{\overline{\rho}}_O$, we can say

$$\overline{\rho}_O = \widehat{\overline{\rho}}_O + \xi_O. \tag{1.53}$$

Equations (1.52) and (1.53) can be written in matrix form as

$$\underbrace{\begin{pmatrix}\overline{\rho}_O \\ \overline{\rho}_{O_2}\end{pmatrix}}_{\overline{\boldsymbol{\rho}}} = \underbrace{\begin{pmatrix}\widehat{\overline{\rho}}_O \\ \widehat{\overline{\rho}}_{O_2}\end{pmatrix}}_{\widehat{\overline{\boldsymbol{\rho}}}} + \underbrace{\begin{pmatrix}1 \\ -\frac{1}{2}\end{pmatrix}}_{\mathbf{D}}\underbrace{(\xi_O)}_{\boldsymbol{\xi}}. \tag{1.54}$$

This gives the dependent variables in terms of a smaller number of transformed dependent variables in a way that satisfies the linear constraints. In vector form,

Eq. (1.54) becomes, similar to the closely related Eq. (5.161),

$$\overline{\rho} = \widehat{\overline{\rho}} + \mathbf{D} \cdot \boldsymbol{\xi}. \tag{1.55}$$

Here, $\mathbf{D}$ is a full rank matrix that spans the same column space as does $\boldsymbol{\nu}$. Note that $\boldsymbol{\nu}$ may or may not be full rank. Because $\mathbf{D}$ spans the same column space as does $\boldsymbol{\nu}$, and we have Eq. (1.35), we must also have in general the equivalent of the upcoming Eq. (5.159):

$$\boldsymbol{\phi} \cdot \mathbf{D} = \mathbf{0}. \tag{1.56}$$

We see here this is true:

$$\underbrace{\begin{pmatrix} 1 & 2 \end{pmatrix}}_{\boldsymbol{\phi}} \underbrace{\begin{pmatrix} 1 \\ -\frac{1}{2} \end{pmatrix}}_{\mathbf{D}} = (0). \tag{1.57}$$

The term $\exp(-\overline{\mathcal{E}}_j/\overline{R}T)$ is a modulating factor to the dynamics. Let us see how this behaves for high and low temperatures. For low- and high-temperature limits we have

$$\lim_{T\to 0} \exp\left(\frac{-\overline{\mathcal{E}}_j}{\overline{R}T}\right) = 0, \quad \lim_{T\to\infty} \exp\left(\frac{-\overline{\mathcal{E}}_j}{\overline{R}T}\right) = 1. \tag{1.58}$$

And last, at intermediate temperature, we have

$$\exp\left(\frac{-\overline{\mathcal{E}}_j}{\overline{R}T}\right) \sim \mathcal{O}(1), \qquad \text{when} \qquad T = \mathcal{O}\left(\frac{\overline{\mathcal{E}}_j}{\overline{R}}\right). \tag{1.59}$$

A sketch of this modulating factor is given in Fig. 1.4. Note that

- For small $T$, the modulation is extreme, and the reaction rate is small.
- For $T \sim \overline{\mathcal{E}}_j/\overline{R}$, the reaction rate is extremely sensitive to temperature.
- For $T \to \infty$, the modulation is unity, and the reaction rate is limited only by the molecular collision frequency.

Reaction 13 has zero activation energy. This suggests that if a collision occurs, the reaction is likely to occur. This reflects that the underlying electronic structure of the O atom is highly receptive to combination with another O atom. In contrast, Reaction 14 has a positive activation energy. This suggests that not all collisions will

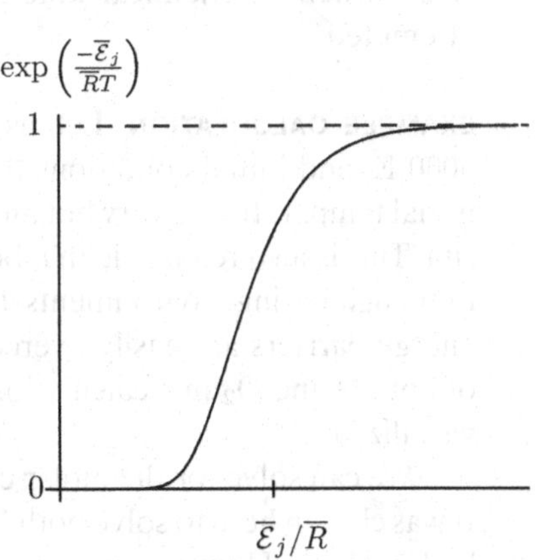


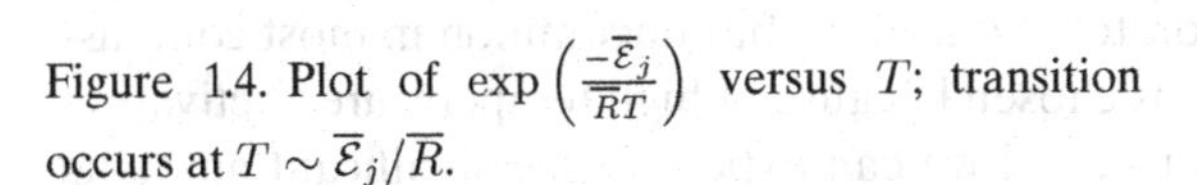
Figure 1.4. Plot of $\exp\left(\frac{-\overline{\mathcal{E}}_j}{\overline{R}T}\right)$ versus $T$; transition occurs at $T \sim \overline{\mathcal{E}}_j/\overline{R}$.

induce dissociation. Because the temperature is a measure of the mean molecular speed, high temperature suggests a high average speed, thus rendering a collision more likely to be sufficiently strong to break the bond holding the $O_2$ molecule together.

Now, $\overline{\rho}_O$ and $\overline{\rho}_{O_2}$ represent molar concentrations that have units of mol/cm$^3$. So, the reaction rates $d\overline{\rho}_O/dt$ and $d\overline{\rho}_{O_2}/dt$ have units of mol/cm$^3$/s. The argument of the exponential is dimensionless. That is,

$$\left[\frac{\overline{\mathcal{E}}_j}{\overline{R}T}\right] = \frac{\text{erg}}{\text{mol}}\frac{\text{mol K}}{\text{erg}}\frac{1}{\text{K}} \Rightarrow \text{dimensionless.} \tag{1.60}$$

Here, the brackets denote the units of a quantity and not molar concentration. Let us use a more informal method than that presented just after Eq. (1.20) to arrive at units for the collision frequency coefficient of Reaction 13, $a_{13}$. We know the units of the rate are mol/cm$^3$/s. Reaction 13 involves three molar species. Because $\beta_{13} = -1$, it also has an extra temperature dependency. The exponential of a dimensionless number is dimensionless, so we need not worry about that. For units to match, we must have

$$\left(\frac{\text{mol}}{\text{cm}^3\ \text{s}}\right) = [a_{13}]\left(\frac{\text{mol}}{\text{cm}^3}\right)\left(\frac{\text{mol}}{\text{cm}^3}\right)\left(\frac{\text{mol}}{\text{cm}^3}\right)\text{K}^{-1}. \tag{1.61}$$

So, the units of $a_{13}$ are

$$[a_{13}] = \left(\frac{\text{mol}}{\text{cm}^3}\right)^{-2}\frac{\text{K}}{\text{s}}. \tag{1.62}$$

For $a_{14}$ we find a different set of units! Following the same procedure, we get

$$\left(\frac{\text{mol}}{\text{cm}^3\ \text{s}}\right) = [a_{14}]\left(\frac{\text{mol}}{\text{cm}^3}\right)\left(\frac{\text{mol}}{\text{cm}^3}\right)\text{K}^{-1}. \tag{1.63}$$

So, the units of $a_{14}$ are

$$[a_{14}] = \left(\frac{\text{mol}}{\text{cm}^3}\right)^{-1}\frac{\text{K}}{\text{s}}. \tag{1.64}$$

This discrepancy in the units of $a_j$, the molecular collision frequency coefficient, is a burden of chemical kinetics, and causes difficulty when classical scaling is attempted.

**EXAMPLE CALCULATION.** Let us consider an example problem. Let us take $T = 5000$ K, and initial conditions $\widehat{\overline{\rho}}_O = 0.001$ mol/cm$^3$ and $\widehat{\overline{\rho}}_{O_2} = 0.001$ mol/cm$^3$. The initial temperature is very hot and is in fact near the temperature of the surface of the sun. This is also realizable in laboratory conditions but uncommon in most combustion engineering environments. It is chosen because at high temperatures, activation energy barriers are easily overcome, and we can expect to see significant amounts of both O and $O_2$ in a calculation. Pedagogically, it renders results to be more easily visualized.

We can solve for the molar concentrations as function of time in variety of ways. It was chosen here to solve both Eqs. (1.44) and (1.45) without the reduction provided by Eq. (1.47). However, we can check after numerical solution to see if Eq. (1.47) is actually satisfied. Substituting numerical values for all the constants, we get

$$-2a_{13}T^{\beta_{13}} \exp\left(\frac{-\overline{\mathcal{E}}_{13}}{\overline{R}T}\right) = -2\left(2.9 \times 10^{7} \left(\frac{\text{mol}}{\text{cm}^3}\right)^{-2} \frac{\text{K}}{\text{s}}\right) \times (5000\ \text{K})^{-1} \exp(0), \tag{1.65}$$

$$= -1.16 \times 10^{14} \left(\frac{\text{mol}}{\text{cm}^3}\right)^{-2} \frac{1}{\text{s}}, \tag{1.66}$$

$$2a_{14}T^{\beta_{14}} \exp\left(\frac{-\overline{\mathcal{E}}_{14}}{\overline{R}T}\right) = 2\left(6.81 \times 10^{18} \left(\frac{\text{mol}}{\text{cm}^3}\right)^{-1} \frac{\text{K}}{\text{s}}\right) (5000\ \text{K})^{-1} \times \exp\left(\frac{-4.9641 \times 10^{12}\ \frac{\text{erg}}{\text{mol}}}{8.314 \times 10^{7}\ \frac{\text{erg}}{\text{mol K}}(5000\ \text{K})}\right), \tag{1.67}$$

$$= 1.77444 \times 10^{10} \left(\frac{\text{mol}}{\text{cm}^3}\right)^{-1} \frac{1}{\text{s}}, \tag{1.68}$$

$$a_{13}T^{\beta_{13}} \exp\left(\frac{-\overline{\mathcal{E}}_{13}}{\overline{R}T}\right) = 5.80 \times 10^{13} \left(\frac{\text{mol}}{\text{cm}^3}\right)^{-2} \frac{1}{\text{s}}, \tag{1.69}$$

$$-a_{14}T^{\beta_{14}} \exp\left(\frac{-\overline{\mathcal{E}}_{14}}{\overline{R}T}\right) = -8.8774 \times 10^{9} \left(\frac{\text{mol}}{\text{cm}^3}\right)^{-1} \frac{1}{\text{s}}. \tag{1.70}$$

Then, the differential equation system, Eqs. (1.23) and (1.24), becomes

$$\frac{d\overline{\rho}_O}{dt} = -(1.16 \times 10^{14})\overline{\rho}_O^2(0.35\overline{\rho}_{O_2} + \overline{\rho}_O) + (1.77444 \times 10^{10})\overline{\rho}_{O_2}(0.35\overline{\rho}_{O_2} + \overline{\rho}_O), \tag{1.71}$$

$$\frac{d\overline{\rho}_{O_2}}{dt} = (5.80 \times 10^{13})\overline{\rho}_O^2(0.35\overline{\rho}_{O_2} + \overline{\rho}_O) - (8.87218 \times 10^{9})\overline{\rho}_{O_2}(0.35\overline{\rho}_{O_2} + \overline{\rho}_O), \tag{1.72}$$

$$\overline{\rho}_O(0) = \overline{\rho}_{O_2}(0) = 0.001\ \text{mol/cm}^3. \tag{1.73}$$

For compact presentation, we omit the units in Eqs. (1.71) and (1.72) that were given earlier. These two nonlinear ordinary differential equations are in a standard form for a variety of numerical software tools. Numerical solution techniques are not the topic of this book, and it will be assumed the reader has access to standard skills and tools.

*Species Concentration versus Time.* A solution was obtained numerically, and a plot of $\overline{\rho}_O(t)$ and $\overline{\rho}_{O_2}(t)$ is given in Fig. 1.5. *The plot is given on a log-log scale, which is often the scale of choice in plotting species concentrations versus time in combustion problems.* The log-log scale illustrates when significant reaction commences and when it is complete. This is especially important in so-called *multiscale problems*, to be studied throughout the book, whose solution features arise at scales of many different orders of magnitude. Significant reaction does not commence until $t \sim 10^{-10}$ s. This can be shown to be similar to the time between molecular collisions. For $t \in [10^{-9}\ \text{s}, 10^{-8}\ \text{s}]$, there is a vigorous reaction. For $t > 10^{-7}$ s, the reaction appears to be equilibrated.

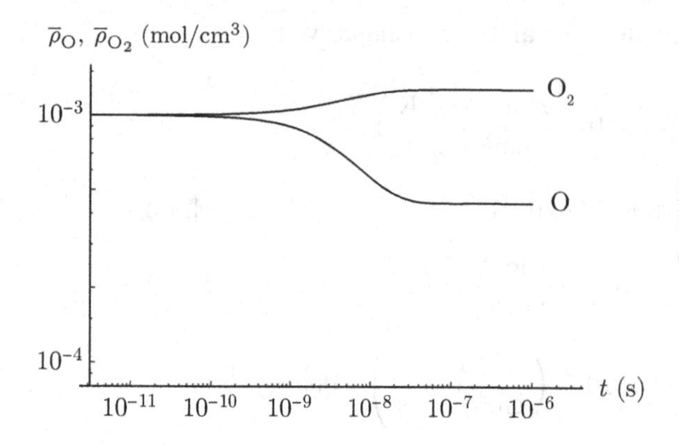


Figure 1.5. Molar concentrations versus time for oxygen dissociation problem; pair of irreversible reactions.

The calculation gives the equilibrium values $\overline{\rho}_O^{eq}$ and $\overline{\rho}_{O_2}^{eq}$ as

$$\lim_{t\to\infty} \overline{\rho}_O = \overline{\rho}_O^{eq} = 0.0004423 \frac{\text{mol}}{\text{cm}^3}, \quad \lim_{t\to\infty} \overline{\rho}_{O_2} = \overline{\rho}_{O_2}^{eq} = 0.00128 \frac{\text{mol}}{\text{cm}^3}. \tag{1.74}$$

Here, the superscript *eq* denotes an equilibrium value. At this high temperature, $O_2$ is preferred over O, but there are significant O molecules present at equilibrium.

*Pressure versus Time.* We next draw upon notions from classical thermodynamics that will be discussed in Chapter 2. We can use the ideal gas law to calculate the pressure, $P$. The ideal gas law for species $i$ is given by the upcoming Eq. (2.45):

$$P_i V = n_i \overline{R} T. \tag{1.75}$$

Here, $P_i$ is the *partial pressure* of molecular species $i$. We also have

$$P_i = \frac{n_i}{V} \overline{R} T. \tag{1.76}$$

By our definition of molecular species concentration, Eq. (1.7), that $\overline{\rho}_i = n_i/V$. So, we also have the ideal gas law as

$$P_i = \overline{\rho}_i \overline{R} T. \tag{1.77}$$

Now, in the Dalton mixture model, to be discussed in Sec. 2.3, all species share the same $T$ and $V$. So, for each species, $V_i = V, T_i = T$. But the mixture pressure is taken to be the sum of the partial pressures, Eq. (2.43):

$$P = \sum_{i=1}^{N} P_i. \tag{1.78}$$

Substituting from Eq. (1.77) into Eq. (1.78), we get

$$P = \sum_{i=1}^{N} \overline{\rho}_i \overline{R} T = \overline{R} T \sum_{i=1}^{N} \overline{\rho}_i. \tag{1.79}$$

For our example, we only have two species, so

$$P = \overline{R} T (\overline{\rho}_O + \overline{\rho}_{O_2}). \tag{1.80}$$

The pressure at the initial state $t = 0$ is

$$P(t = 0) = \overline{R}T(\widehat{\overline{\rho}}_O + \widehat{\overline{\rho}}_{O_2}), \quad (1.81)$$

$$= \left(8.314 \times 10^7 \frac{\text{erg}}{\text{mol K}}\right)(5000\ \text{K})$$

$$\times \left(0.001\ \frac{\text{mol}}{\text{cm}^3} + 0.001\ \frac{\text{mol}}{\text{cm}^3}\right), \quad (1.82)$$

$$= 8.314 \times 10^8\ \frac{\text{dyne}}{\text{cm}^2} = 8.314 \times 10^2\ \text{bar}, \quad (1.83)$$

recalling that 1 bar $= 10^6$ dyne/cm$^2 \approx 1$ atm. This pressure is over 800 atm. It is actually a little too high for good experimental correlation with the underlying data, but we will neglect that for this exercise.

At equilibrium we have more $O_2$ and less O. And we have a different number of molecules, so we expect the pressure to be different. At equilibrium, the pressure is

$$P(t \to \infty) = \lim_{t \to \infty} \overline{R}T(\overline{\rho}_O + \overline{\rho}_{O_2}), \quad (1.84)$$

$$= \left(8.314 \times 10^7 \frac{\text{erg}}{\text{mol K}}\right)(5000\ \text{K})$$

$$\times \left(0.0004423 \frac{\text{mol}}{\text{cm}^3} + 0.00128\ \frac{\text{mol}}{\text{cm}^3}\right), \quad (1.85)$$

$$= 7.155 \times 10^8\ \frac{\text{dyne}}{\text{cm}^2} = 7.155 \times 10^2\ \text{bar}. \quad (1.86)$$

The pressure has dropped because much of the O has recombined to form $O_2$. Thus, there are fewer molecules at equilibrium. The temperature and volume have remained the same. A plot of $P(t)$ is given in Fig. 1.6.

*Dynamical System Form.* Let us next bring to bear well-known methods from the general field of nonlinear dynamical systems to this combustion problem. Background can be found in a wide variety of texts (Arnol'd, 1973; Drazin, 1992; Perko, 2001; Hirsch et al., 2012; Strogatz, 2015; Powers and Sen, 2015). Now, Eqs. (1.71) and (1.72) are of the standard form for an autonomous dynamical system:

$$\frac{d\overline{\boldsymbol{\rho}}}{dt} = \mathbf{f}(\overline{\boldsymbol{\rho}}). \quad (1.87)$$

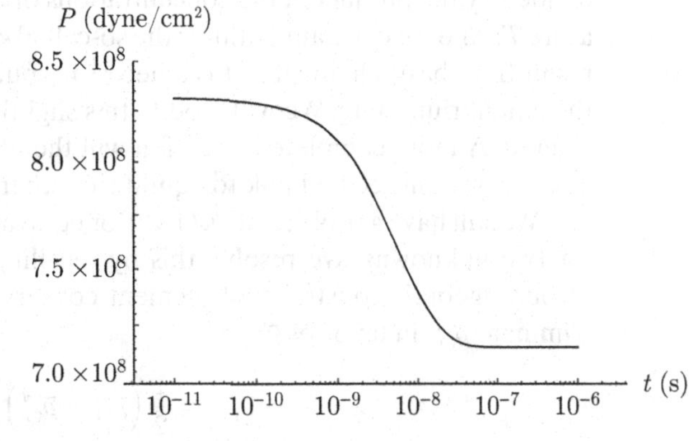


Figure 1.6. Pressure versus time for oxygen dissociation example.

Here, $\overline{\boldsymbol{\rho}}$ is the vector of state variables $(\overline{\rho}_{\mathrm{O}}, \overline{\rho}_{\mathrm{O}_2})^T$, and $\mathbf{f}$ is an algebraic function of the state variables. For the isothermal system, the algebraic function is in fact a polynomial.

*Equilibrium* The dynamical system is in equilibrium when

$$\mathbf{f}(\overline{\boldsymbol{\rho}}) = \mathbf{0}. \tag{1.88}$$

This nonlinear set of algebraic equations can be difficult to solve for large systems. We will later see in Sec. 5.2.3 that for common chemical kinetics systems, there is a guarantee of a unique equilibrium for which all state variables are physical. There are certainly other equilibria for which at least one of the state variables is nonphysical. Such equilibria can be mathematically complicated.

Forming Eq. (1.88) gives us, from Eqs. (1.23) and (1.24),

$$0 = -2a_{13} \exp\left(\frac{-\overline{\mathcal{E}}_{13}}{\overline{R}T}\right) \overline{\rho}_{\mathrm{O}}^{eq} \overline{\rho}_{\mathrm{O}}^{eq} \overline{\rho}_{\mathrm{M}}^{eq} T^{\beta_{13}} + 2a_{14} \exp\left(\frac{-\overline{\mathcal{E}}_{14}}{\overline{R}T}\right) \overline{\rho}_{\mathrm{O}_2}^{eq} \overline{\rho}_{\mathrm{M}}^{eq} T^{\beta_{14}}, \tag{1.89}$$

$$0 = a_{13} T^{\beta_{13}} \exp\left(\frac{-\overline{\mathcal{E}}_{13}}{\overline{R}T}\right) \overline{\rho}_{\mathrm{O}}^{eq} \overline{\rho}_{\mathrm{O}}^{eq} \overline{\rho}_{\mathrm{M}}^{eq} - a_{14} T^{\beta_{14}} \exp\left(\frac{-\overline{\mathcal{E}}_{14}}{\overline{R}T}\right) \overline{\rho}_{\mathrm{O}_2}^{eq} \overline{\rho}_{\mathrm{M}}^{eq}. \tag{1.90}$$

We notice that $\overline{\rho}_{\mathrm{M}}^{eq}$ cancels. This so-called third body concentration will in fact never affect the equilibrium state. It will however influence the dynamics. Removing $\overline{\rho}_{\mathrm{M}}^{eq}$ and slightly rearranging Eqs. (1.89) and (1.90) gives

$$a_{13} T^{\beta_{13}} \exp\left(\frac{-\overline{\mathcal{E}}_{13}}{\overline{R}T}\right) \overline{\rho}_{\mathrm{O}}^{eq} \overline{\rho}_{\mathrm{O}}^{eq} = a_{14} T^{\beta_{14}} \exp\left(\frac{-\overline{\mathcal{E}}_{14}}{\overline{R}T}\right) \overline{\rho}_{\mathrm{O}_2}^{eq}, \tag{1.91}$$

$$a_{13} T^{\beta_{13}} \exp\left(\frac{-\overline{\mathcal{E}}_{13}}{\overline{R}T}\right) \overline{\rho}_{\mathrm{O}}^{eq} \overline{\rho}_{\mathrm{O}}^{eq} = a_{14} T^{\beta_{14}} \exp\left(\frac{-\overline{\mathcal{E}}_{14}}{\overline{R}T}\right) \overline{\rho}_{\mathrm{O}_2}^{eq}. \tag{1.92}$$

These are the same equations! So, we really have two unknowns at the equilibrium, $\overline{\rho}_{\mathrm{O}}^{eq}$ and $\overline{\rho}_{\mathrm{O}_2}^{eq}$ but seemingly only one equation. Rearranging either Eq. (1.91) or (1.92) gives

$$\frac{\prod [\text{products}]}{\prod [\text{reactants}]} = \frac{\overline{\rho}_{\mathrm{O}}^{eq} \overline{\rho}_{\mathrm{O}}^{eq}}{\overline{\rho}_{\mathrm{O}_2}^{eq}} = \frac{a_{14} T^{\beta_{14}} \exp\left(\frac{-\overline{\mathcal{E}}_{14}}{\overline{R}T}\right)}{a_{13} T^{\beta_{13}} \exp\left(\frac{-\overline{\mathcal{E}}_{13}}{\overline{R}T}\right)} = K(T). \tag{1.93}$$

That is, for the net reaction (excluding the inert third body), summarized simply as $O_2 \leftrightharpoons O + O$, at equilibrium the product of the concentrations of the products divided by the product of the concentrations of the reactants is a function of temperature $T$. And for constant $T$, this is the so-called *equilibrium constant*. This is a famous result from basic chemistry. It is a necessary but insufficient condition to determine the equilibrium state. We will modify this slightly when reversible reactions are considered. A more complete discussion will then be given in Sec. 4.4 that will provide an as-of-yet unspecified link to equilibrium thermodynamics.

We still have a problem: Eq. (1.91), or equivalently, Eq. (1.93), is still one equation for two unknowns. We resolve this by recalling we have not yet taken advantage of our algebraic constraint of element conservation, Eq. (1.47). Let us use this to eliminate $\overline{\rho}_{\mathrm{O}_2}^{eq}$ in favor of $\overline{\rho}_{\mathrm{O}}^{eq}$:

$$\overline{\rho}_{\mathrm{O}_2}^{eq} = \frac{1}{2}\left(\widehat{\overline{\rho}}_{\mathrm{O}} - \overline{\rho}_{\mathrm{O}}^{eq}\right) + \widehat{\overline{\rho}}_{\mathrm{O}_2}. \tag{1.94}$$

So, with Eq. (1.94), Eq. (1.91) reduces to

$$a_{13}T^{\beta_{13}} \exp\left(\frac{-\overline{\mathcal{E}}_{13}}{\overline{R}T}\right) \overline{\rho}_{\mathrm{O}}^{eq} \overline{\rho}_{\mathrm{O}}^{eq} = a_{14}T^{\beta_{14}} \exp\left(\frac{-\overline{\mathcal{E}}_{14}}{\overline{R}T}\right) \underbrace{\left(\frac{1}{2}\left(\widehat{\overline{\rho}}_{\mathrm{O}} - \overline{\rho}_{\mathrm{O}}^{eq}\right) + \widehat{\overline{\rho}}_{\mathrm{O}_2}\right)}_{\overline{\rho}_{\mathrm{O}_2}^{eq}}. \tag{1.95}$$

Equation (1.95) is one algebraic equation in one unknown. Its solution gives the equilibrium value $\overline{\rho}_{\mathrm{O}}^{eq}$. It is a quadratic equation for $\overline{\rho}_{\mathrm{O}}^{eq}$. Of its two roots, one will be physical. *The equilibrium state will be a function of the initial conditions.* Mathematically, this is because our system is really best posed as a system of differential-algebraic equations (Ascher and Petzold, 1998; Kee et al., 2003, Chapter 15). Systems that are purely differential equations will have equilibria that are independent of their initial conditions. Most of the literature of mathematical physics focuses on such systems of those. One of the foundational complications of chemical dynamics is the equilibria are functions of the initial conditions, and this renders many common mathematical notions from traditional dynamical system theory to be invalid. Fortunately, after one accounts for the linear constraints of element conservation, one can return to classical notions from traditional dynamical systems theory.

Consider the dynamics of Eq. (1.44) for the evolution of $\overline{\rho}_{\mathrm{O}}$. Equilibrating the right-hand side of this equation gives Eq. (1.89). Using element conservation, Eq. (1.94), to eliminate $\overline{\rho}_{\mathrm{M}}$ and then $\overline{\rho}_{\mathrm{O}_2}$ in Eq. (1.89), then substituting in numerical parameters gives the cubic algebraic equation

$$\underbrace{-9.57000 \times 10^{13}\overline{\rho}_{\mathrm{O}}^3 - 6.82195 \times 10^{10}\overline{\rho}_{\mathrm{O}}^2 + 1.73007 \times 10^{7}\overline{\rho}_{\mathrm{O}} + 13973.7}_{f(\overline{\rho}_{\mathrm{O}})} = 0. \tag{1.96}$$

For compactness, we do not list the units in Eq. (1.96). This equation is cubic because it contains the effect of $\overline{\rho}_{\mathrm{M}}$. This will not affect the equilibrium but will affect the dynamics. We can get an idea of where the roots are by plotting $f(\overline{\rho}_{\mathrm{O}})$, as seen in Fig. 1.7. Zero crossings of $f(\overline{\rho}_{\mathrm{O}})$ in Fig. 1.7 represent equilibria of the system, $\overline{\rho}_{\mathrm{O}}^{eq}$, that

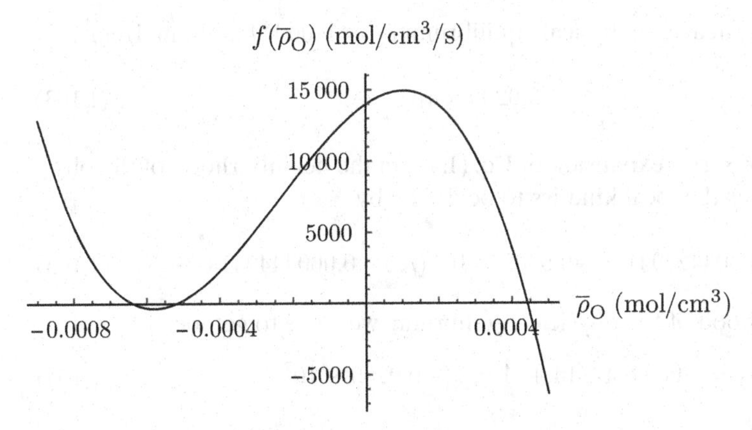


Figure 1.7. Equilibria for oxygen dissociation example; pair of irreversible reactions.

is, locations where $f(\overline{\rho}_{\mathrm{O}}^{eq}) = 0$. The cubic equation has three roots

$$\overline{\rho}_{\mathrm{O}}^{eq} = -0.0006364 \text{ mol/cm}^3, \qquad \text{nonphysical,} \tag{1.97}$$

$$\overline{\rho}_{\mathrm{O}}^{eq} = -0.0005188 \text{ mol/cm}^3, \qquad \text{nonphysical,} \tag{1.98}$$

$$\overline{\rho}_{\mathrm{O}}^{eq} = 0.000442294 \text{ mol/cm}^3, \qquad \text{physical.} \tag{1.99}$$

Values of molar concentration that are negative are obviously nonphysical. The physical root found by our algebraic analysis is effectively identical to that which was identified in the long time limit by our numerical integration of the ordinary differential equations of reaction kinetics; see Eq. (1.74).

*Stability of equilibria* We can get a simple estimate of the stability of the equilibria by considering the slope of $f$ near $f = 0$. Our dynamical system is of the form

$$\frac{d\overline{\rho}_{\mathrm{O}}}{dt} = f(\overline{\rho}_{\mathrm{O}}). \tag{1.100}$$

- Near the first nonphysical root at $\overline{\rho}_{\mathrm{O}}^{eq} = -0.0006364$ mol/cm$^3$, a positive perturbation from equilibrium induces $f < 0$, which induces $d\overline{\rho}_{\mathrm{O}}/dt < 0$, so $\overline{\rho}_{\mathrm{O}}$ returns to its equilibrium. Similarly, a negative perturbation from equilibrium induces $d\overline{\rho}_{\mathrm{O}}/dt > 0$, so the system returns to equilibrium. This nonphysical equilibrium point is *stable.* Stability does not imply physicality!
- We find the nonphysical root at $\overline{\rho}_{\mathrm{O}}^{eq} = -0.0005188$ mol/cm$^3$ is *unstable.*
- We find the physical root at $\overline{\rho}_{\mathrm{O}}^{eq} = 0.000442294$ mol/cm$^3$ is *stable.*

In general, if $f$ crosses zero with a positive slope, the equilibrium is unstable. Otherwise, it is stable.

Consider a formal Taylor[8] series expansion of Eq. (1.100) in the neighborhood of an equilibrium point $\overline{\rho}_{\mathrm{O}}^{eq}$:

$$\frac{d}{dt}\left(\overline{\rho}_{\mathrm{O}} - \overline{\rho}_{\mathrm{O}}^{eq}\right) = \underbrace{f(\overline{\rho}_{\mathrm{O}}^{eq})}_{0} + \left.\frac{df}{d\overline{\rho}_{\mathrm{O}}}\right|_{\overline{\rho}_{\mathrm{O}}=\overline{\rho}_{\mathrm{O}}^{eq}} \left(\overline{\rho}_{\mathrm{O}} - \overline{\rho}_{\mathrm{O}}^{eq}\right) + \cdots \tag{1.101}$$

We find $df/d\overline{\rho}_{\mathrm{O}}$ by differentiating Eq. (1.96) to get

$$\frac{df}{d\overline{\rho}_{\mathrm{O}}} = 1.73007 \times 10^7 - 1.36439 \times 10^{11}\overline{\rho}_{\mathrm{O}} - 2.87100 \times 10^{14}\overline{\rho}_{\mathrm{O}}^2. \tag{1.102}$$

We evaluate $df/d\overline{\rho}_{\mathrm{O}}$ near the physical equilibrium $\overline{\rho}_{\mathrm{O}} = 0.0004442414$ mol/cm$^3$:

$$\frac{df}{d\overline{\rho}_{\mathrm{O}}} = -9.9209 \times 10^7 \; \frac{1}{\mathrm{s}}. \tag{1.103}$$

Thus, the Taylor series expansion of Eq. (1.44) in the neighborhood of the physical equilibrium gives the local kinetics to be driven by

$$\frac{d}{dt}\left(\overline{\rho}_{\mathrm{O}} - 0.000442294\right) = -9.9209 \times 10^7 \left(\overline{\rho}_{\mathrm{O}} - 0.0004442414\right) + \cdots \tag{1.104}$$

So, in the neighborhood of the physical equilibrium, we solve to get locally

$$\overline{\rho}_{\mathrm{O}} = 0.0004442414 + A\exp\left(-9.9209 \times 10^7 t\right). \tag{1.105}$$

[8]Brook Taylor, 1685-1731, English mathematician.

Here, $A$ is an arbitrary constant of integration. The local time constant that governs the times scales of local evolution is $\tau$, where

$$\tau = \frac{1}{9.9209 \times 10^7} = 1.00797 \times 10^{-8} \text{ s.} \tag{1.106}$$

This 10 ns time scale is very fast. It can be shown to be correlated with the mean time between collisions of molecules. Examining Fig. 1.5, we see the simple time scale estimate from linear theory accurately predicts the time where rapid changes occur in molar concentrations.

**EFFECT OF TEMPERATURE.** Let us perform four case studies to see the effect of $T$ on the system's equilibria and its dynamics near equilibrium.

- $T = 3000$ K. Here, we have significantly reduced the temperature, but it is still higher than typically found in ordinary combustion engineering environments. Here, we find

$$\overline{\rho}_{\text{O}}^{eq} = 8.9327 \times 10^{-6} \text{ mol/cm}^3, \quad \tau = 5.43023 \times 10^{-7} \text{ s.} \tag{1.107}$$

  The equilibrium concentration of O dropped by two orders of magnitude relative to $T = 5000$ K, and the time scale of the dynamics near equilibrium slowed by over an order of magnitude.
- $T = 1000$ K. Here, we reduce the temperature more. This temperature is slightly low for common combustion engineering environments. We find

$$\overline{\rho}_{\text{O}}^{eq} = 2.03261 \times 10^{-14} \text{ mol/cm}^3, \quad \tau = 8.07848 \times 10^{1} \text{ s.} \tag{1.108}$$

  The O concentration at equilibrium is greatly diminished to the point of being difficult to detect by standard measurement techniques. And the time scale of combustion has significantly slowed.
- $T = 300$ K. This is near room temperature. We find

$$\overline{\rho}_{\text{O}}^{eq} = 1.13640 \times 10^{-44} \text{ mol/cm}^3, \quad \tau = 4.33485 \times 10^{31} \text{ s.} \tag{1.109}$$

  The O concentration is effectively zero at room temperature, and the relaxation time is too large to validate by experiment.
- $T = 10000$ K. Such high temperature could be achieved in atmospheric reentry:

$$\overline{\rho}_{\text{O}}^{eq} = 2.74801 \times 10^{-3} \text{ mol/cm}^3, \quad \tau = 1.74126 \times 10^{-10} \text{ s.} \tag{1.110}$$

  At this high temperature, O become preferred over $O_2$, and the time scale of reaction becomes extremely small, under a nanosecond. That said, at this elevated temperature, a significant number of electrons are shed via $\text{O} + \text{M} \rightarrow \text{O}^+ + \text{e}^- + \text{M}$, thus generating a so-called *ionized plasma*. Our analysis has neglected the effect of ionization.

### Single Reversible Reaction

The two irreversible reactions studied in the previous section are of a class that is common in combustion modeling. However, the model suffers a defect in that its link to classical equilibrium thermodynamics is missing. A better way to model the same physics and guarantee consistency with classical equilibrium thermodynamics is to model the process as a single *reversible* reaction, with a suitably modified reaction rate term.

**MATHEMATICAL MODEL.**

*Kinetics.* For the reversible O-$O_2$ reaction, consider Reaction 13 from Table 1.2:

$$13\colon O_2 + M \leftrightharpoons O + O + M. \tag{1.111}$$

We have $N = 2$ molecular species in $L = 1$ element reacting in $J = 1$ reaction. Here,

$$a_{13} = 1.85 \times 10^{11} \left(\frac{\text{mol}}{\text{cm}^3}\right)^{-1} (\text{K})^{-0.5}, \ \beta_{13} = 0.5, \ \overline{\mathcal{E}}_{13} = 95560 \ \frac{\text{cal}}{\text{mol}}. \tag{1.112}$$

Units of calories are common in chemistry, but we need to convert to erg:

$$\overline{\mathcal{E}}_{13} = \left(95560 \ \frac{\text{cal}}{\text{mol}}\right)\left(4.186 \ \frac{\text{J}}{\text{cal}}\right)\left(10^7 \ \frac{\text{erg}}{\text{J}}\right) = 4.00014 \times 10^{12} \ \frac{\text{erg}}{\text{mol}}. \tag{1.113}$$

For this reversible reaction, we slightly modify the kinetics equations to

$$\frac{d\overline{\rho}_O}{dt} = \underbrace{\underbrace{2\, a_{13} T^{\beta_{13}} \exp\left(\frac{-\overline{\mathcal{E}}_{13}}{\overline{R}T}\right)}_{k_{13}(T)} \left(\overline{\rho}_{O_2}\overline{\rho}_M - \frac{1}{K_{c,13}}\overline{\rho}_O\overline{\rho}_O\overline{\rho}_M\right)}_{r_{13}}, \tag{1.114}$$

$$\frac{d\overline{\rho}_{O_2}}{dt} = -\underbrace{\underbrace{a_{13} T^{\beta_{13}} \exp\left(\frac{-\overline{\mathcal{E}}_{13}}{\overline{R}T}\right)}_{k_{13}(T)} \left(\overline{\rho}_{O_2}\overline{\rho}_M - \frac{1}{K_{c,13}}\overline{\rho}_O\overline{\rho}_O\overline{\rho}_M\right)}_{r_{13}}. \tag{1.115}$$

All collision efficiency coefficients are unity for this kinetics scheme, so we can use the simple form given by Eq. (1.40) to get

$$\overline{\rho}_M = \overline{\rho}_O + \overline{\rho}_{O_2}. \tag{1.116}$$

Equation (1.115) has its analog in Eq. (5.13). Here, we have used equivalent definitions for $k_{13}(T)$ and $r_{13}$, so that Eqs. (1.114) and (1.115) can be written compactly as

$$\frac{d\overline{\rho}_O}{dt} = 2r_{13}, \qquad \frac{d\overline{\rho}_{O_2}}{dt} = -r_{13}. \tag{1.117}$$

In matrix form, we can simplify to

$$\frac{d}{dt}\begin{pmatrix}\overline{\rho}_O \\ \overline{\rho}_{O_2}\end{pmatrix} = \underbrace{\begin{pmatrix}2 \\ -1\end{pmatrix}}_{\boldsymbol{\nu}}(r_{13}). \tag{1.118}$$

Here, the $N \times J = 2 \times 1$ matrix $\boldsymbol{\nu}$ is

$$\boldsymbol{\nu} = \begin{pmatrix}2 \\ -1\end{pmatrix}. \tag{1.119}$$

Performing row operations, Eq. (1.118) reduces to

$$\frac{d}{dt}\begin{pmatrix}\overline{\rho}_O \\ \overline{\rho}_O + 2\overline{\rho}_{O_2}\end{pmatrix} = \begin{pmatrix}2 \\ 0\end{pmatrix}(r_{13}), \tag{1.120}$$

or

$$\underbrace{\begin{pmatrix} 1 & 0 \\ 1 & 2 \end{pmatrix}}_{\mathbf{L}^{-1}} \frac{d}{dt}\begin{pmatrix} \overline{\rho}_{O} \\ \overline{\rho}_{O_2} \end{pmatrix} = \underbrace{\begin{pmatrix} 2 \\ 0 \end{pmatrix}}_{\mathbf{U}} (r_{13}). \tag{1.121}$$

So, here, the $N \times N = 2 \times 2$ matrix $\mathbf{L}^{-1}$ is

$$\mathbf{L}^{-1} = \begin{pmatrix} 1 & 0 \\ 1 & 2 \end{pmatrix}. \tag{1.122}$$

The $N \times N = 2 \times 2$ permutation matrix $\mathbf{P}$ is the identity matrix. And the $N \times J = 2 \times 1$ upper triangular matrix $\mathbf{U}$ is

$$\mathbf{U} = \begin{pmatrix} 2 \\ 0 \end{pmatrix}. \tag{1.123}$$

Note that $\boldsymbol{\nu} = \mathbf{L} \cdot \mathbf{U}$ or, equivalently, $\mathbf{L}^{-1} \cdot \boldsymbol{\nu} = \mathbf{U}$:

$$\underbrace{\begin{pmatrix} 1 & 0 \\ 1 & 2 \end{pmatrix}}_{\mathbf{L}^{-1}} \underbrace{\begin{pmatrix} 2 \\ -1 \end{pmatrix}}_{\boldsymbol{\nu}} = \underbrace{\begin{pmatrix} 2 \\ 0 \end{pmatrix}}_{\mathbf{U}}. \tag{1.124}$$

Once again the element-species matrix $\boldsymbol{\phi}$ is

$$\boldsymbol{\phi} = (1 \quad 2). \tag{1.125}$$

And we see that $\boldsymbol{\phi} \cdot \boldsymbol{\nu} = \mathbf{0}$ is satisfied:

$$\underbrace{(1 \quad 2)}_{\boldsymbol{\phi}} \underbrace{\begin{pmatrix} 2 \\ -1 \end{pmatrix}}_{\boldsymbol{\nu}} = (0). \tag{1.126}$$

As for the irreversible reactions, the reversible reaction rates are constructed to conserve O atoms. We have from the second row of Eq. (1.121) that

$$\frac{d}{dt}\left(\overline{\rho}_{O} + 2\overline{\rho}_{O_2}\right) = 0. \tag{1.127}$$

Thus, by integrating Eq. (1.127) and applying the initial conditions, we once again find

$$\overline{\rho}_{O} + 2\overline{\rho}_{O_2} = \widehat{\overline{\rho}}_{O} + 2\widehat{\overline{\rho}}_{O_2} = \text{constant}. \tag{1.128}$$

As before, we can say

$$\underbrace{\begin{pmatrix} \overline{\rho}_{O} \\ \overline{\rho}_{O_2} \end{pmatrix}}_{\overline{\boldsymbol{\rho}}} = \underbrace{\begin{pmatrix} \widehat{\overline{\rho}}_{O} \\ \widehat{\overline{\rho}}_{O_2} \end{pmatrix}}_{\widehat{\overline{\boldsymbol{\rho}}}} + \underbrace{\begin{pmatrix} 1 \\ -\frac{1}{2} \end{pmatrix}}_{\mathbf{D}} \underbrace{(\xi_{O})}_{\boldsymbol{\xi}}. \tag{1.129}$$

This gives the dependent variables in terms of a smaller number of transformed variables in a way that satisfies the linear constraints. In vector form, Eq. (1.129) becomes

$$\overline{\boldsymbol{\rho}} = \widehat{\overline{\boldsymbol{\rho}}} + \mathbf{D} \cdot \boldsymbol{\xi}. \tag{1.130}$$

Once again, $\boldsymbol{\phi} \cdot \mathbf{D} = \mathbf{0}$.

*Thermodynamics.* Equations (1.114)–(1.115) are supplemented by an expression for the thermodynamics-based equilibrium constant $K_{c,13}$, that is,

$$K_{c,13} = \left(\frac{P_0}{\overline{R}T}\right)^{\sum_{i=1}^{N} \nu_{i,13}} \exp\left(\frac{-\Delta G_{13}^0}{\overline{R}T}\right), \tag{1.131}$$

$$= \left(\frac{P_0}{\overline{R}T}\right)^{2-1} \exp\left(\frac{-\Delta G_{13}^0}{\overline{R}T}\right), \tag{1.132}$$

$$= \frac{P_0}{\overline{R}T} \exp\left(\frac{-\Delta G_{13}^0}{\overline{R}T}\right). \tag{1.133}$$

Theoretical details are given in Sec. 4.4 for a single reaction and in Sec. 5.2.2 for multiple reactions. Here, $P_0 = 1.01326 \times 10^6$ dyne/cm$^2$ $= 1$ atm is the reference pressure. The net change of Gibbs free energy at the reference pressure for Reaction 13, $\Delta G_{13}^0$, is defined as

$$\Delta G_{13}^0 = 2\overline{g}_{\mathrm{O}}^0 - \overline{g}_{\mathrm{O_2}}^0. \tag{1.134}$$

The Gibbs free energy $G$ is formally defined in Sec. 3.6.2. In short, Eq. (1.134) reflects the fact that we have created two moles of product O for every mol of reactant $O_2$. And with the Gibbs free energy per mol as $\overline{g}_i$, Eq. (1.134) reflects the total change in Gibbs free energy when Reaction 13 occurs. Equation (1.134) is a special case of the more general vector form that will be developed later:

$$\Delta \mathbf{G}^{0^T} = \overline{\mathbf{g}}^{0^T} \cdot \boldsymbol{\nu}. \tag{1.135}$$

For a single reaction, see Eq. (4.206); for multiple reactions, see Eq. (5.5). The superscript $T$ stands for "transpose" and not temperature. Transposing both sides, we also have

$$\Delta \mathbf{G}^0 = \boldsymbol{\nu}^T \cdot \overline{\mathbf{g}}^0. \tag{1.136}$$

Here, $\Delta \mathbf{G}^0$ is a $J \times 1$ vector with the net change of Gibbs free energy for each reaction, and $\overline{\mathbf{g}}^0$ is a $N \times 1$ vector of the Gibbs free energy of each species. Both are evaluated at the reference pressure. For Reaction 13, we have

$$\Delta G_{13}^0 = \underbrace{\left(\overline{g}_{\mathrm{O}}^0 \quad \overline{g}_{\mathrm{O_2}}^0\right)}_{\overline{\mathbf{g}}^{0^T}} \underbrace{\begin{pmatrix} 2 \\ -1 \end{pmatrix}}_{\boldsymbol{\nu}}. \tag{1.137}$$

As can be inferred from the upcoming development of Eq. (3.226), the Gibbs free energy for species $i$ at the reference pressure is defined in terms of the reference pressure enthalpy, $\overline{h}_i^0$, and entropy, $\overline{s}_i^0$, as

$$\overline{g}_i^0 = \overline{h}_i^0 - T\overline{s}_i^0. \tag{1.138}$$

It is common to find $\overline{h}_i^0$ and $\overline{s}_i^0$ in thermodynamic tables given as functions of $T$.

Both Eqs. (1.114) and (1.115) are in equilibrium when

$$\overline{\rho}_{\mathrm{O_2}}^{eq} \overline{\rho}_{\mathrm{M}}^{eq} = \frac{1}{K_{c,13}} \overline{\rho}_{\mathrm{O}}^{eq} \overline{\rho}_{\mathrm{O}}^{eq} \overline{\rho}_{\mathrm{M}}^{eq}. \tag{1.139}$$

We rearrange Eq. (1.139) to find the familiar

$$K_{c,13} = \frac{\overline{\rho}_{\mathrm{O}}^{eq} \overline{\rho}_{\mathrm{O}}^{eq}}{\overline{\rho}_{\mathrm{O_2}}^{eq}} = \frac{\prod \text{[products]}}{\prod \text{[reactants]}}. \tag{1.140}$$

If $K_{c,13} > 1$, the products are preferred. If $K_{c,13} < 1$, the reactants are preferred.

Now, $K_{c,13}$ is a function of $T$ only, so it is known. But Eq. (1.140) once again is one equation in two unknowns. We can use the element conservation constraint, Eq. (1.128), to reduce to one equation and one unknown, valid at equilibrium:

$$K_{c,13} = \frac{\overline{\rho}_O^{eq}\overline{\rho}_O^{eq}}{\widehat{\overline{\rho}}_{O_2} + \frac{1}{2}(\widehat{\overline{\rho}}_O - \overline{\rho}_O^{eq})}. \tag{1.141}$$

Using the element conservation constraint, Eq. (1.128), we can recast the dynamics of our system by modifying Eq. (1.114) into one equation in one unknown:

$$\frac{d\overline{\rho}_O}{dt} = 2a_{13}T^{\beta_{13}} \exp\left(\frac{-\overline{\mathcal{E}}_{13}}{\overline{R}T}\right) \times \left( \underbrace{\left(\widehat{\overline{\rho}}_{O_2} + \frac{1}{2}(\widehat{\overline{\rho}}_O - \overline{\rho}_O)\right)}_{\overline{\rho}_{O_2}} \underbrace{\left(\widehat{\overline{\rho}}_{O_2} + \frac{1}{2}(\widehat{\overline{\rho}}_O - \overline{\rho}_O) + \overline{\rho}_O\right)}_{\overline{\rho}_M} - \frac{1}{K_{c,13}}\overline{\rho}_O\overline{\rho}_O \underbrace{\left(\widehat{\overline{\rho}}_{O_2} + \frac{1}{2}(\widehat{\overline{\rho}}_O - \overline{\rho}_O) + \overline{\rho}_O\right)}_{\overline{\rho}_M} \right). \tag{1.142}$$

**EXAMPLE CALCULATION.** Let us consider the same example as the previous section with $T = 5000$ K. We need numbers for all of the parameters of Eq. (1.142). For O, we find from standard thermodynamic tables[9] at $T = 5000$ K that

$$\overline{h}_O^0 = 3.48903 \times 10^{12} \frac{\text{erg}}{\text{mol}}, \qquad \overline{s}_O^0 = 2.20563 \times 10^9 \frac{\text{erg}}{\text{mol K}}. \tag{1.143}$$

So

$$\overline{g}_O^0 = \left(3.48903 \times 10^{12} \frac{\text{erg}}{\text{mol}}\right) - (5000 \text{ K})\left(2.20563 \times 10^9 \frac{\text{erg}}{\text{mol K}}\right), \tag{1.144}$$

$$= -7.53913 \times 10^{12} \text{ erg/mol}. \tag{1.145}$$

For $O_2$, we find at $T = 5000$ K that

$$\overline{h}_{O_2}^0 = 1.80776 \times 10^{12} \frac{\text{erg}}{\text{mol}}, \qquad \overline{s}_{O_2}^0 = 3.05509 \times 10^9 \frac{\text{erg}}{\text{mol K}}. \tag{1.146}$$

So

$$\overline{g}_{O_2}^0 = \left(1.80776 \times 10^{12} \frac{\text{erg}}{\text{mol}}\right) - (5000 \text{ K})\left(3.05509 \times 10^9 \frac{\text{erg}}{\text{mol K}}\right), \tag{1.147}$$

$$= -1.34677 \times 10^{13} \text{ erg/mol}. \tag{1.148}$$

Thus, by Eq. (1.134), we have

$$\Delta G_{13}^0 = 2(-7.53913 \times 10^{12}) - (-1.34677 \times 10^{13}), \tag{1.149}$$

$$= -1.61056 \times 10^{12} \text{ erg/mol}. \tag{1.150}$$

[9] interpolations of Table A.9 from Borgnakke and Sonntag, 2013, p. 767.

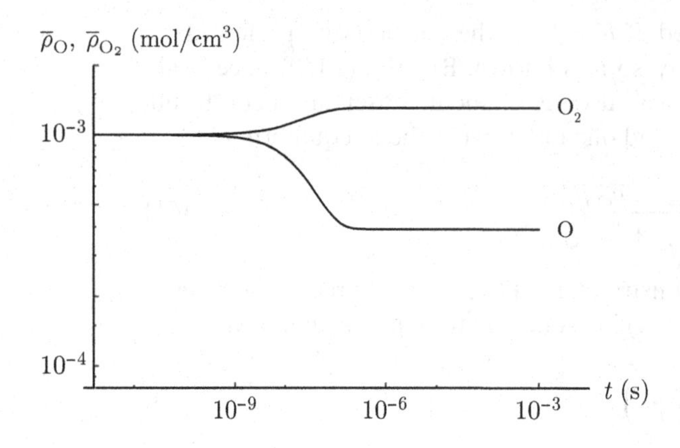


Figure 1.8. Plot of $\overline{\rho}_O(t)$ and $\overline{\rho}_{O_2}(t)$ for oxygen dissociation with reversible reaction.

Thus, by Eq. (1.133), we get for our system

$$K_{c,13} = \frac{1.01326 \times 10^6 \frac{\text{dyne}}{\text{cm}^2}}{\left(8.314 \times 10^7 \frac{\text{erg}}{\text{mol K}}\right)(5000\text{ K})} \times \exp\left(-\left(\frac{-1.61056 \times 10^{12} \frac{\text{erg}}{\text{mol}}}{\left(8.314 \times 10^7 \frac{\text{erg}}{\text{mol K}}\right)(5000\text{ K})}\right)\right), \tag{1.151}$$

$$= 1.17366 \times 10^{-4}\text{ mol/cm}^3. \tag{1.152}$$

Substitution of all numerical parameters into Eq. (1.142) and expansion yields

$$\frac{d\overline{\rho}_O}{dt} = 3897.62 - (2.25725 \times 10^{10})\overline{\rho}_O^2 - (7.37980 \times 10^{12})\overline{\rho}_O^3, \tag{1.153}$$

$$= f(\overline{\rho}_O), \tag{1.154}$$

$$\overline{\rho}_O(0) = 0.001. \tag{1.155}$$

A plot of the time-dependent behavior of $\overline{\rho}_O$ and $\overline{\rho}_{O_2}$ from solution of Eqs. (1.154) and (1.155) is given in Fig. 1.8. The behavior is similar to the predictions given by the pair of irreversible reactions in Fig. 1.4. Here, direct calculation of the equilibrium from time integration reveals

$$\overline{\rho}_O^{eq} = 0.000391265\text{ mol/cm}^3. \tag{1.156}$$

Using Eq. (1.128), we find this corresponds to

$$\overline{\rho}_{O_2}^{eq} = 0.00130437\text{ mol/cm}^3. \tag{1.157}$$

The system begins to undergo significant reaction for $t \sim 10^{-9}$ s and is equilibrated when $t \sim 10^{-7}$ s.

The equilibrium is verified by solving the algebraic equation suggested by Eq. (1.154):

$$f(\overline{\rho}_O) = 3897.62 - (2.25725 \times 10^{10})\overline{\rho}_O^2 - (7.37980 \times 10^{12})\overline{\rho}_O^3 = 0. \tag{1.158}$$

This yields three roots:

$$\overline{\rho}_O^{eq} = -0.003\text{ mol/cm}^3, \qquad \text{nonphysical}, \tag{1.159}$$

$$\overline{\rho}_O^{eq} = -0.000449948\text{ mol/cm}^3, \qquad \text{nonphysical}, \tag{1.160}$$

$$\overline{\rho}_O^{eq} = 0.000391265\text{ mol/cm}^3, \qquad \text{physical}, \tag{1.161}$$

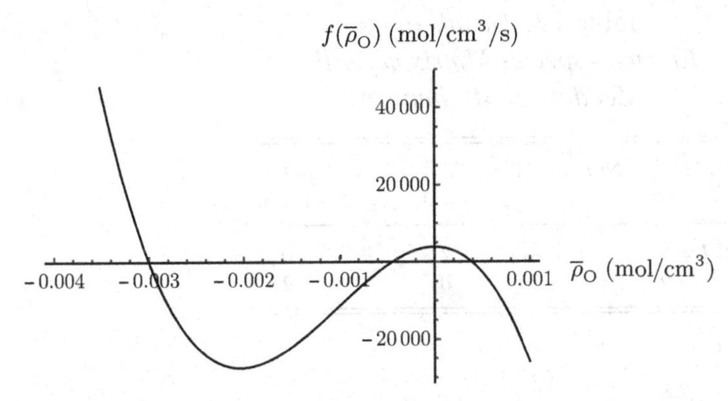


Figure 1.9. Plot of $f(\overline{\rho}_O)$ versus $\overline{\rho}_O$ for oxygen dissociation with reversible reaction.

consistent with the plot given in Fig. 1.9. Linearizing Eq. (1.155) in the neighborhood of the physical equilibrium yields the equation

$$\frac{d}{dt}(\overline{\rho}_O - 0.000391265) = -(2.10529 \times 10^7)(\overline{\rho}_O - 0.000391265) + \cdots \quad (1.162)$$

This has solution

$$\overline{\rho}_O = 0.000391265 + A\exp\left(-2.10529 \times 10^7 t\right). \quad (1.163)$$

Again, $A$ is an arbitrary constant. Obviously the equilibrium is stable. Moreover, the time constant of relaxation to equilibrium is

$$\tau = \frac{1}{2.10529 \times 10^7} = 4.74994 \times 10^{-8} \text{ s}. \quad (1.164)$$

This is consistent with the time scale that comes from integrating the full equation.

### 1.2.2 Zel'dovich Mechanism of NO Production

Let us consider next a more complicated reaction system: that of NO production known as the Zel'dovich[10] mechanism. This is an important model for the production of a major pollutant from combustion processes. It is most important for high-temperature applications. Related calculations and analysis are given by Al-Khateeb et al. (2009).

#### Mathematical model

The model has several versions. One is

$$1: N + NO \leftrightharpoons N_2 + O, \quad (1.165)$$

$$2: N + O_2 \leftrightharpoons NO + O. \quad (1.166)$$

Similar to our model for $O_2$ dissociation, $N_2$ and $O_2$ are preferred at low temperature. As the temperature rises, N and O begin to appear. It is possible when they are mixed for NO to appear as a product.

[10] Yakov Borisovich Zel'dovich (1915–1987), Soviet physicist.

Table 1.3. *Details of the Element-Species Matrix $\phi$ for the Zel'dovich Mechanism*

| | NO $i=1$ | N $i=2$ | $N_2$ $i=3$ | O $i=4$ | $O_2$ $i=5$ |
|---|---|---|---|---|---|
| N, $l=1$ | 1 | 1 | 2 | 0 | 0 |
| O, $l=2$ | 1 | 0 | 0 | 1 | 2 |

**STANDARD MODEL FORM.** Here, we have the reaction of $N = 5$ molecular species with a species molar concentration vector of

$$\overline{\boldsymbol{\rho}} = \begin{pmatrix} \overline{\rho}_{NO} \\ \overline{\rho}_{N} \\ \overline{\rho}_{N_2} \\ \overline{\rho}_{O} \\ \overline{\rho}_{O_2} \end{pmatrix}. \tag{1.167}$$

We have $L = 2$ elements: N ($l = 1$) and O ($l = 2$). The element-species matrix $\phi$ of dimension $L \times N = 2 \times 5$ is

$$\phi = \begin{pmatrix} 1 & 1 & 2 & 0 & 0 \\ 1 & 0 & 0 & 1 & 2 \end{pmatrix}. \tag{1.168}$$

The first row of $\phi$ is for the N atom; the second row is for the O atom. For example, the entry $\phi_{13} = 2$ addresses the row $l = 1$ for the element N and the column $i = 3$ for the species $N_2$. In species 3, we have 2 atoms of element 1. A complete description of the entries of $\phi$ is given in Table 1.3.

We have $J = 2$ reactions. The reaction vector of length $J = 2$ is

$$\mathbf{r} = \begin{pmatrix} r_1 \\ r_2 \end{pmatrix}, \tag{1.169}$$

$$= \begin{pmatrix} \underbrace{a_1 T^{\beta_1} \exp\left(-\frac{T_{a,1}}{T}\right)}_{k_1} \left(\overline{\rho}_N \overline{\rho}_{NO} - \frac{1}{K_{c,1}} \overline{\rho}_{N_2} \overline{\rho}_O\right) \\ \underbrace{a_2 T^{\beta_2} \exp\left(-\frac{T_{a,2}}{T}\right)}_{k_2} \left(\overline{\rho}_N \overline{\rho}_{O_2} - \frac{1}{K_{c,2}} \overline{\rho}_{NO} \overline{\rho}_O\right) \end{pmatrix}, \tag{1.170}$$

$$= \begin{pmatrix} k_1 \left(\overline{\rho}_N \overline{\rho}_{NO} - \dfrac{1}{K_{c,1}} \overline{\rho}_{N_2} \overline{\rho}_O\right) \\ k_2 \left(\overline{\rho}_N \overline{\rho}_{O_2} - \dfrac{1}{K_{c,2}} \overline{\rho}_{NO} \overline{\rho}_O\right) \end{pmatrix}. \tag{1.171}$$

The constants $T_{a,1}$ and $T_{a,2}$ are the so-called *activation temperatures*. These are sometimes used in place of activation energies. They are related by $T_{a,i} = \overline{\mathcal{E}}_i / \overline{R}$. Here, we have reaction rate coefficients of

$$k_1 = a_1 T^{\beta_1} \exp\left(-\frac{T_{a,1}}{T}\right), \quad k_2 = a_2 T^{\beta_2} \exp\left(-\frac{T_{a,2}}{T}\right). \tag{1.172}$$

In matrix form, the model can be written as

$$\frac{d}{dt}\begin{pmatrix} \overline{\rho}_{NO} \\ \overline{\rho}_{N} \\ \overline{\rho}_{N_2} \\ \overline{\rho}_{O} \\ \overline{\rho}_{O_2} \end{pmatrix} = \underbrace{\begin{pmatrix} -1 & 1 \\ -1 & -1 \\ 1 & 0 \\ 1 & 1 \\ 0 & -1 \end{pmatrix}}_{\boldsymbol{\nu}} \begin{pmatrix} r_1 \\ r_2 \end{pmatrix}. \tag{1.173}$$

Here, $\boldsymbol{\nu}$ has dimension $N \times J = 5 \times 2$. The model is of the form of Eq. (1.29):

$$\frac{d\overline{\boldsymbol{\rho}}}{dt} = \boldsymbol{\nu} \cdot \mathbf{r}. \tag{1.174}$$

Let us see how one of these comes about. The first equation from Eq. (1.173) is

$$\frac{d\overline{\rho}_{NO}}{dt} = -r_1 + r_2. \tag{1.175}$$

This reflects the fact that Reaction 1 induces a loss of 1 mol of NO, while Reaction 2 induces a gain of 1 mol of NO. Our stoichiometric constraint on element conservation for each reaction, Eq. (1.35), $\boldsymbol{\phi} \cdot \boldsymbol{\nu} = \mathbf{0}$ holds here:

$$\boldsymbol{\phi} \cdot \boldsymbol{\nu} = \begin{pmatrix} 1 & 1 & 2 & 0 & 0 \\ 1 & 0 & 0 & 1 & 2 \end{pmatrix} \begin{pmatrix} -1 & 1 \\ -1 & -1 \\ 1 & 0 \\ 1 & 1 \\ 0 & -1 \end{pmatrix} = \begin{pmatrix} 0 & 0 \\ 0 & 0 \end{pmatrix}. \tag{1.176}$$

We get 4 zeros because there are 2 reactions each with 2 element constraints.

**REDUCED FORM.** Here, we describe nontraditional, but useful, reductions by using standard techniques from linear algebra to bring the model equations into a reduced form in which all of the linear constraints have been explicitly removed.

Let us perform a series of row operations to Eq. (1.173) to find all of the linear dependencies. Our aim is to convert the matrix $\boldsymbol{\nu}$ into an upper triangular form. The lower left corner of $\boldsymbol{\nu}$ already has a zero, so there is no need to worry about it. Let us replace the fourth equation with the sum of the first and fourth equations to eliminate the 1 in the $(4, 1)$ slot. This gives

$$\frac{d}{dt}\begin{pmatrix} \overline{\rho}_{NO} \\ \overline{\rho}_{N} \\ \overline{\rho}_{N_2} \\ \overline{\rho}_{NO} + \overline{\rho}_{O} \\ \overline{\rho}_{O_2} \end{pmatrix} = \begin{pmatrix} -1 & 1 \\ -1 & -1 \\ 1 & 0 \\ 0 & 2 \\ 0 & -1 \end{pmatrix} \begin{pmatrix} r_1 \\ r_2 \end{pmatrix}. \tag{1.177}$$

Next, replace the third equation with the sum of the first and third equations to get

$$\frac{d}{dt}\begin{pmatrix}\overline{\rho}_{NO}\\ \overline{\rho}_{N}\\ \overline{\rho}_{NO}+\overline{\rho}_{N_2}\\ \overline{\rho}_{NO}+\overline{\rho}_{O}\\ \overline{\rho}_{O_2}\end{pmatrix}=\begin{pmatrix}-1 & 1\\ -1 & -1\\ 0 & 1\\ 0 & 2\\ 0 & -1\end{pmatrix}\begin{pmatrix}r_1\\ r_2\end{pmatrix}. \tag{1.178}$$

Now, multiply the first equation by −1 and add it to the second to get

$$\frac{d}{dt}\begin{pmatrix}\overline{\rho}_{NO}\\ -\overline{\rho}_{NO}+\overline{\rho}_{N}\\ \overline{\rho}_{NO}+\overline{\rho}_{N_2}\\ \overline{\rho}_{NO}+\overline{\rho}_{O}\\ \overline{\rho}_{O_2}\end{pmatrix}=\begin{pmatrix}-1 & 1\\ 0 & -2\\ 0 & 1\\ 0 & 2\\ 0 & -1\end{pmatrix}\begin{pmatrix}r_1\\ r_2\end{pmatrix}. \tag{1.179}$$

Next multiply the fifth equation by −2 and add it to the second to get

$$\frac{d}{dt}\begin{pmatrix}\overline{\rho}_{NO}\\ -\overline{\rho}_{NO}+\overline{\rho}_{N}\\ \overline{\rho}_{NO}+\overline{\rho}_{N_2}\\ \overline{\rho}_{NO}+\overline{\rho}_{O}\\ -\overline{\rho}_{NO}+\overline{\rho}_{N}-2\overline{\rho}_{O_2}\end{pmatrix}=\begin{pmatrix}-1 & 1\\ 0 & -2\\ 0 & 1\\ 0 & 2\\ 0 & 0\end{pmatrix}\begin{pmatrix}r_1\\ r_2\end{pmatrix}. \tag{1.180}$$

Next add the second and fourth equations to get

$$\frac{d}{dt}\begin{pmatrix}\overline{\rho}_{NO}\\ -\overline{\rho}_{NO}+\overline{\rho}_{N}\\ \overline{\rho}_{NO}+\overline{\rho}_{N_2}\\ \overline{\rho}_{N}+\overline{\rho}_{O}\\ -\overline{\rho}_{NO}+\overline{\rho}_{N}-2\overline{\rho}_{O_2}\end{pmatrix}=\begin{pmatrix}-1 & 1\\ 0 & -2\\ 0 & 1\\ 0 & 0\\ 0 & 0\end{pmatrix}\begin{pmatrix}r_1\\ r_2\end{pmatrix}. \tag{1.181}$$

Next multiply the third equation by 2 and add it to the second to get

$$\frac{d}{dt}\begin{pmatrix}\overline{\rho}_{NO}\\ -\overline{\rho}_{NO}+\overline{\rho}_{N}\\ \overline{\rho}_{NO}+\overline{\rho}_{N}+2\overline{\rho}_{N_2}\\ \overline{\rho}_{N}+\overline{\rho}_{O}\\ -\overline{\rho}_{NO}+\overline{\rho}_{N}-2\overline{\rho}_{O_2}\end{pmatrix}=\begin{pmatrix}-1 & 1\\ 0 & -2\\ 0 & 0\\ 0 & 0\\ 0 & 0\end{pmatrix}\begin{pmatrix}r_1\\ r_2\end{pmatrix}. \tag{1.182}$$

Rewritten, this becomes

$$\underbrace{\begin{pmatrix}1 & 0 & 0 & 0 & 0\\ -1 & 1 & 0 & 0 & 0\\ 1 & 1 & 2 & 0 & 0\\ 0 & 1 & 0 & 1 & 0\\ -1 & 1 & 0 & 0 & -2\end{pmatrix}}_{\mathbf{L}^{-1}}\frac{d}{dt}\begin{pmatrix}\overline{\rho}_{NO}\\ \overline{\rho}_{N}\\ \overline{\rho}_{N_2}\\ \overline{\rho}_{O}\\ \overline{\rho}_{O_2}\end{pmatrix}=\underbrace{\begin{pmatrix}-1 & 1\\ 0 & -2\\ 0 & 0\\ 0 & 0\\ 0 & 0\end{pmatrix}}_{\mathbf{U}}\begin{pmatrix}r_1\\ r_2\end{pmatrix}. \tag{1.183}$$

A way to think of this type of row echelon form is that it defines two free variables, those associated with the nonzero pivots of **U**: $\overline{\rho}_{NO}$ and $\overline{\rho}_{N}$. The remaining three variables, $\overline{\rho}_{N_2}$, $\overline{\rho}_{O}$, and $\overline{\rho}_{O_2}$, are bound variables that can be expressed in terms of the free variables. Our set is not unique; had we formed other linear combinations, we could have arrived at a different set that would be as useful as ours.

The last three of Eqs. (1.183) are homogeneous and can be integrated to form

$$\overline{\rho}_{NO} + \overline{\rho}_N + 2\overline{\rho}_{N_2} = C_1, \tag{1.184}$$

$$\overline{\rho}_N + \overline{\rho}_O = C_2, \tag{1.185}$$

$$-\overline{\rho}_{NO} + \overline{\rho}_N - 2\overline{\rho}_{O_2} = C_3. \tag{1.186}$$

The constants $C_1, C_2$, and $C_3$ are determined from the initial conditions. In matrix form,

$$\begin{pmatrix} 1 & 1 & 2 & 0 & 0 \\ 0 & 1 & 0 & 1 & 0 \\ -1 & 1 & 0 & 0 & -2 \end{pmatrix} \begin{pmatrix} \overline{\rho}_{NO} \\ \overline{\rho}_N \\ \overline{\rho}_{N_2} \\ \overline{\rho}_O \\ \overline{\rho}_{O_2} \end{pmatrix} = \begin{pmatrix} C_1 \\ C_2 \\ C_3 \end{pmatrix}. \tag{1.187}$$

Considering the free variables, $\overline{\rho}_{NO}$ and $\overline{\rho}_N$, to be known, we rearrange to get

$$\begin{pmatrix} 2 & 0 & 0 \\ 0 & 1 & 0 \\ 0 & 0 & -2 \end{pmatrix} \begin{pmatrix} \overline{\rho}_{N_2} \\ \overline{\rho}_O \\ \overline{\rho}_{O_2} \end{pmatrix} = \begin{pmatrix} C_1 - \overline{\rho}_{NO} - \overline{\rho}_N \\ C_2 - \overline{\rho}_N \\ C_3 + \overline{\rho}_{NO} - \overline{\rho}_N \end{pmatrix}. \tag{1.188}$$

Solving for the bound variables, we find

$$\begin{pmatrix} \overline{\rho}_{N_2} \\ \overline{\rho}_O \\ \overline{\rho}_{O_2} \end{pmatrix} = \begin{pmatrix} \frac{1}{2}C_1 - \frac{1}{2}\overline{\rho}_{NO} - \frac{1}{2}\overline{\rho}_N \\ C_2 - \overline{\rho}_N \\ -\frac{1}{2}C_3 - \frac{1}{2}\overline{\rho}_{NO} + \frac{1}{2}\overline{\rho}_N \end{pmatrix}. \tag{1.189}$$

We can rewrite this as

$$\begin{pmatrix} \overline{\rho}_{N_2} \\ \overline{\rho}_O \\ \overline{\rho}_{O_2} \end{pmatrix} = \begin{pmatrix} \frac{1}{2}C_1 \\ C_2 \\ -\frac{1}{2}C_3 \end{pmatrix} + \begin{pmatrix} -\frac{1}{2} & -\frac{1}{2} \\ 0 & -1 \\ -\frac{1}{2} & \frac{1}{2} \end{pmatrix} \begin{pmatrix} \overline{\rho}_{NO} \\ \overline{\rho}_N \end{pmatrix}. \tag{1.190}$$

We can get a more elegant form by defining $\tilde{\xi}_{NO} = \overline{\rho}_{NO}$ and $\tilde{\xi}_N = \overline{\rho}_N$. Thus, we can say our state variables have the form

$$\begin{pmatrix} \overline{\rho}_{NO} \\ \overline{\rho}_N \\ \overline{\rho}_{N_2} \\ \overline{\rho}_O \\ \overline{\rho}_{O_2} \end{pmatrix} = \begin{pmatrix} 0 \\ 0 \\ \frac{1}{2}C_1 \\ C_2 \\ -\frac{1}{2}C_3 \end{pmatrix} + \begin{pmatrix} 1 & 0 \\ 0 & 1 \\ -\frac{1}{2} & -\frac{1}{2} \\ 0 & -1 \\ -\frac{1}{2} & \frac{1}{2} \end{pmatrix} \begin{pmatrix} \tilde{\xi}_{NO} \\ \tilde{\xi}_N \end{pmatrix}. \tag{1.191}$$

By translating via $\tilde{\xi}_{NO} = \xi_{NO} + \widehat{\overline{\rho}}_{NO}$ and $\tilde{\xi}_N = \xi_N + \widehat{\overline{\rho}}_N$ and choosing the constants $C_1, C_2$, and $C_3$ appropriately, we can arrive at

$$\underbrace{\begin{pmatrix} \overline{\rho}_{NO} \\ \overline{\rho}_N \\ \overline{\rho}_{N_2} \\ \overline{\rho}_O \\ \overline{\rho}_{O_2} \end{pmatrix}}_{\overline{\boldsymbol{\rho}}} = \underbrace{\begin{pmatrix} \widehat{\overline{\rho}}_{NO} \\ \widehat{\overline{\rho}}_N \\ \widehat{\overline{\rho}}_{N_2} \\ \widehat{\overline{\rho}}_O \\ \widehat{\overline{\rho}}_{O_2} \end{pmatrix}}_{\widehat{\overline{\boldsymbol{\rho}}}} + \underbrace{\begin{pmatrix} 1 & 0 \\ 0 & 1 \\ -\frac{1}{2} & -\frac{1}{2} \\ 0 & -1 \\ -\frac{1}{2} & \frac{1}{2} \end{pmatrix}}_{\mathbf{D}} \underbrace{\begin{pmatrix} \xi_{NO} \\ \xi_N \end{pmatrix}}_{\boldsymbol{\xi}}. \tag{1.192}$$

This takes the form of Eq. (1.55):

$$\overline{\boldsymbol{\rho}} = \widehat{\overline{\boldsymbol{\rho}}} + \mathbf{D} \cdot \boldsymbol{\xi}. \tag{1.193}$$

Here, the matrix $\mathbf{D}$ is of dimension $N \times R = 5 \times 2$. It spans the same column space as does the $N \times J$ matrix $\boldsymbol{\nu}$ that is of rank $R$. In terms of linear algebra, we can consider Eq. (1.193) to be a linear inhomogeneous injection mapping taking $R$-dimensional vectors $\boldsymbol{\xi}$ into an $N$-dimensional space. Here, in fact, $R = J = 2$, so $\mathbf{D}$ has the same dimension as $\boldsymbol{\nu}$. In general it will not. If $\mathbf{c}_1$ and $\mathbf{c}_2$ are the column vectors of $\mathbf{D}$, we see that $-\mathbf{c}_1 - \mathbf{c}_2$ forms the first column vector of $\boldsymbol{\nu}$ and $\mathbf{c}_1 - \mathbf{c}_2$ forms the second column vector of $\boldsymbol{\nu}$. Note that $\boldsymbol{\phi} \cdot \mathbf{D} = \mathbf{0}$:

$$\boldsymbol{\phi} \cdot \mathbf{D} = \begin{pmatrix} 1 & 1 & 2 & 0 & 0 \\ 1 & 0 & 0 & 1 & 2 \end{pmatrix} \begin{pmatrix} 1 & 0 \\ 0 & 1 \\ -\frac{1}{2} & -\frac{1}{2} \\ 0 & -1 \\ -\frac{1}{2} & \frac{1}{2} \end{pmatrix} = \begin{pmatrix} 0 & 0 \\ 0 & 0 \end{pmatrix}. \tag{1.194}$$

Equations (1.184)–(1.186) can also be linearly combined in a way that has more obvious physical relevance. We rewrite the system as three equations in which the first is identical to Eq. (1.184); the second is the difference of Eqs. (1.185) and (1.186); and the third is half of Eq. (1.184) minus half of Eq. (1.186) plus Eq. (1.185):

$$\overline{\rho}_{\mathrm{NO}} + \overline{\rho}_{\mathrm{N}} + 2\overline{\rho}_{\mathrm{N_2}} = C_1, \tag{1.195}$$

$$\overline{\rho}_{\mathrm{O}} + \overline{\rho}_{\mathrm{NO}} + 2\overline{\rho}_{\mathrm{O_2}} = C_2 - C_3, \tag{1.196}$$

$$\overline{\rho}_{\mathrm{NO}} + \overline{\rho}_{\mathrm{N}} + \overline{\rho}_{\mathrm{N_2}} + \overline{\rho}_{\mathrm{O}} + \overline{\rho}_{\mathrm{O_2}} = \frac{1}{2}(C_1 - C_3) + C_2. \tag{1.197}$$

For a constant volume problem, Eq. (1.195) insists that the number of nitrogen elements be constant; Eq. (1.196) demands the number of oxygen elements be constant; and Eq. (1.197) requires the number of moles of molecular species be constant. For general reactions, including the earlier studied oxygen dissociation problem, the number of moles will not be constant. Here, because each reaction considered has two molecules reacting to form two molecules, we are guaranteed the number of moles will be constant. Hence, we get an additional linear constraint beyond the two for element conservation. Because our reaction of ideal gases is isothermal, isochoric and mol-preserving, it will also have constant pressure by the ideal gas law and thus be *isobaric*.

**EXAMPLE CALCULATION.** Let us consider an isothermal reaction at

$$T = 6000 \text{ K}. \tag{1.198}$$

The high temperature is useful in generating easily visualized results. It insures that there will be significant concentrations of all species. Take as an initial condition

$$\widehat{\overline{\rho}}_{\mathrm{NO}} = \widehat{\overline{\rho}}_{\mathrm{N}} = \widehat{\overline{\rho}}_{\mathrm{N_2}} = \widehat{\overline{\rho}}_{\mathrm{O}} = \widehat{\overline{\rho}}_{\mathrm{O_2}} = 1 \times 10^{-6} \text{ mol/cm}^3. \tag{1.199}$$

For this temperature and these concentrations, the pressure, which will remain constant through the reaction, is $P = 2.4942 \times 10^6$ dyne/cm$^2$. This is a little greater than atmospheric.

Kinetic data for this reaction are adopted from Baulch et al. (2005). The data for Reaction 1 are

$$a_1 = 3.5 \times 10^{-11} \left(\frac{\text{molecule}}{\text{cm}^3}\right)^{-1} \frac{1}{\text{s}} \quad \beta_1 = 0, \quad T_{a,1} = 0 \text{ K}. \tag{1.200}$$

We employ *Avogadro's*[11] *number*,

$$\mathcal{N}_A = 6.02 \times 10^{23} \text{ molecule/mol}, \tag{1.201}$$

to convert $a_1$ to molar units:

$$a_1 = \left(3.5 \times 10^{-11} \left(\frac{\text{molecule}}{\text{cm}^3}\right)^{-1} \frac{1}{\text{s}}\right) \left(6.02 \times 10^{23} \frac{\text{molecule}}{\text{mol}}\right), \tag{1.202}$$

$$= 2.107 \times 10^{13} \left(\frac{\text{mol}}{\text{cm}^3}\right)^{-1} \frac{1}{\text{s}}. \tag{1.203}$$

For Reaction 2, we have

$$a_2 = 9.7 \times 10^{-15} \left(\frac{\text{molecule}}{\text{cm}^3}\right)^{-1} \frac{1}{\text{K}^{1.01}\,\text{s}}, \quad \beta_2 = 1.01, \quad T_{a,2} = 3120 \text{ K}. \tag{1.204}$$

We convert $a_2$ to molar units:

$$a_2 = \left(9.7 \times 10^{-15} \left(\frac{\text{molecule}}{\text{cm}^3}\right)^{-1} \frac{1}{\text{s}}\right) \left(6.02 \times 10^{23} \frac{\text{molecule}}{\text{mol}}\right), \tag{1.205}$$

$$= 5.8394 \times 10^{9} \left(\frac{\text{mol}}{\text{cm}^3}\right)^{-1} \frac{1}{\text{s}}, \tag{1.206}$$

Here, the so-called *activation temperature* $T_{a,j}$ for reaction $j$ is really the activation energy scaled by the universal gas constant:

$$T_{a,j} = \overline{\mathcal{E}}_j / \overline{R}. \tag{1.207}$$

These coefficients are written for the forward reactions. Because the forward portion of Reaction 1 is a recombination of two free radicals to form $N_2$, it has no activation energy barrier to overcome, and its activation temperature is 0 K. However, the forward reaction of Reaction 2 dissociates $O_2$. This requires that a significant activation energy barrier be overcome; this is reflected in the activation temperature of 3120 K.

Substituting numbers into Eqs. (1.172), we obtain for the reaction rate coefficients

$$k_1 = (2.107 \times 10^{13})(6000)^{0} \exp\left(\frac{-0}{6000}\right), \tag{1.208}$$

$$= 2.107 \times 10^{13} \left(\frac{\text{mol}}{\text{cm}^3}\right)^{-1} \frac{1}{\text{s}}, \tag{1.209}$$

$$k_2 = (5.8394 \times 10^{9})(6000)^{1.01} \exp\left(\frac{-3120}{6000}\right), \tag{1.210}$$

$$= 2.27231 \times 10^{13} \left(\frac{\text{mol}}{\text{cm}^3}\right)^{-1} \frac{1}{\text{s}}. \tag{1.211}$$

We will also need thermodynamic data. The data here are taken from the `Chemkin` database (Kee et al., 2000). Thermodynamic data for common materials

[11] Lorenzo Romano Amedeo Carlo Bernadette Avogadro di Quaregna e di Cerreto (1776–1856), Italian scientist.

are also found in most thermodynamic texts. For our system at 6000 K, we find

$$\overline{g}^0_{NO} = -1.58757 \times 10^{13} \text{ erg/mol}, \tag{1.212}$$

$$\overline{g}^0_{N} = -7.04286 \times 10^{12} \text{ erg/mol}, \tag{1.213}$$

$$\overline{g}^0_{N_2} = -1.55206 \times 10^{13} \text{ erg/mol}, \tag{1.214}$$

$$\overline{g}^0_{O} = -9.77148 \times 10^{12} \text{ erg/mol}, \tag{1.215}$$

$$\overline{g}^0_{O_2} = -1.65653 \times 10^{13} \text{ erg/mol}. \tag{1.216}$$

We use Eq. (1.135) to find $\Delta G^0_j$ for each reaction:

$$\Delta\mathbf{G}^{0^T} = \overline{\mathbf{g}}^{0^T} \cdot \boldsymbol{\nu}, \tag{1.217}$$

$$\left( \Delta G^0_1 \quad \Delta G^0_2 \right) = \left( \overline{g}^0_{NO} \quad \overline{g}^0_{N} \quad \overline{g}^0_{N_2} \quad \overline{g}^0_{O} \quad \overline{g}^0_{O_2} \right) \begin{pmatrix} -1 & 1 \\ -1 & -1 \\ 1 & 0 \\ 1 & 1 \\ 0 & -1 \end{pmatrix}. \tag{1.218}$$

Thus, for each reaction, we find $\Delta G^0_j$:

$$\Delta G^0_1 = \overline{g}^0_{N_2} + \overline{g}^0_{O} - \overline{g}^0_{N} - \overline{g}^0_{NO}, \tag{1.219}$$

$$= -1.55206 \times 10^{13} - 9.77148 \times 10^{12}$$
$$+ 7.04286 \times 10^{12} + 1.58757 \times 10^{13}, \tag{1.220}$$

$$= -2.37351 \times 10^{12} \text{ erg/mol}, \tag{1.221}$$

$$\Delta G^0_2 = \overline{g}^0_{NO} + \overline{g}^0_{O} - \overline{g}^0_{N} - \overline{g}^0_{O_2}, \tag{1.222}$$

$$= -1.58757 \times 10^{13} - 9.77148 \times 10^{12}$$
$$+ 7.04286 \times 10^{12} + 1.65653 \times 10^{13}, \tag{1.223}$$

$$= -2.03897 \times 10^{12} \text{ erg/mol}. \tag{1.224}$$

At 6000 K, we find the equilibrium constants for the $J = 2$ reactions are

$$K_{c,1} = \left( \frac{P_0}{\overline{R}T} \right)^{\sum_{i=1}^{N} \nu_{i1}} \exp\left( \frac{-\Delta G^0_1}{\overline{R}T} \right), \tag{1.225}$$

$$= \left( \frac{P_0}{\overline{R}T} \right)^{-1-1+1+1+0} \exp\left( \frac{-\Delta G^0_1}{\overline{R}T} \right), \tag{1.226}$$

$$= \exp\left( \frac{-\Delta G^0_1}{\overline{R}T} \right), \tag{1.227}$$

$$= \exp\left( \frac{2.37351 \times 10^{12}}{(8.314 \times 10^7)(6000)} \right), \tag{1.228}$$

$$= 116.52, \tag{1.229}$$

$$K_{c,2} = \left( \frac{P_0}{\overline{R}T} \right)^{\sum_{i=1}^{N} \nu_{i2}} \exp\left( \frac{-\Delta G^0_2}{\overline{R}T} \right), \tag{1.230}$$

$$= \left(\frac{P_0}{\overline{R}T}\right)^{1-1+0+1-1} \exp\left(\frac{-\Delta G_2^0}{\overline{R}T}\right), \quad (1.231)$$

$$= \exp\left(\frac{-\Delta G_2^0}{\overline{R}T}\right), \quad (1.232)$$

$$= \exp\left(\frac{2.03897 \times 10^{12}}{(8.314 \times 10^7)(6000)}\right), \quad (1.233)$$

$$= 59.5861. \quad (1.234)$$

Again, omitting details, we find the two differential equations governing the evolution of the free variables are

$$\frac{d\overline{\rho}_{NO}}{dt} = 0.72331 + \left(2.21806 \times 10^7\right)\overline{\rho}_N + \left(1.14520 \times 10^{13}\right)\overline{\rho}_N^2 - \left(9.4353 \times 10^5\right)\overline{\rho}_{NO} - \left(3.19598 \times 10^{13}\right)\overline{\rho}_N\overline{\rho}_{NO}, \quad (1.235)$$

$$\frac{d\overline{\rho}_N}{dt} = 0.72331 - \left(2.32656 \times 10^7\right)\overline{\rho}_N - \left(1.12711 \times 10^{13}\right)\overline{\rho}_N^2 + \left(5.8187 \times 10^5\right)\overline{\rho}_{NO} - \left(9.9994 \times 10^{12}\right)\overline{\rho}_N\overline{\rho}_{NO}, \quad (1.236)$$

$$\overline{\rho}_{NO}(0) = \overline{\rho}_N(0) = 1 \times 10^{-6} \text{ mol/cm}^3. \quad (1.237)$$

Solving numerically, we obtain the solution shown in Fig. 1.10. The numerics show a relaxation to final concentrations of

$$\lim_{t\to\infty} \overline{\rho}_{NO} = 7.336 \times 10^{-7} \frac{\text{mol}}{\text{cm}^3}, \quad \lim_{t\to\infty} \overline{\rho}_N = 3.708 \times 10^{-8} \frac{\text{mol}}{\text{cm}^3}. \quad (1.238)$$

Equations (1.235)–(1.236) are of the form

$$\frac{d\overline{\rho}_{NO}}{dt} = f_{NO}(\overline{\rho}_{NO}, \overline{\rho}_N), \quad (1.239)$$

$$\frac{d\overline{\rho}_N}{dt} = f_N(\overline{\rho}_{NO}, \overline{\rho}_N). \quad (1.240)$$

At equilibrium, we must have

$$f_{NO}(\overline{\rho}_{NO}, \overline{\rho}_N) = 0, \qquad f_N(\overline{\rho}_{NO}, \overline{\rho}_N) = 0. \quad (1.241)$$

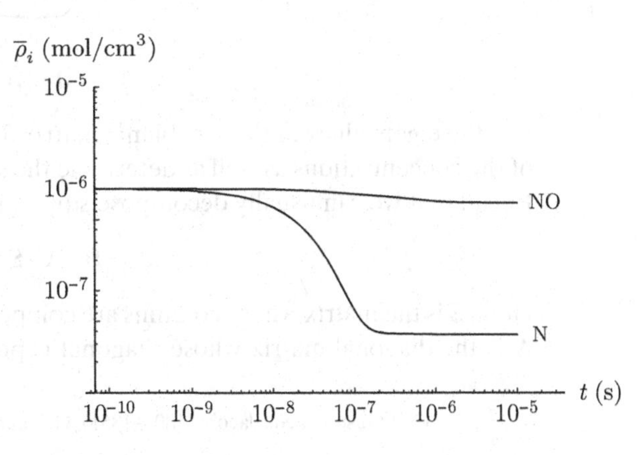


Figure 1.10. NO and N concentrations versus time for $T = 6000$ K, $P = 2.4942 \times 10^6$ dyne/cm² Zel'dovich mechanism.

We find three finite roots to this algebraic problem:

$$(\overline{\rho}_{NO}, \overline{\rho}_{N})_1 = (-1.605 \times 10^{-6}, -3.060 \times 10^{-8}) \text{ mol/cm}^3, \tag{1.242}$$

$$(\overline{\rho}_{NO}, \overline{\rho}_{N})_2 = (-5.173 \times 10^{-8}, -2.048 \times 10^{-6}) \text{ mol/cm}^3, \tag{1.243}$$

$$(\overline{\rho}_{NO}, \overline{\rho}_{N})_3 = (7.336 \times 10^{-7}, 3.708 \times 10^{-8}) \text{ mol/cm}^3. \tag{1.244}$$

Obviously, because of negative concentrations, roots 1 and 2 are nonphysical. Root 3, however, is physical; moreover, it agrees with the equilibrium we found by direct numerical integration of the full nonlinear equations.

We can use local linear analysis in the neighborhood of each equilibrium to rigorously ascertain the stability of each root. Taylor series expansion of Eqs. (1.239) and (1.240) in the neighborhood of an equilibrium point yields

$$\frac{d}{dt}(\overline{\rho}_{NO} - \overline{\rho}_{NO}^{eq}) = \underbrace{f_{NO}|_{eq}}_{0} + \left.\frac{\partial f_{NO}}{\partial \overline{\rho}_{NO}}\right|_{eq}(\overline{\rho}_{NO} - \overline{\rho}_{NO}^{eq}) + \left.\frac{\partial f_{NO}}{\partial \overline{\rho}_{N}}\right|_{eq}(\overline{\rho}_{N} - \overline{\rho}_{N}^{eq}) + \cdots, \tag{1.245}$$

$$\frac{d}{dt}(\overline{\rho}_{N} - \overline{\rho}_{N}^{eq}) = \underbrace{f_{N}|_{eq}}_{0} + \left.\frac{\partial f_{N}}{\partial \overline{\rho}_{NO}}\right|_{eq}(\overline{\rho}_{NO} - \overline{\rho}_{NO}^{eq}) + \left.\frac{\partial f_{N}}{\partial \overline{\rho}_{N}}\right|_{eq}(\overline{\rho}_{N} - \overline{\rho}_{N}^{eq}) + \cdots. \tag{1.246}$$

Evaluation of Eqs. (1.245) and (1.246) near the physical root, root 3, yields the system

$$\frac{d}{dt}\begin{pmatrix} \overline{\rho}_{NO} - 7.336 \times 10^{-7} \\ \overline{\rho}_{N} - 3.708 \times 10^{-8} \end{pmatrix} = \underbrace{\begin{pmatrix} -2.129 \times 10^{6} & -4.155 \times 10^{5} \\ 2.111 \times 10^{5} & -3.144 \times 10^{7} \end{pmatrix}}_{\mathbf{J} = \left.\frac{\partial \mathbf{f}}{\partial \boldsymbol{\rho}}\right|_{eq}} \cdot \begin{pmatrix} \overline{\rho}_{NO} - 7.336 \times 10^{-7} \\ \overline{\rho}_{N} - 3.708 \times 10^{-8} \end{pmatrix}. \tag{1.247}$$

This is of the form

$$\frac{d}{dt}(\overline{\boldsymbol{\rho}} - \overline{\boldsymbol{\rho}}^{eq}) = \underbrace{\left.\frac{\partial \mathbf{f}}{\partial \overline{\boldsymbol{\rho}}}\right|_{eq}}_{\mathbf{J}} \cdot (\overline{\boldsymbol{\rho}} - \overline{\boldsymbol{\rho}}^{eq}), \tag{1.248}$$

$$= \mathbf{J} \cdot (\overline{\boldsymbol{\rho}} - \overline{\boldsymbol{\rho}}^{eq}). \tag{1.249}$$

It is the eigenvalues of the Jacobian[12] matrix $\mathbf{J}$ that give the time scales of evolution of the concentrations as well as determine the stability of the local equilibrium point. Recall that we can usually decompose square matrices via the diagonalization:

$$\mathbf{J} = \mathbf{S} \cdot \boldsymbol{\Lambda} \cdot \mathbf{S}^{-1}. \tag{1.250}$$

Here, $\mathbf{S}$ is the matrix whose columns are composed of the right eigenvectors of $\mathbf{J}$, and $\boldsymbol{\Lambda}$ is the diagonal matrix whose diagonal is populated by the eigenvalues of $\mathbf{J}$. For

[12] After Carl Gustav Jacob Jacobi (1804–1851), German mathematician.

some matrices (typically not those encountered after our removal of linear dependencies), diagonalization is not possible, and one must resort to the so-called near-diagonal Jordan form. This will not be relevant to our discussion but could be easily handled if necessary. We also recall the eigenvector matrix and eigenvalue matrix are defined by the standard eigenvalue problem

$$\mathbf{J} \cdot \mathbf{S} = \mathbf{S} \cdot \mathbf{\Lambda}. \tag{1.251}$$

We also recall that the components $\lambda$ of $\mathbf{\Lambda}$ are found by solving the characteristic polynomial that arises from the equation

$$\det\left(\mathbf{J} - \lambda \mathbf{I}\right) = 0, \tag{1.252}$$

where $\mathbf{I}$ is the identity matrix. Defining $\mathbf{z}$ such that

$$\mathbf{S} \cdot \mathbf{z} \equiv \overline{\rho} - \overline{\rho}^{eq}, \tag{1.253}$$

and using the decomposition Eq. (1.250), Eq. (1.249) can be rewritten to form

$$\frac{d}{dt}(\mathbf{S} \cdot \mathbf{z}) = \underbrace{\mathbf{S} \cdot \mathbf{\Lambda} \cdot \mathbf{S}^{-1}}_{\mathbf{J}} \cdot \underbrace{(\mathbf{S} \cdot \mathbf{z})}_{\overline{\rho}-\overline{\rho}^{eq}}, \tag{1.254}$$

$$\mathbf{S} \cdot \frac{d\mathbf{z}}{dt} = \mathbf{S} \cdot \mathbf{\Lambda} \cdot \mathbf{z}, \tag{1.255}$$

$$\mathbf{S}^{-1} \cdot \mathbf{S} \cdot \frac{d\mathbf{z}}{dt} = \mathbf{S}^{-1} \cdot \mathbf{S} \cdot \mathbf{\Lambda} \cdot \mathbf{z}, \tag{1.256}$$

$$\frac{d\mathbf{z}}{dt} = \mathbf{\Lambda} \cdot \mathbf{z}. \tag{1.257}$$

Equation (1.257) is in diagonal form. This has solution for each component of $\mathbf{z}$ of

$$z_1 = C_1 \exp(\lambda_1 t), \tag{1.258}$$

$$z_2 = C_2 \exp(\lambda_2 t), \tag{1.259}$$

$$\vdots$$

Here, our matrix $\mathbf{J}$, see Eq. (1.247), has two real, negative eigenvalues in the neighborhood of the physical root 3:

$$\lambda_1 = -3.143 \times 10^7 \ 1/\text{s}, \quad \lambda_2 = -2.132 \times 10^6 \ 1/\text{s}. \tag{1.260}$$

Thus, we can conclude that the physical equilibrium is linearly stable. In Sec. 5.5.2, we will prove this result for more general chemical kinetic systems.

The local time constants near equilibrium are given by the reciprocal of the magnitude of the eigenvalues. These are

$$\tau_1 = 1/|\lambda_1| = 3.181 \times 10^{-8} \ \text{s}, \quad \tau_2 = 1/|\lambda_2| = 4.691 \times 10^{-7} \ \text{s}. \tag{1.261}$$

Evolution on these two time scales is predicted in Fig. 1.10. This in fact a multiscale problem. One of the major difficulties in the numerical simulation of combustion problems comes in the effort to capture the effects at all relevant scales. The problem is made more difficult as the breadth of the scales expands. In this problem, the breadth of scales is not particularly challenging. Near equilibrium the ratio of the slowest to the fastest time scale, the *stiffness* ratio $\kappa$, is

$$\kappa = \frac{\tau_2}{\tau_1} = \frac{4.691 \times 10^{-7}\ \text{s}}{3.181 \times 10^{-8}\ \text{s}} = 14.75. \tag{1.262}$$

Many combustion problems can have stiffness ratios over $10^6$. This is more prevalent at lower temperatures. In the neighborhood of root 3, the local solution takes the form

$$\begin{pmatrix} \overline{\rho}_{\text{NO}} - 7.336 \times 10^{-7} \\ \overline{\rho}_{\text{N}} - 3.708 \times 10^{-8} \end{pmatrix} = \underbrace{C_1 \mathbf{s}_1 e^{-3.143\times 10^7 t}}_{\text{mode 1}} + \underbrace{C_2 \mathbf{s}_2 e^{-2.132\times 10^6 t}}_{\text{mode 2}}. \tag{1.263}$$

Here, $\mathbf{s}_1$ and $\mathbf{s}_2$ are the eigenvectors of $\mathbf{J}$, which together form the matrix $\mathbf{S}$. Also, $C_1$ and $C_2$ are arbitrary constants. By inspection, as $t \to \infty$, the terms labeled modes 1 and 2 relax, but mode 1 relaxes faster. So at long time the dynamics are dominated by mode 2. Substituting for the numerical values of $\mathbf{s}_2$, we find at long time

$$\lim_{t\to\infty} \begin{pmatrix} \overline{\rho}_{\text{NO}} - 7.336 \times 10^{-7} \\ \overline{\rho}_{\text{N}} - 3.708 \times 10^{-8} \end{pmatrix} = C_2 \begin{pmatrix} 0.999974 \\ 0.00720323 \end{pmatrix} e^{-2.132\times 10^6 t}. \tag{1.264}$$

We give a sketch of this behavior, not to scale, in Fig. 1.11. Here, we plot the stable equilibrium point, the two eigendirections, along with several trajectories that have initial conditions throughout the domain. Those trajectories rapidly approach the line defined by the *slow eigenvector* associated with mode 2. Once they near that line, they slowly proceed to the physical equilibrium, only reached at infinite time. In that trajectories appear to be attracted to the line whose tangent is the slow eigenvector and moreover in that the linear system spends more time in the neighborhood of this line than away from it, identification of this line may have utility in model reduction. In short, at least for the linear system, we have identified a space of lower dimension to which the dynamics are largely confined. We will exploit this in Chapter 6.

We can do a similar linearization near the initial state, find the local eigenvalues, and find the local time scales. At the initial state here, we find those local time scales

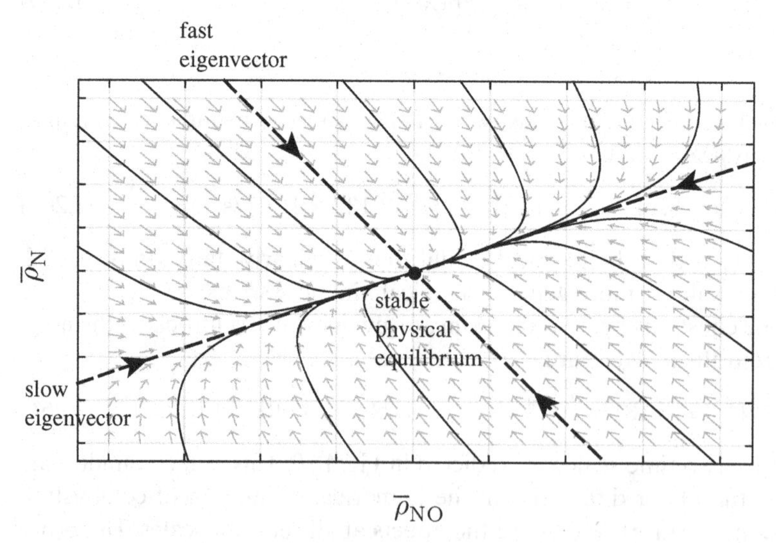


Figure 1.11. Sketch of NO-N phase portraits for the Zel'dovich mechanism near the physical equilibrium.

are

$$\tau_1 = 2.403 \times 10^{-8} \text{ s}, \quad \tau_2 = 2.123 \times 10^{-8} \text{ s}. \tag{1.265}$$

So, initially the stiffness, $\kappa = (2.403 \times 10^{-8} \text{ s})/(2.123 \times 10^{-8} \text{ s}) = 1.13$, is much less, but the time scale itself is small. It is seen from Fig. 1.10 that this initial time scale of $10^{-8}$ s well predicts where significant evolution of species concentrations commences. For $t < 10^{-8}$ s, the model predicts essentially no activity. This can be correlated with the mean time between molecular collisions using more fundamental molecular dynamics theories. It is in fact such a theory on which estimates of the collision frequency coefficients $a_j$ are obtained, as discussed by Vincenti and Kruger (1965).

We briefly consider the nonphysical roots, 1 and 2. A similar eigenvalue analysis of root 1 reveals that the eigenvalues of its local Jacobian matrix are

$$\lambda_1 = -1.193 \times 10^7 \text{ 1/s}, \quad \lambda_2 = 5.434 \times 10^6 \text{ 1/s}. \tag{1.266}$$

Thus, root 1 is a saddle and is unstable. For root 2, we find

$$\lambda_1 = 4.397 \times 10^7 + i7.997 \times 10^6 \frac{1}{\text{s}}, \quad \lambda_2 = 4.397 \times 10^7 - i7.997 \times 10^6 \frac{1}{\text{s}}. \tag{1.267}$$

The eigenvalues are complex with a positive real part. This indicates that the root is an unstable spiral source.

A detailed phase portrait for the full nonlinear system is shown in Fig. 1.12. Here, we see all three finite roots. Their local character of sink, saddle, or spiral source is clearly displayed. Similar to the linear system of Fig. 1.11, we see that trajectories are attracted to a curve labeled SACIM for "Slow Attractive Canonical Invariant Manifold." A part of the SACIM is constructed by the trajectory that originates at root 1 and travels to root 3. The other part is constructed by connecting an equilibrium

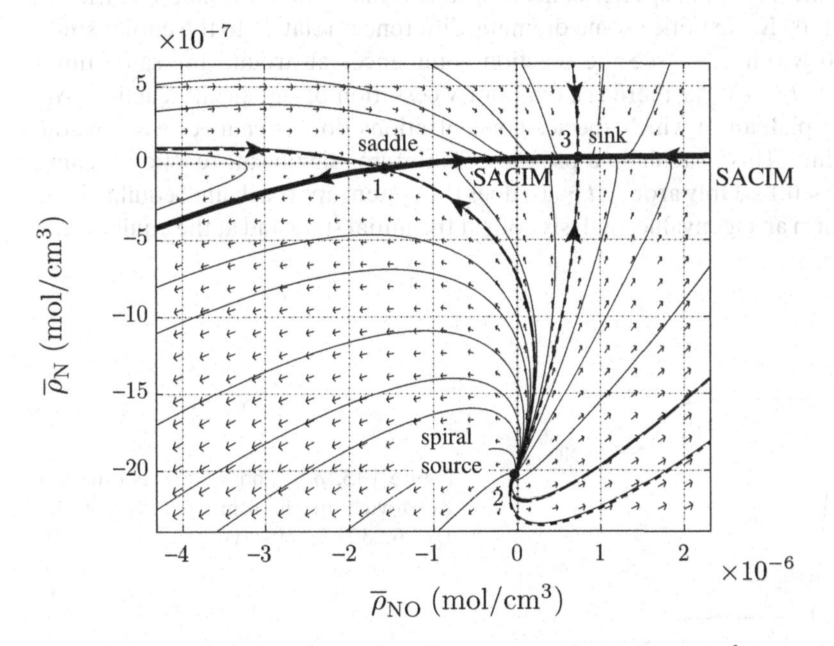


Figure 1.12. NO-N phase portraits for $T = 6000$ K, $P = 2.4942 \times 10^6$ dyne/cm$^2$ Zel'dovich mechanism.

point at infinity into root 3. Details of the full nonlinear analysis are omitted here and are given in Chapter 6. Relevant discussion with related examples is given by Powers et al. (2015). Informally, we call the manifold "slow" because it connects to root 3 along the slow eigenvector at root 3. We call it "attractive" because nearby trajectories appear to be attracted to it. We call it "canonical" because it is constructed by connecting equilibria via trajectories. Such connections are known as *heteroclinic* in the dynamical systems literature. It is invariant because once on such a manifold, the system is such that it can never depart. All trajectories defined by the dynamical system are invariant manifolds. As will be further explained in Chapter 6, some of them are slow, attractive, and canonical. If a SACIM can be identified, it can be useful in both mitigating the effects of stiffness and simplifying the system by studying it in a lower-dimensional space.

### Stiffness, Time Scales, and Numerics

One of the key challenges in computational chemistry is accurately predicting species concentration evolution with time. The problem is made difficult because of the common presence of physical phenomena that evolve on a widely disparate set of time scales. Systems that evolve on a wide range of scales are known as stiff, recognizing a motivating example in mass-spring-damper systems with stiff and nonstiff springs. Here, we will examine the effect of temperature and pressure on time scales and stiffness. We shall also look simplistically at how different numerical approximation methods respond to stiffness.

**EFFECT OF TEMPERATURE.** Let us see how the same Zel'dovich mechanism behaves at lower temperature, $T = 1500$ K; all other parameters, including the initial species concentrations, are the same as the previous high-temperature example. The pressure, however, lowers, and here is $P = 6.23550 \times 10^5$ dyne/cm$^2$. That is close to atmospheric pressure. A plot of species concentrations versus time is given in Fig. 1.13.

At $T = 1500$ K, we notice some dramatic differences relative to the earlier studied $T = 6000$ K. First, we see the reaction commences at around the same time, $t \sim 10^{-8}$ s. For $t \sim 10^{-6}$ s, there is a temporary cessation of significant reaction. We notice a long plateau in which species concentrations do not change over several decades of time. This is actually a pseudo-equilibrium. Significant reaction recommences for $t \sim 0.1$ s. Only around $t \sim 1$ s does the system approach final equilibrium. We can perform an eigenvalue analysis both at the initial state and at the equilibrium

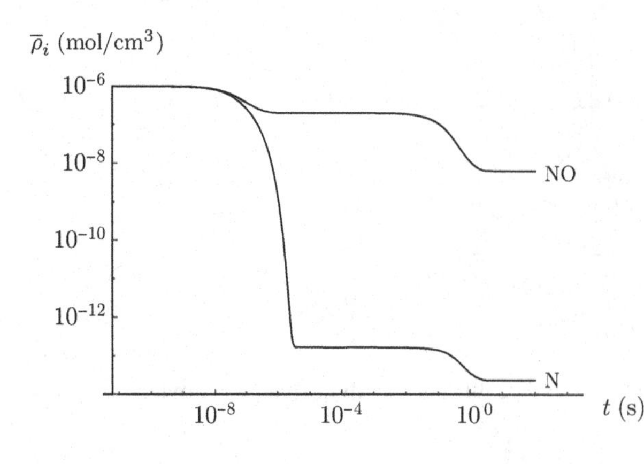


Figure 1.13. $\overline{\rho}_{NO}$ and $\overline{\rho}_{N}$ versus time for Zel'dovich mechanism at $T = 1500$ K, $P = 6.23550 \times 10^5$ dyne/cm$^2$.

state to estimate the time scales of reaction. For this dynamical system that is two ordinary differential equations in two unknowns, we will always find two eigenvalues, and thus two time scales. Let us call them $\tau_1$ and $\tau_2$. Both these scales will evolve with $t$.

At the initial state, we find

$$\tau_1 = 2.37 \times 10^{-8} \text{ s}, \qquad \tau_2 = 4.25 \times 10^{-7} \text{ s}. \tag{1.268}$$

The onset of significant reaction is consistent with the prediction given by $\tau_1$ at the initial state. Moreover, initially, the reaction is not very stiff; the stiffness ratio is $\kappa = 17.9$. At equilibrium, we find

$$\lim_{t\to\infty} \overline{\rho}_{NO} = 6.3 \times 10^{-9} \text{ mol/cm}^3, \qquad \lim_{t\to\infty} \overline{\rho}_{N} = 2.3 \times 10^{-14} \text{ mol/cm}^3, \tag{1.269}$$

and

$$\tau_1 = 7.65 \times 10^{-7} \text{ s}, \qquad \tau_2 = 5.67 \times 10^{-1} \text{ s}. \tag{1.270}$$

The slowest time scale near equilibrium is an excellent indicator of how long the system takes to relax to its final state. Also, near equilibrium, the stiffness ratio is large, $\kappa = \tau_2/\tau_1 \sim 7.4 \times 10^5$. Because $\kappa$ is large, the scales in this problem are widely disparate, and accurate numerical solution becomes challenging.

In summary, we find the effect of lowering temperature while leaving initial concentrations constant:

- Lowers the pressure somewhat, slightly slowing the collision time, and slightly slowing the fastest time scales.
- Slows the slowest time scales many orders of magnitude, stiffening the system, because collisions may not induce reaction with their lower collision speed.

**EFFECT OF INITIAL PRESSURE.** Let us maintain the initial temperature at $T = 1500$ K, but drop the initial concentration of each species to

$$\widehat{\overline{\rho}}_{NO} = \widehat{\overline{\rho}}_{N} = \widehat{\overline{\rho}}_{N_2} = \widehat{\overline{\rho}}_{O_2} = \widehat{\overline{\rho}}_{O} = 10^{-8} \text{ mol/cm}^3. \tag{1.271}$$

With this decrease in number of moles, the pressure now is

$$P = 6.23550 \times 10^3 \text{ dyne/cm}^2. \tag{1.272}$$

This pressure is two orders of magnitude lower than atmospheric. We solve for the species concentration profiles and show the results of numerical prediction in Fig. 1.14. Relative to the high pressure $P = 6.2355 \times 10^5$ dyne/cm$^2$, $T = 1500$ K case, we notice some similarities and dramatic differences. The overall shape of the time profiles of concentration variation is similar. But, we see the reaction commences at a much later time, $t \sim 10^{-6}$ s. For $t \sim 10^{-4}$ s, there is a temporary cessation of significant reaction. We notice a long plateau in which species concentrations do not change over several decades of time. This is again actually a pseudo-equilibrium. Significant reaction recommences for $t \sim 10$ s. Only around $t \sim 100$ s does the system approach final equilibrium. We can perform an eigenvalue analysis both at the initial state and at the equilibrium state to estimate the time scales of reaction.

At the initial state, we find

$$\tau_1 = 2.37 \times 10^{-6} \text{ s}, \qquad \tau_2 = 4.25 \times 10^{-5} \text{ s}. \tag{1.273}$$

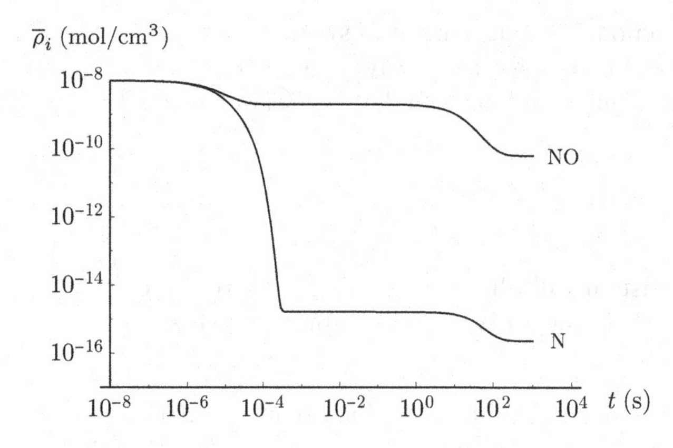


Figure 1.14. $\overline{\rho}_{\text{NO}}$ and $\overline{\rho}_{\text{N}}$ versus time for the Zel'dovich mechanism at $T = 1500$ K, $P = 6.2355 \times 10^3$ dyne/cm$^2$.

The onset of significant reaction is consistent with the prediction given by $\tau_1$ at the initial state. Moreover, initially, the reaction is not very stiff; the stiffness ratio is $\kappa = 17.9$. Interestingly, by decreasing the initial pressure by a factor of $10^2$, we increased the initial time scales by a complementary factor of $10^2$; moreover, we did not alter the stiffness. At equilibrium, we find

$$\lim_{t\to\infty} \overline{\rho}_{\text{NO}} = 6.3 \times 10^{-11} \text{ mol/cm}^3, \quad \lim_{t\to\infty} \overline{\rho}_{\text{N}} = 2.3 \times 10^{-16} \text{ mol/cm}^3, \tag{1.274}$$

and

$$\tau_1 = 7.65 \times 10^{-5} \text{ s}, \qquad \tau_2 = 5.67 \times 10^{1} \text{ s}. \tag{1.275}$$

By decreasing the initial pressure by a factor of $10^2$, we decreased the equilibrium concentrations by a factor of $10^2$ and increased the time scales by a factor of $10^2$, leaving the stiffness ratio unchanged.

In summary, we find the effect of lowering the initial concentrations significantly while leaving temperature constant

- Lowers the pressure significantly, proportionally slowing the collision time, as well as the fastest and slowest time scales.
- Does not affect the stiffness of the system.

**NUMERICAL IMPLICATIONS.** The issue of how to simulate stiff systems of ordinary differential equations, such as presented by our Zel'dovich mechanism, is challenging. Here, a brief summary of some of the issues will be presented. The interested reader should consult the numerical literature for a full discussion (Iserles, 2008; Kee et al., 2003, Chapter 15; Oran and Boris, 2001).

We have seen throughout this section that there are two time scales at work, and they are often disparate. The species evolution is generally characterized by an initial fast transient, followed by a long plateau, then a final relaxation to equilibrium. We noted from the phase plane of Fig. 1.12 that the final relaxation to equilibrium (shown along the curve labeled "SACIM") is an attractive manifold for a wide variety of initial conditions. The relaxation onto the SACIM is fast, and the motion on the SACIM to equilibrium is relatively slow.

Use of common numerical techniques can often mask or obscure the actual dynamics. Numerical methods to solve systems of ordinary differential equations can be broadly categorized as *explicit* or *implicit*. We give a brief synopsis of each class of method. We cast each as a method to solve a system of the form

$$\frac{d\overline{\rho}}{dt} = \mathbf{f}(\overline{\rho}). \tag{1.276}$$

- *Explicit:* The simplest of these methods, the forward Euler[13] method, discretizes Eq. (1.276) as follows:

$$\frac{\overline{\rho}_{n+1} - \overline{\rho}_n}{\Delta t} = \mathbf{f}(\overline{\rho}_n), \tag{1.277}$$

  so that

$$\overline{\rho}_{n+1} = \overline{\rho}_n + \Delta t\, \mathbf{f}(\overline{\rho}_n). \tag{1.278}$$

  Explicit methods
  - Are easy to program, because Eq. (1.278) can be solved explicitly to predict the new value $\overline{\rho}_{n+1}$ in terms of the old values at step $n$.
  - Are often expensive to keep stable as one has to have $\Delta t < \tau_{fastest}$ to achieve numerical stability.
  - Are able to capture all physics and all time scales at great computational expense for stiff problems.
  - Require much computational effort for trajectories near the SACIM, and thus.
  - Are inefficient for some portions of stiff calculations.
- *Implicit:* The simplest of these methods, the backward Euler method, discretizes Eq. (1.276) as follows:

$$\frac{\overline{\rho}_{n+1} - \overline{\rho}_n}{\Delta t} = \mathbf{f}(\overline{\rho}_{n+1}), \tag{1.279}$$

  so that

$$\overline{\rho}_{n+1} = \overline{\rho}_n + \Delta t\, \mathbf{f}(\overline{\rho}_{n+1}). \tag{1.280}$$

  Implicit methods
  - Are more difficult to program because a nonlinear set of algebraic equations, Eq. (1.280), must be solved at every time step with no guarantee of solution.
  - Require potentially significant computational time to advance each time step.
  - Are capable of using large time steps and of remaining numerically stable, although the nonlinear algebraic equations become more difficult to solve as the time step increases.
  - Will miss physics that occur on small time scales $\tau < \Delta t$.
  - Are for many purposes, especially when fine transients are unimportant, better performers than explicit methods.
  - Are potentially highly inaccurate if fine-time-scale transients present in high-frequency physical instabilities are neglected.

A wide variety of software tools exist to solve systems of ordinary differential equations. Most of them use more sophisticated techniques than simple forward and backward Euler methods. One of the most powerful techniques is the use of *error control.* Here, the user specifies how far in time to advance and the error that is able to be tolerated. The algorithm, which is complicated, selects then internal time steps, for either explicit or implicit methods, to achieve a solution within the error tolerance at

[13] Leonhard Euler (1707–1783), Swiss mathematician.

Table 1.4. *Results from Computing Zel'dovich* NO *Production Using Implicit and Explicit Methods with Error Control in the Numerical Integration Algorithm* `dlsode.f`

| $\Delta t$ (s) | Explicit $N_{internal}$ | Explicit $\Delta t_{eff}$ (s) | Implicit $N_{internal}$ | Implicit $\Delta t_{eff}$ (s) |
|---|---|---|---|---|
| $10^2$ | $10^6$ | $10^{-4}$ | $10^0$ | $10^2$ |
| $10^1$ | $10^5$ | $10^{-4}$ | $10^0$ | $10^1$ |
| $10^0$ | $10^4$ | $10^{-4}$ | $10^0$ | $10^0$ |
| $10^{-1}$ | $10^3$ | $10^{-4}$ | $10^0$ | $10^{-1}$ |
| $10^{-2}$ | $10^2$ | $10^{-4}$ | $10^0$ | $10^{-2}$ |
| $10^{-3}$ | $10^1$ | $10^{-4}$ | $10^0$ | $10^{-3}$ |
| $10^{-4}$ | $10^0$ | $10^{-4}$ | $10^0$ | $10^{-4}$ |
| $10^{-5}$ | $10^0$ | $10^{-5}$ | $10^0$ | $10^{-5}$ |
| $10^{-6}$ | $10^0$ | $10^{-6}$ | $10^0$ | $10^{-6}$ |

the specified output time. A well-known public domain algorithm with error control is provided by `lsode.f`, which can be found in the `netlib` repository (Hindmarsh, 1983).

Let us exercise the Zel'dovich mechanism under the conditions simulated in Fig. 1.14, $T = 1500$ K, $P = 6.2355 \times 10^3$ dyne/cm$^2$. Recall in this case that the fastest physical time scale near equilibrium is $\tau_1 = 7.86 \times 10^{-5}$ s $\sim 10^{-4}$ s, and the slowest time scale is $\tau = 3.02 \times 10^1$ s. Let us solve for these conditions using `dlsode.f`, which uses internal time stepping for error control, in both an explicit and implicit mode. We specify a variety of values of $\Delta t$ and report typical values of number of internal time steps selected by `dlsode.f`, and the corresponding effective time step $\Delta t_{eff}$ used for the problem, for both explicit and implicit methods, as reported in Table 1.4.

For the explicit method, if one requests output for large $\Delta t$, the algorithm adjusts the number of unreported internal time steps to keep the error controlled. We see that for large $\Delta t$, the effective internal time step is $10^{-4}$ s, just enough to capture the physical dynamics. When the requested time step is less than $10^{-4}$ s, there is no need to take internal time steps, and we are simply using a time step more fine than necessary to achieve the desired accuracy. In contrast, no internal time stepping is needed for the implicit method.

Obviously if output is requested using $\Delta t > 10^{-4}$ s, the early time dynamics near $t \sim 10^{-4}$ s will be missed. For physically stable systems, codes such as `dlsode.f` will still provide a correct solution at the later times. For physically unstable systems, such as might occur in turbulent flames, it is not clear that one can use large time steps and expect to have fidelity to the underlying equations if a so-called Direct Numerical Simulation (DNS) is desired. The reason is the physical instabilities may evolve on the same time scale as the fine scales that are overlooked by large $\Delta t$.

The errors induced at early times by an implicit method are evident in Fig. 1.15. Here, for the same conditions as modeled to generate Fig. 1.14, namely, $T = 1500$ K, $P = 6.2355 \times 10^3$ dyne/cm$^2$, results are presented for the prediction of a highly resolved explicit method as well as that with an implicit method with a large time step of $\Delta t = 10^{-1}$ s. The predictions of the implicit method are accurate for late time, $t > \Delta t = 10^{-1}$ s, but inaccurate at early time, $t < \Delta t = 10^{-1}$ s. The smooth curves are both interpolations of a finite set of discrete data points.

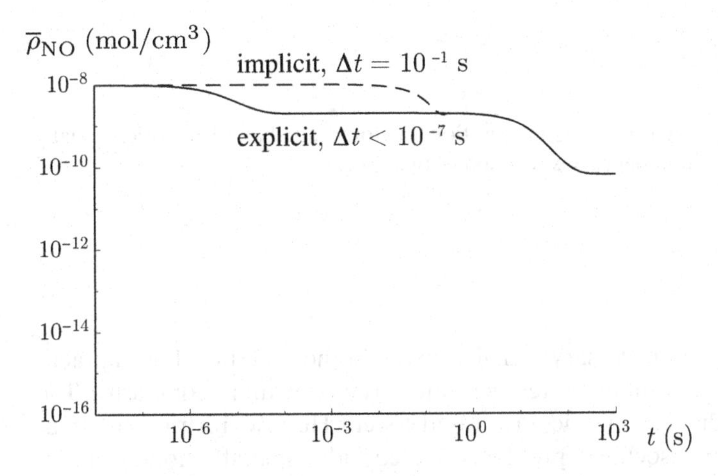


Figure 1.15. $\overline{\rho}_{NO}$ versus time for Zel'dovich mechanism at $T = 1500$ K, $P = 6.2355 \times 10^3$ dyne/cm$^2$ simulated by explicit and implicit numerical methods, demonstrating the loss of accuracy at early time by the implicit method.

As is proved in Sec. 5.2.3, *closed systems of ideal mixtures of ideal gases are guaranteed to have a unique, stable equilibrium that has physical values of all state variables*. Other equilibria may exist mathematically, but they typically induce nonphysical negative values for species concentrations. For such systems, implicit methods using large time steps typically give accurate portrayals of the dynamics on the approach to equilibrium. Such is the case in Fig. 1.14. For such systems the loss of accuracy at early time has no long-term consequences.

For other systems, there may be dangers in using implicit methods with large time steps. Consider that there are many scenarios that induce multiple physical equilibria. Among those are (1) spatially homogeneous open systems that exchange mass with their environment; see Chapter 11, (2) nonideal mixtures, and (3) spatially inhomogeneous systems with advection and diffusion. Consider then that such a system exists with multiple physical equilibria. From a given initial state, a highly resolved calculation will bring one into the correct physical equilibrium. But an implicit calculation with large time steps may misrepresent the early time dynamics in such a fashion to put the solution on the wrong path, so that at late time it relaxes to the wrong physical equilibrium.

If fast scale dynamics are important for a problem of interest or if there is a danger of approaching the wrong physical equilibrium, one should not choose the common strategy in numerical combustion modeling of employing an implicit method with large time steps for reaction kinetics. Such a calculation may be fast and stable, but it will not be accurate at all times. Instead, to capture the fast scale dynamics, it is essential to use a time step slightly smaller than the smallest physical time scale in the problem. This robust strategy unfortunately renders fully resolved solutions to many realistic combustion problems to be prohibitively computationally expensive.

## 1.3 Adiabatic, Isochoric Kinetics

Recall that an *adiabatic* system has no heat transfer across its system boundary. The isothermal systems considered in the previous section were *diabatic*, in that to maintain the constant temperature during reaction, it was necessary for heat to cross the system boundary. Had we considered the first law of thermodynamics, we could have

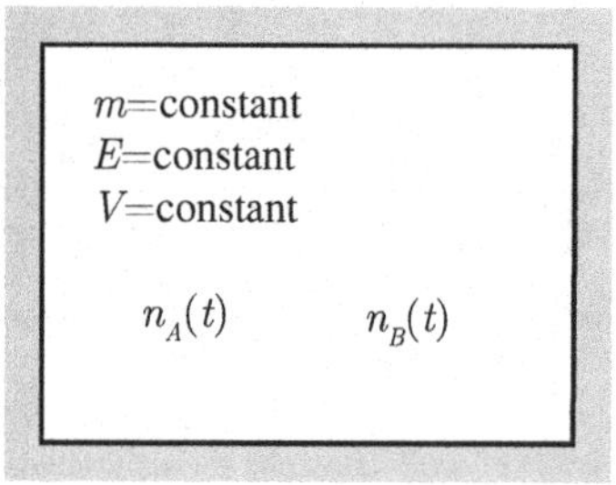

$m$=constant
$E$=constant
$V$=constant

$n_A(t)$ $n_B(t)$

Figure 1.16. Configuration for closed, adiabatic, isochoric, spatially homogeneous reaction of two species.

calculated the heat transfer necessary to maintain the isothermal state. It is of practical engineering relevance to allow for temperature variation within a combustor. The best model for this is adiabatic kinetics in a closed system. Here, we restrict our attention to closed, adiabatic, isochoric problems. We consider spatially homogeneous problems with simple kinetic models, as sketched in Fig. 1.16. We extend this simple model in Chapter 9 to include diffusion for a spatially inhomogeneous problem.

### 1.3.1 Thermal Explosion Theory

There is a simple description known as *thermal explosion theory* that provides a good explanation for how initially slow exothermic reaction induces a sudden temperature rise accompanied by a final relaxation to equilibrium. A related diabatic theory was provided by Semenov (1928, 1959), followed by an adiabatic theory (Kassoy, 1975).

Let us consider a simple *isomerization reaction* in a closed volume

$$A \leftrightharpoons B. \tag{1.281}$$

An isomerization reaction is one in which the molecular structure may rearrange, but the elemental composition, and thus the molecular mass, of the molecule does not change. Let us take $A$ and $B$ to both be calorically perfect ideal gases with identical molecular masses $M_A = M_B = M$ and identical specific heats, $\overline{c}_{vA} = \overline{c}_{vB} = c_v$; $\overline{c}_{PA} = \overline{c}_{PB} = \overline{c}_P$. Here, $\overline{c}_v$ represents the molar specific heat at constant volume, and $\overline{c}_P$ represents the molar specific heat at constant pressure. More discussion of specific heats is given in Chapter 2 and Sec. 3.4. We can consider $A$ and $B$ to be isomers of an identical molecular species. So, we have $N = 2$ species reacting in $J = 1$ reactions. The number of elements $L$ here is irrelevant.

We give in the next section an extensive discussion of how to employ the first law of thermodynamics for a mixture of calorically perfect ideal gases and then to convert a dimensional problem into a more standard dimensionless form. For readers familiar with such analysis, it is possible to go straight to Eqs. (1.335)–(1.337) to focus on the intrinsic dynamics of thermal explosion.

#### One-Step Reversible Kinetics

Let us insist our reaction process be isochoric and adiabatic, and commence with only $A$ present. The reversible reaction kinetics with $\beta = 0$ are

$$\frac{d\overline{\rho}_A}{dt} = -\underbrace{\underbrace{a \exp\left(\frac{-\overline{\mathcal{E}}}{\overline{R}T}\right)}_{k}\left(\overline{\rho}_A - \frac{1}{K_c}\overline{\rho}_B\right)}_{r}, \quad \overline{\rho}_A(0) = \widehat{\overline{\rho}}_A, \tag{1.282}$$

$$\frac{d\overline{\rho}_B}{dt} = \underbrace{\underbrace{a\exp\left(\frac{-\overline{\mathcal{E}}}{\overline{R}T}\right)}_{k}\left(\overline{\rho}_A - \frac{1}{K_c}\overline{\rho}_B\right)}_{r}, \quad \overline{\rho}_B(0) = 0. \tag{1.283}$$

For our alternate compact linear algebra-based form, we see that

$$r = a\exp\left(\frac{-\overline{\mathcal{E}}}{\overline{R}T}\right)\left(\overline{\rho}_A - \frac{1}{K_c}\overline{\rho}_B\right), \tag{1.284}$$

and that

$$\frac{d}{dt}\begin{pmatrix}\overline{\rho}_A\\ \overline{\rho}_B\end{pmatrix} = \begin{pmatrix}-1\\ 1\end{pmatrix}(r). \tag{1.285}$$

Forming the appropriate linear combinations by replacing the second equation with the sum of the two equations yields

$$\frac{d}{dt}\begin{pmatrix}\overline{\rho}_A\\ \overline{\rho}_A + \overline{\rho}_B\end{pmatrix} = \begin{pmatrix}-1\\ 0\end{pmatrix}(r). \tag{1.286}$$

Expanded, this is

$$\begin{pmatrix}1 & 0\\ 1 & 1\end{pmatrix}\frac{d}{dt}\begin{pmatrix}\overline{\rho}_A\\ \overline{\rho}_B\end{pmatrix} = \begin{pmatrix}-1\\ 0\end{pmatrix}(r). \tag{1.287}$$

Integrating the second of Eq. (1.286) yields

$$\frac{d}{dt}(\overline{\rho}_A + \overline{\rho}_B) = 0, \tag{1.288}$$

$$\overline{\rho}_A + \overline{\rho}_B = \widehat{\overline{\rho}}_A, \tag{1.289}$$

$$\overline{\rho}_B = \widehat{\overline{\rho}}_A - \overline{\rho}_A. \tag{1.290}$$

Thus, Eq. (1.282) reduces to

$$\frac{d\overline{\rho}_A}{dt} = -a\exp\left(\frac{-\overline{\mathcal{E}}}{\overline{R}T}\right)\left(\overline{\rho}_A - \frac{1}{K_c}\left(\widehat{\overline{\rho}}_A - \overline{\rho}_A\right)\right). \tag{1.291}$$

Scaling, Eq. (1.291) can be rewritten as

$$\frac{d}{d(at)}\left(\frac{\overline{\rho}_A}{\widehat{\overline{\rho}}_A}\right) = -\exp\left(-\frac{\overline{\mathcal{E}}}{\overline{R}T_0}\frac{1}{T/T_0}\right)\left(\frac{\overline{\rho}_A}{\widehat{\overline{\rho}}_A} - \frac{1}{K_c}\left(1 - \frac{\overline{\rho}_A}{\widehat{\overline{\rho}}_A}\right)\right). \tag{1.292}$$

**First Law of Thermodynamics**

Recall the first law of thermodynamics. Neglecting potential and kinetic energy changes, it can be written as

$$\frac{dE}{dt} = \dot{Q} - \dot{W}. \tag{1.293}$$

Here, $E$ is the extensive internal energy, an extensive property with units of J. The term $\dot{Q}$ represents the rate of heat transfer to the system and $\dot{W}$ represents the rate of pressure-volume work done by the system. Both have units of W. Because we insist the problem is adiabatic, $\dot{Q} = 0$. Because we insist the problem is isochoric, there is no pressure-volume work done, so $\dot{W} = 0$. Thus, we have $dE/dt = 0$, which yields

$$E = E_0. \tag{1.294}$$

We next draw upon concepts from mixture theory that will be fully explored in Chapter 2. With $n_i$ as the number of moles of species $i$, $\overline{e}_i$ as the molar specific internal energy of species $i$, the extensive internal energy for a mixture is

$$E = n_A\overline{e}_A + n_B\overline{e}_B = V\left(\frac{n_A}{V}\overline{e}_A + \frac{n_B}{V}\overline{e}_B\right). \tag{1.295}$$

With $\overline{\rho}_i = n_i/V$ as the molar concentration of species $i$, we then get

$$E = V\left(\overline{\rho}_A\overline{e}_A + \overline{\rho}_B\overline{e}_B\right). \tag{1.296}$$

From the upcoming Eq. (2.204), we replace the molar specific internal energy $\overline{e}_i$ by $\overline{h}_i - P_i/\overline{\rho}_i$, where $\overline{h}_i$ is the molar specific enthalpy of species $i$:

$$E = V\left(\overline{\rho}_A\left(\overline{h}_A - \frac{P_A}{\overline{\rho}_A}\right) + \overline{\rho}_B\left(\overline{h}_B - \frac{P_B}{\overline{\rho}_B}\right)\right). \tag{1.297}$$

We then use the ideal gas law, the upcoming Eq. (2.200), to get

$$E = V\left(\overline{\rho}_A\left(\overline{h}_A - \overline{R}T\right) + \overline{\rho}_B\left(\overline{h}_B - \overline{R}T\right)\right). \tag{1.298}$$

We can then use the fact that $\overline{c}_{Pi}$ is constant for the calorically perfect ideal gases and that we have assumed $\overline{c}_{PA} = \overline{c}_{PB} = \overline{c}_P$ and specialize Eq. (2.204) to eliminate $\overline{h}_i$ to get

$$E = V\left(\overline{\rho}_A\left(\overline{c}_P(T - T_0) + \overline{h}^0_{T_0,A} - \overline{R}T\right) + \overline{\rho}_B\left(\overline{c}_P(T - T_0) + \overline{h}^0_{T_0,B} - \overline{R}T\right)\right). \tag{1.299}$$

Here, $\overline{h}^0_{T_0,i}$ is the enthalpy of species $i$ evaluated at the reference pressure $P_0$ and the reference temperature $T_0$. Then, rearrange to get

$$E = V\left((\overline{\rho}_A + \overline{\rho}_B)(\overline{c}_P(T - T_0) - \overline{R}T) + \overline{\rho}_A\overline{h}^0_{T_0,A} + \overline{\rho}_B\overline{h}^0_{T_0,B}\right), \tag{1.300}$$

$$= V\left((\overline{\rho}_A + \overline{\rho}_B)((\overline{c}_P - \overline{R})T - \overline{c}_PT_0) + \overline{\rho}_A\overline{h}^0_{T_0,A} + \overline{\rho}_B\overline{h}^0_{T_0,B}\right), \tag{1.301}$$

$$= V\left((\overline{\rho}_A + \overline{\rho}_B)((\overline{c}_P - \overline{R})T - (\overline{c}_P - \overline{R} + \overline{R})T_0) + \overline{\rho}_A\overline{h}^0_{T_0,A} + \overline{\rho}_B\overline{h}^0_{T_0,B}\right). \tag{1.302}$$

Next, use Eq. (2.192) to eliminate $\overline{c}_P$ in favor of $\overline{c}_v$ to get

$$E = V\left((\overline{\rho}_A + \overline{\rho}_B)(\overline{c}_vT - (\overline{c}_v + \overline{R})T_0) + \overline{\rho}_A\overline{h}^0_{T_0,A} + \overline{\rho}_B\overline{h}^0_{T_0,B}\right), \tag{1.303}$$

$$= V\left((\overline{\rho}_A + \overline{\rho}_B)\overline{c}_v(T - T_0) + \overline{\rho}_A(\overline{h}^0_{T_0,A} - \overline{R}T_0) + \overline{\rho}_B(\overline{h}^0_{T_0,B} - \overline{R}T_0)\right), \tag{1.304}$$

$$= V\left((\overline{\rho}_A + \overline{\rho}_B)\overline{c}_v(T - T_0) + \overline{\rho}_A\overline{e}^0_{T_0,A} + \overline{\rho}_B\overline{e}^0_{T_0,B}\right). \tag{1.305}$$

Now, at the initial state we have $T = T_0$, so

$$E_0 = V\left(\widehat{\overline{\rho}}_A\overline{e}^0_{T_0,A} + \widehat{\overline{\rho}}_B\overline{e}^0_{T_0,B}\right). \tag{1.306}$$

So, we can say our caloric equation of state is

$$E - E_0 = V\left((\overline{\rho}_A + \overline{\rho}_B)\overline{c}_v(T - T_0) + (\overline{\rho}_A - \widehat{\overline{\rho}}_A)\overline{e}^0_{T_0,A} + (\overline{\rho}_B - \widehat{\overline{\rho}}_B)\overline{e}^0_{T_0,B}\right). \tag{1.307}$$

For general initial conditions, Eq. (1.288) implies $\overline{\rho}_A + \overline{\rho}_B = \widehat{\overline{\rho}_A} + \widehat{\overline{\rho}_B}$, so

$$E - E_0 = V\left((\widehat{\overline{\rho}}_A + \widehat{\overline{\rho}}_B)\overline{c}_v(T - T_0) + (\overline{\rho}_A - \widehat{\overline{\rho}}_A)\overline{e}^0_{T_0,A} + (\overline{\rho}_B - \widehat{\overline{\rho}}_B)\overline{e}^0_{T_0,B}\right). \tag{1.308}$$

As an aside, on a molar basis, we scale Eq. (1.308) to get

$$\overline{e} - \overline{e}_0 = \overline{c}_v(T - T_0) + (y_A - y_{A0})\overline{e}^0_{T_0,A} + (y_B - y_{B0})\overline{e}^0_{T_0,B}, \tag{1.309}$$

where $y_i$ is the *mol fraction* of species $i$; $y_i = n_i/n$. And because we have assumed the molecular masses are the same, $M_A = M_B$, the mol fractions $y_i$ can be shown to equal the mass fractions $Y_i$, and we can write on a mass basis

$$e - e_0 = c_v(T - T_0) + (Y_A - Y_{A0})e^0_{T_0,A} + (Y_B - Y_{B0})e^0_{T_0,B}. \tag{1.310}$$

Our energy conservation relation, Eq. (1.294), gives $E - E_0 = 0$, and Eq. (1.308) becomes

$$0 = V\left((\widehat{\overline{\rho}}_A + \widehat{\overline{\rho}}_B)\overline{c}_v(T - T_0) + (\overline{\rho}_A - \widehat{\overline{\rho}}_A)\overline{e}^0_{T_0,A} + (\overline{\rho}_B - \widehat{\overline{\rho}}_B)\overline{e}^0_{T_0,B}\right). \tag{1.311}$$

Now, we solve for $T$:

$$0 = \overline{c}_v(T - T_0) + \frac{\overline{\rho}_A - \widehat{\overline{\rho}}_A}{\widehat{\overline{\rho}}_A + \widehat{\overline{\rho}}_B}\overline{e}^0_{T_0,A} + \frac{\overline{\rho}_B - \widehat{\overline{\rho}}_B}{\widehat{\overline{\rho}}_A + \widehat{\overline{\rho}}_B}\overline{e}^0_{T_0,B}, \tag{1.312}$$

$$T = T_0 + \frac{\widehat{\overline{\rho}}_A - \overline{\rho}_A}{\widehat{\overline{\rho}}_A + \widehat{\overline{\rho}}_B}\frac{\overline{e}^0_{T_0,A}}{\overline{c}_v} + \frac{\widehat{\overline{\rho}}_B - \overline{\rho}_B}{\widehat{\overline{\rho}}_A + \widehat{\overline{\rho}}_B}\frac{\overline{e}^0_{T_0,B}}{\overline{c}_v}. \tag{1.313}$$

Now, we impose our assumption that $\widehat{\overline{\rho}}_B = 0$, giving also $\overline{\rho}_B = \widehat{\overline{\rho}}_A - \overline{\rho}_A$:

$$T = T_0 + \frac{\widehat{\overline{\rho}}_A - \overline{\rho}_A}{\widehat{\overline{\rho}}_A}\frac{\overline{e}^0_{T_0,A}}{\overline{c}_v} - \frac{\overline{\rho}_B}{\widehat{\overline{\rho}}_A}\frac{\overline{e}^0_{T_0,B}}{\overline{c}_v}, \tag{1.314}$$

$$= T_0 + \frac{\widehat{\overline{\rho}}_A - \overline{\rho}_A}{\widehat{\overline{\rho}}_A}\frac{\overline{e}^0_{T_0,A} - \overline{e}^0_{T_0,B}}{\overline{c}_v}. \tag{1.315}$$

In summary, realizing that $\overline{h}^0_{T_0,A} - \overline{h}^0_{T_0,B} = \overline{e}^0_{T_0,A} - \overline{e}^0_{T_0,B}$, we can write $T$ as a function of $\overline{\rho}_A$:

$$T = T_0 + \frac{(\widehat{\overline{\rho}}_A - \overline{\rho}_A)}{\widehat{\overline{\rho}}_A\overline{c}_v}(\overline{h}^0_{T_0,A} - \overline{h}^0_{T_0,B}). \tag{1.316}$$

We see then that if $\overline{h}^0_{T_0,A} > \overline{h}^0_{T_0,B}$, as $\overline{\rho}_A$ decreases from its initial value of $\widehat{\overline{\rho}}_A$, $T$ will increase. We can scale Eq. (1.316) to form

$$\left(\frac{T}{T_0}\right) = 1 + \left(1 - \frac{\overline{\rho}_A}{\widehat{\overline{\rho}}_A}\right)\left(\frac{\overline{h}^0_{T_0,A} - \overline{h}^0_{T_0,B}}{\overline{c}_v T_0}\right). \tag{1.317}$$

Our caloric state equation, Eq. (1.309), can, for $y_{A0} = 1, y_{B0} = 0$, be rewritten as

$$\overline{e} - \overline{e}_0 = \overline{c}_v(T - T_0) + (y_A - 1)\overline{e}^0_{T_0,A} + y_B\overline{e}^0_{T_0,B}, \tag{1.318}$$

$$= \overline{c}_v(T - T_0) + ((1 - y_B) - 1)\overline{e}^0_{T_0,A} + y_B\overline{e}^0_{T_0,B}, \tag{1.319}$$

$$= \overline{c}_v(T - T_0) - y_B(\overline{e}^0_{T_0,A} - \overline{e}^0_{T_0,B}). \tag{1.320}$$

Similarly, on a mass basis, we can say that

$$e - e_0 = c_v(T - T_0) - Y_B(e^0_{T_0,A} - e^0_{T_0,B}). \tag{1.321}$$

For this problem, we also have

$$K_c = \exp\left(\frac{-\Delta G^0}{\overline{R}T}\right), \tag{1.322}$$

with

$$\Delta G^0 = \overline{g}^0_B - \overline{g}^0_A = (\overline{h}^0_{T_0,B} - \overline{h}^0_{T_0,A}) - T(\overline{s}^0_{T_0,B} - \overline{s}^0_{T_0,A}). \tag{1.323}$$

So

$$K_c = \exp\left(\frac{\overline{h}^0_{T_0,A} - \overline{h}^0_{T_0,B} - T(\overline{s}^0_{T_0,A} - \overline{s}^0_{T_0,B})}{\overline{R}T}\right), \tag{1.324}$$

$$= \exp\left(\frac{\overline{c}_v T_0}{\overline{R}T}\left(\frac{\overline{h}^0_{T_0,A} - \overline{h}^0_{T_0,B} - T(\overline{s}^0_{T_0,A} - \overline{s}^0_{T_0,B})}{\overline{c}_v T_0}\right)\right), \tag{1.325}$$

$$= \exp\left(\frac{1}{\gamma - 1}\frac{1}{\frac{T}{T_0}}\left(\frac{\overline{h}^0_{T_0,A} - \overline{h}^0_{T_0,B}}{\overline{c}_v T_0} - \frac{T}{T_0}\frac{(\overline{s}^0_{T_0,A} - \overline{s}^0_{T_0,B})}{\overline{c}_v}\right)\right). \tag{1.326}$$

Here, we have used the definition of the ratio of specific heats, $\gamma = \overline{c}_P/\overline{c}_v$ along with $\overline{R} = \overline{c}_P - \overline{c}_v$. So, we can solve Eq. (1.291) by first using Eq. (1.326) to eliminate $K_c$ and then Eq. (1.316) to eliminate $T$.

#### Dimensionless Form

Let us try writing dimensionless variables so that our system can be written in a compact dimensionless form. First, let us take the variables of dimensionless time $\tau$, species concentration of reactant $z$, and temperature $\theta$ to be

$$\tau = at, \qquad z = \overline{\rho}_A/\widehat{\overline{\rho}}_A, \qquad \theta = T/T_0. \tag{1.327}$$

Let us take parameters dimensionless heat release $q$, activation energy $\Theta$, and entropy change $\eta$ to be

$$q = \frac{\overline{h}^0_{T_0,A} - \overline{h}^0_{T_0,B}}{\overline{c}_v T_0}, \qquad \Theta = \frac{\overline{\mathcal{E}}}{\overline{R}T_0}, \qquad \eta = \frac{(\overline{s}^0_{T_0,A} - \overline{s}^0_{T_0,B})}{\overline{c}_v}. \tag{1.328}$$

So, our equations, Eqs. (1.292), (1.317), and (1.326), become

$$\frac{dz}{d\tau} = -\exp\left(-\frac{\Theta}{\theta}\right)\left(z - \frac{1}{K_c}(1 - z)\right), \tag{1.329}$$

$$\theta = 1 + (1 - z)q, \tag{1.330}$$

$$K_c = \exp\left(\frac{1}{\gamma - 1}\frac{1}{\theta}(q - \theta\eta)\right). \tag{1.331}$$

It is more common to consider the products. Let us define for general problems

$$\lambda = \frac{\overline{\rho}_B}{\overline{\rho}_A + \overline{\rho}_B} = \frac{\overline{\rho}_B}{\widehat{\overline{\rho}}_A + \widehat{\overline{\rho}}_B}. \tag{1.332}$$

Thus, $\lambda$ is the mol fraction of product, and also the mass fraction because we assume identical molecular masses of $A$ and $B$. For our problem, $\widehat{\overline{\rho}}_B = 0$, so

$$\lambda = \frac{\overline{\rho}_B}{\widehat{\overline{\rho}}_A} = \frac{\widehat{\overline{\rho}}_A - \overline{\rho}_A}{\widehat{\overline{\rho}}_A}. \tag{1.333}$$

Thus,

$$\lambda = 1 - z. \tag{1.334}$$

We can think of $\lambda$ as a reaction progress variable as well. When $\lambda = 0$, we have $\tau = 0$, and the reaction has not begun. Thus, we get

$$\frac{d\lambda}{d\tau} = \exp\left(-\frac{\Theta}{\theta}\right)\left((1-\lambda) - \frac{1}{K_c}\lambda\right), \qquad \lambda(0) = 0, \tag{1.335}$$

$$\theta = 1 + q\lambda, \tag{1.336}$$

$$K_c = \exp\left(\frac{1}{\gamma - 1}\frac{1}{\theta}(q - \theta\eta)\right). \tag{1.337}$$

Equation (1.335) is our spatially homogeneous kinetics equation for the evolution of reaction progress $\lambda$ with time $\tau$. It has exponential dependency on temperature $\theta$. Equation (1.336) is a consequence of the first law of thermodynamics and represents a conservation of the sum of thermal energy, measured by temperature $\theta$ and chemical energy, measured by $(1-\lambda)q$. Rearranged, it can be written as

$$\underbrace{\theta}_{\text{thermal energy}} + \underbrace{(1-\lambda)q}_{\text{chemical energy}} = \underbrace{1+q}_{\text{constant}}. \tag{1.338}$$

Equation (1.337) is an auxiliary equation for the equilibrium constant.

**Example Calculation**

Let us choose some values for the dimensionless parameters:

$$\Theta = 20, \quad \eta = 0, \quad q = 10, \quad \gamma = 7/5. \tag{1.339}$$

With these choices, our model, Eqs. (1.335)–(1.337), reduces to a single nonlinear ordinary differential equation:

$$\frac{d\lambda}{d\tau} = \exp\left(\frac{-20}{1+10\lambda}\right)\left((1-\lambda) - \lambda\exp\left(\frac{-25}{1+10\lambda}\right)\right), \quad \lambda(0) = 0. \tag{1.340}$$

The right side of Eq. (1.340) is at equilibrium for values of $\lambda$ that drive it to zero. Numerical root-finding show this to occur at $\lambda \sim 0.920539$. Near this root, Taylor series expansion shows the dynamics are approximated by

$$\frac{d}{d\tau}(\lambda - 0.920539) = -0.17993(\lambda - 0.920539) + \cdots \tag{1.341}$$

Thus, the local behavior near equilibrium is given by

$$\lambda = 0.920539 + C\exp(-0.17993\,\tau). \tag{1.342}$$

Here, $C$ is some arbitrary constant. Clearly the equilibrium is stable, with a time constant of $1/0.17993 = 5.55773$.

Numerical solution shows the full behavior of the dimensionless species concentration $\lambda(\tau)$ (see Fig. 1.17a). Clearly the product concentration $\lambda$ is small for some

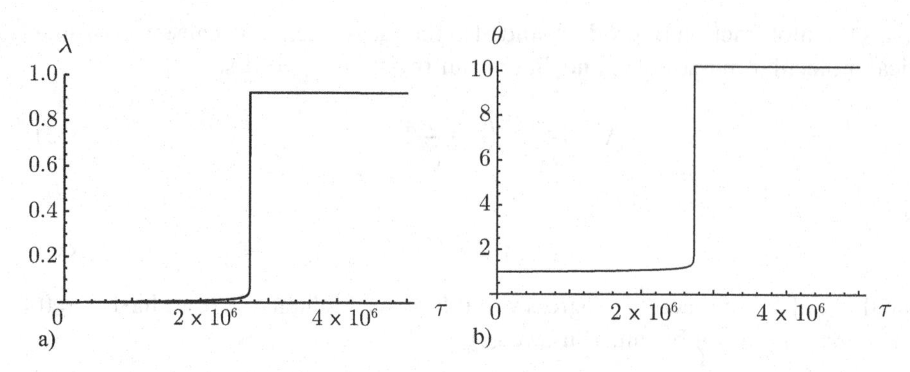


Figure 1.17. Dimensionless plot of (a) reaction product concentration $\lambda$ and (b) temperature $\theta$ versus time $\tau$ for adiabatic isochoric combustion with simple reversible kinetics.

long period of time. At a critical time near $\tau = 2.7 \times 10^6$, there is a so-called thermal explosion with a rapid increase in $\lambda$. *The estimate of the time constant near equilibrium is orders of magnitude less than the explosion time,* $5.55773 \ll 2.7 \times 10^6$. *Thus, linear analysis here is a poor tool to estimate an important physical quantity, the ignition delay time.* Once ignition is commenced, there is a rapid equilibration to the final state. The dimensionless temperature plot is shown in Fig. 1.17b. The temperature plot is similar in behavior to the species concentration plot. At early time, the temperature is low. At a critical time, the ignition delay time, the temperature rapidly rises. This rapid rise, coupled with the exponential sensitivity of reaction rate to temperature, accelerates the formation of product. This process continues until the reverse reaction is activated to the extent it prevents further creation of product.

### High-Activation-Energy Asymptotics

Let us see if we can get an *analytic* prediction of the ignition delay time, $\tau \sim 2.7 \times 10^6$. Such a prediction would be valuable to see how long a reacting material might take to ignite. Prediction of this ignition delay time is a foundational importance for engineering devices that rely on combustion, such as automobile engines, or for safety applications for which ignition is undesirable, such in enclosed mines. Our analysis is similar to that given by Buckmaster and Ludford (1983) and is restricted to the so-called *high-activation-energy limit*:

$$\Theta \gg 1. \tag{1.343}$$

This limit is valid for many important physical systems.

For convenience let us restrict ourselves to $\eta = 0$. In this limit, Eqs. (1.335)–(1.337) reduce to

$$\frac{d\lambda}{d\tau} = \exp\left(-\frac{\Theta}{1+q\lambda}\right)\left((1-\lambda) - \lambda \exp\left(\frac{-q}{(\gamma-1)(1+q\lambda)}\right)\right), \tag{1.344}$$

with $\lambda(0) = 0$. The key trouble in getting an analytic solution to Eq. (1.344) is the presence of $\lambda$ in the denominator of an exponential term. We need to find a way to move it to the numerator. Asymptotic methods provide one such way.

Now, we recall for early time $\lambda \ll 1$. Let us assume $\lambda(\tau)$ takes the form

$$\lambda(\tau) = \epsilon\lambda_1(\tau) + \epsilon^2\lambda_2(\tau) + \epsilon^3\lambda_3(\tau) + \cdots \tag{1.345}$$

Here, we will assume that $0 < \epsilon \ll 1$ and that $\lambda_1(\tau) \sim \mathcal{O}(1)$, $\lambda_2(\tau) \sim \mathcal{O}(1), \ldots$, and we will define the constant $\epsilon$ in terms of physical parameters shortly. With this, we have

$$\frac{1}{1+q\lambda} = \frac{1}{1+\epsilon q\lambda_1 + \epsilon^2 q\lambda_2 + \epsilon^3 q\lambda_3 + \cdots}. \tag{1.346}$$

Long division of the term on the right side yields the approximation

$$\frac{1}{1+q\lambda} = 1 - \epsilon q\lambda_1 + \epsilon^2(q^2\lambda_1^2 - q\lambda_2) + \cdots, \tag{1.347}$$

$$= 1 - \epsilon q\lambda_1 + \mathcal{O}(\epsilon^2). \tag{1.348}$$

So,

$$\exp\left(-\frac{\Theta}{1+q\lambda}\right) \sim \exp\left(-\Theta(1-\epsilon q\lambda_1 + \mathcal{O}(\epsilon^2))\right), \tag{1.349}$$

$$\sim \exp(-\Theta)\exp\left(\epsilon q\Theta\lambda_1 + \mathcal{O}(\epsilon^2)\right). \tag{1.350}$$

We have moved $\lambda$ from the denominator to the numerator of the most important exponential term. Now, let us define $\epsilon$ to be

$$\epsilon \equiv 1/\Theta. \tag{1.351}$$

When we consider the high-activation-energy limit, Eq. (1.343), we thus require that

$$0 < \epsilon \ll 1, \tag{1.352}$$

consistent with our earlier assumption on $\epsilon$. We assume the remaining parameters, $q$ and $\gamma$, are both $\mathcal{O}(1)$ constants. In the high-activation-energy limit, after replacing $\Theta$ with $\epsilon$, Eq. (1.350) becomes

$$\exp\left(-\frac{\Theta}{1+q\lambda}\right) \sim \exp\left(-\frac{1}{\epsilon}\right)\exp\left(q\lambda_1 + \mathcal{O}(\epsilon^2)\right). \tag{1.353}$$

With these assumptions and approximations, Eq. (1.344) can be written as

$$\frac{d}{d\tau}(\epsilon\lambda_1 + \cdots) = \exp\left(-\frac{1}{\epsilon}\right)\exp\left(q\lambda_1 + \mathcal{O}(\epsilon^2)\right)\Bigg((1-\epsilon\lambda_1 - \cdots) - (\epsilon\lambda_1 + \cdots)$$

$$\times \exp\left(\frac{-q}{(\gamma-1)(1+q\epsilon\lambda_1+\cdots)}\right)\Bigg). \tag{1.354}$$

Now, let us rescale time via

$$\tau_* = \frac{1}{\epsilon}\exp\left(\frac{-1}{\epsilon}\right)\tau. \tag{1.355}$$

With this transformation, the chain rule shows how derivatives transform:

$$\frac{d}{d\tau} = \frac{d\tau_*}{d\tau}\frac{d}{d\tau_*} = \frac{1}{\epsilon\exp\left(\frac{1}{\epsilon}\right)}\frac{d}{d\tau_*}. \tag{1.356}$$

With this transformation, Eq. (1.354) becomes

$$\frac{1}{\epsilon\exp\left(\frac{1}{\epsilon}\right)}\frac{d}{d\tau_*}(\epsilon\lambda_1 + \cdots) = \exp\left(-\frac{1}{\epsilon}\right)\exp\left(q\lambda_1 + \mathcal{O}(\epsilon^2)\right)$$

$$\times\Bigg((1-\epsilon\lambda_1 - \cdots) - (\epsilon\lambda_1 + \cdots)$$

$$\times \exp\left(\frac{-q}{(\gamma-1)(1+q\epsilon\lambda_1+\cdots)}\right)\Bigg). \tag{1.357}$$

This simplifies to

$$\frac{d}{d\tau_*}(\lambda_1 + \cdots) = \exp\left(q\lambda_1 + \mathcal{O}(\epsilon^2)\right)\left((1 - \epsilon\lambda_1 - \cdots) - (\epsilon\lambda_1 + \cdots)\right.$$
$$\left.\times \exp\left(\frac{-q}{(\gamma - 1)(1 + q\epsilon\lambda_1 + \cdots)}\right)\right). \tag{1.358}$$

Retaining only $\mathcal{O}(1)$ terms in Eq. (1.358), we get

$$\frac{d\lambda_1}{d\tau_*} = \exp(q\lambda_1). \tag{1.359}$$

This is supplemented by the initial condition $\lambda_1(0) = 0$. Separating variables and solving, we get

$$\exp(-q\lambda_1)\, d\lambda_1 = d\tau_*, \tag{1.360}$$

$$-\frac{1}{q}\exp(-q\lambda_1) = \tau_* + C. \tag{1.361}$$

Applying the initial condition yields

$$-\frac{1}{q} = C. \tag{1.362}$$

So

$$-\frac{1}{q}\exp(-q\lambda_1) = \tau_* - \frac{1}{q}, \tag{1.363}$$

$$\exp(-q\lambda_1) = -q\tau_* + 1, \tag{1.364}$$

$$\exp(-q\lambda_1) = -q\left(\tau_* - \frac{1}{q}\right), \tag{1.365}$$

$$-q\lambda_1 = \ln\left(-q\left(\tau_* - \frac{1}{q}\right)\right), \tag{1.366}$$

$$\lambda_1 = -\frac{1}{q}\ln\left(-q\left(\tau_* - \frac{1}{q}\right)\right). \tag{1.367}$$

For $q = 10$, a plot of $\lambda_1(\tau_*)$ is shown in Fig. 1.18. At a finite $\tau_*$, $\lambda_1$ begins to exhibit unbounded growth. In fact, it is obvious from Eq. (1.367) that as

$$\tau_* \to \frac{1}{q}, \quad \lambda_1 \to \infty. \tag{1.368}$$

That is, there exists a finite time for which $\lambda_1$ violates the assumptions of our asymptotic theory that assumes $\lambda_1 = \mathcal{O}(1)$. We associate this time with the ignition delay time, $\tau_{*i}$:

$$\tau_{*i} = \frac{1}{q}. \tag{1.369}$$

Let us return this to more primitive variables:

$$\frac{1}{\epsilon}\exp\left(\frac{-1}{\epsilon}\right)\tau_i = \frac{1}{q}, \tag{1.370}$$

$$\tau_i = \frac{\epsilon\exp\left(\frac{1}{\epsilon}\right)}{q} = \frac{\exp\Theta}{\Theta q}. \tag{1.371}$$

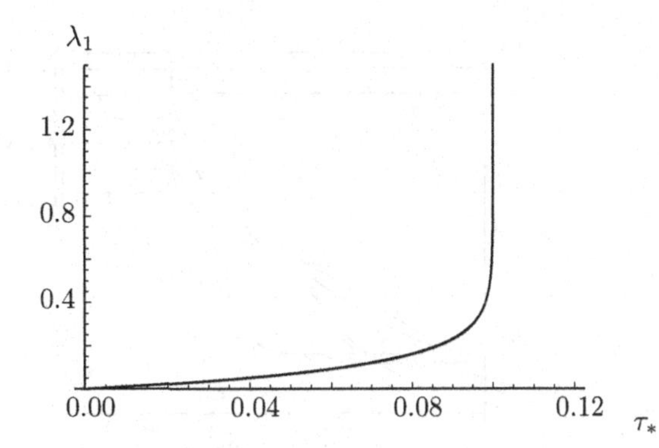


Figure 1.18. $\lambda_1$ versus $\tau_*$ for ignition problem.

For our system with $\Theta = 20$ and $q = 10$, we estimate the dimensionless ignition delay time as

$$\tau_i = \frac{\exp 20}{(20)(10)} = 2.42583 \times 10^6. \tag{1.372}$$

This is a surprisingly good estimate, given the complexity of the problem. Recall the numerical solution showed ignition for $\tau \sim 2.7 \times 10^6$.

In terms of dimensional time, ignition delay time prediction becomes

$$t_i = \frac{\exp \Theta}{a \Theta q}, \tag{1.373}$$

$$= \frac{1}{a}\left(\frac{\overline{R}T_0}{\overline{\mathcal{E}}}\right)\left(\frac{\overline{c}_v T_0}{\overline{h}^0_{T_0,A} - \overline{h}^0_{T_0,B}}\right) \exp\left(\frac{\overline{\mathcal{E}}}{\overline{R}T_0}\right). \tag{1.374}$$

The ignition is suppressed if the ignition delay time is lengthened. That happens when

- The activation energy $\overline{\mathcal{E}}$ is increased, because the exponential sensitivity is stronger than the algebraic sensitivity.
- The energy of combustion $(\overline{h}^0_{T_0,A} - \overline{h}^0_{T_0,B})$ is decreased because it takes longer to react to drive the temperature to a critical value to induce ignition.
- The collision frequency coefficient $a$ is decreased.

### 1.3.2 Detailed $H_2$-Air Kinetics

Here, a physically realistic example that uses detailed kinetics for an adiabatic isochoric system of $H_2$-air is summarized. We mainly present the results and demonstrate they have analog to those of the simple model just considered. Full exposition of the type of model equations needed to generate these results is delayed until Chapter 5. We choose a nonintuitive set of parameters for the problem. Our choices will enable a direct comparison to a detonation of the same mixture via the same reaction mechanism in a later chapter; see Sec. 12.2.8.

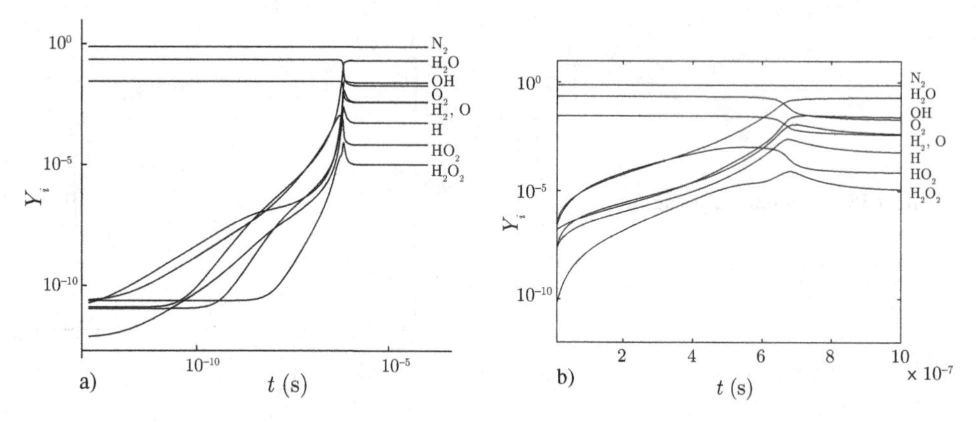


Figure 1.19. Plot of $Y_{H_2}(t), Y_H(t), Y_O(t), Y_{O_2}(t), Y_{OH}(t), Y_{H_2O}(t), Y_{HO_2}(t), Y_{H_2O_2}(t), Y_{N_2}(t)$, for adiabatic, isochoric combustion of a mixture of $2H_2 + O_2 + 3.76N_2$ initially at $T_O = 1542.7$ K, $P_O = 2.8323 \times 10^6$ Pa: (a) global dynamics and (b) dynamics near thermal explosion.

A closed, fixed, adiabatic volume, $V = 0.3061251$ cm$^3$, contains at $t = 0$ s a stoichiometric hydrogen-air mixture of $2 \times 10^{-5}$ mol of $H_2$, $1 \times 10^{-5}$ mol of $O_2$, and $3.76 \times 10^{-5}$ mol of $N_2$ at $P_0 = 2.83230 \times 10^6$ Pa and $T_0 = 1542.7$ K.[14] Thus, the initial molar concentrations are

$$\{\overline{\rho}_{H_2}, \overline{\rho}_{O_2}, \overline{\rho}_{N_2}\} = \{6.533 \times 10^{-5}, 3.267 \times 10^{-5}, 1.228 \times 10^{-4}\} \frac{\text{mol}}{\text{cm}^3}. \quad (1.375)$$

The initial mass fractions, $Y_i$, are calculated using the upcoming Eq. (2.37). That gives $Y_i = M_i\overline{\rho}_i/\rho$; here, $M_i$ is the molecular mass of species $i$. The mass fractions are

$$Y_{H_2} = 0.0285, \qquad Y_{O_2} = 0.226, \qquad Y_{N_2} = 0.745. \quad (1.376)$$

To avoid issues associated with numerical roundoff errors at very early time for species with very small compositions, the minor species were initialized at a small nonzero value near machine precision; each was assigned a value of $10^{-15}$ mol. The minor species all have $\overline{\rho}_i = 10^{-15}/0.3061251 = 3.267 \times 10^{-15}$ mol/cm$^3$. They have correspondingly small initial mass fractions.

We seek the reaction dynamics as the system proceeds from its initial state to its final state. We use the reversible detailed kinetics mechanism of Table 1.2. This problem requires a detailed numerical solution of species equations of the form of Eq. (5.13) coupled with the first law in the form of Eq. (5.27), both to be introduced and fully discussed in Chapter 5. Such a solution was performed by solving the appropriate equations for a mixture of nine interacting species: $H_2$, H, O, $O_2$, OH, $H_2O$, $HO_2$, $H_2O_2$, and $N_2$. The dynamics of the reaction process are reflected in Figs. 1.19 and 1.20.

At early time, $t < 10^{-7}$ s, the pressure, temperature, and major reactant species concentrations ($H_2$, $O_2$, $N_2$) are nearly constant. However, the minor species, for example, OH, $HO_2$, and the major product, $H_2O$, are undergoing rapid growth, albeit with mass fractions whose value remains small. In this time frame, the material is in what is known as the *induction period*.

[14] This temperature and pressure correspond to that of the same ambient mixture of $H_2$, $O_2$, and $N_2$ that was shocked from $1.01325 \times 10^5$ Pa, 298 K, to a value associated with a freely propagating detonation. Relevant comparisons of reaction dynamics are made later, in Sec. 12.2.8.

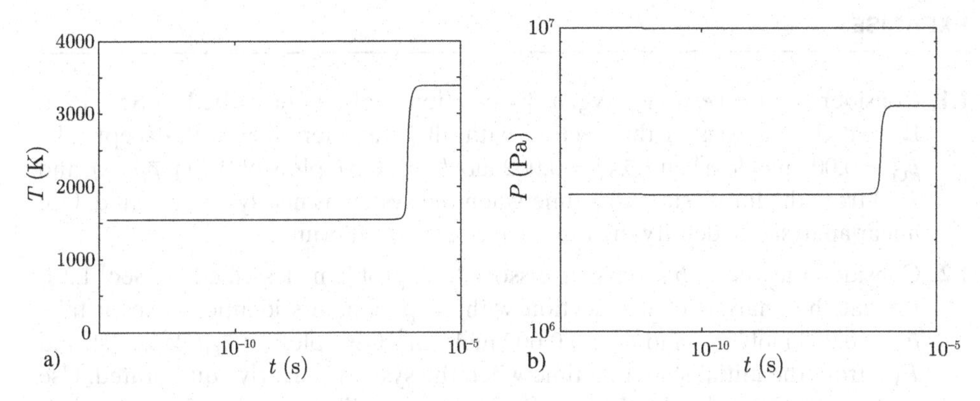


Figure 1.20. Plots of (a) $T(t)$ and (b) $P(t)$ for adiabatic, isochoric combustion of a mixture of $2H_2 + O_2 + 3.76N_2$ initially at $T_O = 1542.7$ K, $P_O = 2.8323 \times 10^6$ Pa.

After a certain critical mass of minor species has accumulated, exothermic recombination of these minor species to form the major product $H_2O$ induces the temperature to rise, which accelerates further the reaction rates. This is manifested in a *thermal explosion.* A common definition of the end of the induction period is the *induction time*, $t = t_{ind}$, the time when $dT/dt$ goes through a maximum. Here, one finds

$$t_{ind} = 6.6 \times 10^{-7} \text{ s.} \tag{1.377}$$

A close-up view of the mass fraction profiles is given in Fig. 1.19b.

At the end of the induction period, there is a final relaxation to equilibrium. The equilibrium mass fractions of each species are

$$Y_{O_2} = 1.85 \times 10^{-2}, \tag{1.378}$$

$$Y_{H} = 5.41 \times 10^{-4}, \tag{1.379}$$

$$Y_{OH} = 2.45 \times 10^{-2}, \tag{1.380}$$

$$Y_{O} = 3.88 \times 10^{-3}, \tag{1.381}$$

$$Y_{H_2} = 3.75 \times 10^{-3}, \tag{1.382}$$

$$Y_{H_2O} = 2.04 \times 10^{-1}, \tag{1.383}$$

$$Y_{HO_2} = 6.84 \times 10^{-5}, \tag{1.384}$$

$$Y_{H_2O_2} = 1.04 \times 10^{-5}, \tag{1.385}$$

$$Y_{N_2} = 7.45 \times 10^{-1}. \tag{1.386}$$

Because our model takes $N_2$ to be inert, its mass fraction remains unchanged. Other than $N_2$, the final products are dominated by $H_2O$. The equilibrium temperature and pressure are 3382.3 K and $5.53 \times 10^6$ Pa, respectively. Both have risen considerably from their respective initial values.

With this, we close this introductory chapter that has highlighted many of the key ideas of this book, especially for spatially homogeneous systems. We turn to a more systematic development of chemical thermodynamics and dynamics in the next several chapters.

## EXERCISES

**1.1.** Consider the *irreversible* oxygen dissociation problem described in Sec. 1.2.1. Repeat the analysis of this section with all parameters identical, except take $\widehat{\overline{\rho}}_O = 0.003\ \text{mol/cm}^3$ and $\widehat{\overline{\rho}}_{O_2} = 0.002\ \text{mol/cm}^3$. Give plots of $\overline{\rho}_O(t), \overline{\rho}_{O_2}(t)$, and $P(t)$ from the initial state to a time when the system is nearly equilibrated. Use linear analysis to identify the time scales near equilibrium.

**1.2.** Consider the *reversible* oxygen dissociation problem described in Sec. 1.2.1. Repeat the analysis of this section with all parameters identical, except take $\widehat{\overline{\rho}}_O = 0.003\ \text{mol/cm}^3$ and $\widehat{\overline{\rho}}_{O_2} = 0.002\ \text{mol/cm}^3$. Give plots of $\overline{\rho}_O(t), \overline{\rho}_{O_2}(t)$, and $P(t)$ from the initial state to a time when the system is nearly equilibrated. Use linear analysis to identify the time scales near equilibrium.

**1.3.** Consider the Zel'dovich mechanism system of Sec. 1.2.2. Repeat the analysis of this section with all parameters identical, except $\widehat{\overline{\rho}}_O = 0\ \text{mol/cm}^3$. Find the concentrations of all species at the physical equilibrium. Plot the molar concentrations of all species as a function of time. Use linear analysis in the neighborhood of the physical equilibrium to identify to local time scales of reaction and the stiffness ratio. Plot a family of trajectories in the $(\overline{\rho}_{NO}, \overline{\rho}_N)$ phase space.

**1.4.** Consider a diabatic extension to the thermal explosion theory of Sec. 1.3.1. In particular, consider that the volume $V$ has a surface area $A$, and that the volume exchanges thermal energy via simple convective heat transfer with surroundings at constant temperature $T_\infty$ at a rate given by

$$\dot{Q} = \mathsf{h}A(T_\infty - T),$$

where $\mathsf{h}$ is a constant convective heat transfer coefficient. Develop a dimensionless system of equations analogous to Eqs. (1.335)–(1.337). Define new dimensionless parameters to account for the ambient temperature and the convective heat transfer. The algebraic energy conservation of Eq. (1.335) will be replaced by a differential equation for temperature evolution. Using values of Eq. (1.339), and your choice of values for parameters capturing the effect of convective heat transfer, show how convective heat transfer can affect the dynamics of combustion. Show numerical results where the effects are small and where they are large. Give a physical explanation for your results.

## References

A. N. Al-Khateeb, J. M. Powers, S. Paolucci, A. J. Sommese, J. A. Diller, J. D. Hauenstein, and J. D. Mengers, 2009, One-dimensional slow invariant manifolds for spatially homogeneous reactive systems, *Journal of Chemical Physics*, 131(2): 024118.

V. I. Arnol'd, 1973, *Ordinary Differential Equations*, MIT Press, Cambridge, MA.

U. M. Ascher and L. R. Petzold, 1998, *Computer Methods for Ordinary Differential Equations and Differential-Algebraic Equations*, SIAM, Philadelphia.

D. L. Baulch et al., 2005, Evaluated kinetic data for combustion modeling: Supplement II, *Journal of Physical and Chemical Reference Data*, 34(3): 757–1397; see pp. 765–766.

C. Borgnakke and R. E. Sonntag, 2013, *Fundamentals of Thermodynamics*, 8th ed., John Wiley, New York.

J. D. Buckmaster and G. S. S. Ludford, 1983, *Lectures on Mathematical Combustion*, SIAM, Philadelphia, Chapter 1.

P. G. Drazin, 1992, *Nonlinear Systems*, Cambridge University Press, Cambridge, UK.

A. C. Hindmarsh, 1983, ODEPACK, a systematized collection of ODE solvers, in *Scientific Computing*, edited by R. S. Stepleman et al., North-Holland, Amsterdam, pp. 55–64. http://www.netlib.org/alliant/ode/prog/lsode.f.

M. W. Hirsch, S. Smale, and R. L. Devaney, 2012, *Differential Equations, Dynamical Systems, and an Introduction to Chaos*, 3rd ed., Academic Press, Oxford, UK.

A. Iserles, 2008, *A First Course in the Numerical Analysis of Differential Equations*, Cambridge University Press, Cambridge, UK.

D. R. Kassoy, 1975, A theory of adiabatic, homogeneous explosion from initiation to completion, *Combustion Science and Technology*, 10(1–2): 27–35.

R. J. Kee et al., 2000, *The* Chemkin *Thermodynamic Data Base*, part of the Chemkin Collection Release 3.6, Reaction Design, San Diego, CA.

R. J. Kee, M. E. Coltrin, and P. Glarborg, 2003, *Chemically Reacting Flow: Theory and Practice*, John Wiley, Hoboken, NJ.

U. Maas and S. B. Pope, 1992, Simplifying chemical kinetics: Intrinsic low-dimensional manifolds in composition space, *Combustion and Flame*, 88(3–4): 239–264.

U. Maas and J. Warnatz, 1988, Ignition processes in hydrogen-oxygen mixtures, *Combustion and Flame*, 74(1): 53–69.

E. S. Oran and J. P. Boris, 2001, *Numerical Simulation of Reactive Flow*, 2nd ed., Cambridge University Press, Cambridge, UK.

L. Perko, 2001, *Differential Equations and Dynamical Systems*, 3rd ed., Springer, New York.

J. M. Powers and S. Paolucci, 2005, Accurate spatial resolution estimates for reactive supersonic flow with detailed chemistry, *AIAA Journal*, 43(5): 1088–1099.

J. M. Powers and M. Sen, 2015, *Mathematical Methods in Engineering*, Cambridge University Press, New York.

J. M. Powers, S. Paolucci, J. D. Mengers, and A. N. Al-Khateeb, 2015, Slow attractive canonical invariant manifolds for reactive systems, *Journal of Mathematical Chemistry*, 53(2): 737–766.

N. N. Semenoff, 1928, Zur Theorie des Verbrennungsprozesses, *Zeitschrift für Physik*, 48(7–8): 571–582.

N. N. Semenoff, 1935, *Chemical Kinetics and Chain Reactions*, Oxford University Press, London.

N. N. Semenov, 1959, *Some Problems in Chemical Kinetics and Reactivity*, Vol. 2, Princeton University Press, Princeton, NJ, Chapter 8.

S. H. Strogatz, 2015, *Nonlinear Dynamics and Chaos*, 2nd ed., Westview Press, Boulder, CO.

W. G. Vincenti and C. H. Kruger, 1965, *Introduction to Physical Gas Dynamics*, John Wiley, New York.

# 2 Gas Mixtures

*The theory of mixtures is more complicated than the theory of a single body but not different in kind.*

– Clifford Ambrose Truesdell III (1919–2000), *Rational Thermodynamics*

One is often faced with mixtures of simple compressible substances, and it the thermodynamics of such mixtures upon which attention is now fixed. Here, a discussion of some of the fundamentals of mixture theory is given. At this stage we focus on inert mixtures, deferring chemically reacting mixtures to a brief reintroduction in Sec. 3.9.2 and full discussion in Chapters 4–12. The thermodynamics of inert mixtures is a challenging topic about which much remains to be learned. We will concentrate on ideal mixtures of ideal gases, for which results are often consistent with intuition. The chemical engineering literature contains a full discussion of the many nuances associated with nonideal mixtures of nonideal materials. Relevant background for this chapter is found in standard undergraduate texts (Tester and Modell, 1997; Smith et al., 2004; Sandler, 2006; Borgnakke and Sonntag, 2013); consequently, this chapter will be review for many readers. Some of our discussion on mixtures is guided by these texts, especially that of Borgnakke and Sonntag, from whom we adopt many notational conventions.

## 2.1 Some General Issues

One of the most important issues in combustion is mixing. Typical gas phase combustion relies upon collisions between molecules of fuel and oxidizer. We give a relevant sketch in Fig. 2.1. On the left, we show a volume segregated by a diaphragm into two chambers. The left chamber is filled with fuel and the right chamber is filled with oxidizer. Because of the segregation, there will obviously be no reaction for the configuration on the left. If, however, the diaphragm is removed, random molecular motions over time will lead the fuel and oxidizer to become fully mixed, as depicted on the right as a reactive mixture. Here, fuel molecules have the opportunity to collide with oxidizer molecules. If the energy of collision is of sufficient magnitude, chemical reaction will occur. This mixing process is one the most important in combustion science, and it is critical that we have a mathematical language of sufficient richness to describe its nature. We focus on treating the mixture as a continuum and not consider

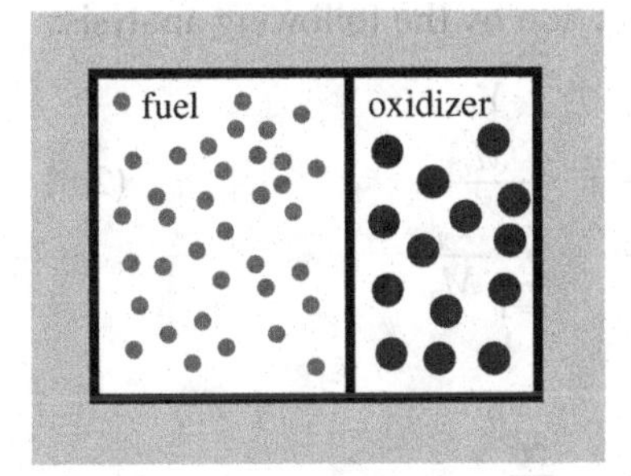

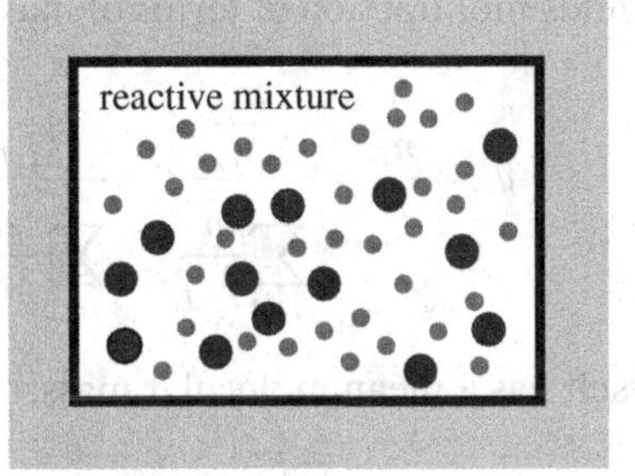


Figure 2.1. Sketch of unmixed and mixed systems of fuel and oxidizer.

individual collisions. We nevertheless recognize that it is molecular collisions on the microscale that ultimately drive the combustion of gas phase mixtures.

Consider then a mixture of $N$ components, each with mass $m_i$ and number of moles $n_i, i = 1, \ldots, N$, so that the total mass $m$ and total number of moles $n$ are

$$m = m_1 + m_2 + m_3 + \cdots + m_N = \sum_{i=1}^{N} m_i, \qquad \text{mass (g)}, \tag{2.1}$$

$$n = n_1 + n_2 + n_3 + \cdots + n_N = \sum_{i=1}^{N} n_i, \qquad \text{number (mol)}. \tag{2.2}$$

Recall 1 mol $= 6.02 \times 10^{23}$ molecules. The mass fraction of component $i$, $Y_i$ is

$$Y_i \equiv m_i/m, \qquad \text{mass fraction, dimensionless.} \tag{2.3}$$

The mol fraction of component $i$ is defined as $y_i$:

$$y_i \equiv n_i/n, \qquad \text{mol fraction, dimensionless.} \tag{2.4}$$

Now, the molecular mass $M_i$ of species $i$ is the mass per unit mol of species $i$. Its units are typically g/mol. Molecular mass is sometimes called "molecular weight," but this is incorrect nomenclature. Mathematically, the definition of $M_i$ is

$$M_i \equiv m_i/n_i, \qquad \text{molecular mass (g/mol)}\,. \tag{2.5}$$

One can write mass fraction in terms of mol fraction by the following analysis:

$$Y_i = \frac{m_i}{m} = \frac{n_i M_i}{m} = \frac{n_i M_i}{\sum_{j=1}^{N} m_j} = \frac{n_i M_i}{\sum_{j=1}^{N} n_j M_j}, = \frac{\dfrac{n_i M_i}{n}}{\dfrac{1}{n}\displaystyle\sum_{j=1}^{N} n_j M_j}, \tag{2.6}$$

$$= \frac{n_i M_i/n}{\displaystyle\sum_{j=1}^{N} n_j M_j/n} = \frac{y_i M_i}{\displaystyle\sum_{j=1}^{N} y_j M_j}. \tag{2.7}$$

Similarly, one finds mol fraction in terms of mass fraction by the following analysis:

$$y_i = \frac{n_i}{n} = \frac{\frac{m_i}{M_i}}{\sum_{j=1}^{N} \frac{m_j}{M_j}} = \frac{\frac{m_i}{M_i m}}{\sum_{j=1}^{N} \frac{m_j}{M_j m}} = \frac{\frac{Y_i}{M_i}}{\sum_{j=1}^{N} \frac{Y_j}{M_j}}. \tag{2.8}$$

The mixture itself has a mean molecular mass:

$$M \equiv \frac{m}{n} = \frac{\sum_{i=1}^{N} m_i}{n} = \sum_{i=1}^{N} \frac{n_i M_i}{n} = \sum_{i=1}^{N} y_i M_i. \tag{2.9}$$

**EXAMPLE 2.1**

Air is often modeled as a mixture in the following molar ratios:

$$O_2 + 3.76N_2. \tag{2.10}$$

This neglects the minor species commonly found in dry atmospheric air. They include typically, in descending order, Ar, $CO_2$, Ne, $O_3$, He, $CH_4$, Kr, $H_2$, and $N_2O$ (Jacob, 1999). The sum of the mol fractions of the minor species is typically less than 0.01. Additionally, a small amount of $H_2O$ is present in air; it can widely vary depending on local humidity. It can have a mol ratio from roughly $10^{-2}$ to as low as $10^{-6}$, similar to the range spanned from Ar to He.

Find the mol fractions, the mass fractions, and the mean molecular mass of the mixture of the major species $O_2$ and $N_2$.

Take $O_2$ to be species 1 and $N_2$ to be species 2. Consider the number of moles of $O_2$ and $N_2$ to be

$$n_1 = 1 \text{ mol}, \qquad n_2 = 3.76 \text{ mol}. \tag{2.11}$$

The molecular mass of $O_2$ is $M_1 = 32$ g/mol. The molecular mass of $N_2$ is $M_2 = 28$ g/mol. The total number of moles is

$$n = 1 \text{ mol} + 3.76 \text{ mol} = 4.76 \text{ mol}. \tag{2.12}$$

So, the mol fractions are

$$y_1 = \frac{1 \text{ mol}}{4.76 \text{ mol}} = 0.2101, \qquad y_2 = \frac{3.76 \text{ mol}}{4.76 \text{ mol}} = 0.7899. \tag{2.13}$$

Note that

$$\sum_{i=1}^{N} y_i = 0.2101 + 0.7899 = 1. \tag{2.14}$$

Now, for the masses, one has

$$m_1 = n_1 M_1 = (1 \text{ mol}) \left(32 \frac{\text{g}}{\text{mol}}\right) = 32 \text{ g}, \tag{2.15}$$

$$m_2 = n_2 M_2 = (3.76 \text{ mol}) \left(28 \frac{\text{g}}{\text{mol}}\right) = 105.28 \text{ g}. \tag{2.16}$$

So, one has

$$m = m_1 + m_2 = 32 \text{ g} + 105.28 \text{ g} = 137.28 \text{ g}. \tag{2.17}$$

The mass fractions then are

$$Y_1 = \frac{m_1}{m} = \frac{32\text{ g}}{137.28\text{ g}} = 0.2331, \quad Y_2 = \frac{m_2}{m} = \frac{105.28\text{ g}}{137.28\text{ g}} = 0.7669. \tag{2.18}$$

Note that

$$\sum_{i=1}^{N} Y_i = 0.2331 + 0.7669 = 1. \tag{2.19}$$

When the molecular masses of component species vary, the mass fractions do not equal the mol fractions. Now, for the mixture molecular mass, one has

$$M = \frac{m}{n} = \frac{137.28\text{ g}}{4.76\text{ mol}} = 28.84\,\frac{\text{g}}{\text{mol}}. \tag{2.20}$$

We can check against Eq. (2.9):

$$M = \sum_{i=1}^{N} y_i M_i = y_1 M_1 + y_2 M_2, \tag{2.21}$$

$$= (0.2101)\left(32\,\frac{\text{g}}{\text{mol}}\right) + (0.7899)\left(28\,\frac{\text{g}}{\text{mol}}\right) = 28.84\,\frac{\text{g}}{\text{mol}}. \tag{2.22}$$

Now, guiding principles for mixtures are not as well established as those for pure substances. Discussion in the scientific literature remains inconclusive (Truesdell, 1984). Often it is possible to model a mixture as a single material. The best example of this is air. It is in fact composed of nitrogen, oxygen, argon, and many other minor species. It is well modeled as a single material with properties that are suitable averages of its components' properties. Such an approach has widely accepted success in predicting actual experimental results for air near atmospheric conditions. For other problems, it is essential to have a theory that addresses the components of the mixture, as well as the mixture itself. Such will be the case for combustion.

## 2.2 Ideal and Nonideal Mixtures

Let us now introduce some concepts from thermodynamics as they apply to mixtures. First one can recall some standard definitions from thermodynamics. An *extensive thermodynamic property* is one that is proportional to the mass of the system. Examples include the mass $m$ itself, the extensive internal energy $E$, and the total volume $V$. An *intensive thermodynamic property* is one that is independent of the mass of the system. Examples in include the temperature $T$, the specific internal energy $e = E/m$, and the specific volume, $v = V/m$. Also recall that a simple compressible substance is a single material that is subject only to work via a pressure force acting through a volume change; other work modes such those due to electric or magnetic forces are not present. A single simple compressible substance can be shown to have its thermodynamic state completely determined by two independent intensive thermodynamic properties. An example is the ideal gas law for a single ideal gas, $P = \rho RT$, where the pressure $P$ is determined by the temperature $T$ and density $\rho$. Here, $R$ is a constant.

In contrast, a general extensive property, such as internal energy $E$, for an $N$-species mixture of simple compressible substances will be taken such that

$$E = E(T, P, n_1, n_2, \ldots, n_N). \tag{2.23}$$

That is to say, it is a function of two intensive properties temperature $T$ and pressure $P$, as well as the number of moles of each component. At this stage, we imagine that $T$ and $P$ are appropriate for the mixture and not necessarily the components of the mixture.

A *partial molar property* is a generalization of an intensive property, and is defined such that it is the *partial derivative of an extensive property with respect to the number of moles, with $T$ and $P$ held constant.* For internal energy, the partial molar internal energy is

$$\overline{e}_i \equiv \left.\frac{\partial E}{\partial n_i}\right|_{T,P,n_j,i\neq j}, \qquad i = 1, \ldots, N. \tag{2.24}$$

Pressure and temperature are held constant because those are convenient variables to control in an experiment. One also has the partial molar volume

$$\overline{\mathrm{v}}_i = \left.\frac{\partial V}{\partial n_i}\right|_{T,P,n_j,i\neq j}, \qquad i = 1, \ldots, N. \tag{2.25}$$

It shall be soon seen that there are other natural ways to think of the volume per mol.

Now, the form of Eq. (2.23) leads us to conclude that one would expect to find

$$\overline{e}_i = \overline{e}_i(T, P, n_1, n_2, \ldots, n_N), \qquad i = 1, \ldots, N, \tag{2.26}$$

$$\overline{\mathrm{v}}_i = \overline{\mathrm{v}}_i(T, P, n_1, n_2, \ldots, n_N), \qquad i = 1, \ldots, N. \tag{2.27}$$

This is the case for what is known as a *nonideal mixture.* An *ideal mixture* is defined as a mixture for which the partial molar properties $\overline{e}_i$ and $\overline{\mathrm{v}}_i$ are not functions of the composition; that is to say

$$\overline{e}_i = \overline{e}_i(T, P), \qquad i = 1, \ldots, N, \qquad \text{if ideal mixture}, \tag{2.28}$$

$$\overline{\mathrm{v}}_i = \overline{\mathrm{v}}_i(T, P), \qquad i = 1, \ldots, N, \qquad \text{if ideal mixture}. \tag{2.29}$$

An ideal mixture also has the feature that

$$\overline{h}_i = \overline{h}_i(T, P), \qquad \text{if ideal mixture}, \tag{2.30}$$

while for a nonideal mixture $\overline{h}_i = \overline{h}_i(T, P, n_1, \ldots, n_N)$. Though not obvious, it will turn out that some partial molar properties of an ideal mixture will depend on composition. For example, the entropy of a constituent of an ideal mixture will be such that

$$\overline{s}_i = \overline{s}_i(T, P, n_1, n_2, \ldots, n_N), \qquad i = 1, \ldots, N, \qquad \text{if ideal mixture}. \tag{2.31}$$

See Sec. 2.3.3 for exposition.

## 2.3 Ideal Mixtures of Ideal Gases

The most straightforward mixture to consider is an ideal mixture of ideal gases. Even here, there are assumptions necessary that remain difficult to validate absolutely.

### 2.3.1 Dalton Model

The most common model for an ideal mixture of ideal gases is the *Dalton*[1] *model.* Key assumptions characterize this model:

- Each constituent shares a common temperature.
- Each constituent occupies the entire volume.
- Each constituent possesses a *partial pressure* that sums to form the total pressure of the mixture.

These features characterize a Dalton model for any gas, ideal or nonideal. One also takes for convenience that

- Each constituent behaves as an ideal gas.
- The mixture behaves as a single ideal gas.

It is more convenient to deal on a molar basis for such a theory. For the Dalton model, additional useful quantities, the species mass concentration $\rho_i$, the mixture mass concentration $\rho$, the species molar concentration $\overline{\rho}_i$, and the mixture molar concentration $\overline{\rho}$, can be defined. These notations for concentrations are useful; however, they are not in common usage. The bar notation, $\bar{\cdot}$, will be reserved for properties that are mol-based rather than mass-based. For the Dalton model, in which each component occupies the same volume, one has

$$V_i = V, \qquad i = 1, \ldots, N, \qquad (\mathrm{cm}^3). \tag{2.32}$$

That is to say the partial volumes are each the mixture volume. Unless necessary for other reasons, we take it to be understood that $i = 1, \ldots, N$, and from here out will often omit writing this explicitly. The mixture mass concentration, also called the density, is simply

$$\rho = m/V, \qquad (\mathrm{g/cm}^3). \tag{2.33}$$

The mixture molar concentration is

$$\overline{\rho} = n/V, \qquad (\mathrm{mol/cm}^3). \tag{2.34}$$

For species $i$, the mass and molar equivalents are

$$\rho_i = m_i/V, \quad (\mathrm{g/cm}^3), \qquad \overline{\rho}_i = n_i/V, \quad (\mathrm{mol/cm}^3). \tag{2.35}$$

One can find a convenient relation between species molar concentration and species mol fraction by the following operations, beginning with the second of Eqs. (2.35):

$$\overline{\rho}_i = \frac{n_i}{V}\frac{n}{n} = \frac{n_i}{n}\frac{n}{V} = y_i\overline{\rho}. \tag{2.36}$$

A similar relation exists between species molar concentration and species mass fraction:

$$\overline{\rho}_i = \frac{n_i}{V}\frac{m}{m}\frac{M_i}{M_i} = \overbrace{\frac{m}{V}}^{\rho}\,\overbrace{\frac{n_i M_i}{m M_i}}^{m_i} = \rho\,\overbrace{\frac{m_i}{m}}^{Y_i}\frac{1}{M_i} = \rho\frac{Y_i}{M_i}. \tag{2.37}$$

The specific volumes, mass and molar, are similar. One takes

$$v = \frac{V}{m}, \qquad \overline{v} = \frac{V}{n}, \qquad v_i = \frac{V}{m_i}, \qquad \overline{v}_i = \frac{V}{n_i}. \tag{2.38}$$

[1] John Dalton (1766–1844), English physicist.

This definition of molar specific volume is *not* the partial molar volume defined in the chemical engineering literature, which takes the form $\overline{v}_i = \partial V/\partial n_i|_{T,P,n_j,i\neq j}$.

### 2.3.2 Thermodynamics of the Dalton Model

The standard laws of thermodynamics are easily extended to the Dalton model. We express the standard axioms as follows:

- *Zeroth law: When two mixtures have equality of temperature with a third mixture, they have equality of temperature.*
- *First law: The rate of change in total energy of a mixture is equal to the difference of the rate of heat added to the mixture and the rate of work done by the mixture*

$$\frac{d}{dt}\left(E + m\frac{\mathbf{u}\cdot\mathbf{u}}{2}\right) = \frac{\delta Q}{dt} - \frac{\delta W}{dt}. \tag{2.39}$$

- *Second law: The rate of change of entropy of a mixture is greater than or equal to the ratio of the rate of heat added to the temperature of the mixture*

$$\frac{dS}{dt} \geq \frac{1}{T}\frac{\delta Q}{dt}. \tag{2.40}$$

- *Third law: Every mixture has a finite positive entropy, but at the absolute zero of temperature, the entropy may become zero, and does so in the case of a pure crystalline substance* (adapted from Lewis and Randall, 1923).

These laws coupled with equations of state for the pressure and internal energy of the mixture form the foundation of the thermodynamics of combustion. We have explicitly introduced time $t$ into our equations describing change of thermodynamic state, contrasting a more traditional approach that confines attention to changes of state without regard to dynamics in time. As will be discussed in detail in Sec. 1.3, $\delta$ denotes a so-called inexact differential. The terms $Q$ and $W$ denote heat transfer into the system and work done by the system, respectively. We will mainly confine our attention to reversible work done by a simple compressible substance, which gives

$$\frac{\delta W}{dt} = P\frac{dV}{dt}. \tag{2.41}$$

That is to say, when we consider work, it will be work done by a pressure force acting through a volume change. We will thus neglect work done by other forces such as electrical or magnetic.

Our first law, Eq. (2.39), has accounted for the possibility of changes in both extensive internal energy $E$ and extensive kinetic energy $m\mathbf{u}\cdot\mathbf{u}/2$. Here, $\mathbf{u}$ is the velocity vector. A significant portion of this book's analyses will consider the case in which the kinetic energy is either zero or unchanging, in which case the first law reduces to

$$\frac{dE}{dt} = \frac{\delta Q}{dt} - \frac{\delta W}{dt}. \tag{2.42}$$

As far as the dynamics of combustion are concerned, we can say the following:

- The zeroth law of thermodynamics plays the same role in reactive mixtures as it does in classical thermodynamics.

- For nonisothermal combustion systems, the first law of thermodynamics plays an essential role in the dynamics of the process. It is explicitly represented as one of the differential equations describing the evolution of the state of the system.
- The second law of thermodynamics places restrictions on the form of the equations governing combustion dynamics; specifically, it requires stability at and in the neighborhood of equilibrium.
- The third law of thermodynamics, developed by Nernst,[2] and is used to guide the construction of state equations for components of mixtures. Its role is often not obvious, but it is essential to have correct values for the equations of state for energy of each mixture component to correctly capture the combustion dynamics. It forces the specific heats to approach zero as the temperature approaches absolute zero.
- The laws of thermodynamics coupled with equations of state are *necessary but insufficient* to describe the dynamics of combustion. They must be supplemented by independent theory for the kinetics of the combustion process.

For the partial pressure of species $i$, one can say for the Dalton model

$$P = \sum_{i=1}^{N} P_i. \tag{2.43}$$

For species $i$, one has

$$P_i V_i = n_i \overline{R} T_i, \tag{2.44}$$

but because $V_i = V$ and $T_i = T$ in the Dalton model, we get

$$P_i V = n_i \overline{R} T, \tag{2.45}$$

$$P_i = \frac{n_i \overline{R} T}{V}, \tag{2.46}$$

$$\underbrace{\sum_{i=1}^{N} P_i}_{P} = \sum_{i=1}^{N} \frac{n_i \overline{R} T}{V}, \tag{2.47}$$

$$P = \frac{\overline{R} T}{V} \underbrace{\sum_{i=1}^{N} n_i}_{n}. \tag{2.48}$$

So, for the mixture, one has

$$PV = n\overline{R}T. \tag{2.49}$$

One could also say

$$P = \frac{n}{V}\overline{R}T = \overline{\rho}\overline{R}T. \tag{2.50}$$

Additionally $\overline{R}$ is the universal gas constant with value

$$\overline{R} = 8.314 \text{ J/(mol K)}. \tag{2.51}$$

[2]Walther Hermann Nernst (1864–1941), German physicist; 1920 Nobel laureate in Chemistry.

Sometimes this is expressed in terms of $k_b$ the *Boltzmann*[3] *constant* and $\mathcal{N}_A$, *Avogadro's*[4] *number*:

$$\overline{R} = k_b \mathcal{N}_A, \quad \mathcal{N}_A = 6.02 \times 10^{23} \, \frac{\text{molecule}}{\text{mol}}, \quad k_b = 1.38 \times 10^{-23} \, \frac{\text{J}}{\text{K molecule}}. \tag{2.52}$$

With the Boltzmann constant and Avogadro's number, the ideal gas law, Eq. (2.49), can be rewritten as

$$PV = \mathcal{N} k_b T, \tag{2.53}$$

where $\mathcal{N} = n\mathcal{N}_A$ is the number of molecules. See Boltzmann, 1995, for background.

**EXAMPLE 2.2**

Compare the molar specific volume with the partial molar volume.

The partial molar volume $\overline{\text{v}}_i$, is given by Eq. (2.25):

$$\overline{\text{v}}_i = \left. \frac{\partial V}{\partial n_i} \right|_{T,P,n_j,i \neq j}. \tag{2.54}$$

For the ideal gas, one has from Eq. (2.48)

$$PV = \overline{R}T \sum_{k=1}^{N} n_k, \tag{2.55}$$

$$V = \frac{\overline{R}T}{P} \sum_{k=1}^{N} n_k, \tag{2.56}$$

$$\left. \frac{\partial V}{\partial n_i} \right|_{T,P,n_j,i \neq j} = \frac{\overline{R}T}{P} \sum_{k=1}^{N} \frac{\partial n_k}{\partial n_i} = \frac{\overline{R}T}{P} \sum_{k=1}^{N} \delta_{ki}, \tag{2.57}$$

$$= \frac{\overline{R}T}{P} \left( \overbrace{\delta_{1i}}^{0} + \overbrace{\delta_{2i}}^{0} + \cdots + \overbrace{\delta_{ii}}^{1} + \cdots + \overbrace{\delta_{Ni}}^{0} \right), \tag{2.58}$$

$$\overline{\text{v}}_i = \frac{\overline{R}T}{P} = \frac{V}{\sum_{k=1}^{N} n_k} = \frac{V}{n}. \tag{2.59}$$

Here, the so-called Kronecker[5] delta has been employed, which is much the same as the identity matrix:

$$\delta_{ki} = \begin{cases} 0, & k \neq i, \\ 1, & k = i. \end{cases} \tag{2.60}$$

Contrast this with the earlier adopted definition of molar specific volume from Eq. (2.38):

$$\overline{v}_i = V/n_i. \tag{2.61}$$

So, why is there a difference? The molar specific volume is a simple definition. One takes the instantaneous volume $V$, which is shared by all species in the Dalton model, and scales it by the instantaneous number of moles of species $i$, and acquires a natural definition of molar specific volume consistent with the notion of a mass specific volume.

[3] Ludwig Boltzmann (1844–1906), Austrian physicist.

[4] Lorenzo Romano Amedeo Carlo Bernadette Avogadro di Quaregna e di Cerreto (1776–1856), Italian scientist.

[5] Leopold Kronecker (1823–1891), German mathematician.

On the other hand, the partial molar volume specifies how the volume changes if the number of moles of species $i$ changes, *while holding $T$ and $P$ and all other species mol numbers constant.* One can imagine that in adding a mol of species $i$, one would find it necessary to change $V$ to guarantee that $P$ remains fixed.

**Binary Mixtures**

Consider now a *binary mixture* of components $A$ and $B$. This is easily extended to a mixture of $N$ components. First, the total number of moles is the sum of the components:

$$n = n_A + n_B. \tag{2.62}$$

Now, write the ideal gas law for each component:

$$P_A V_A = n_A \overline{R} T_A, \quad P_B V_B = n_B \overline{R} T_B. \tag{2.63}$$

But by the assumptions of the Dalton model, $V_A = V_B = V$, and $T_A = T_B = T$, so

$$P_A V = n_A \overline{R} T, \quad P_B V = n_B \overline{R} T. \tag{2.64}$$

One also has for the mixture

$$PV = n\overline{R}T. \tag{2.65}$$

Solving for $n, n_A$ and $n_B$, one finds

$$n = \frac{PV}{\overline{R}T}, \quad n_A = \frac{P_A V}{\overline{R}T}, \quad n_B = \frac{P_B V}{\overline{R}T}. \tag{2.66}$$

Now, $n = n_A + n_B$, so one has

$$\frac{PV}{\overline{R}T} = \frac{P_A V}{\overline{R}T} + \frac{P_B V}{\overline{R}T}, \qquad P = P_A + P_B. \tag{2.67}$$

That is, the total pressure is the sum of the partial pressures. This is a mixture rule for pressure. One can also scale each constituent ideal gas law by the mixture ideal gas law to get

$$\frac{P_A V}{PV} = \frac{n_A \overline{R} T}{n \overline{R} T}, \tag{2.68}$$

$$\frac{P_A}{P} = \frac{n_A}{n} = y_A, \tag{2.69}$$

$$P_A = y_A P. \tag{2.70}$$

Likewise

$$P_B = y_B P. \tag{2.71}$$

Now, one also needs mixture rules for energy, enthalpy, and entropy. The total internal energy $E$ (with units erg) for the binary mixture is taken to be

$$E = me = m_A e_A + m_B e_B, \tag{2.72}$$

$$= m\left(\frac{m_A}{m} e_A + \frac{m_B}{m} e_B\right), \tag{2.73}$$

$$= m\left(Y_A e_A + Y_B e_B\right), \tag{2.74}$$

$$e = Y_A e_A + Y_B e_B. \tag{2.75}$$

For the enthalpy, one has

$$H = mh = m_A h_A + m_B h_B, \tag{2.76}$$

$$= m\left(\frac{m_A}{m} h_A + \frac{m_B}{m} h_B\right) = m\left(Y_A h_A + Y_B h_B\right), \tag{2.77}$$

$$h = Y_A h_A + Y_B h_B. \tag{2.78}$$

It is easy to extend this to a mol fraction basis. One can also obtain a gas constant for the mixture on a mass basis. One can define the mixture gas constant $R$ via

$$PV = n\overline{R}T \equiv mRT, \tag{2.79}$$

$$\frac{PV}{T} \equiv mR = n\overline{R} = (n_A + n_B)\overline{R}, \tag{2.80}$$

$$= \left(\frac{m_A}{M_A} + \frac{m_B}{M_B}\right)\overline{R}, \tag{2.81}$$

$$= \left(m_A \frac{\overline{R}}{M_A} + m_B \frac{\overline{R}}{M_B}\right) = \left(m_A R_A + m_B R_B\right), \tag{2.82}$$

$$R = \left(\frac{m_A}{m} R_A + \frac{m_B}{m} R_B\right) = \left(Y_A R_A + Y_B R_B\right). \tag{2.83}$$

Here, we have taken species gas constants to be defined as

$$R_A = \overline{R}/M_A, \qquad R_B = \overline{R}/M_B. \tag{2.84}$$

The terms $R_A$ and $R_B$ are the typical "gas constants" used in the mechanical engineering community, for which each ideal gas carries its own constant. The units of $R_A$ and $R_B$ are both erg/g/K. For the entropy, one has

$$S = ms = m_A s_A + m_B s_B = m\left(\frac{m_A}{m} s_A + \frac{m_B}{m} s_B\right), \tag{2.85}$$

$$= m\left(Y_A s_A + Y_B s_B\right), \tag{2.86}$$

$$s = Y_A s_A + Y_B s_B. \tag{2.87}$$

In the Dalton model, $s_A$ is evaluated at $T$ and $P_A$, while $s_B$ is evaluated at $T$ and $P_B$. For a calorically perfect ideal gas, one has

$$s_A = \underbrace{s^0_{T_0,A} + c_{PA} \ln\left(\frac{T}{T_0}\right)}_{\equiv s^0_{T,A}} - R_A \ln\left(\frac{P_A}{P_0}\right), \tag{2.88}$$

$$= s^0_{T_0,A} + c_{PA} \ln\left(\frac{T}{T_0}\right) - R_A \ln\left(\frac{y_A P}{P_0}\right). \tag{2.89}$$

Likewise

$$s_B = \underbrace{s^0_{T_0,B} + c_{PB} \ln\left(\frac{T}{T_0}\right)}_{\equiv s^0_{T,B}} - R_B \ln\left(\frac{y_B P}{P_0}\right). \tag{2.90}$$

In general, we can say for calorically perfect ideal gases

$$s_i = \underbrace{s^0_{T_0,i} + c_{Pi} \ln\left(\frac{T}{T_0}\right)}_{\equiv s^0_{T,i}} - R_i \ln\left(\frac{y_i P}{P_0}\right), \qquad i = A, B. \tag{2.91}$$

Here, the 0 denotes some reference state. As a superscript, it typically means that the property is evaluated at a reference pressure. For example, $s^0_{T,A}$ denotes the portion of the entropy of component $A$ that is evaluated at the reference pressure $P_0$ and is allowed to vary with temperature $T$. Furthermore, $s^0_{T_0,A}$ is a constant and represents the entropy evaluated at $T_0$ and $P_0$. It is often found in thermodynamic tables. Note also that $s_A = s_A(T, P, y_A)$ and $s_B = s_B(T, P, y_B)$, so the entropy of a single constituent depends on the composition of the mixture and not just on $T$ and $P$. This contrasts with energy and enthalpy for which $e_A = e_A(T), e_B = e_B(T), h_A = h_A(T), h_B = h_B(T)$ if the mixture is composed of ideal gases. Occasionally, one finds $h^0_A$ and $h^0_B$ used as a notation. This denotes that the enthalpy is evaluated at the reference pressure, and says nothing about temperature. However, if the gas is ideal, the enthalpy is not a function of pressure and $h_A = h^0_A, h_B = h^0_B$.

If one is employing a calorically imperfect ideal gas model, then one has

$$s_i = s^0_{T,i} - R_i \ln\left(\frac{y_i P}{P_0}\right), \qquad i = A, B. \tag{2.92}$$

In this case, one either has a complicated analytic form for $s^0_{T,i}$ as a function of $T$, or it is given in discrete tabular form as a function of $T$.

### Entropy of Mixing

Let us take up an important topic: the entropy of mixing. We know from experience that once components have been mixed, it requires work to separate them. Thus, using standard notions of classical thermodynamics, we believe that all else equal, the entropy of a mixed system should be greater than its unmixed components. We will study through an example the entropy of mixing of two inert ideal gases. The formulation we provide will serve as the foundation for critical concepts in chemical equilibrium to be discussed in Secs. 4.4 and 5.2.

**EXAMPLE 2.3**

Initially calorically perfect ideal gases $A$ and $B$ are segregated within the same volume by a thin frictionless, thermally conducting diaphragm, free to accelerate in response to a net force. See Fig. 2.2. Taking the initial state to be in mechanical and thermal equilibrium, both are at the same initial pressure and temperature, $P_1$ and $T_1$. The total volume is thermally insulated and fixed, so there are no global heat or work exchanges with the environment. The diaphragm is removed, and $A$ and $B$ are allowed to mix. Assume $A$ and $B$ have mass $m_A$ and $m_B$. The gases have distinct molecular masses, $M_A$ and $M_B$. Find the final temperature $T_2$, pressure $P_2$, and the change in entropy.

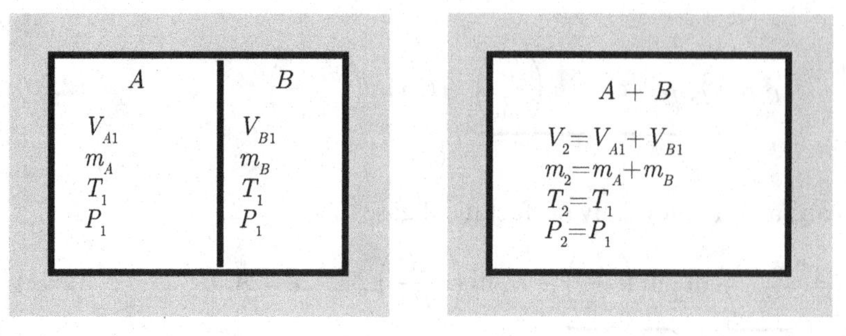


Figure 2.2. Sketch of an adiabatic binary system before and after mixing.

The ideal gas law holds that at the initial state

$$V_{A1} = \frac{m_A R_A T_1}{P_1}, \qquad V_{B1} = \frac{m_B R_B T_1}{P_1}. \tag{2.93}$$

At the final state one has

$$V_2 = V_{A2} = V_{B2} = V_{A1} + V_{B1} = (m_A R_A + m_B R_B)\frac{T_1}{P_1}. \tag{2.94}$$

Mass conservation gives

$$m_2 = m_1 = m_A + m_B. \tag{2.95}$$

One also has the first law of thermodynamics, Eq. (2.42), modified here not to be in rate form:

$$E_2 - E_1 = Q - W. \tag{2.96}$$

Here, the heat transfer $Q = 0$ because the system is adiabatic, and the work $W$ is zero because there is no volume change. So the first law becomes

$$E_2 - E_1 = 0, \tag{2.97}$$

$$E_2 = E_1, \tag{2.98}$$

$$m_2 e_2 = m_A e_{A1} + m_B e_{B1}, \tag{2.99}$$

$$(m_A + m_B)e_2 = m_A e_{A1} + m_B e_{B1}, \tag{2.100}$$

$$0 = m_A(e_{A1} - e_2) + m_B(e_{B1} - e_2), \tag{2.101}$$

$$= m_A c_{vA}(T_1 - T_2) + m_B c_{vB}(T_1 - T_2), \tag{2.102}$$

$$T_2 = \frac{m_A c_{vA} T_1 + m_B c_{vB} T_1}{m_A c_{vA} + m_B c_{vB}}, \tag{2.103}$$

$$= T_1. \tag{2.104}$$

The first law analysis simply confirms that if the two components are at the same initial temperature, the final temperature of the mixture is unchanged from that of the initial state. The final pressure by Dalton's law then is

$$P_2 = P_{A2} + P_{B2} = \frac{m_A R_A T_2}{V_2} + \frac{m_B R_B T_2}{V_2}, \tag{2.105}$$

$$= \frac{m_A R_A T_1}{V_2} + \frac{m_B R_B T_1}{V_2} = \frac{(m_A R_A + m_B R_B)T_1}{V_2}, \tag{2.106}$$

$$= \frac{(m_A R_A + m_B R_B)T_1}{(m_A R_A + m_B R_B)\frac{T_1}{P_1}} = P_1. \tag{2.107}$$

So, the initial and final temperatures and pressures are identical.

Now, the entropy change of gas $A$ is

$$s_{A2} - s_{A1} = c_{PA} \ln\left(\frac{T_{A2}}{T_{A1}}\right) - R_A \ln\left(\frac{P_{A2}}{P_{A1}}\right), \tag{2.108}$$

$$= c_{PA} \ln\left(\frac{T_2}{T_1}\right) - R_A \ln\left(\frac{y_{A2}P_2}{y_{A1}P_1}\right), \tag{2.109}$$

$$= \underbrace{c_{PA} \ln\left(\frac{T_1}{T_1}\right)}_{0} - R_A \ln\left(\frac{y_{A2}P_1}{y_{A1}P_1}\right), \tag{2.110}$$

$$= -R_A \ln\left(\frac{y_{A2}P_1}{(1)P_1}\right) = -R_A \ln y_{A2}. \tag{2.111}$$

Likewise

$$s_{B2} - s_{B1} = -R_B \ln y_{B2}. \tag{2.112}$$

So, the change in entropy of the mixture is

$$\Delta S = m_A(s_{A2} - s_{A1}) + m_B(s_{B2} - s_{B1}), \tag{2.113}$$

$$= -m_A R_A \ln y_{A2} - m_B R_B \ln y_{B2}, \tag{2.114}$$

$$= -\underbrace{(n_A M_A)}_{m_A}\underbrace{\left(\frac{\overline{R}}{M_A}\right)}_{R_A} \ln y_{A2} - \underbrace{(n_B M_B)}_{m_B}\underbrace{\left(\frac{\overline{R}}{M_B}\right)}_{R_B} \ln y_{B2}, \tag{2.115}$$

$$= -\overline{R}(n_A \ln y_{A2} + n_B \ln y_{B2}), \tag{2.116}$$

$$= -\overline{R}\left(\underbrace{n_A \ln\left(\frac{n_A}{n_A+n_B}\right)}_{\leq 0} + \underbrace{n_B \ln\left(\frac{n_B}{n_A+n_B}\right)}_{\leq 0}\right), \tag{2.117}$$

$$\geq 0. \tag{2.118}$$

The mixing process alone induces an entropy increase. Pressure and temperature do not change.

For an $N$-component mixture, mixed in the same fashion such that $P$ and $T$ are constant, Eq. (2.116) extends to

$$\Delta S = -\overline{R}\sum_{k=1}^{N} n_k \ln y_k = -\overline{R}\sum_{k=1}^{N} \frac{n_k}{n} n \ln y_k, \tag{2.119}$$

$$= -\overline{R}\frac{m}{M}\sum_{k=1}^{N} y_k \ln y_k = -Rm \sum_{k=1}^{N} \ln y_k^{y_k}, \tag{2.120}$$

$$= -Rm\left(\ln y_1^{y_1} + \ln y_2^{y_2} + \cdots + \ln y_N^{y_N}\right), \tag{2.121}$$

$$= -Rm \ln\left(y_1^{y_1} y_2^{y_2} \ldots y_N^{y_N}\right) = -Rm \ln\left(\prod_{k=1}^{N} y_k^{y_k}\right), \tag{2.122}$$

$$\frac{\Delta s}{R} = \frac{\Delta \overline{s}}{\overline{R}} = -\ln\left(\prod_{k=1}^{N} y_k^{y_k}\right). \tag{2.123}$$

There is a fundamental dependency of the mixing entropy on the mol fractions. Because $0 \leq y_k \leq 1$, the product is guaranteed to be between 0 and 1. The natural logarithm of such a number is negative, and thus the entropy change for the mixture is guaranteed positive semidefinite.

Now, if one mol of pure $N_2$ is mixed with one mol of pure $O_2$, one certainly expects the resulting mixture to have a higher entropy than the two pure components. The process is not reversible in the sense of classical thermodynamics. Separation into it original constituents could only be achieved with energy input. But what if one mol of pure $N_2$ is mixed with another mol of pure $N_2$? Then, we would expect no increase in entropy. However, if we had the unusual ability to distinguish $N_2$ molecules whose origin was from each respective original chamber, then indeed there would be an entropy of mixing. Increases in entropy thus do correspond to increases in disorder.

The preceding discussion is often taken to define the so-called Gibbs paradox, more specifically the *Gibbs mixing paradox*. Müller and Weiss (2005) note that the Gibbs paradox comes in two flavors: one associated with thermodynamics of mixtures and the other with statistical mechanics. The bulk of the literature on the Gibbs paradox refers to the latter, while our example deals with the former. A nuanced discussion is given by Jaynes (1992). Such nuances are worth appreciating because, as will be seen in Chapters 3–5, it is Dalton's law–motivated formulations of entropy, especially like Eq. (2.92), that (1) are critical in determination of equilibrium conditions of gas phase mixtures (see Secs. 4.4 and 5.2), (2) influence the dynamics of systems away from equilibrium (see Secs. 4.5 and 5.1), and (3) influence many advanced topics in nonequilibrium thermodynamics (see Sec. 5.5).

### Mixtures of Constant Mass Fraction

If the mass fractions, and thus the mol fractions, remain constant during a process, the equations simplify. Inert air at moderate values of temperature and pressure behaves this way. For a calorically perfect ideal gas, one would have

$$e_2 - e_1 = Y_A c_{vA}(T_2 - T_1) + Y_B c_{vB}(T_2 - T_1) = c_v(T_2 - T_1), \tag{2.124}$$

where

$$c_v \equiv Y_A c_{vA} + Y_B c_{vB}. \tag{2.125}$$

Similarly for enthalpy

$$h_2 - h_1 = Y_A c_{PA}(T_2 - T_1) + Y_B c_{PB}(T_2 - T_1) = c_P(T_2 - T_1), \tag{2.126}$$

where

$$c_P \equiv Y_A c_{PA} + Y_B c_{PB}. \tag{2.127}$$

For the entropy

$$s_2 - s_1 = Y_A(s_{A2} - s_{A1}) + Y_B(s_{B2} - s_{B1}), \tag{2.128}$$

$$= Y_A\left(c_{PA} \ln\left(\frac{T_2}{T_1}\right) - R_A \ln\left(\frac{y_A P_2}{y_A P_1}\right)\right)$$
$$+ Y_B\left(c_{PB} \ln\left(\frac{T_2}{T_1}\right) - R_B \ln\left(\frac{y_B P_2}{y_B P_1}\right)\right), \tag{2.129}$$

$$= Y_A \left( c_{PA} \ln \left( \frac{T_2}{T_1} \right) - R_A \ln \left( \frac{P_2}{P_1} \right) \right)$$

$$+ Y_B \left( c_{PB} \ln \left( \frac{T_2}{T_1} \right) - R_B \ln \left( \frac{P_2}{P_1} \right) \right), \tag{2.130}$$

$$= c_P \ln \left( \frac{T_2}{T_1} \right) - R \ln \left( \frac{P_2}{P_1} \right). \tag{2.131}$$

The mixture behaves as a pure substance when the appropriate mixture properties are defined. One can also take for the mixture ratio of specific heats

$$\gamma = c_P / c_v. \tag{2.132}$$

Some intuitive definitions do not hold. For $\gamma_A = c_{PA}/c_{vA}$, $\gamma_B = c_{PB}/c_{vB}$, it can be shown that for the mixture, $\gamma \neq Y_A \gamma_A + Y_B \gamma_B$.

### 2.3.3 Summary of Properties of the Dalton Mixture Model

For convenience, we give here a list of common properties for the Dalton model first on a mass basis, then on a molar basis.

#### Mass Basis

Listed here is a summary of mixture properties for an $N$-component mixture of ideal gases on a mass basis:

$$1 = \sum_{i=1}^{N} Y_i, \tag{2.133}$$

$$M = \sum_{i=1}^{N} y_i M_i, \tag{2.134}$$

$$\rho = \sum_{i=1}^{N} \rho_i, \tag{2.135}$$

$$v = \frac{1}{\sum_{i=1}^{N} \frac{1}{v_i}} = \frac{1}{\rho}, \tag{2.136}$$

$$e = \sum_{i=1}^{N} Y_i e_i, \tag{2.137}$$

$$h = \sum_{i=1}^{N} Y_i h_i, \tag{2.138}$$

$$R = \frac{\overline{R}}{M} = \sum_{i=1}^{N} Y_i R_i = \sum_{i=1}^{N} \frac{y_i \overbrace{M_i R_i}^{\overline{R}}}{\underbrace{\sum_{j=1}^{N} y_j M_j}_{M}} = \frac{\overline{R}}{M} \underbrace{\sum_{i=1}^{N} y_i}_{1}, \tag{2.139}$$

$$c_v = \sum_{i=1}^{N} Y_i c_{vi}, \tag{2.140}$$

$$c_v = c_P - R, \qquad \text{if ideal gas} \tag{2.141}$$

$$c_P = \sum_{i=1}^{N} Y_i c_{Pi}, \tag{2.142}$$

$$\gamma = \frac{c_P}{c_v} = \frac{\sum_{i=1}^{N} Y_i c_{Pi}}{\sum_{i=1}^{N} Y_i c_{vi}}, \tag{2.143}$$

$$s = \sum_{i=1}^{N} Y_i s_i, \tag{2.144}$$

$$Y_i = \frac{y_i M_i}{M}, \tag{2.145}$$

$$P_i = y_i P, \tag{2.146}$$

$$\rho_i = Y_i \rho, \tag{2.147}$$

$$v_i = \frac{v}{Y_i} = \frac{1}{\rho_i}, \tag{2.148}$$

$$V = V_i, \tag{2.149}$$

$$T = T_i, \tag{2.150}$$

$$h_i = h_i^0, \qquad \text{if ideal gas,} \tag{2.151}$$

$$h_i = e_i + \frac{P_i}{\rho_i} = e_i + P_i v_i = \underbrace{e_i + R_i T}_{\text{if ideal gas}}, \tag{2.152}$$

$$h_i = \underbrace{h_{T_0,i}^0 + \int_{T_0}^{T} c_{Pi}(\hat{T})\, d\hat{T}}_{\text{if ideal gas}}, \tag{2.153}$$

$$s_i = \underbrace{\underbrace{s_{T_0,i}^0 + \int_{T_0}^{T} \frac{c_{Pi}(\hat{T})}{\hat{T}}\, d\hat{T}}_{s_{T,i}^0} - R_i \ln\left(\frac{P_i}{P_0}\right)}_{\text{if ideal gas}}, \tag{2.154}$$

$$s_i = \underbrace{s_{T,i}^0 - R_i \ln\left(\frac{y_i P}{P_0}\right)}_{\text{if ideal gas}} = \underbrace{s_{T,i}^0 - R_i \ln\left(\frac{P_i}{P_0}\right)}_{\text{if ideal gas}}, \tag{2.155}$$

$$P_i = \underbrace{\rho_i R_i T = \rho R_i T Y_i = \frac{R_i T}{v_i}}_{\text{if ideal gas}}, \tag{2.156}$$

$$P = \underbrace{\rho R T = \rho \overline{R} T \sum_{i=1}^{N} \frac{Y_i}{M_i} = \frac{RT}{v}}_{\text{if ideal gas}}, \tag{2.157}$$

$$h = \underbrace{\sum_{i=1}^{N} Y_i h_{T_0,i}^0 + \int_{T_0}^{T} c_P(\hat{T})\, d\hat{T}}_{\text{if ideal gas}}, \tag{2.158}$$

$$h = e + \frac{P}{\rho} = e + Pv = \underbrace{e + RT}_{\text{if ideal gas}}, \tag{2.159}$$

$$s = \underbrace{\sum_{i=1}^{N} Y_i s_{T_0,i}^0 + \int_{T_0}^{T} \frac{c_P(\hat{T})}{\hat{T}}\, d\hat{T} - R \ln\left(\frac{P}{P_0}\right) - R \ln\left(\prod_{i=1}^{N} y_i^{y_i}\right)}_{\text{if ideal gas}} \tag{2.160}$$

These relations are not obvious. A few are derived in examples here.

**EXAMPLE 2.4**
Derive the expression $h = e + P/\rho$.

Start from the equation for the constituent $h_i$, multiply by mass fractions, sum over all species, and use properties of mixtures:

$$h_i = e_i + \frac{P_i}{\rho_i}, \tag{2.161}$$

$$Y_i h_i = Y_i e_i + Y_i \frac{P_i}{\rho_i}, \tag{2.162}$$

$$\sum_{i=1}^{N} Y_i h_i = \sum_{i=1}^{N} Y_i e_i + \sum_{i=1}^{N} Y_i \frac{P_i}{\rho_i}, \tag{2.163}$$

$$\underbrace{\sum_{i=1}^{N} Y_i h_i}_{h} = \underbrace{\sum_{i=1}^{N} Y_i e_i}_{e} + \sum_{i=1}^{N} Y_i \frac{\rho_i R_i T}{\rho_i}, \tag{2.164}$$

$$h = e + T \underbrace{\sum_{i=1}^{N} Y_i R_i}_{R} = e + RT = e + \frac{P}{\rho}. \tag{2.165}$$

**EXAMPLE 2.5**
Find the expression for mixture entropy of the ideal gas.

$$s_i = s_{T_0,i}^0 + \int_{T_0}^{T} \frac{c_{Pi}(\hat{T})}{\hat{T}}\, d\hat{T} - R_i \ln\left(\frac{P_i}{P_0}\right), \tag{2.166}$$

$$Y_i s_i = Y_i s_{T_0,i}^0 + Y_i \int_{T_0}^{T} \frac{c_{Pi}(\hat{T})}{\hat{T}}\, d\hat{T} - Y_i R_i \ln\left(\frac{P_i}{P_0}\right), \tag{2.167}$$

$$s = \sum_{i=1}^{N} Y_i s_i = \sum_{i=1}^{N} Y_i s_{T_0,i}^0 + \sum_{i=1}^{N} Y_i \int_{T_0}^{T} \frac{c_{Pi}(\hat{T})}{\hat{T}} d\hat{T} - \sum_{i=1}^{N} Y_i R_i \ln\left(\frac{P_i}{P_0}\right), \tag{2.168}$$

$$= \sum_{i=1}^{N} Y_i s_{T_0,i}^0 + \int_{T_0}^{T} \sum_{i=1}^{N} \frac{Y_i c_{Pi}(\hat{T})}{\hat{T}} d\hat{T} - \sum_{i=1}^{N} Y_i R_i \ln\left(\frac{P_i}{P_0}\right), \tag{2.169}$$

$$= s_{T_0}^0 + \int_{T_0}^{T} \frac{c_P(\hat{T})}{\hat{T}} d\hat{T} - \sum_{i=1}^{N} Y_i R_i \ln\left(\frac{P_i}{P_0}\right). \tag{2.170}$$

All except the last term are natural extensions of the property for a single material. Consider now the last term involving pressure ratios:

$$-\sum_{i=1}^{N} Y_i R_i \ln\left(\frac{P_i}{P_0}\right) = -\left(\sum_{i=1}^{N} Y_i R_i \ln\left(\frac{P_i}{P_0}\right) + R \ln\frac{P}{P_0} - R\ln\frac{P}{P_0}\right), \tag{2.171}$$

$$= -R\left(\sum_{i=1}^{N} Y_i \frac{R_i}{R} \ln\left(\frac{P_i}{P_0}\right) + \ln\frac{P}{P_0} - \ln\frac{P}{P_0}\right), \tag{2.172}$$

$$= -R\left(\sum_{i=1}^{N} Y_i \frac{\overline{R}/M_i}{\sum_{j=1}^{N} Y_j \overline{R}/M_j} \ln\left(\frac{P_i}{P_0}\right) + \ln\frac{P}{P_0} - \ln\frac{P}{P_0}\right), \tag{2.173}$$

$$= -R\left(\sum_{i=1}^{N} \underbrace{\left(\frac{Y_i/M_i}{\sum_{j=1}^{N} Y_j/M_j}\right)}_{y_i} \ln\left(\frac{P_i}{P_0}\right) + \ln\frac{P}{P_0} - \ln\frac{P}{P_0}\right), \tag{2.174}$$

$$= -R\left(\sum_{i=1}^{N} y_i \ln\left(\frac{P_i}{P_0}\right) + \ln\frac{P}{P_0} - \ln\frac{P}{P_0}\right), \tag{2.175}$$

$$= -R\left(\sum_{i=1}^{N} \ln\left(\frac{P_i}{P_0}\right)^{y_i} - \ln\frac{P}{P_0} + \ln\frac{P}{P_0}\right), \tag{2.176}$$

$$= -R\left(\ln\left(\prod_{i=1}^{N}\left(\frac{P_i}{P_0}\right)^{y_i}\right) + \ln\frac{P_o}{P} + \ln\frac{P}{P_0}\right), \tag{2.177}$$

$$= -R\left(\ln\left(\frac{P_0}{P}\prod_{i=1}^{N}\left(\frac{P_i}{P_0}\right)^{y_i}\right) + \ln\frac{P}{P_0}\right), \tag{2.178}$$

$$= -R\left(\ln\left(\frac{P_0}{P^{\sum_{i=1}^{N} y_i}} \frac{1}{P_0^{\sum_{i=1}^{N} y_i}} \prod_{i=1}^{N} (P_i)^{y_i}\right) + \ln\frac{P}{P_0}\right), \tag{2.179}$$

$$= -R\left(\ln\left(\prod_{i=1}^{N}\left(\frac{P_i}{P}\right)^{y_i}\right) + \ln\frac{P}{P_0}\right), \tag{2.180}$$

$$= -R\left(\ln\left(\prod_{i=1}^{N}\left(\frac{y_i P}{P}\right)^{y_i}\right) + \ln\frac{P}{P_0}\right), \tag{2.181}$$

$$= -R\left(\ln\left(\prod_{i=1}^{N} y_i^{y_i}\right) + \ln\frac{P}{P_0}\right). \tag{2.182}$$

So, the mixture entropy becomes

$$s = s^0_{T_0} + \int_{T_0}^{T} \frac{c_P(\hat{T})}{\hat{T}}\,d\hat{T} - R\left(\ln\left(\prod_{i=1}^{N} y_i^{y_i}\right) + \ln\frac{P}{P_0}\right), \tag{2.183}$$

$$= \underbrace{s^0_{T_0} + \int_{T_0}^{T} \frac{c_P(\hat{T})}{\hat{T}}\,d\hat{T} - R\ln\frac{P}{P_0}}_{\text{classical entropy of a single body}} \underbrace{- R\ln\left(\prod_{i=1}^{N} y_i^{y_i}\right)}_{\text{mixture effect}}. \tag{2.184}$$

The first bracketed portion of the mixture entropy based on Dalton's theory of mixtures has indeed the same form as the classical entropy of a single body. But the Dalton theory induces another term, labeled here "mixture effect." While it is possible to redefine the constituent entropy in such a fashion that the mixture entropy in fact takes on the classical form of a single material via $s_i = s^0_{T_0,i} + \int_{T_0}^{T} c_{Pi}(\hat{T})/\hat{T}\,d\hat{T} - R_i \ln(P_i/P_0) + R_i \ln y_i$, this has the disadvantage of predicting *no entropy change* after mixing two pure substances. Such a theory would suggest that this obviously irreversible process is in fact reversible.

### Molar Basis

On a molar basis, one has equivalent relations to those found on a mass basis:

$$1 = \sum_{i=1}^{N} y_i, \tag{2.185}$$

$$\overline{\rho} = \sum_{i=1}^{N} \overline{\rho}_i = \frac{n}{V} = \frac{\rho}{M}, \tag{2.186}$$

$$\overline{v} = \frac{1}{\sum_{i=1}^{N} \frac{1}{\overline{v}_i}} = \frac{V}{n} = \frac{1}{\overline{\rho}} = vM, \tag{2.187}$$

$$\overline{e} = \sum_{i=1}^{N} y_i \overline{e}_i = eM, \tag{2.188}$$

$$\overline{h} = \sum_{i=1}^{N} y_i \overline{h}_i = hM, \tag{2.189}$$

$$\overline{c}_v = \sum_{i=1}^{N} y_i \overline{c}_{vi} = c_v M, \tag{2.190}$$

$$\overline{c}_v = \overline{c}_P - \overline{R}, \tag{2.191}$$

$$\overline{c}_P = \sum_{i=1}^{N} y_i \overline{c}_{Pi} = c_P M, \tag{2.192}$$

$$\gamma = \frac{\overline{c}_P}{\overline{c}_v} = \frac{\sum_{i=1}^{N} y_i \overline{c}_{Pi}}{\sum_{i=1}^{N} y_i \overline{c}_{vi}}, \tag{2.193}$$

$$\overline{s} = \sum_{i=1}^{N} y_i \overline{s}_i = sM, \tag{2.194}$$

$$\overline{\rho}_i = y_i \overline{\rho} = \frac{\rho_i}{M_i}, \tag{2.195}$$

$$\overline{v}_i = \frac{V}{n_i} = \frac{\overline{v}}{y_i} = \frac{1}{\overline{\rho}_i} = v_i M_i, \tag{2.196}$$

$$\overline{\mathrm{v}}_i = \left.\frac{\partial V}{\partial n_i}\right|_{P,T,n_j} = \underbrace{\frac{V}{n} = \overline{v} = vM}_{\text{if ideal gas}}, \tag{2.197}$$

$$P_i = y_i P, \tag{2.198}$$

$$\underbrace{P = \overline{\rho}\overline{R}T = \frac{\overline{R}T}{\overline{v}}}_{\text{if ideal gas}}, \tag{2.199}$$

$$\underbrace{P_i = \overline{\rho}_i \overline{R}T = \frac{\overline{R}T}{\overline{v}_i}}_{\text{if ideal gas}}, \tag{2.200}$$

$$\overline{h} = \overline{e} + \frac{P}{\overline{\rho}} = \overline{e} + P\overline{v} = \underbrace{\overline{e} + \overline{R}T}_{\text{if ideal gas}} = hM, \tag{2.201}$$

$$\overline{h}_i = \overline{h}_i^0, \qquad \text{if ideal gas}, \tag{2.202}$$

$$\overline{h}_i = \overline{e}_i + \frac{P_i}{\overline{\rho}_i} = \overline{e}_i + P_i\overline{v}_i = \overline{e}_i + P\overline{\mathrm{v}}_i = \underbrace{\overline{e}_i + \overline{R}T}_{\text{if ideal gas}} = h_i M_i, \tag{2.203}$$

$$\underbrace{\overline{h}_i = \overline{h}_{T_0,i}^0 + \int_{T_0}^{T} \overline{c}_{Pi}(\hat{T})\, d\hat{T}}_{\text{if ideal gas}} = h_i M_i, \tag{2.204}$$

$$\underbrace{\overline{s}_i = \underbrace{\overline{s}_{T_0,i}^0 + \int_{T_0}^{T} \frac{\overline{c}_{Pi}(\hat{T})}{\hat{T}}\, d\hat{T}}_{\overline{s}_{T,i}^0} - \overline{R}\ln\left(\frac{y_i P}{P_0}\right)}_{\text{if ideal gas}}, \tag{2.205}$$

$$\underbrace{\overline{s}_i = \overline{s}_{T,i}^0 - \overline{R}\ln\left(\frac{y_i P}{P_0}\right)}_{\text{if ideal gas}} = s_i M_i, \tag{2.206}$$

$$\overline{s} = \underbrace{\sum_{i=1}^{N} y_i \overline{s}_{T_0,i}^0 + \int_{T_0}^{T} \frac{\overline{c}_P(\hat{T})}{\hat{T}}\, d\hat{T} - \overline{R}\ln\left(\frac{P}{P_0}\right) - \overline{R}\ln\left(\prod_{i=1}^{N} y_i^{y_i}\right)}_{\text{if ideal gas}} = sM. \tag{2.207}$$

## EXERCISES

**2.1.** A local volume with a sample of air is found to have 2.34 mol of $N_2$, 0.63 mol of $O_2$, 0.0279 mol of Ar, and 0.001095 mol of $CO_2$. With the aid of a periodic table, find the mole and mass fractions of each component, and the mixture gas constant $R$.

**2.2.** A volume with two chambers contains calorically perfect ideal gases. One chamber has volume $V_A$ containing gas $A$ with $n_A$ moles, at temperature $T_A$, with specific heat at constant volume $c_{vA}$. The other chamber has volume $V_B$ containing gas $B$ with $n_B$ moles, at temperature $T_B$, with specific heat at constant volume $c_{vB}$. The two gases are allowed to fully mix while a thermal energy input $Q$ is introduced to the total volume. The surroundings are at $T_0$. At equilibrium, find the final temperature, the final pressure, the total entropy change of the gas mixture, and the total entropy change of the universe.

**2.3.** Derive Eq. (2.207).

## References

L. Boltzmann, 1995, *Lectures on Gas Theory*, Dover, New York.

C. Borgnakke and R. E. Sonntag, 2013, *Fundamentals of Thermodynamics*, 8th ed., John Wiley, New York.

D. J. Jacob, 1999, *Introduction to Atmospheric Chemistry*, Princeton University Press, Princeton, NJ, p. 4.

E. T. Jaynes, 1992, The Gibbs paradox, in *Maximum Entropy and Bayesian Methods*, edited by C. R. Smith, G. J. Erickson, and P. O. Neudorfer, Kluwer, Dordrecht, Netherlands, pp. 1–22.

G. N. Lewis and M. Randall, 1923, *Thermodynamics and the Free Energy of Chemical Substances*, McGraw-Hill, New York, p. 448.

I. Müller and W. Weiss, 2005, *Entropy and Energy*, Springer, New York, pp. 169–178.

S. I. Sandler, 2006, *Chemical, Biochemical, and Engineering Thermodynamics*, 4th ed., John Wiley, New York.

J. M. Smith, H. C. Van Ness, and M. Abbott, 2004, *Introduction to Chemical Engineering Thermodynamics*, 7th ed., McGraw-Hill, New York.

J. W. Tester and M. Modell, 1997, *Thermodynamics and Its Applications*, 3rd ed., Prentice Hall, Upper Saddle River, NJ.

C. A. Truesdell, 1984, *Rational Thermodynamics*, 2nd ed., Springer, New York.

# 3 Mathematical Foundations of Thermodynamics

*Every mathematician knows it is impossible to understand an elementary course in thermodynamics.*[*]

— Vladimir Igorevich Arnol'd (1937–2010), "Contact Geometry: The Geometrical Method of Gibbs's Thermodynamics"

This chapter focuses on the application of mathematical formalism to thermodynamics. A basic understanding of undergraduate engineering thermodynamics is presumed, so many notions will be introduced with little discussion. While the detail will go beyond that covered in typical undergraduate thermodynamics courses, the key mathematical ideas are in fact elementary ones from multivariable calculus, and should be familiar to most readers. We begin with a discussion of single materials, then move on to mixtures. While some of the discussion is seemingly far removed from combustion, most is necessary to frame the second law and equilibrium in binary mixtures, which closes the chapter. This consideration of how to describe equilibrium is the linchpin in connecting dynamics to thermodynamics, and will be built upon throughout the remainder of the book. Throughout the chapter, in keeping with traditional approaches to thermodynamics, there is no introduction of time, and all materials are assumed to be in a state of equilibrium. Understanding of equilibrium-based concepts is essential before moving to the nonequilibrium time-evolving reaction kinetics in Chapter 4. Further background can be found in standard sources (Fermi, 1936; Planck, 1945; Vincenti and Kruger, 1965; Kestin, 1966; Reynolds, 1968; Abbott and Van Ness, 1972; Ziegler, 1977; Callen, 1985; Gyftopoulos and Beretta, 1991; Müller, 2007; Borgnakke and Sonntag, 2013).

## 3.1 Exact Differentials and State Properties

In thermodynamics, one is faced with many equations of similar form as that of the well-known Gibbs equation:[1]

$$de = T\,ds - P\,dv. \tag{3.1}$$

[*] V. I. Arnol'd, 1990, Contact geometry: the geometrical method of Gibbs's thermodynamics, *Proceedings of the Gibbs Symposium*, Yale U., D. Caldi and G. Mostow, eds., American Mathematical Society, p. 163.

[1] This particular equation was certainly given by first by Gibbs (1873). It is called by some authors the "first Gibbs equation." Other authors will commonly introduce its extensive analog that for a single material is $dE = T\,dS - P\,dV$ and call it the "fundamental thermodynamic relation."

This is known to be an exact differential with the consequence that specific internal energy $e$ is a function of the state of the system and not the details of any process that led to the state. As counterexamples, the heat transfer $q$, whose differential is for reversible heat transfer

$$\delta q = T\, ds, \tag{3.2}$$

and the work $w$, whose differential is for reversible pressure-volume work

$$\delta w = P\, dv, \tag{3.3}$$

can be shown to be inexact differentials. We use the notations $d$ and $\delta$ to indicate exact and inexact differentials, respectively. Thus, both the heat transfer and the work depend upon the path of a thermodynamic process. With these definitions of $\delta q$ and $\delta w$, the Gibbs equation gives

$$de = \delta q - \delta w. \tag{3.4}$$

This is also the first law of thermodynamics for systems whose kinetic and potential energy changes are negligible.

**EXAMPLE 3.1**
Show the work is not a state property.

*If* work were a state property, one might expect it to have the form

$$w = w(P, v), \qquad \textit{provisional assumption, to be tested.} \tag{3.5}$$

In such a case, one would have the corresponding exact differential

$$dw = \left.\frac{\partial w}{\partial v}\right|_P dv + \left.\frac{\partial w}{\partial P}\right|_v dP. \tag{3.6}$$

Now, because it is known from elementary mechanics that $dw = P\, dv + 0\, dP$, one deduces that

$$\left.\frac{\partial w}{\partial v}\right|_P = P, \tag{3.7}$$

$$\left.\frac{\partial w}{\partial P}\right|_v = 0. \tag{3.8}$$

Integrating Eq. (3.7), one finds

$$w = Pv + f(P), \tag{3.9}$$

where $f(P)$ is some function of $P$ to be determined. Differentiating Eq. (3.9) with respect to $P$, one gets

$$\left.\frac{\partial w}{\partial P}\right|_v = v + \frac{df(P)}{dP}. \tag{3.10}$$

Now, use Eq. (3.8) to eliminate $\partial w/\partial P|_v$ in Eq. (3.10) so as to obtain

$$\frac{df(P)}{dP} = -v. \tag{3.11}$$

Equation (3.11) cannot be: a function of $P$ cannot be a function of $v$. So, $w$ cannot be a state property:

$$w \neq w(P, v). \tag{3.12}$$

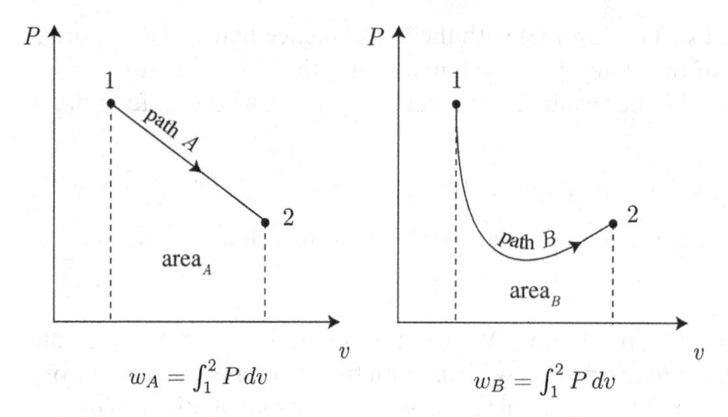


Figure 3.1. Sketch of $w$ from identical end states via different paths, illustrating that $w$ is not a thermodynamic state property.

The path dependence of $w$ is illustrated in Fig. 3.1, where $w$ is determined along two different paths connecting the same end states.

Consider now the more general form

$$\psi_1\,dx_1 + \cdots + \psi_N\,dx_N = \sum_{i=1}^{N} \psi_i\,dx_i. \tag{3.13}$$

Here, $\psi_i$ and $x_i, i = 1, \ldots, N$, may be thermodynamic variables. This form is known in mathematics as a *Pfaff*[2] *differential form.* As formulated, one takes at this stage

- $x_i$: Independent thermodynamic variables.
- $\psi_i$: Thermodynamic variables that are functions of $x_i$.

Now if Eq. (3.13), when set to a differential $dy$, can be integrated to form

$$y = y(x_1, \ldots, x_N), \tag{3.14}$$

the differential is said to be *exact*. In such a case, one has

$$dy = \psi_1\,dx_1 + \cdots + \psi_N\,dx_N = \sum_{i=1}^{N} \psi_i\,dx_i. \tag{3.15}$$

Now, if the algebraic definition of Eq. (3.14) holds, what amounts to the definition of the partial derivative gives the parallel result that

$$dy = \left.\frac{\partial y}{\partial x_1}\right|_{x_j, j\neq 1} dx_1 + \cdots + \left.\frac{\partial y}{\partial x_N}\right|_{x_j, j\neq N} dx_N. \tag{3.16}$$

Now, combining Eqs. (3.15) and (3.16) to eliminate $dy$, one gets

$$\psi_1\,dx_1 + \cdots + \psi_N\,dx_N = \left.\frac{\partial y}{\partial x_1}\right|_{x_j, j\neq 1} dx_1 + \cdots + \left.\frac{\partial y}{\partial x_N}\right|_{x_j, j\neq N} dx_N. \tag{3.17}$$

[2]Johann Friedrich Pfaff (1765–1825), German mathematician.

Rearranging, one gets

$$0 = \left( \left.\frac{\partial y}{\partial x_1}\right|_{x_j, j \neq 1} - \psi_1 \right) dx_1 + \cdots + \left( \left.\frac{\partial y}{\partial x_N}\right|_{x_j, j \neq N} - \psi_N \right) dx_N. \tag{3.18}$$

The variables $x_i, i = 1, \ldots, N$, are independent. Thus, their differentials, $dx_i, i = 1, \ldots, N$, are all independent in Eq. (3.18), and in general nonzero. For equality, one must require that *each* of the coefficients be zero, so

$$\psi_1 = \left.\frac{\partial y}{\partial x_1}\right|_{x_j, j \neq 1}, \quad \ldots, \quad \psi_N = \left.\frac{\partial y}{\partial x_N}\right|_{x_j, j \neq N}. \tag{3.19}$$

When $dy$ is exact, one says that each of the $\psi_i$ and $x_i$ are *conjugate* to each other. From here on, for notational ease, subscripts such as $j \neq 1, j \neq 2, \ldots, j \neq N$ will be ignored in the notation for the partial derivatives. It becomes especially confusing for higher-order derivatives and is fairly obvious for all derivatives.

If $y$ and all its derivatives are continuous and sufficiently differentiable, then one has for all $i = 1, \ldots, N$, and $k = 1, \ldots, N$, that

$$\frac{\partial^2 y}{\partial x_k \partial x_i} = \frac{\partial^2 y}{\partial x_i \partial x_k}. \tag{3.20}$$

That is to say, the partial differentiation operator commutes. Now, from Eq. (3.19), one can infer for general $k$ or $l$ that

$$\psi_k = \left.\frac{\partial y}{\partial x_k}\right|_{x_j}, \qquad \psi_l = \left.\frac{\partial y}{\partial x_l}\right|_{x_j}. \tag{3.21}$$

Taking the partial derivative of the first of Eq. (3.21) with respect to $x_l$ and of the second with respect to $x_k$, one gets

$$\left.\frac{\partial \psi_k}{\partial x_l}\right|_{x_j} = \frac{\partial^2 y}{\partial x_l \partial x_k}, \qquad \left.\frac{\partial \psi_l}{\partial x_k}\right|_{x_j} = \frac{\partial^2 y}{\partial x_k \partial x_l}. \tag{3.22}$$

Because by Eq. (3.20) the order of the mixed second partials does not matter, one deduces from Eq. (3.22) that

$$\left.\frac{\partial \psi_k}{\partial x_l}\right|_{x_j} = \left.\frac{\partial \psi_l}{\partial x_k}\right|_{x_j}. \tag{3.23}$$

This is a necessary and sufficient condition for the exact-ness of Eq. (3.13). It is a generalization of what can be found in most introductory calculus texts for functions of two variables.

For the Gibbs equation, Eq. (3.1), $de = -P\,dv + T\,ds$, one has

$$y \to e, \quad x_1 \to v, \quad x_2 \to s, \quad \psi_1 \to -P, \quad \psi_2 \to T, \tag{3.24}$$

and one expects the natural, or *canonical* form of

$$e = e(v, s). \tag{3.25}$$

Here, $-P$ is conjugate to $v$, and $T$ is conjugate to $s$. Application of the general form of Eq. (3.23) to the Gibbs equation (3.1) gives then

$$\left.\frac{\partial T}{\partial v}\right|_s = -\left.\frac{\partial P}{\partial s}\right|_v. \tag{3.26}$$

Equation (3.26) is known as a *Maxwell*[3] *relation*. Moreover, specialization of Eq. (3.21) to the Gibbs equation (3.1) gives

$$-P = \left.\frac{\partial e}{\partial v}\right|_s, \qquad T = \left.\frac{\partial e}{\partial s}\right|_v. \tag{3.27}$$

If the general differential $dy = \sum_{i=1}^{N} \psi_i \, dx_i$ is exact, one also can show the following:

- The path integral $y_B - y_A = \int_A^B \sum_{i=1}^{N} \psi_i \, dx_i$ is independent of the path of the integral.
- The integral around a closed contour is zero:

$$\oint dy = \oint \sum_{i=1}^{N} \psi_i \, dx_i = 0. \tag{3.28}$$

- The function $y$ can only be determined to within an additive constant. That is, there is no absolute value of $y$; physical significance is only ascribed to differences in $y$. In fact, other means, extraneous to this analysis, can be used to provide absolute specification of key thermodynamic variables. This will be important especially for flows with reaction.

**EXAMPLE 3.2**

Show the heat transfer $q$ is not a state property. Assume all processes are fully reversible and that the only work mode is that of a pressure force acting through a volume change, so $\delta w = -P dv$.

The first law of thermodynamics gives

$$de = \delta q - \delta w, \tag{3.29}$$

$$\delta q = de + \delta w = de + P\, dv. \tag{3.30}$$

Take now the noncanonical, although acceptable, form $e = e(T, v)$. Then, one gets

$$de = \left.\frac{\partial e}{\partial v}\right|_T dv + \left.\frac{\partial e}{\partial T}\right|_v dT. \tag{3.31}$$

So,

$$\delta q = \underbrace{\left.\frac{\partial e}{\partial v}\right|_T dv + \left.\frac{\partial e}{\partial T}\right|_v dT}_{de} + P\, dv, \tag{3.32}$$

$$= \underbrace{\left(\left.\frac{\partial e}{\partial v}\right|_T + P\right)}_{\equiv M} dv + \underbrace{\left.\frac{\partial e}{\partial T}\right|_v}_{\equiv N} dT. \tag{3.33}$$

$$= M\, dv + N\, dT. \tag{3.34}$$

Now, by Eq. (3.23), for $\delta q$ to be exact, one must have

$$\left.\frac{\partial M}{\partial T}\right|_v = \left.\frac{\partial N}{\partial v}\right|_T. \tag{3.35}$$

[3] James Clerk Maxwell (1831–1879), Scottish physicist and mathematician.

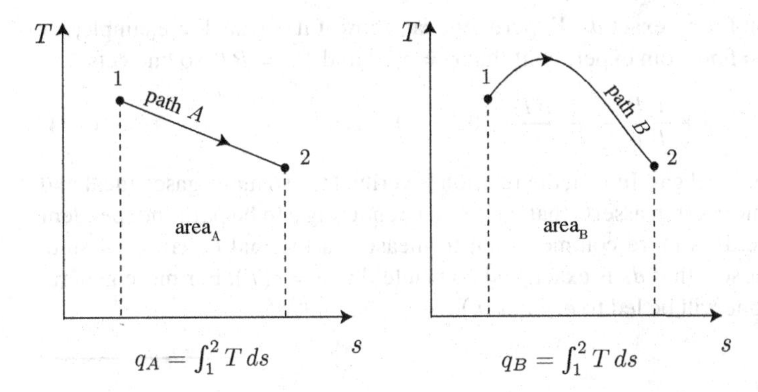


Figure 3.2. Sketch of $q$ from identical end states via different paths, illustrating that $q$ is not a thermodynamic state property.

This reduces to

$$\frac{\partial^2 e}{\partial T \partial v} + \left.\frac{\partial P}{\partial T}\right|_v = \frac{\partial^2 e}{\partial v \partial T}. \tag{3.36}$$

This can only be true if $\partial P/\partial T|_v = 0$. But this is not the case for gases; consider an ideal gas for which $P = \rho RT$. Then $\partial P/\partial T|_v = R/v \neq 0$. So, $\delta q$ is not exact. The path dependence of $q$ is illustrated in Fig. 3.2, where $q$ is determined along two different paths connecting the same end states.

**EXAMPLE 3.3**

Show conditions for $ds$ to be exact in the Gibbs equation, Eq. (3.1).

We have then

$$de = T\,ds - P\,dv, \tag{3.37}$$

$$ds = \frac{de}{T} + \frac{P}{T}\,dv, \tag{3.38}$$

$$= \underbrace{\frac{1}{T}\left(\left.\frac{\partial e}{\partial v}\right|_T dv + \left.\frac{\partial e}{\partial T}\right|_v dT\right)}_{de} + \frac{P}{T}\,dv, \tag{3.39}$$

$$= \underbrace{\left(\frac{1}{T}\left.\frac{\partial e}{\partial v}\right|_T + \frac{P}{T}\right)}_{\equiv M} dv + \underbrace{\frac{1}{T}\left.\frac{\partial e}{\partial T}\right|_v}_{\equiv N} dT. \tag{3.40}$$

Again, invoking Eq. (3.23), one gets

$$\left.\frac{\partial}{\partial T}\right|_v \underbrace{\left(\frac{1}{T}\left.\frac{\partial e}{\partial v}\right|_T + \frac{P}{T}\right)}_{M} = \left.\frac{\partial}{\partial v}\right|_T \underbrace{\left(\frac{1}{T}\left.\frac{\partial e}{\partial T}\right|_v\right)}_{N}, \tag{3.41}$$

$$\frac{1}{T}\frac{\partial^2 e}{\partial T \partial v} - \frac{1}{T^2}\left.\frac{\partial e}{\partial v}\right|_T + \frac{1}{T}\left.\frac{\partial P}{\partial T}\right|_v - \frac{P}{T^2} = \frac{1}{T}\frac{\partial^2 e}{\partial v \partial T}, \tag{3.42}$$

$$-\frac{1}{T^2}\left.\frac{\partial e}{\partial v}\right|_T + \frac{1}{T}\left.\frac{\partial P}{\partial T}\right|_v - \frac{P}{T^2} = 0. \tag{3.43}$$

This is the condition for an exact $ds$. Experiment can show if it is true. For example, for an ideal gas, one can find from experiment that $e = e(T)$ and $Pv = RT$, so one gets

$$0 + \frac{1}{T}\frac{R}{v} - \frac{1}{T^2}\frac{RT}{v} = 0, \qquad 0 = 0. \tag{3.44}$$

So, $ds$ is exact for an ideal gas. In fact, the relation is verified for so many gases, ideal and nonideal, that one more often asserts that $ds$ is exact, rendering $s$ to be path-independent and a state variable. It is more common then to measure a thermal equation of state such as $Pv = RT$, assert that $ds$ is exact, and conclude that $e = e(T)$. For more general equations of state, one will be led to $e = e(T, v)$.

## 3.2 Two Independent Variables

Consider a general implicit function linking three variables, $x$, $y$, $z$:

$$f(x, y, z) = 0. \tag{3.45}$$

In $(x, y, z)$ space, this will represent a surface. If the function can be inverted, it will be possible to write the explicit forms

$$x = x(y, z), \qquad y = y(x, z), \qquad z = z(x, y). \tag{3.46}$$

Differentiating the first two of the Eqs. (3.46) gives

$$dx = \left.\frac{\partial x}{\partial y}\right|_z dy + \left.\frac{\partial x}{\partial z}\right|_y dz, \tag{3.47}$$

$$dy = \left.\frac{\partial y}{\partial x}\right|_z dx + \left.\frac{\partial y}{\partial z}\right|_x dz. \tag{3.48}$$

Now, use Eq. (3.48) to eliminate $dy$ in Eq. (3.47):

$$dx = \left.\frac{\partial x}{\partial y}\right|_z \underbrace{\left( \left.\frac{\partial y}{\partial x}\right|_z dx + \left.\frac{\partial y}{\partial z}\right|_x dz \right)}_{dy} + \left.\frac{\partial x}{\partial z}\right|_y dz, \tag{3.49}$$

$$\left(1 - \left.\frac{\partial x}{\partial y}\right|_z \left.\frac{\partial y}{\partial x}\right|_z\right) dx = \left( \left.\frac{\partial x}{\partial y}\right|_z \left.\frac{\partial y}{\partial z}\right|_x + \left.\frac{\partial x}{\partial z}\right|_y \right) dz, \tag{3.50}$$

$$0\,dx + 0\,dz = \underbrace{\left( \left.\frac{\partial x}{\partial y}\right|_z \left.\frac{\partial y}{\partial x}\right|_z - 1 \right)}_{0} dx + \underbrace{\left( \left.\frac{\partial x}{\partial y}\right|_z \left.\frac{\partial y}{\partial z}\right|_x + \left.\frac{\partial x}{\partial z}\right|_y \right)}_{0} dz. \tag{3.51}$$

Because $x$ and $y$ are independent, so are $dx$ and $dy$, and the coefficients on each in Eq. (3.51) must be zero. Therefore, from the coefficient on $dx$ in Eq. (3.51),

$$\left.\frac{\partial x}{\partial y}\right|_z \left.\frac{\partial y}{\partial x}\right|_z - 1 = 0, \tag{3.52}$$

$$\left.\frac{\partial x}{\partial y}\right|_z = 1 \Big/ \left.\frac{\partial y}{\partial x}\right|_z, \tag{3.53}$$

and also from the coefficient on $dz$ in Eq. (3.51),

$$\left.\frac{\partial x}{\partial y}\right|_z \left.\frac{\partial y}{\partial z}\right|_x + \left.\frac{\partial x}{\partial z}\right|_y = 0, \tag{3.54}$$

$$\left.\frac{\partial x}{\partial z}\right|_y = -\left.\frac{\partial x}{\partial y}\right|_z \left.\frac{\partial y}{\partial z}\right|_x, \tag{3.55}$$

$$\left.\frac{\partial x}{\partial z}\right|_y \left.\frac{\partial y}{\partial x}\right|_z \left.\frac{\partial z}{\partial y}\right|_x = -1. \tag{3.56}$$

If one now divides Eq. (3.47) by a fourth differential, $dw$, one gets

$$\frac{dx}{dw} = \left.\frac{\partial x}{\partial y}\right|_z \frac{dy}{dw} + \left.\frac{\partial x}{\partial z}\right|_y \frac{dz}{dw}. \tag{3.57}$$

Demanding that $z$ be held constant in Eq. (3.57) gives

$$\left.\frac{\partial x}{\partial w}\right|_z = \left.\frac{\partial x}{\partial y}\right|_z \left.\frac{\partial y}{\partial w}\right|_z, \tag{3.58}$$

$$\left.\frac{\partial x}{\partial w}\right|_z \Bigg/ \left.\frac{\partial y}{\partial w}\right|_z = \left.\frac{\partial x}{\partial y}\right|_z, \tag{3.59}$$

$$\left.\frac{\partial x}{\partial w}\right|_z \left.\frac{\partial w}{\partial y}\right|_z = \left.\frac{\partial x}{\partial y}\right|_z. \tag{3.60}$$

If $x = x(y, w)$, one then gets

$$dx = \left.\frac{\partial x}{\partial y}\right|_w dy + \left.\frac{\partial x}{\partial w}\right|_y dw. \tag{3.61}$$

Divide now by $dy$ while holding $z$ constant so

$$\left.\frac{\partial x}{\partial y}\right|_z = \left.\frac{\partial x}{\partial y}\right|_w + \left.\frac{\partial x}{\partial w}\right|_y \left.\frac{\partial w}{\partial y}\right|_z. \tag{3.62}$$

These general operations can be applied to a wide variety of thermodynamic operations.

**EXAMPLE 3.4**

Apply Eq. (3.62) to a standard $(P, v, T)$ system and let

$$\left.\frac{\partial x}{\partial y}\right|_z = \left.\frac{\partial T}{\partial v}\right|_s. \tag{3.63}$$

Give analysis for a calorically perfect ideal gas.

So here we have $T = x$, $v = y$, and $s = z$. Let now $e = w$. So, Eq. (3.62) becomes

$$\left.\frac{\partial T}{\partial v}\right|_s = \left.\frac{\partial T}{\partial v}\right|_e + \left.\frac{\partial T}{\partial e}\right|_v \left.\frac{\partial e}{\partial v}\right|_s. \tag{3.64}$$

Now, by definition the specific heat at constant volume is given as

$$c_v = \left.\frac{\partial e}{\partial T}\right|_v, \quad \text{so} \quad \left.\frac{\partial T}{\partial e}\right|_v = \frac{1}{c_v}. \tag{3.65}$$

Now, by Eq. (3.27), one has $\partial e/\partial v|_s = -P$, so one gets

$$\left.\frac{\partial T}{\partial v}\right|_s = \left.\frac{\partial T}{\partial v}\right|_e - \frac{P}{c_v}. \tag{3.66}$$

For an ideal gas, $e = e(T)$. Inverting, one gets $T = T(e)$, and so $\partial T/\partial v|_e = 0$. Thus,

$$\left.\frac{\partial T}{\partial v}\right|_s = -\frac{P}{c_v}. \tag{3.67}$$

For an isentropic process in a calorically perfect ideal gas, one gets

$$\frac{dT}{dv} = -\frac{P}{c_v} = -\frac{RT}{c_v v}, \tag{3.68}$$

$$\frac{dT}{T} = -\frac{R}{c_v}\frac{dv}{v} = -(\gamma - 1)\frac{dv}{v}, \tag{3.69}$$

$$\ln\frac{T}{T_0} = (\gamma - 1)\ln\frac{v_0}{v}, \tag{3.70}$$

$$\frac{T}{T_0} = \left(\frac{v_0}{v}\right)^{\gamma-1}. \tag{3.71}$$

This is a well-known isentropic relation between $T$ and $v$ valid for calorically perfect ideal gases.

## 3.3 Legendre Transformations

The Gibbs equation (3.1), $de = -P\,dv + T\,ds$, is the fundamental equation of classical thermodynamics of a pure material. It is a canonical form that suggests the most natural set of variables in which to express internal energy $e$ are $s$ and $v$:

$$e = e(v, s). \tag{3.72}$$

However, $v$ and $s$ may not be convenient for a particular problem. There may be other combinations of variables whose canonical form gives a more useful set of independent variables for a particular problem. An example is the enthalpy:

$$h \equiv e + Pv. \tag{3.73}$$

Differentiating the enthalpy gives

$$dh = de + P\,dv + v\,dP. \tag{3.74}$$

Use now Eq. (3.74) to eliminate $de$ in the Gibbs equation, Eq. (3.1), to give

$$\underbrace{dh - P\,dv - v\,dP}_{de} = -P\,dv + T\,ds, \tag{3.75}$$

$$dh = T\,ds + v\,dP. \tag{3.76}$$

Equation (3.76) is called in some sources the "second Gibbs equation," but we will avoid this usage. The canonical variables for $h$ are $s$ and $P$. One then expects

$$h = h(s, P). \tag{3.77}$$

This exercise can be systematized with the Legendre[4] transformation,[5] which defines a set of second-order polynomial combinations of variables. Consider again the exact differential equation (3.15):

$$dy = \psi_1\,dx_1 + \psi_2\,dx_2 + \cdots + \psi_N\,dx_N. \tag{3.78}$$

For $N$ independent variables $x_i$ and $N$ conjugate variables $\psi_i$, by definition, there are $2^N - 1$ Legendre transformed variables:

$$\tau_1 = \tau_1(\psi_1, x_2, x_3, \ldots, x_N) = y - \psi_1 x_1, \tag{3.79}$$

$$\tau_2 = \tau_2(x_1, \psi_2, x_3, \ldots, x_N) = y - \psi_2 x_2, \tag{3.80}$$

$$\vdots$$

$$\tau_N = \tau_N(x_1, x_2, x_3, \ldots, \psi_N) = y - \psi_N x_N, \tag{3.81}$$

$$\tau_{1,2} = \tau_{1,2}(\psi_1, \psi_2, x_3, \ldots, x_N) = y - \psi_1 x_1 - \psi_2 x_2, \tag{3.82}$$

$$\tau_{1,3} = \tau_{1,3}(\psi_1, x_2, \psi_3, \ldots, x_N) = y - \psi_1 x_1 - \psi_3 x_3, \tag{3.83}$$

$$\vdots$$

$$\tau_{1,\ldots,N} = \tau_{1,\ldots,N}(\psi_1, \psi_2, \psi_3, \ldots, \psi_N) = y - \sum_{i=1}^{N} \psi_i x_i. \tag{3.84}$$

Each $\tau$ is a new dependent variable. *Each $\tau$ has the property that when it is known as a function of its N canonical variables, the remaining N variables from the original expression (the $x_i$ and the conjugate $\psi_i$) can be recovered by differentiation of $\tau$.* In general, this is not true for arbitrary transformations. As discussed by Cantwell (2002), the Legendre transformation for thermodynamic systems is a type of *contact transformation* discussed in the theory of Lie groups.

**EXAMPLE 3.5**

Let $y = y(x_1, x_2, x_3)$. With $\psi_i$ thus known to be $\psi_i = \partial y/\partial x_i$, derive expressions for $x_1$, $\psi_2$, and $\psi_3$ in terms of $\psi_1$, $x_2$, and $x_3$. Thus, replace $x_1$ as an independent variable by $\psi_1$. Equivalently, replace $\psi_1$ as a dependent variable by $x_1$.

We have the associated differential form

$$dy = \psi_1\,dx_1 + \psi_2\,dx_2 + \psi_3\,dx_3. \tag{3.85}$$

Because we know $y(x_1, x_2, x_3)$, we know $\psi_i = \partial y/\partial x_i$. Choose now a Legendre transformed variable $\tau_1 \equiv z(\psi_1, x_2, x_3)$:

$$z = y - \psi_1 x_1. \tag{3.86}$$

Then

$$dz = \left.\frac{\partial z}{\partial \psi_1}\right|_{x_1, x_2} d\psi_1 + \left.\frac{\partial z}{\partial x_2}\right|_{\psi_1, x_3} dx_2 + \left.\frac{\partial z}{\partial x_3}\right|_{\psi_1, x_2} dx_3. \tag{3.87}$$

[4]Adrien-Marie Legendre (1752–1833), French mathematician.

[5]Two differentiable functions $f$ and $g$ are said to be Legendre transformations of each other if their first derivatives are inverse functions of each other: $Df = (Dg)^{-1}$. With some effort, not shown here, one can prove that the Legendre transformations of this section satisfy this general condition.

Now, differentiating Eq. (3.86), one also gets

$$dz = dy - \psi_1\, dx_1 - x_1\, d\psi_1. \tag{3.88}$$

Elimination of $dy$ in Eq. (3.88) by using Eq. (3.85) gives

$$dz = \underbrace{\psi_1\, dx_1 + \psi_2\, dx_2 + \psi_3\, dx_3}_{dy} - \psi_1\, dx_1 - x_1\, d\psi_1 = -x_1\, d\psi_1 + \psi_2\, dx_2 + \psi_3\, dx_3. \tag{3.89}$$

Thus, from Eq. (3.87), one gets

$$x_1 = -\left.\frac{\partial z}{\partial \psi_1}\right|_{x_2,x_3}, \qquad \psi_2 = \left.\frac{\partial z}{\partial x_2}\right|_{\psi_1,x_3}, \qquad \psi_3 = \left.\frac{\partial z}{\partial x_3}\right|_{\psi_1,x_2}. \tag{3.90}$$

So, the original expression had three independent variables, $x_1, x_2, x_3$, and three conjugate variables $\psi_1, \psi_2, \psi_3$. Definition of the Legendre function[6] $z$ with canonical variables $\psi_1$, $x_2$, and $x_3$ allowed determination of the remaining variables $x_1$, $\psi_2$, and $\psi_3$ in terms of the canonical variables.

For the Gibbs equation (3.1), $de = -P\, dv + T\, ds$, one has $y = e$, two canonical variables, $x_1 = v$ and $x_2 = s$, and two conjugate variables, $\psi_1 = -P$ and $\psi_2 = T$. Thus, $N = 2$, and one can expect $2^2 - 1 = 3$ Legendre transformations. They are

$$\text{Enthalpy:} \quad \tau_1 = y - \psi_1 x_1 = h = h(P, s) = e + Pv, \tag{3.91}$$

$$\text{Helmholtz free energy:} \quad \tau_2 = y - \psi_2 x_2 = a = a(v, T) = e - Ts, \tag{3.92}$$

$$\text{Gibbs free energy:} \quad \tau_{1,2} = y - \psi_1 x_1 - \psi_2 x_2 = g = g(P, T) = e + Pv - Ts. \tag{3.93}$$

The functions, $e$, $h$, $a$, and $g$, which induce the various canonical and conjugate variable sets, are sometimes described as *thermodynamic potentials* (Callen, 1985). As is discussed in Sec. 5.5.5, such potentials share some, but not all, features with potential fields from classical mechanics. In particular, it is not obvious how to associate with them curl-free vector fields, such as can be done easily in classical mechanics.

It has already been shown for the enthalpy that $dh = T\, ds + v\, dP$, so that the canonical variables are $s$ and $P$. One then also has

$$dh = \left.\frac{\partial h}{\partial s}\right|_P ds + \left.\frac{\partial h}{\partial P}\right|_s dP, \tag{3.94}$$

from which one deduces that

$$T = \left.\frac{\partial h}{\partial s}\right|_P, \qquad v = \left.\frac{\partial h}{\partial P}\right|_s. \tag{3.95}$$

From Eq. (3.95), a second Maxwell relation can be deduced by differentiation of the first with respect to $P$ and the second with respect to $s$:

$$\left.\frac{\partial T}{\partial P}\right|_s = \left.\frac{\partial v}{\partial s}\right|_P. \tag{3.96}$$

[6] We use "Legendre function" in a nontraditional sense here in that it has no relation to the Legendre polynomials.

The relations for Helmholtz[7] and Gibbs free energies each supply additional useful relations including two new Maxwell relations. First, consider the Helmholtz free energy:

$$a = e - Ts, \tag{3.97}$$

$$da = de - T\,ds - s\,dT, \tag{3.98}$$

$$= \underbrace{(-P\,dv + T\,ds)}_{de} - T\,ds - s\,dT, \tag{3.99}$$

$$= -P\,dv - s\,dT. \tag{3.100}$$

So, the canonical variables for $a$ are $v$ and $T$. The conjugate variables are $-P$ and $-s$. Thus

$$da = \left.\frac{\partial a}{\partial v}\right|_T dv + \left.\frac{\partial a}{\partial T}\right|_v dT. \tag{3.101}$$

So, one gets

$$-P = \left.\frac{\partial a}{\partial v}\right|_T, \qquad -s = \left.\frac{\partial a}{\partial T}\right|_v, \tag{3.102}$$

and the consequent Maxwell relation

$$\left.\frac{\partial P}{\partial T}\right|_v = \left.\frac{\partial s}{\partial v}\right|_T. \tag{3.103}$$

For the Gibbs free energy,

$$g = \underbrace{e + Pv}_{h} - Ts = h - Ts, \tag{3.104}$$

$$dg = dh - T\,ds - s\,dT = \underbrace{(T\,ds + v\,dP)}_{dh} - T\,ds - s\,dT, \tag{3.105}$$

$$= v\,dP - s\,dT. \tag{3.106}$$

So, for Gibbs free energy, the canonical variables are $P$ and $T$, while the conjugate variables are $v$ and $-s$. One then has $g = g(P,T)$, which gives

$$dg = \left.\frac{\partial g}{\partial P}\right|_T dP + \left.\frac{\partial g}{\partial T}\right|_P dT. \tag{3.107}$$

So, one finds

$$v = \left.\frac{\partial g}{\partial P}\right|_T, \qquad -s = \left.\frac{\partial g}{\partial T}\right|_P. \tag{3.108}$$

The resulting Maxwell relation is then

$$\left.\frac{\partial v}{\partial T}\right|_P = -\left.\frac{\partial s}{\partial P}\right|_T. \tag{3.109}$$

[7] Hermann von Helmholtz (1821–1894), German physician and physicist.

**EXAMPLE 3.6**

Consider an equation of state in a so-called canonical form. If

$$h(s,P) = c_P T_0 \left(\frac{P}{P_0}\right)^{R/c_P} \exp\left(\frac{s}{c_P}\right) + (h_0 - c_P T_0), \tag{3.110}$$

and $c_P, T_0, R, P_0,$ and $h_0$ are all constants, derive both thermal and caloric state equations.

---

Now, for this material,

$$\left.\frac{\partial h}{\partial s}\right|_P = T_0 \left(\frac{P}{P_0}\right)^{R/c_P} \exp\left(\frac{s}{c_P}\right), \tag{3.111}$$

$$\left.\frac{\partial h}{\partial P}\right|_s = \frac{RT_0}{P_0} \left(\frac{P}{P_0}\right)^{R/c_P - 1} \exp\left(\frac{s}{c_P}\right). \tag{3.112}$$

Now, because, from Eq. (3.95),

$$\left.\frac{\partial h}{\partial s}\right|_P = T, \qquad \left.\frac{\partial h}{\partial P}\right|_s = v, \tag{3.113}$$

one has

$$T = T_0 \left(\frac{P}{P_0}\right)^{R/c_P} \exp\left(\frac{s}{c_P}\right), \tag{3.114}$$

$$v = \frac{RT_0}{P_0} \left(\frac{P}{P_0}\right)^{R/c_P - 1} \exp\left(\frac{s}{c_P}\right). \tag{3.115}$$

Dividing one by the other gives

$$\frac{T}{v} = \frac{P}{R}, \qquad Pv = RT. \tag{3.116}$$

That is the thermal equation of state for an ideal gas. Substituting from Eq. (3.114) into the canonical equation for $h$, Eq. (3.110), one also finds for the caloric equation of state

$$h = c_P T + (h_0 - c_P T_0) = c_P (T - T_0) + h_0, \tag{3.117}$$

which is itself useful and shows we have a calorically perfect gas. Substituting for $T$ and $T_0$,

$$h = c_P \left(\frac{Pv}{R} - \frac{P_0 v_0}{R}\right) + h_0. \tag{3.118}$$

Using $h \equiv e + Pv$, we get

$$\underbrace{e + Pv}_{h} = c_P \left(\frac{Pv}{R} - \frac{P_0 v_0}{R}\right) + \underbrace{e_0 + P_0 v_0}_{h_0}, \tag{3.119}$$

so

$$e = \left(\frac{c_P}{R} - 1\right) Pv - \left(\frac{c_P}{R} - 1\right) P_0 v_0 + e_0. \tag{3.120}$$

Straightforward reduction using $c_P - c_v = R$ and $Pv = RT$ gives

$$e = c_v (T - T_0) + e_0. \tag{3.121}$$

So, *one* canonical equation gives us all the information one needs. Oftentimes, it is difficult to do a single experiment to get the canonical form.

## 3.4 Heat Capacity

Recall that the specific heat at constant volume $c_v$ and specific heat at constant pressure $c_P$ are defined by

$$c_v = \left.\frac{\partial e}{\partial T}\right|_v, \qquad c_P = \left.\frac{\partial h}{\partial T}\right|_P. \tag{3.122}$$

Then, perform operations on the Gibbs equation, Eq. (3.1):

$$de = T\,ds - P\,dv, \tag{3.123}$$

$$\left.\frac{\partial e}{\partial T}\right|_v = T\left.\frac{\partial s}{\partial T}\right|_v, \tag{3.124}$$

$$c_v = T\left.\frac{\partial s}{\partial T}\right|_v. \tag{3.125}$$

Likewise, from Eq. (3.76), we get

$$dh = T\,ds + v\,dP, \tag{3.126}$$

$$\left.\frac{\partial h}{\partial T}\right|_P = T\left.\frac{\partial s}{\partial T}\right|_P, \tag{3.127}$$

$$c_P = T\left.\frac{\partial s}{\partial T}\right|_P. \tag{3.128}$$

One finds further useful relations by operating on the Gibbs equation, Eq. (3.1):

$$de = T\,ds - P\,dv, \tag{3.129}$$

$$\left.\frac{\partial e}{\partial v}\right|_T = T\left.\frac{\partial s}{\partial v}\right|_T - P, \tag{3.130}$$

$$= T\left.\frac{\partial P}{\partial T}\right|_v - P. \tag{3.131}$$

So, one can then say

$$e = e(T, v), \tag{3.132}$$

$$de = \left.\frac{\partial e}{\partial T}\right|_v dT + \left.\frac{\partial e}{\partial v}\right|_T dv = c_v\,dT + \left(T\left.\frac{\partial P}{\partial T}\right|_v - P\right)dv. \tag{3.133}$$

For an ideal gas, one has

$$\left.\frac{\partial e}{\partial v}\right|_T = T\left.\frac{\partial P}{\partial T}\right|_v - P = T\left(\frac{R}{v}\right) - \frac{RT}{v} = 0. \tag{3.134}$$

Consequently, $e$ is not a function of $v$ for an ideal gas, so $e = e(T)$ alone. Because $h = e + Pv$ for an ideal gas reduces to $h = e + RT$, we have for $h$ that

$$h = e(T) + RT = h(T). \tag{3.135}$$

Now, return to general equations of state. With $s = s(T, v)$ or $s = s(T, P)$, one gets

$$ds = \left.\frac{\partial s}{\partial T}\right|_v dT + \left.\frac{\partial s}{\partial v}\right|_T dv, \qquad ds = \left.\frac{\partial s}{\partial T}\right|_P dT + \left.\frac{\partial s}{\partial P}\right|_T dP. \tag{3.136}$$

Now, using Eqs. (3.96), (3.109), (3.125), and (3.128), one gets

$$ds = \frac{c_v}{T}\,dT + \left.\frac{\partial P}{\partial T}\right|_v dv, \qquad ds = \frac{c_P}{T}\,dT - \left.\frac{\partial v}{\partial T}\right|_P dP. \tag{3.137}$$

Subtracting one from the other, one finds

$$0 = \frac{c_v - c_P}{T}\,dT + \left.\frac{\partial P}{\partial T}\right|_v dv + \left.\frac{\partial v}{\partial T}\right|_P dP, \tag{3.138}$$

$$(c_P - c_v)\,dT = T\left.\frac{\partial P}{\partial T}\right|_v dv + T\left.\frac{\partial v}{\partial T}\right|_P dP. \tag{3.139}$$

Now, divide both sides by $dT$ and hold *either* $P$ or $v$ constant. In either case, one gets

$$c_P - c_v = T\left.\frac{\partial P}{\partial T}\right|_v \left.\frac{\partial v}{\partial T}\right|_P. \tag{3.140}$$

**EXAMPLE 3.7**

For an ideal gas, find $c_P - c_v$.

For the ideal gas, $Pv = RT$, one has

$$\left.\frac{\partial P}{\partial T}\right|_v = \frac{R}{v}, \qquad \left.\frac{\partial v}{\partial T}\right|_P = \frac{R}{P}. \tag{3.141}$$

So,

$$c_P - c_v = T\frac{R}{v}\frac{R}{P} = T\frac{R^2}{RT} = R. \tag{3.142}$$

This holds even if the gas is calorically imperfect. That is,

$$c_P(T) - c_v(T) = R. \tag{3.143}$$

For the general ratio of specific heats, one can use Eqs. (3.125) and (3.128) to get

$$\gamma = \frac{c_P}{c_v} = \left(T\left.\frac{\partial s}{\partial T}\right|_P\right) \Big/ \left(T\left.\frac{\partial s}{\partial T}\right|_v\right). \tag{3.144}$$

Then apply Eq. (3.53) to get

$$\gamma = \left.\frac{\partial s}{\partial T}\right|_P \left.\frac{\partial T}{\partial s}\right|_v. \tag{3.145}$$

Then apply Eq. (3.55) to get

$$\gamma = \left(-\left.\frac{\partial s}{\partial P}\right|_T \left.\frac{\partial P}{\partial T}\right|_s\right)\left(-\left.\frac{\partial T}{\partial v}\right|_s \left.\frac{\partial v}{\partial s}\right|_T\right), \tag{3.146}$$

$$= \left(\left.\frac{\partial v}{\partial s}\right|_T \left.\frac{\partial s}{\partial P}\right|_T\right)\left(\left.\frac{\partial P}{\partial T}\right|_s \left.\frac{\partial T}{\partial v}\right|_s\right) = \left.\frac{\partial v}{\partial P}\right|_T \left.\frac{\partial P}{\partial v}\right|_s. \tag{3.147}$$

The first term can be obtained from $P - v - T$ data. The second term is related to the isentropic sound speed of the material, which is also a measurable quantity.

**EXAMPLE 3.8**

For a calorically perfect ideal gas with gas constant $R$ and specific heat at constant volume $c_v$, find expressions for the thermodynamic variables $s, e, h, a,$ and $g$, as functions of $T$ and $P$.

First, get the entropy:

$$de = T\,ds - P\,dv, \tag{3.148}$$

$$T\,ds = de + P\,dv = c_v\,dT + P\,dv, \tag{3.149}$$

$$ds = c_v\frac{dT}{T} + \frac{P}{T}\,dv = c_v\frac{dT}{T} + R\frac{dv}{v}, \tag{3.150}$$

$$\int ds = \int c_v\frac{dT}{T} + \int R\frac{dv}{v}, \tag{3.151}$$

$$s - s_0 = c_v \ln\frac{T}{T_0} + R\ln\frac{v}{v_0}, \tag{3.152}$$

$$\frac{s-s_0}{c_v} = \ln\frac{T}{T_0} + \frac{R}{c_v}\ln\frac{RT/P}{RT_0/P_0} = \ln\left(\frac{T}{T_0}\right) + \ln\left(\frac{T}{T_0}\frac{P_0}{P}\right)^{R/c_v}, \tag{3.153}$$

$$= \ln\left(\frac{T}{T_0}\right)^{1+R/c_v} + \ln\left(\frac{P_0}{P}\right)^{R/c_v} = \ln\left(\frac{T}{T_0}\right)^{\gamma} + \ln\left(\frac{P_0}{P}\right)^{\gamma-1}, \tag{3.154}$$

$$s = s_0 + c_v\ln\left(\frac{T}{T_0}\right)^{\gamma} + c_v\ln\left(\frac{P_0}{P}\right)^{\gamma-1}. \tag{3.155}$$

Now, for the calorically perfect ideal gas, one has

$$e = e_0 + c_v(T - T_0). \tag{3.156}$$

For the enthalpy, one gets

$$h = e + Pv = e + RT = e_0 + c_v(T - T_0) + RT, \tag{3.157}$$

$$= \underbrace{e_0 + RT_0}_{h_0} + c_v(T - T_0) + R(T - T_0), \tag{3.158}$$

$$= h_0 + \underbrace{(c_v + R)}_{c_p}(T - T_0) = h_0 + c_p(T - T_0). \tag{3.159}$$

For the Helmholtz free energy, one gets

$$a = e - Ts, \tag{3.160}$$

$$= e_0 + c_v(T - T_0) - T\left(s_0 + c_v\ln\left(\frac{T}{T_0}\right)^{\gamma} + c_v\ln\left(\frac{P_0}{P}\right)^{\gamma-1}\right). \tag{3.161}$$

For the Gibbs free energy, one gets

$$g = h - Ts, \tag{3.162}$$

$$= h_0 + c_p(T - T_0) - T\left(s_0 + c_v\ln\left(\frac{T}{T_0}\right)^{\gamma} + c_v\ln\left(\frac{P_0}{P}\right)^{\gamma-1}\right). \tag{3.163}$$

## 3.5 Mixtures with Variable Composition

We now turn to a topic of importance for combustion thermodynamics: mixtures with variable composition. Consider then mixtures of $N$ species. The focus here will be on extensive properties and molar properties. We will need to draw upon concepts introduced in Chapter 2. Assume that each species has $n_i$ moles, and the total number of moles is $n = \sum_{i=1}^{N} n_i$. Now, in an analog to the canonical form for a single material of Eq. (3.25), $e = e(v, s)$, one might expect the extensive internal energy to be a function of the extensive entropy, the volume, *and* the number of moles of each species:

$$E = E(V, S, n_i). \tag{3.164}$$

The extensive version of the Gibbs equation, Eq. (3.1), in which *all of the $n_i$ are held constant* is

$$dE = -P\,dV + T\,dS. \tag{3.165}$$

Thus

$$\left.\frac{\partial E}{\partial V}\right|_{S,n_i} = -P, \qquad \left.\frac{\partial E}{\partial S}\right|_{V,n_i} = T. \tag{3.166}$$

In general, because $E = E(S, V, n_i)$, one should expect, for systems in which *the $n_i$ are allowed to change*, that

$$dE = \left.\frac{\partial E}{\partial V}\right|_{S,n_i} dV + \left.\frac{\partial E}{\partial S}\right|_{V,n_i} dS + \sum_{i=1}^{N} \left.\frac{\partial E}{\partial n_i}\right|_{S,V,n_j} dn_i. \tag{3.167}$$

Defining a new thermodynamic property, the *chemical potential* $\overline{\mu}_i$, as

$$\overline{\mu}_i \equiv \left.\frac{\partial E}{\partial n_i}\right|_{S,V,n_j}, \tag{3.168}$$

one has the important *Gibbs equation for multicomponent system*:

$$dE = -P\,dV + T\,dS + \sum_{i=1}^{N} \overline{\mu}_i\,dn_i. \tag{3.169}$$

Obviously, by its definition, $\overline{\mu}_i$ is on a per mol basis, so it is given the appropriate overline notation. In Eq. (3.169), the independent variables and their conjugates are

$$x_1 = V, \qquad \psi_1 = -P, \tag{3.170}$$

$$x_2 = S, \qquad \psi_2 = T, \tag{3.171}$$

$$x_3 = n_1, \qquad \psi_3 = \overline{\mu}_1, \tag{3.172}$$

$$x_4 = n_2, \qquad \psi_4 = \overline{\mu}_2, \tag{3.173}$$

$$\vdots \qquad\qquad \vdots$$

$$x_{N+2} = n_N, \qquad \psi_{N+2} = \overline{\mu}_N. \tag{3.174}$$

Equation (3.169) has $2^{N+2} - 1$ Legendre functions. Three are in wide usage: the extensive analogs to those earlier found. They are

$$H = E + PV, \tag{3.175}$$

$$A = E - TS, \tag{3.176}$$

$$G = E + PV - TS. \tag{3.177}$$

A set of nontraditional but perfectly acceptable additional Legendre functions would be formed from $E - \overline{\mu}_1 n_1$. Another set would be formed from $E + PV - \overline{\mu}_2 n_2$. There are many more, but one in particular is sometimes noted in the literature: the so-called *grand potential*, $\Omega$. The grand potential is defined as

$$\Omega \equiv E - TS - \sum_{i=1}^{N} \overline{\mu}_i n_i. \tag{3.178}$$

Differentiating each defined Legendre function, Eqs. (3.175)–(3.178), and combining with Eq. (3.169), one finds

$$dH = T\,dS + V\,dP + \sum_{i=1}^{N} \overline{\mu}_i\,dn_i, \tag{3.179}$$

$$dA = -S\,dT - P\,dV + \sum_{i=1}^{N} \overline{\mu}_i\,dn_i, \tag{3.180}$$

$$dG = -S\,dT + V\,dP + \sum_{i=1}^{N} \overline{\mu}_i\,dn_i, \tag{3.181}$$

$$d\Omega = -P\,dV - S\,dT - \sum_{i=1}^{N} n_i d\overline{\mu}_i. \tag{3.182}$$

Thus, canonical variables for $H$ are such that $H = H(S, P, n_i)$. One finds a similar set of relations as before from each of the differential forms:

$$T = \left.\frac{\partial E}{\partial S}\right|_{V,n_i} = \left.\frac{\partial H}{\partial S}\right|_{P,n_i}, \tag{3.183}$$

$$P = -\left.\frac{\partial E}{\partial V}\right|_{S,n_i} = -\left.\frac{\partial A}{\partial V}\right|_{T,n_i} = -\left.\frac{\partial \Omega}{\partial V}\right|_{T,\overline{\mu}_i}, \tag{3.184}$$

$$V = \left.\frac{\partial H}{\partial P}\right|_{S,n_i} = \left.\frac{\partial G}{\partial P}\right|_{T,n_i}, \tag{3.185}$$

$$S = -\left.\frac{\partial A}{\partial T}\right|_{V,n_i} = -\left.\frac{\partial G}{\partial T}\right|_{P,n_i} = -\left.\frac{\partial \Omega}{\partial T}\right|_{V,\overline{\mu}_i}, \tag{3.186}$$

$$n_i = -\left.\frac{\partial \Omega}{\partial \overline{\mu}_i}\right|_{V,T,\overline{\mu}_j}, \tag{3.187}$$

$$\overline{\mu}_i = \left.\frac{\partial E}{\partial n_i}\right|_{S,V,n_j} = \left.\frac{\partial H}{\partial n_i}\right|_{S,P,n_j} = \left.\frac{\partial A}{\partial n_i}\right|_{T,V,n_j} = \left.\frac{\partial G}{\partial n_i}\right|_{T,P,n_j}. \tag{3.188}$$

Each of these induces a corresponding Maxwell relation, obtained by cross differentiation. These are

$$\left.\frac{\partial T}{\partial V}\right|_{S,n_i} = -\left.\frac{\partial P}{\partial S}\right|_{V,n_i}, \quad \left.\frac{\partial T}{\partial P}\right|_{S,n_i} = \left.\frac{\partial V}{\partial S}\right|_{P,n_i}, \tag{3.189}$$

$$\left.\frac{\partial P}{\partial T}\right|_{V,n_i} = \left.\frac{\partial S}{\partial V}\right|_{T,n_i}, \quad \left.\frac{\partial V}{\partial T}\right|_{P,n_i} = -\left.\frac{\partial S}{\partial P}\right|_{T,n_i}, \tag{3.190}$$

$$\left.\frac{\partial \overline{\mu}_i}{\partial T}\right|_{P,n_j} = -\left.\frac{\partial S}{\partial n_i}\right|_{V,n_j}, \quad \left.\frac{\partial \overline{\mu}_i}{\partial P}\right|_{T,n_j} = \left.\frac{\partial V}{\partial n_i}\right|_{V,n_j}, \tag{3.191}$$

$$\left.\frac{\partial \overline{\mu}_l}{\partial n_k}\right|_{T,P,n_j} = \left.\frac{\partial \overline{\mu}_k}{\partial n_l}\right|_{T,P,n_j}, \quad \left.\frac{\partial S}{\partial V}\right|_{T,\overline{\mu}_i} = \left.\frac{\partial P}{\partial T}\right|_{V,\overline{\mu}_i}, \tag{3.192}$$

$$\left.\frac{\partial n_i}{\partial \overline{\mu}_k}\right|_{V,T,\overline{\mu}_j,j\neq k} = \left.\frac{\partial n_k}{\partial \overline{\mu}_i}\right|_{V,T,\overline{\mu}_j,j\neq i}. \tag{3.193}$$

## 3.6 Partial Molar Properties

Let us now explore deeper the concept of a partial molar property, introduced in Sec. 2.2.

### 3.6.1 Homogeneous Functions

In mathematics, a *homogeneous function* $f(x_1, \ldots, x_N)$ *of order* $m$ is one such that

$$f(\lambda x_1, \ldots, \lambda x_N) = \lambda^m f(x_1, \ldots, x_N). \tag{3.194}$$

If $m = 1$, one has

$$f(\lambda x_1, \ldots, \lambda x_N) = \lambda f(x_1, \ldots, x_N). \tag{3.195}$$

Thermodynamic functions, typically functions of two or more variables, can be seen to be homogeneous functions.

### 3.6.2 Gibbs Free Energy

Consider an extensive property, such as $G$. One has the canonical functional form

$$G = G(T, P, n_1, n_2, \ldots, n_N). \tag{3.196}$$

One would like to show that if each of the mole numbers $n_i$ is increased by a common factor, say, $\lambda$, with $T$ and $P$ constant, $G$ increases by the same factor $\lambda$:

$$\lambda G(T, P, n_1, n_2, \ldots, n_N) = G(T, P, \lambda n_1, \lambda n_2, \ldots, \lambda n_N). \tag{3.197}$$

This is clearly related to the notion of a homogeneous function with $m = 1$, as discussed in Eq. (3.195). Differentiate both sides of Eq. (3.197) with respect to $\lambda$, while holding $P, T$, and $n_j$ constant, to get

$$G(T, P, n_1, \ldots, n_N)$$

$$= \left.\frac{\partial G}{\partial(\lambda n_1)}\right|_{n_j,P,T} \frac{d(\lambda n_1)}{d\lambda} + \cdots + \left.\frac{\partial G}{\partial(\lambda n_N)}\right|_{n_j,P,T} \frac{d(\lambda n_N)}{d\lambda}, \tag{3.198}$$

$$= \left.\frac{\partial G}{\partial(\lambda n_1)}\right|_{n_j,P,T} n_1 + \cdots + \left.\frac{\partial G}{\partial(\lambda n_N)}\right|_{n_j,P,T} n_N. \tag{3.199}$$

This must hold for all $\lambda$, including $\lambda = 1$, so one requires

$$G(T, P, n_1, \ldots, n_N) = \left.\frac{\partial G}{\partial n_1}\right|_{n_j,P,T} n_1 + \cdots + \left.\frac{\partial G}{\partial n_N}\right|_{n_j,P,T} n_N, \tag{3.200}$$

$$= \sum_{i=1}^{N} \left.\frac{\partial G}{\partial n_i}\right|_{n_j,P,T} n_i. \tag{3.201}$$

Recall now from Sec. 2.2 the definition of a partial molar property, the derivative of an extensive variable with respect to species $n_i$ holding $n_j$, $i \neq j, T$, and $P$ constant. Because the result has units that are per unit mol, an overline superscript is utilized. The partial molar Gibbs free energy of species $i, \overline{g}_i$ is then

$$\overline{g}_i \equiv \left.\frac{\partial G}{\partial n_i}\right|_{n_j,P,T}, \tag{3.202}$$

so that Eq. (3.201) becomes

$$G = \sum_{i=1}^{N} \overline{g}_i n_i. \tag{3.203}$$

Using the definition of chemical potential, Eq. (3.188), one also notes then that

$$G(T, P, n_2, n_2, \ldots, n_N) = \sum_{i=1}^{N} \overline{\mu}_i n_i. \tag{3.204}$$

Consequently, one also sees that the Gibbs free energy per unit mol of species $i$ is the chemical potential of that species:

$$\overline{g}_i = \overline{\mu}_i. \tag{3.205}$$

The $T$ and $P$ dependence of $G$ must lie entirely within $\overline{\mu}_i(T, P, n_j)$, which one notes is also allowed to be a function of $n_j$. Using Eq. (3.203) to eliminate $G$ in Eq. (3.177), one recovers an equation for the energy:

$$E = -PV + TS + \sum_{i=1}^{N} \overline{\mu}_i n_i. \tag{3.206}$$

### 3.6.3 Other Properties

A similar result also holds for any other extensive property such as $V, E, H, A$, or $S$. One can also show that

$$V = \sum_{i=1}^{N} n_i \left.\frac{\partial V}{\partial n_i}\right|_{n_j,A,T}, \qquad E = \sum_{i=1}^{N} n_i \left.\frac{\partial E}{\partial n_i}\right|_{n_j,V,S}, \tag{3.207}$$

$$H = \sum_{i=1}^{N} n_i \left.\frac{\partial H}{\partial n_i}\right|_{n_j,P,S}, \qquad A = \sum_{i=1}^{N} n_i \left.\frac{\partial A}{\partial n_i}\right|_{n_j,T,V}, \tag{3.208}$$

$$S = \sum_{i=1}^{N} n_i \left.\frac{\partial S}{\partial n_i}\right|_{n_j,E,T}. \tag{3.209}$$

These expressions do not formally involve partial molar properties because $P$ and $T$ are not constant. Take now the appropriate partial molar derivatives of $G$ for an ideal mixture of ideal gases to get some useful relations:

$$G = H - TS, \tag{3.210}$$

$$\left.\frac{\partial G}{\partial n_i}\right|_{T,P,n_j} = \left.\frac{\partial H}{\partial n_i}\right|_{T,P,n_j} - T\left.\frac{\partial S}{\partial n_i}\right|_{T,P,n_j}. \tag{3.211}$$

Now, from the definition of an ideal mixture, Eq. (2.30), $\overline{h}_i = \overline{h}_i(T, P)$, so one has

$$H = \sum_{k=1}^{N} n_k \overline{h}_k(T, P), \tag{3.212}$$

$$\left.\frac{\partial H}{\partial n_i}\right|_{T,P,n_j} = \frac{\partial}{\partial n_i}\left(\sum_{k=1}^{N} n_k \overline{h}_k(T, P)\right), \tag{3.213}$$

$$= \sum_{k=1}^{N} \underbrace{\frac{\partial n_k}{\partial n_i}}_{\delta_{ik}} \overline{h}_k(T, P) = \sum_{k=1}^{N} \delta_{ik}\overline{h}_k(T, P) = \overline{h}_i(T, P). \tag{3.214}$$

For an ideal gas, one further has $\overline{h}_i = \overline{h}_i(T)$.

The analysis is more complicated for the entropy, in which

$$S = \sum_{k=1}^{N} n_k \left(\overline{s}_{T,k}^0 - \overline{R}\ln\left(\frac{P_k}{P_0}\right)\right), \tag{3.215}$$

$$= \sum_{k=1}^{N} n_k \left(\overline{s}_{T,k}^0 - \overline{R}\ln\left(\frac{y_k P}{P_0}\right)\right), \tag{3.216}$$

$$= \sum_{k=1}^{N} n_k \left(\overline{s}_{T,k}^0 - \overline{R}\ln\left(\frac{P}{P_0}\right) - \overline{R}\ln\left(\frac{n_k}{\sum_{q=1}^{N} n_q}\right)\right), \tag{3.217}$$

$$= \sum_{k=1}^{N} n_k \left(\overline{s}_{T,k}^0 - \overline{R}\ln\left(\frac{P}{P_0}\right)\right) - \overline{R}\sum_{k=1}^{N} n_k \ln\left(\frac{n_k}{\sum_{q=1}^{N} n_q}\right), \tag{3.218}$$

$$\left.\frac{\partial S}{\partial n_i}\right|_{T,P,n_j} = \frac{\partial}{\partial n_i}\sum_{k=1}^{N} n_k \left(\overline{s}_{T,k}^0 - \overline{R}\ln\left(\frac{P}{P_0}\right)\right) - \overline{R}\left.\frac{\partial}{\partial n_i}\right|_{T,P,n_j}\left(\sum_{k=1}^{N} n_k \ln\left(\frac{n_k}{\sum_{q=1}^{N} n_q}\right)\right), \tag{3.219}$$

$$= \sum_{k=1}^{N} \underbrace{\frac{\partial n_k}{\partial n_i}}_{\delta_{ik}} \left(\overline{s}_{T,k}^0 - \overline{R}\ln\left(\frac{P}{P_0}\right)\right) - \overline{R}\left.\frac{\partial}{\partial n_i}\right|_{T,P,n_j}\left(\sum_{k=1}^{N} n_k \ln\left(\frac{n_k}{\sum_{q=1}^{N} n_q}\right)\right), \tag{3.220}$$

$$= \left(\overline{s}^0_{T,i} - \overline{R}\ln\left(\frac{P}{P_0}\right)\right)$$

$$- \overline{R}\,\frac{\partial}{\partial n_i}\bigg|_{T,P,n_j}\left(\sum_{k=1}^{N} n_k \ln\left(\frac{n_k}{\sum_{q=1}^{N} n_q}\right)\right). \tag{3.221}$$

Evaluation of the final term on the right side requires closer examination and in fact, after tedious but straightforward analysis, yields a simple result that can easily be verified by direct calculation:

$$\frac{\partial}{\partial n_i}\bigg|_{T,P,n_j}\left(\sum_{k=1}^{N} n_k \ln\left(\frac{n_k}{\sum_{q=1}^{N} n_q}\right)\right) = \ln\left(\frac{n_i}{\sum_{q=1}^{N} n_q}\right). \tag{3.222}$$

So, the partial molar entropy is in fact

$$\frac{\partial S}{\partial n_i}\bigg|_{T,P,n_j} = \overline{s}^0_{T,i} - \overline{R}\ln\left(\frac{P}{P_0}\right) - \overline{R}\ln\left(\frac{n_i}{\sum_{q=1}^{N} n_q}\right), \tag{3.223}$$

$$= \overline{s}^0_{T,i} - \overline{R}\ln\left(\frac{P}{P_0}\right) - \overline{R}\ln y_i, \tag{3.224}$$

$$= \overline{s}^0_{T,i} - \overline{R}\ln\left(\frac{P_i}{P_0}\right) = \overline{s}_i. \tag{3.225}$$

Thus, one can in fact claim for the ideal mixture of ideal gases that

$$\overline{g}_i = \overline{h}_i - T\overline{s}_i. \tag{3.226}$$

### 3.6.4 Relation between Mixture and Partial Molar Properties

A simple analysis shows how the partial molar property for an individual species is related to the partial molar property for the mixture. Consider, for example, the Gibbs free energy. The mixture-averaged Gibbs free energy per unit mol is

$$\overline{g} = G/n. \tag{3.227}$$

Now, take a partial molar derivative and analyze to get

$$\frac{\partial \overline{g}}{\partial n_i}\bigg|_{T,P,n_j} = \frac{1}{n}\frac{\partial G}{\partial n_i}\bigg|_{T,P,n_j} - \frac{G}{n^2}\frac{\partial n}{\partial n_i}\bigg|_{T,P,n_j}, \tag{3.228}$$

$$= \frac{1}{n}\frac{\partial G}{\partial n_i}\bigg|_{T,P,n_j} - \frac{G}{n^2}\frac{\partial}{\partial n_i}\bigg|_{T,P,n_j}\left(\sum_{k=1}^{N} n_k\right), \tag{3.229}$$

$$= \frac{1}{n}\frac{\partial G}{\partial n_i}\bigg|_{T,P,n_j} - \frac{G}{n^2}\sum_{k=1}^{N}\delta_{ik}, \tag{3.230}$$

$$= \frac{1}{n}\frac{\partial G}{\partial n_i}\bigg|_{T,P,n_j} - \frac{G}{n^2} = \frac{1}{n}\overline{g}_i - \frac{\overline{g}}{n}. \tag{3.231}$$

Multiplying by $n$ and rearranging, one gets

$$\overline{g}_i = \overline{g} + n\,\frac{\partial \overline{g}}{\partial n_i}\bigg|_{T,P,n_j}. \tag{3.232}$$

A similar result holds for other properties.

## 3.7 Frozen Sound Speed

Let us develop a relation for what is known as the "frozen sound speed," denoted as $c$. This quantity is a thermodynamic property defined by the relationship

$$c^2 \equiv \left.\frac{\partial P}{\partial \rho}\right|_{s,n_i}. \tag{3.233}$$

Later in Sec. 12.1.7 we shall see that this quantity describes the speed of acoustic waves in reactive mixtures, but at this point let us simply treat it as a problem in thermodynamics. The quantity is called "frozen" because the derivative is performed with all of the species mol numbers frozen at a constant value. Because $v = 1/\rho$,

$$c^2 = \left.\frac{\partial P}{\partial \rho}\right|_{s,n_i} = \left.\frac{dv}{d\rho}\frac{\partial P}{\partial v}\right|_{s,n_i} = \left.-\frac{1}{\rho^2}\frac{\partial P}{\partial v}\right|_{s,n_i} = \left.-v^2\frac{\partial P}{\partial v}\right|_{s,n_i}. \tag{3.234}$$

So, we can say

$$(\rho c)^2 = -\left.\frac{\partial P}{\partial v}\right|_{s,n_i}. \tag{3.235}$$

Now, we know that the caloric equation of state for a mixture of ideal gases takes the form, from Eq. (2.137),

$$e = \sum_{i=1}^{N} Y_i e_i(T). \tag{3.236}$$

Employing the ideal gas law, we could say instead that

$$e = \sum_{i=1}^{N} Y_i e_i\left(\frac{P}{\rho R}\right). \tag{3.237}$$

In terms of number of moles, $n_i$, we could say that

$$e = \sum_{i=1}^{N} \underbrace{\frac{M_i n_i}{\rho V}}_{Y_i} e_i\left(\frac{P}{\rho R}\right). \tag{3.238}$$

We can in fact generalize and simply consider caloric equations of state of either of the forms

$$e = e(P, \rho, n_i), \quad \text{or} \quad e = e(P, v, n_i). \tag{3.239}$$

These last two forms can be useful when temperature is unavailable, as can be the case for some materials.

Now, reconsider the Gibbs equation for a multicomponent system, Eq. (3.169):

$$dE = -P\,dV + T\,dS + \sum_{i=1}^{N} \overline{\mu}_i\,dn_i. \tag{3.240}$$

Let us scale Eq. (3.240) by a fixed mass $m$, recalling that $e = E/m$, $v = V/m$, and $s = S/m$, so as to obtain

$$de = -P\,dv + T\,ds + \sum_{i=1}^{N} \frac{\overline{\mu}_i}{m}\,dn_i. \tag{3.241}$$

Considering Eq. (3.241) with $dv = 0$ and $dn_i = 0$, and scaling by $dP$, we can say

$$\left.\frac{\partial e}{\partial P}\right|_{v,n_i} = T \left.\frac{\partial s}{\partial P}\right|_{v,n_i}. \tag{3.242}$$

Considering Eq. (3.241) with $dn_i = 0$ and $dP = 0$, and scaling by $dv$, we can say

$$\left.\frac{\partial e}{\partial v}\right|_{P,n_i} = -P + T \left.\frac{\partial s}{\partial v}\right|_{P,n_i}. \tag{3.243}$$

Applying now Eq. (3.56), we find

$$\left.\frac{\partial P}{\partial v}\right|_{s,n_i} \left.\frac{\partial s}{\partial P}\right|_{v,n_i} \left.\frac{\partial v}{\partial s}\right|_{P,n_i} = -1. \tag{3.244}$$

So, we get

$$\left.\frac{\partial P}{\partial v}\right|_{s,n_i} = -\left.\frac{\partial s}{\partial v}\right|_{P,n_i} \Big/ \left.\frac{\partial s}{\partial P}\right|_{v,n_i}. \tag{3.245}$$

Substituting Eq. (3.245) into Eq. (3.235), we find

$$(\rho c)^2 = \left.\frac{\partial s}{\partial v}\right|_{P,n_i} \Big/ \left.\frac{\partial s}{\partial P}\right|_{v,n_i}. \tag{3.246}$$

Substituting Eqs. (3.242) and (3.243) into Eq. (3.246), we get

$$(\rho c)^2 = \frac{1}{T}\left(P + \left.\frac{\partial e}{\partial v}\right|_{P,n_i}\right) \Big/ \left(\frac{1}{T}\left.\frac{\partial e}{\partial P}\right|_{v,n_i}\right) = \frac{P + \left.\frac{\partial e}{\partial v}\right|_{P,n_i}}{\left.\frac{\partial e}{\partial P}\right|_{v,n_i}}. \tag{3.247}$$

So, the sound speed $c$ is

$$c = v\left(\left(P + \left.\frac{\partial e}{\partial v}\right|_{P,n_i}\right) \Big/ \left.\frac{\partial e}{\partial P}\right|_{v,n_i}\right)^{1/2}. \tag{3.248}$$

Equation (3.248) is useful because we can identify the frozen sound speed from data for $e$, $P$, and $v$ alone. If

$$\left.\frac{\partial e}{\partial v}\right|_{P,n_i} > -P, \qquad \text{and} \qquad \left.\frac{\partial e}{\partial P}\right|_{v,n_i} > 0, \tag{3.249}$$

we will have a real sound speed. Other combinations are possible as well but not observed in nature. It is also common in the literature to consider the so-called equilibrium sound speed. This is the sound speed of the mixture that has relaxed to chemical equilibrium. It will not be necessary for the analysis we present.

**EXAMPLE 3.9**

For an ideal mixture of calorically perfect ideal gases, find the frozen sound speed.

For a calorically perfect ideal gas, we have

$$e_i = e_i(T) = e_0 + c_{vi}(T - T_0) = e_0 + c_{vi}\left(\frac{Pv}{R} - T_0\right), \tag{3.250}$$

and

$$e = \sum_{i=1}^{N} Y_i e_i = \sum_{i=1}^{N} \frac{m_i}{m} e_i = \sum_{i=1}^{N} \frac{n_i M_i}{m} e_i = \sum_{i=1}^{N} \frac{n_i M_i}{m} \left( e_0 + c_{vi} \left( \frac{Pv}{R} - T_0 \right) \right). \quad (3.251)$$

Thus, we get the two relevant partial derivatives

$$\left. \frac{\partial e}{\partial v} \right|_{P,n_i} = \sum_{i=1}^{N} \frac{n_i M_i}{m} c_{vi} \frac{P}{R} = \frac{P}{R} \sum_{i=1}^{N} Y_i c_{vi} = \frac{P c_v}{R}, \quad (3.252)$$

$$\left. \frac{\partial e}{\partial P} \right|_{v,n_i} = \sum_{i=1}^{N} \frac{n_i M_i}{m} c_{vi} \frac{v}{R} = \frac{v}{R} \sum_{i=1}^{N} Y_i c_{vi} = \frac{v c_v}{R}. \quad (3.253)$$

Now, substitute Eqs. (3.252) and (3.253) into Eq. (3.248) to get

$$c = v \sqrt{\frac{P + P c_v / R}{v c_v / R}}. \quad (3.254)$$

Straightforward algebra using $Pv = RT$, $c_P - c_v = R$, and $\gamma = c_P / c_v$ yields

$$c = \sqrt{\gamma R T}. \quad (3.255)$$

This is precisely the formula for the sound speed of a single calorically perfect ideal gas. We interpret $\gamma$ as the correct value for the mixture.

## 3.8 Irreversible Entropy Production

Recall from the second law, $dS \geq \delta Q / T$, that the entropy of a system may go up or down. Specifically, recall that reversible heat transfer with $\delta Q < 0$ will induce $dS < 0$. So the second law does not demand that all entropy changes be positive. But if one removes the effect of reversible heat transfer, there is a portion of entropy change that must be positive semidefinite. We will call that portion the *irreversible entropy production* $\sigma$, with units erg/K and insist that its differential be such that

$$d\sigma \geq 0. \quad (3.256)$$

Let us determine $d\sigma$ for a reacting system. Consider a thermodynamic system *closed to mass exchanges with its surroundings* coming into equilibrium. Allow the system to be exchanging work and heat with its surroundings. Assume the work exchange is restricted to pressure-volume work. Assume the temperature difference between the system and its surroundings is so small that both can be considered to be at temperature $T$. If $\delta Q$ is introduced into the system, then the surroundings suffer a loss of entropy:

$$dS_{surr} = -\delta Q / T. \quad (3.257)$$

The system's entropy $S$ can change via this heat transfer, as well as via other internal irreversible processes, such as internal chemical reaction. Another way to view the second law of thermodynamics is that it requires that the entropy change of our isolated universe to be positive semidefinite:

$$dS + dS_{surr} \geq 0. \quad (3.258)$$

Eliminating $dS_{surr}$, one requires for the system that

$$dS \geq \delta Q / T. \quad (3.259)$$

Consider temporarily the assumption that the work and heat transfer are both reversible. Thus, any irreversible entropy production must be associated with internal chemical reaction. Now, the first law for the entire system gives

$$dE = \delta Q - \delta W = \delta Q - P\,dV, \tag{3.260}$$

$$\delta Q = dE + P\,dV. \tag{3.261}$$

Because the system is closed, there can be no species entering or exiting, and so there is no change $dE$ attributable to $dn_i$. While within the system, the $dn_i$ may not be 0, the *net* contribution to the change in extensive internal energy is zero. A nonzero $dn_i$ within the system simply repartitions a fixed amount of energy from one species to another. Substituting Eq. (3.261) into Eq. (3.259) to eliminate $\delta Q$, one gets

$$dS \geq \frac{1}{T}\underbrace{(dE + P\,dV)}_{\delta Q}, \tag{3.262}$$

$$T\,dS - dE - P\,dV \geq 0, \tag{3.263}$$

$$dE - T\,dS + P\,dV \leq 0. \tag{3.264}$$

Equation (3.264) involves properties only and need not require assumptions of reversibility for processes in its derivation. In special cases, it reduces to simpler forms. For isentropic and isochoric processes, the second law, Eq. (3.264), reduces to

$$dE|_{S,V} \leq 0. \tag{3.265}$$

For isoenergetic and isochoric processes, the second law, Eq. (3.264), reduces to

$$dS|_{E,V} \geq 0. \tag{3.266}$$

For isoenergetic and isentropic processes, the second law, Eq. (3.264), reduces to

$$dV|_{E,S} \leq 0. \tag{3.267}$$

Using Eq. (3.169) to eliminate $dS$ in Eq. (3.266), one can express the second law as

$$\left.\underbrace{\left(\frac{1}{T}dE + \frac{P}{T}dV - \frac{1}{T}\sum_{i=1}^{N}\overline{\mu}_i\,dn_i\right)}_{dS}\right|_{E,V} \geq 0, \tag{3.268}$$

$$\underbrace{-\frac{1}{T}\sum_{i=1}^{N}\overline{\mu}_i\,dn_i}_{dS|_{E,V}\equiv d\sigma} \geq 0. \tag{3.269}$$

The differential irreversible entropy production, $d\sigma$, associated with the internal chemical reaction must be the left side of Eq. (3.269):

$$d\sigma \equiv -\frac{1}{T}\sum_{i=1}^{N}\overline{\mu}_i\,dn_i \geq 0. \tag{3.270}$$

Now, while most standard texts focusing on equilibrium thermodynamics go to great lengths to avoid the introduction of time, to understand combustion dynamics, it really belongs in a discussion describing the approach to equilibrium. One can divide Eq. (3.270) by a *positive* time increment $dt$ to get

$$\dot{\sigma} = \frac{d\sigma}{dt} = -\frac{1}{T}\sum_{i=1}^{N} \overline{\mu}_i \frac{dn_i}{dt} \geq 0. \tag{3.271}$$

Because $T \geq 0$, one can multiply Eq. (3.271) by $-T$ to get

$$-T\frac{d\sigma}{dt} = \sum_{i=1}^{N} \overline{\mu}_i \frac{dn_i}{dt} \leq 0. \tag{3.272}$$

This will hold *if* a model for $dn_i/dt$ is employed that guarantees that the left side of Eq. (3.272) is negative semidefinite. One will expect then for $dn_i/dt$ to be related to the chemical potentials $\overline{\mu}_i$. Elimination of $dE$ in Eq. (3.264) in favor of $dH$ from $dH = dE + P\,dV + V\,dP$ gives

$$\underbrace{dH - P\,dV - V\,dP}_{dE} - T\,dS + P\,dV \leq 0, \tag{3.273}$$

$$dH - V\,dP - T\,dS \leq 0. \tag{3.274}$$

Thus, one finds for isobaric, isentropic equilibration that

$$dH|_{P,S} \leq 0. \tag{3.275}$$

For the Helmholtz and Gibbs free energies, one analogously finds

$$dA|_{T,V} \leq 0, \qquad dG|_{T,P} \leq 0. \tag{3.276}$$

The expression of the second law in terms of $dG$ is especially useful as it may be easy in an experiment to control so that $P$ and $T$ are constant. This is especially true in an isobaric phase change, in which the temperature is guaranteed to be constant as well.

Now, one has Eq. (3.203):

$$G = \sum_{i=1}^{N} n_i \overline{g}_i = \sum_{i=1}^{N} n_i \overline{\mu}_i. \tag{3.277}$$

One also has from Eq. (3.181): $dG = -S\,dT + V\,dP + \sum_{i=1}^{N} \overline{\mu}_i\,dn_i$. We rewrite, holding $T$ and $P$ constant to get

$$dG|_{T,P} = \sum_{i=1}^{N} \overline{\mu}_i\,dn_i. \tag{3.278}$$

Here, the $dn_i$ are associated entirely with internal chemical reactions. Substituting Eq. (3.278) into Eq. (3.276), one gets the important version of the second law:

$$dG|_{T,P} = \sum_{i=1}^{N} \overline{\mu}_i\,dn_i \leq 0. \tag{3.279}$$

One can scale Eq. (3.279) by a positive time increment $dt > 0$ to get

$$\left.\frac{\partial G}{\partial t}\right|_{T,P} = \sum_{i=1}^{N} \overline{\mu}_i \frac{dn_i}{dt} \leq 0. \tag{3.280}$$

While the partial derivative with respect to time usually implies a possible spatial dependency, all we imply here is that $T$ and $P$ are to be held constant, as our spatially homogeneous $G$ may be of the form $G(T, P, n_i, t)$.

## 3.9 Equilibrium in a Two-Component System

A major task of *nonequilibrium thermodynamics* is to find a functional form for $dn_i/dt$ that guarantees satisfaction of the second law, Eq. (3.280), and gives predictions that agree with experiment. This is discussed in more detail in Chapter 4, on thermochemistry. At this point, some simple examples, which for variety will use SI units, will be given in which a naïve but useful functional form for $dn_i/dt$ is posed that leads at least to predictions of the correct equilibrium values. A much better model that gives the correct dynamics in the time domain of the system as it approaches equilibrium is presented in Chapter 4.

### 3.9.1 Phase Equilibrium

Recall that a single material may exist in multiple phases, such as liquid, solid, or vapor. During phase transitions, two or more phases can coexist in a so-called *phase equilibrium*. A material in such a state is often called a multiphase mixture, but because it remains a single material, it is a degenerate case of the general mixtures considered in Chapter 2. Here, consider two examples describing systems in phase equilibrium.

**EXAMPLE 3.10**

Consider an equilibrium two-phase mixture of liquid and vapor $H_2O$ at $T = 100°C$, $x = 0.5$ within a piston-confined cylinder (see Fig. 3.3). Here, $x$ is the so-called *quality*, which is the ratio of the mass of the vapor to the total mass. Use standard steam tables to check if equilibrium properties are satisfied.

In a two-phase gas liquid mixture one can expect liquid and solid molecules to move from one state to the other. In many ways, the phase transition behaves as a chemical reaction. We write it in such a form

$$H_2O_{(l)} \leftrightharpoons H_2O_{(g)}. \tag{3.281}$$

That is, one mol of liquid $(l)$, in the forward phase change, evaporates to form one mol of gas $(g)$. In the reverse phase change, one mol of gas condenses to form one mol of liquid.

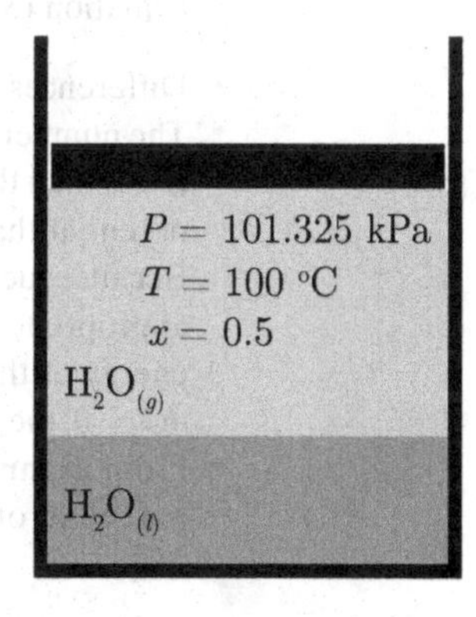


Figure 3.3. Two-phase mixture of liquid and gaseous water in phase equilibrium within a piston-confined cylinder.

Because $T$ is fixed at 100°C and the material is a liquid-gas mixture, the pressure is also fixed at a constant value of $P = 101.325$ kPa.

Equation (3.279) reduces to

$$\overline{\mu}_l dn_l + \overline{\mu}_g dn_g \leq 0. \tag{3.282}$$

Now, for the pure $H_2O$, a loss of moles from one phase must be compensated by the addition to the other. So, one must have

$$dn_l + dn_g = 0. \tag{3.283}$$

This algebraic constraint is essential for the analysis and will take on a larger role in detailed chemical kinetics; see Eq. (4.19) for single-reaction systems or Eq. (5.4) for multiple-reaction systems. We rearrange to say

$$dn_g = -dn_l. \tag{3.284}$$

So, Eq. (3.282), using Eq. (3.284), becomes

$$\overline{\mu}_l \, dn_l - \overline{\mu}_g \, dn_l \leq 0, \qquad dn_l(\overline{\mu}_l - \overline{\mu}_g) \leq 0. \tag{3.285}$$

At this stage of the analysis, it is tempting to use notions grounded in equilibrium thermodynamics and assert that $\overline{\mu}_l = \overline{\mu}_g$, ignoring the fact that the chemical potentials could be different but $dn_l$ could be zero. That approach is not taken here. Instead, let us introduce time. Scale Eq. (3.285) by a positive time increment, $dt \geq 0$ to write the second law as

$$\frac{dn_l}{dt}(\overline{\mu}_l - \overline{\mu}_g) \leq 0. \tag{3.286}$$

Now a popular strategy in continuum mechanics (Truesdell, 1984) is to seek general functional forms that are admissible under the constraints of the second law of thermodynamics. We do not do that here but for argument's sake only choose a convenient, albeit naïve, way to guarantee second law satisfaction by provisionally assuming a kinetics model of

$$\textit{Convenient but naïve kinetics model:} \qquad \frac{dn_l}{dt} = -\kappa(\overline{\mu}_l - \overline{\mu}_g), \qquad \kappa \geq 0. \tag{3.287}$$

Here, $\kappa$ is some positive semidefinite scalar rate constant that dictates the time scale of approach to equilibrium. Equation (3.287) is just a hypothesized model. It has *no experimental validation*; one can imagine that other more complex models exist that both agree with experiment and satisfy the second law. For the purposes of the present argument, however, Eq. (3.287) will suffice. With this assumption, the second law reduces to

$$-\kappa(\overline{\mu}_l - \overline{\mu}_g)^2 \leq 0, \qquad \kappa \geq 0, \tag{3.288}$$

which is always true.

Equation (3.287) has three important consequences:

- Differences in chemical potential drive changes in the number of moles.
- The number of moles of liquid, $n_l$, *increases* when the chemical potential of the liquid is less than that of the gas, $\overline{\mu}_l < \overline{\mu}_g$. That is to say, when the liquid has a lower chemical potential than the gas, the gas is driven toward the phase with the lower potential. Because such a phase change is isobaric and isothermal, the Gibbs free energy is the appropriate variable to consider, and one takes $\overline{\mu} = \overline{g}$. When this is so, the Gibbs free energy of the mixture, $G = n_l\overline{\mu}_l + n_g\overline{\mu}_g$ is being driven to a lower value. So, when $dG = 0$, the system has a minimum $G$.
- From examination of Eq. (3.287), the system is in equilibrium when the chemical potentials of liquid and gas are equal: $\overline{\mu}_l = \overline{\mu}_g$.

The chemical potentials, and hence the molar specific Gibbs free energies, must be the same for each constituent of the binary mixture at the phase equilibrium. That is,

$$\overline{g}_l = \overline{g}_g. \tag{3.289}$$

Now, because both the liquid and gas have the same molecular mass, one also has the mass specific Gibbs free energies equal at phase equilibrium:

$$g_l = g_g. \tag{3.290}$$

This can be verified from standard steam tables,[8] using the definition $g = h - Ts$. We find

$$g_l = h_l - Ts_l = 419.02\,\frac{\text{kJ}}{\text{kg}} - ((100 + 273.15)\text{ K})\left(1.3068\,\frac{\text{kJ}}{\text{kg K}}\right) = -68.6\,\frac{\text{kJ}}{\text{kg}}, \tag{3.291}$$

$$g_g = h_g - Ts_g = 2676.05\,\frac{\text{kJ}}{\text{kg}} - ((100 + 273.15)\text{ K})\left(7.3548\,\frac{\text{kJ}}{\text{kg K}}\right) = -68.4\,\frac{\text{kJ}}{\text{kg}}. \tag{3.292}$$

The two values are essentially the same; the difference is likely due to table inaccuracies.

### 3.9.2 Chemical Equilibrium: Introduction

Next, let us examine a simple reactive mixture of ideal gases. We look at two examples that identify the equilibrium state.

#### Isothermal, Isochoric System

The simplest system to consider is isothermal and isochoric. The isochoric assumption implies there is no work in coming to equilibrium.

**EXAMPLE 3.11**

At high temperatures, collisions between diatomic nitrogen molecules induce the production of monatomic nitrogen molecules. The chemical reaction can be described by the reversible model

$$N_2 + N_2 \leftrightharpoons 2N + N_2. \tag{3.293}$$

Here, one of the $N_2$ molecules behaves as an inert "third" body (really a second body in this scheme). An $N_2$ molecule has to collide with *something*, to induce the reaction. Some authors leave out the third body and write instead $N_2 \leftrightharpoons 2N$, but this does not reflect the true physics as well. The inert third body is especially important when the time scales of reaction are considered. It plays no role in equilibrium chemistry.

Consider 1 kmol of $N_2$ and 0 kmol of N at a pressure of 100 kPa and a temperature of 6000 K. Find the equilibrium concentrations of N and $N_2$ if the equilibration process is *isothermal* and *isochoric*; see Fig. 3.4. Also perform a first and second law analysis with the aid of the ideal gas tables.

The ideal gas law can give the volume:

$$V = \frac{n_{N_2}\overline{R}T}{P_1} = \frac{(1\text{ kmol})\left(8.314\,\frac{\text{kJ}}{\text{kmol K}}\right)(6000\text{ K})}{100\text{ kPa}} = 498.84\text{ m}^3. \tag{3.294}$$

[8] Table B.1 from Borgnakke and Sonntag (2013, p. 777).

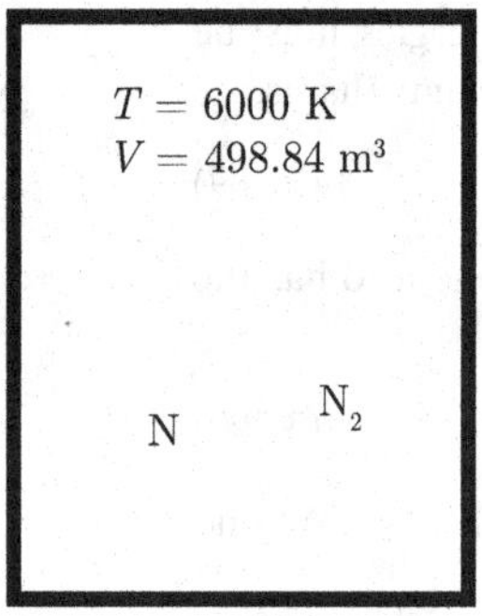

$T = 6000$ K
$V = 498.84$ m$^3$

N $N_2$

Figure 3.4. Isothermal mixture of N and $N_2$ within an isochoric cylinder.

Initially, the mixture is all $N_2$, so its partial pressure is the total pressure, and the initial partial pressure of N is 0. Now, every time an $N_2$ molecule reacts and thus undergoes a negative change, two N molecules are created and thus undergo a positive change. From this, we can infer

$$-dn_{N_2} = \frac{1}{2} dn_N. \tag{3.295}$$

This can be parameterized by a reaction progress variable $\zeta$ defined such that

$$d\zeta = -dn_{N_2}, \qquad d\zeta = \frac{1}{2} dn_N. \tag{3.296}$$

As an aside, one can integrate this, taking $\zeta = 0$ at the initial state to get

$$\zeta = n_{N_2}|_{t=0} - n_{N_2} = -\frac{1}{2}\, n_N|_{t=0} + \frac{1}{2} n_N. \tag{3.297}$$

Thus,

$$n_{N_2} = n_{N_2}|_{t=0} - \zeta, \quad n_N = n_N|_{t=0} + 2\zeta. \tag{3.298}$$

This is the form we will later generalize in Eq. (5.355). One can also eliminate $\zeta$ and recognize that $n_N|_{t=0} = 0$ to get $n_N$ in terms of $n_{N_2}$:

$$n_N = 2\left(n_{N_2}|_{t=0} - n_{N_2}\right). \tag{3.299}$$

Now, for the reaction, one must have, for second law satisfaction from Eq. (3.279), that

$$\overline{\mu}_{N_2}\, dn_{N_2} + \overline{\mu}_N\, dn_N \leq 0, \tag{3.300}$$

$$\overline{\mu}_{N_2}(-d\zeta) + \overline{\mu}_N\,(2\, d\zeta) \leq 0, \tag{3.301}$$

$$\left(-\overline{\mu}_{N_2} + 2\overline{\mu}_N\right) d\zeta \leq 0, \tag{3.302}$$

$$\left(-\overline{\mu}_{N_2} + 2\overline{\mu}_N\right) \frac{d\zeta}{dt} \leq 0. \tag{3.303}$$

Similar to the previous example, to satisfy the second law, one can usefully, but naïvely, hypothesize that the *nonequilibrium* reaction kinetics are given by

*Convenient but naïve kinetics model:* $$\frac{d\zeta}{dt} = -k(-\overline{\mu}_{N_2} + 2\overline{\mu}_N), \qquad k \geq 0. \tag{3.304}$$

There are other ways to guarantee second law satisfaction. In fact, a more complicated model is known to fit data well, and will be studied in Chapter 4; realistic reaction kinetics for precisely this system will be considered in an example in Sec. 4.5.1. For the present purposes, this naïve model will suffice. With this assumption, the second law reduces to

$$-k(-\overline{\mu}_{N_2} + 2\overline{\mu}_N)^2 \leq 0, \qquad k \geq 0, \tag{3.305}$$

which is always true. Obviously, the reaction ceases when $d\zeta/dt = 0$, which holds only when

$$2\overline{\mu}_N = \overline{\mu}_{N_2}. \tag{3.306}$$

Away from equilibrium, for the reaction to go forward, one must expect $d\zeta/dt > 0$, and then one must have

$$-\overline{\mu}_{N_2} + 2\overline{\mu}_N \leq 0, \tag{3.307}$$

$$2\overline{\mu}_N \leq \overline{\mu}_{N_2}. \tag{3.308}$$

The chemical potentials are the molar specific Gibbs free energies; thus, for the reaction to go forward,

$$2\overline{g}_N \leq \overline{g}_{N_2}. \tag{3.309}$$

Substituting using the definitions of Gibbs free energy and entropy, see Eqs. (3.226) and (2.206), one gets

$$2\left(\overline{h}_N - T\overline{s}_N\right) \leq \overline{h}_{N_2} - T\overline{s}_{N_2}, \tag{3.310}$$

$$2\left(\overline{h}_N - T\left(\overline{s}^{\,0}_{T,N} - \overline{R}\ln\left(\frac{y_N P}{P_0}\right)\right)\right) \leq \overline{h}_{N_2} - T\left(\overline{s}^{\,0}_{T,N_2} - \overline{R}\ln\left(\frac{y_{N_2} P}{P_0}\right)\right), \tag{3.311}$$

$$2\left(\overline{h}_N - T\,\overline{s}^{\,0}_{T,N}\right) - \left(\overline{h}_{N_2} - T\,\overline{s}^{\,0}_{T,N_2}\right) \leq -2\overline{R}T\ln\left(\frac{y_N P}{P_0}\right) + \overline{R}T\ln\left(\frac{y_{N_2} P}{P_0}\right), \tag{3.312}$$

$$-2\left(\overline{h}_N - T\,\overline{s}^{\,0}_{T,N}\right) + \left(\overline{h}_{N_2} - T\,\overline{s}^{\,0}_{T,N_2}\right) \geq 2\overline{R}T\ln\left(\frac{y_N P}{P_0}\right) - \overline{R}T\ln\left(\frac{y_{N_2} P}{P_0}\right), \tag{3.313}$$

$$\geq \overline{R}T\ln\left(\frac{y_N^2 P^2}{P_0^2}\frac{P_0}{P y_{N_2}}\right), \tag{3.314}$$

$$\geq \overline{R}T\ln\left(\frac{y_N^2}{y_{N_2}}\frac{P}{P_0}\right). \tag{3.315}$$

At the initial state, one has $y_N = 0$, so the right-hand side approaches $-\infty$, and the inequality holds. At equilibrium, one has equality:

$$-2\left(\overline{h}_N - T\,\overline{s}^{\,0}_{T,N}\right) + \left(\overline{h}_{N_2} - T\,\overline{s}^{\,0}_{T,N_2}\right) = \overline{R}T\ln\left(\frac{y_N^2}{y_{N_2}}\frac{P}{P_0}\right). \tag{3.316}$$

Taking numerical values from standard thermodynamic tables,[9]

$$-2\left(5.9727\times10^5\,\frac{\text{kJ}}{\text{kmol}}-(6000\text{ K})\left(216.926\,\frac{\text{kJ}}{\text{kg K}}\right)\right)$$
$$+\left(2.05848\times10^5\,\frac{\text{kJ}}{\text{kmol}}-(6000\text{ K})\left(292.984\,\frac{\text{kJ}}{\text{kg K}}\right)\right)=\left(8.314\,\frac{\text{kJ}}{\text{kmol K}}\right)$$
$$\times(6000\text{ K})\ln\left(\frac{y_{\text{N}}^2}{y_{\text{N}_2}}\frac{P}{P_0}\right). \tag{3.317}$$

This yields

$$\underbrace{-2.87635}_{\equiv\ln K_P}=\ln\left(\frac{y_{\text{N}}^2}{y_{\text{N}_2}}\frac{P}{P_0}\right), \tag{3.318}$$

$$\underbrace{0.0563399}_{\equiv K_P}=\frac{y_{\text{N}}^2}{y_{\text{N}_2}}\frac{P}{P_0}, \tag{3.319}$$

$$=\frac{\left(\frac{n_{\text{N}}}{n_{\text{N}}+n_{\text{N}_2}}\right)^2}{\left(\frac{n_{\text{N}_2}}{n_{\text{N}}+n_{\text{N}_2}}\right)}(n_{\text{N}}+n_{\text{N}_2})\frac{\overline{R}T}{P_0V}, \tag{3.320}$$

$$=\frac{n_{\text{N}}^2}{n_{\text{N}_2}}\frac{\overline{R}T}{P_0V}, \tag{3.321}$$

$$=\frac{\left(2\left(n_{\text{N}_2}|_{t=0}-n_{\text{N}_2}\right)\right)^2}{n_{\text{N}_2}}\frac{\overline{R}T}{P_0V}, \tag{3.322}$$

$$=\frac{\left(2\left(1\text{ kmol}-n_{\text{N}_2}\right)\right)^2}{n_{\text{N}_2}}\frac{(8.314)(6000)}{(100)(498.84)}. \tag{3.323}$$

This is a quadratic equation for $n_{\text{N}_2}$. It has two roots,

$$n_{\text{N}_2}=0.888154\text{ kmol, (physical)},\qquad n_{\text{N}_2}=1.12593\text{ kmol, (nonphysical)}. \tag{3.324}$$

The second root generates more $N_2$ than at the start and also yields nonphysically negative $n_{\text{N}}=-0.25186$ kmol. So, at equilibrium, the physical root is

$$n_{\text{N}}=2(1-n_{\text{N}_2})=2(1-0.888154)=0.223693\text{ kmol}. \tag{3.325}$$

The diatomic species is preferred.

In the preceding analysis, the term $K_P$ was introduced. This is the so-called equilibrium constant that is really a function of temperature. It will be described in more detail in Sec. 4.4, but it is commonly tabulated for some reactions. Its tabular value can be derived from the more fundamental quantities shown in this example. Standard thermodynamic tables[10] give for this reaction at 6000 K the value of $\ln K_P=-2.876$. Note that $K_P$ is fundamentally defined in terms of thermodynamic properties for a system that may or may not be at chemical equilibrium. Only at chemical equilibrium can it can further be related to mol fraction and pressure ratios.

[9]Table A.9 from Borgnakke and Sonntag (2013, p. 766).
[10]Table A.11 from Borgnakke and Sonntag (2013, p. 773).

The pressure at equilibrium is

$$P_2 = \frac{(n_{N_2} + n_N)\overline{R}T}{V}, \tag{3.326}$$

$$= \frac{(0.888154 \text{ kmol} + 0.223693 \text{ kmol})\left(8.314 \frac{\text{kJ}}{\text{kmol K}}\right)(6000 \text{ K})}{498.84}, \tag{3.327}$$

$$= 111.185 \text{ kPa}. \tag{3.328}$$

The pressure has increased because there are more molecules with the volume and temperature being equal. The molar concentrations $\overline{\rho}_i$ at equilibrium, are

$$\overline{\rho}_N = \frac{0.223693 \text{ kmol}}{498.84 \text{ m}^3} = 4.48426 \times 10^{-4} \frac{\text{kmol}}{\text{m}^3} = 4.48426 \times 10^{-7} \frac{\text{mol}}{\text{cm}^3}, \tag{3.329}$$

$$\overline{\rho}_{N_2} = \frac{0.888154 \text{ kmol}}{498.84 \text{ m}^3} = 1.78044 \times 10^{-3} \frac{\text{kmol}}{\text{m}^3} = 1.78044 \times 10^{-6} \frac{\text{mol}}{\text{cm}^3}. \tag{3.330}$$

Now, consider the heat transfer. One knows for the isochoric process that $Q = E_2 - E_1$. The initial energy is given by

$$E_1 = n_{N_2}\overline{e}_{N_2} = n_{N_2}(\overline{h}_{N_2} - \overline{R}T), \tag{3.331}$$

$$= (1 \text{ kmol})\left(2.05848 \times 10^5 \frac{\text{kJ}}{\text{kmol}} - \left(8.314 \frac{\text{kJ}}{\text{kmol K}}\right)(6000 \text{ K})\right), \tag{3.332}$$

$$= 1.55964 \times 10^5 \text{ kJ}. \tag{3.333}$$

The energy at the final state is

$$E_2 = n_{N_2}\overline{e}_{N_2} + n_N\overline{e}_N = n_{N_2}(\overline{h}_{N_2} - \overline{R}T) + n_N(\overline{h}_N - \overline{R}T), \tag{3.334}$$

$$= (0.888154 \text{ kmol})\left(2.05848 \times 10^5 \frac{\text{kJ}}{\text{kmol}} - \left(8.314 \frac{\text{kJ}}{\text{kmol K}}\right)(6000 \text{ K})\right)$$
$$+ (0.223693 \text{ kmol})\left(5.9727 \times 10^5 \frac{\text{kJ}}{\text{kmol}} - \left(8.314 \frac{\text{kJ}}{\text{kmol K}}\right)(6000 \text{ K})\right), \tag{3.335}$$

$$= 2.60966 \times 10^5 \text{ kJ}. \tag{3.336}$$

So

$$Q = E_2 - E_1 = 2.60966 \times 10^5 \text{ kJ} - 1.555964 \times 10^5 \text{ kJ} = 1.05002 \times 10^5 \text{ kJ}. \tag{3.337}$$

Heat needed to be added to keep the system at the constant temperature. This is because the nitrogen dissociation process is *endothermic*. Breaking the bonds of $N_2$ required an energy input.

One can check for second law satisfaction in two ways. Fundamentally, one can demand that Eq. (3.259), $dS \geq \delta Q/T$, be satisfied for the process, giving

$$S_2 - S_1 \geq \int_1^2 \frac{\delta Q}{T}. \tag{3.338}$$

For this isothermal process, this reduces to

$$S_2 - S_1 \geq \frac{Q}{T}, \tag{3.339}$$

$$(n_{N_2}\overline{s}_{N_2} + n_N\overline{s}_N)|_2 - (n_{N_2}\overline{s}_{N_2} + n_N\overline{s}_N)|_1 \geq \frac{Q}{T}, \tag{3.340}$$

$$\left(n_{N_2}\left(\overline{s}^0_{T,N_2} - \overline{R}\ln\left(\frac{y_{N_2}P}{P_0}\right)\right) + n_N\left(\overline{s}^0_{T,N} - \overline{R}\ln\left(\frac{y_N P}{P_0}\right)\right)\right)\Big|_2 - \left(n_{N_2}\left(\overline{s}^0_{T,N_2} - \overline{R}\ln\left(\frac{y_{N_2}P}{P_0}\right)\right) + n_N\left(\overline{s}^0_{T,N} - \overline{R}\ln\left(\frac{y_N P}{P_0}\right)\right)\right)\Big|_1 \geq \frac{Q}{T}, \quad (3.341)$$

$$\left(n_{N_2}\left(\overline{s}^0_{T,N_2} - \overline{R}\ln\left(\frac{P_{N_2}}{P_0}\right)\right) + n_N\left(\overline{s}^0_{T,N} - \overline{R}\ln\left(\frac{P_N}{P_0}\right)\right)\right)\Big|_2 - \left(n_{N_2}\left(\overline{s}^0_{T,N_2} - \overline{R}\ln\left(\frac{P_{N_2}}{P_0}\right)\right) + n_N\left(\overline{s}^0_{T,N} - \overline{R}\ln\left(\frac{P_N}{P_0}\right)\right)\right)\Big|_1 \geq \frac{Q}{T}, \quad (3.342)$$

$$\left(n_{N_2}\left(\overline{s}^0_{T,N_2} - \overline{R}\ln\left(\frac{n_{N_2}\overline{R}T}{P_0 V}\right)\right) + n_N\left(\overline{s}^0_{T,N} - \overline{R}\ln\left(\frac{n_N\overline{R}T}{P_0 V}\right)\right)\right)\Big|_2 - \left(n_{N_2}\left(\overline{s}^0_{T,N_2} - \overline{R}\ln\left(\frac{n_{N_2}\overline{R}T}{P_0 V}\right)\right) + \underbrace{n_N}_{0}\left(\overline{s}^0_{T,N} - \overline{R}\ln\left(\frac{n_N\overline{R}T}{P_0 V}\right)\right)\right)\Big|_1 \geq \frac{Q}{T}. \quad (3.343)$$

Now, at the initial state, $n_N = 0$ kmol, and $\overline{R}T/P_0/V$ has a constant value of

$$\frac{\overline{R}T}{P_0 V} = \frac{\left(8.314\ \frac{\text{kJ}}{\text{kmol K}}\right)(6000\ \text{K})}{(100\ \text{kPa})(498.84\ \text{m}^3)} = 1\ \frac{1}{\text{kmol}}, \quad (3.344)$$

so Eq. (3.343) reduces to

$$\left(n_{N_2}\left(\overline{s}^0_{T,N_2} - \overline{R}\ln\left(\frac{n_{N_2}}{1\ \text{kmol}}\right)\right) + n_N\left(\overline{s}^0_{T,N} - \overline{R}\ln\left(\frac{n_N}{1\ \text{kmol}}\right)\right)\right)\Big|_2 - \left(n_{N_2}\left(\overline{s}^0_{T,N_2} - \overline{R}\ln\left(\frac{n_{N_2}}{1\ \text{kmol}}\right)\right)\right)\Big|_1 \geq \frac{Q}{T}. \quad (3.345)$$

Substituting numbers, we get

$$((0.888143)(292.984 - 8.314\ln(0.88143)) + (0.223714)(216.926 - 8.314\ln(0.223714)))|_2 - ((1)(292.984 - 8.314\ln(1)))|_1 \geq \frac{105002}{6000},$$

$$19.4181\ \frac{\text{kJ}}{\text{K}} \geq 17.5004\ \frac{\text{kJ}}{\text{K}}. \quad (3.346)$$

Indeed, the second law is satisfied. Moreover, the irreversible entropy production of the chemical reaction is $\sigma = 19.4181 - 17.5004 = +1.91772$ kJ/K.

For the isochoric, isothermal process, it is also appropriate to use Eq. (3.276), $dA|_{T,V} \leq 0$, to check for second law satisfaction. This turns out to give an identical result. Because by Eq. (3.176), $A = U - TS$, $A_2 - A_1 = (E_2 - T_2S_2) - (E_1 - T_1S_1)$. Because the process is isothermal, $A_2 - A_1 = E_2 - E_1 - T(S_2 - S_1)$. For $A_2 - A_1 \leq 0$, one must demand $E_2 - E_1 - T(S_2 - S_1) \leq 0$, or $E_2 - E_1 \leq T(S_2 - S_1)$, or $S_2 - S_1 \geq (E_2 - E_1)/T$. Because $Q = E_2 - E_1$ for this isochoric process, one then recovers $S_2 - S_1 \geq Q/T$.

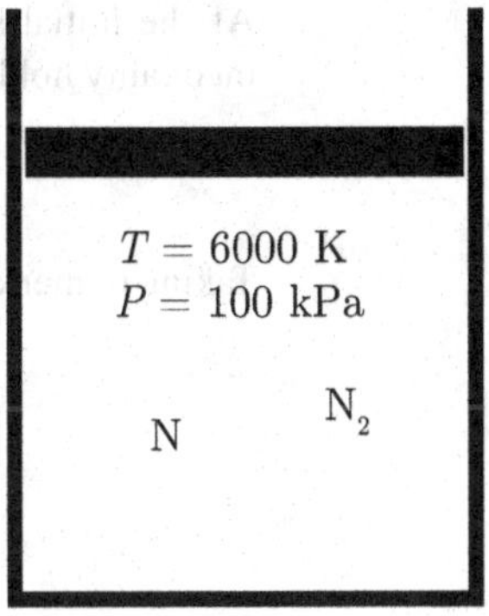

$T = 6000$ K
$P = 100$ kPa
N $N_2$

Figure 3.5. Isothermal mixture of N and $N_2$ within an isobaric piston-confined cylinder.

## Isothermal, Isobaric System

Allowing for isobaric rather than isochoric equilibration introduces a small variation in the analysis.

**EXAMPLE 3.12**

Consider the same reaction,

$$N_2 + N_2 \leftrightharpoons 2N + N_2, \tag{3.347}$$

for an *isobaric* and *isothermal* process. That is, consider 1 kmol of $N_2$ and 0 kmol of N at a pressure of 100 kPa and a temperature of 6000 K; see Fig. 3.5. Using the tables, find the equilibrium concentrations of N and $N_2$ if the equilibration process is isothermal and isobaric.

The initial volume is the same as from the previous example, $V_1 = 498.84\ \mathrm{m}^3$. The volume will change in this isobaric process. Initially, the mixture is all $N_2$, so its partial pressure is the total pressure, and the initial partial pressure of N is 0.

A few other key results are identical to the previous example, $n_{\rm N} = 2\left(n_{\rm N_2}|_{t=0} - n_{\rm N_2}\right)$, and $2\overline{g}_{\rm N} \le \overline{g}_{\rm N_2}$. Substituting using the definitions of Gibbs free energy, one gets

$$2\left(\overline{h}_{\rm N} - T\,\overline{s}_{\rm N}\right) \le \overline{h}_{\rm N_2} - T\,\overline{s}_{\rm N_2}, \tag{3.348}$$

$$2\left(\overline{h}_{\rm N} - T\left(\overline{s}^{\,0}_{T,\rm N} - \overline{R}\ln\left(\frac{y_{\rm N}P}{P_0}\right)\right)\right) \le \overline{h}_{\rm N_2} - T\left(\overline{s}^{\,0}_{T,\rm N_2} - \overline{R}\ln\left(\frac{y_{\rm N_2}P}{P_0}\right)\right), \tag{3.349}$$

$$2\left(\overline{h}_{\rm N} - T\,\overline{s}^{\,0}_{T,\rm N}\right) - \left(\overline{h}_{\rm N_2} - T\,\overline{s}^{\,0}_{T,\rm N_2}\right) \le -2\overline{R}T\ln\left(\frac{y_{\rm N}P}{P_0}\right) + \overline{R}T\ln\left(\frac{y_{\rm N_2}P}{P_0}\right), \tag{3.350}$$

$$-2\left(\overline{h}_{\rm N} - T\,\overline{s}^{\,0}_{T,\rm N}\right) + \left(\overline{h}_{\rm N_2} - T\,\overline{s}^{\,0}_{T,\rm N_2}\right) \ge 2\overline{R}T\ln\left(\frac{y_{\rm N}P}{P_0}\right) - \overline{R}T\ln\left(\frac{y_{\rm N_2}P}{P_0}\right), \tag{3.351}$$

$$\ge \overline{R}T\ln\left(\frac{y_{\rm N}^2P^2}{P_0^2}\frac{P_0}{Py_{\rm N_2}}\right). \tag{3.352}$$

In this case $P_0 = P$, so one gets

$$-2\left(\overline{h}_{\rm N} - T\,\overline{s}^{\,0}_{T,\rm N}\right) + \left(\overline{h}_{\rm N_2} - T\,\overline{s}^{\,0}_{T,\rm N_2}\right) \ge \overline{R}T\ln\left(\frac{y_{\rm N}^2}{y_{\rm N_2}}\right). \tag{3.353}$$

At the initial state, one has $y_{\rm N} = 0$, so the right-hand side approaches $-\infty$, and the inequality holds. At equilibrium, one has equality:

$$-2\left(\overline{h}_{\rm N} - T\,\overline{s}^{0}_{T,{\rm N}}\right) + \left(\overline{h}_{{\rm N}_2} - T\,\overline{s}^{0}_{T,{\rm N}_2}\right) = \overline{R}T\ln\left(\frac{y_{\rm N}^2}{y_{{\rm N}_2}}\right). \tag{3.354}$$

Taking numerical values from standard thermodynamic tables,[11]

$$\begin{aligned}
&-2\left(5.9727\times10^5\,\frac{\rm kJ}{\rm kmol} - (6000\ {\rm K})\left(216.926\,\frac{\rm kJ}{\rm kg\ K}\right)\right)\\
&+\left(2.05848\times10^5\,\frac{\rm kJ}{\rm kmol} - (6000\ {\rm K})\left(292.984\,\frac{\rm kJ}{\rm kg\ K}\right)\right)\\
&=\left(8.314\,\frac{\rm kJ}{\rm kmol\ K}\right)(6000\ {\rm K})\ln\left(\frac{y_{\rm N}^2}{y_{{\rm N}_2}}\right).
\end{aligned} \tag{3.355}$$

This yields

$$\underbrace{-2.87635}_{\equiv\ln K_P} = \ln\left(\frac{y_{\rm N}^2}{y_{{\rm N}_2}}\right), \tag{3.356}$$

$$\underbrace{0.0563399}_{\equiv K_P} = \frac{y_{\rm N}^2}{y_{{\rm N}_2}} = \frac{\left(\frac{n_{\rm N}}{n_{\rm N}+n_{{\rm N}_2}}\right)^2}{\left(\frac{n_{{\rm N}_2}}{n_{\rm N}+n_{{\rm N}_2}}\right)} = \frac{n_{\rm N}^2}{n_{{\rm N}_2}(n_{\rm N}+n_{{\rm N}_2})}, \tag{3.357}$$

$$= \frac{\left(2\left(n_{{\rm N}_2}|_{t=0} - n_{{\rm N}_2}\right)\right)^2}{n_{{\rm N}_2}\left(2\left(n_{{\rm N}_2}|_{t=0} - n_{{\rm N}_2}\right) + n_{{\rm N}_2}\right)}, \tag{3.358}$$

$$= \frac{\left(2\left(1\ {\rm kmol} - n_{{\rm N}_2}\right)\right)^2}{n_{{\rm N}_2}\left(2\left(1\ {\rm kmol} - n_{{\rm N}_2}\right) + n_{{\rm N}_2}\right)}. \tag{3.359}$$

This is a quadratic equation for $n_{{\rm N}_2}$. It has two roots:

$$n_{{\rm N}_2} = 0.882147\ {\rm kmol},\ \text{(physical)},\quad n_{{\rm N}_2} = 1.11785\ {\rm kmol},\ \text{(nonphysical)}. \tag{3.360}$$

The second root generates more $N_2$ than at the start and also yields nonphysically negative $n_{\rm N} = -0.235706$ kmol. So, at equilibrium, the physical root is

$$n_{\rm N} = 2(1 - n_{{\rm N}_2}) = 2(1 - 0.882147) = 0.235706\ {\rm kmol}. \tag{3.361}$$

Again, the diatomic species is preferred. As the temperature is raised, one could show that the monatomic species would come to dominate.

The volume at equilibrium is

$$V_2 = \frac{(n_{{\rm N}_2} + n_{\rm N})\overline{R}T}{P}, \tag{3.362}$$

$$= \frac{(0.882147\ {\rm kmol} + 0.235706\ {\rm kmol})\left(8.314\,\frac{\rm kJ}{\rm kmol\ K}\right)(6000\ {\rm K})}{100\ {\rm kPa}}, \tag{3.363}$$

$$= 557.630\ {\rm m}^3. \tag{3.364}$$

The volume has increased because there are more molecules with the pressure and temperature being equal. The molar concentrations $\overline{\rho}_i$ at equilibrium are

$$\overline{\rho}_{\rm N} = \frac{0.235706\ {\rm kmol}}{556.630\ {\rm m}^3} = 4.22693\times10^{-4}\,\frac{\rm kmol}{\rm m^3} = 4.22693\times10^{-7}\,\frac{\rm mol}{\rm cm^3}, \tag{3.365}$$

$$\overline{\rho}_{{\rm N}_2} = \frac{0.882147\ {\rm kmol}}{556.630\ {\rm m}^3} = 1.58196\times10^{-3}\,\frac{\rm kmol}{\rm m^3} = 1.58196\times10^{-6}\,\frac{\rm mol}{\rm cm^3}. \tag{3.366}$$

[11] Table A.9 from Borgnakke and Sonntag (2013, p. 766).

The molar concentrations are a little smaller than for the isochoric case, mainly because the volume is larger at equilibrium. Now, consider the heat transfer. One knows for the isobaric process that $Q = H_2 - H_1$. The initial enthalpy is given by

$$H_1 = n_{N_2}\overline{h}_{N_2} = (1 \text{ kmol})\left(2.05848 \times 10^5 \frac{\text{kJ}}{\text{kmol}}\right) = 2.05848 \times 10^5 \text{ kJ}. \tag{3.367}$$

The enthalpy at the final state is

$$H_2 = n_{N_2}\overline{h}_{N_2} + n_N\overline{h}_N, \tag{3.368}$$

$$= (0.882147 \text{ kmol})\left(2.05848 \times 10^5 \frac{\text{kJ}}{\text{kmol}}\right)$$

$$+ (0.235706 \text{ kmol})\left(5.9727 \times 10^5 \frac{\text{kJ}}{\text{kmol}}\right) = 3.22368 \times 10^5 \text{ kJ}. \tag{3.369}$$

So,

$$Q = H_2 - H_1 = 3.22389 \times 10^5 \text{ kJ} - 2.05848 \times 10^5 \text{ kJ} = 1.16520 \times 10^5 \text{ kJ}. \tag{3.370}$$

Heat needed to be added to keep the system at the constant temperature. This is because the nitrogen dissociation process is endothermic. Relative to the isochoric process, *more* heat had to be added to maintain the temperature. This to counter the cooling effect of the expansion. It is a straightforward exercise to show that the second law is satisfied for this process. Last, the actual time dynamics of this problem with realistic kinetic rates are evaluated in Sec. 4.5.2.

The previous examples, in which a functional form of a progress variable's time variation, $d\zeta/dt$, was postulated to satisfy the second law, gave a condition for equilibrium. This can be generalized for $N$ species so as to require at equilibrium that

$$\underbrace{\sum_{i=1}^{N} \overline{\mu}_i \nu_i}_{\equiv -\overline{\alpha}} = 0. \tag{3.371}$$

Here, $\nu_i$ represents the *net* number of moles of species $i$ generated in the forward reaction. This is discussed in detail in Sec. 4.4. In vector form, we would say

$$\overline{\boldsymbol{\mu}}^T \cdot \boldsymbol{\nu} = 0. \tag{3.372}$$

So, in the phase equilibrium example, Eq. (3.371) becomes

$$\overline{\mu}_l(-1) + \overline{\mu}_g(1) = 0, \quad \text{with} \quad \boldsymbol{\nu} = \begin{pmatrix} -1 \\ 1 \end{pmatrix}. \tag{3.373}$$

In the nitrogen chemistry example, Eq. (3.371) becomes

$$\overline{\mu}_{N_2}(-1) + \overline{\mu}_N(2) = 0, \quad \text{with} \quad \boldsymbol{\nu} = \begin{pmatrix} -1 \\ 2 \end{pmatrix}. \tag{3.374}$$

The negation of the term on the left side of Eq (3.371) is sometimes defined as the *chemical affinity* for a single reaction, $\overline{\alpha}$:

$$\overline{\alpha} \equiv -\sum_{i=1}^{N} \overline{\mu}_i \nu_i = -\overline{\boldsymbol{\mu}} \cdot \boldsymbol{\nu}. \tag{3.375}$$

A multiple reaction extension is given in Sec. 5.1.

We finally note that while the examples of this section all introduced time $t$ in a loose sense that at least maintained the correct time directionality of the equilibration processes, we did not solve for the time-dependent variation of state variables. Among other things, Chapter 4 introduces physically based kinetics so that the time dynamics of thermodynamic state variables can be predicted.

## EXERCISES

**3.1.** Consider a material that obeys a van der Waals[12] thermal equation of state:

$$P = \frac{RT}{v - b} - \frac{a}{v^2}.$$

If $c_v$ is a constant, show that the caloric state equation must take the form

$$e(T, v) = e_0 + c_v(T - T_0) + a\left(\frac{1}{v_0} - \frac{1}{v}\right).$$

Find $c_P - c_v$ and show that it is not constant. Find an expression for the frozen sound speed.

**3.2.** Derive Eqs. (3.187) and (3.193).

**3.3.** Consider a modification of the phase equilibrium example problem of Sec. 3.9.1. With all else equal, take that the initial quality is $x = 0$ and that thermal energy is added isobarically from a surroundings at $T = 200°\text{C}$ until $x = 0.5$. Find the work done in the process, the heat added, the irreversible entropy production of the system, and the entropy change of the universe.

**3.4.** Repeat the isothermal, isochoric example of Sec. 3.9.2 under the same conditions, but replace nitrogen by oxygen: $O_2 + O_2 \leftrightharpoons 2O + O_2$. Consult a standard thermodynamics text for appropriate material properties.

**3.5.** Repeat the isothermal, isobaric example of Sec. 3.9.2 under the same conditions, but replace nitrogen by oxygen: $O_2 + O_2 \leftrightharpoons 2O + O_2$. Consult a standard thermodynamics text for appropriate material properties.

## References

M. M. Abbott and H. C. Van Ness, 1972, *Thermodynamics*, Schaum's Outline Series in Engineering, McGraw-Hill, New York, Chapter 3.

C. Borgnakke and R. E. Sonntag, 2013, *Fundamentals of Thermodynamics*, 8th ed., John Wiley, New York, Chapters 11–14.

H. B. Callen, 1985, *Thermodynamics and an Introduction to Thermostatistics*, 2nd ed., John Wiley, New York.

B. J. Cantwell, 2002, *Introduction to Symmetry Analysis*, Cambridge University Press, Cambridge, UK.

E. Fermi, 1936, *Thermodynamics*, Dover, New York.

J. W. Gibbs, 1873, Graphical methods in the thermodynamics of fluids, *Transactions of the Connecticut Academy of Arts and Sciences*, 2: 309–342.

E. P. Gyftopoulos and G. P. Beretta, 1991, *Thermodynamics: Foundations and Applications*, Macmillan, New York.

J. Kestin, 1966, *A Course in Thermodynamics*, Blaisdell, Waltham, MA.

[12] Johannes Diderik van der Waals (1837–1923), Dutch physicist; 1910 Nobel laureate in Physics.

I. Müller, 2007, *A History of Thermodynamics: The Doctrine of Energy and Entropy*, Springer, New York.
M. Planck, 1945, *Treatise on Thermodynamics*, Dover, New York.
W. C. Reynolds, 1968, *Thermodynamics*, 2nd ed., McGraw-Hill, New York.
C. A. Truesdell, 1984, *Rational Thermodynamics*, 2nd ed., Springer, New York.
W. G. Vincenti and C. H. Kruger, 1965, *Introduction to Physical Gas Dynamics*, John Wiley, New York, Chapter 3.
H. Ziegler, 1977, *An Introduction to Thermomechanics*, North-Holland, Amsterdam, Chapter 4.

# 4 Thermochemistry of a Single Reaction

*These four bodies are fire, air, water, earth. Fire occupies the highest place among them all.*
– Aristotle (384–322 BC), *Meteorology*, Book 1, § 2

Building on the foundations of Chapters 2 and 3, this chapter gives a detailed development of the thermodynamics and dynamics of spatially homogeneous chemically reacting gaseous mixtures evolving over time in a single chemical reaction. The main goal is to develop a mathematical framework for predicting the time-dependent behavior of $N$ gaseous species composed of $L$ elements reacting in $J = 1$ reaction. We consider first standard notions from equilibrium thermochemistry such as stoichiometry, the first law of thermodynamics, enthalpies of formation, adiabatic flame temperatures, and additional aspects of chemical equilibrium. We then present a model of gas phase kinetics. We spend considerable effort on formulating stoichiometry in the formal language of linear algebra. While not often done, this formulation will have benefits in enabling an efficient general model formulation at the end of the chapter and will be even more useful when multiple reactions are considered in Chapter 5. As much as possible, we use conventional notation; however, because our approach to stoichiometry is both fundamentally important and atypical in the combustion literature, some of the key ideas will be formulated with a unique nomenclature. Several sources can be consulted as references on combustion chemistry (Abbott and Van Ness, 1972; Strehlow, 1984; Strahle, 1993; Kondepudi and Prigogine, 1998; Kuo, 2005; Annamalai and Puri, 2006; Warnatz et al., 2006; McAllister et al., 2011; Ragland and Bryden, 2011; Turns, 2011; Borgnakke and Sonntag, 2013; Date, 2014; Glassman et al., 2014). Physical chemistry underpinnings of chemical kinetics are given by many, including Berry et al. (2000), and Laidler (1965). Our approach to linear algebra is most influenced by Strang (2005).

## 4.1 Molecular Mass

The molecular mass of a molecule is a straightforward notion from chemistry. One simply sums the product of the number of atoms and each atom's atomic mass to form the molecular mass. If one defines $L$ as the number of elements present in species $i$, $\phi_{li}$ as the number of moles of atomic element $l$ in species $i$, and $\mathcal{M}_l$ as the atomic mass of element $l$, then the molecular mass $M_i$ of species $i$ is

$$M_i = \sum_{l=1}^{L} \mathcal{M}_l \phi_{li}. \tag{4.1}$$

In vector form, one would say

$$\mathbf{M}^T = \mathcal{M}^T \cdot \phi, \qquad \text{or} \qquad \mathbf{M} = \phi^T \cdot \mathcal{M}. \tag{4.2}$$

Here, superscript $T$ denotes the transpose, $\mathbf{M}$ is the vector of length $N$ containing the molecular masses, $\mathcal{M}$ is the vector of length $L$ containing the atomic element masses, and $\phi$ is the element-species matrix of dimension $L \times N$ containing the number of moles of each element in each species. Generally, $\phi$ is full rank. If $N > L$, $\phi$ has rank $L$. If $N < L$, $\phi$ has rank $N$. In any problem involving an actual chemical reaction, one will find $N \geq L$, and in most cases $N > L$. In isolated problems not involving a reaction, one may have $N < L$. In any case, in terms of linear algebra's nomenclature, $\mathbf{M}$ lies in the column space of $\phi^T$; that space is well known to be the row space of $\phi$. The element-species matrix is the same as was introduced in Sec. 1.2.

**EXAMPLE 4.1**

Find the molecular mass of $H_2O$.

Here, one has two elements H and O, so $L = 2$, and one species, so $N = 1$; thus, $N < L$. Take $i = 1$ for species $H_2O$. Take $l = 1$ for element H. Take $l = 2$ for element O. From the periodic table, one gets $\mathcal{M}_1 = 1$ g/mol for H, $\mathcal{M}_2 = 16$ g/mol for O. For element 1, there are 2 atoms, so $\phi_{11} = 2$. For element 2, there is 1 atom so $\phi_{21} = 1$. So, the molecular mass of species 1, $H_2O$, is

$$(M_1) = (\mathcal{M}_1 \quad \mathcal{M}_2) \begin{pmatrix} \phi_{11} \\ \phi_{21} \end{pmatrix} = \mathcal{M}_1\phi_{11} + \mathcal{M}_2\phi_{21}, \tag{4.3}$$

$$= \left(1 \frac{\text{g}}{\text{mol}}\right)(2) + \left(16 \frac{\text{g}}{\text{mol}}\right)(1) = 18 \frac{\text{g}}{\text{mol}}. \tag{4.4}$$

This trivial problem is sketched in Fig. 4.1.

**EXAMPLE 4.2**

Find the molecular masses of the two species octane $C_8H_{18}$ and carbon dioxide $CO_2$.

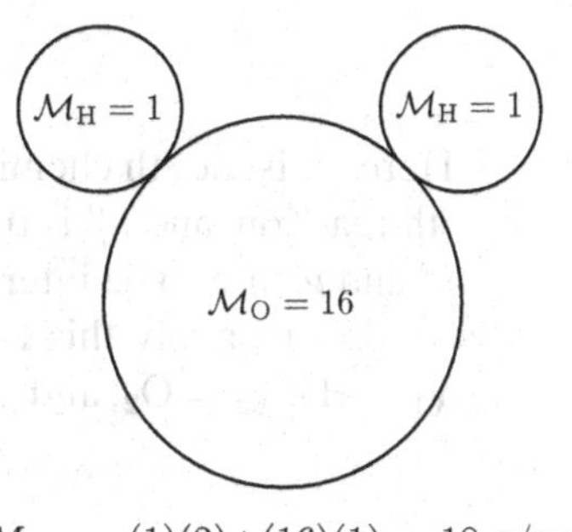


Figure 4.1. Molecular mass of water from its components.

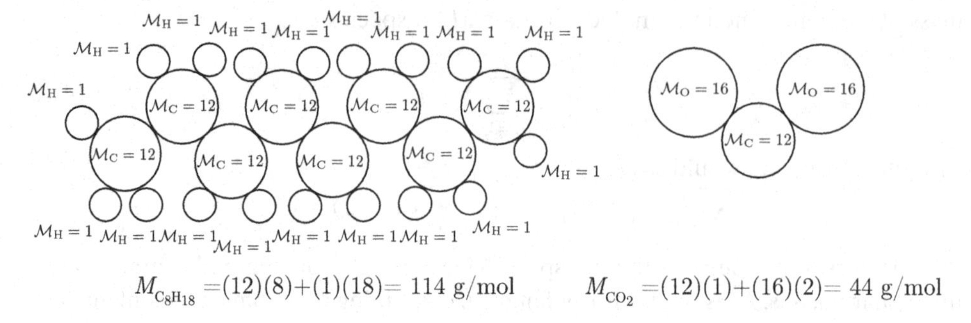


Figure 4.2. Molecular masses of octane and carbon dioxide from their components.

Here, the vector-matrix notation is more useful. One has $N = 2$ species and takes $i = 1$ for $C_8H_{18}$ and $i = 2$ for $CO_2$. One also has $L = 3$ elements and takes $l = 1$ for C, $l = 2$ for H, and $l = 3$ for O. Now, for each element, one has C: $\mathcal{M}_1 = 12$ g/mol, H: $\mathcal{M}_2 = 1$ g/mol, O: $\mathcal{M}_3 = 16$ g/mol. The molecular masses are then given by

$$( M_1 \;\; M_2 ) = ( \mathcal{M}_1 \;\; \mathcal{M}_2 \;\; \mathcal{M}_3 ) \begin{pmatrix} \phi_{11} & \phi_{12} \\ \phi_{21} & \phi_{22} \\ \phi_{31} & \phi_{32} \end{pmatrix} = ( 12 \;\; 1 \;\; 16 ) \begin{pmatrix} 8 & 1 \\ 18 & 0 \\ 0 & 2 \end{pmatrix} = ( 114 \;\; 44 ). \tag{4.5}$$

That is, for $C_8H_{18}$, the molecular mass is $M_1 = 114$ g/mol. For $CO_2$, the molecular mass is $M_2 = 44$ g/mol. This problem is sketched in Fig. 4.2.

## 4.2 Stoichiometry

### 4.2.1 General Development

Stoichiometry represents a mass and number balance on each element in a chemical reaction. For example, in the simple global reaction

$$2H_2 + O_2 \leftrightharpoons 2H_2O, \tag{4.6}$$

one has four H atoms and two O atoms on both the reactant and product sides. Both the mass and the number of H and O are balanced. Here, such *stoichiometry* is systematized.

Consider a general reaction with $N$ species. This reaction can be represented by

$$\sum_{i=1}^{N} \nu_i' \chi_i \leftrightharpoons \sum_{i=1}^{N} \nu_i'' \chi_i. \tag{4.7}$$

Here, $\chi_i$ is the $i$th chemical species, $\nu_i'$ is the *forward stoichiometric coefficient* of the $i$th reaction, and $\nu_i''$ is the *reverse stoichiometric coefficient* of the $i$th reaction. Both $\nu_i'$ and $\nu_i''$ are to be interpreted as pure dimensionless numbers.

Let us apply this to Eq. (4.6). There, one has $N = 3$ species. One might take $\chi_1 = H_2$, $\chi_2 = O_2$, and $\chi_3 = H_2O$. The reaction is written in more general form as

$$(2)\chi_1 + (1)\chi_2 + (0)\chi_3 \leftrightharpoons (0)\chi_1 + (0)\chi_2 + (2)\chi_3, \tag{4.8}$$

$$(2)H_2 + (1)O_2 + (0)H_2O \leftrightharpoons (0)H_2 + (0)O_2 + (2)H_2O. \tag{4.9}$$

Here, one has

$$\nu_1' = 2, \qquad \nu_1'' = 0, \tag{4.10}$$

$$\nu_2' = 1, \qquad \nu_2'' = 0, \tag{4.11}$$

$$\nu_3' = 0, \qquad \nu_3'' = 2. \tag{4.12}$$

It is common and useful to define another pure dimensionless number, the *net stoichiometric coefficients* for species $i$, $\nu_i$. Here, $\nu_i$ represents the net production of number if the reaction goes forward. It is given by

$$\nu_i = \nu_i'' - \nu_i'. \tag{4.13}$$

For the reaction $2H_2 + O_2 \leftrightharpoons 2H_2O$, one has

$$\nu_1 = \nu_1'' - \nu_1' = 0 - 2 = -2, \tag{4.14}$$

$$\nu_2 = \nu_2'' - \nu_2' = 0 - 1 = -1, \tag{4.15}$$

$$\nu_3 = \nu_3'' - \nu_3' = 2 - 0 = 2. \tag{4.16}$$

With these definitions, it is possible to summarize a chemical reaction, in both index and vector notation, as

$$\sum_{i=1}^{N} \nu_i \chi_i = 0, \qquad \boldsymbol{\nu}^T \cdot \boldsymbol{\chi} = 0. \tag{4.17}$$

For the reaction of Eq. (4.6), one can write the nontraditional form

$$-2H_2 - O_2 + 2H_2O = 0. \tag{4.18}$$

It remains to enforce a stoichiometric balance systematically. This is achieved if, for each element $l = 1, \dots, L$, one has the following equality, given in index and vector notation:

$$\sum_{i=1}^{N} \phi_{li}\nu_i = 0, \qquad l = 1, \dots, L, \quad \boldsymbol{\phi} \cdot \boldsymbol{\nu} = \mathbf{0}. \tag{4.19}$$

That is to say, for each element, the sum of the product of the net species production and the number of elements in the species must be zero. For this single reaction, $\boldsymbol{\nu}$ is the *net stoichiometric vector* of dimension $N \times 1$ whose component $\nu_i$ contains the net stoichiometric coefficients of species $i$. One may recall from linear algebra that this demands that $\boldsymbol{\nu}$ lie in the *right null space* of the $L \times N$ element-species matrix $\boldsymbol{\phi}$.

**EXAMPLE 4.3**

Show stoichiometric balance is achieved for $-2H_2 - O_2 + 2H_2O = 0$ in the context of the formalism of linear algebra.

Here, again, the number of elements $L = 2$, and one can take $l = 1$ for H and $l = 2$ for O. Also the number of species $N = 3$, and one takes $i = 1$ for $H_2$, $i = 2$ for $O_2$, and $i = 3$ for $H_2O$. A complete description of the entries of $\phi$ is given in Table 4.1. In matrix

Table 4.1. *Details of the Element-Species Matrix $\phi$ for Hydrogen Combustion*

| | $H_2$ $i=1$ | $O_2$ $i=2$ | $H_2O$ $i=3$ |
|---|---|---|---|
| $H, l=1$ | 2 | 0 | 2 |
| $O, l=2$ | 0 | 2 | 1 |

form, then, $\boldsymbol{\phi} \cdot \boldsymbol{\nu} = \mathbf{0}$ gives

$$\begin{pmatrix} 2 & 0 & 2 \\ 0 & 2 & 1 \end{pmatrix} \begin{pmatrix} \nu_1 \\ \nu_2 \\ \nu_3 \end{pmatrix} = \begin{pmatrix} 0 \\ 0 \end{pmatrix}. \tag{4.20}$$

This is two equations in three unknowns. Thus, it is formally underconstrained. Certainly the trivial solution $\nu_1 = \nu_2 = \nu_3 = 0$ will satisfy, but one seeks nontrivial solutions. Assume that $\nu_3$ has a known value $\nu_3 = \xi$. Then, the system reduces to

$$\begin{pmatrix} 2 & 0 \\ 0 & 2 \end{pmatrix} \begin{pmatrix} \nu_1 \\ \nu_2 \end{pmatrix} = \begin{pmatrix} -2\xi \\ -\xi \end{pmatrix}. \tag{4.21}$$

The inversion here is easy, and one finds $\nu_1 = -\xi, \nu_2 = -\xi/2$. Or, in vector form,

$$\begin{pmatrix} \nu_1 \\ \nu_2 \\ \nu_3 \end{pmatrix} = \begin{pmatrix} -\xi \\ -\frac{1}{2}\xi \\ \xi \end{pmatrix} = \xi \begin{pmatrix} -1 \\ -\frac{1}{2} \\ 1 \end{pmatrix}, \qquad \xi \in \mathbb{R}^1. \tag{4.22}$$

Again, this amounts to saying that the solution vector $(\nu_1, \nu_2, \nu_3)^T$ lies in the right null space of $\boldsymbol{\phi}$. Here, $\xi$ is any real scalar. If one takes $\xi = 2$, one gets

$$\begin{pmatrix} \nu_1 \\ \nu_2 \\ \nu_3 \end{pmatrix} = \begin{pmatrix} -2 \\ -1 \\ 2 \end{pmatrix}, \tag{4.23}$$

This simply corresponds to

$$-2H_2 - O_2 + 2H_2O = 0. \tag{4.24}$$

If one takes $\xi = -4$, one still achieves stoichiometric balance, with

$$\begin{pmatrix} \nu_1 \\ \nu_2 \\ \nu_3 \end{pmatrix} = \begin{pmatrix} 4 \\ 2 \\ -4 \end{pmatrix}, \tag{4.25}$$

which corresponds to the equally valid

$$4H_2 + 2O_2 - 4H_2O = 0. \tag{4.26}$$

In summary, *the net stoichiometric coefficients are nonunique but partially constrained by mass conservation.* Which set is chosen is to some extent arbitrary and often based on traditional conventions from chemistry.

There is a small issue with units here, which will be seen to be difficult to reconcile. In practice, it will have little to no importance. In the previous example, one might be tempted to ascribe units of moles to $\nu_i$. Later, it will be seen that in classical

Table 4.2. *Details of the Element-Species Matrix $\phi$ for Ethane Combustion*

| | $C_2H_6$ $i=1$ | $O_2$ $i=2$ | $CO_2$ $i=3$ | $H_2O$ $i=4$ |
|---|---|---|---|---|
| $C, l=1$ | 2 | 0 | 1 | 0 |
| $H, l=2$ | 6 | 0 | 0 | 2 |
| $O, l=3$ | 0 | 2 | 2 | 1 |

reaction kinetics, $\nu_i$ is best interpreted as a pure dimensionless number, consistent with the definition of this section. So, in the context of the previous example, one would then take $\xi$ to be dimensionless as well, which is perfectly acceptable for the example. In later problems, it will be more useful to give $\xi$ the units of moles. Multiplication of $\xi$ by any scalar, for example, mol/$(6.02 \times 10^{23}$ molecule), still yields an acceptable result.

**EXAMPLE 4.4**

Balance an equation for hypothesized ethane combustion

$$\nu_1' C_2H_6 + \nu_2' O_2 \leftrightharpoons \nu_3'' CO_2 + \nu_4'' H_2O. \tag{4.27}$$

One could also say in terms of the net stoichiometric coefficients that

$$\nu_1 C_2H_6 + \nu_2 O_2 + \nu_3 CO_2 + \nu_4 H_2O = 0. \tag{4.28}$$

We take $\chi_1 = C_2H_6, \chi_2 = O_2, \chi_3\ CO_2, \chi_4 = H_2O$. So, there are $N = 4$ species. There are also $L = 3$ elements: $l = 1 : C, l = 2 : H, l = 3 : O$. A complete description of the entries of $\phi$ is given in Table 4.2. So, the stoichiometry equation, $\phi \cdot \nu = \mathbf{0}$, is given by

$$\begin{pmatrix} 2 & 0 & 1 & 0 \\ 6 & 0 & 0 & 2 \\ 0 & 2 & 2 & 1 \end{pmatrix} \begin{pmatrix} \nu_1 \\ \nu_2 \\ \nu_3 \\ \nu_4 \end{pmatrix} = \begin{pmatrix} 0 \\ 0 \\ 0 \end{pmatrix}. \tag{4.29}$$

Here, there are three equations in four unknowns, so the system is underconstrained. There are many ways to address this problem. Here, choose the robust way of casting the system into *row echelon form*. This is easily achieved by Gaussian[1] elimination. Row echelon form is such that zeros appear below the diagonal of the matrix. The lower left corner has a zero already, so that is useful. Now, multiply the top equation by 3 and subtract the result from the second to get

$$\begin{pmatrix} 2 & 0 & 1 & 0 \\ 0 & 0 & -3 & 2 \\ 0 & 2 & 2 & 1 \end{pmatrix} \begin{pmatrix} \nu_1 \\ \nu_2 \\ \nu_3 \\ \nu_4 \end{pmatrix} = \begin{pmatrix} 0 \\ 0 \\ 0 \end{pmatrix}. \tag{4.30}$$

[1]Johann Carl Friedrich Gauss (1777–1855), German mathematician.

Next, switch the last two equations to get

$$\begin{pmatrix} 2 & 0 & 1 & 0 \\ 0 & 2 & 2 & 1 \\ 0 & 0 & -3 & 2 \end{pmatrix} \begin{pmatrix} \nu_1 \\ \nu_2 \\ \nu_3 \\ \nu_4 \end{pmatrix} = \begin{pmatrix} 0 \\ 0 \\ 0 \end{pmatrix}. \tag{4.31}$$

Now, divide the first by 2, the second by 2, and the third by $-3$ to get unity in the diagonal:

$$\begin{pmatrix} 1 & 0 & \frac{1}{2} & 0 \\ 0 & 1 & 1 & \frac{1}{2} \\ 0 & 0 & 1 & -\frac{2}{3} \end{pmatrix} \begin{pmatrix} \nu_1 \\ \nu_2 \\ \nu_3 \\ \nu_4 \end{pmatrix} = \begin{pmatrix} 0 \\ 0 \\ 0 \end{pmatrix}. \tag{4.32}$$

So-called *bound variables* have nonzero coefficients on the diagonal, so one can take the bound variables to be $\nu_1, \nu_2$, and $\nu_3$. The remaining variables are *free variables*. Here, one takes the free variable to be $\nu_4$. So, set $\nu_4 = \xi$, and rewrite the system as

$$\begin{pmatrix} 1 & 0 & \frac{1}{2} \\ 0 & 1 & 1 \\ 0 & 0 & 1 \end{pmatrix} \begin{pmatrix} \nu_1 \\ \nu_2 \\ \nu_3 \end{pmatrix} = \begin{pmatrix} 0 \\ -\frac{1}{2}\xi \\ \frac{2}{3}\xi \end{pmatrix}. \tag{4.33}$$

Solving, one finds

$$\begin{pmatrix} \nu_1 \\ \nu_2 \\ \nu_3 \\ \nu_4 \end{pmatrix} = \begin{pmatrix} -\frac{1}{3}\xi \\ -\frac{7}{6}\xi \\ \frac{2}{3}\xi \\ \xi \end{pmatrix} = \xi \begin{pmatrix} -\frac{1}{3} \\ -\frac{7}{6} \\ \frac{2}{3} \\ 1 \end{pmatrix}, \qquad \xi \in \mathbb{R}^1. \tag{4.34}$$

Again one finds a nonunique solution in the right null space of $\phi$. If one chooses $\xi = 6$, one gets

$$\begin{pmatrix} \nu_1 \\ \nu_2 \\ \nu_3 \\ \nu_4 \end{pmatrix} = \begin{pmatrix} -2 \\ -7 \\ 4 \\ 6 \end{pmatrix}, \tag{4.35}$$

which corresponds to the stoichiometrically balanced reaction

$$2C_2H_6 + 7O_2 \leftrightharpoons 4CO_2 + 6H_2O. \tag{4.36}$$

In this example, $\xi$ is dimensionless.

**EXAMPLE 4.5**

Consider stoichiometric balance for a propane oxidation reaction that may produce carbon monoxide and hydroxyl in addition to carbon dioxide and water.

The hypothesized reaction takes the form

$$\nu_1' C_3H_8 + \nu_2' O_2 \leftrightharpoons \nu_3'' CO_2 + \nu_4'' CO + \nu_5'' H_2O + \nu_6'' OH. \tag{4.37}$$

In terms of net stoichiometric coefficients, this becomes

$$\nu_1 C_3H_8 + \nu_2 O_2 + \nu_3 CO_2 + \nu_4 CO + \nu_5 H_2O + \nu_6 OH = 0. \tag{4.38}$$

Table 4.3. *Details of the Element-Species Matrix $\phi$ for Propane Oxidation*

| | $C_3H_8$ $i=1$ | $O_2$ $i=2$ | $CO_2$ $i=3$ | $CO$ $i=4$ | $H_2O$ $i=5$ | $OH$ $i=6$ |
|---|---|---|---|---|---|---|
| C, $l=1$ | 3 | 0 | 1 | 1 | 0 | 0 |
| H, $l=2$ | 8 | 0 | 0 | 0 | 2 | 1 |
| O, $l=3$ | 0 | 2 | 2 | 1 | 1 | 1 |

There are $N = 6$ species and $L = 3$ elements. A complete description of the entries of $\phi$ is given in Table 4.3. The equation $\phi \cdot \boldsymbol{\nu} = \mathbf{0}$ then becomes

$$\begin{pmatrix} 3 & 0 & 1 & 1 & 0 & 0 \\ 8 & 0 & 0 & 0 & 2 & 1 \\ 0 & 2 & 2 & 1 & 1 & 1 \end{pmatrix} \begin{pmatrix} \nu_1 \\ \nu_2 \\ \nu_3 \\ \nu_4 \\ \nu_5 \\ \nu_6 \end{pmatrix} = \begin{pmatrix} 0 \\ 0 \\ 0 \end{pmatrix}. \tag{4.39}$$

Multiplying the first equation by $-8/3$ and adding it to the second gives

$$\begin{pmatrix} 3 & 0 & 1 & 1 & 0 & 0 \\ 0 & 0 & -\frac{8}{3} & -\frac{8}{3} & 2 & 1 \\ 0 & 2 & 2 & 1 & 1 & 1 \end{pmatrix} \begin{pmatrix} \nu_1 \\ \nu_2 \\ \nu_3 \\ \nu_4 \\ \nu_5 \\ \nu_6 \end{pmatrix} = \begin{pmatrix} 0 \\ 0 \\ 0 \end{pmatrix}. \tag{4.40}$$

Trading the second and third rows gives

$$\begin{pmatrix} 3 & 0 & 1 & 1 & 0 & 0 \\ 0 & 2 & 2 & 1 & 1 & 1 \\ 0 & 0 & -\frac{8}{3} & -\frac{8}{3} & 2 & 1 \end{pmatrix} \begin{pmatrix} \nu_1 \\ \nu_2 \\ \nu_3 \\ \nu_4 \\ \nu_5 \\ \nu_6 \end{pmatrix} = \begin{pmatrix} 0 \\ 0 \\ 0 \end{pmatrix}. \tag{4.41}$$

Dividing the first row by 3, the second by 2, and the third by $-8/3$ gives

$$\begin{pmatrix} 1 & 0 & \frac{1}{3} & \frac{1}{3} & 0 & 0 \\ 0 & 1 & 1 & \frac{1}{2} & \frac{1}{2} & \frac{1}{2} \\ 0 & 0 & 1 & 1 & -\frac{3}{4} & -\frac{3}{8} \end{pmatrix} \begin{pmatrix} \nu_1 \\ \nu_2 \\ \nu_3 \\ \nu_4 \\ \nu_5 \\ \nu_6 \end{pmatrix} = \begin{pmatrix} 0 \\ 0 \\ 0 \end{pmatrix}. \tag{4.42}$$

Take bound variables to be $\nu_1$, $\nu_2$, and $\nu_3$ and free variables to be $\nu_4$, $\nu_5$, and $\nu_6$. So, set $\nu_4 = \xi_1$, $\nu_5 = \xi_2$, and $\nu_6 = \xi_3$, and get

$$\begin{pmatrix} 1 & 0 & \frac{1}{3} \\ 0 & 1 & 1 \\ 0 & 0 & 1 \end{pmatrix} \begin{pmatrix} \nu_1 \\ \nu_2 \\ \nu_3 \end{pmatrix} = \begin{pmatrix} -\frac{\xi_1}{3} \\ -\frac{\xi_1}{2} - \frac{\xi_2}{2} - \frac{\xi_3}{2} \\ -\xi_1 + \frac{3}{4}\xi_2 + \frac{3}{8}\xi_3 \end{pmatrix}. \tag{4.43}$$

Solving, one finds

$$\begin{pmatrix} \nu_1 \\ \nu_2 \\ \nu_3 \end{pmatrix} = \begin{pmatrix} \frac{1}{8}(-2\xi_2 - \xi_3) \\ \frac{1}{8}(4\xi_1 - 10\xi_2 - 7\xi_3) \\ \frac{1}{8}(-8\xi_1 + 6\xi_2 + 3\xi_3) \end{pmatrix}. \tag{4.44}$$

For all the coefficients, one then has

$$\begin{pmatrix} \nu_1 \\ \nu_2 \\ \nu_3 \\ \nu_4 \\ \nu_5 \\ \nu_6 \end{pmatrix} = \begin{pmatrix} \frac{1}{8}(-2\xi_2 - \xi_3) \\ \frac{1}{8}(4\xi_1 - 10\xi_2 - 7\xi_3) \\ \frac{1}{8}(-8\xi_1 + 6\xi_2 + 3\xi_3) \\ \xi_1 \\ \xi_2 \\ \xi_3 \end{pmatrix} = \frac{\xi_1}{8}\begin{pmatrix} 0 \\ 4 \\ -8 \\ 8 \\ 0 \\ 0 \end{pmatrix} + \frac{\xi_2}{8}\begin{pmatrix} -2 \\ -10 \\ 6 \\ 0 \\ 8 \\ 0 \end{pmatrix} + \frac{\xi_3}{8}\begin{pmatrix} -1 \\ -7 \\ 3 \\ 0 \\ 0 \\ 8 \end{pmatrix}. \tag{4.45}$$

Here, one finds *three* independent vectors in the right null space. To simplify the notation, take $\hat{\xi}_1 = \xi_1/8$, $\hat{\xi}_2 = \xi_2/8$, and $\hat{\xi}_3 = \xi_3/8$. Then,

$$\underbrace{\begin{pmatrix} \nu_1 \\ \nu_2 \\ \nu_3 \\ \nu_4 \\ \nu_5 \\ \nu_6 \end{pmatrix}}_{\boldsymbol{\nu}} = \hat{\xi}_1\begin{pmatrix} 0 \\ 4 \\ -8 \\ 8 \\ 0 \\ 0 \end{pmatrix} + \hat{\xi}_2\begin{pmatrix} -2 \\ -10 \\ 6 \\ 0 \\ 8 \\ 0 \end{pmatrix} + \hat{\xi}_3\begin{pmatrix} -1 \\ -7 \\ 3 \\ 0 \\ 0 \\ 8 \end{pmatrix} = \underbrace{\begin{pmatrix} 0 & -2 & -1 \\ 4 & -10 & -7 \\ -8 & 6 & 3 \\ 8 & 0 & 0 \\ 0 & 8 & 0 \\ 0 & 0 & 8 \end{pmatrix}}_{\mathbf{D}} \underbrace{\begin{pmatrix} \hat{\xi}_1 \\ \hat{\xi}_2 \\ \hat{\xi}_3 \end{pmatrix}}_{\boldsymbol{\xi}}. \tag{4.46}$$

The most general reaction that can achieve a stoichiometric balance is given by

$$(-2\hat{\xi}_2 - \hat{\xi}_3)C_3H_8 + (4\hat{\xi}_1 - 10\hat{\xi}_2 - 7\hat{\xi}_3)O_2 \\ + (-8\hat{\xi}_1 + 6\hat{\xi}_2 + 3\hat{\xi}_3)CO_2 + 8\hat{\xi}_1 CO + 8\hat{\xi}_2 H_2O + 8\hat{\xi}_3 OH = 0. \tag{4.47}$$

Rearranging, one gets

$$(2\hat{\xi}_2 + \hat{\xi}_3)C_3H_8 + (-4\hat{\xi}_1 + 10\hat{\xi}_2 + 7\hat{\xi}_3)O_2 \leftrightharpoons (-8\hat{\xi}_1 + 6\hat{\xi}_2 + 3\hat{\xi}_3)CO_2 \\ + 8\hat{\xi}_1 CO + 8\hat{\xi}_2 H_2O + 8\hat{\xi}_3 OH. \tag{4.48}$$

This will be balanced for all $\hat{\xi}_1, \hat{\xi}_2$, and $\hat{\xi}_3$. The values that are actually achieved in practice depend on the thermodynamics of the problem. Stoichiometry only provides constraints.

A slightly more familiar form is found by taking $\hat{\xi}_2 = 1/2$ and rearranging, giving

$$(1 + \hat{\xi}_3)C_3H_8 + (5 - 4\hat{\xi}_1 + 7\hat{\xi}_3)O_2 \leftrightharpoons (3 - 8\hat{\xi}_1 + 3\hat{\xi}_3)CO_2 \\ + 4H_2O + 8\hat{\xi}_1 CO + 8\hat{\xi}_3 OH. \tag{4.49}$$

Often the production of CO and OH will be small. If there is no production of CO or OH, $\hat{\xi}_1 = \hat{\xi}_3 = 0$, and one recovers the familiar balance of

$$C_3H_8 + 5O_2 \leftrightharpoons 3CO_2 + 4H_2O. \tag{4.50}$$

Stoichiometry alone admits unusual solutions. For instance, if $\hat{\xi}_1 = 100$, $\hat{\xi}_2 = 1/2$, and $\hat{\xi}_3 = 1$, one has

$$2C_3H_8 + 794CO_2 \leftrightharpoons 388O_2 + 4H_2O + 800CO + 8OH. \tag{4.51}$$

This reaction is certainly admitted by stoichiometry but is not observed in nature. To determine precisely which of the infinitely many possible final states are realized requires a consideration of the equilibrium condition from Eq. (3.371): $\sum_{i=1}^{N} \nu_i \overline{\mu}_i = 0$.

Looked at in another way, we can think of three independent classes of reactions admitted by the stoichiometry, one for each of the linearly independent null space vectors.

Taking first $\hat{\xi}_1 = 1/4, \hat{\xi}_2 = 0, \hat{\xi}_3 = 0$, one gets, after rearrangement,

$$2CO + O_2 \leftrightharpoons 2CO_2 \tag{4.52}$$

as one class of reaction admitted by stoichiometry. Taking next $\hat{\xi}_1 = 0, \hat{\xi}_2 = 1/2, \hat{\xi}_3 = 0$, one gets

$$C_3H_8 + 5O_2 \leftrightharpoons 3CO_2 + 4H_2O \tag{4.53}$$

as the second class admitted by stoichiometry. The third class is given by taking $\hat{\xi}_1 = 0$, $\hat{\xi}_2 = 0, \hat{\xi}_3 = 1$, and is

$$C_3H_8 + 7O_2 \leftrightharpoons 3CO_2 + 8OH. \tag{4.54}$$

These three classes are simply mathematical conveniences. To predict what is observed in nature requires introduction of more principles.

In general, one can expect to find the stoichiometric coefficients for $N$ species composed of $L$ elements to be of the following form (written in index and vector notation):

$$\nu_i = \sum_{k=1}^{N-L} \mathcal{D}_{ik}\xi_k, \qquad i = 1, \ldots, N, \qquad \boldsymbol{\nu} = \mathbf{D} \cdot \boldsymbol{\xi}. \tag{4.55}$$

Here, $\mathcal{D}_{ik}$ is a dimensionless component of a full rank matrix of dimension $N \times (N - L)$; its rank must then be $N - L$, and $\xi_k$ is a dimensionless component of a vector of parameters of length $N - L$. The matrix whose components are $\mathcal{D}_{ik}$ is constructed by populating its columns with vectors that lie in the right null space of $\phi$. Multiplication of $\xi_k$ by any constant gives another set of $\nu_i$, and mass conservation for each element is still satisfied.

### 4.2.2 Fuel-Air Mixtures

Let us discuss some notions from stoichiometry of practical importance for internal combustion engines. In combustion with air, one often models air as a simple mixture of diatomic oxygen and inert diatomic nitrogen:

$$\text{Air:} \qquad \nu'_{air}(O_2 + 3.76N_2). \tag{4.56}$$

The *air-fuel ratio*, $\mathcal{A}$, and its reciprocal, the *fuel-air ratio*, $\mathcal{F}$, can be defined on a mass and mol basis:

$$\mathcal{A}_{mass} = \frac{m_{air}}{m_{fuel}}, \qquad \mathcal{A}_{mol} = \frac{n_{air}}{n_{fuel}}. \tag{4.57}$$

Via the molecular masses, one has

$$\mathcal{A}_{mass} = \frac{m_{air}}{m_{fuel}} = \frac{n_{air}M_{air}}{n_{fuel}M_{fuel}} = \mathcal{A}_{mol}\frac{M_{air}}{M_{fuel}}. \tag{4.58}$$

If there is not enough air to burn all the fuel, the mixture is said to be *rich*. If there is excess air, the mixture is said to be *lean*. If there is just enough, the mixture is said to be *stoichiometric*. The *equivalence ratio*, $\Phi$, is defined as the actual ($act$) fuel-air ratio scaled by the stoichiometric ($st$) fuel-air ratio:

$$\Phi \equiv \frac{\mathcal{F}_{act}}{\mathcal{F}_{st}} = \frac{\mathcal{A}_{st}}{\mathcal{A}_{act}}. \tag{4.59}$$

The ratio $\Phi$ is the same whether $\mathcal{F}$s are taken on a mass or mol basis, because the ratio of molecular masses cancels.

**EXAMPLE 4.6**

Calculate the stoichiometry of the combustion of methane with air with an equivalence ratio of $\Phi = 0.5$. If the pressure is 0.1 MPa, find the dew point of the products.

First calculate the coefficients for stoichiometric combustion:

$$\nu_1' CH_4 + \nu_2'(O_2 + 3.76N_2) \leftrightharpoons \nu_3'' CO_2 + \nu_4'' H_2O + \nu_5'' N_2. \tag{4.60}$$

We can also say in terms of the net stoichiometric coefficients that

$$\nu_1 CH_4 + \nu_2 O_2 + \nu_3 CO_2 + \nu_4 H_2O + (\nu_5 + 3.76\nu_2)N_2 = 0. \tag{4.61}$$

Here, one has $N = 5$ species and $L = 4$ elements. Adopting a slightly more intuitive procedure for variety, one easily deduces a conservation equation for each element:

$$\nu_1 + \nu_3 = 0, \quad C, \tag{4.62}$$

$$4\nu_1 + 2\nu_4 = 0, \quad H, \tag{4.63}$$

$$2\nu_2 + 2\nu_3 + \nu_4 = 0, \quad O, \tag{4.64}$$

$$3.76\nu_2 + \nu_5 = 0, \quad N. \tag{4.65}$$

In matrix form, this becomes

$$\begin{pmatrix} 1 & 0 & 1 & 0 & 0 \\ 4 & 0 & 0 & 2 & 0 \\ 0 & 2 & 2 & 1 & 0 \\ 0 & 3.76 & 0 & 0 & 1 \end{pmatrix} \begin{pmatrix} \nu_1 \\ \nu_2 \\ \nu_3 \\ \nu_4 \\ \nu_5 \end{pmatrix} = \begin{pmatrix} 0 \\ 0 \\ 0 \\ 0 \end{pmatrix}. \tag{4.66}$$

Now, one might expect to have one free variable, because one has five unknowns in four equations. While casting the equation in row echelon form is guaranteed to yield a proper solution, one can often use intuition to get a solution more rapidly. One certainly expects that $CH_4$ will need to be present for the reaction to take place. One might also expect to find an answer if there is one mol of $CH_4$. So take $\nu_1 = -1$. Realize that one could also get a physically valid answer by assuming $\nu_1$ to be equal to *any* scalar. With $\nu_1 = -1$, one gets

$$\begin{pmatrix} 0 & 1 & 0 & 0 \\ 0 & 0 & 2 & 0 \\ 2 & 2 & 1 & 0 \\ 3.76 & 0 & 0 & 1 \end{pmatrix} \begin{pmatrix} \nu_2 \\ \nu_3 \\ \nu_4 \\ \nu_5 \end{pmatrix} = \begin{pmatrix} 1 \\ 4 \\ 0 \\ 0 \end{pmatrix}. \tag{4.67}$$

One easily finds the unique inverse does exist and that the solution is

$$\begin{pmatrix} \nu_2 \\ \nu_3 \\ \nu_4 \\ \nu_5 \end{pmatrix} = \begin{pmatrix} -2 \\ 1 \\ 2 \\ 7.52 \end{pmatrix}. \tag{4.68}$$

If there had been more than one free variable, the $4 \times 4$ matrix would have been singular, and no unique inverse would have existed. In any case, the reaction under stoichiometric conditions is

$$-CH_4 - 2O_2 + CO_2 + 2H_2O + (7.52 + (3.76)(-2))N_2 = 0. \tag{4.69}$$

In more traditional form, it is

$$CH_4 + 2(O_2 + 3.76N_2) \leftrightharpoons CO_2 + 2H_2O + 7.52N_2. \tag{4.70}$$

For the stoichiometric reaction, the fuel-air ratio on a mol basis is

$$\mathcal{F}_{st} = \frac{1}{2+2(3.76)} = 0.1050. \tag{4.71}$$

Now, $\Phi = 0.5$, so

$$\mathcal{F}_{act} = \Phi \mathcal{F}_{st} = (0.5)(0.1050) = 0.0525. \tag{4.72}$$

By inspection, one can write the *actual* reaction equation as

$$CH_4 + 4(O_2 + 3.76N_2) \leftrightharpoons CO_2 + 2H_2O + 2O_2 + 15.04N_2. \tag{4.73}$$

Check:

$$\mathcal{F}_{act} = \frac{1}{4+4(3.76)} = 0.0525. \tag{4.74}$$

For the dew point of the products, it is known from elementary thermodynamics that one needs the partial pressure of $H_2O$. The mol fraction of $H_2O$ in the products is

$$y_{H_2O} = \frac{2}{1+2+2+15.04} = 0.0499. \tag{4.75}$$

So, the partial pressure of $H_2O$ is

$$P_{H_2O} = y_{H_2O}P = 0.0499(100 \text{ kPa}) = 4.99 \text{ kPa}. \tag{4.76}$$

From the steam tables (Borgnakke and Sonntag, 2013, Table B.1.2, p. 780) the saturation temperature at this pressure is $T_{sat} = 32.88°C$. This is also the dew point temperature. If the products cool to this temperature in an exhaust device, the water could condense in the apparatus.

## 4.3 First Law Analysis of Reacting Systems

The first law of thermodynamics yields important results for reacting systems.

### 4.3.1 Enthalpy of Formation

The enthalpy of formation is the enthalpy that is required to form a molecule from its constituents at $T = 298$ K and $P = 0.1$ MPa.[2] Consider the irreversible reaction

$$C + O_2 \rightarrow CO_2. \tag{4.77}$$

To maintain the process at constant temperature, it is found that heat transfer to the volume is necessary. For the steady constant pressure process, one has

$$E_2 - E_1 = Q - W = Q - \int_1^2 P\,dV = Q - P(V_2 - V_1), \tag{4.78}$$

$$Q = E_2 - E_1 + P(V_2 - V_1) = (E_2 + PV_2) - (E_1 + PV_1), \tag{4.79}$$

$$= H_2 - H_1, \tag{4.80}$$

$$= H_{products} - H_{reactants} = \sum_{products} n_i\overline{h}_i - \sum_{reactants} n_i\overline{h}_i. \tag{4.81}$$

[2]Taking the reference pressure to be 0.1 MPa is common in the engineering literature (e.g., Borgnakke and Sontag, 2013). It is more common in the chemistry literature to take the similar value of 1 atm = 0.101325 MPa. This induces slightly different values for enthalpies of formation.

In this reaction, one measures that $Q = -393522$ kJ for the reaction of 1 kmol of C and $O_2$. That is, the reaction liberates thermal energy to the environment. So, measuring the heat transfer can give a measure of the enthalpy difference between reactants and products. Assign a value of enthalpy of zero to elements in their standard state at the reference state. Thus, C and $O_2$ both have enthalpies of 0 kJ/kmol at $T = 298$ K, $P = 0.1$ MPa. This enthalpy is designated, for species $i$,

$$\overline{h}^0_{f,i} = \overline{h}^0_{T_0,i}, \tag{4.82}$$

and is called the *enthalpy of formation.* Recall that the superscript 0 refers only to the property being evaluated at the reference pressure. So, the energy balance for the products and reactants, here both at the standard state, becomes

$$Q = n_{CO_2}\overline{h}^0_{f,CO_2} - n_C\overline{h}^0_{f,C} - n_{O_2}\overline{h}^0_{f,O_2}, \tag{4.83}$$

$$-393522 \text{ kJ} = (1 \text{ kmol})\overline{h}^0_{f,CO_2} - (1 \text{ kmol})\left(0 \frac{\text{kJ}}{\text{kmol}}\right) -(1 \text{ kmol})\left(0 \frac{\text{kJ}}{\text{kmol}}\right). \tag{4.84}$$

Thus, the enthalpy of formation of $CO_2$ is $\overline{h}^0_{f,CO_2} = -393522$ kJ/kmol, because the reaction involved creation of 1 kmol of $CO_2$.

Often values of enthalpy are tabulated in the forms of enthalpy differences $\Delta\overline{h}_i$. These are defined such that

$$\overline{h}_i = \overline{h}^0_{f,i} + \underbrace{(\overline{h}_i - \overline{h}^0_{f,i})}_{\Delta\overline{h}_i} = \overline{h}^0_{f,i} + \Delta\overline{h}_i. \tag{4.85}$$

Last, for an ideal gas the enthalpy is a function of temperature only and so does not depend on the reference pressure; hence

$$\overline{h}_i = \overline{h}^0_i, \qquad \Delta\overline{h}_i = \Delta\overline{h}^0_i, \qquad \text{if ideal gas.} \tag{4.86}$$

**EXAMPLE 4.7**

Consider the following irreversible reaction in a piston-confined cylinder:

$$CH_4 + 2O_2 \rightarrow CO_2 + 2H_2O(g). \tag{4.87}$$

The entire process is at $P = 100$ kPa; the process commences at $T = 298$ K. Following reaction, the mixture cools back to the same temperature via heat transfer with the surroundings. Assume the stoichiometric coefficients represent the number of kilomoles of each species. Figure 4.3 gives a sketch.

The first law holds that

$$Q = \sum_{products} n_i\overline{h}_i - \sum_{reactants} n_i\overline{h}_i. \tag{4.88}$$

All components are at their reference states. Standard tables (Borgnakke and Sonntag, 2013, Table A.10, p. 772) give properties, and one finds

$$Q = n_{CO_2}\overline{h}_{CO_2} + n_{H_2O}\overline{h}_{H_2O} - n_{CH_4}\overline{h}_{CH_4} - n_{O_2}\overline{h}_{O_2}, \tag{4.89}$$

$$= (1 \text{ kmol})\left(-393522 \frac{\text{kJ}}{\text{kmol}}\right) + (2 \text{ kmol})\left(-241826 \frac{\text{kJ}}{\text{kmol}}\right) - (1 \text{ kmol})\left(-74873 \frac{\text{kJ}}{\text{kmol}}\right) - (2 \text{ kmol})\left(0 \frac{\text{kJ}}{\text{kmol}}\right) = -802301 \text{ kJ}. \tag{4.90}$$

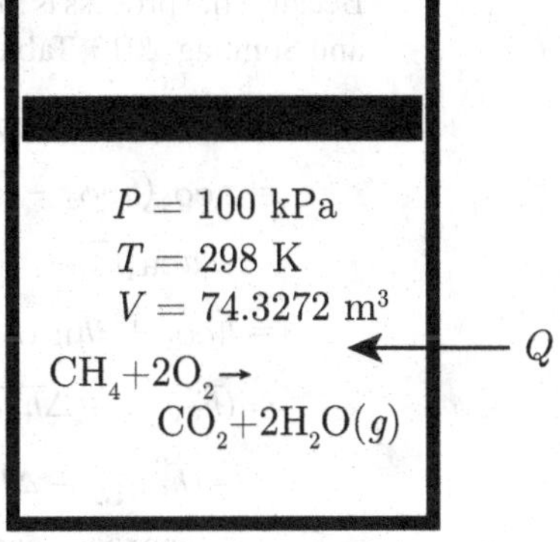


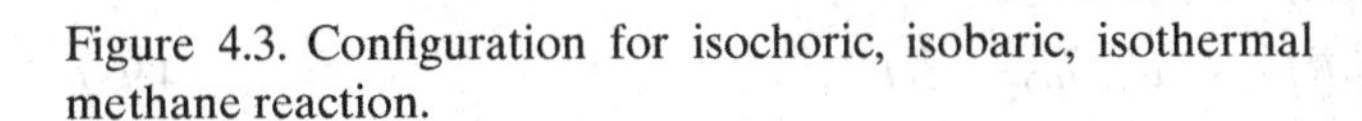
Figure 4.3. Configuration for isochoric, isobaric, isothermal methane reaction.

Consistent with the conventions of engineering thermodynamics, Figure 4.3 depicts $Q$ into the system as being positive. Because we find the $Q$ here to be negative, thermal energy must leave the system after chemical reaction to maintain the temperature at 298 K.

In this problem, the total number of moles remains constant, so the volume also remains constant:

$$V = \frac{n\overline{R}T}{P} = \frac{(3 \text{ kmol})\left(8.314 \frac{\text{kJ}}{\text{kmol K}}\right)(298 \text{ K})}{100 \text{ kPa}} = 74.3272 \text{ m}^3. \tag{4.91}$$

A more detailed analysis is required in the likely case in which the system is not at the reference state.

**EXAMPLE 4.8**

A mixture of 1 kmol of methane and 2 kmol of oxygen initially at 298 K and 101.325 kPa burns completely in a closed, rigid container. Heat transfer occurs until the final temperature is 1000 K. Figure 4.4 gives a sketch. Find the heat transfer and the final pressure.

The combustion is stoichiometric. Assume that no small concentration species are generated. The global reaction is given by

$$CH_4 + 2O_2 \rightarrow CO_2 + 2H_2O. \tag{4.92}$$

The first law analysis for the closed system is slightly different:

$$E_2 - E_1 = Q - W. \tag{4.93}$$

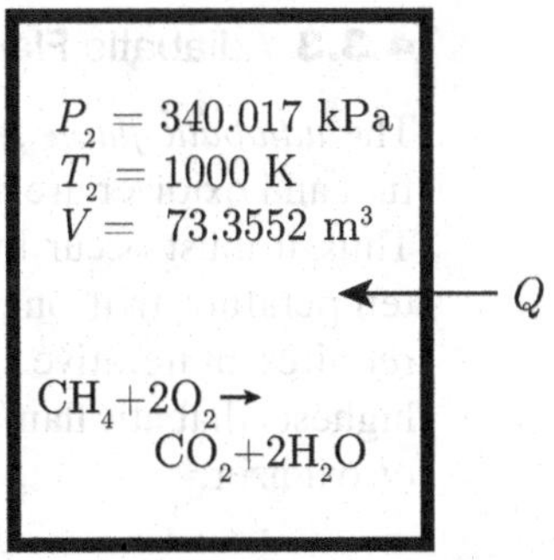

$P_2$ = 340.017 kPa
$T_2$ = 1000 K
$V$ = 73.3552 m³
$Q$
$CH_4+2O_2 \rightarrow CO_2+2H_2O$

Figure 4.4. Final configuration for isochoric methane reaction.

Because the process is isochoric, $W = 0$. So using analysis and tabular values (Borgnakke and Sonntag, 2013, Tables A.9 and A.10, pp. 767–772), we find

$$Q = E_2 - E_1 = n_{CO_2}\overline{e}_{CO_2} + n_{H_2O}\overline{e}_{H_2O} - n_{CH_4}\overline{e}_{CH_4} - n_{O_2}\overline{e}_{O_2}, \tag{4.94}$$

$$= n_{CO_2}(\overline{h}_{CO_2} - \overline{R}T_2) + n_{H_2O}(\overline{h}_{H_2O} - \overline{R}T_2)$$
$$- n_{CH_4}(\overline{h}_{CH_4} - \overline{R}T_1) - n_{O_2}(\overline{h}_{O_2} - \overline{R}T_1), \tag{4.95}$$

$$= \overline{h}_{CO_2} + 2\overline{h}_{H_2O} - \overline{h}_{CH_4} - 2\overline{h}_{O_2} - 3\overline{R}(T_2 - T_1), \tag{4.96}$$

$$= (\overline{h}^0_{CO_2,f} + \Delta\overline{h}_{CO_2}) + 2(\overline{h}^0_{H_2O,f} + \Delta\overline{h}_{H_2O})$$
$$- (\overline{h}^0_{CH_4,f} + \Delta\overline{h}_{CH_4}) - 2(\overline{h}^0_{O_2,f} + \Delta\overline{h}_{O_2}) - 3\overline{R}(T_2 - T_1), \tag{4.97}$$

$$= (-393522 + 33397) + 2(-241826 + 26000) - (-74873 + 0) - 2(0 + 0)$$
$$- 3(8.314)(1000 - 298) = -734413 \text{ kJ}. \tag{4.98}$$

Thus, thermal energy has left the system. To find the final pressure, one has

$$V_1 = \frac{(n_{CH_4} + n_{O_2})\overline{R}T_1}{P_1}, \tag{4.99}$$

$$= \frac{(1 \text{ kmol} + 2 \text{ kmol})\left(8.314 \frac{\text{kJ}}{\text{kg K}}\right)(298 \text{ K})}{101.325 \text{ kPa}} = 73.3552 \text{ m}^3. \tag{4.100}$$

Now, $V_2 = V_1$, so

$$P_2 = \frac{(n_{CO_2} + n_{H_2O})\overline{R}T_2}{V_2}, \tag{4.101}$$

$$= \frac{(1 \text{ kmol} + 2 \text{ kmol})\left(8.314 \frac{\text{kJ}}{\text{kg K}}\right)(1000 \text{ K})}{73.3552 \text{ m}^3} = 340.017 \text{ kPa}. \tag{4.102}$$

The pressure increased in the reaction. This is entirely attributable to the temperature rise, as the number of moles and the volume remained constant here.

### 4.3.2 Enthalpy and Internal Energy of Combustion

The *enthalpy of combustion* is the difference between the enthalpy of products and reactants when complete combustion occurs. It is also known as the *heating value* or the *heat of reaction*. The *internal energy of combustion* is related and is the difference between the internal energy of products and reactants when complete combustion occurs at a given volume and temperature. The term *higher heating value* refers to the energy of combustion when liquid water is in the products. *Lower heating value* refers to the energy of combustion when water vapor is in the product.

### 4.3.3 Adiabatic Flame Temperature

The *adiabatic flame temperature* refers to the temperature that is achieved when a fuel and oxidizer are combined with no loss of energy due to work or heat transfers. Thus, it must occur in a closed, insulated, fixed volume. It is generally the highest temperature that one can expect to achieve in a combustion process. It generally requires an iterative solution. Of all mixtures, stoichiometric mixtures will yield the highest adiabatic flame temperatures because there is no need to heat the excess fuel or oxidizer.

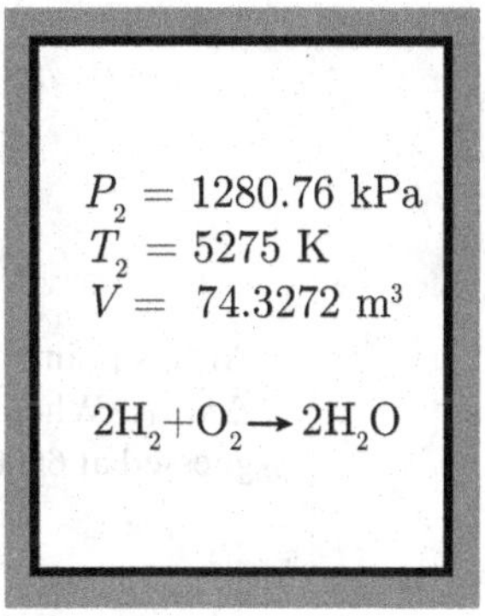

$P_2 = 1280.76$ kPa
$T_2 = 5275$ K
$V = 74.3272$ m$^3$

$2H_2 + O_2 \rightarrow 2H_2O$

Figure 4.5. Final configuration for isochoric, adiabatic hydrogen-oxygen reaction, undiluted and initially cold.

Here, four examples will be presented to illustrate the following points:

- The adiabatic flame temperature can be over 5000 K for ordinary mixtures.
- Dilution of the mixture with an inert gas lowers the adiabatic flame temperature. The same effect would happen in rich and lean mixtures.
- Preheating the mixture, such as one might find in the compression stroke of an internal combustion engine, increases the adiabatic flame temperature.
- Consideration of minor species lowers the adiabatic flame temperature.

Many thermodynamic properties will be needed in these examples; all are taken from Borgnakke and Sonntag (2013, Table A.9, pp. 767–770).

## Undiluted, Cold Mixture

**EXAMPLE 4.9**

A closed, fixed, adiabatic volume contains a stoichiometric mixture of 2 kmol of $H_2$ and 1 kmol of $O_2$ at 100 kPa and 298 K. Figure 4.5 gives a sketch of the configuration. Find the adiabatic flame temperature assuming the irreversible reaction

$$2H_2 + O_2 \rightarrow 2H_2O. \tag{4.103}$$

The volume is given by

$$V = \frac{(n_{H_2} + n_{O_2})\overline{R}T_1}{P_1}, \tag{4.104}$$

$$= \frac{(2 \text{ kmol} + 1 \text{ kmol})\left(8.314 \frac{\text{kJ}}{\text{kmol K}}\right)(298 \text{ K})}{100 \text{ kPa}} = 74.3272 \text{ m}^3. \tag{4.105}$$

The first law gives

$$E_2 - E_1 = Q - W, \tag{4.106}$$

$$E_2 - E_1 = 0, \tag{4.107}$$

$$n_{H_2O}\overline{e}_{H_2O} - n_{H_2}\overline{e}_{H_2} - n_{O_2}\overline{e}_{O_2} = 0, \tag{4.108}$$

$$n_{H_2O}(\overline{h}_{H_2O} - \overline{R}T_2) - n_{H_2}(\overline{h}_{H_2} - \overline{R}T_1) - n_{O_2}(\overline{h}_{O_2} - \overline{R}T_1) = 0, \tag{4.109}$$

$$2\overline{h}_{H_2O} - 2\underbrace{\overline{h}_{H_2}}_{0} - \underbrace{\overline{h}_{O_2}}_{0} + \overline{R}(-2T_2 + 3T_1) = 0, \tag{4.110}$$

$$2\overline{h}_{H_2O} + (8.314)((-2)T_2 + (3)(298)) = 0, \tag{4.111}$$

$$\overline{h}_{H_2O} - 8.314T_2 + 3716.4 = 0, \tag{4.112}$$

$$\overline{h}^0_{f,H_2O} + \Delta\overline{h}_{H_2O} - 8.314T_2 + 3716.4 = 0, \tag{4.113}$$

$$-241826 + \Delta\overline{h}_{H_2O} - 8.314T_2 + 3716.4 = 0, \tag{4.114}$$

$$-238110 + \Delta\overline{h}_{H_2O} - 8.314T_2 = 0. \tag{4.115}$$

At this point, one begins an iteration process, guessing a value of $T_2$ and an associated $\Delta\overline{h}_{H_2O}$. When $T_2$ is guessed at 5600 K, the left side becomes $-6507.04$. When $T_2$ is guessed at 6000 K, the left side becomes 14301.4. One can interpolate to arrive at

$$T_2 = 5725 \text{ K}. \tag{4.116}$$

This is an extremely high temperature. At such temperatures, in fact, one can expect other species to co-exist in the equilibrium state in large quantities. These may include H, OH, O, $HO_2$, and $H_2O_2$, among others. The final pressure is given by

$$P_2 = \frac{n_{H_2O}\overline{R}T_2}{V} = \frac{(2 \text{ kmol})\left(8.314 \frac{\text{kJ}}{\text{kmol K}}\right)(5725 \text{ K})}{74.3272 \text{ m}^3} = 1280.76 \text{ kPa}. \tag{4.117}$$

The final concentration of $H_2O$ is

$$\overline{\rho}_{H_2O} = \frac{2 \text{ kmol}}{74.3272 \text{ m}^3} = 2.69 \times 10^{-2} \frac{\text{kmol}}{\text{m}^3}. \tag{4.118}$$

## Dilute, Cold Mixture

**EXAMPLE 4.10**

Consider a variant on the previous example in which the mixture is diluted with an inert, taken here to be $N_2$. A closed, fixed, adiabatic volume contains a stoichiometric mixture of 2 kmol of $H_2$, 1 kmol of $O_2$, and 8 kmol of $N_2$ at 100 kPa and 298 K. Figure 4.6 gives a sketch. Find the adiabatic flame temperature and the final pressure, assuming the irreversible reaction

$$2H_2 + O_2 + 8N_2 \rightarrow 2H_2O + 8N_2. \tag{4.119}$$

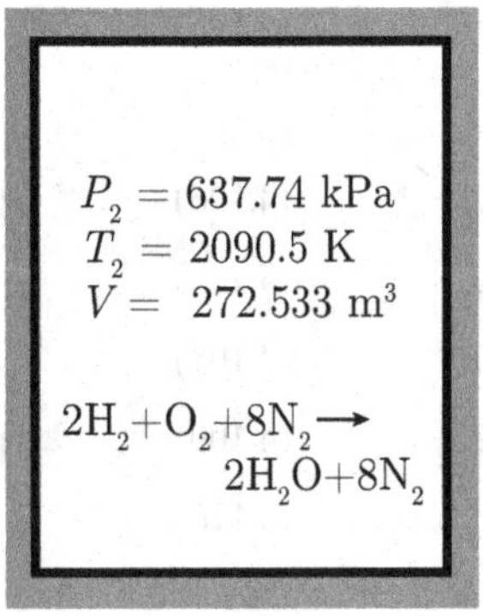

$P_2 = 637.74$ kPa
$T_2 = 2090.5$ K
$V = 272.533$ m$^3$

$2H_2+O_2+8N_2 \rightarrow$
$2H_2O+8N_2$

Figure 4.6. Final configuration for isochoric, adiabatic hydrogen-oxygen reaction diluted with nitrogen, initially cold.

The volume is given by

$$V = \frac{(n_{H_2} + n_{O_2} + n_{N_2})\overline{R}T_1}{P_1}, \tag{4.120}$$

$$= \frac{(2+1+8)(\text{kmol})\left(8.314 \frac{\text{kJ}}{\text{kmol K}}\right)(298\text{ K})}{100\text{ kPa}} = 272.533\text{ m}^3. \tag{4.121}$$

The first law gives

$$E_2 - E_1 = Q - W, \tag{4.122}$$

$$E_2 - E_1 = 0, \tag{4.123}$$

$$n_{H_2O}\overline{e}_{H_2O} - n_{H_2}\overline{e}_{H_2} - n_{O_2}\overline{e}_{O_2} + n_{N_2}(\overline{e}_{N_{2}2} - \overline{e}_{N_{2}1}) = 0, \tag{4.124}$$

$$n_{H_2O}(\overline{h}_{H_2O} - \overline{R}T_2) - n_{H_2}(\overline{h}_{H_2} - \overline{R}T_1) - n_{O_2}(\overline{h}_{O_2} - \overline{R}T_1) + n_{N_2}((\overline{h}_{N_{2}2} - \overline{R}T_2) - (\overline{h}_{N_{2}1} - \overline{R}T_1)) = 0, \tag{4.125}$$

$$2\overline{h}_{H_2O} - 2\underbrace{\overline{h}_{H_2}}_{0} - \underbrace{\overline{h}_{O_2}}_{0} + \overline{R}(-10T_2 + 11T_1) + 8(\underbrace{\overline{h}_{N_{2}2}}_{\Delta\overline{h}_{N_2}} - \underbrace{\overline{h}_{N_{2}1}}_{0}) = 0, \tag{4.126}$$

$$2\overline{h}_{H_2O} + (8.314)(-10T_2 + (11)(298)) + 8\Delta\overline{h}_{N_{2}2} = 0, \tag{4.127}$$

$$2\overline{h}_{H_2O} - 83.14T_2 + 27253.3 + 8\Delta\overline{h}_{N_{2}2} = 0, \tag{4.128}$$

$$2\overline{h}^0_{f,H_2O} + 2\Delta\overline{h}_{H_2O} - 83.14T_2 + 27253.3 + 8\Delta\overline{h}_{N_{2}2} = 0, \tag{4.129}$$

$$2(-241826) + 2\Delta\overline{h}_{H_2O} - 83.14T_2 + 27253.3 + 8\Delta\overline{h}_{N_{2}2} = 0, \tag{4.130}$$

$$-456399 + 2\Delta\overline{h}_{H_2O} - 83.14T_2 + 8\Delta\overline{h}_{N_{2}2} = 0. \tag{4.131}$$

At this point, one begins an iteration process, guessing a value of $T_2$ and an associated $\Delta\overline{h}_{H_2O}$. When $T_2$ is guessed at 2000 K, the left side becomes $-28006.7$. When $T_2$ is guessed at 2200 K, the left side becomes 33895.3. Interpolate then to arrive at

$$T_2 = 2090.5\text{ K}. \tag{4.132}$$

The inert diluent significantly lowers the adiabatic flame temperature. This is because $N_2$ serves as a heat sink for the energy of reaction. If the mixture were at nonstoichiometric conditions, the excess species would also serve as a heat sink, and the adiabatic flame temperature would be lower than that of the stoichiometric mixture. The final pressure is given by

$$P_2 = \frac{(n_{H_2O} + n_{N_2})\overline{R}T_2}{V}, \tag{4.133}$$

$$= \frac{(2\text{ kmol} + 8\text{ kmol})\left(8.314 \frac{\text{kJ}}{\text{kmol K}}\right)(2090.5\text{ K})}{272.533\text{ m}^3} = 637.74\text{ kPa}. \tag{4.134}$$

The final concentrations of $H_2O$ and $N_2$ are

$$\overline{\rho}_{H_2O} = \frac{2\text{ kmol}}{272.533\text{ m}^3} = 7.34 \times 10^{-3}\frac{\text{kmol}}{\text{m}^3}, \tag{4.135}$$

$$\overline{\rho}_{N_2} = \frac{8\text{ kmol}}{272.533\text{ m}^3} = 2.94 \times 10^{-2}\frac{\text{kmol}}{\text{m}^3}. \tag{4.136}$$

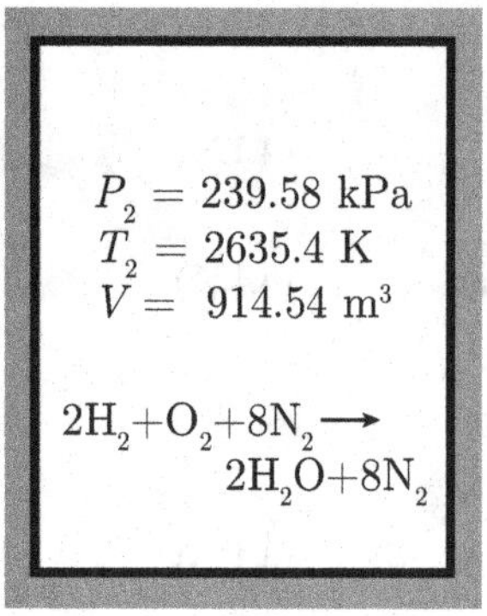

$P_2 = 239.58$ kPa
$T_2 = 2635.4$ K
$V = 914.54$ m$^3$

$2H_2+O_2+8N_2 \rightarrow$
$2H_2O+8N_2$

Figure 4.7. Final configuration for isochoric, adiabatic hydrogen-oxygen reaction diluted with nitrogen and preheated to $T_1 = 1000$ K.

### Dilute, Preheated Mixture

**EXAMPLE 4.11**

Consider a variant on the previous example in which the diluted mixture is preheated to 1000 K. One can achieve this via an isentropic compression of the cold mixture, such as might occur in an internal combustion engine. To simplify, the temperature of the mixture will be increased, while the pressure will be maintained. A closed, fixed, adiabatic volume contains a stoichiometric mixture of 2 kmol of $H_2$, 1 kmol of $O_2$, and 8 kmol of $N_2$ at 100 kPa and 1000 K. Figure 4.7 gives a sketch. Find the adiabatic flame temperature and the final pressure, assuming the irreversible reaction

$$2H_2 + O_2 + 8N_2 \rightarrow 2H_2O + 8N_2. \tag{4.137}$$

The volume is given by

$$V = \frac{(n_{H_2} + n_{O_2} + n_{N_2})\overline{R}T_1}{P_1}, \tag{4.138}$$

$$= \frac{(2+1+8)(\text{kmol})\left(8.314\ \frac{\text{kJ}}{\text{kmol K}}\right)(1000\ \text{K})}{100\ \text{kPa}} = 914.54\ \text{m}^3. \tag{4.139}$$

The first law gives

$$E_2 - E_1 = Q - W, \tag{4.140}$$

$$E_2 - E_1 = 0, \tag{4.141}$$

$$n_{H_2O}\overline{e}_{H_2O} - n_{H_2}\overline{e}_{H_2} - n_{O_2}\overline{e}_{O_2} + n_{N_2}(\overline{e}_{N_2 2} - \overline{e}_{N_2 1}) = 0, \tag{4.142}$$

$$n_{H_2O}(\overline{h}_{H_2O} - \overline{R}T_2) - n_{H_2}(\overline{h}_{H_2} - \overline{R}T_1) - n_{O_2}(\overline{h}_{O_2} - \overline{R}T_1) + n_{N_2}((\overline{h}_{N_2 2} - \overline{R}T_2) - (\overline{h}_{N_2 1} - \overline{R}T_1)) = 0, \tag{4.143}$$

$$2\overline{h}_{H_2O} - 2\overline{h}_{H_2} - \overline{h}_{O_2} + \overline{R}(-10T_2 + 11T_1) + 8(\overline{h}_{N_2 2} - \overline{h}_{N_2 1}) = 0, \tag{4.144}$$

$$2(-241826 + \Delta\overline{h}_{H_2O}) - 2(20663) - 22703 + (8.314)(-10T_2 + (11)(1000)) + 8\Delta\overline{h}_{N_2 2} - 8(21463) = 0, \tag{4.145}$$

$$2\Delta\overline{h}_{H_2O} - 83.14T_2 - 627931 + 8\Delta\overline{h}_{N_2 2} = 0. \tag{4.146}$$

At this point, one begins an iteration process, guessing a value of $T_2$ and an associated $\Delta\overline{h}_{H_2O}$. When $T_2$ is guessed at 2600 K, the left side becomes $-11351$. When $T_2$ is guessed at 2800 K, the left side becomes 52787. Interpolate then to arrive at

$$T_2 = 2635.4 \text{ K}. \tag{4.147}$$

The preheating raised the adiabatic flame temperature. The preheating was by 1000 K − 298 K = 702 K. The new adiabatic flame temperature is only 2635.4 K − 2090.5 K = 544.9 K greater. The final pressure is given by

$$P_2 = \frac{(n_{H_2O} + n_{N_2})\overline{R}T_2}{V}, \tag{4.148}$$

$$= \frac{(2 \text{ kmol} + 8 \text{ kmol})\left(8.314 \frac{\text{kJ}}{\text{kmol K}}\right)(2635.4 \text{ K})}{914.54 \text{ m}^3} = 239.58 \text{ kPa}. \tag{4.149}$$

The final concentrations of $H_2O$ and $N_2$ are

$$\overline{\rho}_{H_2O} = \frac{2 \text{ kmol}}{914.54 \text{ m}^3} = 2.19 \times 10^{-3} \frac{\text{kmol}}{\text{m}^3}, \tag{4.150}$$

$$\overline{\rho}_{N_2} = \frac{8 \text{ kmol}}{914.54 \text{ m}^3} = 8.75 \times 10^{-3} \frac{\text{kmol}}{\text{m}^3}. \tag{4.151}$$

## Dilute, Preheated Mixture with Minor Species

**EXAMPLE 4.12**

Consider a variant on the previous example. Here, allow for minor species to be present at equilibrium. A closed, fixed, adiabatic volume contains a stoichiometric mixture of 2 kmol of $H_2$, 1 kmol of $O_2$, and 8 kmol of $N_2$ at 100 kPa and 1000 K. Figure 4.8 gives a sketch of the configuration. Find the adiabatic flame temperature and the final pressure, assuming reversible reactions.

Here, the details of the analysis are omitted, but the result is given that is the consequence of a calculation involving detailed reactions rates involving equations of the form to be studied in Sec. 5.1. One can also solve an optimization problem to minimize the Gibbs free energy of a wide variety of products to get the same answer; for a general discussion, see Sec. 5.2.1. In this case, the equilibrium temperature and pressure are found to be

$$T = 2484.8 \text{ K}, \qquad P = 227.89 \text{ kPa}. \tag{4.152}$$

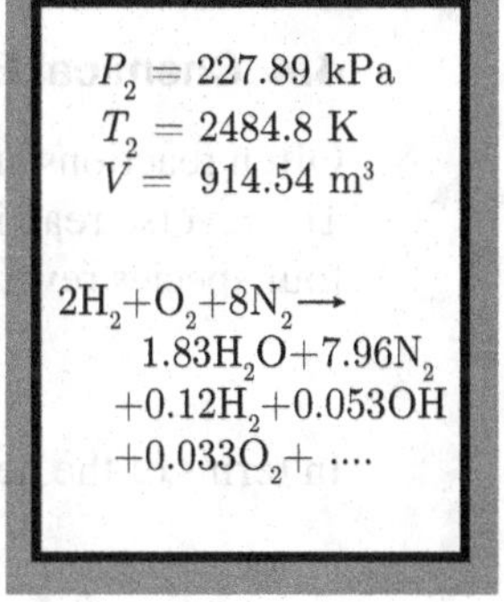

$P_2 = 227.89$ kPa
$T_2 = 2484.8$ K
$V = 914.54$ m$^3$

$2H_2 + O_2 + 8N_2 \rightarrow 1.83H_2O + 7.96N_2 + 0.12H_2 + 0.053OH + 0.033O_2 + \ldots$

Figure 4.8. Final configuration for isochoric, adiabatic hydrogen-oxygen reaction diluted with nitrogen and preheated to $T_1 = 1000$ K; minor species predicted via a detailed chemical equilibrium calculation as from Sec. 5.2.1.

Equilibrium species concentrations are found to be

$$\text{minor product} \quad \overline{\rho}_{H_2} = 1.3 \times 10^{-4}\ \text{kmol/m}^3, \tag{4.153}$$

$$\text{minor product} \quad \overline{\rho}_{H} = 1.9 \times 10^{-5}\ \text{kmol/m}^3, \tag{4.154}$$

$$\text{minor product} \quad \overline{\rho}_{O} = 5.7 \times 10^{-6}\ \text{kmol/m}^3, \tag{4.155}$$

$$\text{minor product} \quad \overline{\rho}_{O_2} = 3.6 \times 10^{-5}\ \text{kmol/m}^3, \tag{4.156}$$

$$\text{minor product} \quad \overline{\rho}_{OH} = 5.9 \times 10^{-5}\ \text{kmol/m}^3, \tag{4.157}$$

$$\textbf{major}\ \text{product} \quad \overline{\rho}_{H_2O} = 2.0 \times 10^{-3}\ \text{kmol/m}^3, \tag{4.158}$$

$$\text{trace product} \quad \overline{\rho}_{HO_2} = 1.1 \times 10^{-8}\ \text{kmol/m}^3, \tag{4.159}$$

$$\text{trace product} \quad \overline{\rho}_{H_2O_2} = 1.2 \times 10^{-9}\ \text{kmol/m}^3, \tag{4.160}$$

$$\text{trace product} \quad \overline{\rho}_{N} = 1.7 \times 10^{-9}\ \text{kmol/m}^3, \tag{4.161}$$

$$\text{trace product} \quad \overline{\rho}_{NH} = 3.7 \times 10^{-10}\ \text{kmol/m}^3, \tag{4.162}$$

$$\text{trace product} \quad \overline{\rho}_{NH_2} = 1.5 \times 10^{-10}\ \text{kmol/m}^3, \tag{4.163}$$

$$\text{trace product} \quad \overline{\rho}_{NH_3} = 3.1 \times 10^{-10}\ \text{kmol/m}^3, \tag{4.164}$$

$$\text{trace product} \quad \overline{\rho}_{NNH} = 1.0 \times 10^{-10}\ \text{kmol/m}^3, \tag{4.165}$$

$$\text{minor product} \quad \overline{\rho}_{NO} = 3.1 \times 10^{-6}\ \text{kmol/m}^3, \tag{4.166}$$

$$\text{trace product} \quad \overline{\rho}_{NO_2} = 5.3 \times 10^{-9}\ \text{kmol/m}^3, \tag{4.167}$$

$$\text{trace product} \quad \overline{\rho}_{N_2O} = 2.6 \times 10^{-9}\ \text{kmol/m}^3, \tag{4.168}$$

$$\text{trace product} \quad \overline{\rho}_{HNO} = 1.7 \times 10^{-9}\ \text{kmol/m}^3, \tag{4.169}$$

$$\textbf{major}\ \text{product} \quad \overline{\rho}_{N_2} = 8.7 \times 10^{-3}\ \text{kmol/m}^3. \tag{4.170}$$

The concentrations of the major products went down when the minor species were considered. The adiabatic flame temperature also went down by a significant amount: 2635 K − 2484.8 K = 150.2 K. Some thermal energy was necessary to break the bonds that induce the presence of minor species. Multiplying the final concentrations by the volume allows us to write what amounts to the global reaction for these conditions. Retaining only a few of the species, the global reaction is

$$\begin{aligned} 2H_2 + O_2 + 8N_2 &\rightarrow 1.83H_2O + 7.96N_2 \\ &+ 0.12H_2 + 0.053OH + 0.033O_2 + 0.017H + 0.005O + \cdots \end{aligned} \tag{4.171}$$

## 4.4 Chemical Equilibrium

Often reactions are not simply unidirectional, as alluded to in the previous example. The reverse reaction, especially at high temperature, can be important. Consider the four species reversible reaction

$$\nu_1'\chi_1 + \nu_2'\chi_2 \leftrightharpoons \nu_3''\chi_3 + \nu_4''\chi_4. \tag{4.172}$$

In terms of the net stoichiometric coefficients, this becomes

$$\nu_1\chi_1 + \nu_2\chi_2 + \nu_3\chi_3 + \nu_4\chi_4 = 0. \tag{4.173}$$

One can define a variable $\zeta$, the *reaction progress*. When $t = 0$, one takes $\zeta = 0$. Now, as the reaction goes forward, one takes $d\zeta > 0$. And a forward reaction will decrease the number of moles of $\chi_1$ and $\chi_2$ while increasing the number of moles of $\chi_3$ and $\chi_4$. This will occur in ratios dictated by the net stoichiometric coefficients:

$$dn_1 = -\nu_1' \, d\zeta, \quad dn_2 = -\nu_2' \, d\zeta, \quad dn_3 = +\nu_3'' \, d\zeta, \quad dn_4 = +\nu_4'' \, d\zeta. \qquad (4.174)$$

If $n_i$ is taken to have units of mol, $\nu_i'$, and $\nu_i''$ are taken as dimensionless, then $\zeta$ must have units of moles. In terms of the net stoichiometric coefficients, one has

$$dn_1 = \nu_1 \, d\zeta, \quad dn_2 = \nu_2 \, d\zeta, \quad dn_3 = \nu_3 \, d\zeta, \quad dn_4 = \nu_4 \, d\zeta. \qquad (4.175)$$

Assume that at $t = 0$, one has

$$n_1|_{t=0} = n_{10}, \quad n_2|_{t=0} = n_{20}, \quad n_3|_{t=0} = n_{30}, \quad n_4|_{t=0} = n_{40}. \qquad (4.176)$$

Then, after integrating Eqs. (4.175), one finds

$$n_1 = \nu_1\zeta + n_{10}, \quad n_2 = \nu_2\zeta + n_{20}, \quad n_3 = \nu_3\zeta + n_{30}, \quad n_4 = \nu_4\zeta + n_{40}. \qquad (4.177)$$

One can also eliminate $\zeta$ in a variety of fashions and parameterize the reaction by one of the species mol numbers. Choosing, for example, $n_1$ as a parameter, one gets

$$\zeta = \frac{n_1 - n_{10}}{\nu_1}. \qquad (4.178)$$

Eliminating $\zeta$, one finds all other mol numbers in terms of $n_1$:

$$n_2 = \nu_2 \frac{n_1 - n_{10}}{\nu_1} + n_{20}, \qquad (4.179)$$

$$n_3 = \nu_3 \frac{n_1 - n_{10}}{\nu_1} + n_{30}, \qquad (4.180)$$

$$n_4 = \nu_4 \frac{n_1 - n_{10}}{\nu_1} + n_{40}. \qquad (4.181)$$

Written another way, one has

$$\frac{n_1 - n_{10}}{\nu_1} = \frac{n_2 - n_{20}}{\nu_2} = \frac{n_3 - n_{30}}{\nu_3} = \frac{n_4 - n_{40}}{\nu_4} = \zeta. \qquad (4.182)$$

For an $N$-species reaction, $\sum_{i=1}^{N} \nu_i \chi_i = 0$, one can generalize to say

$$dn_i = \nu_i \, d\zeta, \qquad (4.183)$$

$$n_i = \nu_i\zeta + n_{i0}, \qquad (4.184)$$

$$\frac{n_i - n_{i0}}{\nu_i} = \zeta. \qquad (4.185)$$

Note that

$$\frac{dn_i}{d\zeta} = \nu_i. \qquad (4.186)$$

From Chapter 3, one manifestation of the second law is Eq. (3.279):

$$dG|_{T,P} = \sum_{i=1}^{N} \overline{\mu}_i \, dn_i \le 0. \qquad (4.187)$$

Now, one can eliminate $dn_i$ in Eq. (4.187) by use of Eq. (4.183) to get

$$dG|_{T,P} = \sum_{i=1}^{N} \overline{\mu}_i \nu_i \, d\zeta \leq 0, \tag{4.188}$$

$$\left.\frac{\partial G}{\partial \zeta}\right|_{T,P} = \sum_{i=1}^{N} \overline{\mu}_i \nu_i = -\overline{\alpha} \leq 0. \tag{4.189}$$

Then, for the reaction to go forward, one must require that the chemical affinity, defined earlier in Eq. (3.375), be positive, $\overline{\alpha} \geq 0$. One also knows from Chapter 3 that the irreversible entropy production takes the form of Eq. (3.270):

$$d\sigma = -\frac{1}{T}\sum_{i=1}^{N} \overline{\mu}_i \, dn_i = -\frac{1}{T} d\zeta \sum_{i=1}^{N} \overline{\mu}_i \nu_i \geq 0, \tag{4.190}$$

$$\dot{\sigma} = \frac{d\sigma}{dt} = -\frac{1}{T}\frac{d\zeta}{dt}\sum_{i=1}^{N} \overline{\mu}_i \nu_i \geq 0. \tag{4.191}$$

In terms of the chemical affinity, $\overline{\alpha} = -\sum_{i=1}^{N} \overline{\mu}_i \nu_i$, Eq. (4.191) can be written as

$$\dot{\sigma} = \frac{1}{T}\frac{d\zeta}{dt}\overline{\alpha} \geq 0. \tag{4.192}$$

Now, one straightforward, albeit naïve, way to guarantee positive semidefiniteness of the irreversible entropy production and thus satisfaction of the second law is to construct the chemical kinetic rate equation so that

$$\textit{Provisional, naïve kinetics model:} \qquad \frac{d\zeta}{dt} = -k \sum_{i=1}^{N} \overline{\mu}_i \nu_i = k\overline{\alpha}, \quad k \geq 0. \tag{4.193}$$

This provisional assumption of convenience will be supplanted later in Sec. 4.5 by a form that agrees well with experiment. Here, $k$ is a positive semidefinite scalar. In general, it is a function of temperature, $k = k(T)$, so that reactions proceed rapidly at high temperature and slowly at low temperature. Then, certainly the reaction progress variable $\zeta$ will cease to change when the equilibrium condition

$$\sum_{i=1}^{N} \overline{\mu}_i \nu_i = 0 \tag{4.194}$$

is met. This is equivalent to requiring

$$\overline{\alpha} = 0 \tag{4.195}$$

at equilibrium.

Now, while Eq. (4.194) is a compact form of the equilibrium condition, it is not the most commonly used form. One can perform the following analysis to obtain the form in most common usage. Start by employing Eq. (3.205) equating the chemical potential with the Gibbs free energy per unit mol for each species $i$: $\overline{\mu}_i = \overline{g}_i$. Then, employ the definition of Gibbs free energy for an ideal gas and carry out a set of operations

$$\sum_{i=1}^{N} \overline{g}_i \nu_i = \sum_{i=1}^{N} (\overline{h}_i - T\overline{s}_i)\nu_i = 0 \qquad \text{at equilibrium.} \tag{4.196}$$

For the ideal gas, one can substitute for $\overline{h}_i(T)$ and $\overline{s}_i(T,P)$ and write the equilibrium condition as

$$\sum_{i=1}^{N}\underbrace{\left(\underbrace{\overline{h}^0_{T_0,i}+\underbrace{\int_{T_0}^{T}\overline{c}_{Pi}(\hat{T})\,d\hat{T}}_{\Delta\overline{h}^0_{T,i}}}_{\overline{h}^0_{T,i}=\overline{h}_{T,i}}-T\underbrace{\left(\underbrace{\overline{s}^0_{T_0,i}+\int_{T_0}^{T}\frac{\overline{c}_{Pi}(\hat{T})}{\hat{T}}d\hat{T}}_{\overline{s}^0_{T,i}}-\overline{R}\ln\left(\frac{y_iP}{P_0}\right)\right)}_{\overline{s}_{T,i}}\right)}_{\overline{g}_i}\nu_i=0. \quad (4.197)$$

Now, writing the equilibrium condition in terms of the enthalpies and entropies referred to the standard pressure, one gets

$$\sum_{i=1}^{N}\left(\overline{h}^0_{T,i}-T\left(\overline{s}^0_{T,i}-\overline{R}\ln\left(\frac{y_iP}{P_0}\right)\right)\right)\nu_i=0. \quad (4.198)$$

Let us rearrange to get terms at the reference pressure on the left side:

$$\sum_{i=1}^{N}\underbrace{\left(\overline{h}^0_{T,i}-T\overline{s}^0_{T,i}\right)}_{\overline{g}^0_{T,i}=\overline{\mu}^0_{T,i}}\nu_i=-\sum_{i=1}^{N}\overline{R}T\nu_i\ln\left(\frac{y_iP}{P_0}\right), \quad (4.199)$$

$$-\underbrace{\sum_{i=1}^{N}\overline{g}^0_{T,i}\nu_i}_{\equiv\Delta G^0}=\overline{R}T\sum_{i=1}^{N}\ln\left(\frac{y_iP}{P_0}\right)^{\nu_i}, \quad (4.200)$$

$$-\frac{\Delta G^0}{\overline{R}T}=\sum_{i=1}^{N}\ln\left(\frac{y_iP}{P_0}\right)^{\nu_i}=\ln\left(\prod_{i=1}^{N}\left(\frac{y_iP}{P_0}\right)^{\nu_i}\right), \quad (4.201)$$

$$\underbrace{\exp\left(-\frac{\Delta G^0}{\overline{R}T}\right)}_{\equiv K_P}=\prod_{i=1}^{N}\left(\frac{y_iP}{P_0}\right)^{\nu_i}, \quad (4.202)$$

$$K_P=\prod_{i=1}^{N}\left(\frac{y_iP}{P_0}\right)^{\nu_i}=\left(\frac{P}{P_0}\right)^{\sum_{i=1}^{N}\nu_i}\prod_{i=1}^{n}y_i^{\nu_i}. \quad (4.203)$$

From Eq. (4.203), we recognize the common result

$$K_P=\prod_{i=1}^{N}\left(\frac{P_i}{P_0}\right)^{\nu_i} \quad \text{at equilibrium.} \quad (4.204)$$

Here, $K_P$ is what is known as the pressure-based *equilibrium constant.* It is dimensionless. Despite its name, it is not a constant. It is defined in terms of thermodynamic properties, and for the ideal gas is a function of $T$ only:

$$K_P\equiv\exp\left(-\frac{\Delta G^0}{\overline{R}T}\right), \quad \text{generally valid.} \quad (4.205)$$

*Only at equilibrium* does the property $K_P$ also equal the product of the partial pressures as in Eq. (4.204). The subscript $P$ comes about because it is related to the product of the ratio of the partial pressure to the reference pressure raised to the net stoichiometric coefficient. Also, the net change in Gibbs free energy of the reaction at the reference pressure, $\Delta G^0$, which is a function of $T$ only, has been defined, much as in Sec. 1.2.1, as

$$\Delta G^0 \equiv \sum_{i=1}^{N} \overline{g}^0_{T,i} \nu_i. \tag{4.206}$$

The term $\Delta G^0$ has units of kJ/kmol; it traditionally does not get an overbar. If $\Delta G^0 > 0$, one has $K_P \in (0, 1)$, and reactants are favored over products. If $\Delta G^0 < 0$, one gets $K_P > 1$, and products are favored over reactants. One can also define $\Delta G^0$ in terms of the chemical affinity, referred to the reference pressure, as

$$\Delta G^0 = -\overline{\alpha}^0. \tag{4.207}$$

One can also define another convenient thermodynamic property, which, for an ideal gas, is a function of $T$ alone, the equilibrium constant $K_c$:

$$K_c \equiv \left(\frac{P_0}{\overline{R}T}\right)^{\sum_{i=1}^{N} \nu_i} \exp\left(-\frac{\Delta G^0}{\overline{R}T}\right), \quad \text{generally valid.} \tag{4.208}$$

This property is dimensional, and the units depend on the stoichiometry of the reaction. The units of $K_c$ will be $(\text{mol/cm}^3)^{\sum_{i=1}^{N} \nu_i}$.

The equilibrium condition, Eq. (4.204), is often written in terms of molar concentrations and $K_c$. This can be achieved by the operations, valid only at equilibrium,

$$K_P = \prod_{i=1}^{N} \left(\frac{\overline{\rho}_i \overline{R}T}{P_0}\right)^{\nu_i}, \tag{4.209}$$

$$\exp\left(\frac{-\Delta G^0}{\overline{R}T}\right) = \left(\frac{\overline{R}T}{P_0}\right)^{\sum_{i=1}^{N} \nu_i} \prod_{i=1}^{N} \overline{\rho}_i^{\nu_i}, \tag{4.210}$$

$$\underbrace{\left(\frac{P_0}{\overline{R}T}\right)^{\sum_{i=1}^{N} \nu_i} \exp\left(\frac{-\Delta G^0}{\overline{R}T}\right)}_{\equiv K_c} = \prod_{i=1}^{N} \overline{\rho}_i^{\nu_i}, \tag{4.211}$$

$$K_c = \prod_{i=1}^{N} \overline{\rho}_i^{\nu_i}, \quad \text{at equilibrium.} \tag{4.212}$$

*Only at equilibrium* does the property $K_c$ also equal the product of the molar species concentrations, as in Eq. (4.212).

## 4.5 Chemical Kinetics of a Single Isothermal Reaction

In the same fashion in ordinary mechanics that an understanding of statics enables an understanding of dynamics, an understanding of chemical equilibrium is necessary to understand the more challenging topic of chemical kinetics. Chemical kinetics describes the time evolution of systems that may have an initial state far from equilibrium; it typically describes the path of such systems to an equilibrium state.

Here, gas phase kinetics of ideal gas mixtures that obey Dalton's law will be studied. Important topics such as catalysis and solid or liquid reactions will not be considered. This section will be restricted to strictly *isothermal* systems. This simplifies the analysis greatly. It is straightforward to extend the analysis of this system to non-isothermal systems. One must then make further appeal to the energy equation to get an equation for temperature evolution. This extension is made in Sec. 4.6.3.

The general form for spatially homogeneous evolution of species in a closed system of variable volume is taken to be

$$\frac{d}{dt}\left(\frac{\overline{\rho}_i}{\rho}\right) = \frac{\dot{\omega}_i}{\rho}. \tag{4.213}$$

Multiplying both sides of Eq. (4.213) by molecular mass $M_i$ and using Eq. (2.37) to exchange $\overline{\rho}_i$ for mass fraction $Y_i$ then gives the alternate form

$$\frac{dY_i}{dt} = \frac{\dot{\omega}_i M_i}{\rho}. \tag{4.214}$$

### 4.5.1 Isochoric Systems

Consider the evolution of species concentration in a system that is isothermal, isochoric and spatially homogeneous. The system is undergoing a single chemical reaction involving $N$ species of the familiar form of Eq. (4.17):

$$\sum_{i=1}^{N} \nu_i \chi_i = 0. \tag{4.215}$$

Because the density is constant for the closed isochoric system, Eq. (4.213) reduces to

$$\frac{d\overline{\rho}_i}{dt} = \dot{\omega}_i. \tag{4.216}$$

Then, experiment, as well as a more fundamental molecular collision theory, shows that the evolution of species concentration $i$ is given by

$$\frac{d\overline{\rho}_i}{dt} = \overbrace{\nu_i \underbrace{\underbrace{aT^\beta \exp\left(\frac{-\mathcal{E}}{\overline{R}T}\right)}_{\equiv k(T)} \underbrace{\left(\prod_{k=1}^{N} \overline{\rho}_k^{\nu_k'}\right)}_{\text{forward reaction}} \left(1 - \underbrace{\frac{1}{K_c}\prod_{k=1}^{N} \overline{\rho}_k^{\nu_k}}_{\text{reverse reaction}}\right)}_{\equiv r}}^{\equiv \dot{\omega}_i}, \quad \text{isochoric system.} \tag{4.217}$$

This relation actually holds for isochoric, nonisothermal systems as well. Those will not be considered in detail yet. Here, some new variables are defined as follows:

- $a$: A kinetic rate constant called the *collision frequency coefficient*. Its units will depend on the actual reaction and could involve various combinations of length, time, and temperature. It is constructed so that $d\overline{\rho}_i/dt$ has units of mol/cm$^3$/s; this requires it to have units of $(\text{mol/cm}^3)^{(1-\nu_M'-\sum_{k=1}^{N}\nu_k')}/\text{s/K}^\beta$. Here, $\nu_M'$ is a coefficient that is present if an inert third body participates in the reaction.

- $\beta$: A dimensionless parameter whose value is set by experiments, sometimes combined with guiding theory, to account for weak temperature dependency of reaction rates.
- $\overline{\mathcal{E}}$: The *activation energy*. It has units of cal/mol, though others are often used, and is fit by both experiment and fundamental theory to account for the often strong temperature dependency of reaction.

In Eq. (4.217) molar concentrations are raised to the $\nu_k'$ and $\nu_k$ powers. As it does not make sense to raise a physical quantity to a power with units, this is a reason one traditionally interprets the values of $\nu_k$, $\nu_k'$, as well as $\nu_k''$ to be dimensionless pure numbers. They are also interpreted in a standard fashion: the smallest integer values that actually correspond to the underlying molecular collision that has been modeled. While stoichiometric balance can be achieved by a variety of $\nu_k$ values, the kinetic rates are linked to one particular set that is defined by the community.

Equation (4.217) is written in such a way that the species molar production rate *increases* when

- The net number of moles generated in the reaction, measured by $\nu_i$ increases.
- The temperature increases; here, the sensitivity may be high, as one observes in nature.
- The species concentrations of species involved in the forward reaction increase; this embodies the principle that the *collision-based* reaction rates are enhanced when there are more molecules to collide.
- The species concentrations of species involved in the reverse reaction decrease.

Here, three intermediate variables that are in common usage have been defined. First one takes the reaction rate to be

$$r \equiv \underbrace{aT^{\beta} \exp\left(\frac{-\overline{\mathcal{E}}}{\overline{R}T}\right)}_{\equiv k(T)} \underbrace{\left(\prod_{k=1}^{N} \overline{\rho}_k^{\nu_k'}\right)}_{\text{forward reaction}} \left(1 - \underbrace{\frac{1}{K_c} \prod_{k=1}^{N} \overline{\rho}_k^{\nu_k}}_{\text{reverse reaction}}\right), \tag{4.218}$$

$$= \underbrace{aT^{\beta} \exp\left(\frac{-\overline{\mathcal{E}}}{\overline{R}T}\right)}_{\equiv k(T),\ \text{Arrhenius rate}} \underbrace{\left(\underbrace{\prod_{k=1}^{N} \overline{\rho}_k^{\nu_k'}}_{\text{forward reaction}} - \underbrace{\frac{1}{K_c} \prod_{k=1}^{N} \overline{\rho}_k^{\nu_k''}}_{\text{reverse reaction}}\right)}_{\text{law of mass action}}. \tag{4.219}$$

The reaction rate $r$ has units of $\text{mol/cm}^3\text{/s}$.

The temperature dependency of the reaction rate is embodied in $k(T)$. The reaction rate coefficient $k(T)$ is defined by what is known as an *Arrhenius*[3] *rate law*:

$$k(T) \equiv aT^{\beta} \exp\left(\frac{-\overline{\mathcal{E}}}{\overline{R}T}\right). \tag{4.220}$$

[3]Svante August Arrhenius (1859–1927), Swedish physicist; 1903 Nobel laureate in Chemistry.

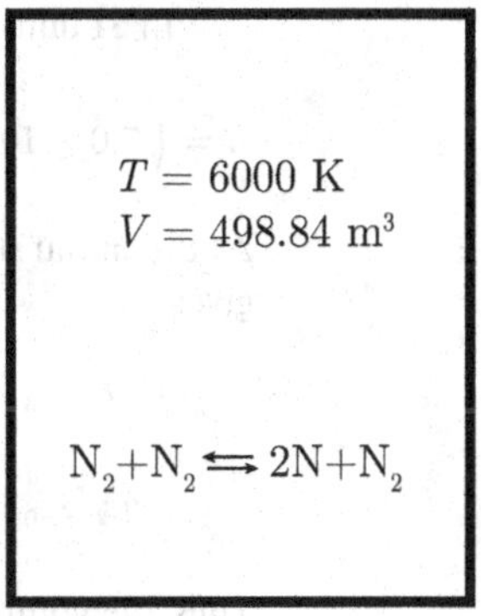

$T = 6000$ K
$V = 498.84$ m$^3$

$N_2 + N_2 \leftrightharpoons 2N + N_2$

Figure 4.9. Configuration for isochoric, isothermal nitrogen dissociation reaction.

This equation was advocated by van't Hoff (1884);[4] Arrhenius (1889) gave a physical justification. The units of $k(T)$ actually depend on the reaction. This precludes a clean nondimensionalization. The units must be $(\mathrm{mol/cm^3})^{(1-\nu'_M-\sum_{k=1}^{N}\nu'_k)}/\mathrm{s}$.

In terms of reaction progress, one can also take

$$r = \frac{1}{V}\frac{d\zeta}{dt}. \tag{4.221}$$

The factor of $1/V$ is necessary because $r$ has units of molar concentration per time and $\zeta$ has units of mol. The overriding importance of the temperature sensitivity is illustrated as part of the next example. The remainder of the expression involving the products of the species concentrations is the defining characteristic of systems that obey the *law of mass action*. Though the history is complex, most attribute the law of mass action to Waage[5] and Guldberg[6] (1864). Last, the overall species molar production rate of species $i$, often written as $\dot{\omega}_i$, is defined as

$$\dot{\omega}_i \equiv \nu_i r. \tag{4.222}$$

As $\nu_i$ is considered to be dimensionless, the units of $\dot{\omega}_i$ must be mol/cm$^3$/s.

**EXAMPLE 4.13**

Study the nitrogen dissociation problem considered in an earlier example that was confined to equilibrium analysis (see Sec. 3.9.2), in which at $t = 0$ s, 1 kmol of $N_2$ exists at $P = 100$ kPa and $T = 6000$ K. Take as before the reaction to be *isothermal* and *isochoric*. Consider again the elementary nitrogen dissociation reaction

$$N_2 + N_2 \leftrightharpoons 2N + N_2, \tag{4.223}$$

which has kinetic rate parameters (Park, 1990)

$$a = 7.0 \times 10^{21}\ \frac{\mathrm{cm^3\ K^{1.6}}}{\mathrm{mol\ s}}, \quad \beta = -1.6, \quad \overline{\mathcal{E}} = 941145\ \frac{\mathrm{kJ}}{\mathrm{kmol}}. \tag{4.224}$$

Park actually reports a so-called activation temperature $T_a = \overline{\mathcal{E}}/\overline{R} = 113200$ K. We have converted it to activation energy. Figure 4.9 gives a sketch of the configuration.

[4]Jacobus Henricus van't Hoff (1852–1922), Dutch chemist; 1901 Nobel laureate in Chemistry.
[5]Peter Waage (1833–1900), Norwegian chemist.
[6]Cato Maximilian Guldberg (1836–1902), Norwegian mathematician and chemist.

In SI units, $a$ is expressed as

$$a = \left(7.0 \times 10^{21}\, \frac{\text{cm}^3\ \text{K}^{1.6}}{\text{mol s}}\right) \left(\frac{1\ \text{m}}{100\ \text{cm}}\right)^3 \left(\frac{1000\ \text{mol}}{\text{kmol}}\right) = 7.0 \times 10^{18}\, \frac{\text{m}^3\ \text{K}^{1.6}}{\text{kmol s}}. \quad (4.225)$$

At the initial state, the material is all $N_2$, so $P_{N_2} = P = 100$ kPa. The ideal gas law then gives

$$P|_{t=0} = P_{N_2}|_{t=0} = \overline{\rho}_{N_2}\big|_{t=0} \overline{R}T, \quad (4.226)$$

$$\overline{\rho}_{N_2}\big|_{t=0} = \frac{P|_{t=0}}{\overline{R}T} = \frac{100\ \text{kPa}}{\left(8.314\, \frac{\text{kJ}}{\text{kmol K}}\right)(6000\ \text{K})} = 2.00465 \times 10^{-3}\, \frac{\text{kmol}}{\text{m}^3}. \quad (4.227)$$

Thus, the volume, constant for all time in the isochoric process, is

$$V = \frac{n_{N_2}|_{t=0}}{\overline{\rho}_{N_2}\big|_{t=0}} = \frac{1\ \text{kmol}}{2.00465 \times 10^{-3}\, \frac{\text{kmol}}{\text{m}^3}} = 4.9884 \times 10^2\ \text{m}^3. \quad (4.228)$$

Now, the stoichiometry of the reaction is such that

$$-dn_{N_2} = \frac{1}{2} dn_N, \quad (4.229)$$

$$-(n_{N_2} - \underbrace{n_{N_2}|_{t=0}}_{1\ \text{kmol}}) = \frac{1}{2}(n_N - \underbrace{n_N|_{t=0}}_{0}), \quad (4.230)$$

$$n_N = 2(1\ \text{kmol} - n_{N_2}), \quad (4.231)$$

$$\frac{n_N}{V} = 2\left(\frac{1\ \text{kmol}}{V} - \frac{n_{N_2}}{V}\right), \quad (4.232)$$

$$\overline{\rho}_N = 2\left(\frac{1\ \text{kmol}}{4.9884 \times 10^2\ \text{m}^3} - \overline{\rho}_{N_2}\right), \quad (4.233)$$

$$= 2\left(2.00465 \times 10^{-3}\, \frac{\text{kmol}}{\text{m}^3} - \overline{\rho}_{N_2}\right). \quad (4.234)$$

Now, the general equation for kinetics of a single reaction, Eq. (4.217), reduces for $N_2$ molar concentration to

$$\frac{d\overline{\rho}_{N_2}}{dt} = \nu_{N_2} a T^\beta \exp\left(\frac{-\mathcal{E}}{\overline{R}T}\right) (\overline{\rho}_{N_2})^{\nu'_{N_2}} (\overline{\rho}_N)^{\nu'_N} \left(1 - \frac{1}{K_c} (\overline{\rho}_{N_2})^{\nu_{N_2}} (\overline{\rho}_N)^{\nu_N}\right). \quad (4.235)$$

Realizing that $\nu'_{N_2} = 2$, $\nu'_N = 0$, $\nu_{N_2} = -1$, and $\nu_N = 2$, one gets

$$\frac{d\overline{\rho}_{N_2}}{dt} = \underbrace{-\ aT^\beta \exp\left(\frac{-\mathcal{E}}{\overline{R}T}\right)}_{k(T)} \overline{\rho}_{N_2}^2 \left(1 - \frac{1}{K_c} \frac{\overline{\rho}_N^2}{\overline{\rho}_{N_2}}\right). \quad (4.236)$$

Examine the primary temperature dependency of the reaction:

$$k(T) = aT^\beta \exp\left(\frac{-\mathcal{E}}{\overline{R}T}\right), \quad (4.237)$$

$$= \left(7.0 \times 10^{18}\, \frac{\text{m}^3\text{K}^{1.6}}{\text{kmol s}}\right) T^{-1.6} \exp\left(\frac{-941145\, \frac{\text{kJ}}{\text{kmol}}}{\left(8.314\, \frac{\text{kJ}}{\text{kmol K}}\right) T}\right), \quad (4.238)$$

$$= \frac{7.0 \times 10^{18}}{T^{1.6}} \exp\left(\frac{-1.13200 \times 10^5}{T}\right). \quad (4.239)$$

Figure 4.10 gives a plot of $k(T)$ that shows its strong dependency on temperature. For this

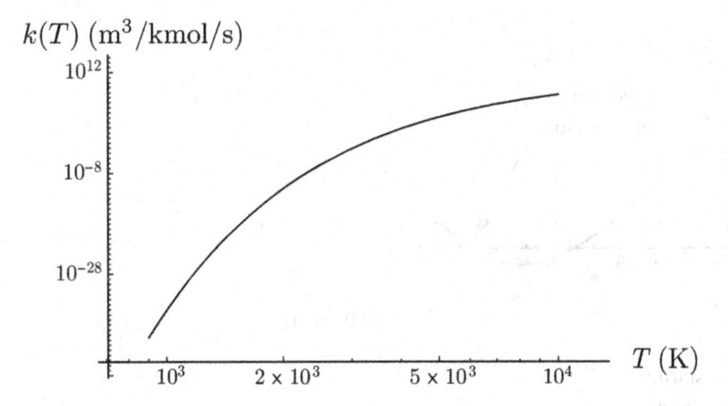


Figure 4.10. Reaction rate coefficient $k(T)$ for nitrogen dissociation example.

problem, $T = 6000$ K, so

$$k(6000) = \frac{7.0 \times 10^{18}}{6000^{1.6}} \exp\left(\frac{-1.13200 \times 10^5}{6000}\right) = 40398.5 \frac{\text{m}^3}{\text{kmol s}}. \tag{4.240}$$

Now, the equilibrium constant $K_c$ is needed. Recall Eq. (4.208):

$$K_c = \left(\frac{P_0}{\overline{R}T}\right)^{\sum_{i=1}^{N} \nu_i} \exp\left(\frac{-\Delta G^0}{\overline{R}T}\right). \tag{4.241}$$

For this system, because $\sum_{i=1}^{N} \nu_i = 1$, this reduces to

$$K_c = \left(\frac{P_0}{\overline{R}T}\right) \exp\left(\frac{-(2\overline{g}_{\text{N}}^0 - \overline{g}_{\text{N}_2}^0)}{\overline{R}T}\right), \tag{4.242}$$

$$= \left(\frac{P_0}{\overline{R}T}\right) \exp\left(\frac{-(2(\overline{h}_{\text{N}}^0 - T\overline{s}_{T,N}^0) - (\overline{h}_{\text{N}_2}^0 - T\overline{s}_{T,\text{N}_2}^0))}{\overline{R}T}\right), \tag{4.243}$$

$$= \left(\frac{100}{(8.314)(6000)}\right) \times \exp\left(\frac{-(2(597270 - (6000)216.926) - (205848 - (6000)292.984))}{(8.314)(6000)}\right), \tag{4.244}$$

$$= 0.000112942 \frac{\text{kmol}}{\text{m}^3}. \tag{4.245}$$

The differential equation for $N_2$ evolution is then given by

$$\frac{d\overline{\rho}_{\text{N}_2}}{dt} = -\left(40398.5 \frac{\text{m}^3}{\text{kmol}}\right) \overline{\rho}_{\text{N}_2}^2 \times \left(1 - \frac{1}{0.000112942 \frac{\text{kmol}}{\text{m}^3}} \frac{\left(2\left(2.00465 \times 10^{-3} \frac{\text{kmol}}{\text{m}^3} - \overline{\rho}_{\text{N}_2}\right)\right)^2}{\overline{\rho}_{\text{N}_2}}\right), \tag{4.246}$$

$$= f(\overline{\rho}_{\text{N}_2}). \tag{4.247}$$

The system is at equilibrium when $f(\overline{\rho}_{\text{N}_2}) = 0$. This is an algebraic function of $\overline{\rho}_{\text{N}_2}$ only and can be plotted. Figure 4.11 gives a plot of $f(\overline{\rho}_{\text{N}_2})$ and shows that it has three potential

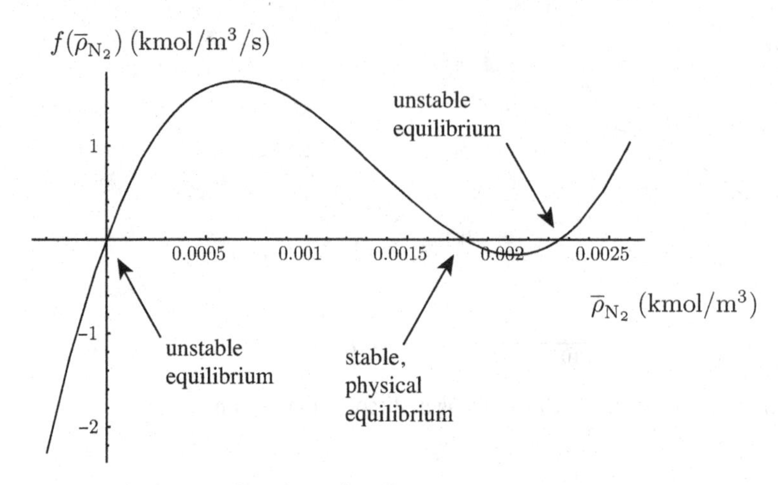


Figure 4.11. Forcing function, $f(\overline{\rho}_{N_2})$, that drives changes of $\overline{\rho}_{N_2}$ as a function of $\overline{\rho}_{N_2}$ in isothermal, isochoric problem.

equilibria. Finding the equilibria requires solving

$$0 = -\left(40398.5\ \frac{\mathrm{m}^3}{\mathrm{kmol}}\right)\overline{\rho}_{N_2}^2 \times \left(1 - \frac{1}{0.000112942\ \frac{\mathrm{kmol}}{\mathrm{m}^3}}\frac{\left(2\left(2.00465\times 10^{-3}\ \frac{\mathrm{kmol}}{\mathrm{m}^3}\right) - \overline{\rho}_{N_2}\right)^2}{\overline{\rho}_{N_2}}\right). \tag{4.248}$$

The three roots are

$$\overline{\rho}_{N_2} = \underbrace{0\ \frac{\mathrm{kmol}}{\mathrm{m}^3}}_{\text{unstable}},\quad \underbrace{0.00178044\ \frac{\mathrm{kmol}}{\mathrm{m}^3}}_{\text{stable}}, \underbrace{0.00225710\ \frac{\mathrm{kmol}}{\mathrm{m}^3}}_{\text{unstable}}. \tag{4.249}$$

By inspection of the topology of Fig. 4.11, the only stable root is 0.00178044 kmol/m$^3$. This root agrees precisely with the equilibrium value found in an earlier example for the same problem conditions; see Eq. (3.330). Small perturbations from this equilibrium induce the forcing function to supply dynamics that restore the system to its original equilibrium state. Small perturbations from the unstable equilibria induce nonrestoring dynamics. For this root, one can then determine that the stable equilibrium value of $\overline{\rho}_N = 0.000448426$ kmol/m$^3$.

One can examine this stability more formally. Define an equilibrium concentration $\overline{\rho}_{N_2}^{eq}$ such that

$$f(\overline{\rho}_{N_2}^{eq}) = 0. \tag{4.250}$$

Now, perform a Taylor series of $f(\overline{\rho}_{N_2})$ about $\overline{\rho}_{N_2} = \overline{\rho}_{N_2}^{eq}$:

$$f(\overline{\rho}_{N_2}) \sim \underbrace{f(\overline{\rho}_{N_2}^{eq})}_{0} + \left.\frac{df}{d\overline{\rho}_{N_2}}\right|_{\overline{\rho}_{N_2}=\overline{\rho}_{N_2}^{eq}}(\overline{\rho}_{N_2} - \overline{\rho}_{N_2}^{eq}) + \frac{1}{2}\frac{d^2 f}{d\overline{\rho}_{N_2}^2}(\overline{\rho}_{N_2} - \overline{\rho}_{N_2}^{eq})^2 + \cdots \tag{4.251}$$

Now, the first term of the Taylor series is zero by construction. Next, neglect all higher-order terms as small so that the approximation becomes

$$f(\overline{\rho}_{N_2}) \sim \left.\frac{df}{d\overline{\rho}_{N_2}}\right|_{\overline{\rho}_{N_2}=\overline{\rho}_{N_2}^{eq}}(\overline{\rho}_{N_2} - \overline{\rho}_{N_2}^{eq}). \tag{4.252}$$

Thus, near equilibrium, one can write

$$\frac{d\overline{\rho}_{N_2}}{dt} \sim \left.\frac{df}{d\overline{\rho}_{N_2}}\right|_{\overline{\rho}_{N_2}=\overline{\rho}^{eq}_{N_2}} (\overline{\rho}_{N_2} - \overline{\rho}^{eq}_{N_2}). \tag{4.253}$$

Because the derivative of a constant is zero, one can also write the equation as

$$\frac{d}{dt}(\overline{\rho}_{N_2} - \overline{\rho}^{eq}_{N_2}) \sim \left.\frac{df}{d\overline{\rho}_{N_2}}\right|_{\overline{\rho}_{N_2}=\overline{\rho}^{eq}_{N_2}} (\overline{\rho}_{N_2} - \overline{\rho}^{eq}_{N_2}). \tag{4.254}$$

This has a solution, valid near the equilibrium point, of

$$(\overline{\rho}_{N_2} - \overline{\rho}^{eq}_{N_2}) = C \exp\left( \left.\frac{df}{d\overline{\rho}_{N_2}}\right|_{\overline{\rho}_{N_2}=\overline{\rho}^{eq}_{N_2}} t \right), \tag{4.255}$$

$$\overline{\rho}_{N_2} = \overline{\rho}^{eq}_{N_2} + C \exp\left( \left.\frac{df}{d\overline{\rho}_{N_2}}\right|_{\overline{\rho}_{N_2}=\overline{\rho}^{eq}_{N_2}} t \right). \tag{4.256}$$

Here, $C$ is some constant whose value is not important for this discussion. If the slope of $f$ is positive, that is,

$$\left.\frac{df}{d\overline{\rho}_{N_2}}\right|_{\overline{\rho}_{N_2}=\overline{\rho}^{eq}_{N_2}} > 0, \qquad \text{unstable,} \tag{4.257}$$

the equilibrium will be *unstable*. That is, a perturbation will grow without bound as $t \to \infty$. If the slope is zero,

$$\left.\frac{df}{d\overline{\rho}_{N_2}}\right|_{\overline{\rho}_{N_2}=\overline{\rho}^{eq}_{N_2}} = 0, \qquad \text{neutrally stable,} \tag{4.258}$$

the solution is stable in that there is no unbounded growth and moreover is known as *neutrally stable*. If the slope is negative,

$$\left.\frac{df}{d\overline{\rho}_{N_2}}\right|_{\overline{\rho}_{N_2}=\overline{\rho}^{eq}_{N_2}} < 0, \qquad \text{asymptotically stable,} \tag{4.259}$$

the solution is stable in that there is no unbounded growth and moreover is known as *asymptotically stable*.

A solution via numerical integration is found for Eq. (4.246). The solution for $\overline{\rho}_{N_2}$, along with $\overline{\rho}_N$, is plotted in Fig. 4.12. Linearization of Eq. (4.246) about the equilibrium

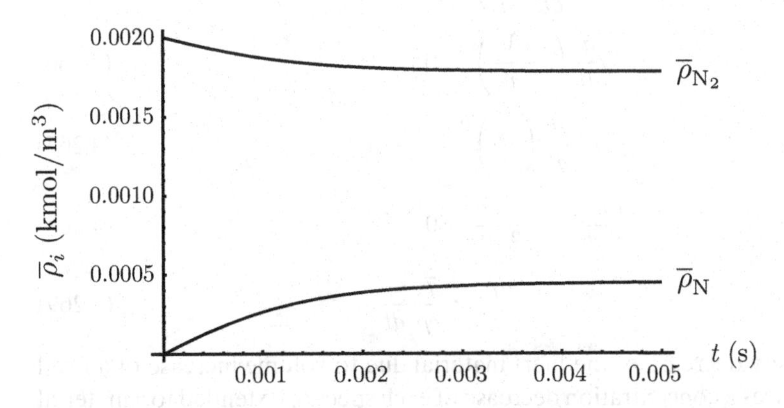


Figure 4.12. $\overline{\rho}_{N_2}(t)$ and $\overline{\rho}_N(t)$ in isothermal, isochoric nitrogen dissociation problem.

state gives rise to the locally linearly valid

$$\frac{d}{dt}\left(\overline{\rho}_{N_2} - 0.00178044\right) = -1214.25(\overline{\rho}_{N_2} - 0.00178044) + \cdots \tag{4.260}$$

This has local asymptotically stable solution

$$\overline{\rho}_{N_2} = 0.00178044 + C\exp\left(-1214.25t\right). \tag{4.261}$$

Here, $C$ is some integration constant whose value is irrelevant for this analysis. The time scale of relaxation $\tau$ is the time when the argument of the exponential is $-1$, that is,

$$\tau = \frac{1}{1214.25\ \text{s}^{-1}} = 8.24 \times 10^{-4}\ \text{s}. \tag{4.262}$$

One usually finds this time scale to have high sensitivity to temperature, with high temperatures giving fast time constants and thus fast reactions.

The equilibrium values agree exactly with those found in the earlier example. Here, the kinetics provide the details of *how much time it takes* to achieve equilibrium. This is one of the key questions of nonequilibrium thermodynamics.

### 4.5.2 Isobaric Systems

The analysis of the previous section is important as it is easily extended to both (1) a formulation of the partial differential equations describing spatially inhomogeneous combustion (see Chapter 7) and (2) a computational grid with fixed volume elements in reactive fluid flow problems. However, there is another important spatially homogeneous problem in which the formulation needs slight modification: an *isobaric* reaction for which $P$ is equal to a constant. Again, in this section, only *closed isothermal* conditions will be considered.

In an isobaric problem, there can be volume change. Consider first the problem of isobaric expansion of a closed inert mixture. In such a mixture, the total number of moles of each species must be constant, so one gets

$$\frac{dn_i}{dt} = 0, \qquad \text{inert, closed isobaric mixture.} \tag{4.263}$$

Now, carry out a sequence of operations, realizing the total mass $m$ is also constant:

$$\frac{1}{m}\frac{d}{dt}\left(n_i\right) = 0, \tag{4.264}$$

$$\frac{d}{dt}\left(\frac{n_i}{m}\right) = 0, \tag{4.265}$$

$$\frac{d}{dt}\left(\frac{n_i}{V}\frac{V}{m}\right) = 0, \tag{4.266}$$

$$\frac{d}{dt}\left(\frac{\overline{\rho}_i}{\rho}\right) = 0, \tag{4.267}$$

$$\frac{1}{\rho}\frac{d\overline{\rho}_i}{dt} - \frac{\overline{\rho}_i}{\rho^2}\frac{d\rho}{dt} = 0, \tag{4.268}$$

$$\frac{d\overline{\rho}_i}{dt} = \frac{\overline{\rho}_i}{\rho}\frac{d\rho}{dt}. \tag{4.269}$$

So, a global density decrease of the inert material due to volume increase of a fixed mass system induces a concentration decrease of each species. Extended to a material with a single reaction rate $r$, one could say either

$$\frac{d\overline{\rho}_i}{dt} = \nu_i r + \frac{\overline{\rho}_i}{\rho}\frac{d\rho}{dt}, \qquad \text{or} \tag{4.270}$$

$$\frac{d}{dt}\left(\frac{\overline{\rho}_i}{\rho}\right) = \frac{1}{\rho}\nu_i r = \frac{\dot{\omega}_i}{\rho}. \tag{4.271}$$

Equation (4.271) is consistent with Eq. (4.213) and is actually valid for general systems with variable density, temperature, and pressure. However, in this section, it is required that pressure and temperature be constant. Now, differentiate with respect to time the isobaric, isothermal, ideal gas law to get the density derivative:

$$P = \sum_{i=1}^{N} \overline{\rho}_i \overline{R} T, \tag{4.272}$$

$$0 = \sum_{i=1}^{N} \overline{R} T \frac{d\overline{\rho}_i}{dt}, \tag{4.273}$$

$$= \sum_{i=1}^{N} \frac{d\overline{\rho}_i}{dt} = \sum_{i=1}^{N}\left(\nu_i r + \frac{\overline{\rho}_i}{\rho}\frac{d\rho}{dt}\right) = r\sum_{i=1}^{N}\nu_i + \frac{1}{\rho}\frac{d\rho}{dt}\sum_{i=1}^{N}\overline{\rho}_i, \tag{4.274}$$

$$\frac{d\rho}{dt} = \frac{-r\sum_{i=1}^{N}\nu_i}{\sum_{i=1}^{N}\frac{\overline{\rho}_i}{\rho}} = -\frac{\rho r\sum_{i=1}^{N}\nu_i}{\sum_{i=1}^{N}\overline{\rho}_i} = -\frac{\rho r\sum_{i=1}^{N}\nu_i}{\frac{P}{\overline{R}T}}, \tag{4.275}$$

$$= -\frac{\rho \overline{R} T r\sum_{i=1}^{N}\nu_i}{P} = -\frac{\rho \overline{R} T r\sum_{k=1}^{N}\nu_k}{P}. \tag{4.276}$$

If there is no net number change in the reaction, $\sum_{k=1}^{N}\nu_k = 0$, the isobaric, isothermal reaction also guarantees there would be no density or volume change. It is convenient to define the net number change in the elementary reaction as $\Delta\mathsf{n}$:

$$\Delta\mathsf{n} \equiv \sum_{k=1}^{N}\nu_k. \tag{4.277}$$

Here, $\Delta\mathsf{n}$ is taken to be a dimensionless pure number. It is associated with the number change in the elementary reaction and not the actual mol change in a physical system; it is, however, proportional to the actual mol change.

Now, use Eq. (4.276) to eliminate the density derivative in Eq. (4.270) to get

$$\frac{d\overline{\rho}_i}{dt} = \nu_i r - \frac{\overline{\rho}_i}{\rho}\frac{\rho \overline{R} T r \Delta\mathsf{n}}{P}, \tag{4.278}$$

$$= r\left(\underbrace{\nu_i}_{\text{reaction}} - \underbrace{\frac{\overline{\rho}_i \overline{R} T}{P}\Delta\mathsf{n}}_{\text{expansion}}\right) = r\left(\underbrace{\nu_i}_{\text{reaction}} - \underbrace{y_i\Delta\mathsf{n}}_{\text{expansion}}\right). \tag{4.279}$$

There are two terms dictating the rate change. The first, a reaction effect, is precisely the same term that drove the isochoric reaction. The second is due to the fact that the volume can change if the number of moles change, and this induces an intrinsic change in concentration. The term $\overline{\rho}_i \overline{R} T / P = y_i$ is the mol fraction.

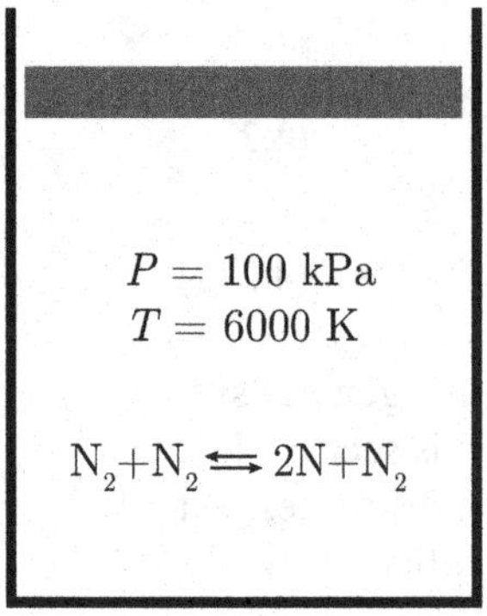

$P = 100$ kPa
$T = 6000$ K

$N_2 + N_2 \leftrightarrows 2N + N_2$

Figure 4.13. Configuration for isobaric, isothermal nitrogen dissociation reaction.

**EXAMPLE 4.14**

Study a variant of the nitrogen dissociation problem considered in an earlier example in which at $t = 0$ s, 1 kmol of $N_2$ exists at $P = 100$ kPa, and $T = 6000$ K. In this case, take the reaction to be *isothermal* and *isobaric*. Consider again the elementary nitrogen dissociation reaction

$$N_2 + N_2 \leftrightarrows 2N + N_2, \tag{4.280}$$

which has kinetic rate parameters of the previous example problem. Figure 4.13 gives a sketch of the configuration.

As in the previous example, the initial $N_2$ concentration and volume is

$$\overline{\rho}_{N_2}\big|_{t=0} = 2.00465 \times 10^{-3} \frac{\text{kmol}}{\text{m}^3}, \qquad V\big|_{t=0} = 4.9884 \times 10^2 \text{ m}^3. \tag{4.281}$$

In this isobaric process, one always has $P = 100$ kPa. Now, in general,

$$P = \overline{R}T(\overline{\rho}_{N_2} + \overline{\rho}_N), \tag{4.282}$$

therefore, one can write $\overline{\rho}_N$ in terms of $\overline{\rho}_{N_2}$:

$$\overline{\rho}_N = \frac{P}{\overline{R}T} - \overline{\rho}_{N_2}, \tag{4.283}$$

$$= \frac{100 \text{ kPa}}{\left(8.314 \frac{\text{kJ}}{\text{kmol K}}\right)(6000 \text{ K})} - \overline{\rho}_{N_2} = \left(2.00465 \times 10^{-3} \frac{\text{kmol}}{\text{m}^3}\right) - \overline{\rho}_{N_2}. \tag{4.284}$$

Then, the equations for kinetics of a single isobaric isothermal reaction, Eq. (4.279), in conjunction with Eq. (4.218), reduce for $N_2$ molar concentration to

$$\frac{d\overline{\rho}_{N_2}}{dt} = \underbrace{\left(aT^\beta \exp\left(\frac{-\mathcal{E}}{\overline{R}T}\right)(\overline{\rho}_{N_2})^{\nu'_{N_2}}(\overline{\rho}_N)^{\nu'_N}\left(1 - \frac{1}{K_c}(\overline{\rho}_{N_2})^{\nu_{N_2}}(\overline{\rho}_N)^{\nu_N}\right)\right)}_{r}$$
$$\times \left(\nu_{N_2} - \frac{\overline{\rho}_{N_2}\overline{R}T}{P}(\nu_{N_2} + \nu_N)\right). \tag{4.285}$$

Realizing that $\nu'_{N_2} = 2$, $\nu'_N = 0$, $\nu_{N_2} = -1$, and $\nu_N = 2$, one gets

$$\frac{d\overline{\rho}_{N_2}}{dt} = \underbrace{aT^\beta \exp\left(\frac{-\mathcal{E}}{\overline{R}T}\right)}_{k(T)}\overline{\rho}_{N_2}^2\left(1 - \frac{1}{K_c}\frac{\overline{\rho}_N^2}{\overline{\rho}_{N_2}}\right)\left(-1 - \frac{\overline{\rho}_{N_2}\overline{R}T}{P}\right). \tag{4.286}$$

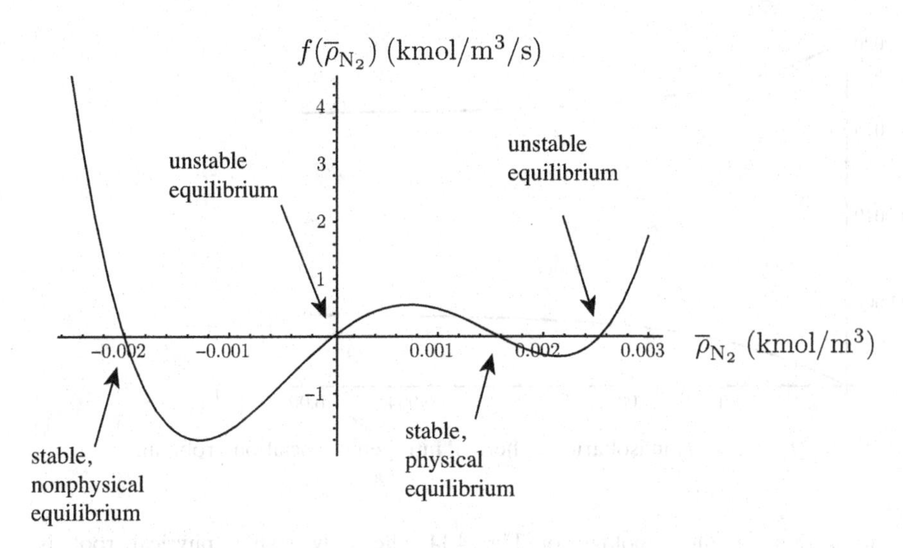


Figure 4.14. Forcing function, $f(\overline{\rho}_{N_2})$, that drives changes of $\overline{\rho}_{N_2}$ as a function of $\overline{\rho}_{N_2}$ in isothermal, isobaric problem.

The temperature dependency of the reaction is unchanged from the previous example:

$$k(6000) = 40398.5 \frac{\text{m}^3}{\text{kmol s}}. \tag{4.287}$$

The value of $K_c$ is also unchanged from the previous example, with $K_c = 0.000112942$ kmol/m$^3$. The differential equation for $N_2$ evolution is then given by

$$\begin{aligned}\frac{d\overline{\rho}_{N_2}}{dt} &= \left(40398.5 \frac{\text{m}^3}{\text{kmol}}\right)\overline{\rho}_{N_2}^2 \\ &\times \left(1 - \frac{1}{0.000112942 \frac{\text{kmol}}{\text{m}^3}} \frac{\left(\left(2.00465 \times 10^{-3} \frac{\text{kmol}}{\text{m}^3}\right) - \overline{\rho}_{N_2}\right)^2}{\overline{\rho}_{N_2}}\right) \\ &\times \left(-1 - \frac{\overline{\rho}_{N_2}\left(8.314 \frac{\text{kJ}}{\text{kmol K}}\right)(6000 \text{ K})}{100 \text{ kPa}}\right), \qquad (4.288)\\ &\equiv f(\overline{\rho}_{N_2}). \qquad (4.289)\end{aligned}$$

The system is at equilibrium when $f(\overline{\rho}_{N_2}) = 0$. This is an algebraic function of $\overline{\rho}_{N_2}$ only and can be plotted. Figure 4.14 gives a plot of $f(\overline{\rho}_{N_2})$ and shows that it has four equilibrium points. Solving for the equilibria requires solving

$$\begin{aligned}0 = \left(40398.5 \frac{\text{m}^3}{\text{kmol}}\right)\overline{\rho}_{N_2}^2 \left(1 - \frac{1}{0.000112942 \frac{\text{kmol}}{\text{m}^3}} \frac{\left(\left(2.00465 \times 10^{-3} \frac{\text{kmol}}{\text{m}^3}\right) - \overline{\rho}_{N_2}\right)^2}{\overline{\rho}_{N_2}}\right) \\ \times \left(-1 - \frac{\overline{\rho}_{N_2}\left(8.314 \frac{\text{kJ}}{\text{kmol K}}\right)(6000 \text{ K})}{100 \text{ kPa}}\right). \qquad (4.290)\end{aligned}$$

The four roots are

$$\begin{aligned}\overline{\rho}_{N_2} = \underbrace{-0.00200465 \frac{\text{kmol}}{\text{m}^3}}_{\text{stable, nonphysical}}, \underbrace{0 \frac{\text{kmol}}{\text{m}^3}}_{\text{unstable}}, \\ \underbrace{0.00158196 \frac{\text{kmol}}{\text{m}^3}}_{\text{stable, physical}}, \underbrace{0.00254029 \frac{\text{kmol}}{\text{m}^3}}_{\text{unstable}}. \qquad (4.291)\end{aligned}$$

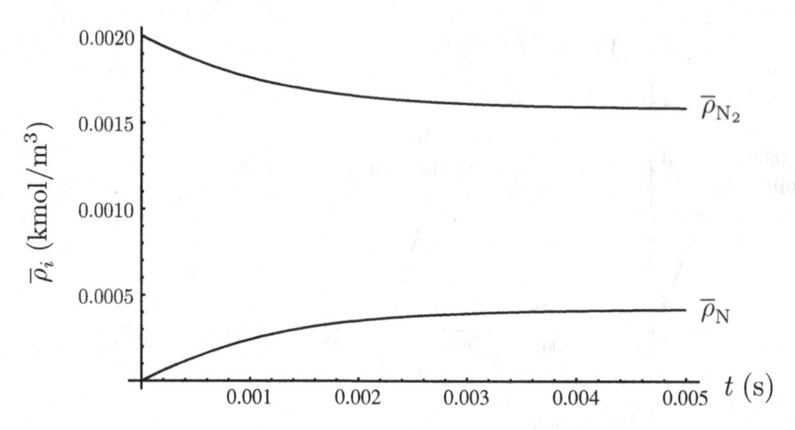


Figure 4.15. $\overline{\rho}_{N_2}(t)$ and $\overline{\rho}_N(t)$ in isobaric, isothermal nitrogen dissociation problem.

By inspection of the topology of Fig. 4.14, the only stable, physical root is $0.00158196\ \text{kmol/m}^3$. Its value agrees with that found in earlier analysis; see Eq. (3.366). Small perturbations from this equilibrium induce the forcing function to supply dynamics that restore the system to its original equilibrium state. Small perturbations from the unstable equilibria induce nonrestoring dynamics. For this root, one can then determine that the stable equilibrium value of $\overline{\rho}_N = 0.000422693\ \text{kmol/m}^3$.

A numerical solution is found for Eq. (4.290). The solution for $\overline{\rho}_{N_2}$, along with $\overline{\rho}_N$, is plotted in Fig. 4.15. Linearization of Eq. (4.290) about the equilibrium gives rise to the locally linearly valid

$$\frac{d}{dt}\left(\overline{\rho}_{N_2} - 0.00158196\right) = -970.208(\overline{\rho}_{N_2} - 0.00158196) + \cdots \tag{4.292}$$

This has local solution

$$\overline{\rho}_{N_2} = 0.00158196 + C\exp\left(-970.208t\right). \tag{4.293}$$

Again, $C$ is an irrelevant integration constant. The time scale of relaxation $\tau$ is the time when the argument of the exponential is $-1$, that is,

$$\tau = \frac{1}{970.208\ \text{s}^{-1}} = 1.03 \times 10^{-3}\ \text{s}. \tag{4.294}$$

The time constant for the isobaric combustion is about a factor 1.25 greater than for isochoric combustion under the otherwise identical conditions. The equilibrium values agree exactly with those found in the earlier example. Again, the kinetics provide the details of *how much time it takes* to achieve equilibrium.

## 4.6 Some Conservation and Evolution Equations

Here, a few useful global conservation and evolution equations are presented.

### 4.6.1 Total Mass Conservation: Isochoric Reaction

One can show that the isochoric reaction rate model, Eq. (4.217), satisfies the principle of mixture mass conservation. Begin with Eq. (4.217) in a compact form, using the definition of the reaction rate $r$, Eq. (4.219), and perform the following operations:

$$\frac{d\overline{\rho}_i}{dt} = \nu_i r, \tag{4.295}$$

$$\frac{d}{dt}\left(\frac{\rho Y_i}{M_i}\right) = \nu_i r, \tag{4.296}$$

$$\frac{d}{dt}(\rho Y_i) = \nu_i M_i r = \nu_i \underbrace{\sum_{l=1}^{L} \mathcal{M}_l \phi_{li}\, r}_{M_i} = \sum_{l=1}^{L} \mathcal{M}_l \phi_{li} \nu_i r, \tag{4.297}$$

$$\sum_{i=1}^{N} \frac{d}{dt}(\rho Y_i) = \sum_{i=1}^{N}\sum_{l=1}^{L} \mathcal{M}_l \phi_{li} \nu_i r, \tag{4.298}$$

$$\frac{d}{dt}\left(\rho \underbrace{\sum_{i=1}^{N} Y_i}_{1}\right) = \sum_{l=1}^{L}\sum_{i=1}^{N} \mathcal{M}_l \phi_{li} \nu_i r, \tag{4.299}$$

$$\frac{d\rho}{dt} = r \sum_{l=1}^{L} \mathcal{M}_l \underbrace{\sum_{i=1}^{N} \phi_{li} \nu_i}_{0} = 0, \tag{4.300}$$

$$\rho = \text{constant}. \tag{4.301}$$

So, $\rho =$ is a constant for the closed isochoric system. The term $\sum_{i=1}^{N} \phi_{li}\nu_i = 0$ because of Eq. (4.19). *The use of the linear algebra formalism for both molecular mass and the element-species matrix was essential to verify this obvious notion.*

### 4.6.2 Element Mass Conservation: Isochoric Reaction

Through a similar series of operations, one can show that the mass of each element, $l = 1, \ldots, L$, is conserved in this reaction, which is chemical, not nuclear. Once again, begin with Eq. (4.219) and perform a set of operations

$$\frac{d\overline{\rho}_i}{dt} = \nu_i r, \tag{4.302}$$

$$\phi_{li}\frac{d\overline{\rho}_i}{dt} = \phi_{li}\nu_i r, \qquad l = 1, \ldots, L, \tag{4.303}$$

$$\frac{d}{dt}(\phi_{li}\overline{\rho}_i) = r\phi_{li}\nu_i, \qquad l = 1, \ldots, L, \tag{4.304}$$

$$\sum_{i=1}^{N} \frac{d}{dt}(\phi_{li}\overline{\rho}_i) = \sum_{i=1}^{N} r\phi_{li}\nu_i, \qquad l = 1, \ldots, L, \tag{4.305}$$

$$\frac{d}{dt}\left(\sum_{i=1}^{N} \phi_{li}\overline{\rho}_i\right) = r \underbrace{\sum_{i=1}^{N} \phi_{li}\nu_i}_{0} = 0, \qquad l = 1, \ldots, L. \tag{4.306}$$

The term $\sum_{i=1}^{N} \phi_{li}\overline{\rho}_i$ represents the number of moles of element $l$ per unit volume, by the following analysis:

$$\sum_{i=1}^{N} \phi_{li}\overline{\rho}_i = \sum_{i=1}^{N} \frac{\text{moles element } l}{\text{moles species } i} \frac{\text{moles species } i}{\text{volume}}, \tag{4.307}$$

$$= \frac{\text{moles element } l}{\text{volume}} \equiv \overline{\rho}_l{}^e. \tag{4.308}$$

Here, the *element mol density*, $\overline{\rho}_l{}^e$, for element $l$ has been defined. So, the element concentration for each element remains constant in a constant volume reaction process:

$$\frac{d\overline{\rho}_l{}^e}{dt} = 0, \qquad l = 1, \ldots, L, \tag{4.309}$$

$$\overline{\rho}_l{}^e = \text{constant}, \qquad l = 1, \ldots, L. \tag{4.310}$$

One can also multiply by the element mass, $\mathcal{M}_l$, to get the *element mass density*, $\rho_l^e$:

$$\rho_l^e \equiv \mathcal{M}_l \overline{\rho}_l{}^e, \qquad l = 1, \ldots, L. \tag{4.311}$$

Because $\mathcal{M}_l$ is a constant, one can incorporate this definition into Eq. (4.309) to get

$$\frac{d\rho_l^e}{dt} = 0, \qquad l = 1, \ldots, L, \tag{4.312}$$

$$\rho_l^e = \text{constant}, \qquad l = 1, \ldots, L. \tag{4.313}$$

The element mass density remains constant in the constant volume reaction. One could also simply say that because the elements' density is constant, and the mixture is simply a sum of the elements, the mixture density is conserved as well.

### 4.6.3 Energy Conservation: Adiabatic, Isochoric Reaction

Consider a simple application of the first law of thermodynamics to reaction kinetics: that of a closed, adiabatic, isochoric combustion process in a mixture of ideal gases. One may be interested in the rate of temperature change. First, because the system is closed, there can be no mass change, and because the system is isochoric, the total volume is a nonzero constant; hence,

$$\frac{dm}{dt} = \frac{d}{dt}(\rho V) = V\frac{d\rho}{dt} = 0, \tag{4.314}$$

$$\frac{d\rho}{dt} = 0. \tag{4.315}$$

For such a process, the first law of thermodynamics is

$$\frac{dE}{dt} = \dot{Q} - \dot{W}. \tag{4.316}$$

But there is no enthalpy or work in the adiabatic isochoric process, so one gets

$$\frac{dE}{dt} = \frac{d}{dt}(me) = m\frac{de}{dt} + e\underbrace{\frac{dm}{dt}}_{0} = 0, \tag{4.317}$$

$$\frac{de}{dt} = 0. \tag{4.318}$$

Thus, for the mixture of ideal gases, $e(T, \overline{\rho}_1, \ldots, \overline{\rho}_N) = e_0$. One can see how reaction rates affect temperature changes by expanding the derivative in Eq. (4.318):

$$\frac{d}{dt}\left(\sum_{i=1}^{N} Y_i e_i\right) = \sum_{i=1}^{N} \frac{d}{dt}(Y_i e_i) = 0, \tag{4.319}$$

$$\sum_{i=1}^{N}\left(Y_i \frac{de_i}{dt} + e_i \frac{dY_i}{dt}\right) = 0, \tag{4.320}$$

$$\sum_{i=1}^{N}\left(Y_i \underbrace{\frac{de_i}{dT}}_{c_{vi}} \frac{dT}{dt} + e_i \frac{dY_i}{dt}\right) = 0, \tag{4.321}$$

$$\sum_{i=1}^{N}\left(Y_i c_{vi} \frac{dT}{dt} + e_i \frac{dY_i}{dt}\right) = 0, \tag{4.322}$$

$$\frac{dT}{dt}\underbrace{\sum_{i=1}^{N} Y_i c_{vi}}_{c_v} = -\sum_{i=1}^{N} e_i \frac{dY_i}{dt}, \tag{4.323}$$

$$c_v \frac{dT}{dt} = -\sum_{i=1}^{N} e_i \frac{d}{dt}\left(\frac{M_i \overline{\rho}_i}{\rho}\right), \tag{4.324}$$

$$\rho c_v \frac{dT}{dt} = -\sum_{i=1}^{N} e_i M_i \frac{d\overline{\rho}_i}{dt}, \tag{4.325}$$

$$= -\sum_{i=1}^{N} e_i M_i \nu_i r, \tag{4.326}$$

$$\frac{dT}{dt} = -\frac{r \sum_{i=1}^{N} \nu_i \overline{e}_i}{\rho c_v}. \tag{4.327}$$

If one defines the net energy change of the reaction as

$$\Delta E \equiv \sum_{i=1}^{N} \nu_i \overline{e}_i, \tag{4.328}$$

one then gets

$$\frac{dT}{dt} = -\frac{r \Delta E}{\rho c_v}. \tag{4.329}$$

If the reaction is going forward, so $r > 0$, and that is a direction in which the net molar energy change is negative, then the temperature will rise.

### 4.6.4 Energy Conservation: Adiabatic, Isobaric Reaction

Solving for the reaction dynamics in an adiabatic, isobaric system requires some nonobvious manipulations. First, the first law of thermodynamics says that $dE = \delta Q - \delta W$. Because the process is adiabatic, one has $\delta Q = 0$, so $dE + P\,dV = 0$. Because it is isobaric, one gets $d(E + PV) = 0$, or $dH = 0$. So, the total enthalpy is constant. Then

$$\frac{d}{dt}H = \frac{d}{dt}(mh) = 0, \tag{4.330}$$

$$\frac{dh}{dt} = 0, \tag{4.331}$$

$$\frac{d}{dt}\left(\sum_{i=1}^{N} Y_i h_i\right) = \sum_{i=1}^{N} \frac{d}{dt}(Y_i h_i) = 0, \tag{4.332}$$

$$\sum_{i=1}^{N} Y_i \frac{dh_i}{dt} + h_i \frac{dY_i}{dt} = \sum_{i=1}^{N} Y_i \underbrace{\frac{dh_i}{dT}}_{c_{Pi}} \frac{dT}{dt} + h_i \frac{dY_i}{dt} = 0, \tag{4.333}$$

$$\sum_{i=1}^{N} Y_i c_{Pi} \frac{dT}{dt} + \sum_{i=1}^{N} h_i \frac{dY_i}{dt} = \frac{dT}{dt} \underbrace{\sum_{i=1}^{N} Y_i c_{Pi}}_{c_P} + \sum_{i=1}^{N} h_i \frac{dY_i}{dt} = 0, \tag{4.334}$$

$$c_P \frac{dT}{dt} + \sum_{i=1}^{N} h_i \frac{d}{dt}\left(\frac{\overline{\rho}_i M_i}{\rho}\right) = c_P \frac{dT}{dt} + \sum_{i=1}^{N} h_i M_i \frac{d}{dt}\left(\frac{\overline{\rho}_i}{\rho}\right) = 0. \tag{4.335}$$

Now, use Eq. (4.271) to eliminate the term in Eq. (4.335) involving molar concentration derivatives to get

$$c_P \frac{dT}{dt} + \sum_{i=1}^{N} \overline{h}_i \frac{\nu_i r}{\rho} = 0, \tag{4.336}$$

$$\frac{dT}{dt} = -\frac{r \sum_{i=1}^{N} \overline{h}_i \nu_i}{\rho c_P}. \tag{4.337}$$

So, the temperature derivative is known as an algebraic function. If one defines the net enthalpy change as

$$\Delta H \equiv \sum_{i=1}^{N} \overline{h}_i \nu_i, \tag{4.338}$$

one finds that Eq. (4.337) transforms to

$$\frac{dT}{dt} = -\frac{r \Delta H}{\rho c_P}, \tag{4.339}$$

or

$$\rho c_P \frac{dT}{dt} = -r \Delta H. \tag{4.340}$$

Equation (4.340) is in a form that can easily be compared to a form to be derived later when we account for multiple reactions, variable pressure, advection, and diffusion; see Eq. (7.97).

Now, differentiate the isobaric ideal gas law to get the density derivative:

$$P = \sum_{i=1}^{N} \overline{\rho}_i \overline{R} T, \tag{4.341}$$

$$0 = \sum_{i=1}^{N} \overline{\rho}_i \overline{R} \frac{dT}{dt} + \sum_{i=1}^{N} \overline{R} T \frac{d\overline{\rho}_i}{dt}, \tag{4.342}$$

$$= \frac{dT}{dt} \sum_{i=1}^{N} \overline{\rho}_i + \sum_{i=1}^{N} T \left( \nu_i r + \frac{\overline{\rho}_i}{\rho} \frac{d\rho}{dt} \right), \tag{4.343}$$

$$= \frac{1}{T} \frac{dT}{dt} \sum_{i=1}^{N} \overline{\rho}_i + r \sum_{i=1}^{N} \nu_i + \frac{1}{\rho} \frac{d\rho}{dt} \sum_{i=1}^{N} \overline{\rho}_i, \tag{4.344}$$

$$\frac{d\rho}{dt} = \frac{-\frac{1}{T} \frac{dT}{dt} \sum_{i=1}^{N} \overline{\rho}_i - r \sum_{i=1}^{N} \nu_i}{\sum_{i=1}^{N} \frac{\overline{\rho}_i}{\rho}}. \tag{4.345}$$

One takes $dT/dt$ from Eq. (4.337) to get

$$\frac{d\rho}{dt} = \frac{\frac{1}{T} \frac{r \sum_{i=1}^{N} \overline{h}_i \nu_i}{\rho c_P} \sum_{i=1}^{N} \overline{\rho}_i - r \sum_{i=1}^{N} \nu_i}{\sum_{i=1}^{N} \frac{\overline{\rho}_i}{\rho}}. \tag{4.346}$$

Now, recall from Eqs. (2.186) and (2.192) that $\overline{\rho} = \rho/M$ and $\overline{c}_P = c_P M$, so $\overline{\rho}\,\overline{c}_P = \rho c_P$. Then Eq. (4.346) can be reduced slightly:

$$\frac{d\rho}{dt} = r\rho \frac{\frac{\sum_{i=1}^{N} \overline{h}_i \nu_i}{\overline{c}_P T} \overbrace{\sum_{i=1}^{N} \frac{\overline{\rho}_i}{\overline{\rho}}}^{1} - \sum_{i=1}^{N} \nu_i}{\sum_{i=1}^{N} \overline{\rho}_i}, \tag{4.347}$$

$$= r\rho \frac{\sum_{i=1}^{N} \frac{\overline{h}_i \nu_i}{\overline{c}_P T} - \sum_{i=1}^{N} \nu_i}{\sum_{i=1}^{N} \overline{\rho}_i} = r\rho \frac{\sum_{i=1}^{N} \nu_i \left( \frac{\overline{h}_i}{\overline{c}_P T} - 1 \right)}{\frac{P}{\overline{R} T}}, \tag{4.348}$$

$$= r \frac{\rho \overline{R} T}{P} \sum_{i=1}^{N} \nu_i \left( \frac{\overline{h}_i}{\overline{c}_P T} - 1 \right) = r M \sum_{i=1}^{N} \nu_i \left( \frac{\overline{h}_i}{\overline{c}_P T} - 1 \right), \tag{4.349}$$

where $M$ is the mean molecular mass. For exothermic reaction $\sum_{i=1}^{N} \nu_i \overline{h}_i < 0$, so exothermic reaction induces a density decrease as the increased temperature at constant pressure causes the volume to increase. Then, using Eq. (4.349) to eliminate the density derivative in Eq. (4.270), and changing the dummy index from $i$ to $k$, one gets an explicit expression for concentration evolution:

$$\frac{d\overline{\rho}_i}{dt} = \nu_i r + \frac{\overline{\rho}_i}{\rho} r M \sum_{k=1}^{N} \nu_k \left( \frac{\overline{h}_k}{\overline{c}_P T} - 1 \right), \tag{4.350}$$

$$= r \left( \nu_i + \underbrace{\frac{\overline{\rho}_i}{\rho} M}_{y_i} \sum_{k=1}^{N} \nu_k \left( \frac{\overline{h}_k}{\overline{c}_P T} - 1 \right) \right), \tag{4.351}$$

$$= r \left( \nu_i + y_i \sum_{k=1}^{N} \nu_k \left( \frac{\overline{h}_k}{\overline{c}_P T} - 1 \right) \right). \tag{4.352}$$

Defining the change of enthalpy of the reaction as $\Delta H \equiv \sum_{k=1}^{N} \nu_k \overline{h}_k$, and the change of number of the reaction as $\Delta \mathsf{n} \equiv \sum_{k=1}^{N} \nu_k$, one can also say

$$\frac{d\overline{\rho}_i}{dt} = r \left( \nu_i + y_i \left( \frac{\Delta H}{\overline{c}_P T} - \Delta \mathsf{n} \right) \right). \tag{4.353}$$

Exothermic reaction, $\Delta H < 0$, and net number increases, $\Delta \mathsf{n} > 0$, both tend to decrease the molar concentrations of the species in the isobaric reaction.

Last, the evolution of the adiabatic, isobaric system, can be described by the simultaneous, coupled ordinary differential equations: Eqs. (4.337), (4.346), and (4.352). These require numerical solution in general. One could also employ a more fundamental treatment as a differential-algebraic system involving $H = H_1$, $P = P_1 = \overline{R} T \sum_{i=1}^{N} \overline{\rho}_i$ and Eq. (4.270).

### 4.6.5 Irreversible Entropy Production: Clausius-Duhem Inequality

Now, consider whether the kinetics law that has been posed actually satisfies the second law. Consider again Eq. (3.269). There is an algebraic relation on its left side. If it can be shown that this algebraic relation is positive semidefinite, then the second law is satisfied, and the algebraic relation is known as a Clausius[7]-Duhem[8] inequality.

Now, take Eq. (3.269) and perform some straightforward operations on it:

$$dS|_{E,V} = \underbrace{-\frac{1}{T} \sum_{i=1}^{N} \overline{\mu}_i \, dn_i}_{d\sigma} \geq 0, \tag{4.354}$$

$$\left. \frac{dS}{dt} \right|_{E,V} = \dot{\sigma} = -\frac{V}{T} \sum_{i=1}^{N} \overline{\mu}_i \frac{dn_i}{dt} \frac{1}{V} \geq 0, \tag{4.355}$$

$$= -\frac{V}{T} \sum_{i=1}^{N} \overline{\mu}_i \frac{d\overline{\rho}_i}{dt} \geq 0. \tag{4.356}$$

[7] Rudolf Julius Emanuel Clausius (1822–1888), German physicist.
[8] Pierre Maurice Marie Duhem (1861–1916), French physicist.

Now employ Eq. (4.217) to get

$$\dot{\sigma} = -\frac{V}{T}\sum_{i=1}^{N}\overline{\mu}_i\nu_i \underbrace{\times aT^{\beta}\exp\left(\frac{-\mathcal{E}}{\overline{R}T}\right)}_{k(T)}\left(\prod_{k=1}^{N}\overline{\rho}_k^{\nu'_k}\right)\left(1-\frac{1}{K_c}\prod_{k=1}^{N}\overline{\rho}_k^{\nu_k}\right) \geq 0, \tag{4.357}$$

$$= -\frac{V}{T}\sum_{i=1}^{N}\overline{\mu}_i\nu_i k(T)\left(\prod_{k=1}^{N}\overline{\rho}_k^{\nu'_k}\right)\left(1-\frac{1}{K_c}\prod_{k=1}^{N}\overline{\rho}_k^{\nu_k}\right) \geq 0, \tag{4.358}$$

$$= -\frac{V}{T}k(T)\left(\prod_{k=1}^{N}\overline{\rho}_k^{\nu'_k}\right)\left(1-\frac{1}{K_c}\prod_{k=1}^{N}\overline{\rho}_k^{\nu_k}\right)\underbrace{\left(\sum_{i=1}^{N}\overline{\mu}_i\nu_i\right)}_{-\overline{\alpha}} \geq 0. \tag{4.359}$$

Change the dummy index from $k$ back to $i$:

$$\dot{\sigma} = \frac{V}{T}k(T)\left(\prod_{i=1}^{N}\overline{\rho}_i^{\nu'_i}\right)\left(1-\frac{1}{K_c}\prod_{i=1}^{N}\overline{\rho}_i^{\nu_i}\right)\overline{\alpha} \geq 0, \tag{4.360}$$

$$= \frac{V}{T}r\overline{\alpha} = \frac{\overline{\alpha}}{T}\frac{d\zeta}{dt} \geq 0. \tag{4.361}$$

Consider now the chemical affinity $\overline{\alpha}$ term in Eq. (4.359) and expand it so that it has a more useful form:

$$\overline{\alpha} = -\sum_{i=1}^{N}\overline{\mu}_i\nu_i = -\sum_{i=1}^{N}\overline{g}_i\nu_i, \tag{4.362}$$

$$= -\sum_{i=1}^{N}\left(\overline{g}^0_{T,i} + \overline{R}T\ln\left(\frac{P_i}{P_0}\right)\right)\nu_i, \tag{4.363}$$

$$= -\underbrace{\sum_{i=1}^{N}\overline{g}^0_{T,i}\nu_i}_{\Delta G^0} - \overline{R}T\sum_{i=1}^{N}\ln\left(\frac{P_i}{P_0}\right)^{\nu_i}, \tag{4.364}$$

$$= \overline{R}T\left(\underbrace{\frac{-\Delta G^0}{\overline{R}T}}_{\ln K_P} - \sum_{i=1}^{N}\ln\left(\frac{P_i}{P_0}\right)^{\nu_i}\right), \tag{4.365}$$

$$= \overline{R}T\left(\ln K_P - \ln\prod_{i=1}^{N}\left(\frac{P_i}{P_0}\right)^{\nu_i}\right), \tag{4.366}$$

$$= -\overline{R}T\left(\ln\frac{1}{K_P} + \ln\prod_{i=1}^{N}\left(\frac{P_i}{P_0}\right)^{\nu_i}\right), \tag{4.367}$$

$$= -\overline{R}T\ln\left(\frac{1}{K_P}\prod_{i=1}^{N}\left(\frac{P_i}{P_0}\right)^{\nu_i}\right), \tag{4.368}$$

$$= -\overline{R}T \ln \left( \frac{\left(\frac{P_0}{\overline{R}T}\right)^{\sum_{i=1}^{N} \nu_i}}{K_c} \prod_{i=1}^{N} \left(\frac{\overline{\rho}_i \overline{R}T}{P_0}\right)^{\nu_i} \right), \tag{4.369}$$

$$= -\overline{R}T \ln \left( \frac{1}{K_c} \prod_{i=1}^{N} \overline{\rho}_i^{\nu_i} \right). \tag{4.370}$$

Equation (4.370) is the common definition of affinity. Sometimes, it is written as

$$\overline{\alpha} = \overline{R}T \ln \left( \frac{K_c}{\prod_{i=1}^{N} \overline{\rho}_i^{\nu_i}} \right). \tag{4.371}$$

Another form can be found by employing the definition of $K_c$ from Eq. (4.208) to get

$$\overline{\alpha} = -\overline{R}T \ln \left( \left(\frac{P_0}{\overline{R}T}\right)^{-\sum_{i=1}^{N} \nu_i} \exp\left(\frac{\Delta G^0}{\overline{R}T}\right) \prod_{i=1}^{N} \overline{\rho}_i^{\nu_i} \right), \tag{4.372}$$

$$= -\overline{R}T \left( \frac{\Delta G^0}{\overline{R}T} + \ln \left( \left(\frac{P_0}{\overline{R}T}\right)^{-\sum_{i=1}^{N} \nu_i} \prod_{i=1}^{N} \overline{\rho}_i^{\nu_i} \right) \right), \tag{4.373}$$

$$= -\Delta G^0 - \overline{R}T \ln \left( \left(\frac{P_0}{\overline{R}T}\right)^{-\sum_{i=1}^{N} \nu_i} \prod_{i=1}^{N} \overline{\rho}_i^{\nu_i} \right). \tag{4.374}$$

To see clearly that the irreversible entropy production rate is positive semidefinite, substitute Eq. (4.370) into Eq. (4.359) to get

$$\dot{\sigma} = \frac{V}{T} k(T) \left( \prod_{i=1}^{N} \overline{\rho}_i^{\nu_i'} \right) \left( 1 - \frac{1}{K_c} \prod_{i=1}^{N} \overline{\rho}_i^{\nu_i} \right) \left( -\overline{R}T \ln \left( \frac{1}{K_c} \prod_{i=1}^{N} \overline{\rho}_i^{\nu_i} \right) \right) \geq 0, \tag{4.375}$$

$$= -\overline{R} V k(T) \left( \prod_{i=1}^{N} \overline{\rho}_i^{\nu_i'} \right) \left( 1 - \frac{1}{K_c} \prod_{i=1}^{N} \overline{\rho}_i^{\nu_i} \right) \ln \left( \frac{1}{K_c} \prod_{i=1}^{N} \overline{\rho}_i^{\nu_i} \right) \geq 0. \tag{4.376}$$

Define forward and reverse reaction coefficients, $\mathcal{R}'$ and $\mathcal{R}''$, respectively, as

$$\mathcal{R}' \equiv k(T) \prod_{i=1}^{N} \overline{\rho}_i^{\nu_i'}, \qquad \mathcal{R}'' \equiv \frac{k(T)}{K_c} \prod_{i=1}^{N} \overline{\rho}_i^{\nu_i''}. \tag{4.377}$$

Both $\mathcal{R}'$ and $\mathcal{R}''$ have units of mol/cm$^3$/s. It is easy to see that

$$r = \mathcal{R}' - \mathcal{R}''. \tag{4.378}$$

Because $k(T) > 0$, $K_c > 0$, and $\overline{\rho}_i \geq 0$, both $\mathcal{R}' \geq 0$ and $\mathcal{R}'' \geq 0$. Because $\nu_i = \nu_i'' - \nu_i'$, one finds that

$$\frac{1}{K_c} \prod_{i=1}^{N} \overline{\rho}_i^{\nu_i} = \frac{1}{K_c} \frac{k(T)}{k(T)} \prod_{i=1}^{N} \overline{\rho}_i^{\nu_i'' - \nu_i'} = \frac{\mathcal{R}''}{\mathcal{R}'}. \tag{4.379}$$

Then, Eq. (4.376) reduces to

$$\dot{\sigma} = \left.\frac{dS}{dt}\right|_{E,V} = -\overline{R}V\mathcal{R}'\left(1 - \frac{\mathcal{R}''}{\mathcal{R}'}\right)\ln\left(\frac{\mathcal{R}''}{\mathcal{R}'}\right) \geq 0, \tag{4.380}$$

$$= \overline{R}V\left(\mathcal{R}' - \mathcal{R}''\right)\ln\left(\frac{\mathcal{R}'}{\mathcal{R}''}\right) \geq 0. \tag{4.381}$$

Obviously, if the forward rate is greater than the reverse rate $\mathcal{R}' - \mathcal{R}'' > 0$, $\ln(\mathcal{R}'/\mathcal{R}'') > 0$, and the entropy production is positive. If the forward rate is less than the reverse rate, $\mathcal{R}' - \mathcal{R}'' < 0$, $\ln(\mathcal{R}'/\mathcal{R}'') < 0$, and the entropy production is still positive. The production rate is zero when $\mathcal{R}' = \mathcal{R}''$. We have made no requirement that the system be at or near equilibrium. *Thus our selected reaction kinetics law is guaranteed to satisfy the second law of thermodynamics, globally.*

Note that the chemical affinity $\overline{\alpha}$ can be written as

$$\overline{\alpha} = \overline{R}T\ln\left(\frac{\mathcal{R}'}{\mathcal{R}''}\right). \tag{4.382}$$

And so when the forward reaction rate exceeds the reverse, the chemical affinity is positive. It is zero at equilibrium, when the forward reaction rate equals the reverse.

## 4.7 Simple One-Step Kinetics

A common model in theoretical combustion is that of so-called simple one-step kinetics. Such a model, in which the molecular mass does not change, is quantitatively appropriate only for isomerization reactions. However, as a pedagogical tool as well as a qualitative model to capture gross features of real chemistry, it can be valuable. We already employed this model in Sec. 1.3.1 to study thermal explosion theory. In later chapters, this model will be the primary kinetics used to study reaction-diffusion in Chapter 9, advection-reaction-diffusion in Chapter 10, and advection-reaction in Chapter 12. The effects of advection and diffusion are sufficiently complicated that having a simple qualitatively correct reaction model is valuable.

Consider the reversible reaction

$$A \rightleftharpoons B, \tag{4.383}$$

where chemical species $A$ and $B$ have identical molecular masses $M_A = M_B = M$. Consider further the case in which at the initial state, $n_0$, moles of $A$ only are present. Also take the reaction to be isochoric and isothermal. These assumptions can easily be relaxed for more general cases. Specializing, then, Eq. (4.184) for this case, one has

$$n_A = \underbrace{\nu_A}_{-1}\zeta + \underbrace{n_{A0}}_{n_0}, \tag{4.384}$$

$$n_B = \underbrace{\nu_B}_{1}\zeta + \underbrace{n_{B0}}_{0}. \tag{4.385}$$

Thus,

$$n_A = -\zeta + n_0, \quad n_B = \zeta. \tag{4.386}$$

Scale by $n_0$ and define the dimensionless reaction progress as $\lambda \equiv \zeta/n_0$ to get

$$\underbrace{\frac{n_A}{n_0}}_{y_A} = -\lambda + 1, \qquad \underbrace{\frac{n_B}{n_0}}_{y_B} = \lambda. \tag{4.387}$$

In terms of the mol fractions, then, one has

$$y_A = 1 - \lambda, \qquad y_B = \lambda. \tag{4.388}$$

The reaction kinetics for each species reduce to

$$\frac{d\overline{\rho}_A}{dt} = -r, \qquad \overline{\rho}_A(0) = \frac{n_0}{V} \equiv \overline{\rho}_0, \tag{4.389}$$

$$\frac{d\overline{\rho}_B}{dt} = r, \qquad \overline{\rho}_B(0) = 0. \tag{4.390}$$

Addition of Eqs. (4.389) and (4.390) gives

$$\frac{d}{dt}\left(\overline{\rho}_A + \overline{\rho}_B\right) = 0, \tag{4.391}$$

$$\overline{\rho}_A + \overline{\rho}_B = \overline{\rho}_0, \tag{4.392}$$

$$\underbrace{\frac{\overline{\rho}_A}{\overline{\rho}_0}}_{y_A} + \underbrace{\frac{\overline{\rho}_B}{\overline{\rho}_0}}_{y_B} = 1, \tag{4.393}$$

$$y_A + y_B = 1. \tag{4.394}$$

The reaction rate $r$ is then

$$r = k\overline{\rho}_A\left(1 - \frac{1}{K_c}\frac{\overline{\rho}_B}{\overline{\rho}_A}\right) = k\overline{\rho}_0\frac{\overline{\rho}_A}{\overline{\rho}_0}\left(1 - \frac{1}{K_c}\frac{\overline{\rho}_B/\overline{\rho}_0}{\overline{\rho}_A/\overline{\rho}_0}\right), \tag{4.395}$$

$$= k\overline{\rho}_0 y_A\left(1 - \frac{1}{K_c}\frac{y_B}{y_A}\right) = k\overline{\rho}_0(1-\lambda)\left(1 - \frac{1}{K_c}\frac{\lambda}{1-\lambda}\right). \tag{4.396}$$

By Eq. (4.221), $r = (1/V)d\zeta/dt = (1/V)d(n_0\lambda)/dt = (n_0/V)d\lambda/dt = \overline{\rho}_0 d\lambda/dt$. So, the reaction dynamics can be described by a single equation in a single unknown:

$$\overline{\rho}_0\frac{d\lambda}{dt} = k\overline{\rho}_0(1-\lambda)\left(1 - \frac{1}{K_c}\frac{\lambda}{1-\lambda}\right), \tag{4.397}$$

$$\frac{d\lambda}{dt} = k(1-\lambda)\left(1 - \frac{1}{K_c}\frac{\lambda}{1-\lambda}\right). \tag{4.398}$$

Equation (4.398) is in equilibrium when

$$\lambda = \frac{1}{1 + \frac{1}{K_c}} \sim 1 - \frac{1}{K_c} + \cdots \tag{4.399}$$

As $K_c \to \infty$, the equilibrium value of $\lambda \to 1$. In this limit, the reaction is irreversible. That is, the species $B$ is preferred over $A$. Equation (4.398) has exact solution

$$\lambda = \frac{1 - \exp\left(-k\left(1 + \frac{1}{K_c}\right)t\right)}{1 + \frac{1}{K_c}}. \tag{4.400}$$

For $k > 0, K_c > 0$, the equilibrium is stable. The time constant of relaxation $\tau$ is

$$\tau = \frac{1}{k\left(1 + \frac{1}{K_c}\right)}. \tag{4.401}$$

For the isothermal, isochoric system, one can also consider the second law and irreversible entropy production in terms of the Helmholtz free energy. Combine then Eq. (3.276), $dA|_{T,V} \leq 0$, with Eq. (3.180), $dA = -S\,dT - P\,dV + \sum_{i=1}^{N} \overline{\mu}_i dn_i$:

$$dA|_{T,V} = \left(-S\,dT - P\,dV + \sum_{i=1}^{N} \overline{\mu}_i dn_i\right)\Bigg|_{T,V} \leq 0. \tag{4.402}$$

Taking time derivatives, one finds

$$\left.\frac{dA}{dt}\right|_{T,V} = \sum_{i=1}^{N} \overline{\mu}_i \frac{dn_i}{dt} \leq 0, \tag{4.403}$$

$$-\frac{1}{T}\left.\frac{dA}{dt}\right|_{T,V} = -\frac{V}{T}\sum_{i=1}^{N} \overline{\mu}_i \frac{d\overline{\rho}_i}{dt} \geq 0. \tag{4.404}$$

This is exactly the same form as Eq. (4.376), which can be directly substituted into Eq. (4.404) to give

$$\underbrace{-\frac{1}{T}\left.\frac{dA}{dt}\right|_{T,V}}_{\dot{\sigma}} = -\overline{R}Vk(T)\left(\prod_{i=1}^{N} \overline{\rho}_i^{\nu_i'}\right)\left(1 - \frac{1}{K_c}\prod_{i=1}^{N} \overline{\rho}_i^{\nu_i}\right) \ln\left(\frac{1}{K_c}\prod_{i=1}^{N} \overline{\rho}_i^{\nu_i}\right) \geq 0. \tag{4.405}$$

For the assumptions of this section, Eq. (4.405) reduces to

$$\dot{\sigma} = -\overline{R}k\overline{\rho}_0 V(1-\lambda)\left(1 - \frac{1}{K_c}\frac{\lambda}{1-\lambda}\right)\ln\left(\frac{1}{K_c}\frac{\lambda}{1-\lambda}\right) \geq 0, \tag{4.406}$$

$$= \overline{R}Vk\overline{\rho}_0\left((1-\lambda) - \frac{\lambda}{K_c}\right)\ln\left(\frac{1-\lambda}{\frac{\lambda}{K_c}}\right) \geq 0. \tag{4.407}$$

This takes the form of Eq. (4.381) when we recognize that

$$\mathcal{R}' = k\overline{\rho}_0(1-\lambda), \qquad \mathcal{R}'' = \frac{k}{K_c}\overline{\rho}_0\lambda. \tag{4.408}$$

This completes our development of single-step kinetics. In the next chapter, we give the multistep extensions that are necessary to understand modern continuum combustion chemistry.

## EXERCISES

**4.1.** Find values of $\phi$ and $\nu$ associated with Reaction 1 of the mechanism of Table 1.2.

**4.2.** Find the most general way to balance the reaction

$$\nu_1' H_2 + \nu_2' O_2 \leftrightharpoons \nu_3'' H_2O + \nu_4'' OH.$$

**4.3.** Find the most general way to balance the reaction

$$\nu_1' C_2H_2 + \nu_2' O_2 \leftrightharpoons \nu_3'' H_2O + \nu_4'' CO_2 + \nu_5'' OH + \nu_6'' O + \nu_7'' H.$$

**4.4.** An isochoric, adiabatic chamber contains 1 kmol of $CH_4$ and 2 kmol of $O_2$ at 600 K and 100 kPa. Find the adiabatic flame temperature if the reaction is

$$CH_4 + 2O_2 \rightarrow CO_2 + 2H_2O.$$

Consult a thermodynamics text for tabular values of thermodynamic properties.

**4.5.** Consider a problem under identical isochoric, isothermal conditions as that studied in the example of Sec. 4.5.1, except hydrogen dissociation is considered instead of nitrogen:

$$H_2 + M \leftrightharpoons H + H + M.$$

Kinetic parameters are given by Reaction 12 of Table 1.2. Thermodynamic properties can be found in elementary thermodynamics texts. Find the physical equilibrium state and the local time scales near equilibrium. Numerically integrate and give plots of $\overline{\rho}_H(t)$ and $\overline{\rho}_{H_2}(t)$.

**4.6.** Consider a problem under identical isobaric, isothermal conditions as that studied in the example of Sec. 4.5.1, except hydrogen dissociation is considered instead of nitrogen:

$$H_2 + M \leftrightharpoons H + H + M.$$

Kinetic parameters are given by Reaction 12 of Table 1.2. Thermodynamic properties can be found in elementary thermodynamics texts. Find the physical equilibrium state and the local time scales near equilibrium. Numerically integrate and give plots of $\overline{\rho}_H(t)$ and $\overline{\rho}_{H_2}(t)$.

**4.7.** Extend the analysis of Sec. 4.6.3 to a diabatic system for which there is convective heat exchange with the surroundings at the rate given by

$$\dot{Q} = \mathsf{h}A(T_\infty - T),$$

where $\mathsf{h}$ is the constant convective heat transfer coefficient, $A$ is the surface area, and $T_\infty$ is the constant temperature of the surroundings. Find the analog to the expression for $dT/dt$ given in Eq. (4.329) for the diabatic system.

**4.8.** Extend the analysis of Sec. 4.7 to the simple two-step system

$$A \leftrightharpoons B, \qquad B \leftrightharpoons C.$$

Make an analogous set of assumptions, for example, $M_A = M_B = M_C = M$. Find an expression for $\dot{\sigma}$ similar to that given for the one-step system in Eq. (4.407).

## References

M. M. Abbott and H. C. Van Ness, 1972, *Thermodynamics*, Schaum's Outline Series in Engineering, McGraw-Hill, New York, Chapter 7.

K. Annamalai and I. K. Puri, 2006, *Combustion Science and Engineering*, CRC Press, Boca Raton, FL.

S. A. Arrhenius, 1889, Über die Reaktiongeschwindigkeit bei der Inversion von Rohzucker durch Säuren, *Zeitschrift für physikalische Chemie*, 4: 226–248.

R. S. Berry, S. A. Rice, and J. Ross, 2000, *Physical Chemistry*, 2nd ed., Oxford University Press, New York.

C. Borgnakke and R. E. Sonntag, 2013, *Fundamentals of Thermodynamics*, 8th ed., John Wiley, New York, Chapters 13 and 14.

A. W. Date, 2014, *Analytic Combustion: With Thermodynamics, Chemical Kinetics and Mass Transfer*, Cambridge University Press, New York.

I. Glassman, R. A. Yetter, and N. G. Glumac, 2014, *Combustion*, 5th ed., Academic Press, New York, Chapters 1, 2.

D. Kondepudi and I. Prigogine, 1998, *Modern Thermodynamics: From Heat Engines to Dissipative Structures*, John Wiley, New York, Chapters 4, 5, 7, and 9.

K. K. Kuo, 2005, *Principles of Combustion*, 2nd ed., John Wiley, New York, Chapters 1, 2.

K. J. Laidler, 1965, *Chemical Kinetics*, McGraw-Hill, New York.

S. McAllister, J.-Y. Chen, and A. C. Fernandez-Pello, 2011, *Fundamentals of Combustion Processes*, Springer, New York.

C. Park, 1990, *Nonequilibrium Hypersonic Aerothermodynamics*, John Wiley, New York, p. 326.

K. W. Ragland and K. M. Bryden, 2011, *Combustion Engineering*, 2nd ed., CRC Press, Boca Raton, FL.

W. C. Strahle, 1993, *An Introduction to Combustion*, Gordon and Breach, Amsterdam.

G. Strang, 2005, *Linear Algebra and Its Applications*, 4th ed., Cengage Learning, Stamford, CT.

R. A. Strehlow, 1984, *Combustion Fundamentals*, McGraw-Hill, New York, Chapters 2, 4, 6.

S. R. Turns, 2011, *An Introduction to Combustion*, 3rd ed., McGraw-Hill, Boston, Chapter 2.

J. H. van't Hoff, 1884, *Etudes de Dynamique Chimique*, Frederik Muller, Amsterdam.

P. Waage and C. M. Guldberg, 1864, Studies concerning affinity, *Forhandlinger: Videnskabs-Selskabet i Christiania*, 35. English translation: *Journal of Chemical Education*, 63(12): 1044–1047.

J. Warnatz, U. Maas, and R. W. Dibble, 2006, *Combustion*, 4th ed., Springer, New York, Chapter 4.

# 5 Thermochemistry of Multiple Reactions

*Double, double, toile and trouble;*
*Fire burne, and Cauldron bubble.*

— William Shakespeare (1564–1616), *Macbeth*

This chapter will extend notions associated with the thermodynamics of a single chemical reaction from Chapter 4 to systems in which many reactions occur simultaneously. Additional background is in some standard sources (Buckmaster and Ludford, 1982; Williams, 1985; Liñan and Williams, 1993; Kee et al., 2003; Kuo, 2005; Law, 2006; Turns, 2011). Many of the extensions are immediately obvious and will not need significant discussion. Others benefit from detailed exposition. We will then take on the important problem of identification of chemical equilibrium for general ideal gas mixtures. We will show that two common methods, (1) minimization of Gibbs free energy of the mixture and (2) equilibration of all chemical reactions, nearly always give identical results. We then give an extended discussion of a proof of uniqueness of chemical equilibrium, an important notion for nonlinear chemical dynamics. A simple linear example is used to illustrate the dynamics of multistep kinetics quantitatively; exact solutions are obtained for both reversible and irreversible mechanisms. We next show how, depending on whether species or reactions are dominant in their number, two distinct ways of concisely summarizing detailed reaction kinetics models exist. Both of these methods account explicitly for linear algebraic constraints from element conservation. We close with a discussion of topics often considered in more detail in many works on what is called "thermodynamics of irreversible processes" or "nonequilibrium thermodynamics." We connect ideas associated with so-called Onsager[1] reciprocity with irreversible entropy production (Hirschfelder et al., 1954; Reynolds, 1968; Gyarmati, 1970; Woods, 1975; Lavenda, 1978; de Groot and Mazur, 1984; Kondepudi and Prigogine, 1998; Müller and Ruggeri, 1998; Bejan, 2006; Terao, 2007).

## 5.1 Summary of Multiple Reaction Extensions

Consider here the reaction of $N$ species, composed of $L$ elements, in $J$ reactions. This section focuses on the most common case in which $J \geq (N - L)$. That is usually the

[1] Lars Onsager (1903–1976), Norwegian-American physicist; 1968 Nobel laureate in Chemistry.

case in large chemical kinetic systems in use in engineering models. While much of the analysis will only require $J > 0$, certain results will depend on $J \geq (N - L)$. It is not difficult to study the complementary case where $0 < J < (N - L)$, and we will briefly consider that case in Sec. 5.4.2.

The molecular mass of species $i$ is given by Eq. (4.1):

$$M_i = \sum_{l=1}^{L} \mathcal{M}_l \phi_{li}, \qquad i = 1, \ldots, N. \tag{5.1}$$

However, each reaction, $j = 1, \ldots, J$, has a stoichiometric coefficient. The $j$th reaction can be summarized in the following ways:

$$\sum_{i=1}^{N} \chi_i \nu'_{ij} \rightleftharpoons \sum_{i=1}^{N} \chi_i \nu''_{ij}, \quad \text{or} \quad \sum_{i=1}^{N} \chi_i \nu_{ij} = 0, \quad j = 1, \ldots, J. \tag{5.2}$$

We have taken the net stoichiometric coefficient $\nu_{ij}$ to be defined by

$$\nu_{ij} = \nu''_{ij} - \nu'_{ij}, \qquad i = 1, \ldots N, \; j = 1, \ldots, J. \tag{5.3}$$

Stoichiometry for the $j$th reaction and $l$th element is given by the extension of Eq. (4.19):

$$\sum_{i=1}^{N} \phi_{li} \nu_{ij} = 0, \qquad l = 1, \ldots, L, \; j = 1, \ldots, J. \tag{5.4}$$

The net change in Gibbs free energy and equilibrium constants of the $j$th reaction are defined by extensions of Eqs. (4.206), (4.205), and (4.208):

$$\Delta G_j^0 \equiv \sum_{i=1}^{N} \overline{g}_{T,i}^0 \nu_{ij}, \qquad j = 1, \ldots, J, \tag{5.5}$$

$$K_{P,j} \equiv \exp\left(\frac{-\Delta G_j^0}{\overline{R}T}\right), \qquad j = 1, \ldots, J, \tag{5.6}$$

$$K_{c,j} \equiv \left(\frac{P_0}{\overline{R}T}\right)^{\sum_{i=1}^{N} \nu_{ij}} \exp\left(\frac{-\Delta G_j^0}{\overline{R}T}\right), \qquad j = 1, \ldots, J. \tag{5.7}$$

The dimension and rank of the matrix composed of $\nu_{ij}$ may be such that a set of $N$ values of $\overline{g}_{T,i}^0$ maps to a unique set of $J$ values of $\Delta G_j^0$, or it could be such that some of the $\Delta G_j^0$ are linearly dependent. In previous examples, that has been the case: the Zel'dovich model of Sec. 1.2.2 had $J = 2$ reactions and two independent values of $\Delta G_1^0$ and $\Delta G_2^0$. However, linear dependencies will exist if $\boldsymbol{\nu}$ is less than full rank, as will be seen in Sec. 5.3.

The equilibrium of the $j$th reaction is given by the extension of Eq. (4.194) or the extension of Eq. (4.196):

$$\sum_{i=1}^{N} \overline{\mu}_i \nu_{ij} = 0, \quad \text{or} \quad \sum_{i=1}^{N} \overline{g}_i \nu_{ij} = 0, \qquad j = 1, \ldots, J. \tag{5.8}$$

The multireaction extensions of Eq. (3.375) for chemical affinity is

$$\overline{\alpha}_j = -\sum_{i=1}^{N} \overline{\mu}_i \nu_{ij}, \qquad j = 1, \ldots, J. \tag{5.9}$$

The multireaction extension of the chemical affinity definition of Eq. (4.382) is

$$\overline{\alpha}_j = \overline{R}T \ln\left(\frac{\mathcal{R}'_j}{\mathcal{R}''_j}\right), \qquad j = 1, \ldots, J. \tag{5.10}$$

In terms of the chemical affinity of each reaction, the equilibrium condition is simply the extension of Eq. (4.195):

$$\overline{\alpha}_j = 0, \qquad j = 1, \ldots, J. \tag{5.11}$$

At equilibrium, then the equilibrium constraints can be shown to reduce to the extensions of Eq. (4.204) or (4.212):

$$K_{P,j} = \prod_{i=1}^{N}\left(\frac{P_i}{P_0}\right)^{\nu_{ij}}, \quad K_{c,j} = \prod_{i=1}^{N}\overline{\rho}_i^{\,\nu_{ij}}, \qquad j = 1, \ldots, J. \tag{5.12}$$

For isochoric reaction, the evolution of species concentration $i$ due to the combined effect of $J$ reactions is given by the extension of Eq. (4.217):

$$\frac{d\overline{\rho}_i}{dt} = \overbrace{\sum_{j=1}^{J}\nu_{ij}\underbrace{\underbrace{a_j T^{\beta_j}\exp\left(\frac{-\overline{\mathcal{E}}_j}{\overline{R}T}\right)}_{k_j(T)}\underbrace{\left(\prod_{k=1}^{N}\overline{\rho}_k^{\,\nu'_{kj}}\right)}_{\text{forward reaction}}\left(1 - \underbrace{\frac{1}{K_{c,j}}\prod_{k=1}^{N}\overline{\rho}_k^{\,\nu_{kj}}}_{\text{reverse reaction}}\right)}_{r_j=(1/V)\,d\zeta_j/dt}}^{\dot{\omega}_i},$$

$$i = 1, \ldots, N. \tag{5.13}$$

The extension to isobaric reactions, not given here, is straightforward, and follows the same analysis as for a single reaction. Again, three intermediate variables that are in common usage have been defined. First, one takes the reaction rate of the $j$th reaction to be the extension of Eq. (4.218):

$$r_j \equiv \underbrace{a_j T^{\beta_j}\exp\left(\frac{-\overline{\mathcal{E}}_j}{\overline{R}T}\right)}_{k_j(T)}\underbrace{\left(\prod_{k=1}^{N}\overline{\rho}_k^{\,\nu'_{kj}}\right)}_{\text{forward reaction}}\left(1 - \underbrace{\frac{1}{K_{c,j}}\prod_{k=1}^{N}\overline{\rho}_k^{\,\nu_{kj}}}_{\text{reverse reaction}}\right),$$

$$j = 1, \ldots, J, \tag{5.14}$$

or the extension of Eq. (4.219):

$$r_j = \underbrace{a_j T^{\beta_j}\exp\left(\frac{-\overline{\mathcal{E}}_j}{\overline{R}T}\right)}_{k_j(T),\ \text{Arrhenius rate}}\underbrace{\left(\underbrace{\prod_{k=1}^{N}\overline{\rho}_k^{\,\nu'_{kj}}}_{\text{forward reaction}} - \underbrace{\frac{1}{K_{c,j}}\prod_{k=1}^{N}\overline{\rho}_k^{\,\nu''_{kj}}}_{\text{reverse reaction}}\right)}_{\text{law of mass action}},$$

$$j = 1, \ldots, J, \tag{5.15}$$

$$= \frac{1}{V}\frac{d\zeta_j}{dt}. \tag{5.16}$$

Here, $\zeta_j$ is the reaction progress variable for the $j$th reaction. Each reaction has a temperature-dependent reaction rate coefficient $k_j(T)$ that is an extension of Eq. (4.220):

$$k_j(T) \equiv a_j T^{\beta_j} \exp\left(\frac{-\overline{\mathcal{E}}_j}{\overline{R}T}\right), \qquad j = 1, \ldots, J. \tag{5.17}$$

The species molar production rate is given by $\dot{\omega}_i$, defined as an extension of Eq. (4.222):

$$\dot{\omega}_i \equiv \sum_{j=1}^{J} \nu_{ij} r_j, \qquad i = 1, \ldots, N. \tag{5.18}$$

So, we can summarize the isochoric Eq. (5.13) as

$$\frac{d\overline{\rho}_i}{dt} = \dot{\omega}_i. \tag{5.19}$$

It will be useful to cast this in terms of mass fraction. Using the definition Eq. (2.37), Eq. (5.19) can be rewritten as

$$\frac{d}{dt}\left(\frac{\rho Y_i}{M_i}\right) = \dot{\omega}_i. \tag{5.20}$$

Because $M_i$ is a constant, Eq. (5.20) can be recast as

$$\frac{d}{dt}(\rho Y_i) = M_i \dot{\omega}_i. \tag{5.21}$$

The multireaction extension of Eq. (4.183) for mol change in terms of progress variables is

$$dn_i = \sum_{j=1}^{J} \nu_{ij}\, d\zeta_j, \qquad i = 1, \ldots, N. \tag{5.22}$$

One also has Eq. (4.187):

$$dG|_{T,P} = \sum_{i=1}^{N} \overline{\mu}_i\, dn_i. \tag{5.23}$$

Let us see how $G$ changes for the $j$th reaction. We will use Eq. (5.22) to assist.

$$dG|_{T,P} = \sum_{i=1}^{N} \overline{\mu}_i \sum_{k=1}^{J} \nu_{ik}\, d\zeta_k, \tag{5.24}$$

$$\left.\frac{\partial G}{\partial \zeta_j}\right|_{\zeta_p} = \sum_{i=1}^{N} \overline{\mu}_i \sum_{k=1}^{J} \nu_{ik} \frac{\partial \zeta_k}{\partial \zeta_j} = \sum_{i=1}^{N} \overline{\mu}_i \sum_{j=1}^{J} \nu_{ik}\delta_{kj} = \sum_{i=1}^{N} \overline{\mu}_i \nu_{ij}, \tag{5.25}$$

$$= -\overline{\alpha}_j, \qquad j = 1, \ldots, J. \tag{5.26}$$

Positive chemical affinity induces $G$ to decrease as the reaction moves forward. For a set of adiabatic, isochoric reactions, one can show the extension of Eq. (4.329) is

$$\frac{dT}{dt} = -\frac{\sum_{j=1}^{J} r_j \Delta E_j}{\rho c_v}, \tag{5.27}$$

where the energy change for a reaction $\Delta E_j$ is defined as the extension of Eq. (4.328):

$$\Delta E_j = \sum_{i=1}^{N} \overline{e}_i \nu_{ij}, \qquad j = 1, \ldots, J. \tag{5.28}$$

For adiabatic, isobaric reactions, one can show the extension of Eq. (4.339) to be

$$\frac{dT}{dt} = -\frac{\sum_{j=1}^{J} r_j \Delta H_j}{\rho c_P}, \tag{5.29}$$

where the enthalpy change for a reaction $\Delta H_j$ is defined as the extension of Eq. (4.338):

$$\Delta H_j = \sum_{i=1}^{N} \overline{h}_i \nu_{ij}, \qquad j = 1, \ldots, J. \tag{5.30}$$

Moreover, the density and species concentration derivatives for an adiabatic, isobaric set can be shown to be the extensions of Eqs. (4.349) and (4.353):

$$\frac{d\rho}{dt} = M \sum_{j=1}^{J} r_j \sum_{i=1}^{N} \nu_{ij} \left( \frac{\overline{h}_i}{c_P T} - 1 \right), \tag{5.31}$$

$$\frac{d\overline{\rho}_i}{dt} = \sum_{j=1}^{J} r_j \left( \nu_{ij} + y_i \left( \frac{\Delta H_j}{\overline{c}_P T} - \Delta \mathsf{n}_j \right) \right), \tag{5.32}$$

where the net mol change for the $j$th reaction $\Delta \mathsf{n}_j$ is

$$\Delta \mathsf{n}_j = \sum_{k=1}^{N} \nu_{kj}. \tag{5.33}$$

In a similar fashion to that shown for a single reaction, one can sum over all reactions and prove that mixture mass is conserved, element mass and number are conserved.

**EXAMPLE 5.1**

Show that element mass and number are conserved for the multireaction formulation.

Start with Eq. (5.13) and expand as follows:

$$\frac{d\overline{\rho}_i}{dt} = \sum_{j=1}^{J} \nu_{ij} r_j, \tag{5.34}$$

$$\phi_{li} \frac{d\overline{\rho}_i}{dt} = \phi_{li} \sum_{j=1}^{J} \nu_{ij} r_j, \tag{5.35}$$

$$\frac{d}{dt} \left( \phi_{li} \overline{\rho}_i \right) = \sum_{j=1}^{J} \phi_{li} \nu_{ij} r_j, \tag{5.36}$$

$$\sum_{i=1}^{N} \frac{d}{dt} \left( \phi_{li} \overline{\rho}_i \right) = \sum_{i=1}^{N} \sum_{j=1}^{J} \phi_{li} \nu_{ij} r_j, \tag{5.37}$$

$$\underbrace{\frac{d}{dt}\left(\sum_{i=1}^{N}\phi_{li}\overline{\rho}_i\right)}_{\overline{\rho}_l^{\,e}} = \sum_{j=1}^{J}\sum_{i=1}^{N}\phi_{li}\nu_{ij}r_j, \tag{5.38}$$

$$\frac{d\overline{\rho}_l^{\,e}}{dt} = \sum_{j=1}^{J} r_j \underbrace{\sum_{i=1}^{N}\phi_{li}\nu_{ij}}_{0} = 0, \qquad l = 1,\dots,L, \tag{5.39}$$

$$\frac{d}{dt}\left(\mathcal{M}_l\overline{\rho}_l^{\,e}\right) = 0, \qquad l = 1,\dots,L, \tag{5.40}$$

$$\frac{d\rho_l^{\,e}}{dt} = 0, \qquad l = 1,\dots,L. \tag{5.41}$$

It is also straightforward to show that the mixture density is conserved:

$$\frac{d\rho}{dt} = 0. \tag{5.42}$$

The proof of the Clausius-Duhem inequality is an extension of the single reaction result. Start with Eq. (4.354) and operate much as for a single reaction model:

$$dS|_{E,V} = \underbrace{-\frac{1}{T}\sum_{i=1}^{N}\overline{\mu}_i\,dn_i}_{d\sigma} \geq 0, \tag{5.43}$$

$$\dot{\sigma} = \left.\frac{dS}{dt}\right|_{E,V} = -\frac{V}{T}\sum_{i=1}^{N}\overline{\mu}_i\frac{dn_i}{dt}\frac{1}{V} = -\frac{V}{T}\sum_{i=1}^{N}\overline{\mu}_i\frac{d\overline{\rho}_i}{dt} \geq 0, \tag{5.44}$$

$$= -\frac{V}{T}\sum_{i=1}^{N}\overline{\mu}_i\sum_{j=1}^{J}\nu_{ij}r_j = -\frac{V}{T}\sum_{i=1}^{N}\sum_{j=1}^{J}\overline{\mu}_i\nu_{ij}r_j \geq 0. \tag{5.45}$$

Continuing, we see

$$\dot{\sigma} = -\frac{V}{T}\sum_{j=1}^{J}r_j\sum_{i=1}^{N}\overline{\mu}_i\nu_{ij} \geq 0, \tag{5.46}$$

$$= -\frac{V}{T}\sum_{j=1}^{J}k_j\prod_{i=1}^{N}\overline{\rho}_i^{\,\nu'_{ij}}\left(1-\frac{1}{K_{c,j}}\prod_{i=1}^{N}\overline{\rho}_i^{\,\nu_{ij}}\right)\sum_{i=1}^{N}\overline{\mu}_i\nu_{ij} \geq 0, \tag{5.47}$$

$$= -\frac{V}{T}\sum_{j=1}^{J}k_j\prod_{i=1}^{N}\overline{\rho}_i^{\,\nu'_{ij}}\left(1-\frac{1}{K_{c,j}}\prod_{i=1}^{N}\overline{\rho}_i^{\,\nu_{ij}}\right) \times\left(\overline{R}T\ln\left(\frac{1}{K_{c,j}}\prod_{i=1}^{N}\overline{\rho}_i^{\,\nu_{ij}}\right)\right) \geq 0, \tag{5.48}$$

$$= -\overline{R}V\sum_{j=1}^{J}k_j\prod_{i=1}^{N}\overline{\rho}_i^{\,\nu'_{ij}}\left(1-\frac{1}{K_{c,j}}\prod_{i=1}^{N}\overline{\rho}_i^{\,\nu_{ij}}\right)\times\ln\left(\frac{1}{K_{c,j}}\prod_{i=1}^{N}\overline{\rho}_i^{\,\nu_{ij}}\right) \geq 0. \tag{5.49}$$

Equation (5.46) can also be written in terms of the affinities (see Eq. (5.9)) and reaction progress variables (see Eq. (5.16)) as

$$\dot{\sigma} = \left.\frac{dS}{dt}\right|_{E,V} = \frac{1}{T}\sum_{j=1}^{J}\overline{\alpha}_j\frac{d\zeta_j}{dt} \geq 0. \tag{5.50}$$

Similar to the argument for a single reaction, if one defines

$$\mathcal{R}'_j = k_j\prod_{i=1}^{N}\overline{\rho}_i^{\nu'_{ij}}, \qquad \mathcal{R}''_j = \frac{k_j}{K_{c,j}}\prod_{i=1}^{N}\overline{\rho}_i^{\nu''_{ij}}, \tag{5.51}$$

then it is easy to show that

$$r_j = \mathcal{R}'_j - \mathcal{R}''_j, \tag{5.52}$$

and we get the equivalent of Eq. (4.381):

$$\dot{\sigma} = \left.\frac{dS}{dt}\right|_{E,V} = \overline{R}V\sum_{j=1}^{J}(\mathcal{R}'_j - \mathcal{R}''_j)\ln\left(\frac{\mathcal{R}'_j}{\mathcal{R}''_j}\right) \geq 0. \tag{5.53}$$

Because $k_j(T) > 0, \overline{R} > 0$, and $V \geq 0$, and each term in the summation combines to be positive semidefinite, one sees that the Clausius-Duhem inequality is guaranteed to be satisfied for multiple reactions.

Now, let us interpret this analysis with some of the language and notation similar to that of Prigogine's[2] influential monograph (Prigogine, 1967). We will use a "P" to denote equations found in Prigogine's text. First, Eq. (5.46) can also be written in terms of the affinities (see Eq. (5.9)) as

$$\dot{\sigma} = \left.\frac{dS}{dt}\right|_{E,V} = V\mathbf{r}^T\cdot\left(\frac{\overline{\boldsymbol{\alpha}}}{T}\right). \tag{5.54}$$

We can take Prigogine's so-called generalized force and flux to be $\mathbf{x}$ and $\mathbf{j}$ with

$$\mathbf{x} = \frac{\overline{\boldsymbol{\alpha}}}{T}, \qquad \mathbf{j} = V\mathbf{r}. \tag{5.55}$$

Therefore, Eq. (5.54) can be recast as

$$\dot{\sigma} = \left.\frac{dS}{dt}\right|_{E,V} = \mathbf{j}^T\cdot\mathbf{x}. \tag{5.56}$$

Equations (5.55) can be compared with Eqs. (P4.2). We know that at equilibrium both $\mathbf{j} = \mathbf{0}$ and $\mathbf{x} = \mathbf{0}$ from earlier analysis: when all $\overline{\alpha}_j$ are zero, $\mathbf{x} = \mathbf{0}$, and by Eq. (5.10), we must have $\mathcal{R}'_j = \mathcal{R}''_j$, and thus because by Eq. (5.52), $r_j = \mathcal{R}'_j - \mathcal{R}''_j, r_j = 0$. When all $r_j$ are zero, $\mathbf{j} = \mathbf{0}$. Moreover, Eq. (5.18) shows all $\dot{\omega}_i$ are zero; thus, Eq. (5.19) is in equilibrium. This proof did not rely on any linearization near equilibrium. We shall return to this topic in Sec. 5.5, including a discussion of the influential Onsager approach, which relies upon a linearization near equilibrium.

[2]Ilya Prigogine (1917–2003), Russian-born Belgian chemist; 1977 Nobel laureate in Chemistry.

## 5.2 Equilibrium Conditions

For multicomponent mixtures undergoing multiple reactions, determining the equilibrium condition is more difficult than for a single reaction. There are two primary approaches; they are essentially equivalent. The most straightforward method requires formal minimization of the Gibbs free energy of the mixture. Nearly always, the same equilibrium is found by the second method: equilibrating each of the actual reactions in the mechanism.

### 5.2.1 Minimization of G via Lagrange Multipliers

Let us first describe determination of equilibrium by minimization of $G$. Recall Eq. (3.276), $dG|_{T,P} \leq 0$. Recall also Eq. (3.277), $G = \sum_{i=1}^{N} \overline{g}_i n_i$. Because $\overline{\mu}_i = \overline{g}_i = \partial G/\partial n_i|_{P,T,n_j}$, one also has $G = \sum_{i=1}^{N} \overline{\mu}_i n_i$. From Eq. (3.278), $dG|_{T,P} = \sum_{i=1}^{N} \overline{\mu}_i \, dn_i$. Now, one must also demand for a system coming to equilibrium that the element numbers are conserved. This can be achieved by requiring

$$\sum_{i=1}^{N} \phi_{li}(n_{i0} - n_i) = 0, \qquad l = 1, \ldots, L. \tag{5.57}$$

Recall $n_{i0}$ is the initial number of moles of species $i$, and $\phi_{li}$ is the number of moles of element $l$ in species $i$. Equation (5.57) thus provides $L$ linear constraints. One can use the method of constrained optimization given by the method of Lagrange[3] multipliers to extremize $G$ subject to the constraints of element conservation. The extremum will be a minimum; this will not be proved, but will be demonstrated. Define a set of $L$ Lagrange multipliers $\lambda_l$. Next define an augmented Gibbs free energy function $G^*$, which is simply $G$ plus the product of the Lagrange multipliers and the constraints:

$$G^* = G + \sum_{l=1}^{L} \lambda_l \sum_{i=1}^{N} \phi_{li}(n_{i0} - n_i). \tag{5.58}$$

When the constraints are satisfied, $G^* = G$, so assuming the constraints can be satisfied, extremizing $G$ is equivalent to extremizing $G^*$. To extremize $G^*$, take its derivative with respect to $n_i$, with $P, T$ and $n_j$ constant and set it to zero for each species:

$$\left.\frac{\partial G^*}{\partial n_i}\right|_{T,P,n_j} = \underbrace{\left.\frac{\partial G}{\partial n_i}\right|_{T,P,n_j}}_{\overline{\mu}_i} - \sum_{l=1}^{L} \lambda_l \phi_{li} = 0, \qquad i = 1, \ldots, N. \tag{5.59}$$

With the definition of $\overline{\mu}_i$, see Eq. (3.188) and Sec. 3.6, one gets

$$\overline{\mu}_i - \sum_{l=1}^{L} \lambda_l \phi_{li} = 0, \qquad i = 1, \ldots, N. \tag{5.60}$$

We recall that $\overline{\mu}_i = \overline{g}_i = \overline{h}_i - T\overline{s}_i$ from Eq. (3.226). Now use Eq. (2.206) to segregate the chemical potential into that at a reference pressure and the deviation from the reference pressure, recalling that $\overline{h}_i = \overline{h}_i^0$:

$$\overline{g}_i = \overline{\mu}_i = \overline{h}_i^0 - T\left(\overline{s}_i^0 - \overline{R}\ln\left(\frac{P_i}{P_0}\right)\right) = (\overline{h}_i^0 - T\overline{s}_i^0) - \overline{R}T\ln\left(\frac{P_i}{P_0}\right). \tag{5.61}$$

[3]Joseph-Louis Lagrange (1736–1813), Italian-French mathematician.

Now, with $\overline{\mu}_i^0 \equiv \overline{h}_i^0 - T\overline{s}_i^0$, we get

$$\overline{\mu}_i = \overline{\mu}_i^0 - \overline{R}T \ln\left(\frac{P_i}{P_0}\right). \tag{5.62}$$

Then one can expand Eq. (5.60) so as to get

$$\underbrace{\overline{\mu}_{T,i}^0 + \overline{R}T \ln\left(\frac{P_i}{P_0}\right)}_{\overline{\mu}_i} - \sum_{l=1}^{L} \lambda_l \phi_{li} = 0, \qquad i = 1, \ldots, N, \tag{5.63}$$

$$\overline{\mu}_{T,i}^0 + \overline{R}T \ln\left(\underbrace{\left(\frac{n_i P}{\sum_{k=1}^{N} n_k}\right)}_{P_i} \frac{1}{P_0}\right) - \sum_{l=1}^{L} \lambda_l \phi_{li} = 0, \qquad i = 1, \ldots, N. \tag{5.64}$$

Recalling that $\sum_{k=1}^{N} n_k = n$, in summary then, one has $N + L$ equations

$$\overline{\mu}_{T,i}^0 + \overline{R}T \ln\left(\frac{n_i}{n}\frac{P}{P_0}\right) - \sum_{l=1}^{L} \lambda_l \phi_{li} = 0, \qquad i = 1, \ldots, N, \tag{5.65}$$

$$\sum_{i=1}^{N} \phi_{li}(n_{io} - n_i) = 0, \qquad l = 1, \ldots, L, \tag{5.66}$$

in $N + L$ unknowns: $n_i, i = 1, \ldots, N,\ \lambda_l, l = 1, \ldots, L$.

We will next consider a series of examples. For each of these thermodynamic properties are taken from standard tables (Borgnakke and Sonntag, 2013). Units for chemical potentials and enthalpies are all kJ/kmol. Units for entropy are all kJ/kmol/K. Units for temperature are all K.

**EXAMPLE 5.2**

Consider the isothermal, isobaric example from Sec. 3.9.2, in which

$$N_2 + N_2 \leftrightharpoons 2N + N_2. \tag{5.67}$$

Take $T = 6000$ K and $P = 100$ kPa. Initially one has 1 kmol of $N_2$ and 0 kmol of N. Use the extremization of Gibbs free energy to find the equilibrium composition.

First, find the chemical potentials at the reference pressure of each of the possible constituents.

$$\overline{\mu}_{T,i}^0 = \overline{g}_i^0 = \overline{h}_i^0 - T\overline{s}_i^0 = (\overline{h}_{298,i}^0 + \Delta\overline{h}_i^0) - T\overline{s}_i^0. \tag{5.68}$$

For each species, one then finds

$$\overline{\mu}_{N_2}^0 = 0 + 205848 - (6000)(292.984) = -1552056 \text{ kJ/kmol}, \tag{5.69}$$

$$\overline{\mu}_{N}^0 = 472680 + 124590 - (6000)(216.926) = -704286 \text{ kJ/kmol}. \tag{5.70}$$

To each of these one must add

$$\overline{R}T \ln\left(\frac{n_i P}{n P_0}\right)$$

to get the full chemical potential. Now, $P = P_0 = 100$ kPa for this problem. One also has $\overline{R}T = 8.314(6000) = 49884$ kJ/kmol. So, the chemical potentials are

$$\overline{\mu}_{N_2} = -1552056 + 49884 \ln \left( \frac{n_{N_2}}{n_N + n_{N_2}} \right), \tag{5.71}$$

$$\overline{\mu}_{N} = -704286 + 49884 \ln \left( \frac{n_{N}}{n_N + n_{N_2}} \right). \tag{5.72}$$

Then, one adds the Lagrange multiplier term and then considers element conservation to get the following coupled set of nonlinearalgebraic equations:

$$-1552056 + 49884 \ln \left( \frac{n_{N_2}}{n_N + n_{N_2}} \right) - 2\lambda_N = 0, \tag{5.73}$$

$$-704286 + 49884 \ln \left( \frac{n_{N}}{n_N + n_{N_2}} \right) - \lambda_N = 0, \tag{5.74}$$

$$n_N + 2n_{N_2} = 2. \tag{5.75}$$

These nonlinear equations are solved numerically to get

$$n_{N_2} = 0.882147 \text{ kmol}, \tag{5.76}$$

$$n_{N} = 0.235706 \text{ kmol}, \tag{5.77}$$

$$\lambda_{N} = -781934 \text{ kJ/kmol}. \tag{5.78}$$

These agree exactly with results found in the earlier example problem as seen in Eqs. (3.360) and (3.361).

**EXAMPLE 5.3**

Consider a mixture of 2 kmol of $H_2$ and 1 kmol of $O_2$ at $T = 3000$ K and $P = 100$ kPa. Assuming an isobaric and isothermal equilibration process with the products consisting of $H_2$, $O_2$, $H_2O$, OH, H, and O, find the equilibrium concentrations. Consider the same mixture at $T = 298$ K and $T = 1000$ K.

The first task is to find the chemical potentials of each species at the reference pressure and $T = 3000$ K. Here, one can use the standard tables along with the general equation

$$\overline{\mu}^0_{T,i} = \overline{g}^0_i = \overline{h}^0_i - T\overline{s}^0_i = (\overline{h}^0_{298,i} + \Delta\overline{h}^0_i) - T\overline{s}^0_i. \tag{5.79}$$

For each species, one then finds

$$\overline{\mu}^0_{H_2} = 0 + 88724 - 3000(202.989) = -520242 \text{ kJ/kmol}, \tag{5.80}$$

$$\overline{\mu}^0_{O_2} = 0 + 98013 - 3000(284.466) = -755385 \text{ kJ/kmol}, \tag{5.81}$$

$$\overline{\mu}^0_{H_2O} = -241826 + 126548 - 3000(286.504) = -974790 \text{ kJ/kmol}, \tag{5.82}$$

$$\overline{\mu}^0_{OH} = 38987 + 89585 - 3000(256.825) = -641903 \text{ kJ/kmol}, \tag{5.83}$$

$$\overline{\mu}^0_{H} = 217999 + 56161 - 3000(162.707) = -213961 \text{ kJ/kmol}, \tag{5.84}$$

$$\overline{\mu}^0_{O} = 249170 + 56574 - 3000(209.705) = -323371 \text{ kJ/kmol}. \tag{5.85}$$

To each of these one must add

$$\overline{R}T\ln\left(\frac{n_iP}{nP_0}\right)$$

to get the full chemical potential. Now, $P = P_0 = 100$ kPa for this problem. One also has $\overline{R}T = 8.314(3000) = 24942$ kJ/kmol. So, the chemical potentials are

$$\overline{\mu}_{H_2} = -520243 + 24942\ln\left(\frac{n_{H_2}}{\sum_{i=1}^{5} n_i}\right), \tag{5.86}$$

$$\overline{\mu}_{O_2} = -755385 + 24942\ln\left(\frac{n_{O_2}}{\sum_{i=1}^{5} n_i}\right), \tag{5.87}$$

$$\overline{\mu}_{H_2O} = -974790 + 24942\ln\left(\frac{n_{H_2O}}{\sum_{i=1}^{5} n_i}\right), \tag{5.88}$$

$$\overline{\mu}_{OH} = -641903 + 24942\ln\left(\frac{n_{OH}}{\sum_{i=1}^{5} n_i}\right), \tag{5.89}$$

$$\overline{\mu}_{H} = -213961 + 24942\ln\left(\frac{n_{H}}{\sum_{i=1}^{5} n_i}\right), \tag{5.90}$$

$$\overline{\mu}_{O} = -323371 + 24942\ln\left(\frac{n_{O}}{\sum_{i=1}^{5} n_i}\right). \tag{5.91}$$

Then, one adds on the Lagrange multipliers and then considers element conservation to get the following coupled set of nonlinear equations:

$$0 = -520243 + 24942\ln\left(\frac{n_{H_2}}{\sum_{i=1}^{5} n_i}\right) - 2\lambda_H, \tag{5.92}$$

$$0 = -755385 + 24942\ln\left(\frac{n_{O_2}}{\sum_{i=1}^{5} n_i}\right) - 2\lambda_O, \tag{5.93}$$

$$0 = -974790 + 24942\ln\left(\frac{n_{H_2O}}{\sum_{i=1}^{5} n_i}\right) - 2\lambda_H - \lambda_O, \tag{5.94}$$

$$0 = -641903 + 24942\ln\left(\frac{n_{OH}}{\sum_{i=1}^{5} n_i}\right) - \lambda_H - \lambda_O, \tag{5.95}$$

$$0 = -213961 + 24942\ln\left(\frac{n_{H}}{\sum_{i=1}^{5} n_i}\right) - \lambda_H, \tag{5.96}$$

$$0 = -323371 + 24942\ln\left(\frac{n_{O}}{\sum_{i=1}^{5} n_i}\right) - \lambda_O, \tag{5.97}$$

$$4 = 2n_{H_2} + 2n_{H_2O} + n_{OH} + n_H, \tag{5.98}$$

$$2 = 2n_{O_2} + n_{H_2O} + n_{OH} + n_O. \tag{5.99}$$

These nonlinear algebraic equations can be solved numerically via a Newton[4]-Raphson[5] technique. The equations are sensitive to the initial guess, and one can use intuition to help guide the selection. For example, one might expect to have $n_{H_2O}$

[4]Isaac Newton (1643–1727), English polymath.

[5]Joseph Raphson (c. 1648–c. 1715), English mathematician.

somewhere near 2 kmol. Application of the Newton-Raphson iteration yields

$$n_{\mathrm{H_2}} = 3.19 \times 10^{-1} \text{ kmol}, \tag{5.100}$$

$$n_{\mathrm{O_2}} = 1.10 \times 10^{-1} \text{ kmol}, \tag{5.101}$$

$$n_{\mathrm{H_2O}} = 1.50 \times 10^{0} \text{ kmol}, \tag{5.102}$$

$$n_{\mathrm{OH}} = 2.20 \times 10^{-1} \text{ kmol}, \tag{5.103}$$

$$n_{\mathrm{H}} = 1.36 \times 10^{-1} \text{ kmol}, \tag{5.104}$$

$$n_{\mathrm{O}} = 5.74 \times 10^{-2} \text{ kmol}, \tag{5.105}$$

$$\lambda_{\mathrm{H}} = -2.85 \times 10^{5} \text{ kJ/kmol}, \tag{5.106}$$

$$\lambda_{\mathrm{O}} = -4.16 \times 10^{5} \text{ kJ/kmol}. \tag{5.107}$$

At this relatively high value of temperature, all species considered have a relatively major presence. That is, there are no truly minor species.

Unless a good guess is provided, it may be difficult to find a solution for this set of nonlinear equations. Straightforward algebra allows the equations to be recast in a form that sometimes converges more rapidly:

$$\frac{n_{\mathrm{H_2}}}{\sum_{i=1}^{5} n_i} = \exp\left(\frac{520243}{24942}\right)\left(\exp\left(\frac{\lambda_{\mathrm{H}}}{24942}\right)\right)^2, \tag{5.108}$$

$$\frac{n_{\mathrm{O_2}}}{\sum_{i=1}^{5} n_i} = \exp\left(\frac{755385}{24942}\right)\left(\exp\left(\frac{\lambda_{\mathrm{O}}}{24942}\right)\right)^2, \tag{5.109}$$

$$\frac{n_{\mathrm{H_2O}}}{\sum_{i=1}^{5} n_i} = \exp\left(\frac{974790}{24942}\right)\exp\left(\frac{\lambda_{\mathrm{O}}}{24942}\right)\left(\exp\left(\frac{\lambda_{\mathrm{H}}}{24942}\right)\right)^2, \tag{5.110}$$

$$\frac{n_{\mathrm{OH}}}{\sum_{i=1}^{5} n_i} = \exp\left(\frac{641903}{24942}\right)\exp\left(\frac{\lambda_{\mathrm{O}}}{24942}\right)\exp\left(\frac{\lambda_{\mathrm{H}}}{24942}\right), \tag{5.111}$$

$$\frac{n_{\mathrm{H}}}{\sum_{i=1}^{5} n_i} = \exp\left(\frac{213961}{24942}\right)\exp\left(\frac{\lambda_{\mathrm{H}}}{24942}\right), \tag{5.112}$$

$$\frac{n_{\mathrm{O}}}{\sum_{i=1}^{5} n_i} = \exp\left(\frac{323371}{24942}\right)\exp\left(\frac{\lambda_{\mathrm{O}}}{24942}\right), \tag{5.113}$$

$$4 = 2n_{\mathrm{H_2}} + 2n_{\mathrm{H_2O}} + n_{\mathrm{OH}} + n_{\mathrm{H}}, \tag{5.114}$$

$$2 = 2n_{\mathrm{O_2}} + n_{\mathrm{H_2O}} + n_{\mathrm{OH}} + n_{\mathrm{O}}. \tag{5.115}$$

Then, solve these considering $n_i$, $\exp(\lambda_{\mathrm{O}}/24942)$, and $\exp(\lambda_{\mathrm{H}}/24942)$ as unknowns. The same result is recovered, but a broader range of initial guesses converges to the correct solution.

One can verify that this choice extremizes $G$ by direct computation; moreover, this will show that the extremum is actually a minimum. In so doing, one must exercise care to see that element conservation is retained. As an example, perturb the equilibrium solution for $n_{\mathrm{H_2}}$ and $n_{\mathrm{H}}$ such that

$$n_{\mathrm{H_2}} = 3.19 \times 10^{-1} + \xi, \qquad n_{\mathrm{H}} = 1.36 \times 10^{-1} - 2\xi. \tag{5.116}$$

Leave all other species mol numbers the same. In this way, when $\xi = 0$, one has the original equilibrium solution. For $\xi \neq 0$, the solution moves off the equilibrium value in such a way that elements are conserved. Then, one has $G = \sum_{i=1}^{N} \overline{\mu}_i n_i = G(\xi)$. The difference $G(\xi) - G(0)$ is plotted in Fig. 5.1. When $\xi = 0$, there is no deviation from the value predicted by the Newton-Raphson iteration. Clearly when $\xi = 0$, $G(\xi) - G(0)$, takes on a minimum value, and so then does $G(\xi)$. So, the procedure works.

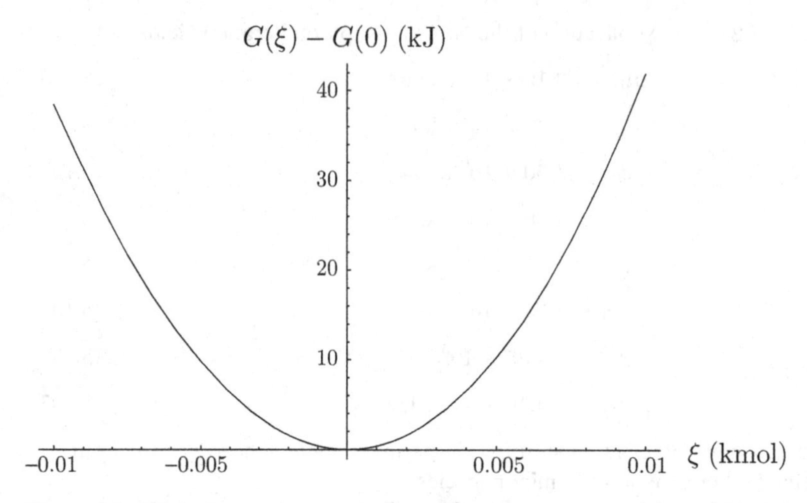


Figure 5.1. Gibbs free energy variation with mixture composition maintaining element conservation for mixture of $H_2$, $O_2$, $H_2O$, OH, H, and O at $T = 3000$ K, $P = 100$ kPa.

At the lower temperature, $T = 298$ K, application of the same procedure yields

$$n_{H_2} = 4.88 \times 10^{-27} \text{ kmol}, \tag{5.117}$$

$$n_{O_2} = 2.44 \times 10^{-27} \text{ kmol}, \tag{5.118}$$

$$n_{H_2O} = 2.00 \times 10^{0} \text{ kmol}, \tag{5.119}$$

$$n_{OH} = 2.22 \times 10^{-29} \text{ kmol}, \tag{5.120}$$

$$n_{H} = 2.29 \times 10^{-49} \text{ kmol}, \tag{5.121}$$

$$n_{O} = 1.67 \times 10^{-54} \text{ kmol}, \tag{5.122}$$

$$\lambda_{H} = -9.54 \times 10^{4} \text{ kJ/kmol}, \tag{5.123}$$

$$\lambda_{O} = -1.07 \times 10^{5} \text{ kJ/kmol}. \tag{5.124}$$

Clearly at the low temperature, $T = 298$ K, there is one dominant product, $H_2$. All other species have many orders of magnitude fewer number.

At the intermediate temperature, $T = 1000$ K, application of the same procedure shows the minor species become slightly more prominent:

$$n_{H_2} = 4.99 \times 10^{-7} \text{ kmol}, \tag{5.125}$$

$$n_{O_2} = 2.44 \times 10^{-7} \text{ kmol}, \tag{5.126}$$

$$n_{H_2O} = 2.00 \times 10^{0} \text{ kmol}, \tag{5.127}$$

$$n_{OH} = 2.09 \times 10^{-8} \text{ kmol}, \tag{5.128}$$

$$n_{H} = 2.26 \times 10^{-12} \text{ kmol}, \tag{5.129}$$

$$n_{O} = 1.10 \times 10^{-13} \text{ kmol}, \tag{5.130}$$

$$\lambda_{H} = -1.36 \times 10^{5} \text{ kJ/kmol}, \tag{5.131}$$

$$\lambda_{O} = -1.77 \times 10^{5} \text{ kJ/kmol}. \tag{5.132}$$

At $T = 1000$ K, $H_2O$ still dominates, but one finds larger numbers of minor species, especially OH.

### 5.2.2 Equilibration of All Reactions

In another equivalent method, if one commences with a multireaction model, one can require each reaction to be in equilibrium. This leads to a set of algebraic equations for $r_j = 0$, that from Eq. (5.15) leads to

$$K_{c,j} = \left(\frac{P_0}{\overline{R}T}\right)^{\sum_{i=1}^{N}\nu_{ij}} \exp\left(\frac{-\Delta G_j^0}{\overline{R}T}\right) = \prod_{k=1}^{N} \overline{\rho}_k^{\nu_{kj}}, \qquad j = 1,\ldots,J. \quad (5.133)$$

With some effort it can be shown that not all of the $J$ equations are linearly independent. Moreover, they do not possess a unique solution for the equilibrium. However, for closed systems, only one of the solutions is physical, as will be shown in the following Sec. 5.2.3. The others typically involve nonphysical, negative concentrations.

That said, Eqs. (5.133) are entirely consistent with the predictions of the $N + L$ equations that arise from extremization of Gibbs free energy while enforcing element number constraints. This can be shown by beginning with Eq. (5.64), rewritten in terms of molar concentrations, and performing the following sequence of operations:

$$\overline{\mu}_{T,i}^0 + \overline{R}T \ln\left(\frac{n_i/V}{\sum_{k=1}^{N} n_k/V}\frac{P}{P_0}\right) - \sum_{l=1}^{L}\lambda_l\phi_{li} = 0, \qquad i = 1,\ldots,N, \quad (5.134)$$

$$\overline{\mu}_{T,i}^0 + \overline{R}T \ln\left(\frac{\overline{\rho}_i}{\sum_{k=1}^{N} \overline{\rho}_k}\frac{P}{P_0}\right) - \sum_{l=1}^{L}\lambda_l\phi_{li} = 0, \qquad i = 1,\ldots,N, \quad (5.135)$$

$$\overline{\mu}_{T,i}^0 + \overline{R}T \ln\left(\frac{\overline{\rho}_i}{\overline{\rho}}\frac{P}{P_0}\right) - \sum_{l=1}^{L}\lambda_l\phi_{li} = 0, \qquad i = 1,\ldots,N, \quad (5.136)$$

$$\overline{\mu}_{T,i}^0 + \overline{R}T \ln\left(\overline{\rho}_i\frac{\overline{R}T}{P_0}\right) - \sum_{l=1}^{L}\lambda_l\phi_{li} = 0, \qquad i = 1,\ldots,N, \quad (5.137)$$

$$\nu_{ij}\overline{\mu}_{T,i}^0 + \nu_{ij}\overline{R}T \ln\left(\overline{\rho}_i\frac{\overline{R}T}{P_0}\right) - \nu_{ij}\sum_{l=1}^{L}\lambda_l\phi_{li} = 0, \qquad i = 1,\ldots,N, \quad j = 1,\ldots,J. \quad (5.138)$$

Continuing, we see

$$\underbrace{\sum_{i=1}^{N}\nu_{ij}\overline{\mu}_{T,i}^0}_{\Delta G_j^0} + \sum_{i=1}^{N}\nu_{ij}\overline{R}T \ln\left(\overline{\rho}_i\frac{\overline{R}T}{P_0}\right) - \sum_{i=1}^{N}\nu_{ij}\sum_{l=1}^{L}\lambda_l\phi_{li} = 0, \qquad j = 1,\ldots,J, \quad (5.139)$$

$$\Delta G_j^0 + \overline{R}T\sum_{i=1}^{N}\nu_{ij}\ln\left(\overline{\rho}_i\frac{\overline{R}T}{P_0}\right) - \sum_{l=1}^{L}\lambda_l\underbrace{\sum_{i=1}^{N}\phi_{li}\nu_{ij}}_{0} = 0, \qquad j = 1,\ldots,J. \quad (5.140)$$

Simplifying, we get

$$\Delta G_j^0 + \overline{R}T\sum_{i=1}^{N}\nu_{ij}\ln\left(\overline{\rho}_i\frac{\overline{R}T}{P_0}\right) = 0, \qquad j = 1,\ldots,J. \quad (5.141)$$

Here, the stoichiometry for each reaction has been employed to remove the effect of the Lagrange multipliers. Continue to find

$$\sum_{i=1}^{N} \ln\left(\overline{\rho}_i \frac{\overline{R}T}{P_0}\right)^{\nu_{ij}} = -\frac{\Delta G_j^0}{\overline{R}T}, \qquad j = 1, \ldots, J, \tag{5.142}$$

$$\exp\left(\sum_{i=1}^{N} \ln\left(\overline{\rho}_i \frac{\overline{R}T}{P_0}\right)^{\nu_{ij}}\right) = \exp\left(-\frac{\Delta G_j^0}{\overline{R}T}\right), \qquad j = 1, \ldots, J, \tag{5.143}$$

$$\prod_{i=1}^{N} \left(\overline{\rho}_i \frac{\overline{R}T}{P_0}\right)^{\nu_{ij}} = \exp\left(-\frac{\Delta G_j^0}{\overline{R}T}\right), \qquad j = 1, \ldots, J, \tag{5.144}$$

$$\left(\frac{\overline{R}T}{P_0}\right)^{\sum_{i=1}^{N}\nu_{ij}} \prod_{i=1}^{N} \overline{\rho}_i^{\nu_{ij}} = \exp\left(-\frac{\Delta G_j^0}{\overline{R}T}\right), \qquad j = 1, \ldots, J, \tag{5.145}$$

$$\prod_{i=1}^{N} \overline{\rho}_i^{\nu_{ij}} = \underbrace{\left(\frac{P_0}{\overline{R}T}\right)^{\sum_{i=1}^{N}\nu_{ij}} \exp\left(-\frac{\Delta G_j^0}{\overline{R}T}\right)}_{K_{c,j}}, \qquad j = 1, \ldots, J, \tag{5.146}$$

$$\prod_{i=1}^{N} \overline{\rho}_i^{\nu_{ij}} = K_{c,j}, \qquad j = 1, \ldots, J. \tag{5.147}$$

Extremization of $G$ is consistent with equilibrating each of the $J$ reactions.

### 5.2.3 Zel'dovich's Uniqueness Proof

Here, a detailed proof is given for the global uniqueness and stability of the chemical equilibrium point. We follow a procedure given in a little known paper by Zel'dovich (1938). The proof follows the basic outline of Zel'dovich, but the notation will be consistent with the present development. We adapt large portions of the text given originally by Powers and Paolucci (2008).[6] There citation is given for conditions under which uniqueness may not exist. Among those features that can induce nonunique physically acceptable equilibria are open systems and nonideal mixtures. Some complementary mathematical details in the context of network theory are given by Feinberg (1995a, 1995b).

The uniqueness or nonuniqueness of a physical equilibrium is critical in studies of the nonlinear dynamics of a chemically reacting system. As studied in the examples of Sec. 1.2, it is also essential to determine the stability of each equilibrium. If cases exist where multiple physical stable equilibria exist, the full nonlinear dynamics of nonequilibrium thermodynamics becomes important in determining which stable equilibria is approached from a given initial condition. We do not address this important problem, though others have.

The key result is summarized as follows. *In the physically accessible region of composition space for a closed, spatially homogeneous, ideal mixture of ideal gases, there exists a unique equilibrium point in each of four cases: (1) isothermal, isochoric,*

[6]Reproduced with permission from J. M. Powers and S. Paolucci, Uniqueness of chemical equilibria in ideal mixtures of ideal gases, *American Journal of Physics*, 76(9): 848–855, Copyright 2008, American Association of Physics Teachers.

*(2) isothermal, isobaric, (3) adiabatic, isochoric, and (4) adiabatic, isobaric. The equilibrium is globally stable within the entire physical domain of composition space.* We give full details mainly because they are difficult to find elsewhere and the result is important. It is possible to accept the proof and proceed straight to Sec. 5.3, where relevant numerical examples are given.

### Isothermal, Isochoric Case

Consider an ideal mixture of ideal gases in a closed fixed volume $V$ at fixed temperature $T$. For such a system, the canonical equilibration relation is given by Eq. (3.276), $dA|_{T,V} \leq 0$. So, $A$ must be always decreasing until it reaches a minimum. Consider then $A$. First, combining Eqs. (3.176) and (3.177), one finds

$$A = -PV + G. \tag{5.148}$$

Now, from Eq. (3.277) one can eliminate $G$ to get

$$A = -PV + \sum_{i=1}^{N} n_i \overline{\mu}_i. \tag{5.149}$$

From the ideal gas law, $PV = n\overline{R}T$, and again with $n = \sum_{i=1}^{N} n_i$, one gets

$$A = -n\overline{R}T + \sum_{i=1}^{N} n_i \overline{\mu}_i, \tag{5.150}$$

$$= -\sum_{i=1}^{N} n_i \overline{R}T + \sum_{i=1}^{N} n_i \left( \overline{\mu}_{T,i}^0 + \overline{R}T \ln\left(\frac{P_i}{P_0}\right)\right), \tag{5.151}$$

$$= \overline{R}T \sum_{i=1}^{N} n_i \left( \frac{\overline{\mu}_{T,i}^0}{\overline{R}T} - 1 + \ln\left(\frac{n_i P}{n P_0}\right)\right), \tag{5.152}$$

$$= \overline{R}T \sum_{i=1}^{N} n_i \left( \frac{\overline{\mu}_{T,i}^0}{\overline{R}T} - 1 + \ln\left(n_i \frac{\overline{R}T}{P_0 V}\right)\right). \tag{5.153}$$

For convenience, define, for this isothermal, isochoric problem $n^0$, the total number of moles at the reference pressure, which, for this isothermal, isochoric problem, is a constant:

$$n^0 \equiv \frac{P_0 V}{\overline{R}T}. \tag{5.154}$$

So

$$A = \overline{R}T \sum_{i=1}^{N} n_i \left( \frac{\overline{\mu}_{T,i}^0}{\overline{R}T} - 1 + \ln\left(\frac{n_i}{n^0}\right)\right). \tag{5.155}$$

Now, recall that atomic element conservation, Eq. (5.57), demands that

$$\sum_{i=1}^{N} \phi_{li}(n_{io} - n_i) = 0, \qquad l = 1, \ldots, L. \tag{5.156}$$

As defined earlier in Sec. 1.2.1, $\phi_{li}$ is the number of atoms of element $l$ in species $i$; note that $\phi_{li} \geq 0$. It is described by a $L \times N$ nonsquare element-species matrix, $\phi$, typically of full rank, $L$. Defining the initial number of moles of element

$l$, $\varepsilon_l = \sum_{i=1}^{N} \phi_{li} n_{i0}$, one can rewrite Eq. (5.156) as

$$\sum_{i=1}^{N} \phi_{li} n_i = \varepsilon_l, \qquad l = 1, \ldots, L. \tag{5.157}$$

Equation (5.157) is generally underconstrained, and one can find solutions of the form

$$\begin{pmatrix} n_1 \\ n_2 \\ \vdots \\ n_N \end{pmatrix} = \begin{pmatrix} n_{10} \\ n_{20} \\ \vdots \\ n_{N0} \end{pmatrix} + \begin{pmatrix} \mathcal{D}_{11} \\ \mathcal{D}_{21} \\ \vdots \\ \mathcal{D}_{N\,1} \end{pmatrix} \xi_1 + \begin{pmatrix} \mathcal{D}_{12} \\ \mathcal{D}_{22} \\ \vdots \\ \mathcal{D}_{N\,2} \end{pmatrix} \xi_2 + \cdots + \begin{pmatrix} \mathcal{D}_{1\,N-L} \\ \mathcal{D}_{2\,N-L} \\ \vdots \\ \mathcal{D}_{N\,N-L} \end{pmatrix} \xi_{N-L}. \tag{5.158}$$

Here, $\mathcal{D}_{ik}$ represents a dimensionless component of a matrix of dimension $N \times (N - L)$. Here, in contrast to earlier analysis, $\xi_i$ is interpreted as having the dimensions of mol. Each of the $N - L$ column vectors of the matrix whose components are $\mathcal{D}_{ik}$ has length $N$ and lies in the right null space of the matrix whose components are $\phi_{li}$. That is,

$$\sum_{i=1}^{N} \phi_{li} \mathcal{D}_{ik} = 0, \qquad k = 1, \ldots, N - L, \qquad l = 1, \ldots, L. \tag{5.159}$$

There is an exception to this rule when the left null space of $\nu_{ij}$ is of higher dimension than the right null space of $\phi_{li}$. In such a case, there are quantities conserved in addition to the elements as a consequence of the form of the reaction law. The most common exception occurs when each of the reactions also conserves the number of molecules. In such a case, there will be $N - L - 1$ free variables, rather than $N - L$. One can robustly form $\mathcal{D}_{ij}$ from the set of independent column space vectors of $\nu_{ij}$. These vectors are included in the right null space of $\phi_{li}$ because $\sum_{i=1}^{N} \phi_{li} \nu_{ij} = 0$.

One can have $\mathcal{D}_{ik} \in (-\infty, \infty)$. Each of these is a function of $\phi_{li}$. In matrix form, one can say

$$\begin{pmatrix} n_1 \\ n_2 \\ \vdots \\ n_N \end{pmatrix} = \begin{pmatrix} n_{10} \\ n_{20} \\ \vdots \\ n_{N0} \end{pmatrix} + \begin{pmatrix} \mathcal{D}_{11} & \mathcal{D}_{12} & \ldots & \mathcal{D}_{1\,N-L} \\ \mathcal{D}_{21} & \mathcal{D}_{22} & \ldots & \mathcal{D}_{2\,N-L} \\ \vdots & \vdots & \vdots & \vdots \\ \mathcal{D}_{N1} & \mathcal{D}_{N2} & \ldots & \mathcal{D}_{N\,N-L} \end{pmatrix} \begin{pmatrix} \xi_1 \\ \xi_2 \\ \vdots \\ \xi_{N-L} \end{pmatrix}. \tag{5.160}$$

In index form, this becomes, in an equation scaled slightly differently than the earlier studied Eq. (1.55),

$$n_i = n_{i0} + \sum_{k=1}^{N-L} \mathcal{D}_{ik} \xi_k, \qquad i = 1, \ldots, N. \tag{5.161}$$

It is also easy to show that the $N - L$ column space vectors in $\mathcal{D}_{ik}$ are linear combinations of $N - L$ column space vectors that span the column space of the rank-deficient $N \times J$ components of $\nu_{ij}$. The $N$ values of $n_i$ are uniquely determined once $N - L$ values of $\xi_k$ are specified. That is, a set of independent $\xi_k$, $k = 1, \ldots, N - L$, is sufficient to describe the system. This insures the initial element concentrations are always maintained. One can also develop a $J$-reaction generalization of the single reaction Eq. (4.55). Let $\xi_{kj}$, $k = 1, \ldots, N - L$, $j = 1, \ldots, J$, be the extension of $\xi_k$. Then, the appropriate generalization of Eq. (4.55) in index and Gibbs's boldfaced notation is

$$\nu_{ij} = \sum_{k=1}^{N-L} \mathcal{D}_{ik}\xi_{kj}, \qquad \boldsymbol{\nu} = \mathbf{D} \cdot \boldsymbol{\Xi}. \tag{5.162}$$

One can find the matrix $\boldsymbol{\Xi}$ by the following operations:

$$\mathbf{D} \cdot \boldsymbol{\Xi} = \boldsymbol{\nu}, \tag{5.163}$$

$$\mathbf{D}^T \cdot \mathbf{D} \cdot \boldsymbol{\Xi} = \mathbf{D}^T \cdot \boldsymbol{\nu}, \tag{5.164}$$

$$\boldsymbol{\Xi} = \left(\mathbf{D}^T \cdot \mathbf{D}\right)^{-1} \cdot \mathbf{D}^T \cdot \boldsymbol{\nu}. \tag{5.165}$$

Returning to the primary exercise, one can form the partial derivative of Eq. (5.161):

$$\left.\frac{\partial n_i}{\partial \xi_p}\right|_{\xi_j} = \sum_{k=1}^{N-L} \mathcal{D}_{ik} \left.\frac{\partial \xi_k}{\partial \xi_p}\right|_{\xi_j}, \quad i = 1, \ldots, N; p = 1, \ldots, N-L, \tag{5.166}$$

$$= \sum_{k=1}^{N-L} \mathcal{D}_{ik}\delta_{kp} = \mathcal{D}_{ip}, \quad i = 1, \ldots, N; p = 1, \ldots, N-L. \tag{5.167}$$

Next, return to consideration of Eq. (5.153). It is sought to minimize $A$ while holding $T$ and $V$ constant. The only available variables are $n_i$, $i = 1, \ldots, N$. These are not fully independent, but they are known in terms of the independent $\xi_p$, $p = 1, \ldots, N-L$. So, one can find an extremum of $A$ by differentiating it with respect to each of the $\xi_p$ and setting each derivative to zero:

$$\left.\frac{\partial A}{\partial \xi_p}\right|_{\xi_j,T,V} = \overline{R}T \sum_{i=1}^{N} \left( \left.\frac{\partial n_i}{\partial \xi_p}\right|_{\xi_j} \left(\frac{\overline{\mu}_{T,i}^0}{\overline{R}T} - 1\right) + \left.\frac{\partial n_i}{\partial \xi_p}\right|_{\xi_j} \ln\left(\frac{n_i}{n^0}\right) + n_i \frac{\partial}{\partial \xi_p} \ln\left(\frac{n_i}{n^0}\right)\right) = 0, \quad p = 1, \ldots, N-L, \tag{5.168}$$

$$= \overline{R}T \sum_{i=1}^{N} \left( \left.\frac{\partial n_i}{\partial \xi_p}\right|_{\xi_j} \left(\frac{\overline{\mu}_{T,i}^0}{\overline{R}T} - 1\right) + \left.\frac{\partial n_i}{\partial \xi_p}\right|_{\xi_j} \ln\left(\frac{n_i}{n^0}\right) + n_i \sum_{q=1}^{N} \frac{\partial n_q}{\partial \xi_p} \frac{\partial}{\partial n_q} \ln\left(\frac{n_i}{n^0}\right)\right) = 0, \quad p = 1, \ldots, N-L, \tag{5.169}$$

$$= \overline{R}T \sum_{i=1}^{N} \left( \mathcal{D}_{ip} \left(\frac{\overline{\mu}_{T,i}^0}{\overline{R}T} - 1\right) + \mathcal{D}_{ip} \ln\left(\frac{n_i}{n^0}\right) + n_i \sum_{q=1}^{N} \mathcal{D}_{qp} \frac{1}{n_i} \delta_{iq}\right) = 0, \quad p = 1, \ldots, N-L, \tag{5.170}$$

$$= \overline{R}T \sum_{i=1}^{N} \left( \mathcal{D}_{ip} \left(\frac{\overline{\mu}_{T,i}^0}{\overline{R}T} - 1\right) + \mathcal{D}_{ip} \ln\left(\frac{n_i}{n^0}\right) + \mathcal{D}_{ip}\right) = 0, \quad p = 1, \ldots, N-L, \tag{5.171}$$

$$= \overline{R}T \sum_{i=1}^{N} \mathcal{D}_{ip} \left(\frac{\overline{\mu}_{T,i}^0}{\overline{R}T} + \ln\left(\frac{n_i}{n^0}\right)\right) = 0, \quad p = 1, \ldots, N-L. \tag{5.172}$$

Following Zel'dovich, one then rearranges Eq. (5.172) to define the equations for equilibrium:

$$\sum_{i=1}^{N} \mathcal{D}_{ip} \ln\left(\frac{n_i}{n^0}\right) = -\sum_{i=1}^{N} \frac{\mathcal{D}_{ip}\overline{\mu}_{T,i}^0}{\overline{R}T}, \quad p = 1, \ldots, N-L, \tag{5.173}$$

$$\sum_{i=1}^{N} \ln\left(\frac{n_i}{n^0}\right)^{\mathcal{D}_{ip}} = -\sum_{i=1}^{N} \frac{\mathcal{D}_{ip}\overline{\mu}_{T,i}^0}{\overline{R}T}, \quad p = 1, \ldots, N-L, \tag{5.174}$$

$$\ln \prod_{i=1}^{N} \left(\frac{n_i}{n^0}\right)^{\mathcal{D}_{ip}} = -\sum_{i=1}^{N} \frac{\mathcal{D}_{ip}\overline{\mu}_{T,i}^0}{\overline{R}T}, \quad p = 1, \ldots, N-L, \tag{5.175}$$

$$\ln \prod_{i=1}^{N} \left(\left(n_{i0} + \sum_{k=1}^{N-L} \mathcal{D}_{ik}\xi_k\right) \frac{1}{n^0}\right)^{\mathcal{D}_{ip}} = -\sum_{i=1}^{N} \frac{\mathcal{D}_{ip}\overline{\mu}_{T,i}^0}{\overline{R}T}, \quad p = 1, \ldots, N-L. \tag{5.176}$$

Equation (5.176) forms $N - L$ equations in the $N - L$ unknown values of $\xi_p$ and can be solved by Newton-Raphson iteration. As of yet, no proof exists that this is a unique solution. Nor is it certain whether or not $A$ is maximized or minimized at such a solution point.

One also notices that the method of Zel'dovich is consistent with a more rudimentary form. Rearrange Eq. (5.172) to get at equilibrium

$$\left.\frac{\partial A}{\partial \xi_p}\right|_{\xi_j,T,V} = \sum_{i=1}^{N} \mathcal{D}_{ip}\left(\overline{\mu}_{T,i}^0 + \overline{R}T \ln\left(\frac{n_i}{n^0}\right)\right) = 0, \quad p = 1, \ldots, N-L, \tag{5.177}$$

$$= \sum_{i=1}^{N} \mathcal{D}_{ip}\left(\overline{\mu}_{T,i}^0 + \overline{R}T \ln\left(\frac{n_i P}{n P_0}\right)\right) = 0, \tag{5.178}$$

$$= \sum_{i=1}^{N} \mathcal{D}_{ip}\left(\overline{\mu}_{T,i}^0 + \overline{R}T \ln\left(\frac{P_i}{P_0}\right)\right) = \sum_{i=1}^{N} \overline{\mu}_i \mathcal{D}_{ip} = 0. \tag{5.179}$$

Now, Eq. (5.179) is easily found via another method. Recall Eq. (3.180), force it to zero so as to extremize $A$, and then operate on it in an isochoric, isothermal limit, taking its derivative with respect to $\xi_p, p = 1, \ldots, N - L$:

$$dA = -SdT - PdV + \sum_{i=1}^{N} \overline{\mu}_i \, dn_i = 0, \tag{5.180}$$

$$dA|_{T,V} = \sum_{i=1}^{N} \overline{\mu}_i \, dn_i = 0, \tag{5.181}$$

$$\left.\frac{\partial A}{\partial \xi_p}\right|_{\xi_j,T,V} = \sum_{i=1}^{N} \overline{\mu}_i \frac{\partial n_i}{\partial \xi_p} = \sum_{i=1}^{N} \overline{\mu}_i \mathcal{D}_{ip} = 0. \tag{5.182}$$

One can determine whether extreme solutions of Eq. (5.176) are maxima or minima by examining the second derivative, given by differentiating Eq. (5.172). We find the Hessian[7] of $A$, $\mathbf{H}_A$, to be

$$\mathbf{H}_A = \frac{\partial^2 A}{\partial \xi_j \partial \xi_p} = \overline{R}T \sum_{i=1}^{N} \mathcal{D}_{ip} \frac{\partial}{\partial \xi_j} \ln\left(\frac{n_i}{n^0}\right), \tag{5.183}$$

$$= \overline{R}T \sum_{i=1}^{N} \mathcal{D}_{ip} \left( \frac{\partial}{\partial \xi_j} \ln n_i - \frac{\partial}{\partial \xi_j} \ln n^0 \right), \tag{5.184}$$

$$= \overline{R}T \sum_{i=1}^{N} \mathcal{D}_{ip} \frac{1}{n_i} \frac{\partial n_i}{\partial \xi_j} = \overline{R}T \sum_{i=1}^{N} \frac{\mathcal{D}_{ip}\mathcal{D}_{ij}}{n_i}. \tag{5.185}$$

Here, $j = 1, \ldots, N - L$. Scaling each of the rows of $\mathcal{D}_{ip}$ by a constant does not affect the rank. So, we can say that the $N \times (N - L)$ matrix whose entries are $\mathcal{D}_{ip}/\sqrt{n_i}$ has rank $N - L$, and consequently the Hessian $\mathbf{H}_A$, of dimension $(N - L) \times (N - L)$, has full rank $N - L$, and is symmetric. It is easy to show by means of *singular value decomposition* (Strang, 2005), or other methods, that the eigenvalues of a full rank symmetric square matrix are all real and nonzero.

Now, consider whether $\partial^2 A/\partial \xi_j \partial \xi_p$ is positive definite. By definition, it is positive definite if for an arbitrary vector $z_i$ of length $N - L$ with nonzero norm that the term $\Upsilon_{TV}$, defined next, be always positive for nonzero $z_i$:

$$\Upsilon_{TV} = \sum_{j=1}^{N-L} \sum_{p=1}^{N-L} \frac{\partial^2 A}{\partial \xi_j \partial \xi_p} z_j z_p > 0. \tag{5.186}$$

Substitute Eq. (5.185) into Eq. (5.186) to find

$$\Upsilon_{TV} = \sum_{j=1}^{N-L} \sum_{p=1}^{N-L} \overline{R}T \sum_{i=1}^{N} \frac{\mathcal{D}_{ip}\mathcal{D}_{ij}}{n_i} z_j z_p, \tag{5.187}$$

$$= \overline{R}T \sum_{i=1}^{N} \frac{1}{n_i} \sum_{j=1}^{N-L} \sum_{p=1}^{N-L} \mathcal{D}_{ip}\mathcal{D}_{ij} z_j z_p, \tag{5.188}$$

$$= \overline{R}T \sum_{i=1}^{N} \frac{1}{n_i} \sum_{j=1}^{N-L} \mathcal{D}_{ij} z_j \sum_{p=1}^{N-L} \mathcal{D}_{ip} z_p, \tag{5.189}$$

$$= \overline{R}T \sum_{i=1}^{N} \frac{1}{n_i} \left( \sum_{j=1}^{N-L} \mathcal{D}_{ij} z_j \right) \left( \sum_{p=1}^{N-L} \mathcal{D}_{ip} z_p \right). \tag{5.190}$$

Define now

$$y_i \equiv \sum_{j=1}^{N-L} \mathcal{D}_{ij} z_j, \qquad i = 1, \ldots, N. \tag{5.191}$$

[7] After Ludwig Otto Hesse (1811–1874), German mathematician.

This yields

$$\Upsilon_{TV} = \overline{R}T \sum_{i=1}^{N} \frac{y_i^2}{n_i}. \tag{5.192}$$

Now, to restrict the domain to physically accessible space, one is only concerned with $n_i \geq 0, T > 0$, so for arbitrary nonzero $y_i$, one finds $\Upsilon_{TV} > 0$, so one concludes that the second mixed partial of $A$ is positive definite *globally*. Our analysis here did not rely upon any linearization near equilibrium, assuring its global validity.

We next consider the behavior of $A$ near a generic point in the physical space $\hat{\boldsymbol{\xi}}$ where $A = \hat{A}$. If we let $d\boldsymbol{\xi}$ represent an differential deviation of $\boldsymbol{\xi}$ from $\hat{\boldsymbol{\xi}}$, we can represent the Helmholtz free energy by the Taylor series

$$A(\boldsymbol{\xi}) = \hat{A} + d\boldsymbol{\xi}^T \cdot \nabla A + \frac{1}{2} d\boldsymbol{\xi}^T \cdot \mathbf{H}_A \cdot d\boldsymbol{\xi} + \cdots, \tag{5.193}$$

where $\nabla A$ and $\mathbf{H}_A$ are evaluated at $\hat{\boldsymbol{\xi}}$. Now, if $\hat{\boldsymbol{\xi}} = \boldsymbol{\xi}^{eq}$, where *eq* denotes an equilibrium point, $\nabla A = \mathbf{0}$, $\hat{A} = A^{eq}$, and

$$A(\boldsymbol{\xi}) - A^{eq} = \frac{1}{2} d\boldsymbol{\xi}^T \cdot \mathbf{H}_A \cdot d\boldsymbol{\xi} + \cdots. \tag{5.194}$$

Because $\mathbf{H}_A$ is positive definite in the entire physical domain, any isolated critical point will be a minimum. If more than one isolated minimum point of $A$ were to exist in the domain interior, a maximum would also have to exist in the interior, but maxima are not allowed by the global positive definite nature of $\mathbf{H}_A$. Subsequently, any extremum that exists away from the boundary of the physical region must be a minimum, and the minimum is global.

Global positive definiteness of $\mathbf{H}_A$ alone does not rule out the possibility of nonisolated multiple equilibria, as seen by the following analysis. Because it is symmetric, $\mathbf{H}_A$ can be orthogonally decomposed into $\mathbf{H}_A = \mathbf{Q}^T \cdot \mathbf{\Lambda} \cdot \mathbf{Q}$, where $\mathbf{Q}$ is an orthogonal matrix whose columns are the normalized eigenvectors of $\mathbf{H}_A$. Note that $\mathbf{Q}^T = \mathbf{Q}^{-1}$. Also $\mathbf{\Lambda}$ is a diagonal matrix with real eigenvalues on its diagonal. We can effect a volume-preserving rotation of axes by taking the transformation $d\tilde{\boldsymbol{\xi}} = \mathbf{Q} \cdot d\boldsymbol{\xi}$; thus, $d\boldsymbol{\xi} = \mathbf{Q}^T \cdot d\tilde{\boldsymbol{\xi}}$. Hence,

$$A - A^{eq} = \frac{1}{2} (\mathbf{Q}^T \cdot d\tilde{\boldsymbol{\xi}})^T \cdot \mathbf{H}_A \cdot \mathbf{Q}^T \cdot d\tilde{\boldsymbol{\xi}}, \tag{5.195}$$

$$= \frac{1}{2} d\tilde{\boldsymbol{\xi}}^T \cdot \mathbf{Q} \cdot \mathbf{Q}^T \cdot \mathbf{\Lambda} \cdot \mathbf{Q} \cdot \mathbf{Q}^T \cdot d\tilde{\boldsymbol{\xi}} = \frac{1}{2} d\tilde{\boldsymbol{\xi}}^T \cdot \mathbf{\Lambda} \cdot d\tilde{\boldsymbol{\xi}}. \tag{5.196}$$

Application of these transformations gives in the near-equilibrium quadratic form

$$A - A^{eq} = \frac{1}{2} \sum_{p=1}^{N-L} \lambda_p (d\tilde{\xi}_p)^2. \tag{5.197}$$

For $A$ to be a unique minimum, $\lambda_p > 0$. If one or more of the $\lambda_p = 0$, then the minimum could be realized on a line or higher-dimensional plane, depending on how many zeros are present. The positive definite nature along with symmetry of $\mathbf{H}_A$ guarantees that $\lambda_p > 0$. Had the symmetric $\mathbf{H}_A$ been positive semidefinite, at least one of the $\lambda_p$ could have been zero, and uniqueness would have been lost. For our problem the unique global minimum that exists in the interior will exist at a unique point.

One must check the boundary of the physical region to see if it can form an extremum. Near a physical boundary given by $n_q = 0$, one finds that $A$ behaves as

$$\lim_{n_q \to 0} A \sim \overline{R}T \sum_{i=1}^{N} n_i \left( \frac{\overline{\mu}_{T,i}}{\overline{R}T} - 1 + \ln\left(\frac{n_i}{n^0}\right)\right) \to \text{finite.} \tag{5.198}$$

The behavior of $A$ itself is finite because $\lim_{n_q \to 0} n_q \ln n_q = 0$, and the remaining terms in the summation are nonzero and finite.

The analysis of the behavior of the derivative of $A$ on a boundary of $n_q = 0$ is more complex. We require that $n_q \geq 0$ for all $N$ species. The hyperplanes given by $n_q = 0$ define a closed physical boundary in reduced composition space. We require that changes in $n_q$ that originate from the surface $n_q = 0$ be positive. For such curves we thus require that near $n_q = 0$, perturbations $d\xi_k$ be such that

$$dn_q = \sum_{k=1}^{N-L} \mathcal{D}_{qk}\, d\xi_k > 0. \tag{5.199}$$

We next examine changes in $A$ near boundaries given by $n_q = 0$. We restrict attention to changes that give rise to $dn_q > 0$. We employ Eq. (5.177) and find that

$$dA = \sum_{p=1}^{N-L} d\xi_p \left.\frac{\partial A}{\partial \xi_p}\right|_{T,V,\xi_{j \neq p}} = \sum_{i=1}^{N} \left(\overline{\mu}_i^0 + \overline{R}T \ln\left(\frac{n_i}{n^0}\right)\right) \sum_{p=1}^{N-L} \mathcal{D}_{ip}\, d\xi_p. \tag{5.200}$$

On the boundary given by $n_q = 0$, the dominant term in the sum is for $i = q$, so

$$\lim_{n_q \to 0} dA = \overline{R}T \ln\left(\frac{n_q}{n^0}\right) \underbrace{\sum_{p=1}^{N-L} \mathcal{D}_{qp}\, d\xi_p}_{>0} \to -\infty. \tag{5.201}$$

The term identified by the brace is positive because of Eq. (5.199). Because $\overline{R}$ and $T > 0$, as $n_q$ moves away from zero into the physical region, changes in $A$ are large and negative. So, the physical boundary can be a local maximum, but never a local minimum in $A$. Hence, the only admissible equilibrium is the unique minimum of $A$ found from Eq. (5.176); this equilibrium is found at a unique point in reduced composition space.

Last, let us determine if we can identify an appropriate Lyapunov[8] function for the isothermal isochoric system. Let us define the deviation of $A$ and $\xi_i$ from equilibrium as $\tilde{A}$ and $\tilde{\xi}_i$:

$$\tilde{A} = A - A^{eq}, \qquad \tilde{\xi}_i = \xi_i - \xi_i^{eq}. \tag{5.202}$$

For $\tilde{A}$ to be a Lyapunov function, it must be such that

- $\tilde{A}|_{\tilde{\xi}_i=0} = 0.$
- $\tilde{A}$ is positive definite, so $\tilde{A}|_{\tilde{\xi}_i \neq 0} > 0.$
- $d\tilde{A}/dt|_{\tilde{\xi}_i=0} = 0.$
- $d\tilde{A}/dt$ is negative definite, so $d\tilde{A}/dt|_{\tilde{\xi}_i \neq 0} < 0.$

Such a function will have global stability in that it is everywhere decreasing, except at a single point. And one it reaches that point, it is guaranteed to remain. By

[8]Aleksandr Mikhailovich Lyapunov (1857–1918), Russian mathematician.

construction $\tilde{A} = 0$ at equilibrium, where $\tilde{\xi}_i = 0$. We have already seen that the equilibrium point is guaranteed to be a global minimum of $A$ within the physical domain, so by construction $\tilde{A}$ is positive definite in the physical domain. Moreover the second law, Eq. (3.276), holds that $dA|_{T,V} \leq 0$. For $dt > 0$, the second law holds $dA/dt|_{T,V} \leq 0$. Because $A^{eq}$ is constant, we can also express the second law as $d\tilde{A}/dt|_{T,V} \leq 0$. So because $\tilde{A} = 0$ when $\tilde{\xi}_i = 0$, because $d\tilde{A}/dt|_{T,V} < 0$ wherever $\tilde{\xi}_i \neq 0$, and $d\tilde{A}/dt|_{T,V} = 0$ when $\tilde{\xi}_i = 0$, we can say that *the deviation of the Helmholtz free energy from its equilibrium value, $\tilde{A}$, is a Lyapunov function for isothermal, isochoric combustion.*

### Isothermal, Isobaric Case

Next, consider the related case in which $T$ and $P$ are constant. For such a system, the canonical equilibration relation given by Eq. (3.276) holds that $dG|_{T,P} \leq 0$. So, $G$ must be decreasing until it reaches a minimum. So consider $G$, which, from Eq. (3.277), is

$$G = \sum_{i=1}^{N} n_i \overline{\mu}_i = \sum_{i=1}^{N} n_i \left( \overline{\mu}_{T,i}^0 + \overline{R}T \ln\left(\frac{P_i}{P_0}\right)\right), \tag{5.203}$$

$$= \sum_{i=1}^{N} n_i \left( \overline{\mu}_{T,i}^0 + \overline{R}T \ln\left(\frac{n_i P}{n P_0}\right)\right), \tag{5.204}$$

$$= \overline{R}T \sum_{i=1}^{N} \left( n_i \left( \frac{\overline{\mu}_{T,i}^0}{\overline{R}T} + \ln\left(\frac{P}{P_0}\right)\right) + n_i \ln\left(\frac{n_i}{\sum_{k=1}^{N} n_k}\right)\right). \tag{5.205}$$

Differentiate with respect to each of the independent variables $\xi_p$ for $p = 1, \ldots, N - L$:

$$\frac{\partial G}{\partial \xi_p} = \overline{R}T \sum_{i=1}^{N} \left( \frac{\partial n_i}{\partial \xi_p} \left( \frac{\overline{\mu}_{T,i}^0}{\overline{R}T} + \ln\left(\frac{P}{P_0}\right)\right) + \frac{\partial}{\partial \xi_p}\left( n_i \ln\left(\frac{n_i}{\sum_{k=1}^{N} n_k}\right)\right)\right), \tag{5.206}$$

$$= \overline{R}T \sum_{i=1}^{N} \left( \frac{\partial n_i}{\partial \xi_p} \left( \frac{\overline{\mu}_{T,i}^0}{\overline{R}T} + \ln\left(\frac{P}{P_0}\right)\right) + \sum_{q=1}^{N} \frac{\partial n_q}{\partial \xi_p} \frac{\partial}{\partial n_q}\left( n_i \ln\left(\frac{n_i}{\sum_{k=1}^{N} n_k}\right)\right)\right), \tag{5.207}$$

$$= \overline{R}T \sum_{i=1}^{N} \left( \mathcal{D}_{ip} \left( \frac{\overline{\mu}_{T,i}^0}{\overline{R}T} + \ln\left(\frac{P}{P_0}\right)\right) \right.$$
$$\left. + \sum_{q=1}^{N} \mathcal{D}_{qp} \left( \frac{-n_i}{\sum_{k=1}^{N} n_k} + \delta_{iq}\left(1 + \ln\left(\frac{n_i}{\sum_{k=1}^{N} n_k}\right)\right)\right)\right), \tag{5.208}$$

$$= \overline{R}T \sum_{i=1}^{N} \left( \mathcal{D}_{ip} \left( \frac{\overline{\mu}_{T,i}^0}{\overline{R}T} + \ln\left(\frac{P}{P_0}\right) + 1 + \ln\left(\frac{n_i}{n}\right)\right) - \frac{n_i}{n} \sum_{q=1}^{N} \mathcal{D}_{qp}\right), \tag{5.209}$$

$$= \overline{R}T \sum_{i=1}^{N} \left( \mathcal{D}_{ip} \left( \frac{\overline{\mu}_{T,i}^0}{\overline{R}T} + \ln\left(\frac{P}{P_0}\right) + 1 + \ln\left(\frac{n_i}{n}\right)\right)\right) - \overline{R}T \sum_{i=1}^{N} \frac{n_i}{n} \sum_{q=1}^{N} \mathcal{D}_{qp}, \tag{5.210}$$

$$= \overline{R}T\sum_{i=1}^{N}\left(\mathcal{D}_{ip}\left(\frac{\overline{\mu}_{T,i}^{0}}{\overline{R}T}+\ln\left(\frac{P}{P_0}\right)+1+\ln\left(\frac{n_i}{n}\right)\right)\right)-\overline{R}T\sum_{q=1}^{N}\mathcal{D}_{qp}\sum_{i=1}^{N}\frac{n_i}{n}, \tag{5.211}$$

$$= \overline{R}T\sum_{i=1}^{N}\left(\mathcal{D}_{ip}\left(\frac{\overline{\mu}_{T,i}^{0}}{\overline{R}T}+\ln\left(\frac{P}{P_0}\right)+1+\ln\left(\frac{n_i}{n}\right)\right)\right)-\overline{R}T\sum_{q=1}^{N}\mathcal{D}_{qp}, \tag{5.212}$$

$$= \overline{R}T\sum_{i=1}^{N}\left(\mathcal{D}_{ip}\left(\frac{\overline{\mu}_{T,i}^{0}}{\overline{R}T}+\ln\left(\frac{P}{P_0}\right)+1+\ln\left(\frac{n_i}{n}\right)\right)\right)-\overline{R}T\sum_{i=1}^{N}\mathcal{D}_{ip}, \tag{5.213}$$

$$= \overline{R}T\sum_{i=1}^{N}\left(\mathcal{D}_{ip}\left(\frac{\overline{\mu}_{T,i}^{0}}{\overline{R}T}+\ln\left(\frac{P}{P_0}\right)+\ln\left(\frac{n_i}{n}\right)\right)\right), \tag{5.214}$$

$$= \sum_{i=1}^{N}\mathcal{D}_{ip}\left(\overline{\mu}_{T,i}^{0}+\overline{R}T\ln\left(\frac{n_iP}{nP_0}\right)\right), \tag{5.215}$$

$$= \sum_{i=1}^{N}\mathcal{D}_{ip}\left(\overline{\mu}_{T,i}^{0}+\overline{R}T\ln\left(\frac{n_iP}{nP_0}\right)\right), \tag{5.216}$$

$$= \sum_{i=1}^{N}\overline{\mu}_i\mathcal{D}_{ip}, \qquad p=1,\ldots,N-L. \tag{5.217}$$

Note this simple result is entirely consistent with a result that could have been deduced by commencing with the alternative Eq. (3.279), $dG|_{T,P}=\sum_{i=1}^{N}\overline{\mu}_i\,dn_i$. Had this simplification been taken, one could readily deduce that

$$\left.\frac{\partial G}{\partial \xi_p}\right|_{\xi_j}=\sum_{i=1}^{N}\overline{\mu}_i\frac{\partial n_i}{\partial \xi_p}=\sum_{i=1}^{N}\overline{\mu}_i\mathcal{D}_{ip}, \qquad p=1,\ldots,N-L. \tag{5.218}$$

Now, to equilibrate, one sets the derivatives to zero to get

$$\left.\frac{\partial G}{\partial \xi_p}\right|_{\xi_j}=\sum_{i=1}^{N}\mathcal{D}_{ip}\left(\overline{\mu}_{T,i}^{0}+\overline{R}T\ln\left(\frac{n_iP}{nP_0}\right)\right)=0, \qquad p=1,\ldots,N-L, \tag{5.219}$$

$$\sum_{i=1}^{N}\ln\left(\frac{n_iP}{nP_0}\right)^{\mathcal{D}_{ip}}=-\sum_{i=1}^{N}\mathcal{D}_{ip}\frac{\overline{\mu}_{T,i}^{0}}{\overline{R}T}, \tag{5.220}$$

$$\ln\prod_{i=1}^{N}\left(\frac{n_iP}{nP_0}\right)^{\mathcal{D}_{ip}}=-\sum_{i=1}^{N}\mathcal{D}_{ip}\frac{\overline{\mu}_{T,i}^{0}}{\overline{R}T}. \tag{5.221}$$

Rearrange and use Eq. (5.161) to get

$$\ln\prod_{i=1}^{N}\left(\frac{\left(n_{i0}+\sum_{k=1}^{N-L}\mathcal{D}_{ik}\xi_k\right)P}{\left(\sum_{q=1}^{N}\left(n_{q0}+\sum_{k=1}^{N-L}\mathcal{D}_{qk}\xi_k\right)\right)P_0}\right)^{\mathcal{D}_{ip}}=-\sum_{i=1}^{N}\mathcal{D}_{ip}\frac{\overline{\mu}_{T,i}^{0}}{\overline{R}T},$$

$$p=1,\ldots,N-L. \tag{5.222}$$

These $N-L$ nonlinear algebraic equations can be solved for the $N-L$ unknown values of $\xi_p$ via an iterative technique.

We can follow the outline of our previous analysis to show that this equilibrium is unique in the physically accessible region of composition space. By differentiating Eq. (5.218) it is seen that the Hessian of $G$, $\mathbf{H}_G = \partial^2 G/\partial\xi_j \partial\xi_p$, is

$$\frac{\partial^2 G}{\partial\xi_j \partial\xi_p} = \overline{R}T \sum_{i=1}^{N} \mathcal{D}_{ip} \frac{\partial}{\partial\xi_j} \ln\left(\frac{n_i}{n}\right), \tag{5.223}$$

$$= \overline{R}T \left( \sum_{i=1}^{N} \mathcal{D}_{ip} \frac{\partial}{\partial\xi_j} \ln n_i - \sum_{i=1}^{N} \mathcal{D}_{ip} \frac{\partial}{\partial\xi_j} \ln n \right), \tag{5.224}$$

$$= \overline{R}T \left( \sum_{i=1}^{N} \mathcal{D}_{ip} \frac{1}{n_i} \frac{\partial n_i}{\partial\xi_j} - \sum_{i=1}^{N} \mathcal{D}_{ip} \frac{1}{n} \frac{\partial n}{\partial\xi_j} \right), \tag{5.225}$$

$$= \overline{R}T \left( \sum_{i=1}^{N} \mathcal{D}_{ip} \frac{1}{n_i} \frac{\partial n_i}{\partial\xi_j} - \sum_{i=1}^{N} \mathcal{D}_{ip} \frac{1}{n} \frac{\partial}{\partial\xi_j} \left( \sum_{q=1}^{N} n_q \right) \right), \tag{5.226}$$

$$= \overline{R}T \left( \sum_{i=1}^{N} \mathcal{D}_{ip} \frac{1}{n_i} \frac{\partial n_i}{\partial\xi_j} - \sum_{i=1}^{N} \left( \sum_{q=1}^{N} \mathcal{D}_{ip} \frac{1}{n} \frac{\partial n_q}{\partial\xi_j} \right) \right), \tag{5.227}$$

$$= \overline{R}T \left( \sum_{i=1}^{N} \frac{\mathcal{D}_{ip}\mathcal{D}_{ij}}{n_i} - \frac{1}{n} \sum_{i=1}^{N} \sum_{q=1}^{N} \mathcal{D}_{ip}\mathcal{D}_{qj} \right). \tag{5.228}$$

Next consider the sum

$$\Upsilon_{TP} = \sum_{j=1}^{N-L} \sum_{p=1}^{N-L} \frac{\partial^2 G}{\partial\xi_j \partial\xi_p} z_j z_p, \tag{5.229}$$

$$= \overline{R}T \sum_{j=1}^{N-L} \sum_{p=1}^{N-L} \left( \sum_{i=1}^{N} \frac{\mathcal{D}_{ip}\mathcal{D}_{ij}}{n_i} - \frac{1}{n} \sum_{i=1}^{N} \sum_{q=1}^{N} \mathcal{D}_{ip}\mathcal{D}_{qj} \right) z_j z_p. \tag{5.230}$$

We use Eq. (5.191) to define $y_i$ and following a long series of calculations, reduce Eq. (5.230), to the positive definite form

$$\Upsilon_{TP} = \frac{\overline{R}T}{n} \sum_{i=1}^{N} \sum_{j=i+1}^{N} \left( \sqrt{\frac{n_j}{n_i}} y_i - \sqrt{\frac{n_i}{n_j}} y_j \right)^2 > 0. \tag{5.231}$$

It is easily verified by direct expansion that Eqs. (5.230) and (5.231) are equivalent. Thus we have a result that will be important in upcoming analysis of Sec. 5.5.2: *the Hessian matrix* $\mathbf{H}_G$ *is globally positive definite.*

On the boundary $n_i = 0$, and as for the isothermal-isochoric case, it can be shown that $dG \to -\infty$ for changes with $dn_i > 0$. Thus, the boundary has no local minimum, and we can conclude that $G$ is minimized in the interior and the minimum is unique. Similarly, it is easy to show *the deviation of the Gibbs free energy from its equilibrium value,* $\tilde{G}$, *is a Lyapunov function for isothermal, isobaric combustion.*

**Adiabatic, Isochoric Case**

One can extend Zel'dovich's proof to other sets of conditions. For example, consider a case that is isochoric and isoenergetic. This corresponds to a chemical reaction in a fixed volume that is thermally insulated. In this case, one operates on Eq. (3.276):

$$\underbrace{dE}_{0} = -P \underbrace{dV}_{0} + T\,dS + \sum_{i=1}^{N} \overline{\mu}_i\,dn_i, \tag{5.232}$$

$$0 = T\,dS + \sum_{i=1}^{N} \overline{\mu}_i\,dn_i, \tag{5.233}$$

$$dS = -\frac{1}{T}\sum_{i=1}^{N} \overline{\mu}_i\,dn_i = -\frac{1}{T}\sum_{i=1}^{N} \overline{\mu}_i \sum_{k=1}^{N-L} \mathcal{D}_{ik}\,d\xi_k, \tag{5.234}$$

$$= -\frac{1}{T}\sum_{i=1}^{N}\sum_{k=1}^{N-L} \overline{\mu}_i \mathcal{D}_{ik}\,d\xi_k, \tag{5.235}$$

$$\frac{\partial S}{\partial \xi_j} = -\frac{1}{T}\sum_{i=1}^{N}\sum_{k=1}^{N-L} \overline{\mu}_i \mathcal{D}_{ik}\frac{\partial \xi_k}{\partial \xi_j} = -\frac{1}{T}\sum_{i=1}^{N}\sum_{k=1}^{N-L} \overline{\mu}_i \mathcal{D}_{ik}\delta_{kj}, \tag{5.236}$$

$$= -\frac{1}{T}\sum_{i=1}^{N} \overline{\mu}_i \mathcal{D}_{ij}, \tag{5.237}$$

$$= -\frac{1}{T}\sum_{i=1}^{N} \left(\overline{\mu}_{T,i}^{0} + \overline{R}T\ln\left(\frac{P_i}{P_0}\right)\right)\mathcal{D}_{ij}, \tag{5.238}$$

$$= -\frac{1}{T}\sum_{i=1}^{N} \left(\overline{\mu}_{T,i}^{0} + \overline{R}T\ln\left(\frac{n_i P}{n P_0}\right)\right)\mathcal{D}_{ij}, \tag{5.239}$$

$$= -\frac{1}{T}\sum_{i=1}^{N} \left(\overline{\mu}_{T,i}^{0} + \overline{R}T\ln\left(\frac{n_i \overline{R} T T_0}{P_0 V T_0}\right)\right)\mathcal{D}_{ij}, \tag{5.240}$$

$$= -\frac{1}{T}\sum_{i=1}^{N} \left(\overline{\mu}_{T,i}^{0} + \overline{R}T\ln\left(\frac{T}{T_0}\right) + \overline{R}T\ln\left(\frac{n_i \overline{R} T_0}{P_0 V}\right)\right) \times \mathcal{D}_{ij}, \tag{5.241}$$

$$= -\sum_{i=1}^{N} \left(\underbrace{\frac{\overline{\mu}_{T,i}^{0}}{T} + \overline{R}\ln\left(\frac{T}{T_0}\right)}_{\psi_i(T)} + \overline{R}\ln\left(\frac{n_i \overline{R} T_0}{P_0 V}\right)\right)\mathcal{D}_{ij}, \tag{5.242}$$

$$= -\sum_{i=1}^{N} \left(\psi_i(T) + \overline{R}\ln\left(\frac{n_i \overline{R} T_0}{P_0 V}\right)\right)\mathcal{D}_{ij}, \tag{5.243}$$

$$\frac{\partial^2 S}{\partial \xi_k \partial \xi_j} = -\sum_{i=1}^{N} \left(\frac{\partial \psi_i(T)}{\partial T}\frac{\partial T}{\partial \xi_k} + \frac{\overline{R}}{n_i}\frac{\partial n_i}{\partial \xi_k}\right)\mathcal{D}_{ij}, \tag{5.244}$$

$$= -\sum_{i=1}^{N} \left(\frac{\partial \psi_i(T)}{\partial T}\frac{\partial T}{\partial \xi_k} + \frac{\overline{R}}{n_i}\mathcal{D}_{ik}\right)\mathcal{D}_{ij}. \tag{5.245}$$

Now, consider $dT$ for the adiabatic system:

$$E = E_0 = \sum_{q=1}^{N} n_q \overline{e}_q(T), \tag{5.246}$$

$$dE = 0 = \sum_{q=1}^{N} \left( n_q \underbrace{\frac{\partial \overline{e}_q}{\partial T}}_{\overline{c}_{vq}} dT + \overline{e}_q \, dn_q \right), \tag{5.247}$$

$$0 = \sum_{q=1}^{N} \left( n_q \overline{c}_{vq} \, dT + \overline{e}_q \, dn_q \right), \tag{5.248}$$

$$dT = -\frac{\sum_{q=1}^{N} \overline{e}_q \, dn_q}{\sum_{q=1}^{N} n_q \overline{c}_{vq}} = -\frac{1}{n\overline{c}_v} \sum_{q=1}^{N} \overline{e}_q \sum_{p=1}^{N-L} \mathcal{D}_{qp} \, d\xi_p, \tag{5.249}$$

$$= -\frac{1}{n\overline{c}_v} \sum_{q=1}^{N} \sum_{p=1}^{N-L} \overline{e}_q \mathcal{D}_{qp} \, d\xi_p, \tag{5.250}$$

$$\frac{\partial T}{\partial \xi_k} = -\frac{1}{n\overline{c}_v} \sum_{q=1}^{N} \sum_{p=1}^{N-L} \overline{e}_q \mathcal{D}_{qp} \frac{\partial \xi_p}{\partial \xi_k} = -\frac{1}{n\overline{c}_v} \sum_{q=1}^{N} \sum_{p=1}^{N-L} \overline{e}_q \mathcal{D}_{qp} \delta_{pk}, \tag{5.251}$$

$$= -\frac{1}{n\overline{c}_v} \sum_{q=1}^{N} \overline{e}_q \mathcal{D}_{qk}. \tag{5.252}$$

Now, return to Eq. (5.245), using Eq. (5.252) to expand:

$$\frac{\partial^2 S}{\partial \xi_k \partial \xi_j} = \sum_{i=1}^{N} \left( \frac{\partial \psi_i(T)}{\partial T} \frac{1}{n\overline{c}_v} \sum_{q=1}^{N} \overline{e}_q \mathcal{D}_{qk} - \frac{\overline{R}}{n_i} \mathcal{D}_{ik} \right) \mathcal{D}_{ij}, \tag{5.253}$$

$$= \sum_{i=1}^{N} \frac{\partial \psi_i(T)}{\partial T} \frac{1}{n\overline{c}_v} \sum_{q=1}^{N} \overline{e}_q \mathcal{D}_{qk} \mathcal{D}_{ij} - \sum_{i=1}^{N} \frac{\overline{R}}{n_i} \mathcal{D}_{ik} \mathcal{D}_{ij}, \tag{5.254}$$

$$= \frac{1}{n\overline{c}_v} \sum_{i=1}^{N} \sum_{q=1}^{N} \frac{\partial \psi_i(T)}{\partial T} \overline{e}_q \mathcal{D}_{qk} \mathcal{D}_{ij} - \overline{R} \sum_{i=1}^{N} \frac{\mathcal{D}_{ik} \mathcal{D}_{ij}}{n_i}. \tag{5.255}$$

Now, consider the temperature derivative of $\psi_i(T)$, where $\psi_i(T)$ is defined in Eq. (5.242):

$$\psi_i \equiv \frac{\overline{\mu}_{T,i}^0}{T} + \overline{R} \ln \left( \frac{T}{T_0} \right), \tag{5.256}$$

$$\frac{d\psi_i}{dT} = -\frac{\overline{\mu}_{T,i}^0}{T^2} + \frac{1}{T} \frac{d\overline{\mu}_{T,i}^0}{dT} + \frac{\overline{R}}{T}. \tag{5.257}$$

Now

$$\overline{\mu}^0_{T,i} = \overline{h}^0_{T,i} - T\overline{s}^0_{T,i}, \tag{5.258}$$

$$= \underbrace{\overline{h}^0_{T_0,i} + \int_{T_0}^{T} \overline{c}_{Pi}(\hat{T})\, d\hat{T}}_{\overline{h}^0_{T,i}} - T\underbrace{\left(\overline{s}^0_{T_0,i} + \int_{T_0}^{T} \frac{\overline{c}_{Pi}(\hat{T})}{\hat{T}}\, d\hat{T}\right)}_{\overline{s}^0_{T,i}}, \tag{5.259}$$

$$\frac{d\overline{\mu}^0_{T,i}}{dT} = \overline{c}_{Pi} - \overline{s}^0_{T_0,i} - \int_{T_0}^{T} \frac{\overline{c}_{Pi}(\hat{T})}{\hat{T}}\, d\hat{T} - \overline{c}_{Pi}, \tag{5.260}$$

$$= -\overline{s}^0_{T_0,i} - \int_{T_0}^{T} \frac{\overline{c}_{Pi}(\hat{T})}{\hat{T}}\, d\hat{T} = -\overline{s}^0_{T,i}. \tag{5.261}$$

Note that Eq. (5.261) is a special case of the Gibbs equation given by the first of Eqs. (3.191). With this, one finds that Eq. (5.257) reduces to

$$\frac{d\psi_i}{dT} = -\frac{\overline{\mu}^0_{T,i}}{T^2} - \frac{\overline{s}^0_{T,i}}{T} + \frac{\overline{R}}{T}, \tag{5.262}$$

$$= -\frac{1}{T^2}\left(\overline{\mu}^0_{T,i} + T\overline{s}^0_{T,i} - \overline{R}T\right) = -\frac{1}{T^2}\left(\overline{g}^0_{T,i} + T\overline{s}^0_{T,i} - \overline{R}T\right), \tag{5.263}$$

$$= -\frac{1}{T^2}\left(\underbrace{\overline{h}^0_{T,i} - T\overline{s}^0_{T,i}}_{\overline{g}^0_{T,i}} + T\overline{s}^0_{T,i} - \overline{R}T\right), \tag{5.264}$$

$$= -\frac{1}{T^2}\left(\overline{h}^0_{T,i} - \overline{R}T\right), \tag{5.265}$$

$$= -\frac{1}{T^2}\left(\overline{e}_i + P_i\overline{v}_i - \overline{R}T\right) = -\frac{1}{T^2}\left(\overline{e}_i + \overline{R}T - \overline{R}T\right), \tag{5.266}$$

$$= -\frac{1}{T^2}\overline{e}_i. \tag{5.267}$$

So, substituting Eq. (5.267) into Eq. (5.255), one gets

$$\frac{\partial^2 S}{\partial \xi_k \partial \xi_j} = -\frac{1}{n\overline{c}_v T^2}\sum_{i=1}^{N}\sum_{q=1}^{N} \overline{e}_i\overline{e}_q \mathcal{D}_{qk}\mathcal{D}_{ij} - \overline{R}\sum_{i=1}^{N}\frac{\mathcal{D}_{ik}\mathcal{D}_{ij}}{n_i}. \tag{5.268}$$

Next, as before, consider the sum

$$\sum_{k=1}^{N-L}\sum_{j=1}^{N-L}\frac{\partial^2 S}{\partial \xi_k \partial \xi_j} z_k z_j = -\frac{1}{n\overline{c}_v T^2}\sum_{k=1}^{N-L}\sum_{j=1}^{N-L}\sum_{i=1}^{N}\sum_{q=1}^{N} \overline{e}_i\overline{e}_q \mathcal{D}_{qk}\mathcal{D}_{ij} z_k z_j$$

$$-\overline{R}\sum_{k=1}^{N-L}\sum_{j=1}^{N-L}\sum_{i=1}^{N}\frac{\mathcal{D}_{ik}\mathcal{D}_{ij}}{n_i} z_k z_j, \tag{5.269}$$

$$= -\frac{1}{n\overline{c}_v T^2}\sum_{i=1}^{N}\sum_{q=1}^{N}\sum_{k=1}^{N-L}\sum_{j=1}^{N-L} \overline{e}_q \mathcal{D}_{qk} z_k \overline{e}_i \mathcal{D}_{ij} z_j$$

$$-\overline{R}\sum_{i=1}^{N}\sum_{k=1}^{N-L}\sum_{j=1}^{N-L}\frac{\mathcal{D}_{ik} z_k \mathcal{D}_{ij} z_j}{n_i}, \tag{5.270}$$

$$= -\frac{1}{n\overline{c}_v T^2} \sum_{i=1}^{N} \sum_{q=1}^{N} \left( \sum_{k=1}^{N-L} \overline{e}_q \mathcal{D}_{qk} z_k \right) \left( \sum_{j=1}^{N-L} \overline{e}_i \mathcal{D}_{ij} z_j \right)$$

$$-\overline{R} \sum_{i=1}^{N} \frac{1}{n_i} \left( \sum_{k=1}^{N-L} \mathcal{D}_{ik} z_k \right) \left( \sum_{j=1}^{N-L} \mathcal{D}_{ij} z_j \right), \tag{5.271}$$

$$= -\frac{1}{n\overline{c}_v T^2} \left( \sum_{i=1}^{N} \sum_{j=1}^{N-L} \overline{e}_i \mathcal{D}_{ij} z_j \right) \left( \sum_{q=1}^{N} \sum_{k=1}^{N-L} \overline{e}_q \mathcal{D}_{qk} z_k \right)$$

$$-\overline{R} \sum_{i=1}^{N} \frac{1}{n_i} \left( \sum_{k=1}^{N-L} \mathcal{D}_{ik} z_k \right) \left( \sum_{j=1}^{N-L} \mathcal{D}_{ij} z_j \right), \tag{5.272}$$

$$= -\frac{1}{n\overline{c}_v T^2} \left( \sum_{i=1}^{N} \sum_{j=1}^{N-L} \overline{e}_i \mathcal{D}_{ij} z_j \right)^2$$

$$-\overline{R} \sum_{i=1}^{N} \frac{1}{n_i} \left( \sum_{k=1}^{N-L} \mathcal{D}_{ik} z_k \right)^2, \tag{5.273}$$

$$= -\frac{1}{n\overline{c}_v T^2} \left( \sum_{i=1}^{N} \overline{e}_i y_i \right)^2 - \overline{R} \sum_{i=1}^{N} \frac{y_i^2}{n_i}. \tag{5.274}$$

Because $\overline{c}_v > 0, T > 0, \overline{R} > 0, n_i > 0$, and the other terms are perfect squares, it is obvious that the second partial derivative of $S < 0$; consequently, critical points of $S$ represent a maximum. Again near the boundary of the physical region, $S \sim -n_i \ln n_i$, so $\lim_{n_i \to 0} S \to 0$. From Eq. (5.252), there is no formal restriction on the slope at the boundary. However, if a critical point is to exist in the physical domain in which the second derivative is guaranteed negative, the slope at the boundary must be positive everywhere. This combines to guarantee that if a critical point exists in the physically accessible region of composition space, it is unique. Moreover, it is not difficult to show that *the negative deviation of the entropy from its equilibrium value, $-\tilde{S}$, is a Lyapunov function for adiabatic, isochoric combustion.*

#### Adiabatic, Isobaric Case

A similar proof holds for the adiabatic-isobaric case. The appropriate Legendre transformation is $H = E + PV$. We omit the details, which are similar to those of previous sections, and find a term that must be negative definite, $\Upsilon_{HP}$:

$$\Upsilon_{HP} = \sum_{k=1}^{N-L} \sum_{j=1}^{N-L} \frac{\partial^2 S}{\partial \xi_k \partial \xi_j} z_k z_j. \tag{5.275}$$

Detailed analysis gives

$$\Upsilon_{HP} = -\frac{1}{n\overline{c}_P T^2} \left( \sum_{i=1}^{N} \overline{h}_i y_i \right)^2 - \frac{\overline{R}}{n} \sum_{i=1}^{N} \sum_{j=i+1}^{N} \left( \sqrt{\frac{n_j}{n_i}} y_i - \sqrt{\frac{n_i}{n_j}} y_j \right)^2. \tag{5.276}$$

Because $\overline{c}_P > 0$ and $n_i \geq 0$, the term involving $\overline{h}_i y_i$ is a perfect square, and the term multiplying $\overline{R}$ is positive definite for the same reasons as discussed before. Hence, $\Upsilon_{HP} < 0$, and the Hessian matrix is negative definite. It can be shown that *the negative deviation of the entropy from its equilibrium value, $-\tilde{S}$, is a Lyapunov function for adiabatic, isobaric combustion.*

## 5.3 Simple Three-Step Kinetics

Let us turn now to kinetics of multiple reactions. We begin with two simple linear examples, the first with reversible kinetics, the second with irreversible. They will illustrate important features of the kinetics of spatially homogeneous closed systems that will be generalized for the remainder of the chapter.

### 5.3.1 Reversible Kinetics

Consider the $N = 3, J = 3$ set of idealized reversible reactions that are an extension of those considered in Sec. 4.7:

$$1: \; A \leftrightharpoons B, \tag{5.277}$$

$$2: \; B \leftrightharpoons C, \tag{5.278}$$

$$3: \; C \leftrightharpoons A. \tag{5.279}$$

Here, chemical species $A$, $B$, and $C$ have identical molecular masses, $M_A = M_B = M_C$. We take the reaction to be isochoric and isothermal. We then have the following evolution equations and initial conditions:

$$\frac{d\overline{\rho}_A}{dt} = -r_1 + r_3, \qquad \overline{\rho}_A(0) = \overline{\rho}_{A0}, \tag{5.280}$$

$$\frac{d\overline{\rho}_B}{dt} = r_1 - r_2, \qquad \overline{\rho}_B(0) = \overline{\rho}_{B0}, \tag{5.281}$$

$$\frac{d\overline{\rho}_C}{dt} = r_2 - r_3, \qquad \overline{\rho}_C(0) = \overline{\rho}_{C0}. \tag{5.282}$$

Summing the three and integrating, we get

$$\frac{d}{dt}\left(\overline{\rho}_A + \overline{\rho}_B + \overline{\rho}_C\right) = 0, \tag{5.283}$$

$$\overline{\rho}_A + \overline{\rho}_B + \overline{\rho}_C = \overline{\rho}_{A0} + \overline{\rho}_{B0} + \overline{\rho}_{C0}. \tag{5.284}$$

Thus, we can say

$$\overline{\rho}_C = \overline{\rho}_{A0} + \overline{\rho}_{B0} + \overline{\rho}_{C0} - \overline{\rho}_A - \overline{\rho}_B. \tag{5.285}$$

Now let us focus only on evolution equations for $A$ and $B$, substituting our known forms of reaction rate $r$ for reversible reactions:

$$\frac{d\overline{\rho}_A}{dt} = -k_1\overline{\rho}_A\left(1 - \frac{1}{K_{c,1}}\frac{\overline{\rho}_B}{\overline{\rho}_A}\right) + k_3\overline{\rho}_C\left(1 - \frac{1}{K_{c,3}}\frac{\overline{\rho}_A}{\overline{\rho}_C}\right), \tag{5.286}$$

$$\frac{d\overline{\rho}_B}{dt} = k_1\overline{\rho}_A\left(1 - \frac{1}{K_{c,1}}\frac{\overline{\rho}_B}{\overline{\rho}_A}\right) - k_2\overline{\rho}_B\left(1 - \frac{1}{K_{c,2}}\frac{\overline{\rho}_C}{\overline{\rho}_B}\right). \tag{5.287}$$

We expand these to get

$$\frac{d\overline{\rho}_A}{dt} = -\left(k_1 + \frac{k_3}{K_{c,3}}\right)\overline{\rho}_A + \frac{k_1}{K_{c,1}}\overline{\rho}_B + k_3\overline{\rho}_C, \tag{5.288}$$

$$\frac{d\overline{\rho}_B}{dt} = k_1\overline{\rho}_A - \left(k_2 + \frac{k_1}{K_{c,1}}\right)\overline{\rho}_B + \frac{k_2}{K_{c,2}}\overline{\rho}_C. \tag{5.289}$$

Eliminating $\overline{\rho}_C$ with Eq. (5.285), we get

$$\frac{d\overline{\rho}_A}{dt} = -\left(k_1 + \frac{k_3}{K_{c,3}}\right)\overline{\rho}_A + \frac{k_1}{K_{c,1}}\overline{\rho}_B + k_3\left(\overline{\rho}_{A0} + \overline{\rho}_{B0} + \overline{\rho}_{C0} - \overline{\rho}_A - \overline{\rho}_B\right), \tag{5.290}$$

$$\frac{d\overline{\rho}_B}{dt} = k_1\overline{\rho}_A - \left(k_2 + \frac{k_1}{K_{c,1}}\right)\overline{\rho}_B + \frac{k_2}{K_{c,2}}\left(\overline{\rho}_{A0} + \overline{\rho}_{B0} + \overline{\rho}_{C0} - \overline{\rho}_A - \overline{\rho}_B\right). \tag{5.291}$$

Now Eq. (5.5) for this system is the linear system

$$\underbrace{\left(\Delta G_1^0 \quad \Delta G_2^0 \quad \Delta G_3^0\right)}_{\Delta\mathbf{G}^0} = \underbrace{\left(\overline{g}_A^0 \quad \overline{g}_B^0 \quad \overline{g}_C^0\right)}_{\overline{\mathbf{g}}^0} \underbrace{\begin{pmatrix} -1 & 0 & 1 \\ 1 & -1 & 0 \\ 0 & 1 & -1 \end{pmatrix}}_{\boldsymbol{\nu}}. \tag{5.292}$$

Because the rank of $\boldsymbol{\nu}$ is easily seen to be two, a set of $\overline{\mathbf{g}}^0$ does not uniquely determine $\Delta\mathbf{G}^0$. This is because an arbitrary three-dimensional $\overline{\mathbf{g}}^0$ is mapped by the rank two $\boldsymbol{\nu}$ into a two-dimensional subspace. There will exist a linear dependency between the three $\Delta G_j^0$ values. Because of this, there will exist a functional dependency between the $K_{c,j}$ values.

We can rewrite by first transposing to get

$$\begin{pmatrix} \Delta G_1^0 \\ \Delta G_2^0 \\ \Delta G_3^0 \end{pmatrix} = \begin{pmatrix} -1 & 1 & 0 \\ 0 & -1 & 1 \\ 1 & 0 & -1 \end{pmatrix} \begin{pmatrix} \overline{g}_A^0 \\ \overline{g}_B^0 \\ \overline{g}_C^0 \end{pmatrix}. \tag{5.293}$$

Row reduction leads us to

$$\begin{pmatrix} \Delta G_1^0 \\ \Delta G_2^0 \\ \Delta G_1^0 + \Delta G_2^0 + \Delta G_3^0 \end{pmatrix} = \begin{pmatrix} -1 & 1 & 0 \\ 0 & -1 & 1 \\ 0 & 0 & 0 \end{pmatrix} \begin{pmatrix} \overline{g}_A^0 \\ \overline{g}_B^0 \\ \overline{g}_C^0 \end{pmatrix}. \tag{5.294}$$

Thus, our $\Delta\mathbf{G}^0$ must have

$$-\Delta G_3^0 = \Delta G_1^0 + \Delta G_2^0. \tag{5.295}$$

Because $\sum_{i=1}^{N} \nu_{ij} = 0$ for each $j$, we can use Eq. (5.7) along with Eq. (5.295) to get

$$K_{c,1} = \exp\left(\frac{-\Delta G_1^0}{\overline{R}T}\right), \tag{5.296}$$

$$K_{c,2} = \exp\left(\frac{-\Delta G_2^0}{\overline{R}T}\right), \tag{5.297}$$

$$K_{c,3} = \exp\left(\frac{-\Delta G_3^0}{\overline{R}T}\right) = \exp\left(\frac{\Delta G_1^0 + \Delta G_2^0}{\overline{R}T}\right). \tag{5.298}$$

By inspection, we see that

$$K_{c,3} = \frac{1}{K_{c,1}K_{c,2}}. \tag{5.299}$$

This is used to rewrite Eqs. (5.290) and (5.291) as

$$\frac{d\overline{\rho}_A}{dt} = -\left(k_1 + k_3\left(1 + K_{c,1}K_{c,2}\right)\right)\overline{\rho}_A - \left(k_3 - \frac{k_1}{K_{c,1}}\right)\overline{\rho}_B + k_3\left(\overline{\rho}_{A0} + \overline{\rho}_{B0} + \overline{\rho}_{C0}\right), \tag{5.300}$$

$$\frac{d\overline{\rho}_B}{dt} = \left(k_1 - \frac{k_2}{K_{c,2}}\right)\overline{\rho}_A - \left(\frac{k_1}{K_{c,1}} + k_2\left(1 + \frac{1}{K_{c,2}}\right)\right)\overline{\rho}_B + \frac{k_2}{K_{c,2}}\left(\overline{\rho}_{A0} + \overline{\rho}_{B0} + \overline{\rho}_{C0}\right). \tag{5.301}$$

This is a linear system of two ordinary differential equations in two unknowns. The unique finite equilibrium is easily determined to be

$$\overline{\rho}_A^{eq} = \left(\overline{\rho}_{A0} + \overline{\rho}_{B0} + \overline{\rho}_{C0}\right)\frac{1}{1 + K_{c,1} + K_{c,1}K_{c,2}}, \tag{5.302}$$

$$\overline{\rho}_B^{eq} = \left(\overline{\rho}_{A0} + \overline{\rho}_{B0} + \overline{\rho}_{C0}\right)\frac{K_{c,1}}{1 + K_{c,1} + K_{c,1}K_{c,2}}. \tag{5.303}$$

Using Eq. (5.285), we then find

$$\overline{\rho}_C^{eq} = \left(\overline{\rho}_{A0} + \overline{\rho}_{B0} + \overline{\rho}_{C0}\right)\frac{K_{c,1}K_{c,2}}{1 + K_{c,1} + K_{c,1}K_{c,2}}. \tag{5.304}$$

We see by inspection that *all the equilibrium concentrations are positive and related only to the initial concentrations and the equilibrium constants. Kinetic parameters do not influence the equilibrium.* We also see that $\overline{\rho}_i^{eq}/\sum_{i=1}^{N}\overline{\rho}_{i0} < 1$ for each species.

The Jacobian for the linear system is constant and is

$$\mathbf{J} = \begin{pmatrix} -k_1 - k_3(1 + K_{c,1}K_{c,2}) & \frac{k_1}{K_{c,1}} - k_3 \\ k_1 - \frac{k_2}{K_{c,2}} & -\frac{k_1}{K_{c,1}} - k_2\left(1 - \frac{1}{K_{c,2}}\right) \end{pmatrix}. \tag{5.305}$$

Let us evaluate numerically when $k_1 = 1$, $k_2 = 2$, $k_3 = 1$, $K_{c,1} = 2$, $K_{c,2} = 2$, $\overline{\rho}_{A0} = \overline{\rho}_{B0} = \overline{\rho}_{C0} = 1/3$. For simplicity, we neglect units. Our system becomes

$$\frac{d\overline{\rho}_A}{dt} = -6\overline{\rho}_A - \frac{1}{2}\overline{\rho}_B + 1, \qquad \overline{\rho}_A(0) = \frac{1}{3}, \tag{5.306}$$

$$\frac{d\overline{\rho}_B}{dt} = -\frac{7}{2}\overline{\rho}_B + 1, \qquad \overline{\rho}_B(0) = \frac{1}{3}. \tag{5.307}$$

This has a unique finite equilibrium at

$$\overline{\rho}_A^{eq} = 1/7, \quad \overline{\rho}_B^{eq} = 2/7, \quad \overline{\rho}_C^{eq} = 4/7. \tag{5.308}$$

The eigenvalues of the Jacobian are

$$\lambda_1 = -6, \qquad \lambda_2 = -7/2. \tag{5.309}$$

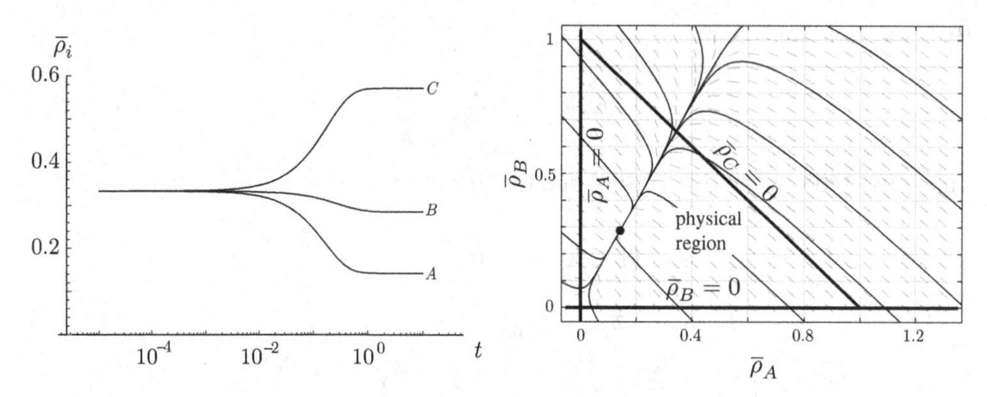


Figure 5.2. Species concentrations versus time and phase plane portrait in simple three-step linear reversible kinetics problem.

The eigenvalues are real and negative. For reversible kinetics in spatially homogeneous closed systems, Sec. 5.5.2 will show this to be generally the case. The magnitudes of the reciprocals of the eigenvalues are the time scales of reaction:

$$\tau_1 = 1/6 = 0.166667, \qquad \tau_2 = 2/7 = 0.285714. \tag{5.310}$$

The exact solution is

$$\overline{\rho}_A(t) = \frac{1}{7} + \frac{1}{5}e^{-6t} - \frac{1}{105}e^{-7t/2}, \tag{5.311}$$

$$\overline{\rho}_B(t) = \frac{2}{7} + \frac{1}{21}e^{-7t/2}, \tag{5.312}$$

$$\overline{\rho}_C(t) = \frac{4}{7} - \frac{1}{5}e^{-6t} - \frac{4}{105}e^{-7t/2}. \tag{5.313}$$

We plot this solution along with the phase plane portrait in Fig. 5.2. Also shown in the phase plane portrait is the physical equilibrium and the boundaries for which $\overline{\rho}_i = 0$. Within this triangle, all species concentrations are positive. It is evident that the equilibrium is a sink. Moreover, it appears that trajectories are attracted to a slow manifold. That slow manifold can be shown to coincide with the eigenvector associated with the least negative eigenvalue.

### 5.3.2 Irreversible Kinetics

Now let us consider the analogous problem but with irreversible kinetics. Because our second law analysis relies upon the presence of reversible kinetics near equilibrium, we can expect different results when reverse reactions are not considered. Consider then the $N = 3$, $J = 3$ set of idealized irreversible reactions that are an extension of those considered in Sec. 4.7:

$$1: \ A \rightarrow B, \tag{5.314}$$

$$2: \ B \rightarrow C, \tag{5.315}$$

$$3: \ C \rightarrow A. \tag{5.316}$$

Again chemical species $A$, $B$, and $C$ have identical molecular masses, $M_A = M_B = M_C$. And again, we take the reaction to be isochoric and isothermal. We then have

$$\frac{d\overline{\rho}_A}{dt} = -r_1 + r_3, \qquad \overline{\rho}_A(0) = \overline{\rho}_{A0}, \tag{5.317}$$

$$\frac{d\overline{\rho}_B}{dt} = r_1 - r_2, \qquad \overline{\rho}_B(0) = \overline{\rho}_{B0}, \tag{5.318}$$

$$\frac{d\overline{\rho}_C}{dt} = r_2 - r_3, \qquad \overline{\rho}_C(0) = \overline{\rho}_{C0}. \tag{5.319}$$

Summing and integrating, we get an identical result to that of the previous example:

$$\frac{d}{dt}\left(\overline{\rho}_A + \overline{\rho}_B + \overline{\rho}_C\right) = 0, \tag{5.320}$$

$$\overline{\rho}_A + \overline{\rho}_B + \overline{\rho}_C = \overline{\rho}_{A0} + \overline{\rho}_{B0} + \overline{\rho}_{C0}. \tag{5.321}$$

Thus, we can say

$$\overline{\rho}_C = \overline{\rho}_{A0} + \overline{\rho}_{B0} + \overline{\rho}_{C0} - \overline{\rho}_A - \overline{\rho}_B. \tag{5.322}$$

Now let us focus only on evolution equations for $A$ and $B$, substituting for $r_j$ for irreversible reactions:

$$\frac{d\overline{\rho}_A}{dt} = -k_1\overline{\rho}_A + k_3\overline{\rho}_C, \tag{5.323}$$

$$\frac{d\overline{\rho}_B}{dt} = k_1\overline{\rho}_A - k_2\overline{\rho}_B. \tag{5.324}$$

Now rewrite using Eq. (5.322) to eliminate $\overline{\rho}_C$:

$$\frac{d\overline{\rho}_A}{dt} = -k_1\overline{\rho}_A + k_3\left(\overline{\rho}_{A0} + \overline{\rho}_{B0} + \overline{\rho}_{C0} - \overline{\rho}_A - \overline{\rho}_B\right), \tag{5.325}$$

$$\frac{d\overline{\rho}_B}{dt} = k_1\overline{\rho}_A - k_2\overline{\rho}_B. \tag{5.326}$$

These equations are linear and have a unique equilibrium. Solving for it, and using Eq. (5.322) to get the equilibrium value of $\overline{\rho}_C$, we find

$$\overline{\rho}_A^{eq} = \frac{\overline{\rho}_{A0} + \overline{\rho}_{B0} + \overline{\rho}_{C0}}{k_1} \frac{1}{\frac{1}{k_1} + \frac{1}{k_2} + \frac{1}{k_3}}, \tag{5.327}$$

$$\overline{\rho}_B^{eq} = \frac{\overline{\rho}_{A0} + \overline{\rho}_{B0} + \overline{\rho}_{C0}}{k_2} \frac{1}{\frac{1}{k_1} + \frac{1}{k_2} + \frac{1}{k_3}}, \tag{5.328}$$

$$\overline{\rho}_C^{eq} = \frac{\overline{\rho}_{A0} + \overline{\rho}_{B0} + \overline{\rho}_{C0}}{k_3} \frac{1}{\frac{1}{k_1} + \frac{1}{k_2} + \frac{1}{k_3}}. \tag{5.329}$$

Taking $\overline{\rho}_0 \equiv (\overline{\rho}_{A0} + \overline{\rho}_{B0} + \overline{\rho}_{C0})/3$, the arithmetic mean initial concentration, and $k_h \equiv 3/(1/k_1 + 1/k_2 + 1/k_3)$, the harmonic mean reaction rate coefficient, we get

$$\overline{\rho}_A^{eq} = \overline{\rho}_0 \frac{k_h}{k_1}, \quad \overline{\rho}_B^{eq} = \overline{\rho}_0 \frac{k_h}{k_2}, \quad \overline{\rho}_C^{eq} = \overline{\rho}_0 \frac{k_h}{k_3}. \tag{5.330}$$

In stark contrast to reversible kinetics, the equilibrium point in this irreversible model depends on the reaction rate coefficients, $k_j$. All of the equilibria are positive and real, with the equilibrium concentrations scaled by the total initial concentration bounded between zero and one. The constant Jacobian is

$$\mathbf{J} = \begin{pmatrix} -k_1 - k_3 & -k_3 \\ k_1 & -k_2 \end{pmatrix}. \tag{5.331}$$

In contrast to reversible kinetics, this Jacobian may possess complex eigenvalues, inducing oscillatory components to the dynamics. We see this when we numerically evaluate the system using as much as possible the same numerical values as for the previous reversible example. Take then $k_1 = 1, k_2 = 2, k_3 = 1, \overline{\rho}_{A0} = 1/3, \overline{\rho}_{B0} = 1/3$, $\overline{\rho}_{C0} = 1/3$. The differential system becomes

$$\frac{d\overline{\rho}_A}{dt} = -2\overline{\rho}_A - \overline{\rho}_B + 1, \tag{5.332}$$

$$\frac{d\overline{\rho}_B}{dt} = \overline{\rho}_A - 2\overline{\rho}_B. \tag{5.333}$$

This has a unique finite equilibrium at

$$\overline{\rho}_A^{eq} = 2/5, \quad \overline{\rho}_B^{eq} = 1/5, \quad \overline{\rho}_C^{eq} = 2/5. \tag{5.334}$$

The eigenvalues of the Jacobian are

$$\lambda = -2 \pm i. \tag{5.335}$$

So this will induce an overdamped oscillation on the approach to a stable equilibrium. The exact solution is

$$\overline{\rho}_A(t) = \frac{2}{5} + e^{-2t}\left(-\frac{1}{15}\cos t - \frac{2}{15}\sin t\right), \tag{5.336}$$

$$\overline{\rho}_B(t) = \frac{1}{5} + e^{-2t}\left(\frac{2}{15}\cos t - \frac{1}{15}\sin t\right), \tag{5.337}$$

$$\overline{\rho}_C(t) = \frac{2}{5} + e^{-2t}\left(-\frac{1}{15}\cos t + \frac{1}{5}\sin t\right). \tag{5.338}$$

We plot this solution and the phase plane behavior in Fig. 5.3. The overdamped nature of the solution prevents the oscillatory behavior from being clearly displayed in Fig. 5.3. However, it is more evident in the phase plane. Here, one can see the spiral nature of the approach of trajectories to equilibrium. Also shown are the boundaries for which $\overline{\rho}_i = 0$. Within this triangle, all species concentrations are positive. It

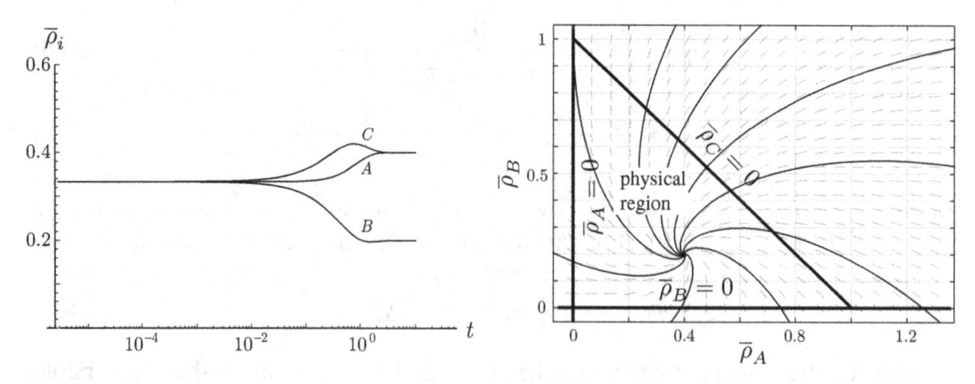


Figure 5.3. Species concentrations versus time and phase plane in simple three-step linear irreversible kinetics problem.

is evident that the equilibrium is stable. Now for closed spatially homogeneous ideal mixtures of reacting gases, oscillatory behavior is not observed. This induces many to always model reversible reactions. For other configurations, especially open systems, oscillatory behavior can be observed, and there is a large literature devoted to chemical oscillations (e.g., Kuramoto, 1984).

## 5.4 Concise Reaction Rate Law Formulations

We now turn to more general systems. One can employ notions developed in the Zel'dovich uniqueness proof, Sec. 5.2.3, to obtain a more efficient representation of the reaction rate law for multiple reactions. There are two important cases: (1) reactions dominate species: $J \geq (N - L)$ (this is most common for large chemical kinetic systems) and (2) species dominate reactions: $J < (N - L)$ (this is common for simple chemistry models).

First let us do some preliminary analysis before moving on to cases 1 and 2. The species molar production rate is given by Eq. (5.13). It reduces to

$$\frac{d\overline{\rho}_i}{dt} = \frac{1}{V}\sum_{j=1}^{J} \nu_{ij}\frac{d\zeta_j}{dt}, \qquad i = 1, \ldots, N. \tag{5.339}$$

Now, differentiating Eq. (5.161), one obtains

$$dn_i = \sum_{k=1}^{N-L} \mathcal{D}_{ik}\, d\xi_k, \qquad i = 1, \ldots, N. \tag{5.340}$$

Comparing then Eq. (5.340) to Eq. (5.22), one sees that

$$\sum_{j=1}^{J} \nu_{ij}\, d\zeta_j = \sum_{k=1}^{N-L} \mathcal{D}_{ik}\, d\xi_k, \qquad i = 1, \ldots, N, \tag{5.341}$$

$$\frac{1}{V}\sum_{j=1}^{J} \nu_{ij}\, d\zeta_j = \frac{1}{V}\sum_{k=1}^{N-L} \mathcal{D}_{ik}\, d\xi_k, \qquad i = 1, \ldots, N. \tag{5.342}$$

### 5.4.1 Reactions Dominant over Species

Consider the most common case in which $J \geq (N - L)$. The species molar production rate is

$$\frac{d\overline{\rho}_i}{dt} = \frac{1}{V}\sum_{k=1}^{N-L} \mathcal{D}_{ik}\frac{d\xi_k}{dt} = \sum_{j=1}^{J} \nu_{ij} r_j, \qquad i = 1, \ldots, N. \tag{5.343}$$

One would like to invert and solve directly for $d\xi_k/dt$. However, $\mathcal{D}_{ik}$ is nonsquare and has no inverse. But because $\sum_{i=1}^{N} \phi_{li}\mathcal{D}_{ip} = 0$, and $\sum_{i=1}^{N} \phi_{li}\nu_{ij} = 0$, $L$ of the equations $N$ equations in Eq. (5.343) are redundant.

At this point, it is more convenient to go to Gibbs's boldfaced notation, where there is an obvious correspondence between the bold symbols and the indicial counterparts. We start by writing Eq. (5.343) as

$$\frac{d\overline{\boldsymbol{\rho}}}{dt} = \frac{1}{V}\mathbf{D}\cdot\frac{d\boldsymbol{\xi}}{dt} = \boldsymbol{\nu}\cdot\mathbf{r}, \tag{5.344}$$

$$\mathbf{D}^T\cdot\mathbf{D}\cdot\frac{d\boldsymbol{\xi}}{dt} = V\mathbf{D}^T\cdot\boldsymbol{\nu}\cdot\mathbf{r}, \tag{5.345}$$

$$\frac{d\boldsymbol{\xi}}{dt} = V(\mathbf{D}^T\cdot\mathbf{D})^{-1}\cdot\mathbf{D}^T\cdot\boldsymbol{\nu}\cdot\mathbf{r}. \tag{5.346}$$

Here, $\boldsymbol{\nu}$ is the matrix of net stoichiometric coefficients, of dimension $N \times J$. Because of the $L$ linear dependencies, there is no loss of information in this projection. This system of $N - L$ equations is the smallest number of differential equations that can be solved for a general system in which $J > (N - L)$. As an aside, note that $(\mathbf{D}^T\cdot\mathbf{D})^{-1}$ does not exist for *all* rectangular matrices. But if $\mathbf{D}$ is full rank and $J > (N - L)$, it is guaranteed to exist.

For a properly posed system of ordinary differential equations, we need the derivatives of the variables as functions of those variables. Thus we write Eq. (5.346) as

$$\frac{d\boldsymbol{\xi}}{dt} = V(\mathbf{D}^T\cdot\mathbf{D})^{-1}\cdot\mathbf{D}^T\cdot\boldsymbol{\nu}\cdot\mathbf{r}(\overline{\boldsymbol{\rho}}). \tag{5.347}$$

If we scale Eq. (5.161) by volume $V$, we get the scaled equivalent of Eq. (1.55):

$$\overline{\boldsymbol{\rho}} = \widehat{\overline{\boldsymbol{\rho}}} + \frac{1}{V}\mathbf{D}\cdot\boldsymbol{\xi}. \tag{5.348}$$

We can then rewrite our reaction kinetics equation in proper dynamical system form:

$$\frac{d\boldsymbol{\xi}}{dt} = V(\mathbf{D}^T\cdot\mathbf{D})^{-1}\cdot\mathbf{D}^T\cdot\boldsymbol{\nu}\cdot\mathbf{r}\left(\widehat{\overline{\boldsymbol{\rho}}} + \frac{1}{V}\mathbf{D}\cdot\boldsymbol{\xi}\right) \equiv \mathbf{f}(\boldsymbol{\xi}). \tag{5.349}$$

Last, one recovers the original system when forming

$$\mathbf{D}\cdot\frac{d\boldsymbol{\xi}}{dt} = V\underbrace{\mathbf{D}\cdot(\mathbf{D}^T\cdot\mathbf{D})^{-1}\cdot\mathbf{D}^T}_{\mathbf{P}}\cdot\boldsymbol{\nu}\cdot\mathbf{r}. \tag{5.350}$$

Here, the $N \times N$ projection matrix $\mathbf{P}$ is symmetric, has norm of unity, has rank of $N - L$, has $N - L$ eigenvalues of value unity, and $L$ eigenvalues of value zero. And, while application of a general projection matrix to $\boldsymbol{\nu}\cdot\mathbf{r}$ filters some of the information in $\boldsymbol{\nu}\cdot\mathbf{r}$, because the $N \times (N - L)$ matrix $\mathbf{D}$ spans the same column space as the $N \times J$ matrix $\boldsymbol{\nu}$, no information is lost in Eq. (5.350) relative to the original Eq. (5.344). Mathematically, one can say

$$\mathbf{P}\cdot\boldsymbol{\nu} = \boldsymbol{\nu}, \tag{5.351}$$

and the projection operator $\mathbf{P}$ maps $\boldsymbol{\nu}$ onto itself. For any given mechanism, it is possible to form the matrix $\mathbf{D}$. With that in hand, one can then form a smaller set of $N - L$ ordinary differential equations to solve. The remaining $L$ concentrations are then available from the linear constraints.

### 5.4.2 Species Dominant over Reactions

Next consider the case in which $J < (N - L)$. This often arises in models of simple chemistry, for example one- or two-step kinetics. The fundamental reaction dynamics are most concisely governed by the $J$ equations that form

$$\frac{1}{V}\frac{d\boldsymbol{\zeta}}{dt} = \mathbf{r}. \tag{5.352}$$

However, $\mathbf{r}$ is a function of the concentrations; one must therefore recover $\overline{\boldsymbol{\rho}}$ as a function of reaction progress $\boldsymbol{\zeta}$. In vector form, Eq. (5.339) is written as

$$\frac{d\overline{\boldsymbol{\rho}}}{dt} = \frac{1}{V}\boldsymbol{\nu}\cdot\frac{d\boldsymbol{\zeta}}{dt}. \tag{5.353}$$

Take as an initial condition that the reaction progress is zero at $t = 0$ and that there are an appropriate set of initial conditions on the species concentrations $\overline{\boldsymbol{\rho}}$:

$$\boldsymbol{\zeta} = \mathbf{0}, \quad \overline{\boldsymbol{\rho}} = \overline{\boldsymbol{\rho}}_0, \qquad t = 0. \tag{5.354}$$

Then, because $\boldsymbol{\nu}$ is a constant, Eq. (5.353) is easily integrated. After applying the initial conditions, Eq. (5.354), one gets

$$\overline{\boldsymbol{\rho}} = \overline{\boldsymbol{\rho}}_0 + \frac{1}{V}\boldsymbol{\nu}\cdot\boldsymbol{\zeta}. \tag{5.355}$$

Again for a properly posed system of ordinary differential equations, we need the derivatives of the state variables as functions of those state variables. Thus we write Eq. (5.352) as follows, using the fact we know how $\mathbf{r}$ depends on $\overline{\boldsymbol{\rho}}$:

$$\frac{d\boldsymbol{\zeta}}{dt} = V\mathbf{r}(\overline{\boldsymbol{\rho}}). \tag{5.356}$$

We then use Eq. (5.355) to get the proper dynamical systems form:

$$\frac{d\boldsymbol{\zeta}}{dt} = V\mathbf{r}\left(\overline{\boldsymbol{\rho}}_0 + \frac{1}{V}\boldsymbol{\nu}\cdot\boldsymbol{\zeta}\right) \equiv \mathbf{g}(\boldsymbol{\zeta}). \tag{5.357}$$

Last, if $J = (N - L)$, either approach is equally concise.

### 5.4.3 Linear Mapping Features

As an aside, let us consider the linear mapping of Eq. (5.348) in terms of some standard notions from linear algebra. We recall the physical origins of this mapping are in the conservation of elements that is guaranteed in chemical reactions. Let us for clarity add the dimensions of the matrices in Eq. (5.348) to rewrite it as

$$\overline{\boldsymbol{\rho}}_{N\times 1} = \widehat{\overline{\boldsymbol{\rho}}}_{N\times 1} + \frac{1}{V}\mathbf{D}_{N\times(N-L)}\cdot\boldsymbol{\xi}_{(N-L)\times 1}. \tag{5.358}$$

This represents a linear inhomogeneous mapping of a vector of length $N - L$ into a higher-dimensional space with dimension $N$; mathematically one can say $\mathbb{R}^{N-L} \to \mathbb{R}^{N}$. The mapping is known as *injective* in that for every $\boldsymbol{\xi}$ there is a $\overline{\boldsymbol{\rho}}$, but there is not a $\boldsymbol{\xi}$ for every $\overline{\boldsymbol{\rho}}$. The matrix $\mathbf{D}$ deforms $\boldsymbol{\xi}$ and then it is translated by $\widehat{\overline{\boldsymbol{\rho}}}_{N\times 1}$. The fact that $\widehat{\overline{\boldsymbol{\rho}}}_{N\times 1} \neq \mathbf{0}$ renders the transformation inhomogeneous.

Let us find a geometrical interpretation of this linear mapping. It is well known from linear algebra that all matrices possess a singular value decomposition (Strang, 2005). For $\mathbf{D}$, that decomposition is

$$\mathbf{D}_{N\times(N-L)} = \mathbf{Q}_{N\times N}\cdot\boldsymbol{\Sigma}_{N\times(N-L)}\cdot\mathbf{Q}^T_{(N-L)\times(N-L)}. \tag{5.359}$$

Here, $\mathbf{Q}_{N\times N}$ and $\mathbf{Q}_{(N-L)\times(N-L)}$ are so-called orthogonal matrices. They are nonunique, and can be constructed so that they effect a pure rotation in a space of dimension either $N$ or $N - L$. The action of a rotation matrix on a vector is to purely rotate it, but not stretch it. The rectangular matrix $\mathbf{\Sigma}$ may have positive semidefinite entries known as *singular values* on its main diagonal, and has zeros for all other entries. The columns of $\mathbf{Q}_{N\times N}$ are the normalized eigenvectors of $\mathbf{D}\cdot\mathbf{D}^T$. The columns of $\mathbf{Q}_{(N-L)\times(N-L)}$ are the normalized eigenvectors of $\mathbf{D}^T\cdot\mathbf{D}$. The singular values within $\mathbf{\Sigma}$ are the positive square roots of the eigenvalues of either $\mathbf{D}\cdot\mathbf{D}^T$ or $\mathbf{D}^T\cdot\mathbf{D}$. The matrix $\mathbf{\Sigma}$ stretches vectors and brings them into a higher dimension space. For condensed notation take

$$\mathbf{Q}_{N\times N} \equiv \mathbf{Q}_{\overline{\rho}}, \qquad \mathbf{Q}_{(N-L)\times(N-L)} \equiv \mathbf{Q}_{\xi}. \tag{5.360}$$

Then we can recast Eq. (5.348) as

$$\overline{\boldsymbol{\rho}} = \widehat{\overline{\boldsymbol{\rho}}} + \frac{1}{V}\mathbf{Q}_{\overline{\rho}}\cdot\mathbf{\Sigma}\cdot\mathbf{Q}_{\xi}^T\cdot\boldsymbol{\xi}. \tag{5.361}$$

The transformation is then interpreted geometrically as the following procedure:

- Use $\mathbf{Q}_{\xi}^T$ to rotate the vector $\boldsymbol{\xi}$ into an appropriate configuration and remain in a space of dimension $N - L$.
- Use $\mathbf{\Sigma}$ to variably stretch the components of this vector and expand the dimension of the space.
- Use $\mathbf{Q}_{\overline{\rho}}$ to rotate the vector in the higher dimension space.
- Use $\widehat{\overline{\boldsymbol{\rho}}}$ to translate the vector to $\overline{\boldsymbol{\rho}}$.

**EXAMPLE 5.4**

Consider the mapping of Eq. (1.192), altered here to include $V$, and consider it in terms of the singular value decomposition.

For this example we have $N = 5$ species and $L = 2$ atoms, but because of the additional feature of molecule conservation, there is an extra constraint. Thus, we should replace $L$ by $L + 1$ everywhere in our matrix dimensioning. We first give the modified equation

$$\underbrace{\begin{pmatrix} \overline{\rho}_{\mathrm{NO}} \\ \overline{\rho}_{\mathrm{N}} \\ \overline{\rho}_{\mathrm{N_2}} \\ \overline{\rho}_{\mathrm{O}} \\ \overline{\rho}_{\mathrm{O_2}} \end{pmatrix}}_{\overline{\boldsymbol{\rho}}} = \underbrace{\begin{pmatrix} \widehat{\overline{\rho}}_{\mathrm{NO}} \\ \widehat{\overline{\rho}}_{\mathrm{N}} \\ \widehat{\overline{\rho}}_{\mathrm{N_2}} \\ \widehat{\overline{\rho}}_{\mathrm{O}} \\ \widehat{\overline{\rho}}_{\mathrm{O_2}} \end{pmatrix}}_{\widehat{\overline{\boldsymbol{\rho}}}} + \frac{1}{V}\underbrace{\begin{pmatrix} 1 & 0 \\ 0 & 1 \\ -\frac{1}{2} & -\frac{1}{2} \\ 0 & -1 \\ -\frac{1}{2} & \frac{1}{2} \end{pmatrix}}_{\mathbf{D}} \underbrace{\begin{pmatrix} \xi_{\mathrm{NO}} \\ \xi_{\mathrm{N}} \end{pmatrix}}_{\boldsymbol{\xi}}. \tag{5.362}$$

Now, leaving out details, the singular value decomposition of $\mathbf{D}$ can be shown to be

$$\mathbf{D} = \underbrace{\begin{pmatrix} 0 & \sqrt{\frac{2}{3}} & \frac{1}{\sqrt{6}} & \frac{1}{\sqrt{66}} & \sqrt{\frac{5}{33}} \\ \sqrt{\frac{2}{5}} & 0 & -\frac{1}{\sqrt{6}} & \frac{5}{\sqrt{66}} & \sqrt{\frac{3}{55}} \\ -\frac{1}{\sqrt{10}} & -\frac{1}{\sqrt{6}} & 0 & 0 & \sqrt{\frac{11}{15}} \\ -\sqrt{\frac{2}{5}} & 0 & 0 & \sqrt{\frac{6}{11}} & -\sqrt{\frac{3}{55}} \\ \frac{1}{\sqrt{10}} & -\frac{1}{\sqrt{6}} & \sqrt{\frac{2}{3}} & \sqrt{\frac{2}{33}} & -\frac{1}{\sqrt{165}} \end{pmatrix}}_{\mathbf{Q}_{\overline{\rho}}} \underbrace{\begin{pmatrix} \sqrt{\frac{5}{2}} & 0 \\ 0 & \sqrt{\frac{3}{2}} \\ 0 & 0 \\ 0 & 0 \\ 0 & 0 \end{pmatrix}}_{\mathbf{\Sigma}} \underbrace{\begin{pmatrix} 1 & 0 \\ 0 & 1 \end{pmatrix}}_{\mathbf{Q}_{\xi}^T}. \tag{5.363}$$

Because we worked in Chapter 1 to render the top two rows of $\mathbf{D}$ to be diagonal with unit values, the rotation matrix $\mathbf{Q}_\xi$ is in fact the identity matrix. Had we not done row reductions to achieve this form, we would have found a more general $\mathbf{Q}_\xi$. The effects of the actions of $\mathbf{\Sigma}$ and $\mathbf{Q}_\xi^T$ on $(\xi_{NO}, \xi_N)^T$ are

$$\mathbf{\Sigma} \cdot (\mathbf{Q}_\xi^T \cdot \boldsymbol{\xi}) = \begin{pmatrix} \sqrt{\frac{5}{2}}\xi_{NO} \\ \sqrt{\frac{3}{2}}\xi_N \\ 0 \\ 0 \\ 0 \end{pmatrix}. \tag{5.364}$$

In general this vector will have nonzero entries in the number that is the dimension of $\boldsymbol{\xi}$; here that is $N - L - 1 = 2$; usually it will be $N - L$. Rotation of this $N = 5$-dimensional vector by $\mathbf{Q}_{\overline{\rho}}$ then reveals what the deviation of $\overline{\rho}$ from its initial value is for the chosen value of $\boldsymbol{\xi}$, yielding

$$\mathbf{Q}_\xi \cdot (\mathbf{\Sigma} \cdot \mathbf{Q}_\xi^T \cdot \boldsymbol{\xi}) = \begin{pmatrix} \xi_{NO} \\ \xi_N \\ -\frac{\xi_N}{2} - \frac{\xi_{NO}}{2} \\ -\xi_N \\ \frac{\xi_N}{2} - \frac{\xi_{NO}}{2} \end{pmatrix}. \tag{5.365}$$

Translating by adding $\widehat{\overline{\rho}}$ then gives the actual value of $\overline{\rho}$.

## 5.5 Irreversible Entropy Production

This section considers the production of entropy due to irreversible chemical reactions. It is closely related to discussion of the Clausius-Duhem inequality, Eq. (5.53); it builds on general analysis given in Sec. 3.8 and for single reaction systems in Sec. 4.6.5. We first consider the irreversible production of entropy in the context of what is known as *Onsager reciprocity*. This notion relies on a linearization of the reaction rate laws in the neighborhood of equilibrium. Because the Clausius-Duhem inequality is already known to hold for the full nonlinear regime, we find only that Onsager's analysis simply confirms this notion in a more restricted context. We are able to use the Onsager analysis to lead us to an important conclusion regarding the existence of real eigenvalues and nonorthogonal eigenvectors of species production near equilibrium. We then go on to give a more general analysis, and finish with an illustration based on (1) the Zel'dovich mechanism for the example problem of Sec. 1.2.2 and (2) an extended Zel'dovich mechanism including $N_2$ and $O_2$ dissociation. The original Zel'dovich mechanism induces a strictly diagonal Onsager matrix, and thus does not fully capture the essence of the analysis; the extended Zel'dovich mechanism induces a full Onsager matrix.

### 5.5.1 Onsager Reciprocity

There is an elegant result from physical chemistry that speaks to how systems behave near equilibrium. It has applications beyond physical chemistry into mechanics in general. It will be illustrated here for multiple reaction systems. Developed by Onsager (1931a, 1931b), it is known as the reciprocity principle. Where it holds, one can use its results to assist in showing that systems approach equilibrium in a nonoscillatory manner.[9]

Let us begin by using Eq. (5.52) to rewrite Eq. (5.16) as

$$\frac{1}{V}\frac{d\zeta_j}{dt} = r_j = \mathcal{R}'_j - \mathcal{R}''_j = \mathcal{R}'_j\left(1 - \frac{\mathcal{R}''_j}{\mathcal{R}'_j}\right), \qquad j = 1, \ldots, J. \tag{5.366}$$

The definition of chemical affinity, Eq. (5.10), gives

$$\exp\left(\frac{-\overline{\alpha}_j}{\overline{R}T}\right) = \frac{\mathcal{R}''_j}{\mathcal{R}'_j}, \qquad j = 1, \ldots, J. \tag{5.367}$$

Therefore, one can say

$$\frac{1}{V}\frac{d\zeta_j}{dt} = r_j = \mathcal{R}'_j\left(1 - \exp\left(\frac{-\overline{\alpha}_j}{\overline{R}T}\right)\right), \qquad j = 1, \ldots, J. \tag{5.368}$$

Now, as each reaction comes to equilibrium, one finds that $\overline{\alpha}_j \to 0$, so a Taylor series expansion of $r_j$ yields

$$\frac{1}{V}\frac{d\zeta_j}{dt} = r_j \sim \mathcal{R}'_j\left(1 - \left(1 - \frac{\overline{\alpha}_j}{\overline{R}T} + \cdots\right)\right), \quad j = 1, \ldots, J, \tag{5.369}$$

$$\sim \mathcal{R}'_j\frac{\overline{\alpha}_j}{\overline{R}T}, \qquad j = 1, \ldots, J. \tag{5.370}$$

Note that $\mathcal{R}'_j > 0$, while $\overline{\alpha}_j$ can be positive or negative. At equilibrium, we are guaranteed $\mathcal{R}'_j > 0$ with $\overline{\alpha}_j = 0$. There is no summation on $j$. Take now the matrix $\mathbf{R}'$ to be the diagonal matrix with $\mathcal{R}_j$ populating its diagonal:

$$\mathbf{R}' \equiv \begin{pmatrix} \mathcal{R}'_1 & 0 & \ldots & 0 \\ 0 & \mathcal{R}'_2 & \ldots & 0 \\ \vdots & \vdots & \ddots & \vdots \\ 0 & 0 & 0 & \mathcal{R}'_J \end{pmatrix}. \tag{5.371}$$

Adopting Gibbs's vector notation, we find that

$$\mathbf{r} = \frac{1}{\overline{R}T}\mathbf{R}' \cdot \overline{\alpha}, \qquad \text{near equilibrium.} \tag{5.372}$$

The irreversible entropy production is given for multicomponent systems by Eq. (5.50):

$$\dot{\sigma} = \left.\frac{dS}{dt}\right|_{E,V} = \frac{V}{T}\sum_{j=1}^{J}\overline{\alpha}_j r_j = \frac{V}{T}\overline{\alpha}^T \cdot \mathbf{r} \geq 0. \tag{5.373}$$

Now, consider the definition of chemical affinity, Eq. (5.9):

$$\overline{\alpha} = -\boldsymbol{\nu}^T \cdot \overline{\mu}. \tag{5.374}$$

Now, $\boldsymbol{\nu}^T$ is of dimension $J \times N$ with rank typically $N - L$. Because $\boldsymbol{\nu}^T$ is typically not of full rank, one finds only $N - L$ of the components of $\overline{\alpha}$ to be linearly independent. When one recalls that $\boldsymbol{\nu}^T$ maps vectors $\overline{\mu}$ into the column space of $\boldsymbol{\nu}^T$, one recognizes

[9] One recognizes, of course, that many systems in nature do approach equilibrium in an oscillatory manner; unforced and underdamped mass-spring dampers are the paradigm example.

that $\overline{\alpha}$ can be represented as

$$\overline{\alpha} = \mathbf{C} \cdot \hat{\overline{\alpha}}. \tag{5.375}$$

Here, $\mathbf{C}$ is a $J \times (N - L)$-dimensional matrix of full rank, $N - L$, whose $N - L$ columns are populated by the linearly independent vectors that form the column space of $\boldsymbol{\nu}^T$, and $\hat{\overline{\alpha}}$ is a column vector of dimension $(N - L) \times 1$. If $J \geq N - L$, one can explicitly solve for $\hat{\overline{\alpha}}$, starting by operating on both sides of Eq. (5.375) by $\mathbf{C}^T$:

$$\mathbf{C}^T \cdot \overline{\alpha} = \mathbf{C}^T \cdot \mathbf{C} \cdot \hat{\overline{\alpha}}, \tag{5.376}$$

$$\mathbf{C}^T \cdot \mathbf{C} \cdot \hat{\overline{\alpha}} = \mathbf{C}^T \cdot \overline{\alpha}, \tag{5.377}$$

$$\hat{\overline{\alpha}} = \left(\mathbf{C}^T \cdot \mathbf{C}\right)^{-1} \cdot \mathbf{C}^T \cdot \overline{\alpha}, \tag{5.378}$$

$$= -\left(\mathbf{C}^T \cdot \mathbf{C}\right)^{-1} \cdot \mathbf{C}^T \cdot \boldsymbol{\nu}^T \cdot \overline{\boldsymbol{\mu}}, \tag{5.379}$$

$$\overline{\alpha} = \mathbf{C} \cdot \hat{\overline{\alpha}} = -\underbrace{\mathbf{C} \cdot \left(\mathbf{C}^T \cdot \mathbf{C}\right)^{-1} \cdot \mathbf{C}^T}_{\mathbf{B}} \cdot \boldsymbol{\nu}^T \cdot \overline{\boldsymbol{\mu}}. \tag{5.380}$$

Here, in the recomposition of $\overline{\alpha}$, one can employ the $J \times J$ symmetric projection matrix $\mathbf{B}$, which has $N - L$ eigenvalues of unity and $J - (N - L)$ eigenvalues of zero. The matrix $\mathbf{B}$ has rank $N - L$ and is thus not full rank.

Substitute Eqs. (5.372) and (5.375) into Eq. (5.373) to get near equilibrium

$$\dot{\sigma} = \left.\frac{dS}{dt}\right|_{E,V} = \frac{V}{T}\overbrace{\left(\mathbf{C} \cdot \hat{\overline{\alpha}}\right)^T}^{\overline{\alpha}^T} \cdot \underbrace{\frac{1}{\overline{R}T}\mathbf{R}' \cdot \overbrace{\mathbf{C} \cdot \hat{\overline{\alpha}}}^{\overline{\alpha}}}_{\mathbf{r}} \geq 0, \tag{5.381}$$

$$= \frac{V}{\overline{R}}\left(\frac{\hat{\overline{\alpha}}^T}{T}\right) \cdot \underbrace{\mathbf{C}^T \cdot \mathbf{R}' \cdot \mathbf{C}}_{\mathbf{L}} \cdot \left(\frac{\hat{\overline{\alpha}}}{T}\right) \geq 0. \tag{5.382}$$

Because each of the entries of the diagonal $\mathbf{R}'$ are guaranteed positive semidefinite in the physical region of composition space, the irreversible entropy production rate near equilibrium is also positive semidefinite. If some of the rate constants or species concentrations are identically zero, then some of the $\mathcal{R}'_j$ could be also be zero, inducing the positive semidefinite nature of $\mathbf{R}'$. The constant square matrix $\mathbf{L}$, of dimension $(N - L) \times (N - L)$, is given by

$$\mathbf{L} \equiv \mathbf{C}^T \cdot \mathbf{R}' \cdot \mathbf{C}. \tag{5.383}$$

The matrix $\mathbf{L}$ has rank $N - L$ and is thus full rank. Because $\mathbf{R}'$ is diagonal with positive semidefinite elements, $\mathbf{L}$ is symmetric positive semidefinite. Thus, its eigenvalues are guaranteed to be nonnegative and real. Off-diagonal elements of $\mathbf{L}$ can be negative, but that the matrix itself remains positive semidefinite. Onsager reciprocity simply demands that near equilibrium, the linearized version of the combination of the thermodynamic "forces" (here the chemical affinity $\overline{\alpha}$) and "fluxes" (here the reaction rate $\mathbf{r}$) be positive semidefinite. Upon linearization, one should always be able to find a positive semidefinite matrix associated with the dynamics of the approach to equilibrium. Here, that matrix is $\mathbf{L}$, and it has the desired properties.

One can also formulate an alternative version of Onsager reciprocity using the projection matrix **B**, which, from Eq. (5.380), is

$$\mathbf{B} \equiv \mathbf{C} \cdot \left(\mathbf{C}^T \cdot \mathbf{C}\right)^{-1} \cdot \mathbf{C}^T. \tag{5.384}$$

With a series of straightforward substitutions, it can be shown that the irreversible entropy production rate given by Eq. (5.382) reduces to

$$\dot{\sigma} = \left.\frac{dS}{dt}\right|_{E,V} = \frac{V}{\overline{R}}\left(\frac{\overline{\boldsymbol{\alpha}}^T}{T}\right) \cdot \underbrace{\mathbf{B}^T \cdot \mathbf{R}' \cdot \mathbf{B}}_{\mathsf{L}} \cdot \left(\frac{\overline{\boldsymbol{\alpha}}}{T}\right) \geq 0. \tag{5.385}$$

Here, an alternative symmetric positive semidefinite matrix $\mathsf{L}$, of dimension $J \times J$ and rank $N - L$, has been defined as

$$\mathsf{L} \equiv \mathbf{B}^T \cdot \mathbf{R}' \cdot \mathbf{B}. \tag{5.386}$$

One can also express the entropy generation directly in terms of the chemical potential rather than the chemical affinity by defining the $J \times N$ matrix **S** as

$$\mathbf{S} \equiv \mathbf{B} \cdot \boldsymbol{\nu}^T = \mathbf{C} \cdot (\mathbf{C}^T \cdot \mathbf{C})^{-1} \cdot \mathbf{C} \cdot \boldsymbol{\nu}^T. \tag{5.387}$$

The matrix **S** has rank $N - L$ and thus is not full rank. With a series of straightforward substitutions, it can be shown that the irreversible entropy production rate given by Eq. (5.385) reduces to

$$\dot{\sigma} = \left.\frac{dS}{dt}\right|_{E,V} = \frac{V}{\overline{R}}\left(\frac{\overline{\boldsymbol{\mu}}^T}{T}\right) \cdot \underbrace{\mathbf{S}^T \cdot \mathbf{R}' \cdot \mathbf{S}}_{\mathcal{L}} \cdot \left(\frac{\overline{\boldsymbol{\mu}}}{T}\right) \geq 0. \tag{5.388}$$

Here, an alternative symmetric positive semidefinite matrix $\mathcal{L}$, of dimension $N \times N$ and rank $N - L$, has been defined as

$$\mathcal{L} \equiv \mathbf{S}^T \cdot \mathbf{R}' \cdot \mathbf{S}. \tag{5.389}$$

**EXAMPLE 5.5**

Find the matrices associated with Onsager reciprocity for the reaction mechanism given by

$$H_2 + O_2 \leftrightharpoons 2OH, \tag{5.390}$$

$$H_2 + OH \leftrightharpoons H + H_2O, \tag{5.391}$$

$$H + O_2 \leftrightharpoons O + OH, \tag{5.392}$$

$$H_2 + O \leftrightharpoons H + OH, \tag{5.393}$$

$$H + H \leftrightharpoons H_2, \tag{5.394}$$

$$2OH \leftrightharpoons O + H_2O, \tag{5.395}$$

$$H_2 \leftrightharpoons H + H, \tag{5.396}$$

$$O_2 \leftrightharpoons O + O, \tag{5.397}$$

$$H + OH \leftrightharpoons H_2O. \tag{5.398}$$

Here, there are $N = 6$ species $(H, H_2, O, O_2, OH, H_2O)$, composed of $L = 2$ elements $(H, O)$, reacting in $J = 9$ reactions. Take species $i = 1$ as H, $i = 2$ as $H_2, \ldots, i = N = 6$ as

$H_2O$. Take element $l = 1$ as H and element $l = L = 2$ as O. The full rank element-species matrix $\phi$, of dimension $L \times N = 2 \times 6$ and rank $L = 2$, is

$$\phi = \begin{pmatrix} 1 & 2 & 0 & 0 & 1 & 2 \\ 0 & 0 & 1 & 2 & 1 & 1 \end{pmatrix}. \tag{5.399}$$

The rank-deficient matrix of net stoichiometric coefficients $\nu$, of dimension $N \times J = 6 \times 9$ and rank $N - L = 4$ is

$$\nu = \begin{pmatrix} 0 & 1 & -1 & 1 & -2 & 0 & 2 & 0 & -1 \\ -1 & -1 & 0 & -1 & 1 & 0 & -1 & 0 & 0 \\ 0 & 0 & 1 & -1 & 0 & 1 & 0 & 2 & 0 \\ -1 & 0 & -1 & 0 & 0 & 0 & 0 & -1 & 0 \\ 2 & -1 & 1 & 1 & 0 & -2 & 0 & 0 & -1 \\ 0 & 1 & 0 & 0 & 0 & 1 & 0 & 0 & 1 \end{pmatrix}. \tag{5.400}$$

Check for stoichiometric balance:

$$\phi \cdot \nu = \begin{pmatrix} 1 & 2 & 0 & 0 & 1 & 2 \\ 0 & 0 & 1 & 2 & 1 & 1 \end{pmatrix} \cdot \begin{pmatrix} 0 & 1 & -1 & 1 & -2 & 0 & 2 & 0 & -1 \\ -1 & -1 & 0 & -1 & 1 & 0 & -1 & 0 & 0 \\ 0 & 0 & 1 & -1 & 0 & 1 & 0 & 2 & 0 \\ -1 & 0 & -1 & 0 & 0 & 0 & 0 & -1 & 0 \\ 2 & -1 & 1 & 1 & 0 & -2 & 0 & 0 & -1 \\ 0 & 1 & 0 & 0 & 0 & 1 & 0 & 0 & 1 \end{pmatrix}, \tag{5.401}$$

$$= \begin{pmatrix} 0 & 0 & 0 & 0 & 0 & 0 & 0 & 0 & 0 \\ 0 & 0 & 0 & 0 & 0 & 0 & 0 & 0 & 0 \end{pmatrix}. \tag{5.402}$$

So, the number and mass of every element is conserved in every reaction; every vector in the column space of $\nu$ is in the right null space of $\phi$.

The detailed version of the reaction kinetics law is given by

$$\frac{d\overline{\rho}}{dt} = \nu \cdot \mathbf{r}, \tag{5.403}$$

$$= \begin{pmatrix} 0 & 1 & -1 & 1 & -2 & 0 & 2 & 0 & -1 \\ -1 & -1 & 0 & -1 & 1 & 0 & -1 & 0 & 0 \\ 0 & 0 & 1 & -1 & 0 & 1 & 0 & 2 & 0 \\ -1 & 0 & -1 & 0 & 0 & 0 & 0 & -1 & 0 \\ 2 & -1 & 1 & 1 & 0 & -2 & 0 & 0 & -1 \\ 0 & 1 & 0 & 0 & 0 & 1 & 0 & 0 & 1 \end{pmatrix} \begin{pmatrix} r_1 \\ r_2 \\ r_3 \\ r_4 \\ r_5 \\ r_6 \\ r_7 \\ r_8 \\ r_9 \end{pmatrix}, \tag{5.404}$$

$$= \begin{pmatrix} r_2 - r_3 + r_4 - 2r_5 + 2r_7 - r_9 \\ -r_1 - r_2 - r_4 + r_5 - r_7 \\ r_3 - r_4 + r_6 + 2r_8 \\ -r_1 - r_3 - r_8 \\ 2r_1 - r_2 + r_3 + r_4 - 2r_6 - r_9 \\ r_2 + r_6 + r_9 \end{pmatrix}. \tag{5.405}$$

The full rank matrix $\mathbf{D}$, of dimension $N \times (N - L) = 6 \times 4$ and rank $N - L = 4$, is composed of vectors in the right null space of $\phi$. It is nonunique, as linear combinations of right null space vectors suffice. It is equivalently composed by casting the $N - L$ linearly independent vectors of the column space of $\nu$ in its columns. Recall that some of the columns of $\nu$ are linearly dependent. In the present example, the first $N - L = 4$ column vectors of $\boldsymbol{\nu}$ happen to be linearly dependent, and thus will not suffice. Other sets are not; the last $N - L = 4$ column vectors of $\boldsymbol{\nu}$ happen to be linearly independent and thus suffice for the present purposes. Take then

$$\mathbf{D} = \begin{pmatrix} 0 & 2 & 0 & -1 \\ 0 & -1 & 0 & 0 \\ 1 & 0 & 2 & 0 \\ 0 & 0 & -1 & 0 \\ -2 & 0 & 0 & -1 \\ 1 & 0 & 0 & 1 \end{pmatrix}. \tag{5.406}$$

It is easily verified by direct substitution that $\mathbf{D}$ is in the right null space of $\boldsymbol{\phi}$:

$$\boldsymbol{\phi} \cdot \mathbf{D} = \begin{pmatrix} 1 & 2 & 0 & 0 & 1 & 2 \\ 0 & 0 & 1 & 2 & 1 & 1 \end{pmatrix} \begin{pmatrix} 0 & 2 & 0 & -1 \\ 0 & -1 & 0 & 0 \\ 1 & 0 & 2 & 0 \\ 0 & 0 & -1 & 0 \\ -2 & 0 & 0 & -1 \\ 1 & 0 & 0 & 1 \end{pmatrix} = \begin{pmatrix} 0 & 0 & 0 & 0 \\ 0 & 0 & 0 & 0 \end{pmatrix}. \tag{5.407}$$

Because $\boldsymbol{\phi} \cdot \boldsymbol{\nu} = \mathbf{0}$ and $\boldsymbol{\phi} \cdot \mathbf{D} = \mathbf{0}$, one concludes that the column spaces of both $\boldsymbol{\nu}$ and $\mathbf{D}$ are one and the same. The nonunique concise version of the reaction kinetics law is given by

$$\frac{d\boldsymbol{\xi}}{dt} = V(\mathbf{D}^T \cdot \mathbf{D})^{-1} \cdot \mathbf{D}^T \cdot \boldsymbol{\nu} \cdot \mathbf{r}, \tag{5.408}$$

$$= V \left( \begin{pmatrix} 0 & 0 & 1 & 0 & -2 & 1 \\ 2 & -1 & 0 & 0 & 0 & 0 \\ 0 & 0 & 2 & -1 & 0 & 0 \\ -1 & 0 & 0 & 0 & -1 & 1 \end{pmatrix} \begin{pmatrix} 0 & 2 & 0 & -1 \\ 0 & -1 & 0 & 0 \\ 1 & 0 & 2 & 0 \\ 0 & 0 & -1 & 0 \\ -2 & 0 & 0 & -1 \\ 1 & 0 & 0 & 1 \end{pmatrix} \right)^{-1}$$

$$\cdot \begin{pmatrix} 0 & 0 & 1 & 0 & -2 & 1 \\ 2 & -1 & 0 & 0 & 0 & 0 \\ 0 & 0 & 2 & -1 & 0 & 0 \\ -1 & 0 & 0 & 0 & -1 & 1 \end{pmatrix}$$

$$\cdot \begin{pmatrix} 0 & 1 & -1 & 1 & -2 & 0 & 2 & 0 & -1 \\ -1 & -1 & 0 & -1 & 1 & 0 & -1 & 0 & 0 \\ 0 & 0 & 1 & -1 & 0 & 1 & 0 & 2 & 0 \\ -1 & 0 & -1 & 0 & 0 & 0 & 0 & -1 & 0 \\ 2 & -1 & 1 & 1 & 0 & -2 & 0 & 0 & -1 \\ 0 & 1 & 0 & 0 & 0 & 1 & 0 & 0 & 1 \end{pmatrix} \begin{pmatrix} r_1 \\ r_2 \\ r_3 \\ r_4 \\ r_5 \\ r_6 \\ r_7 \\ r_8 \\ r_9 \end{pmatrix}, \tag{5.409}$$

$$= V \begin{pmatrix} -2r_1 - r_3 - r_4 + r_6 \\ r_1 + r_2 + r_4 - r_5 + r_7 \\ r_1 + r_3 + r_8 \\ 2r_1 + r_2 + r_3 + r_4 + r_9 \end{pmatrix}. \tag{5.410}$$

The rank-deficient projection matrix $\mathbf{P}$, of dimension $N \times N = 6 \times 6$ and rank $N - L = 4$, is

$$\mathbf{P} = \mathbf{D} \cdot \left(\mathbf{D}^T \cdot \mathbf{D}\right)^{-1} \cdot \mathbf{D}^T, \tag{5.411}$$

$$= \begin{pmatrix} \frac{54}{61} & \frac{-14}{61} & \frac{3}{61} & \frac{6}{61} & \frac{-4}{61} & \frac{-11}{61} \\ \frac{-14}{61} & \frac{33}{61} & \frac{6}{61} & \frac{12}{61} & \frac{-8}{61} & \frac{-22}{61} \\ \frac{3}{61} & \frac{6}{61} & \frac{51}{61} & \frac{-20}{61} & \frac{-7}{61} & \frac{-4}{61} \\ \frac{6}{61} & \frac{12}{61} & \frac{-20}{61} & \frac{21}{61} & \frac{-14}{61} & \frac{-8}{61} \\ \frac{-4}{61} & \frac{-8}{61} & \frac{-7}{61} & \frac{-14}{61} & \frac{50}{61} & \frac{-15}{61} \\ \frac{-11}{61} & \frac{-22}{61} & \frac{-4}{61} & \frac{-8}{61} & \frac{-15}{61} & \frac{35}{61} \end{pmatrix}. \tag{5.412}$$

The projection matrix $\mathbf{P}$ has $N - L = 4$ eigenvalues of unity and $L = 2$ eigenvalues of zero.

The chemical affinity vector $\overline{\boldsymbol{\alpha}}$, of dimension $J \times 1 = 9 \times 1$, is given by

$$\overline{\boldsymbol{\alpha}} = -\boldsymbol{\nu}^T \cdot \overline{\boldsymbol{\mu}}, \tag{5.413}$$

$$= - \begin{pmatrix} 0 & 1 & -1 & 1 & -2 & 0 & 2 & 0 & -1 \\ -1 & -1 & 0 & -1 & 1 & 0 & -1 & 0 & 0 \\ 0 & 0 & 1 & -1 & 0 & 1 & 0 & 2 & 0 \\ -1 & 0 & -1 & 0 & 0 & 0 & 0 & -1 & 0 \\ 2 & -1 & 1 & 1 & 0 & -2 & 0 & 0 & -1 \\ 0 & 1 & 0 & 0 & 0 & 1 & 0 & 0 & 1 \end{pmatrix}^T \begin{pmatrix} \overline{\mu}_1 \\ \overline{\mu}_2 \\ \overline{\mu}_3 \\ \overline{\mu}_4 \\ \overline{\mu}_5 \\ \overline{\mu}_6 \end{pmatrix}, \tag{5.414}$$

$$= - \begin{pmatrix} 0 & -1 & 0 & -1 & 2 & 0 \\ 1 & -1 & 0 & 0 & -1 & 1 \\ -1 & 0 & 1 & -1 & 1 & 0 \\ 1 & -1 & -1 & 0 & 1 & 0 \\ -2 & 1 & 0 & 0 & 0 & 0 \\ 0 & 0 & 1 & 0 & -2 & 1 \\ 2 & -1 & 0 & 0 & 0 & 0 \\ 0 & 0 & 2 & -1 & 0 & 0 \\ -1 & 0 & 0 & 0 & -1 & 1 \end{pmatrix} \begin{pmatrix} \overline{\mu}_1 \\ \overline{\mu}_2 \\ \overline{\mu}_3 \\ \overline{\mu}_4 \\ \overline{\mu}_5 \\ \overline{\mu}_6 \end{pmatrix}, \tag{5.415}$$

$$= \begin{pmatrix} \overline{\mu}_2 + \overline{\mu}_4 - 2\overline{\mu}_5 \\ -\overline{\mu}_1 + \overline{\mu}_2 + \overline{\mu}_5 - \overline{\mu}_6 \\ \overline{\mu}_1 - \overline{\mu}_3 + \overline{\mu}_4 - \overline{\mu}_5 \\ -\overline{\mu}_1 + \overline{\mu}_2 + \overline{\mu}_3 - \overline{\mu}_5 \\ 2\overline{\mu}_1 - \overline{\mu}_2 \\ -\overline{\mu}_3 + 2\overline{\mu}_5 - \overline{\mu}_6 \\ -2\overline{\mu}_1 + \overline{\mu}_2 \\ -2\overline{\mu}_3 + \overline{\mu}_4 \\ \overline{\mu}_1 + \overline{\mu}_5 - \overline{\mu}_6 \end{pmatrix}. \tag{5.416}$$

The full rank matrix $\mathbf{C}$, of dimension $J \times (N - L) = 9 \times 4$ and rank $N - L = 4$, is composed of the set of $N - L = 4$ linearly independent column space vectors of $\boldsymbol{\nu}^T$; thus, they also comprise the $N - L = 4$ linearly independent row space vectors of $\boldsymbol{\nu}$. It does not matter which four are chosen, so long as they are linearly independent. In this case, the first four column vectors of $\boldsymbol{\nu}^T$ suffice:

$$\mathbf{C} = \begin{pmatrix} 0 & -1 & 0 & -1 \\ 1 & -1 & 0 & 0 \\ -1 & 0 & 1 & -1 \\ 1 & -1 & -1 & 0 \\ -2 & 1 & 0 & 0 \\ 0 & 0 & 1 & 0 \\ 2 & -1 & 0 & 0 \\ 0 & 0 & 2 & -1 \\ -1 & 0 & 0 & 0 \end{pmatrix}. \tag{5.417}$$

When $J > N - L$, not all of the components of $\overline{\boldsymbol{\alpha}}$ are linearly independent. In this case, one can form the reduced chemical affinity vector, $\hat{\overline{\boldsymbol{\alpha}}}$, of dimension $(N - L) \times 1 = 4 \times 1$ via

$$\hat{\overline{\boldsymbol{\alpha}}} = (\mathbf{C}^T \cdot \mathbf{C})^{-1} \cdot \mathbf{C}^T \cdot \overline{\boldsymbol{\alpha}}, \tag{5.418}$$

$$= \left( \begin{pmatrix} 0 & 1 & -1 & 1 & -2 & 0 & 2 & 0 & -1 \\ -1 & -1 & 0 & -1 & 1 & 0 & -1 & 0 & 0 \\ 0 & 0 & 1 & -1 & 0 & 1 & 0 & 2 & 0 \\ -1 & 0 & -1 & 0 & 0 & 0 & 0 & -1 & 0 \end{pmatrix} \begin{pmatrix} 0 & -1 & 0 & -1 \\ 1 & -1 & 0 & 0 \\ -1 & 0 & 1 & -1 \\ 1 & -1 & -1 & 0 \\ -2 & 1 & 0 & 0 \\ 0 & 0 & 1 & 0 \\ 2 & -1 & 0 & 0 \\ 0 & 0 & 2 & -1 \\ -1 & 0 & 0 & 0 \end{pmatrix} \right)^{-1}$$

$$\cdot \begin{pmatrix} 0 & 1 & -1 & 1 & -2 & 0 & 2 & 0 & -1 \\ -1 & -1 & 0 & -1 & 1 & 0 & -1 & 0 & 0 \\ 0 & 0 & 1 & -1 & 0 & 1 & 0 & 2 & 0 \\ -1 & 0 & -1 & 0 & 0 & 0 & 0 & -1 & 0 \end{pmatrix} \begin{pmatrix} \overline{\mu}_2 + \overline{\mu}_4 - 2\overline{\mu}_5 \\ -\overline{\mu}_1 + \overline{\mu}_2 + \overline{\mu}_5 - \overline{\mu}_6 \\ \overline{\mu}_1 - \overline{\mu}_3 + \overline{\mu}_4 - \overline{\mu}_5 \\ -\overline{\mu}_1 + \overline{\mu}_2 + \overline{\mu}_3 - \overline{\mu}_5 \\ 2\overline{\mu}_1 - \overline{\mu}_2 \\ -\overline{\mu}_3 + 2\overline{\mu}_5 - \overline{\mu}_6 \\ -2\overline{\mu}_1 + \overline{\mu}_2 \\ -2\overline{\mu}_3 + \overline{\mu}_4 \\ \overline{\mu}_1 + \overline{\mu}_5 - \overline{\mu}_6 \end{pmatrix}, \tag{5.419}$$

$$= \begin{pmatrix} -\overline{\mu}_1 - \overline{\mu}_5 + \overline{\mu}_6 \\ -\overline{\mu}_2 - 2\overline{\mu}_5 + 2\overline{\mu}_6 \\ -\overline{\mu}_3 + 2\overline{\mu}_5 - \overline{\mu}_6 \\ -\overline{\mu}_4 + 4\overline{\mu}_5 - 2\overline{\mu}_6 \end{pmatrix}. \tag{5.420}$$

The rank-deficient projection matrix $\mathbf{B}$, of dimension $J \times J = 9 \times 9$ and rank $N - L = 4$, is

$$\mathbf{B} = \mathbf{C} \cdot (\mathbf{C}^T \cdot \mathbf{C})^{-1} \cdot \mathbf{C}^T, \tag{5.421}$$

$$= \begin{pmatrix} \frac{57}{94} & \frac{9}{94} & \frac{31}{94} & \frac{26}{94} & \frac{-1}{94} & \frac{-17}{94} & \frac{1}{94} & \frac{6}{94} & \frac{8}{94} \\ \frac{9}{94} & \frac{41}{94} & \frac{-5}{94} & \frac{14}{94} & \frac{-15}{94} & \frac{27}{94} & \frac{15}{94} & \frac{-4}{94} & \frac{26}{94} \\ \frac{31}{94} & \frac{-5}{94} & \frac{35}{94} & \frac{-4}{94} & \frac{11}{94} & \frac{-1}{94} & \frac{-11}{94} & \frac{28}{94} & \frac{6}{94} \\ \frac{26}{94} & \frac{14}{94} & \frac{-4}{94} & \frac{30}{94} & \frac{-12}{94} & \frac{16}{94} & \frac{12}{94} & \frac{-22}{94} & \frac{2}{94} \\ \frac{-1}{94} & \frac{-15}{94} & \frac{11}{94} & \frac{-12}{94} & \frac{33}{94} & \frac{-3}{94} & \frac{-33}{94} & \frac{-10}{94} & \frac{18}{94} \\ \frac{-17}{94} & \frac{27}{94} & \frac{-1}{94} & \frac{-16}{94} & \frac{-3}{94} & \frac{43}{94} & \frac{3}{94} & \frac{18}{94} & \frac{24}{94} \\ \frac{1}{94} & \frac{15}{94} & \frac{-11}{94} & \frac{12}{94} & \frac{-33}{94} & \frac{3}{94} & \frac{33}{94} & \frac{10}{94} & \frac{-18}{94} \\ \frac{6}{94} & \frac{-4}{94} & \frac{28}{94} & \frac{-22}{94} & \frac{-10}{94} & \frac{18}{94} & \frac{10}{94} & \frac{60}{94} & \frac{-14}{94} \\ \frac{8}{94} & \frac{26}{94} & \frac{6}{94} & \frac{2}{94} & \frac{18}{94} & \frac{24}{94} & \frac{-18}{94} & \frac{-14}{94} & \frac{44}{94} \end{pmatrix}. \tag{5.422}$$

The projection matrix $\mathbf{B}$ has a set of $J = 9$ eigenvalues, $N - L = 4$ of which are unity, and $J - (N - L) = 5$ of which are zero. One can recover $\overline{\alpha}$ by the operation $\overline{\alpha} = \mathbf{C} \cdot \hat{\overline{\alpha}} = -\mathbf{B} \cdot \boldsymbol{\nu}^T \cdot \overline{\boldsymbol{\mu}}$.

The square full rank Onsager matrix $\mathbf{L}$, of dimension $(N - L) \times (N - L) = 4 \times 4$ and rank $N - L = 4$, is given by

$$\mathbf{L} = \mathbf{C}^T \cdot \mathbf{R} \cdot \mathbf{C}, \tag{5.423}$$

$$= \begin{pmatrix} 0 & 1 & -1 & 1 & -2 & 0 & 2 & 0 & -1 \\ -1 & -1 & 0 & -1 & 1 & 0 & -1 & 0 & 0 \\ 0 & 0 & 1 & -1 & 0 & 1 & 0 & 2 & 0 \\ -1 & 0 & -1 & 0 & 0 & 0 & 0 & -1 & 0 \end{pmatrix}$$

$$\cdot \begin{pmatrix} \mathcal{R}'_1 & 0 & 0 & 0 & 0 & 0 & 0 & 0 & 0 \\ 0 & \mathcal{R}'_2 & 0 & 0 & 0 & 0 & 0 & 0 & 0 \\ 0 & 0 & \mathcal{R}'_3 & 0 & 0 & 0 & 0 & 0 & 0 \\ 0 & 0 & 0 & \mathcal{R}'_4 & 0 & 0 & 0 & 0 & 0 \\ 0 & 0 & 0 & 0 & \mathcal{R}'_5 & 0 & 0 & 0 & 0 \\ 0 & 0 & 0 & 0 & 0 & \mathcal{R}'_6 & 0 & 0 & 0 \\ 0 & 0 & 0 & 0 & 0 & 0 & \mathcal{R}'_7 & 0 & 0 \\ 0 & 0 & 0 & 0 & 0 & 0 & 0 & \mathcal{R}'_8 & 0 \\ 0 & 0 & 0 & 0 & 0 & 0 & 0 & 0 & \mathcal{R}'_9 \end{pmatrix}$$

$$\cdot \begin{pmatrix} 0 & -1 & 0 & -1 \\ 1 & -1 & 0 & 0 \\ -1 & 0 & 1 & -1 \\ 1 & -1 & -1 & 0 \\ -2 & 1 & 0 & 0 \\ 0 & 0 & 1 & 0 \\ 2 & -1 & 0 & 0 \\ 0 & 0 & 2 & -1 \\ -1 & 0 & 0 & 0 \end{pmatrix}, \tag{5.424}$$

$$= \begin{pmatrix} \mathcal{R}'_2 + \mathcal{R}'_3 + \mathcal{R}'_4 + 4\mathcal{R}'_5 + 4\mathcal{R}'_7 + \mathcal{R}'_9 & -\mathcal{R}'_2 - \mathcal{R}'_4 - 2\mathcal{R}'_5 - 2\mathcal{R}'_7 & -\mathcal{R}'_3 - \mathcal{R}'_4 & \mathcal{R}'_3 \\ -\mathcal{R}'_2 - \mathcal{R}'_4 - 2\mathcal{R}'_5 - 2\mathcal{R}'_7 & \mathcal{R}'_1 + \mathcal{R}'_2 + \mathcal{R}'_4 + \mathcal{R}'_5 + \mathcal{R}'_7 & \mathcal{R}'_4 & \mathcal{R}'_1 \\ -\mathcal{R}'_3 - \mathcal{R}'_4 & \mathcal{R}'_4 & \mathcal{R}'_3 + \mathcal{R}'_4 + \mathcal{R}'_6 + 4\mathcal{R}'_8 & -\mathcal{R}'_3 - 2\mathcal{R}'_8 \\ \mathcal{R}'_3 & \mathcal{R}'_1 & -\mathcal{R}'_3 - 2\mathcal{R}'_8 & \mathcal{R}'_1 + \mathcal{R}'_3 + \mathcal{R}'_8 \end{pmatrix}. \tag{5.425}$$

Obviously $\mathbf{L}$ (shown here in split form because of its large size) is symmetric, and thus has all real eigenvalues. It is also positive semidefinite. With a similar effort, one can obtain the alternate rank-deficient square Onsager matrices $\mathsf{L}$ and $\mathcal{L}$. Recall that $\mathbf{L}$, $\mathsf{L}$, and $\mathcal{L}$ each have rank $N - L$, while $\mathbf{L}$ has the smallest dimension, $(N - L) \times (N - L)$, and so forms the most efficient Onsager matrix.

### 5.5.2 Eigenvalues at Equilibrium

We can use our previous results to prove an important result: *for closed, spatially homogeneous systems of ideal mixtures of ideal gases obeying mass action kinetics with reversible reactions, the eigenvalues at equilibrium will be real and negative.* We confine discussion to isothermal, isochoric reaction, but results could be extended. It is known that $G$ and $\dot{\sigma}$ reach minima at equilibrium. Let us see what else can be learned in the approach to chemical equilibrium.

Consider the gradient of the irreversible entropy production rate in the space of species progress variables $\xi_k$. First, recall Eq. (5.218) for the gradient of Gibbs free energy with respect to the independent species progress variables, $\partial G/\partial \xi_k = \sum_{i=1}^{N} \overline{\mu}_i \mathcal{D}_{ik}$. Now, recalling Eq. (3.271), take the irreversible entropy production rate, $\dot{\sigma}$, as

$$\dot{\sigma} = \frac{d\sigma}{dt} = -\frac{1}{T}\sum_{i=1}^{N} \overline{\mu}_i \frac{dn_i}{dt}. \tag{5.426}$$

Take now the time derivative of Eq. (5.161) to get

$$\frac{dn_i}{dt} = \sum_{k=1}^{N-L} \mathcal{D}_{ik}\frac{d\xi_k}{dt}. \tag{5.427}$$

Substitute from Eq. (5.427) into Eq. (5.426) to get

$$\dot{\sigma} = -\frac{1}{T}\sum_{i=1}^{N} \overline{\mu}_i \sum_{k=1}^{N-L} \mathcal{D}_{ik}\frac{d\xi_k}{dt} = -\frac{1}{T}\sum_{k=1}^{N-L} \frac{d\xi_k}{dt} \underbrace{\sum_{i=1}^{N} \overline{\mu}_i \mathcal{D}_{ik}}_{\frac{\partial G}{\partial \xi_k}}, \tag{5.428}$$

$$= -\frac{1}{T}\sum_{k=1}^{N-L} \frac{d\xi_k}{dt}\frac{\partial G}{\partial \xi_k}. \tag{5.429}$$

Now, Eq. (5.346) gives an explicit algebraic formula for $d\xi_k/dt$. Define then the constitutive function $\hat{\omega}_k$:

$$\hat{\omega}_k(\xi_1, \ldots, \xi_{N-L}) \equiv \frac{d\xi_k}{dt} = V(\mathbf{D}^T \cdot \mathbf{D})^{-1} \cdot \mathbf{D}^T \cdot \boldsymbol{\nu} \cdot \mathbf{r}. \tag{5.430}$$

So, the irreversible entropy production rate is

$$\dot{\sigma} = -\frac{1}{T}\sum_{k=1}^{N-L} \hat{\omega}_k \frac{\partial G}{\partial \xi_k}. \tag{5.431}$$

The gradient of this field in the space defined by reaction coordinate $\xi_p$ is given by

$$\frac{\partial \dot{\sigma}}{\partial \xi_p} = -\frac{1}{T}\sum_{k=1}^{N-L}\left(\frac{\partial \hat{\omega}_k}{\partial \xi_p}\frac{\partial G}{\partial \xi_k} + \hat{\omega}_k \frac{\partial^2 G}{\partial \xi_p \partial \xi_k}\right). \tag{5.432}$$

The Hessian of this field is given by

$$\frac{\partial^2 \dot{\sigma}}{\partial \xi_l \partial \xi_p} = -\frac{1}{T}\sum_{k=1}^{N-L}\left(\frac{\partial^2 \hat{\omega}_k}{\partial \xi_l \partial \xi_p}\frac{\partial G}{\partial \xi_k} + \frac{\partial \hat{\omega}_k}{\partial \xi_p}\frac{\partial^2 G}{\partial \xi_l \partial \xi_k} + \frac{\partial \hat{\omega}_k}{\partial \xi_l}\frac{\partial^2 G}{\partial \xi_p \partial \xi_k} + \hat{\omega}_k \frac{\partial^3 G}{\partial \xi_l \partial \xi_p \partial \xi_k}\right). \tag{5.433}$$

Now, at equilibrium, $\xi_k = \xi_k^{eq}$, we have $\hat{\omega}_k = 0$ as well as $\partial G/\partial \xi_k = 0$. Thus

$$\dot{\sigma}\big|_{\xi_k=\xi_k^{eq}} = 0, \tag{5.434}$$

$$\left.\frac{\partial \dot{\sigma}}{\partial \xi_k}\right|_{\xi_k=\xi_k^{eq}} = 0, \tag{5.435}$$

$$\left.\frac{\partial^2 \dot{\sigma}}{\partial \xi_l \partial \xi_p}\right|_{\xi_k=\xi_k^{eq}} = -\frac{1}{T}\sum_{k=1}^{N-L}\left.\left(\frac{\partial \hat{\omega}_k}{\partial \xi_p}\frac{\partial^2 G}{\partial \xi_l \partial \xi_k} + \frac{\partial \hat{\omega}_k}{\partial \xi_l}\frac{\partial^2 G}{\partial \xi_p \partial \xi_k}\right)\right|_{\xi_k=\xi_k^{eq}}. \tag{5.436}$$

Because Hessian matrices are symmetric, we can exchange $p$ and $l$ as needed. This induces the simplification

$$\left.\frac{\partial^2 \dot{\sigma}}{\partial \xi_l \partial \xi_p}\right|_{\xi_k=\xi_k^{eq}} = -\frac{2}{T}\sum_{k=1}^{N-L}\left.\frac{\partial \hat{\omega}_k}{\partial \xi_p}\frac{\partial^2 G}{\partial \xi_l \partial \xi_k}\right|_{\xi_k=\xi_k^{eq}}. \tag{5.437}$$

Equation (5.437) gives a formula for the Hessian of $\dot{\sigma}$ at equilibrium. Now this Hessian of $\dot{\sigma}$ is obviously symmetric, but it is possible to show that it is positive semidefinite. Consider, for instance, Eq. (5.382), a result that is valid in the neighborhood of equilibrium:

$$\dot{\sigma} = \frac{V}{\overline{\overline{R}}}\left(\frac{\hat{\boldsymbol{\alpha}}}{T}\right)^T \cdot \mathbf{L} \cdot \left(\frac{\hat{\boldsymbol{\alpha}}}{T}\right), \quad \text{valid only near equilibrium.} \tag{5.438}$$

Recall that by construction of Eq. (5.383), $\mathbf{L}$ is positive semidefinite. Now, near equilibrium, $\hat{\boldsymbol{\alpha}}$ has a Taylor series expansion

$$\hat{\boldsymbol{\alpha}} = \underbrace{\hat{\boldsymbol{\alpha}}\big|_{eq}}_{\mathbf{0}} + \left.\frac{\partial \hat{\boldsymbol{\alpha}}}{\partial \boldsymbol{\xi}}\right|_{eq} \cdot (\boldsymbol{\xi} - \boldsymbol{\xi}|_{eq}) + \cdots. \tag{5.439}$$

Recall at equilibrium that $\hat{\boldsymbol{\alpha}} = \mathbf{0}$. Let us also define the Jacobian of $\hat{\boldsymbol{\alpha}}$ as $\mathbf{J}_\alpha$. Thus, Eq. (5.438) can be rewritten as

$$\dot{\sigma} = \frac{V}{\overline{\overline{R}}T^2}\left(\mathbf{J}_\alpha \cdot (\boldsymbol{\xi} - \boldsymbol{\xi}|_{eq})\right)^T \cdot \mathbf{L} \cdot \left(\mathbf{J}_\alpha \cdot (\boldsymbol{\xi} - \boldsymbol{\xi}|_{eq})\right), \tag{5.440}$$

$$= \frac{V}{\overline{\overline{R}}T^2}(\boldsymbol{\xi} - \boldsymbol{\xi}|_{eq})^T \cdot \mathbf{J}_\alpha^T \cdot \mathbf{L} \cdot \mathbf{J}_\alpha \cdot (\boldsymbol{\xi} - \boldsymbol{\xi}|_{eq}). \tag{5.441}$$

By inspection, we see that the Hessian of $\dot{\sigma}$ is

$$\mathbf{H}_{\dot{\sigma}} = \frac{2V}{\overline{\overline{R}}T^2}\mathbf{J}_\alpha^T \cdot \mathbf{L} \cdot \mathbf{J}_\alpha, \quad \text{valid only near equilibrium.} \tag{5.442}$$

The eigenvalues of $\mathbf{L}$ are guaranteed positive and real. Moreover, $\mathbf{L}$ has the same eigenvalues as the similar matrix $\mathbf{J}_\alpha^T \cdot \mathbf{L} \cdot \mathbf{J}_\alpha$, so $\mathbf{H}_{\dot{\sigma}}$ is positive semidefinite in the neighborhood of equilibrium.

Returning to the formulation that does not rely upon the Onsager approach, we can write Eq. (5.437) in Gibbs's boldfaced notation as

$$\mathbf{H}_{\dot{\sigma}} = -\frac{2}{T}\mathbf{H}_G \cdot \mathbf{J}, \quad \text{valid only near equilibrium,} \tag{5.443}$$

where $\mathbf{J}$ is the Jacobian matrix of $\hat{\omega}_k$. So knowing that $\mathbf{H}_{\dot{\sigma}}$ is symmetric and positive semidefinite, and $\mathbf{H}_G$ is symmetric and positive definite, we ask what may be concluded about $\mathbf{J}$ near equilibrium. Certainly one can invert to find

$$\mathbf{J} = -\frac{T}{2}\mathbf{H}_G^{-1} \cdot \mathbf{H}_{\dot{\sigma}}. \tag{5.444}$$

Now for the symmetric and positive definite $\mathbf{H}_G$, it is easy to show that $\mathbf{H}_G^{-1}$ is symmetric and positive definite as well. We do this by first diagonally decomposing $\mathbf{H}_G$ as

$$\mathbf{H}_G = \mathbf{Q} \cdot \mathbf{\Lambda}_G \cdot \mathbf{Q}^T. \tag{5.445}$$

Here, $\mathbf{\Lambda}_G$ is diagonal with positive entries, and $\mathbf{Q}$ is an orthogonal matrix. Then, we get

$$\mathbf{H}_G^{-1} = \mathbf{Q} \cdot \mathbf{\Lambda}_G^{-1} \cdot \mathbf{Q}^T, \tag{5.446}$$

with $\mathbf{\Lambda}_G^{-1}$ also being diagonal with positive entries, each of which are the reciprocals of entries in $\mathbf{\Lambda}_G$, all of which are nonzero. Now recall that any symmetric positive definite matrix has a Cholesky decomposition; we take that to be

$$\mathbf{H}_G^{-1} = \mathbf{U}^T \cdot \mathbf{U}. \tag{5.447}$$

Here, $\mathbf{U}$ is an upper triangular matrix with real positive eigenvalues. It can be shown that the eigenvalues of the symmetric positive semidefinite $\mathbf{U} \cdot \mathbf{H}_{\dot{\sigma}} \cdot \mathbf{U}^T$ are identical to those of the asymmetric $\mathbf{U}^T \cdot \mathbf{U} \cdot \mathbf{H}_{\dot{\sigma}} = \mathbf{H}_G^{-1} \cdot \mathbf{H}_{\dot{\sigma}}$. Because we multiply by the negative scalar $-T/2$, we conclude that $\mathbf{J} = -(T/2)\mathbf{H}_G^{-1} \cdot \mathbf{H}_{\dot{\sigma}}$ is guaranteed to have real and negative semidefinite eigenvalues near equilibrium. So for our positive definite $\mathbf{H}_{\dot{\sigma}}$, the eigenvalues of $\mathbf{J}$ are guaranteed to be real and negative.

This important result yields two important properties relevant to the chemical dynamics of closed systems of ideal mixtures of ideal gases:

- *Because the eigenvalues of the asymmetric* $\mathbf{J}$ *are guaranteed real near chemical equilibrium associated with reversible kinetics, there can be no oscillations in the neighborhood of such an equilibrium.*
- *Because the eigenvalues of* $\mathbf{J}$ *are guaranteed negative semidefinite and real near chemical equilibrium, that equilibrium is stable.*

The eigenvectors of $\mathbf{J}$ give the directions of fast and slow modes. Because $\mathbf{J}$ is asymmetric, its eigenvectors are most generally nonorthogonal. Near equilibrium, the dynamics will relax to the slow mode, and the motion toward equilibrium will be along the eigenvector associated with the slowest time scale. Now, in the unlikely circumstance that $\mathbf{H}_G$ were the identity matrix, one would have the eigenvalues of $\mathbf{H}_{\dot{\sigma}}$ equal to the product of $-2/T$ and the eigenvalues of $\mathbf{H}_G$. So, they would be positive, as expected. Moreover, the eigenvectors of $\mathbf{H}_{\dot{\sigma}}$ would be identical to those of $\mathbf{J}$, so in this unusual case, the slow dynamics could be inferred from examining the slowest descent down contours of $\dot{\sigma}$. Essentially the same conclusion would be reached if $\mathbf{H}_G$ had a diagonalization with equal eigenvalues on its diagonal. This would correspond to reactions proceeding at the same rate near equilibrium. However, in the usual case, the eigenvalues of $\mathbf{H}_G$ are nonuniform. Thus, the action of $\mathbf{H}_G$ on $\mathbf{J}$ is to stretch it nonuniformly in such a fashion that $\mathbf{H}_{\dot{\sigma}}$ does not share the same eigenvalues or eigenvectors. Thus, irreversible entropy production cannot be used to directly infer such important chemical dynamics. But in the ordinary case, the eigenvectors of $\mathbf{J}$ will not align with either the eigenvectors of $\mathbf{H}_G$ nor of $\mathbf{H}_{\dot{\sigma}}$; thus, steepest descents down contours of either $G$ or $\dot{\sigma}$ will not provide information on the dynamics of the system.

Now, consider the behavior of $\dot{\sigma}$ in the neighborhood of an equilibrium point. In the neighborhood of a general point $\xi_k = \hat{\xi}_k$, $\dot{\sigma}$ has a Taylor series expansion

$$\dot{\sigma} = \dot{\sigma}\big|_{\xi_k=\hat{\xi}_k} + \sum_{k=1}^{N-L} \frac{\partial \dot{\sigma}}{\partial \xi_k}\bigg|_{\xi_k=\hat{\xi}_k} \left(\xi_k - \hat{\xi}_k\right)$$

$$+\frac{1}{2}\sum_{l=1}^{N-L}\sum_{p=1}^{N-L} \left(\xi_l - \hat{\xi}_l\right) \frac{\partial^2 \dot{\sigma}}{\partial \xi_l \partial \xi_p}\bigg|_{\xi_k=\hat{\xi}_k} \left(\xi_p - \hat{\xi}_p\right) + \cdots \tag{5.448}$$

Near equilibrium, the first two terms of this Taylor series are zero, and $\dot{\sigma}$ has the behavior

$$\dot{\sigma} = \frac{1}{2}\sum_{l=1}^{N-L}\sum_{p=1}^{N-L} (\xi_l - \xi_l^{eq}) \frac{\partial^2 \dot{\sigma}}{\partial \xi_l \partial \xi_p}\bigg|_{\xi_k=\xi_k^{eq}} (\xi_p - \xi_p^{eq}) + \cdots \tag{5.449}$$

Substituting from Eq. (5.437), we find near equilibrium that

$$\dot{\sigma} = -\frac{1}{T}\sum_{l=1}^{N-L}\sum_{p=1}^{N-L}\sum_{k=1}^{N-L} (\xi_l - \xi_l^{eq}) \frac{\partial \hat{\omega}_k}{\partial \xi_p}\bigg|_{\xi_k=\xi_k^{eq}} \frac{\partial^2 G}{\partial \xi_l \partial \xi_k}\bigg|_{\xi_k=\xi_k^{eq}} (\xi_p - \xi_p^{eq}) + \cdots \tag{5.450}$$

As the literature in irreversible thermodynamics has much discussion about extremization of $\dot{\sigma}$, let us study whether $\dot{\sigma}$ is a Lyapunov function. Our analysis will be inconclusive. Loosely interpreted, we have already shown in Sec. 5.2.3 that $\sigma$ is related to a Lyapunov function, though there it was cast in terms of $A$, $G$, or $S$, as appropriate for the selected constraints. Because $\dot{\sigma}$ is also extreme at equilibrium, it is natural to ask if it too may be a Lyapunov function.

We can show that $\dot{\sigma} > 0$, $\xi_p \neq \xi_p^{eq}$, and $\dot{\sigma} = 0$, $\xi_p = \xi_p^{eq}$. Now, to determine whether or not the Lyapunov function exists, we must study $d\dot{\sigma}/dt$ and determine if it is negative semidefinite globally. We expand it as

$$\frac{d\dot{\sigma}}{dt} = \sum_{p=1}^{N-L} \frac{\partial \dot{\sigma}}{\partial \xi_p}\frac{d\xi_p}{dt} = \sum_{p=1}^{N-L} \frac{\partial \dot{\sigma}}{\partial \xi_p}\hat{\omega}_p, \tag{5.451}$$

$$= \sum_{p=1}^{N-L} \left( -\frac{1}{T}\sum_{k=1}^{N-L} \left( \frac{\partial \hat{\omega}_k}{\partial \xi_p}\frac{\partial G}{\partial \xi_k} + \hat{\omega}_k \frac{\partial^2 G}{\partial \xi_p \partial \xi_k} \right)\right) \hat{\omega}_p, \tag{5.452}$$

$$= -\frac{1}{T}\sum_{p=1}^{N-L}\sum_{k=1}^{N-L} \left( \frac{\partial \hat{\omega}_k}{\partial \xi_p}\frac{\partial G}{\partial \xi_k}\hat{\omega}_p + \hat{\omega}_p \frac{\partial^2 G}{\partial \xi_p \partial \xi_k}\hat{\omega}_k \right). \tag{5.453}$$

At the equilibrium state, $\hat{\omega}_k = 0$, so obviously $d\dot{\sigma}/dt = 0$ at equilibrium. This is a necessary condition for $\dot{\sigma}$ to be a Lyapunov function. Away from equilibrium, we know from our uniqueness analysis that the term $\partial^2 G/\partial\xi_p\partial\xi_k$ is positive definite; so this term contributes to rendering $d\dot{\sigma}/dt < 0$. But the other term does not transparently contribute, so we can draw no conclusion away from equilibrium.

Let us examine the gradient of $d\dot{\sigma}/dt$. It is

$$\frac{\partial}{\partial\xi_m}\frac{d\dot{\sigma}}{dt} = -\frac{1}{T}\sum_{p=1}^{N-L}\sum_{k=1}^{N-L}\left(\frac{\partial^2\hat{\omega}_k}{\partial\xi_m\partial\xi_p}\frac{\partial G}{\partial\xi_k}\hat{\omega}_p + \frac{\partial\hat{\omega}_k}{\partial\xi_p}\frac{\partial^2 G}{\partial\xi_m\partial\xi_k}\hat{\omega}_p + \frac{\partial\hat{\omega}_k}{\partial\xi_p}\frac{\partial G}{\partial\xi_k}\frac{\partial\hat{\omega}_p}{\partial\xi_m}\right.$$
$$\left. + \frac{\partial\hat{\omega}_p}{\partial\xi_m}\frac{\partial^2 G}{\partial\xi_p\partial\xi_k}\hat{\omega}_k + \hat{\omega}_p\frac{\partial^3 G}{\partial\xi_m\partial\xi_p\partial\xi_k}\hat{\omega}_k + \hat{\omega}_p\frac{\partial^2 G}{\partial\xi_p\partial\xi_k}\frac{\partial\hat{\omega}_k}{\partial\xi_m}\right). \tag{5.454}$$

At equilibrium, all terms in the gradient of $d\dot{\sigma}/dt$ are zero, so it is a critical point.

Let us next study the Hessian of $d\dot{\sigma}/dt$ to ascertain the nature of this critical point. Because there are so many terms, let us only write those terms that will be nonzero at equilibrium. In this limit, the Hessian is

$$\left.\frac{\partial^2}{\partial\xi_n\partial\xi_m}\frac{d\dot{\sigma}}{dt}\right|_{eq} = -\frac{1}{T}\sum_{p=1}^{N-L}\sum_{k=1}^{N-L}\left(\frac{\partial\hat{\omega}_k}{\partial\xi_p}\frac{\partial^2 G}{\partial\xi_m\partial\xi_k}\frac{\partial\hat{\omega}_p}{\partial\xi_n} + \frac{\partial\hat{\omega}_k}{\partial\xi_p}\frac{\partial^2 G}{\partial\xi_n\partial\xi_k}\frac{\partial\hat{\omega}_p}{\partial\xi_m}\right.$$
$$\left. + \frac{\partial\hat{\omega}_p}{\partial\xi_m}\frac{\partial^2 G}{\partial\xi_p\partial\xi_k}\frac{\partial\hat{\omega}_k}{\partial\xi_n} + \frac{\partial\hat{\omega}_p}{\partial\xi_n}\frac{\partial^2 G}{\partial\xi_p\partial\xi_k}\frac{\partial\hat{\omega}_k}{\partial\xi_m}\right). \tag{5.455}$$

With some effort, it may be possible to show the total sum is negative semidefinite. This is because each term involves a Hessian of $G$. This would guarantee that $d\dot{\sigma}/dt < 0$ in the neighborhood of equilibrium. If so, $\dot{\sigma}$ is a Lyapunov function in the neighborhood of equilibrium. Far from equilibrium, it is not clear whether $\dot{\sigma}$ is a Lyapunov function.

### 5.5.3 Zel'dovich Mechanism Example

Let us now consider a numerical example, building on the analysis of Sec. 1.2.2; many numerical details have already been presented and will not be repeated here. For the conditions simulated in Sec. 1.2.2, we plot $dS/dt|_{E,V}/V = \dot{\sigma}/V$ as a function of time as predicted by the full theory of Eq. (5.54), repeated here in slightly modified form

$$\left.\frac{1}{V}\frac{dS}{dt}\right|_{E,V} = \frac{\dot{\sigma}}{V} = \mathbf{r}^T\cdot\left(\frac{\overline{\alpha}}{T}\right), \tag{5.456}$$

in Fig. 5.4. Clearly, $\dot{\sigma}/V$ is always positive and approaches zero as the system approaches equilibrium. We have made no appeal to notions of Onsager nor have we confined ourselves to the neighborhood of equilibrium. The function $\dot{\sigma}/V$ is nonlinear, positive, and decreases to zero by its construction. That is consistent with the Clausius-Duhem analysis shown earlier in Eq. (5.53).

Now, let us consider the system of Sec. 1.2.2 in the language of Onsager reciprocity. We see from Eq. (1.173) that

$$\boldsymbol{\nu}^T = \begin{pmatrix} -1 & -1 & 1 & 1 & 0 \\ 1 & -1 & 0 & 1 & -1 \end{pmatrix}. \tag{5.457}$$

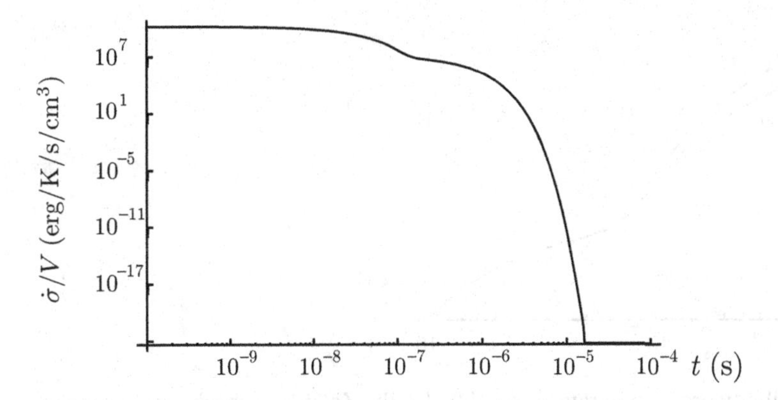


Figure 5.4. Irreversibility production rate versus time for the Zel'dovich mechanism problem of Sec. 1.2.2.

It is easily shown that $\boldsymbol{\nu}^T$ is of full rank, $R = 2$, and thus has two linearly independent column vectors. Following the discussion of Eq. (5.375), let us simply choose the first two column vectors of $\boldsymbol{\nu}^T$ to form the matrix $\mathbf{C}$:

$$\mathbf{C} = \begin{pmatrix} -1 & -1 \\ 1 & -1 \end{pmatrix}. \tag{5.458}$$

Then, by Eq. (5.384), we find the projection matrix $\mathbf{B}$ to be

$$\mathbf{B} = \mathbf{C} \cdot \left(\mathbf{C}^T \cdot \mathbf{C}\right)^{-1} \cdot \mathbf{C}^T, \tag{5.459}$$

$$= \begin{pmatrix} -1 & -1 \\ 1 & -1 \end{pmatrix} \left( \begin{pmatrix} -1 & 1 \\ -1 & -1 \end{pmatrix} \begin{pmatrix} -1 & -1 \\ 1 & -1 \end{pmatrix} \right)^{-1} \begin{pmatrix} -1 & 1 \\ -1 & -1 \end{pmatrix}, \tag{5.460}$$

$$= \begin{pmatrix} 1 & 0 \\ 0 & 1 \end{pmatrix}. \tag{5.461}$$

Remarkably, we recover the identity matrix: $\mathbf{B} = \mathbf{I}$. For more general reaction mechanisms that have $\boldsymbol{\nu}^T$ of less than full rank, this will not be true.

So, from Eq. (5.386), we find one of the Onsager matrices to reduce to

$$\mathsf{L} = \mathbf{B}^T \cdot \mathbf{R}' \cdot \mathbf{B} = \mathbf{I}^T \cdot \mathbf{R}' \cdot \mathbf{I} = \mathbf{R}' = \begin{pmatrix} \mathcal{R}'_1 & 0 \\ 0 & \mathcal{R}'_2 \end{pmatrix}. \tag{5.462}$$

Because each $\mathcal{R}_j$, $j = 1, \ldots, J$, is positive semidefinite, the matrix $\mathbf{R}'$ is positive semidefinite. We can then cast the irreversible entropy production of Eq. (5.385) as

$$\dot{\sigma} = \left.\frac{dS}{dt}\right|_{E,V} = \underbrace{\frac{V}{\overline{\overline{R}}} \left(\frac{\overline{\boldsymbol{\alpha}}^T}{T}\right) \cdot \mathsf{L}}_{\sim \mathbf{j}^T} \cdot \underbrace{\left(\frac{\overline{\boldsymbol{\alpha}}}{T}\right)}_{\mathbf{x}}. \tag{5.463}$$

Comparing Eqs. (5.463) and (5.56), we see in the Onsager limit that

$$\mathbf{j}^T \sim \frac{V}{\overline{\overline{R}}} \left(\frac{\overline{\boldsymbol{\alpha}}^T}{T}\right) \cdot \mathsf{L} = \frac{V}{\overline{\overline{R}}} \mathbf{x}^T \cdot \mathsf{L}. \tag{5.464}$$

Transposing both sides, we can also say

$$\mathbf{j} \sim \frac{V}{\overline{\overline{R}}} \mathsf{L}^T \cdot \left(\frac{\overline{\boldsymbol{\alpha}}}{T}\right) = \frac{V}{\overline{\overline{R}}} \mathsf{L}^T \cdot \mathbf{x}. \tag{5.465}$$

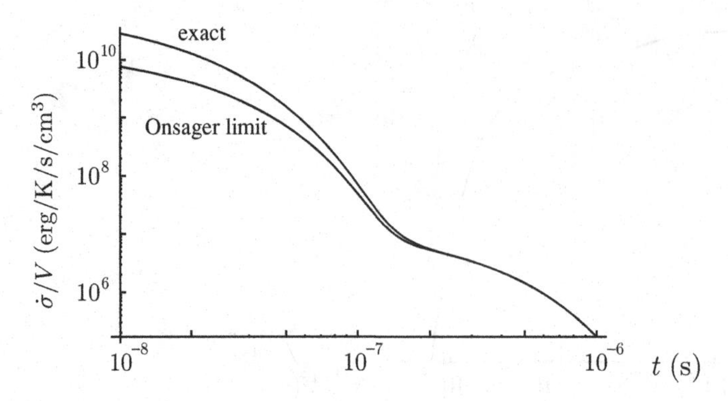


Figure 5.5. Irreversibility production rate versus time for the Zel'dovich mechanism problem of Sec. 1.2.2 for the full nonlinear model and the Onsager limit.

We also can recast in the standard form

$$\left.\frac{dS}{dt}\right|_{E,V} = \dot{\sigma} = \frac{V}{\overline{\overline{R}}}\mathbf{x}^T \cdot \mathsf{L} \cdot \mathbf{x}. \tag{5.466}$$

For the conditions simulated in Sec. 1.2.2, we plot $\dot{\sigma}/V$ as a function of time for both the full nonlinear model as well as the prediction in the Onsager limit of Eq. (5.463) in Fig. 5.5. Clearly at early time the two theories predict different values, but the Onsager limit becomes a better approximation as equilibrium is approached.

Numerically, direct substitution of values at equilibrium shows that

$$\mathsf{L} = \begin{pmatrix} 0.573127 & 0 \\ 0 & 0.549134 \end{pmatrix}. \tag{5.467}$$

The units of the entries of $\mathsf{L}$ are mol/cm$^3$/s.

Now at equilibrium, Eq. (5.11) tells us the chemical affinity of each reaction must be zero; thus, at equilibrium, we must have $\mathbf{x} = \mathbf{0}$. This then induces the fluxes $\mathbf{j} = \mathbf{0}$ at equilibrium. Near equilibrium, it is possible to perform Prigogine's extremization exercise surrounding his Eq. (P6.8). Recognizing that the conditions of our problem, which have $\mathsf{L}_{12} = \mathsf{L}_{21} = 0$, will not hold for more general problems, we slightly extend our analysis to consider the more general case. For such a case, we have near equilibrium that

$$\begin{pmatrix} j_1 \\ j_2 \end{pmatrix} = \frac{V}{\overline{\overline{R}}} \begin{pmatrix} \mathsf{L}_{11} & \mathsf{L}_{21} \\ \mathsf{L}_{12} & \mathsf{L}_{22} \end{pmatrix} \begin{pmatrix} x_1 \\ x_2 \end{pmatrix}. \tag{5.468}$$

Thus,

$$j_1 = \frac{V}{\overline{\overline{R}}}\left(\mathsf{L}_{11}x_1 + \mathsf{L}_{21}x_2\right), \qquad j_2 = \frac{V}{\overline{\overline{R}}}\left(\mathsf{L}_{12}x_1 + \mathsf{L}_{22}x_2\right). \tag{5.469}$$

The irreversible entropy production becomes

$$\dot{\sigma} = \frac{V}{\overline{\overline{R}}}\left(\mathsf{L}_{11}x_1^2 + \left(\mathsf{L}_{12} + \mathsf{L}_{21}\right)x_1x_2 + \mathsf{L}_{22}x_2^2\right). \tag{5.470}$$

Let us extremize $\dot{\sigma}$. We then require at a critical point

$$\frac{\partial \dot{\sigma}}{\partial x_1} = \frac{V}{\overline{R}} \left(2\mathsf{L}_{11} x_1 + (\mathsf{L}_{12} + \mathsf{L}_{21}) x_2\right) = 0, \quad (5.471)$$

$$\frac{\partial \dot{\sigma}}{\partial x_2} = \frac{V}{\overline{R}} \left((\mathsf{L}_{12} + \mathsf{L}_{21}) x_1 + 2\mathsf{L}_{22} x_2\right) = 0. \quad (5.472)$$

Now by its method of construction, $\mathsf{L}$ must be symmetric; thus, $\mathsf{L}_{12} = \mathsf{L}_{21}$, and we can thus state at an extreme point that

$$\frac{\partial \dot{\sigma}}{\partial x_1} = \frac{2V}{\overline{R}} \left(\mathsf{L}_{11} x_1 + \mathsf{L}_{21} x_2\right) = 2 j_1 = 0, \quad (5.473)$$

$$\frac{\partial \dot{\sigma}}{\partial x_2} = \frac{2V}{\overline{R}} \left(\mathsf{L}_{12} x_1 + \mathsf{L}_{22} x_2\right) = 2 j_2 = 0. \quad (5.474)$$

Thus, $j_1 = j_2 = 0$ at an extremum of $\dot{\sigma}$, consistent with Eq. (P6.6).

Far from equilibrium, the formulation of $\dot{\sigma}$ does not lend itself to such transparent analysis. One can however easily formulate $\dot{\sigma}$ globally in terms of reduced species variables $\xi_k$ as in Eq. (5.431). While the formulation is straightforward, its mathematical form is lengthy, and will not be repeated here. Following the analysis after Eq. (5.431), one can form the gradient and Hessian of $\dot{\sigma}$, and note that at chemical equilibrium that $\dot{\sigma} = 0$ and $\nabla \dot{\sigma} = 0$. Moreover, from Eq. (5.442), we see that the Hessian of $\dot{\sigma}$ is positive semidefinite, so that in fact $\dot{\sigma}$ is a minimum at equilibrium.

For our Zel'dovich mechanism example problem, the Hessian matrix, $\mathbf{H}_{\dot{\sigma}}$, evaluated at chemical equilibrium is from Eq. (5.443):

$$\mathbf{H}_{\dot{\sigma}} = \left. \frac{\partial^2 \dot{\sigma}}{\partial \xi_l \partial \xi_p} \right|_{\xi_k = \xi_k^{eq}} = \begin{pmatrix} 6.81073 \times 10^{20} & -1.06440 \times 10^{21} \\ -1.06440 \times 10^{21} & 1.46438 \times 10^{23} \end{pmatrix}. \quad (5.475)$$

The units of each of the entries of the Hessian are erg/K/s/mol$^2$. The eigenvalues of $\mathbf{H}_{\dot{\sigma}}$ are $1.464 \times 10^{23}$ and $6.733 \times 10^{20}$, each with units erg/K/s/mol$^2$, and each positive, consistent with its positive semidefinite nature.

For the same example, the Hessian matrix $\mathbf{H}_G$, evaluated at chemical equilibrium is

$$\mathbf{H}_G = \left. \frac{\partial^2 G}{\partial \xi_l \partial \xi_p} \right|_{\xi_k = \xi_k^{eq}} = \begin{pmatrix} 9.48582 \times 10^{17} & -1.14112 \times 10^{17} \\ -1.14112 \times 10^{17} & 1.39760 \times 10^{19} \end{pmatrix}. \quad (5.476)$$

The units of the entries of $\mathbf{H}_G$ are erg/mol$^2$. The matrix has eigenvalues of $1.39770 \times 10^{19}$, $9.47583 \times 10^{17}$, each with units of erg/mol$^2$, and each positive, consistent with its positive definite nature.

We recall from Eq. (1.247) that for this example, in the neighborhood of chemical equilibrium, we have the Jacobian $\mathbf{J}$ of the chemical dynamics given by

$$\mathbf{J} = \begin{pmatrix} -2.12858 \times 10^6 & -4.15485 \times 10^5 \\ 2.11099 \times 10^5 & -3.14369 \times 10^7 \end{pmatrix}. \quad (5.477)$$

The entries of $\mathbf{J}$ and its eigenvalues each have units of s$^{-1}$. It is easily verified by direct substitution that Eq. (5.443) holds here: $\mathbf{H}_{\dot{\sigma}} = -(2/T)\mathbf{H}_G \cdot \mathbf{J}$. Moreover, with the Cholesky decomposition of $\mathbf{H}_G^{-1}$ as

$$\mathbf{U} = \begin{pmatrix} 1.02725 \times 10^{-9} & 8.38735 \times 10^{-12} \\ 0 & 2.67491 \times 10^{-10} \end{pmatrix}, \quad (5.478)$$

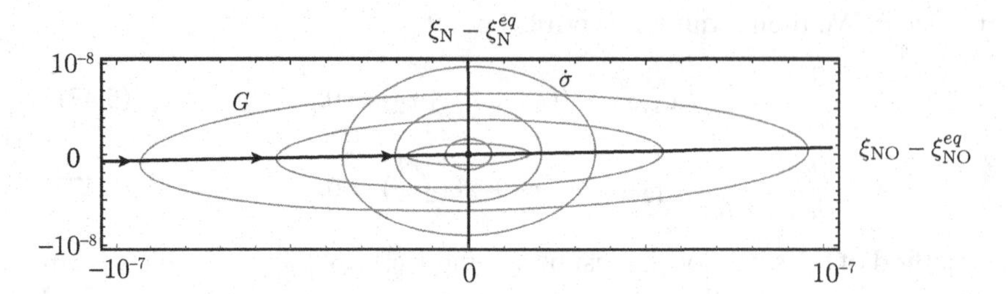


Figure 5.6. Contours of $G$ and $\dot{\sigma}$ along with a solution trajectory in the neighborhood of equilibrium for the Zel'dovich mechanism example problem of Sec. 1.2.2.

we see the eigenvalues of the negative semidefinite $-(2/T)\mathbf{U} \cdot \mathbf{H}_{\dot{\sigma}} \cdot \mathbf{U}^T$ are identical to the eigenvalues of $\mathbf{J}$ from Eqs. (1.260):

$$\lambda_1 = -3.143 \times 10^7 \text{ 1/s}, \qquad \lambda_1 = -2.132 \times 10^6 \text{ 1/s}. \tag{5.479}$$

Consider the dynamics of the approach to equilibrium. In the neighborhood of the stable equilibrium, it is straightforward to show that *the solution trajectory is directed toward the equilibrium in a direction given by the eigenvector of* $\mathbf{J}$ *that is associated with the slowest time scale. For our example, this eigenvector is closely aligned with the eigenvectors of both* $\mathbf{H}_G$ *and* $\mathbf{H}_{\dot{\sigma}}$*, but is not identical to them. Thus, consistent with the results of Al-Khateeb et al. (2009), it is correct to use eigenvectors of* $\mathbf{J}$ *to study slow dynamics near equilibrium. Use of extreme descent contours of* $G$ *or* $\dot{\sigma}$ *introduces error.*

We show contours of $G$ and $\dot{\sigma}$ along with the trajectory to equilibrium in Fig. 5.6. On the scale shown, it *appears* that the solution trajectory to equilibrium is along the principal axes of the ellipses that form the contours of both $G$ and $\dot{\sigma}$. Close examination in the rescaled space of Fig. 5.7 reveals small deviation. Direct computation of the angles of (1) the appropriate slow eigenvector of $\mathbf{J}$, (2) the appropriate eigenvector of $\mathbf{H}_{\dot{\sigma}}$, and (3) the appropriate eigenvector of $\mathbf{H}_G$ shows

$$\mathbf{J}\text{: } 0.413°, \quad \mathbf{H}_{\dot{\sigma}}\text{: } 0.418°, \quad \mathbf{H}_G\text{: } 0.502°. \tag{5.480}$$

The proximity of these angles suggest there may be an asymptotic theory that could be exploited in which the trajectory of the approach to equilibrium is approximated by extreme descents down contours of $G$ or $\dot{\sigma}$. The actual path to equilibrium does not coincide with extreme descent paths.

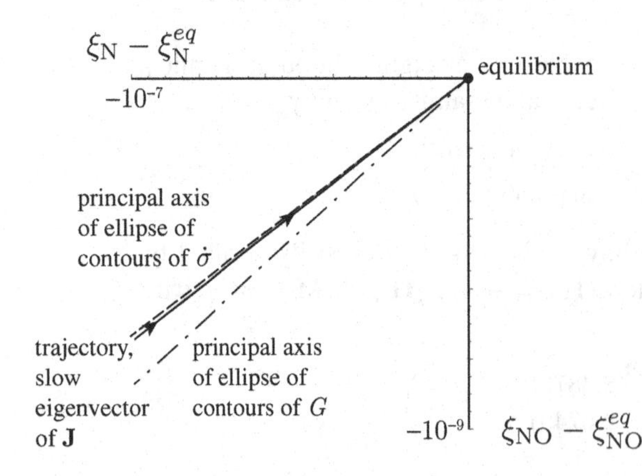


Figure 5.7. Principal axes of ellipses of contours of $G$ and $\dot{\sigma}$ along with a solution trajectory in the neighborhood of equilibrium for the Zel'dovich mechanism example problem of Sec. 1.2.2.

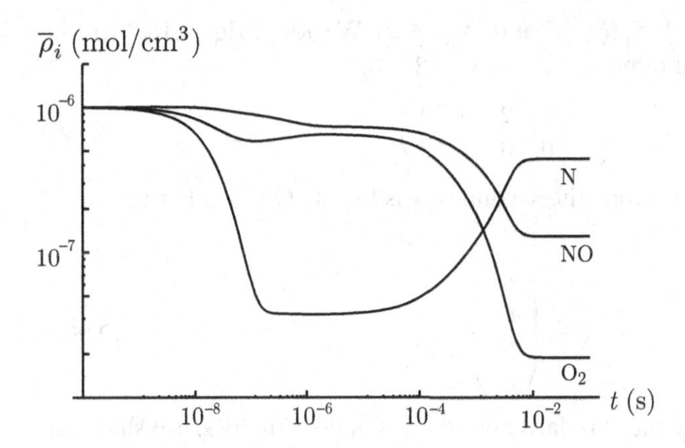


Figure 5.8. NO, N, and $O_2$ concentrations versus time for $T = 6000$ K, extended Zel'dovich mechanism.

### 5.5.4 Extended Zel'dovich Mechanism Example

The example of the previous section induced a diagonal Onsager matrix because the matrix $\boldsymbol{\nu}$ was full rank. Let us extend the kinetics mechanism so as to induce a $\boldsymbol{\nu}$ that is less than full rank, and thus induce a full Onsager matrix. To do so, we can add $N_2$ and $O_2$ dissociation mechanisms to our kinetics laws. We take our extended Zel'dovich reaction mechanism to be

$$N + NO \leftrightharpoons N_2 + O, \tag{5.481}$$

$$N + O_2 \leftrightharpoons NO + O, \tag{5.482}$$

$$N_2 + N_2 \leftrightharpoons 2N + N_2, \tag{5.483}$$

$$O_2 + O_2 \leftrightharpoons 2O + O_2. \tag{5.484}$$

For this system we have $N = 5$ species composed of $L = 2$ elements reacting in $J = 4$ reactions. We take the reaction constants for the first two reactions from Sec. 1.2.2; the $N_2$ and $O_2$ dissociation kinetic rates are those given earlier in Secs. 4.5.1 and 1.2.1, respectively.

Applying the extended Zel'dovich model for the conditions simulated in Sec. 1.2.2, we plot $\overline{\rho}_{NO}$, $\overline{\rho}_N$ and $\overline{\rho}_{O_2}$ as a function of time in Fig. 5.8. The results for NO and N are similar to those for the ordinary Zel'dovich mechanism, see Fig. 1.10, at early time. But at late time there is an additional equilibration with more complicated dynamics, with the relative concentrations of NO and N switching places in importance.

We summarize some of the key vectors and matrices involved in the analysis. We adopt Eq. (1.167) for our vector of $N = 5$ concentrations:

$$\overline{\boldsymbol{\rho}} = \begin{pmatrix} \overline{\rho}_{NO} \\ \overline{\rho}_N \\ \overline{\rho}_{N_2} \\ \overline{\rho}_O \\ \overline{\rho}_{O_2} \end{pmatrix}. \tag{5.485}$$

We have $L = 2$ elements, with N, $(l = 1)$ and O, $(l = 2)$. We adopt Eq. (1.168) for the element-species matrix $\boldsymbol{\phi}$, of dimension $L \times N = 2 \times 5$,

$$\boldsymbol{\phi} = \begin{pmatrix} 1 & 1 & 2 & 0 & 0 \\ 1 & 0 & 0 & 1 & 2 \end{pmatrix}. \tag{5.486}$$

The first row of $\boldsymbol{\phi}$ is for the N atom; the second row is for the O atom. For the $J = 4$ reactions, we take

$$\mathbf{r} = \begin{pmatrix} r_1 \\ r_2 \\ r_3 \\ r_4 \end{pmatrix}, \tag{5.487}$$

where the rates are given by the standard law of mass action kinetics, not shown in detail here. The matrix $\boldsymbol{\nu}$, of dimension $N \times J = 5 \times 4$ is new for this problem and is

$$\boldsymbol{\nu} = \begin{pmatrix} -1 & 1 & 0 & 0 \\ -1 & -1 & 2 & 0 \\ 1 & 0 & -1 & 0 \\ 1 & 1 & 0 & 2 \\ 0 & -1 & 0 & -1 \end{pmatrix}. \tag{5.488}$$

The matrix $\boldsymbol{\nu}$ has rank $R = N - L = 3$. It is not full rank in this case, in contrast to the previous example. Our stoichiometric constraint on element conservation for each reaction, Eq. (1.35), $\boldsymbol{\phi} \cdot \boldsymbol{\nu} = \mathbf{0}$, holds here:

$$\boldsymbol{\phi} \cdot \boldsymbol{\nu} = \begin{pmatrix} 1 & 1 & 2 & 0 & 0 \\ 1 & 0 & 0 & 1 & 2 \end{pmatrix} \begin{pmatrix} -1 & 1 & 0 & 0 \\ -1 & -1 & 2 & 0 \\ 1 & 0 & -1 & 0 \\ 1 & 1 & 0 & 2 \\ 0 & -1 & 0 & -1 \end{pmatrix} = \begin{pmatrix} 0 & 0 & 0 & 0 \\ 0 & 0 & 0 & 0 \end{pmatrix}. \tag{5.489}$$

With a similar analysis to that given in Sec. 1.2.2, we find that

$$\underbrace{\begin{pmatrix} \overline{\rho}_{NO} \\ \overline{\rho}_{N} \\ \overline{\rho}_{N_2} \\ \overline{\rho}_{O} \\ \overline{\rho}_{O_2} \end{pmatrix}}_{\overline{\boldsymbol{\rho}}} = \underbrace{\begin{pmatrix} \widehat{\overline{\rho}}_{NO} \\ \widehat{\overline{\rho}}_{N} \\ \widehat{\overline{\rho}}_{N_2} \\ \widehat{\overline{\rho}}_{O} \\ \widehat{\overline{\rho}}_{O_2} \end{pmatrix}}_{\widehat{\overline{\boldsymbol{\rho}}}} + \underbrace{\begin{pmatrix} 1 & 0 & 0 \\ 0 & 1 & 0 \\ -\frac{1}{2} & -\frac{1}{2} & 0 \\ -1 & 0 & -2 \\ 0 & 0 & 1 \end{pmatrix}}_{\mathbf{D}} \underbrace{\begin{pmatrix} \xi_{NO} \\ \xi_{N} \\ \xi_{O_2} \end{pmatrix}}_{\boldsymbol{\xi}}. \tag{5.490}$$

This takes the form of Eq. (1.55):

$$\overline{\boldsymbol{\rho}} = \widehat{\overline{\boldsymbol{\rho}}} + \mathbf{D} \cdot \boldsymbol{\xi}. \tag{5.491}$$

Here, the matrix $\mathbf{D}$ is of dimension $N \times R = 5 \times 3$. It spans the same column space as does the $N \times J$ matrix $\boldsymbol{\nu}$ that is of rank $R$. Note that $\boldsymbol{\phi} \cdot \mathbf{D} = \mathbf{0}$:

$$\boldsymbol{\phi} \cdot \mathbf{D} = \begin{pmatrix} 1 & 1 & 2 & 0 & 0 \\ 1 & 0 & 0 & 1 & 2 \end{pmatrix} \begin{pmatrix} 1 & 0 & 0 \\ 0 & 1 & 0 \\ -\frac{1}{2} & -\frac{1}{2} & 0 \\ -1 & 0 & -2 \\ 0 & 0 & 1 \end{pmatrix} = \begin{pmatrix} 0 & 0 & 0 \\ 0 & 0 & 0 \end{pmatrix}. \tag{5.492}$$

Now let us move to the Onsager analysis. The matrix $\mathbf{C}$ is of dimension $J \times (N - L) = 4 \times 3$ and has in its columns $R = N - L = 3$ linearly independent columns of $\boldsymbol{\nu}^T$. We easily see that $\boldsymbol{\nu}^T$ is

$$\boldsymbol{\nu}^T = \begin{pmatrix} -1 & -1 & 1 & 1 & 0 \\ 1 & -1 & 0 & 1 & -1 \\ 0 & 2 & -1 & 0 & 0 \\ 0 & 0 & 0 & 2 & -1 \end{pmatrix}. \quad (5.493)$$

We can take the first, second, and fourth columns of $\boldsymbol{\nu}^T$ to form $\mathbf{C}$:

$$\mathbf{C} = \begin{pmatrix} -1 & -1 & 1 \\ 1 & -1 & 1 \\ 0 & 2 & 0 \\ 0 & 0 & 2 \end{pmatrix}. \quad (5.494)$$

Then, by Eq. (5.384), we find the $J \times J = 4 \times 4$ projection matrix $\mathbf{B}$ to be

$$\mathbf{B} = \mathbf{C} \cdot \left(\mathbf{C}^T \cdot \mathbf{C}\right)^{-1} \cdot \mathbf{C}^T, \quad (5.495)$$

$$= \begin{pmatrix} \frac{3}{4} & -\frac{1}{4} & -\frac{1}{4} & \frac{1}{4} \\ -\frac{1}{4} & \frac{3}{4} & -\frac{1}{4} & \frac{1}{4} \\ -\frac{1}{4} & -\frac{1}{4} & \frac{3}{4} & \frac{1}{4} \\ \frac{1}{4} & \frac{1}{4} & \frac{1}{4} & \frac{3}{4} \end{pmatrix}. \quad (5.496)$$

The matrix $\mathbf{B}$ has rank $N - L = 3$ and is thus not full rank. It has eigenvalues of $1, 1, 1, 0$. The Onsager matrix $\mathsf{L}$ is given by

$$\mathsf{L} = \mathbf{B}^T \cdot \mathbf{R}' \cdot \mathbf{B}, \quad (5.497)$$

$$= \frac{1}{16} \begin{pmatrix} 9\mathcal{R}_1' + \mathcal{R}_2' + \mathcal{R}_3' + \mathcal{R}_4' & -3\mathcal{R}_1' - 3\mathcal{R}_2' + \mathcal{R}_3' + \mathcal{R}_4' & -3\mathcal{R}_1' + \mathcal{R}_2' - 3\mathcal{R}_3' + \mathcal{R}_4' & 3\mathcal{R}_1' - \mathcal{R}_2' - \mathcal{R}_3' + 3\mathcal{R}_4' \\ -3\mathcal{R}_1' - 3\mathcal{R}_2' + \mathcal{R}_3' + \mathcal{R}_4' & \mathcal{R}_1' + 9\mathcal{R}_2' + \mathcal{R}_3' + \mathcal{R}_4' & \mathcal{R}_1' - 3\mathcal{R}_2' - 3\mathcal{R}_3' + \mathcal{R}_4' & -\mathcal{R}_1' + 3\mathcal{R}_2' - \mathcal{R}_3' + 3\mathcal{R}_4' \\ -3\mathcal{R}_1' + \mathcal{R}_2' - 3\mathcal{R}_3' + \mathcal{R}_4' & \mathcal{R}_1' - 3\mathcal{R}_2' - 3\mathcal{R}_3' + \mathcal{R}_4' & \mathcal{R}_1' + \mathcal{R}_2' + 9\mathcal{R}_3' + \mathcal{R}_4' & -\mathcal{R}_1' - \mathcal{R}_2' + 3\mathcal{R}_3' + 3\mathcal{R}_4' \\ 3\mathcal{R}_1' - \mathcal{R}_2' - \mathcal{R}_3' + 3\mathcal{R}_4' & -\mathcal{R}_1' + 3\mathcal{R}_2' - \mathcal{R}_3' + 3\mathcal{R}_4' & -\mathcal{R}_1' - \mathcal{R}_2' + 3\mathcal{R}_3' + 3\mathcal{R}_4' & \mathcal{R}_1' - \mathcal{R}_2' + \mathcal{R}_3' + 9\mathcal{R}_4' \end{pmatrix}. \quad (5.498)$$

One can estimate the behavior of $\dot{\sigma}$ with both the full nonlinear model and its Onsager approximation. As expected, the Onsager approximation works well near equilibrium. Meaningful graphical display of the comparison of the two approximations is difficult due to numerical precision issues. As expected, the Hessian matrix of $\dot{\sigma}$ is symmetric and positive semidefinite at equilibrium with eigenvalues of $9.245 \times 10^{22}, 1.50 \times 10^{22}$ and $4.025 \times 10^{17}$.

For variety, let us work with a different Onsager matrix. Let us first get the reduced chemical affinity by Eq. (5.378):

$$\hat{\overline{\alpha}} = \left(\mathbf{C}^T \cdot \mathbf{C}\right)^{-1} \cdot \mathbf{C}^T \cdot \overline{\alpha}, \tag{5.499}$$

$$\begin{pmatrix} \hat{\overline{\alpha}}_1 \\ \hat{\overline{\alpha}}_2 \\ \hat{\overline{\alpha}}_3 \end{pmatrix} = \begin{pmatrix} -\frac{1}{2} & \frac{1}{2} & 0 & 0 \\ -\frac{1}{8} & -\frac{1}{8} & \frac{3}{8} & \frac{1}{8} \\ \frac{1}{8} & \frac{1}{8} & \frac{1}{8} & \frac{3}{8} \end{pmatrix} \begin{pmatrix} \overline{\alpha}_1 \\ \overline{\alpha}_2 \\ \overline{\alpha}_3 \\ \overline{\alpha}_4 \end{pmatrix}, \tag{5.500}$$

$$= \begin{pmatrix} -\frac{\overline{\alpha}_1}{2} + \frac{\overline{\alpha}_2}{2} \\ -\frac{\overline{\alpha}_1}{8} - \frac{\overline{\alpha}_2}{8} + \frac{3\overline{\alpha}_3}{8} + \frac{\overline{\alpha}_4}{8} \\ \frac{\overline{\alpha}_1}{8} + \frac{\overline{\alpha}_2}{8} + \frac{\overline{\alpha}_3}{8} + \frac{3\overline{\alpha}_4}{8} \end{pmatrix}. \tag{5.501}$$

The $(N-L) \times (N-L) = 3 \times 3$ Onsager matrix $\mathbf{L}$ is given by Eq. (5.383):

$$\mathbf{L} = \mathbf{C}^T \cdot \mathbf{R}' \cdot \mathbf{C}, \tag{5.502}$$

$$= \begin{pmatrix} -1 & 1 & 0 & 0 \\ -1 & -1 & 2 & 0 \\ 1 & 1 & 0 & 2 \end{pmatrix} \begin{pmatrix} \mathcal{R}'_1 & 0 & 0 & 0 \\ 0 & \mathcal{R}'_2 & 0 & 0 \\ 0 & 0 & \mathcal{R}'_3 & 0 \\ 0 & 0 & 0 & \mathcal{R}'_4 \end{pmatrix} \begin{pmatrix} -1 & -1 & 1 \\ 1 & -1 & 1 \\ 0 & 2 & 0 \\ 0 & 0 & 2 \end{pmatrix}, \tag{5.503}$$

$$= \begin{pmatrix} \mathcal{R}'_1 + \mathcal{R}'_2 & \mathcal{R}'_1 - \mathcal{R}'_2 & -\mathcal{R}'_1 + \mathcal{R}'_2 \\ \mathcal{R}'_1 - \mathcal{R}'_2 & \mathcal{R}'_1 + \mathcal{R}'_2 + 4\mathcal{R}'_3 & -\mathcal{R}'_1 - \mathcal{R}'_2 \\ -\mathcal{R}'_1 + \mathcal{R}'_2 & -\mathcal{R}'_1 - \mathcal{R}'_2 & \mathcal{R}'_1 + \mathcal{R}'_2 + 4\mathcal{R}'_4 \end{pmatrix}. \tag{5.504}$$

Then, we can write Eq. (5.382) as

$$\left.\frac{dS}{dt}\right|_{E,V} = \frac{V}{T} \begin{pmatrix} \hat{\overline{\alpha}}_1 & \hat{\overline{\alpha}}_2 & \hat{\overline{\alpha}}_3 \end{pmatrix} \cdot \begin{pmatrix} \mathcal{R}'_1 + \mathcal{R}'_2 & \mathcal{R}'_1 - \mathcal{R}'_2 & -\mathcal{R}'_1 + \mathcal{R}'_2 \\ \mathcal{R}'_1 - \mathcal{R}'_2 & \mathcal{R}'_1 + \mathcal{R}'_2 + 4\mathcal{R}'_3 & -\mathcal{R}'_1 - \mathcal{R}'_2 \\ -\mathcal{R}'_1 + \mathcal{R}'_2 & -\mathcal{R}'_1 - \mathcal{R}'_2 & \mathcal{R}'_1 + \mathcal{R}'_2 + 4\mathcal{R}'_4 \end{pmatrix} \begin{pmatrix} \hat{\overline{\alpha}}_1 \\ \hat{\overline{\alpha}}_2 \\ \hat{\overline{\alpha}}_3 \end{pmatrix}. \tag{5.505}$$

## 5.5.5 On Potentials, Entropy, and Dynamics

The notion of a scalar potential field has long history and great utility in mechanics. Typically the gradient of the potential field gives a vector field that describes the motion of a physical entity in the field. Among many, Hubbard and Hubbard (2009) nicely discuss the relevant mathematical physics. In a three-dimensional potential field, the corresponding vector field will be curl-free. For example, three-dimensional steady heat transfer in an isotropic material with constant thermal conductivity is described by $\nabla^2 T = 0$. Here, $T(x, y, z)$ is actually a potential field. Its gradient, $\nabla T$, is parallel to the vector of thermal energy flow given by Fourier's law, $\mathbf{j}^q = -\mathsf{k}\nabla T$, where $\mathbf{j}^q$ is the heat flux vector and $\mathsf{k}$ the thermal conductivity. Thermal energy moves from high potential to low potential. Moreover the induced vector field $\mathbf{j}^q(x, y, z)$ is curl-free, $\nabla \times \mathbf{j}^q = \mathbf{0}$, and divergence-free $\nabla \cdot \mathbf{j}^q = 0$. Analogous potential fields arise in incompressible irrotational fluid dynamics, electrodynamics, and other fields where dynamics is important.

In combustion dynamics, we have seen the analogy is not clear. Indeed, use of the term "chemical potential" is suggestive, but is not a classical potential in the sense just described. As was introduced in Sec. 3.9.1, differences in chemical potential induce chemical dynamics. And our actual mass action Arrhenius kinetics with reverse reactions depend heavily on chemical potentials to induce the dynamics. However, the classical chemical potential is difficult to reconcile with a potential field as known in mechanics. While it certainly would be desirable to have a scalar field whose gradient defines chemical dynamics, it remains to be identified for common gas phase combustion models. Certainly the Lyapunov functions such as $\tilde{G}$ defined earlier in this chapter are suggestive of potential functions as nonzero values of $\tilde{G}$ induce chemical dynamics that bring $\tilde{G}$ to zero. But as we saw in this chapter, their gradients do not identically yield the chemical dynamics as a true potential would. There is certainly much discussion in the literature about potentials, and we refer here to two of the sources that contain richer exposition (Edelen, 1973; Goddard, 2014).

We have also spent considerable effort delineating notions of production of irreversible entropy because it is associated with an ongoing debate on whether additional principles beyond the traditional laws of thermodynamics reflect nature. One of the more influential contentions is given by Prigogine (1967, p. 75), who notes,

> We shall see that stationary states may be characterized by an extremum principle that states that in the stationary state, the entropy production has its minimum value compatible with some auxiliary conditions to be specified in each case.

Notions such as these have spurred much activity in a variety of disciplines, including combustion dynamics, with the goal of using new principles to better codify observed behaviors (Ziegler, 1983; Lengyel, 1988; Pekar, 2005; Kojima, 2014). One must note, however, that these notions have also attracted skepticism. For example, Müller and Weiss (2005, p. 182) present a simple counterexample from heat transfer, from which they conclude, "Therefore we can safely discard the principle of maximum entropy source: It contradicts the *first law of thermodynamics*." Jaynes (1980) gives relevant discussion, which is extended in detail by Grandy (2008), who notes,[10]

> These observations do not in themselves seem to support the notion of a *general* extremum principle for dissipation rates in steady-state processes, either linear or nonlinear, near or far from equilibrium.

Grandy later concludes,

> After all is said and done, the issue with which we began this discussion—the role of additional principles in nonequilibrium thermodynamics—does not appear to be much closer to a definite resolution. In those cases where we are able to apply a principle of minimum (or maximum) rate of energy dissipation, it seems that we already have sufficient tools in the conservation laws and constitutive equations. Although the conservation laws are exact, the constitutive relations are of necessity approximated to some degree, so it may be

[10] W. T. Grandy, Jr., 2008, *Entropy and the Time Evolution of Macroscopic Systems*, pp. 170–172, from Ch. 12: Entropy Production and Dissipation Rates, by permission of Oxford University Press.

> that the choice of which route to follow very much depends on the details of any specific problem. Rather than focus on some general and improbable principle, it may be more reasonable to search for general conditions or classes of problems for which extremum rate principles might be both valid and productive (p. 172).

Consistent with Grandy, our analysis does not explicitly rely on invocation of additional extremum principles beyond the ordinary laws of thermodynamics, though one cannot ignore the possibility that such principles may be in play at a deeper level in the construction of constitutive models. But at this stage, we are not prepared to accept as proven that additional laws of nature based on extremizing irreversible entropy production rates have been established in continuum models.

## EXERCISES

**5.1.** Find the extension of Eq. (5.13) to isobaric reaction.

**5.2.** Show a detailed analysis that develops Eqs. (5.29)–(5.32).

**5.3.** Consider a mixture of 1 kmol of $H_2$ and 1 kmol of $O_2$ at $T = 3000$ K and $P = 100$ kPa. Assuming an isobaric and isothermal equilibration process with the products consisting of $H_2$, $O_2$, $H_2O$, OH, H, and O, use the methods of Sec. 5.2.1 to find the equilibrium concentrations. Consider the same mixture at $T = 298$ K and $T = 1000$ K. Consult a thermodynamics text as needed for property values.

**5.4.** Show details of Zel'dovich's uniqueness proof leading to Eq. (5.276).

**5.5.** Study the dynamics of a problem identical to that considered in Sec. 5.3.1, except take the reaction mechanism to be

$$1: \; A \leftrightharpoons B,$$

$$2: \; B \leftrightharpoons C,$$

$$3: \; C \rightarrow A.$$

Determine and plot the exact solution for species concentrations as functions of time for the same numerical parametric values given in Sec. 5.3.1.

**5.6.** For the reaction mechanism of Table 1.2, form Eq. (5.348). Find the associated projection matrix **P** and demonstrate the validity of Eq. (5.351).

**5.7.** For the reaction mechanism of Table 1.2, form any of the matrices associated with Onsager reciprocity presented in Sec. 5.5.1.

**5.8.** Study the extended Zel'dovich mechanism of Sec. 5.5.4. First, formulate the differential equations and generate the plot of Fig. 5.8. Then, in the neighborhood of equilibrium, calculate $\mathbf{J}$, $\mathbf{H}_{\dot{\sigma}}$, and $\mathbf{H}_G$ and demonstrate that Eq. (5.443) holds.

## References

A. N. Al-Khateeb, J. M. Powers, S. Paolucci, A. J. Sommese, J. A. Diller, J. D. Hauenstein, and J. D. Mengers, 2009, One-dimensional slow invariant manifolds for spatially homogeneous reactive systems, *Journal of Chemical Physics*, 131(2): 024118.

A. Bejan, 2006, *Advanced Engineering Thermodynamics*, 3rd ed., John Wiley, Hoboken, NJ.

C. Borgnakke and R. E. Sonntag, 2013, *Fundamentals of Thermodynamics*, 8th ed., John Wiley, New York, Table A.9, p. 766.

J. D. Buckmaster and G. S. S. Ludford, 1982, *Theory of Laminar Flames*, Cambridge University Press, Cambridge, UK, Chapter 1.

S. R. de Groot and P. Mazur, 1984, *Non-Equilibrium Thermodynamics*, Dover, New York.

D. G. B. Edelen, 1973, Existence of symmetry relations and dissipation potentials, *Archive for Rational Mechanics and Analysis*, 51(3): 218–227.

M. Feinberg, 1995a, The existence and uniqueness of steady states for a class of chemical reaction networks, *Archive for Rational Mechanics and Analysis*, 132(4): 311–370.

M. Feinberg, 1995b, Multiple steady states for chemical reaction networks of deficiency one, *Archive for Rational Mechanics and Analysis*, 132(4): 371–406.

J. D. Goddard, 2014, Dissipation potentials for reaction-diffusion systems, *Industrial and Engineering Chemistry Research*, 54(16): 4078–4083.

W. T. Grandy, 2008, *Entropy and the Time Evolution of Macroscopic Systems*, Oxford University Press, Oxford, UK, pp. 170–172.

I. Gyarmati, 1970, *Non-equilibrium Thermodynamics: Field Theory and Variational Principles*, Springer, New York.

J. O. Hirschfelder, C. F. Curtis, and R. B. Bird, 1954, *Molecular Theory of Gases and Liquids*, John Wiley, New York, Chapter 8.

J. H. Hubbard and B. B. Hubbard, 2009, *Vector Calculus, Linear Algebra, and Differential Forms: A Unified Approach*, 4th ed., Matrix Editions, Ithaca, NY.

E. T. Jaynes, 1980, The minimum entropy production principle, *Annual Review of Physical Chemistry*, 31: 579–601.

R. J. Kee, M. E. Coltrin, and P. Glarborg, 2003, *Chemically Reacting Flow: Theory and Practice*, John Wiley, Hoboken, NJ, Chapter 9.

S. Kojima, 2014, Reaction kinetics path based on entropy production rate and its relevance to low-dimensional manifolds, *Entropy*, 16(6): 2904–2943.

D. Kondepudi and I. Prigogine, 1998, *Modern Thermodynamics: From Heat Engines to Dissipative Structures*, John Wiley, New York, Chapters 15–19.

K. K. Kuo, 2005, *Principles of Combustion*, 2nd ed., John Wiley, New York, Chapters 1 and 2.

Y. Kuramoto, 1984, *Chemical Oscillations, Waves, and Turbulence*, Dover, Mineola, NY.

B. H. Lavenda, 1978, *Thermodynamics of Irreversible Processes*, John Wiley, New York.

C. K. Law, 2006, *Combustion Physics*, Cambridge University Press, Cambridge, UK, Chapter 2.

S. Lengyel, 1988, Deduction of the Guldberg-Waage mass action law from Gyarmati's governing principle of dissipative processes, *Journal of Chemical Physics*, 88(3): 1617–1621.

A. Liñan and F. A. Williams, 1993, *Fundamental Aspects of Combustion*, Oxford University Press, Oxford, UK.

I. Müller and T. Ruggeri, 1998, *Rational Extended Thermodynamics*, 2nd ed., Springer, New York.

I. Müller and W. Weiss, 2005, *Entropy and Energy*, Springer, New York, p. 182.

L. Onsager, 1931a, Reciprocal relations in irreversible process, I., *Physical Review*, 37(4): 405–426.

L. Onsager, 1931b, Reciprocal relations in irreversible process, II., *Physical Review*, 38(12): 2265–2279.

M. Pekar, 2005, Thermodynamics and the foundations of mass-action kinetics, *Progress in Reaction Kinetics and Mechanism*, 30(1-2): 3–113.

J. M. Powers and S. Paolucci, 2008, Uniqueness of chemical equilibria in ideal mixtures of ideal gases, *American Journal of Physics*, 76(9): 848–855.

I. Prigogine, 1967, *Introduction to Thermodynamics of Irreversible Processes*, Interscience, New York.

W. C. Reynolds, 1968, *Thermodynamics*, 2nd ed., McGraw-Hill, New York.

G. Strang, 2005, *Linear Algebra and Its Applications*, 4th ed., Cengage Learning, Stamford, CT.

K. Terao, 2007, *Irreversible Phenomena: Ignitions, Combustion, and Detonation Waves*, Springer, Berlin.

S. R. Turns, 2011, *An Introduction to Combustion*, 3rd ed., McGraw-Hill, Boston. Chapters 4–6.

F. A. Williams, 1985, *Combustion Theory*, 2nd ed., Benjamin-Cummings, Menlo Park, CA, Chapter 1.

L. C. Woods, 1975, *The Thermodynamics of Fluid Systems*, Oxford University Press, Oxford, UK.

Ya. B. Zel'dovich, 1938, A proof of the uniqueness of the solution of the equations for the law of mass action, *Zhurnal Fizicheskoi Khimii*, 11: 685–687.

H. Ziegler, 1983, Chemical reactions and the principle of maximal rate of entropy production, *Zeitschrift für angewandte Mathematik und Physik*, 34(6): 832–844.

# 6 Nonlinear Dynamics of Reduced Kinetics

*It may be hoped that the complication of physical phenomena likewise hides from us some simple cause still unknown.**

— Henri Poincaré (1854–1912), *The Value of Science*

In this chapter, we introduce a topic that brings the interplay of thermodynamics and dynamics of combustion into clearer focus: reduction of complex combustion systems to simpler systems. The topic of model reduction is vast and transcends combustion. Even within combustion, the literature is extensive. A summary of the many traditional methods of kinetics reduction is found in reviews (Griffiths, 1996; Lu and Law, 2009). Despite decades of effort, there remains no systematic, robust way of reduction in such a way that guarantees optimal fidelity with the underlying detailed model. This chapter will consider one approach to reduction that exploits the fact that common combustion models are dynamical systems and thus are amenable to the same mathematical analysis strategies used in broader applications in dynamics. We will draw directly and extensively from two papers for this discussion, appropriately updated. The first (Powers, et al., 2015)[1] will focus on spatially homogeneous systems. The second (Mengers and Powers, 2013) will give our first glimpse at the effects of spatial inhomogeneity in the form of diffusion. While the problem we consider is simple, the significant complications of diffusion on dynamics will be apparent and serve as a useful transition to the remainder of the book, for which spatial inhomogeneity will be a dominant feature. Indeed, there is much more that could be said about reduction; the intention here is to highlight issues exposed by one approach that likely extends to others. The approach chosen here is not particularly common. However, it is mathematically well defined, is clearly a reduction, and has useful pedagogical value. Our conclusion will be pessimistic: rigorous reduction is possible for simple systems, but is dauntingly difficult for realistic systems.

It is widely recognized that combustion problems possess features that can only be captured by mathematical models that are both nonlinear and reflect the problems' multiscale nature. In some cases, the disparity of scales and effects of nonlinearity are not severe, and a so-called direct numerical modeling approach, which

*H. Poincaré, 1907, *The Value of Science*, trans. G. B. Halsted, The Science Press, New York, 1907, p. 87.
[1] With kind permission from Springer Science+Business Media: *Journal of Mathematical Chemistry*, Slow attractive canonical invariant manifolds for reactive systems, 53, 2015, 737–766, J. M. Powers, S. Paolucci, J. D. Mengers, A. N. Al-Khateeb, text and Figs. 1–4, 11–21, 2015.

employs a large number of degrees of freedom, is viable. In other cases, the disparity of scales is too large for even the most powerful computers. This necessitates the employment of additional reduction strategies in which the model is simplified in such a fashion that the essence of the solution can be captured, while reducing the required computational resources. Such reductions nearly always entail the filtering of detail. For nonlinear multiscale problems, one should ensure that such a filter has retained as much of the original nature of the full solution as possible; colloquially, one wishes to retain the "signal" and filter the "noise." However, determining what is signal and what is noise is not always straightforward.

One class of reduction that has attracted much attention in both the applied mathematics and combustion chemistry communities is based on so-called manifold methods. The best known examples are known as intrinsic low-dimensional manifolds (ILDMs) (Maas and Pope, 1992) and computationally singular perturbation (CSP) (Lam and Goussis, 1989). The essential idea is as follows. The dynamics of a combustion system can nearly always be modeled as an evolution of a trajectory within a finite- or infinite-dimensional phase space. For many systems, there is a rapid relaxation of a trajectory onto a manifold of lower dimension than the original space. On this lower-dimensional manifold, it is often the case that only slow dynamics occur, and these are often the dynamics that are most physically relevant. If one can identify a priori such lower-dimensional manifolds, it may be possible to project the trajectories of combustion dynamics onto such a manifold so that fine time scale events are "filtered," thus reducing the computational time necessary for solving a challenging and otherwise multiscale problem. Such a manifold has been colloquially named a *slow manifold.* As an aside, we will be concerned only with so-called *smooth manifolds*, which possess sufficient continuity and differentiability for our purposes.

A visualization of a common manifold reduction is sketched in Fig. 6.1a. Shown are two equilibria: one a sink that corresponds to a physical equilibrium in a closed fixed-mass combustion system, and the other a saddle with one unstable mode. The saddle is a nonphysical equilibrium in a closed spatially homogeneous combustion system. It is often possible to identify a heteroclinic connection between the saddle and the sink. Because such a connection is a trajectory, it is also an *invariant manifold* (IM) of the system (Perko, 2001). There are other ways to construct IMs (cf. Roussel and Fraser, 2001; Gorban and Karlin, 2003). Because it is a canonical trajectory, due to its connection to the single unstable mode of the saddle, we call it a canonical invariant manifold (CIM). It is well known that the CIM attracts nearby trajectories near both the saddle and sink equilibria. Figure 6.1a also depicts trajectories far from

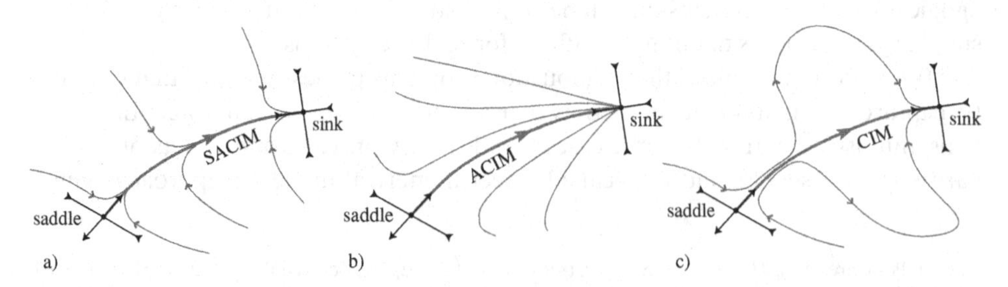


Figure 6.1. Sketches of (a) SACIM, (b) nonslow ACIM, and (c) nonattractive CIM. Adapted from Powers et al. (2015), with permission.

the equilibria that are attracted to the CIM; in combustion problems such behavior is often realized. If it is, we define the CIM to be an attractive canonical invariant manifold (ACIM). If it is further the case that relaxation to the ACIM occurs much faster than the motion on the ACIM, we call the ACIM a slow attractive canonical invariant manifold (SACIM). Much of manifold-based reduction in combustion chemistry relies on the assumption that the manifolds employed are in fact SACIMs.

Unfortunately, classical gas phase combustion models offer no guarantee that manifolds identified as a reduction have all the desirable properties found in a SACIM. Such models, which employ the law of mass action and Arrhenius kinetics, are at most guaranteed to have a unique sink equilibrium in the region of phase space where all species mass fractions are positive. First while a CIM may be attractive, there is no guarantee that it is slow; such a scenario is depicted in Fig. 6.1b. Moreover, far from the sink, there is no guarantee that trajectories nearby to the CIM are attracted to it, thus rendering the scenario of Fig. 6.1c a possibility. In this scenario, a trajectory that is initially near the CIM may actually be repelled from the CIM before ultimately returning to the sink equilibrium. Such a scenario would render any reduction that was based on projection onto a CIM to have large inaccuracies and would predict nonphysical system behavior.

We will describe diagnostic tools that allow one to ascertain whether a given CIM is actually a SACIM. If so, it would have utility in rational reduction of reactive dynamics. Our main result is that CIMs need not be SACIMs, though many are, and that users of reduced chemistry should take precautions to see that any manifolds they employ do not suppress relevant chemical dynamics. We will take advantage of local linearization tools that have been previously developed (Adrover et al., 2007) for the deformation of trajectories near a CIM. These useful stretching-based diagnostics are necessary but insufficient to determine whether a CIM is a SACIM. To remedy this deficiency, we develop the notion of rotation of trajectories for three-dimensional systems and show how the competition between stretching and rotation rates can play a determining role in SACIM diagnosis.

Many of these tools are similar to those used in analyzing the kinematics of translation, stretching, and rotation of compressible fluid particles in motion. While some of our analysis will apply to arbitrarily large systems, most of our discussion where rotation is relevant will be limited to systems that evolve in a phase space with dimension of three. This is because the notion of a generalized rotation in higher dimensions introduces significant complication. Indeed, realistic combustion systems are typically of higher dimension; nevertheless, we will demonstrate that even for the low-dimensional systems, consideration of rotation yields new insights into chemical dynamics.

The plan of the chapter is as follows. We first give a general mathematical background of CIM construction via heteroclinic connections. This is accompanied by a discussion of stretching- and rotation-based diagnostics. We then illustrate the diagnostic tools by applying them to four spatially homogeneous problems: (i) a two-dimensional combustion-inspired model problem, (2) a three-dimensional combustion-inspired model problem, (3) the Zel'dovich mechanism of nitric oxide production, and (4) hydrogen-air kinetics. "Dimension" here refers to composition space and not physical space. The first problem consists of a system that evolves in a two-dimensional chemical composition space; rotation is not a concern. It is sufficiently simple that many closed form analytical results are possible; moreover, it has been well studied in the literature. The second problem consists of a system that evolves in a three-dimensional chemical composition space, rendering rotation to be a relevant concern. The third, the Zel'dovich model, is studied under the same conditions considered in Sec. 1.2.2. The Zel'dovich model evolves in a two-dimensional chemical composition space; thus, while stretching-based diagnostics are relevant, rotation is not. The fourth, the hydrogen-air model, is restricted to a small number of species and reactions to better illustrate the geometrical features of the chemical dynamics. It evolves in a three-dimensional chemical composition space, again rendering rotation to have potential relevance. It is extracted from a larger model that describes dynamics of a more realistic hydrogen-air system. We then turn to a simple linear reaction-diffusion model, and use a Galerkin[2] method to cast the partial differential equation system as an infinite-dimensional ordinary differential equation system. We then reduce this system by retaining a small number of modes and study its dynamics.

## 6.1 Mathematical Background

We describe our framework for reduction of spatially homogeneous reactive systems.

### 6.1.1 Nonlinear Problem

We restrict discussion to a spatially homogeneous ideal mixture of $N$ ideal gases in a closed vessel with volume $V$. The volume may or may not be constant. The constant total mixture mass is $m$. The number of moles of each species $n_i, i = 1, \ldots, N$, evolves in time $t$ due to chemical reactions. The equation for reaction kinetics of for a closed system is given by Eq. (4.213):

$$\frac{d}{dt}\left(\frac{\overline{\rho}_i}{\rho}\right) = \frac{\dot{\omega}_i}{\rho}. \tag{6.1}$$

Defining the specific number of species $i$, $z_i$ as

$$z_i = \frac{n_i}{m} = \frac{n_i}{V}\frac{V}{m} = \frac{\overline{\rho}_i}{\rho}, \qquad i = 1, \ldots, N, \tag{6.2}$$

we can rewrite Eq. (6.1) as

$$\frac{dz_i}{dt} = \frac{\dot{\omega}_i}{\rho}. \tag{6.3}$$

The formulation in terms of $z_i$ instead of $\overline{\rho}_i$ is slightly more convenient for closed systems that may be either isochoric or isobaric. It is also common in some of the literature, though by no means is it required. At any given $t$ only $N' \leq N$ of the $z_i$ are linearly independent, as physical restrictions provided by principles such as element conservation for chemical reactions provide $N - N'$ linear constraints. Sometimes reaction mechanisms are such that other linear constraints exist. For example, a mechanism in which each elementary reaction has the property that the number of molecules is unchanged will have an additional linear constraint due to conservation of number of molecules. Much of the following analysis applies to $N' \geq 1$; as such, we leave $N'$ general at this point. However, we will later confine attention to systems that have $N' = 2$ or 3, as it is those systems that best illustrate our results.

[2] Boris Gigoryevich Galerkin (1871–1945), Russian-Soviet engineer.

Time-dependent spatially homogeneous reaction of this mixture is described by a system of $N'$ ordinary differential equations of the form

$$\frac{d\mathbf{z}}{dt} = \mathbf{f}(\mathbf{z}), \qquad \mathbf{z}(0) = \mathbf{z}_0, \qquad \{\mathbf{z}, \mathbf{z}_0, \mathbf{f}\} \in \mathbb{R}^{N'}. \tag{6.4}$$

Here, the independent variable is $t$. The dependent variables $z_i$, $i = 1, \ldots, N'$, are embodied in $\mathbf{z}$, with $\mathbf{z}_0$ as their set of initial values. The remaining $N - N'$ values of $z_i$ can be determined by the linear constraints. One way to develop the linear constraints is to specify a provisional set of initial conditions for $z_i$, $i = 1, \ldots, N$, which can later be relaxed so as to allow $\mathbf{z}_0$ to have parametric variation. The law of mass action with Arrhenius kinetics is represented within the nonlinear algebraic function $\mathbf{f}$.

Equilibrium is attained at points for which

$$\mathbf{f}(\mathbf{z}) = \mathbf{0}, \tag{6.5}$$

and in general one can expect multiple equilibria within $\mathbb{R}^{N'}$. However, as discussed in Sec. 5.2.3, a unique stable equilibrium exists when $\mathbf{z}$ is restricted to values that are physically realizable. We have shown in Sec. 5.5.2 that the eigenvalues of the Jacobian

$$\mathbf{J} = \nabla \mathbf{f}, \tag{6.6}$$

where $\nabla = \partial/\partial z_i, i = 1, \ldots, N'$, are guaranteed real and negative at this equilibrium. The physically realizable region of $\mathbb{R}^{N'}$ is that portion for which $z_i \geq 0, i = 1, \ldots, N$, the boundary of which is a convex polytope. We define the physically realizable region as $\mathbb{S}$, its boundary as $\partial\mathbb{S}$, and the nonphysical region as $\mathbb{S}'$. No species can have a negative number of moles within $\mathbb{S}$. Physical systems are further constrained by the second law of thermodynamics, which is manifested in additional restrictions on the functional form of $\mathbf{f}$. As described in Sec. 5.2.3, $\mathbf{f}$ must be constructed such that within $\mathbb{S}$, there exists a scalar function of the state variables $\mathbf{z}$ whose value changes monotonically with $t$ through the action of $\mathbf{f}$ until it reaches an extreme value at the physical equilibrium. Depending on the particular physical scenario, that scalar function could be the entropy, the Helmholtz free energy, or the Gibbs free energy. We have seen in Sec. 5.2.3 that it is straightforward to formulate this scalar so that it is a Lyapunov function for the system within $\mathbb{S}$. Typically these scalar functions are singular on $\partial\mathbb{S}$.

Although the state variables $\mathbf{z}$ are obviously nonphysical in $\mathbb{S}'$, they are often otherwise mathematically well behaved. In $\mathbb{S}'$, one typically finds a set of nonphysical equilibria with local dynamics that can include all combinations of modes: stable, unstable, and oscillatory. For equilibria with one or more unstable modes, it is possible for there to exist a heteroclinic connection between a nonphysical equilibrium and the physical equilibrium. Typically, such orbits are easy to identify by numerical integration, with values of $\mathbf{z}$ encountering no singularities (even while the second law-motivated scalar function takes on a singular value as the trajectory passes through $\partial\mathbb{S}$).

Our procedure is to focus attention on saddles with one unstable mode and to numerically integrate the system starting from a perturbation from the saddle in the eigen-direction of the unstable mode pointing toward the physical equilibrium. We call a heteroclinic trajectory connecting a nonphysical equilibrium to the physical equilibrium a CIM. Because it is often the case that a CIM is the trajectory to which the slowest dynamics of the system are confined as well as the trajectory that attracts nearby trajectories, it has the potential to be the ideal candidate for a reduced model of chemical kinetics. Because it is a trajectory, it meets well-known criteria for IMs. We focus attention on one-dimensional CIMs that originate from saddle points with one unstable mode. This guarantees that the CIM is attractive in the neighborhood of each equilibrium; far from equilibrium there is no guarantee of attraction to the CIM, as illustrated in Fig. 6.1b. Higher-dimensional CIMs could be considered, though the topology becomes more challenging. So, given a one-dimensional CIM, which is not difficult to identify, one would like to quantifiably ascertain whether such a CIM is both slow and attractive. This is more difficult, and we describe in detail in the following sections how to use a set of well-defined diagnostic tools based on linear analysis to make this determination for a general CIM.

Before proceeding to linear tools, one recognizes that a somewhat useful nonlinear diagnostic tool for attractiveness of a CIM is the time evolution of the distance separating the CIM and a local trajectory that originates near the CIM as both progress toward the stable equilibrium point. We take this distance as

$$s(t) = \min_{\tau \in (-\infty, \infty)} ||\mathbf{z}_{CIM}(\tau) - \mathbf{z}(t)||. \tag{6.7}$$

Here, $\mathbf{z}_{CIM}$ represents the CIM that is known parametrically as a function of $\tau$. At a given time $t$ the coordinates of a nearby trajectory $\mathbf{z}(t)$ are known, and $s(t)$ is the minimum distance from $\mathbf{z}(t)$ to any point on the entire CIM. A necessary but insufficient condition for an attractive CIM is that $s(t)/s(0) \ll 1$ as $t \to \infty$. This condition is in fact weak in that any trajectory originating within the basin of attraction of the sink equilibrium will satisfy Eq. (6.7). Loosely speaking, for an attractive CIM, we would like $s(t)$ for a trajectory such as that sketched in Fig. 6.1a to rapidly decrease to near zero on its approach to the CIM and then slowly approach zero as it moves to the sink along the CIM.

### 6.1.2 Local Linear Analysis

To attempt to quantify the notion of slowness and attractiveness just discussed, we will in this section pose two useful ansatzs based on local linear analysis, which will be tested *a posteriori* in a later section on simple examples with $N'$ of either 2 or 3. Let us then consider Eq. (6.4) and require that $\mathbf{z}_0$ be a point on a CIM, far from equilibrium. Then, local linearization in the neighborhood of $\mathbf{z}_0$ gives

$$\frac{d}{dt}(\mathbf{z} - \mathbf{z}_0) = \mathbf{f}(\mathbf{z}_0) + \mathbf{J}|_{\mathbf{z}_0} \cdot (\mathbf{z} - \mathbf{z}_0) + \cdots. \tag{6.8}$$

One recognizes that $\mathbf{f}(\mathbf{z}_0)$ is a constant vector of dimension $N'$, and $\mathbf{J}|_{\mathbf{z}_0}$ is a constant Jacobian matrix of dimension $N' \times N'$. Now let us define the symmetric and antisymmetric parts of $\mathbf{J}$ as $\mathbf{J}_s$ and $\mathbf{J}_a$, respectively:

$$\mathbf{J}_s = \frac{\mathbf{J} + \mathbf{J}^T}{2}, \qquad \mathbf{J}_a = \frac{\mathbf{J} - \mathbf{J}^T}{2}, \tag{6.9}$$

which allows us to recast Eq. (6.8) as

$$\frac{d}{dt}(\mathbf{z} - \mathbf{z}_0) = \underbrace{\mathbf{f}(\mathbf{z}_0)}_{\text{translation}} + \underbrace{\mathbf{J}_s|_{\mathbf{z}_0} \cdot (\mathbf{z} - \mathbf{z}_0)}_{\text{stretch}} + \underbrace{\mathbf{J}_a|_{\mathbf{z}_0} \cdot (\mathbf{z} - \mathbf{z}_0)}_{\text{rotation}} + \cdots. \tag{6.10}$$

Equation (6.10) describes the motion of points in the neighborhood of the CIM. For $N' = 3$, this motion has a geometric interpretation that is easily visualized.

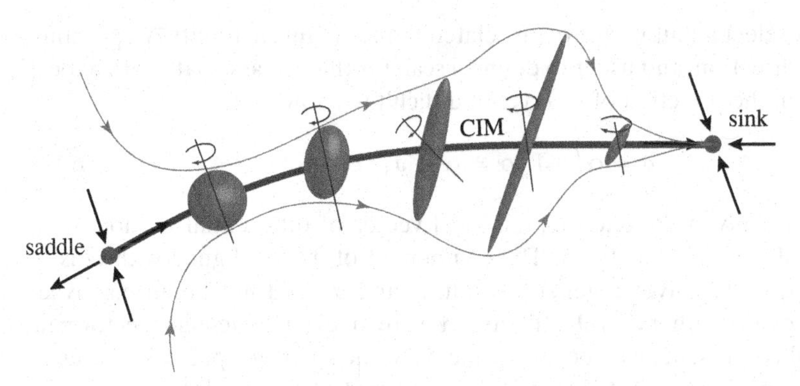


Figure 6.2. Sketch of volume and trajectory modulation near a CIM. Adapted from Powers et al. (2015), with permission.

Because this motion is analogous to that of a three-dimensional fluid particle moving in a known velocity field, one can draw upon standard notions from differential geometry and continuum kinematics (Aris, 1962); we draw upon many of these concepts and present results without detailed proof. As sketched in Fig. 6.2, one might imagine a set of points initially confined within a small sphere that is close to a non-physical saddle equilibrium point. These points evolve under the action of $\mathbf{f}$, until they are brought into the physical sink equilibrium point. Let us imagine that the volume defined by these points is monotonically shrinking with time. We are thus considering systems for which

$$\nabla \cdot \mathbf{f} < 0. \tag{6.11}$$

In the dynamical systems literature, such a system is known as dissipative. In the chemistry literature, "dissipative" often has a different connotation, implying satisfaction of a Clausius-Duhem inequality. Within $\mathbb{S}$, chemistry demands the second type of dissipation, but requires the first only in the neighborhood of the physical equilibrium. As it is common for reactive systems to satisfy Eq. (6.11) in large portions of $\mathbb{R}^{N'}$, we restrict discussion to systems satisfying Eq. (6.11). We shall see that even for such CIMs, it remains an open question as to whether they are additionally slow and attractive.

The decomposition of motion given in Eq. (6.10) allows one to associate $\mathbf{f}(\mathbf{z}_0)$ with translation of the volume along the CIM. The term involving $\mathbf{J}_s$ can be associated with stretching of the volume; moreover, because of the symmetry of $\mathbf{J}_s$, one can define a local orthonormal set of basis vectors as well as principal axes of stretch. The term involving $\mathbf{J}_a$ can, for $N' = 3$, be associated with rotation as a solid body about a central axis. For $N' = 3$, the dual vector associated with $\mathbf{J}_a$ is aligned with the axis of rotation, and its magnitude gives the rotation rate. For $N' > 3$, translation and stretching are not difficult to imagine, but rotation is not easily generalized.

Taking $\upsilon$ as the small volume initially near the saddle equilibrium, one can easily show that the local relative volumetric stretching rate is given by

$$\overline{\ln \upsilon}^{\,\cdot} = \frac{1}{\upsilon}\frac{d\upsilon}{dt} = \operatorname{tr}\mathbf{J} = \operatorname{tr}\mathbf{J}_s = \nabla \cdot \mathbf{f}, \tag{6.12}$$

which we consider to be negative. This "volume" $\upsilon$ has nothing to do with the physical property specific volume. It is a volume in chemical composition space, not physical geometric space.

Let us now select a unit vector $\boldsymbol{\alpha}$ (unrelated to the chemical affinity $\overline{\boldsymbol{\alpha}}$), pointing in an arbitrary direction, and use it to define a scalar $\sigma$ that we associate with a local stretching rate in the direction of $\boldsymbol{\alpha}$. It is not difficult to show that

$$\sigma = \boldsymbol{\alpha}^T \cdot \mathbf{J} \cdot \boldsymbol{\alpha} = \boldsymbol{\alpha}^T \cdot \mathbf{J}_s \cdot \boldsymbol{\alpha}. \tag{6.13}$$

Note that for any given direction $\boldsymbol{\alpha}$, the local vector of differential motion at $\mathbf{z}_0$ attributable to $\mathbf{J}_s$ is given by $\mathbf{J}_s \cdot \boldsymbol{\alpha}$. The component of $\mathbf{J}_s \cdot \boldsymbol{\alpha}$ aligned with $\boldsymbol{\alpha}$ is in fact $\sigma$. Also, Eq. (6.13) gives a general result; $\boldsymbol{\alpha}$ and $\sigma$ need not be an eigenvector/eigenvalue pair of either $\mathbf{J}$ or $\mathbf{J}_s$. If $\sigma$ and $\boldsymbol{\alpha}$ were an eigenvalue/eigenvector pair of $\mathbf{J}_s$, they would represent an associated principal value and principal axis of stretch.

Now let us consider a CIM that is known parametrically as a function of $t$ via numerical integration, $\mathbf{z}_{CIM}$. It is then straightforward to determine the unit tangent vector to the CIM, $\boldsymbol{\alpha}_t$, to be

$$\boldsymbol{\alpha}_t = \frac{\mathbf{f}(\mathbf{z}_{CIM})}{|\mathbf{f}(\mathbf{z}_{CIM})|}. \tag{6.14}$$

The unit tangent is uniquely defined and points in the direction of motion; however, $\boldsymbol{\alpha}_t$ need not be a principal axis of stretch.

We next consider the unit vectors orthogonal to the direction of motion. For $N' = 1$, there can be no motion orthogonal to $\boldsymbol{\alpha}_t$. For $N' = 2$, there are two unit vectors orthogonal to $\boldsymbol{\alpha}_t$, and one can be selected. For $N' \geq 3$, there exists an infinite number of unit vectors that are orthogonal to $\boldsymbol{\alpha}_t$. Through algorithms such as the Gram[3]-Schmidt[4] procedure, we can select $N' - 1$ of them, labeled $\boldsymbol{\alpha}_{n,i}$, $i = 1, \ldots, N' - 1$, which are mutually orthonormal and, in combination with $\boldsymbol{\alpha}_t$, span the $N'$-dimensional space. Note that the $\boldsymbol{\alpha}_{n,i}$ are not unique.

With $\boldsymbol{\alpha}_t$ and $\boldsymbol{\alpha}_{n,i}$, one then has the tangential stretching rate $\sigma_t$:

$$\sigma_t = \boldsymbol{\alpha}_t^T \cdot \mathbf{J}_s \cdot \boldsymbol{\alpha}_t, \tag{6.15}$$

and the normal stretching rates $\sigma_{n,i}$:

$$\sigma_{n,i} = \boldsymbol{\alpha}_{n,i}^T \cdot \mathbf{J}_s \cdot \boldsymbol{\alpha}_{n,i}, \qquad i = 1, \ldots, N' - 1. \tag{6.16}$$

Because the $\boldsymbol{\alpha}_{n,i}$ are not yet uniquely determined, we have not yet determined the extreme values of $\sigma_{n,i}$. These will be of relevance.

At a generic point on a CIM, each individual stretching rate can be either positive, negative, or zero. It is also easy to show that the sum of all stretching rates yields the relative volumetric stretching rate:

$$\dot{\overline{\ln v}} = \sigma_t + \sum_{i=1}^{N'-1} \sigma_{n,i}, \tag{6.17}$$

and because we are focusing on systems for which volumetric stretching is negative, any positive stretching in a given direction must be more than counterbalanced by negative stretching in other directions.

[3] Jørgan Pedersen Gram (1850–1916), Danish mathematician.
[4] Erhard Schmidt (1876–1959), German mathematician.

### 6.1.3 Diagnostics in the Normal Plane

It will be useful to isolate attention to the plane whose unit normal is $\boldsymbol{\alpha}_t$; we call this the normal plane. To this end, we first form the $N' \times (N'-1)$ matrix $\mathbf{Q}_n$ whose columns are populated with the $N'-1$ unit normals, $\boldsymbol{\alpha}_{n,i}$:

$$\mathbf{Q}_n = \begin{pmatrix} \vdots & \vdots & \vdots & \vdots \\ \boldsymbol{\alpha}_{n,1} & \boldsymbol{\alpha}_{n,2} & \vdots & \boldsymbol{\alpha}_{n,N'-1} \\ \vdots & \vdots & \vdots & \vdots \end{pmatrix}. \tag{6.18}$$

We then project the $N' \times N'$ Jacobian $\mathbf{J}$ onto the normal plane to form the $(N'-1) \times (N'-1)$ Jacobian $\mathbf{J}_n$:

$$\mathbf{J}_n = \mathbf{Q}_n^T \cdot \mathbf{J} \cdot \mathbf{Q}_n. \tag{6.19}$$

As $\mathbf{J} = \mathbf{J}_s + \mathbf{J}_a$, we also can easily show that $\mathbf{J}_n = \mathbf{J}_{ns} + \mathbf{J}_{na}$, where

$$\mathbf{J}_{ns} = \mathbf{Q}_n^T \cdot \mathbf{J}_s \cdot \mathbf{Q}_n, \qquad \mathbf{J}_{na} = \mathbf{Q}_n^T \cdot \mathbf{J}_a \cdot \mathbf{Q}_n. \tag{6.20}$$

It is clear that $\mathbf{J}_{ns}$ is symmetric and $\mathbf{J}_{na}$ is antisymmetric.

#### Stretching

To identify the magnitudes and directions of extremal normal stretching, one can consider $\mathbf{J}_{ns}$ and calculate its $N'-1$ eigenvalues $\sigma_{n,i}$ and eigenvectors $\boldsymbol{\beta}_{n,i}$. Because of symmetry, these eigenvalues are purely real. It can be verified by applying a standard optimization procedure to Eq. (6.13), namely, to select $\boldsymbol{\alpha}$ such that one extremizes $\sigma = \boldsymbol{\alpha}^T \cdot \mathbf{J}_s \cdot \boldsymbol{\alpha}$, subject to $\boldsymbol{\alpha}^T \cdot \boldsymbol{\alpha} = 1$ and $\boldsymbol{\alpha}^T \cdot \boldsymbol{\alpha}_t = 0$, that the eigenvalues of $\mathbf{J}_{ns}$ are the extreme values of $\sigma$ from Eq. (6.13) in the plane whose unit normal vector is $\boldsymbol{\alpha}_t$. One can find the directions of extreme normal stretch via $\boldsymbol{\alpha}_{n,i} = (\mathbf{Q}_n^T)^+ \cdot \boldsymbol{\beta}_{n,i}$; here the "+" denotes the Moore[5]-Penrose[6] inverse (Strang, 2005). Because $\boldsymbol{\alpha}_{n,i}$, $i = 1, \dots, N'-1$, span the same space as the column vectors of $\mathbf{Q}_n$, there is no loss of information in the Moore-Penrose projection. One also has $\mathbf{Q}_n^T \cdot \boldsymbol{\alpha}_{n,i} = \boldsymbol{\beta}_{n,i}$. Direct application of the optimization procedure yields the identical directions $\boldsymbol{\alpha}_{n,i}$ associated with the extreme values of normal stretchings $\sigma_{n,i}$.

#### Rotation

We next discuss the portion of the motion in the normal plane attributable to $\mathbf{J}_a$, which for $N' = 3$ will be interpreted as being related to rotation, that is, nondeforming. We first recognize that for any given direction $\boldsymbol{\alpha}$, the local vector of differential motion at $\mathbf{z}_0$ attributable to $\mathbf{J}_a$ is given by $\mathbf{J}_a \cdot \boldsymbol{\alpha}$. Part of this vector will be parallel with $\boldsymbol{\alpha}_t$, and part will be normal. The part that is normal is $\mathbf{J}_a \cdot \boldsymbol{\alpha} - (\boldsymbol{\alpha}_t^T \cdot \mathbf{J}_a \cdot \boldsymbol{\alpha})\boldsymbol{\alpha}_t$.

We then wish to identify the $\boldsymbol{\alpha}$ that is associated with the maximum magnitude of this vector in the normal plane. Thus our optimization problem is to select $\boldsymbol{\alpha}$ so as to maximize $\left|\left|\mathbf{J}_a \cdot \boldsymbol{\alpha} - (\boldsymbol{\alpha}_t^T \cdot \mathbf{J}_a \cdot \boldsymbol{\alpha})\boldsymbol{\alpha}_t\right|\right|$, subject to $\boldsymbol{\alpha}^T \cdot \boldsymbol{\alpha} = 1$, and $\boldsymbol{\alpha}^T \cdot \boldsymbol{\alpha}_t = 0$. It will be useful to take

$$||\mathbf{J}_{na}|| \equiv \omega. \tag{6.21}$$

[5] Eliakim Hastings Moore (1862–1932), American mathematician.
[6] Roger Penrose (1931–), English mathematical physicist.

For $N' = 1, 2$, it is easy to show that $||\mathbf{J}_a \cdot \boldsymbol{\alpha} - (\boldsymbol{\alpha}_t^T \cdot \mathbf{J}_a \cdot \boldsymbol{\alpha})\boldsymbol{\alpha}_t|| = ||\mathbf{J}_{na}|| = \omega = 0$ for all suitably constrained $\boldsymbol{\alpha}$. For $N' = 3$, it can be shown that *all* $\boldsymbol{\alpha}$ satisfying the constraints induce $||\mathbf{J}_a \cdot \boldsymbol{\alpha} - (\boldsymbol{\alpha}_t^T \cdot \mathbf{J}_a \cdot \boldsymbol{\alpha})\boldsymbol{\alpha}_t|| = ||\mathbf{J}_{na}|| = \omega$. For $N' = 3$, we interpret $\omega$ as the magnitude of the relevant component of rotational velocity in the normal plane. For $N' = 3$, indeed each $\boldsymbol{\alpha}$ induces a unique vector $\mathbf{J}_a \cdot \boldsymbol{\alpha} - (\boldsymbol{\alpha}_t^T \cdot \mathbf{J}_a \cdot \boldsymbol{\alpha})\boldsymbol{\alpha}_t$; however, each of these has identical magnitude. For $N' > 3$, the value of $||\mathbf{J}_a \cdot \boldsymbol{\alpha} - (\boldsymbol{\alpha}_t^T \cdot \mathbf{J}_a \cdot \boldsymbol{\alpha})\boldsymbol{\alpha}_t||$ varies with suitably constrained $\boldsymbol{\alpha}$. However, its maximum value is in fact given by $\omega$. For $N' > 3$, the geometric interpretation of $\omega$ is unclear. Equipped with these tools, we next turn to how to use them to diagnose the attractiveness and slowness of a candidate CIM.

### Attractiveness

Loosely defined, an attractive CIM is one for which trajectories originating from points near the CIM are brought toward the CIM by the action of $\mathbf{f}$. Certainly for points on the CIM for which all possible normal stretching rates are negative: $\sigma_{n,i} < 0; i = 1, \ldots, N' - 1$, trajectories that originate in the near neighborhood of the CIM will be carried toward the CIM. This important result holds for all $N' > 1$. Conversely, if all possible normal stretching rates are positive, the CIM is not attractive. For $N' = 1$ there is no possibility of normal stretching, and the notion of an attractive manifold is irrelevant. If $N' = 2$, there is only one normal stretching rate, and one has sufficient information to determine attractiveness.

We next consider the interesting case where some of the $\sigma_{n,i}$ may be positive and some negative. For $N' = 3$, a CIM with one positive and one negative normal stretching rate can still be attractive in the presence of rotation in the normal plane of sufficiently large magnitude and a negative volumetric stretching rate. One can imagine a trajectory originating at a point near the CIM where $\sigma_{n,i} > 0$ for some $i$. Such a trajectory could be initially repelled from the CIM. However, if that trajectory is simultaneously being rotated with sufficient rapidity through the action of nonzero $\mathbf{J}_{na}$, one can imagine it being rapidly rotated out of the region of positive normal stretching into a region of negative normal stretching. If the rotation rate is sufficiently more rapid than the positive normal stretching rate, the trajectory will be modified so that it spends more time in regions of negative normal stretching than positive normal stretching, because the overall volumetric stretching rate is negative. Thus, even though a particle could experience a small transient growth away from the CIM, it is rapidly restored by rotation to return to the CIM. Mathematically, if for $N' = 3$, one of the extreme values of $\sigma_{n,i}$ is positive, we take as our *first ansatz* that the CIM is attractive if the ratio $\mu\ (\geq 0)$ is

$$\mu \equiv \frac{\omega}{\max_i \sigma_{n,i}} > 1, \qquad N' = 3. \tag{6.22}$$

This plausible hypothesis is tested in upcoming examples. Note that $\mu$ can vary along the CIM. In practice $\mu \gtrsim 1$ may be sufficient to realize attractiveness. For $\mu \lesssim 1$ or $\mu < 1$, a trajectory could be carried far from the CIM. If the trajectory originates within $\mathbb{S}$, it will ultimately return to the physical sink. If it originates within $\mathbb{S}'$, there is no guarantee of connection to the physical sink. For systems with $N' > 3$ that have both positive and negative $\sigma_{n,i}$, it is difficult to characterize rotation, and thus difficult to arrive at a simple criteria for attractiveness.

**Slowness**

For a given attractive CIM (ACIM), we now ask if the ACIM is also slow, rendering it a SACIM. For the ACIM to be slow, one can insist that normal stretching be faster than tangential stretching on the ACIM. It is the value of $\sigma_{n,i}$ with the smallest magnitude that must be compared to the tangential stretching rate. And thus we take as a *second ansatz* that the criteria for a SACIM must be that the ACIM exist and possess

$$\eta \equiv \frac{\min_i |\sigma_{n,i}|}{|\sigma_t|} > 1. \tag{6.23}$$

This too is tested in upcoming examples. Note that $\eta$ can vary along the ACIM. Large $\eta \gg 1$ will correspond to trajectories moving normally toward the SACIM, followed by a region of high trajectory curvature, where the trajectory then aligns to be nearly tangent to the SACIM. As $\eta > 1$ reduces, such a trajectory will encounter weaker curvature in its relaxation to the SACIM. For $\eta < 1$, trajectory motion will be toward the equilibrium point, but not strongly aligned with the ACIM. In practice $\eta \gtrsim 1$ may be sufficient for a SACIM. ACIMs with $\eta \lesssim 1$ or $\eta < 1$ will have nearby trajectories whose dynamics are as slow or slower than those on the ACIM, thus rendering such an ACIM to be of little or no value in a rational reduction. If $N' = 2$, there is only one normal direction, and $\eta = \kappa$, the stiffness ratio.

One recognizes that any trajectory nearby an ACIM will formally coincide with the ACIM only at the sink equilibrium. A SACIM will have nearby trajectories that are rapidly pulled normally toward it, while an ACIM that is not a SACIM has nearby trajectories that evolve nearly parallel to it. This is sketched in Fig. 6.1b. A SACIM is one that locally attracts nearby trajectories to an arbitrarily close distance of the CIM on a time scale that is much faster than that of the dynamics on the CIM.

### 6.1.4 Algorithmic Diagnostic Procedure

We summarize an algorithmic diagnostic procedure for identification of a SACIM for a system of the form of Eq. (6.4) as follows:

- Identify all equilibria by solving Eq. (6.5). This requires some means to solve nonlinear algebraic equations.
- Determine the Jacobian $\mathbf{J}$ from Eq. (6.6), along with $\mathbf{J}_s$ and $\mathbf{J}_a$ from Eqs. (6.9).
- Evaluate $\mathbf{J}$ near each equilibrium to determine its local dynamical character as a sink, saddle, source, and so on.
- Integrate from the neighborhood of all saddles within $\mathbb{S}'$ with one unstable mode to identify any and all heteroclinic connections to the unique sink within $\mathbb{S}$ so as to determine a one-dimensional CIM, $\mathbf{z}_{CIM}$, which is a candidate ACIM.
- Determine the unit tangent $\boldsymbol{\alpha}_t$ along the CIM from Eq. (6.14).
- Determine the tangential stretching rate $\sigma_t$ along the CIM from Eq. (6.15).
- Use a Gram-Schmidt procedure to identify $N' - 1$ unit normal vectors, thus forming the orthonormal basis $\{\boldsymbol{\alpha}_t, \boldsymbol{\alpha}_{n,1}, \ldots, \boldsymbol{\alpha}_{n,N'-1}\}$.
- Form $\mathbf{Q}_n$ from Eq. (6.18).
- Form $\mathbf{J}_{ns}$ and $\mathbf{J}_{na}$ from Eq. (6.20).
- Identify the extremal values of normal stretching and their associated directions from calculation of the eigenvalues and eigenvectors of $\mathbf{J}_{ns}$.
- Identify $\omega$ from $||\mathbf{J}_{na}||$ and associate it with rotation for $N' = 3$.

- If all extremal normal stretching is negative, the CIM is an ACIM; use Eq. (6.23) to determine $\eta$ to discern if and where the ACIM is a SACIM.
- For $N' = 3$, if one extremal normal stretching is positive and the other negative, from Eqs. (6.22) and (6.23), determine $\mu$ and $\eta$ along the CIM so as to discern if and where the CIM is an ACIM and, moreover, a SACIM.

To concisely characterize the system behavior, one can from Eq. (6.7) determine $s(t)/s(0)$ for several trajectories to demonstrate the attractiveness of the CIM.

## 6.2 Reduction of Model Systems

We present results for two model systems relevant to combustion in this section. The first has a simple mathematical form and shares certain features with physically derived combustion models. It has $N' = 2$, and thus rotation will be irrelevant. The second also has a simple mathematical form and also shares certain features with physically derived combustion models. It has $N' = 3$, and the magnitude of rotation will be shown to be critical in diagnosing the manifold as a CIM, ACIM, or SACIM.

### 6.2.1 Two-Dimensional Phase Space

Consider the system, of the form of Eq. (6.8), with $N' = 2$, taking the parameter $\gamma > 1$:

$$\frac{dz_1}{dt} = -z_1, \qquad \frac{dz_2}{dt} = -\gamma z_2 + \frac{(\gamma - 1)z_1 + \gamma z_1^2}{(1 + z_1)^2}. \tag{6.24}$$

This problem was introduced by Davis and Skodje (1999). It has been widely used (e.g., Singh et al., 2002), as a model problem to test a variety of reduction methods. Its key advantage is that is possesses a closed-form analytic solution, thus enabling simple exposition of many features of reduction. Unusually, it allows the results of many common reductions to be written in closed form, allowing straightforward exposition of the errors induced by reduction.

The problem shares many, but not all, features with practical combustion problems. Shared features include possession of (1) nonlinearity, (2) a unique stable sink equilibrium, and (3) a local volumetric deformation rate that is strictly negative on the CIM as well as throughout the domain of positive $z_i$. In contrast with combustion systems, the system has no clearly identified (1) additional linear constraints, so one cannot discuss the convex polytope $\mathbb{S}$, or (2) entropy-based scalar Lyapunov function.

With arbitrary initial conditions $z_1(0) = z_{1o}$, $z_2(0) = z_{2o}$, it is easy to show that the exact solution is

$$z_1(t) = z_{1o}e^{-t}, \tag{6.25}$$

$$z_2(t) = \left(z_{2o} - \frac{z_{1o}}{1 + z_{1o}}\right) e^{-\gamma t} + \frac{z_{1o}e^{-t}}{1 + z_{1o}e^{-t}}. \tag{6.26}$$

By inspection, as $t \to \infty$, $(z_1, z_2) \to (0, 0)$, and this equilibrium is stable. Combining $z_1(t)$ and $z_2(t)$ so as to eliminate $t$, we obtain a solution in the phase plane:

$$z_2 = \left(z_{2o} - \frac{z_{1o}}{1 + z_{1o}}\right)\left(\frac{z_1}{z_{1o}}\right)^{\gamma} + \frac{z_1}{1 + z_1}. \tag{6.27}$$

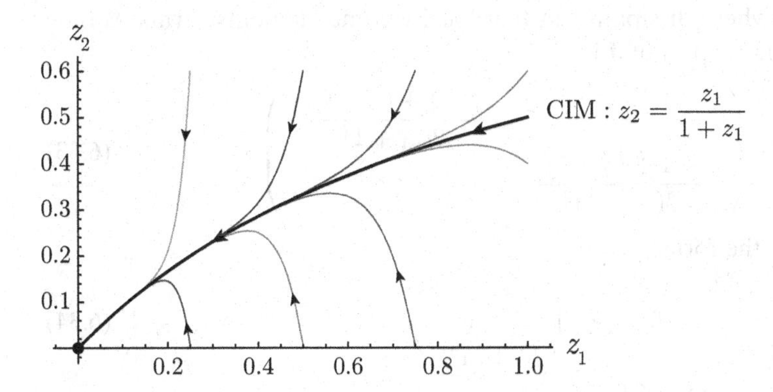


Figure 6.3. Davis-Skodje manifold and nearby trajectories for $\gamma = 10$.

Because $\gamma > 1$, it is easy to see that systems beginning at $(z_{1o}, z_{2o})$ first rapidly approach the curve:

$$z_2 = \frac{z_1}{1 + z_1}. \tag{6.28}$$

Solution trajectories then proceed nearly along this curve until they reach the stable equilibrium at the origin. If the initial conditions are such that $z_{2o} = z_{1o}/(1 + z_{1o})$, the system begins and stays on the CIM.

Davis and Skodje discuss how when the system is transformed appropriately, $z_2 = z_1/(1 + z_1)$ represents a heteroclinic connection from a saddle point at infinity to the stable fixed point at $(z_1, z_2) = (0, 0)$. Because it is a heteroclinic trajectory connecting equilibria, Eq. (6.28) is a CIM. Because it is attractive, it is an ACIM. We plot the ACIM and several nearby trajectories in Fig. 6.3 for $\gamma = 10$. Let us use our diagnostics to see if it is a SACIM. The vector field in phase space is given by

$$\mathbf{f} = \begin{pmatrix} -z_1 \\ -\gamma z_2 + \dfrac{(\gamma - 1)z_1 + \gamma z_1^2}{(1 + z_1)^2} \end{pmatrix}. \tag{6.29}$$

The exact Jacobian is

$$\mathbf{J} = \nabla \mathbf{f} = \begin{pmatrix} -1 & 0 \\ \dfrac{\gamma - 1 + (\gamma + 1)z_1}{(1 + z_1)^3} & -\gamma \end{pmatrix}. \tag{6.30}$$

Its eigenvalues throughout the domain are

$$\lambda_1 = -1, \qquad \lambda_2 = -\gamma. \tag{6.31}$$

There is only one finite equilibrium point; by inspection, it is at the origin $(0, 0)$. Because the eigenvalues of $\mathbf{J}$ are everywhere negative, including the origin, the origin is linearly stable. The relative volumetric stretching rate is

$$\overline{\ln v}^{\cdot} = \nabla \cdot \mathbf{f} = -1 - \gamma < 0. \tag{6.32}$$

The system is everywhere dissipative in that local volume elements shrink as time increases. The symmetric part of $\mathbf{J}$ is

$$\mathbf{J}_s = \begin{pmatrix} -1 & \dfrac{\gamma - 1 + (\gamma+1)z_1}{2(z_1+1)^3} \\ \dfrac{\gamma - 1 + (\gamma+1)z_1}{2(z_1+1)^3} & -\gamma \end{pmatrix}. \tag{6.33}$$

On the CIM, $\mathbf{f}$ takes the form

$$\mathbf{f}_{CIM} = \begin{pmatrix} -z_1 \\ \dfrac{-z_1}{(1+z_1)^2} \end{pmatrix}. \tag{6.34}$$

The unit tangent vector to the CIM is

$$\boldsymbol{\alpha}_t = \frac{\mathbf{f}_{CIM}}{|\mathbf{f}_{CIM}|} = \frac{1}{\sqrt{\dfrac{z_1^2}{(1+z_1)^4} + z_1^2}} \begin{pmatrix} -z_1 \\ \dfrac{-z_1}{(1+z_1)^2} \end{pmatrix}. \tag{6.35}$$

Now, knowing from Eq. (6.25) that $z_1(t) = z_{1o}e^{-t}$, we can evaluate $\mathbf{J}_s$ and $\boldsymbol{\alpha}_t$ as functions of $t$. Then we can use Eq. (6.15) to form $\sigma_t = \boldsymbol{\alpha}_t^T \cdot \mathbf{J}_s \cdot \boldsymbol{\alpha}_t$ as a function of $t$. That function is too lengthy to write here, but it is available in closed form. Knowing $\sigma_t(t)$, for this special problem, we are able to directly use Eq. (6.17) to calculate $\sigma_n$:

$$\sigma_n(t) = \nabla \cdot \mathbf{f} - \sigma_t(t) = -1 - \gamma - \sigma_t(t). \tag{6.36}$$

Knowing $\sigma_t(t)$ and $\sigma_n(t)$, one can then use Eq. (6.23) to determine $\eta$ and learn if the ACIM is a SACIM. We take $z_{1o} = 1$, and calculate $\sigma_t(t), \sigma_n(t), \eta(t)$ and plot the results in Fig. 6.4. We see that the normal and tangential stretching rates are both negative in the domain studied, and that normal stretching dominates tangential, allowing us to conclude the ACIM is a SACIM. The ratio $\eta \to \kappa = \gamma$ as $t \to \infty$.

In the time domain, the SACIM is described by a limit of Eqs. (6.25) and (6.26):

$$z_1(t) = z_{1o}e^{-t}, \qquad z_2(t) = \frac{z_{1o}e^{-t}}{1 + z_{1o}e^{-t}}. \tag{6.37}$$

Significantly, on the SACIM there is only a single slow time scale $\tau = 1$. Away from the SACIM, two time scales are in play, $\tau = 1, \tau = 1/\gamma$.

Let us describe how a numerical calculation might proceed in the context of an explicit numerical method. If one were to simulate the full system, Eqs. (6.25) and

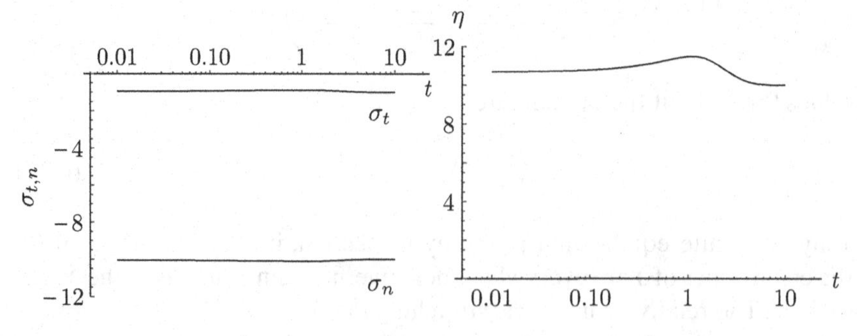


Figure 6.4. For the Davis-Skodje system with $\gamma = 10$, (a) the tangential and normal stretching rates along the ACIM and (b) the ratio of normal to tangential stretching.

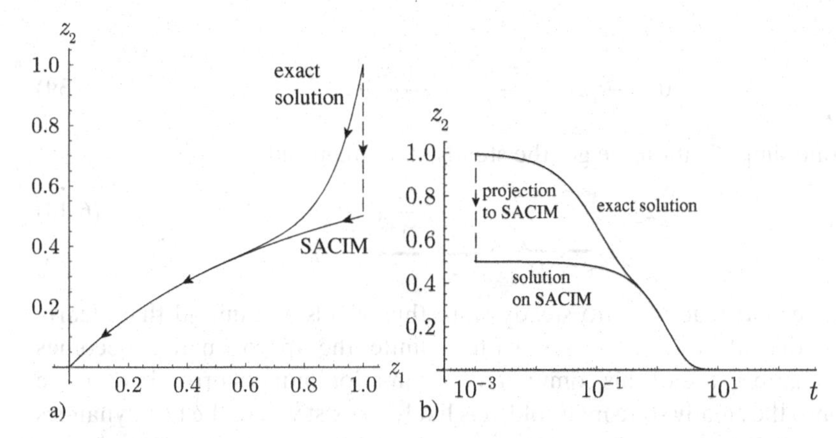


Figure 6.5. For the Davis-Skodje system with $\gamma = 10, (z_{1o}, z_{2o}) = (1, 1)$, the exact and approximate solutions in (a) the phase plane and (b) the time domain.

(6.26), one would be forced to use a time step $\Delta t < 1/\gamma$ to capture the full dynamics and retain numerical stability. If one takes advantage of the SACIM, one could solve numerically the differential-algebraic system

$$\frac{dz_1}{dt} = -z_1, \qquad 0 = z_2 - \frac{z_1}{1+z_1}. \tag{6.38}$$

Importantly, an initial condition is lost when the differential equation is replaced with the algebraic relation, however good that algebraic relation is. Loss of the initial condition also loses information about the dynamics of relaxation onto the SACIM. Such is the nature of reduction. Some information must be sacrificed. So for a given set of initial conditions, one must first project them onto the SACIM in zero time. There is no a priori guarantee that any projection will be optimal. Once one is projected onto the SACIM, the differential equation can be integrated, and one can capture the slow dynamics of the system as it relaxes to equilibrium.

Taking $\gamma = 10$, $(z_{1o}, z_{2o}) = (1, 1)$, let us solve the full stiff Davis-Skodje system, Eqs. (6.25) and (6.26), and the reduced nonstiff Eqs. (6.38). Let us choose one of the simplest projections for the reduced system. Take $(z_{1o}, z_{2o}) = (1, 1)$ and project it onto the SACIM. Choose the projection such that $z_{1o}$ is held constant, and project to the point $(1, 1/2)$. We give plots of the phase space and $z_2(t)$ for the exact and approximate solutions in Fig. 6.5. Clearly, there is error at early time because of the projection and the neglect of fast time scale events. At late time, the approximation is accurate.

Additional error is introduced when manifolds that are not SACIMs are used to approximate the solution. We compare the SACIM with two common methods of determining reduced kinetics manifolds: the so-called *steady state* assumption and the ILDM. In the steady state assumption, the user selects a reaction to bring into steady state by setting one of the time derivatives to zero, yielding an algebraic constraint. It is not always clear which reaction to select. If we bring the first reaction into steady state, we get the steady state manifold $z_1 = 0$. The second equation reduces then to $dz_2/dt = -\gamma z_2$, and has solution $z_2 = z_{2o}e^{-\gamma t}$. This assumption does not appear to capture significant dynamics. The first variable $z_1$ is driven to zero, and the second evolves on a fast time scale. If instead, we bring the second reaction into steady state, we get

$$0 = -\gamma z_2 + \frac{(\gamma - 1)z_1 + \gamma z_1^2}{(1 + z_1)^2}. \tag{6.39}$$

After algebraic simplification, we get the steady state manifold

$$z_2 = \underbrace{\frac{z_1}{1 + z_1}}_{\text{SACIM}} - \underbrace{\frac{z_1}{\gamma(1 + z_1)^2}}_{\text{error}}. \tag{6.40}$$

Bringing the second reaction into steady state thus yields a manifold that clearly approximates the SACIM. As $\gamma \to \infty$ with $z_1$ finite, the approximation becomes exact, but for finite $\gamma$, use of this common approximation induces error. If we force the system onto the steady state manifold, it is likely to exist where the fast dynamics are active in the full system, inducing the need for a fine time scale if the kinetics are to be resolved.

Another method of manifold estimation is the ILDM method. The essence of the method is to identify the set of points on which motion in the fast direction defined by the eigenvector associated with the fastest reaction mode is suppressed. For our Davis-Skodje system, there are always two eigenvectors, $\mathbf{v}_1$ and $\mathbf{v}_2$. These form a basis. By forming the inverse of the matrix of basis vectors, a reciprocal basis can be found. These vectors are $\tilde{\mathbf{v}}_1$ and $\tilde{\mathbf{v}}_2$. Detailed analysis reveals that we are interested in the reciprocal basis vectors. Because $\lambda_1 = -1$ is slow and $\lambda_2 = -\gamma$ is fast, we seek the reciprocal vector $\mathbf{v}_2$ associated with $\lambda_2$. With it, the ILDM is the set of points for which

$$\tilde{\mathbf{v}}_2^T \cdot \mathbf{f} = 0. \tag{6.41}$$

The eigenvectors of the Jacobian of the Davis-Skodje system are easily calculated to be

$$\mathbf{v}_1 = \begin{pmatrix} 1 \\ \dfrac{\gamma - 1 + (\gamma + 1)z_1}{(\gamma - 1)(1 + z_1)^3} \end{pmatrix}, \qquad \mathbf{v}_2 = \begin{pmatrix} 0 \\ 1 \end{pmatrix}. \tag{6.42}$$

These form a matrix of basis vectors

$$\mathbf{V} = \begin{pmatrix} 1 & 0 \\ \dfrac{\gamma - 1 + (\gamma + 1)z_1}{(\gamma - 1)(1 + z_1)^3} & 1 \end{pmatrix}. \tag{6.43}$$

The inverse is

$$\mathbf{V}^{-1} = \tilde{\mathbf{V}} = \begin{pmatrix} 1 & 0 \\ -\dfrac{\gamma - 1 + (\gamma + 1)z_1}{(\gamma - 1)(1 + z_1)^3} & 1 \end{pmatrix}. \tag{6.44}$$

The reciprocal basis vectors are the row vectors of $\tilde{\mathbf{V}}$:

$$\tilde{\mathbf{v}}_1 = \begin{pmatrix} 1 \\ 0 \end{pmatrix}, \qquad \tilde{\mathbf{v}}_2 = \begin{pmatrix} -\dfrac{\gamma - 1 + (\gamma + 1)z_1}{(\gamma - 1)(1 + z_1)^3} \\ 1 \end{pmatrix}. \tag{6.45}$$

The ILDM is formed by taking $\tilde{\mathbf{v}}_2^T \cdot \mathbf{f} = 0$, giving

$$\begin{pmatrix} -\dfrac{\gamma - 1 + (\gamma + 1)z_1}{(\gamma - 1)(1 + z_1)^3} & 1 \end{pmatrix} \begin{pmatrix} -z_1 \\ -\gamma z_2 + \dfrac{(\gamma - 1)z_1 + \gamma z_1^2}{(1 + z_1)^2} \end{pmatrix} = 0. \tag{6.46}$$

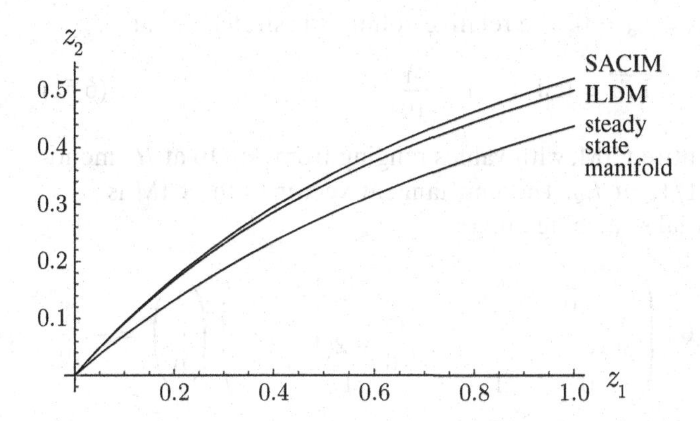


Figure 6.6. Davis-Skodje SACIM, steady state manifold, and ILDM for $\gamma = 4$.

Algebraic analysis shows this is equivalent to

$$z_2 = \underbrace{\frac{z_1}{1+z_1}}_{\text{SACIM}} + \underbrace{\frac{2z_1^2}{\gamma(\gamma-1)(1+z_1)^3}}_{\text{error}}. \tag{6.47}$$

Similar to the steady state manifold, the ILDM is a deviation from the SACIM with the deviation going to zero as $\gamma \to \infty$. For this problem, the ILDM is a better approximation to the SACIM than is the steady state manifold. This is seen in Fig. 6.6 for $\gamma = 4$.

### 6.2.2 Three-Dimensional Phase Space

Consider now the system first studied by Mengers (2012), and Powers et al. (2015), of the form of Eq. (6.8), with $N' = 3$:

$$\frac{dz_1}{dt} = \frac{1}{20}(1 - z_1^2), \tag{6.48}$$

$$\frac{dz_2}{dt} = -2z_2 - \frac{35}{16}z_3 + 2(1 - z_1^2)z_3, \tag{6.49}$$

$$\frac{dz_3}{dt} = z_2 + z_3. \tag{6.50}$$

Similar to the Davis-Skodje system, this problem shares many, but not all, features with practical combustion problems. This system has two finite roots, $R_1$ at $\mathbf{z} = (-1, 0, 0)^T$ and $R_2$ at $\mathbf{z} = (1, 0, 0)^T$. The Jacobian

$$\mathbf{J} = \begin{pmatrix} -\frac{z_1}{10} & 0 & 0 \\ -4z_1z_3 & -2 & -\frac{35}{16} + 2(1 - z_1^2) \\ 0 & 1 & 1 \end{pmatrix} \tag{6.51}$$

has eigenvalues $\boldsymbol{\lambda} = \{1/10, -1/4, -3/4\}$ at $R_1$ and $\boldsymbol{\lambda} = \{-1/10, -1/4, -3/4\}$ at $R_2$. Thus, $R_1$ is a saddle with one unstable mode, and $R_2$ is a sink, analogous to a physical equilibrium in a reactive system. There is a CIM defined by the heteroclinic orbit that connects $R_1$ to $R_2$; by inspection, it is seen that the CIM is confined to the $z_1$

axis with $z_1 \in [-1, 1]$ and $z_2 = z_3 = 0$. The relative volumetric stretching rate is

$$\dot{\overline{\ln v}} = \operatorname{tr} \mathbf{J} = -1 - \frac{z_1}{10} \tag{6.52}$$

and is negative along the entire CIM, with values ranging from $-9/10$ at $R_1$ monotonically decreasing to $-11/10$ at $R_2$. The unit tangent vector to the CIM is $\boldsymbol{\alpha}_t = (1, 0, 0)^T$, yielding a tangential stretching rate of

$$\sigma_t = \boldsymbol{\alpha}_t^T \cdot \mathbf{J} \cdot \boldsymbol{\alpha}_t = (1 \quad 0 \quad 0) \begin{pmatrix} -\frac{z_1}{10} & 0 & 0 \\ -4z_1 z_3 & -2 & -\frac{35}{16} + 2(1 - z_1^2) \\ 0 & 1 & 1 \end{pmatrix} \begin{pmatrix} 1 \\ 0 \\ 0 \end{pmatrix} = -\frac{z_1}{10}. \tag{6.53}$$

On the CIM, we thus find that $\sigma_t \sim 1/10$ near $R_1$ and $\sigma_t \sim -1/10$ near the physical equilibrium $R_2$. On the CIM, we have

$$\mathbf{J} = \begin{pmatrix} -\frac{z_1}{10} & 0 & 0 \\ 0 & -2 & -\frac{35}{16} + 2(1 - z_1^2) \\ 0 & 1 & 1 \end{pmatrix}, \tag{6.54}$$

and

$$\mathbf{J}_s = \begin{pmatrix} -\frac{z_1}{10} & 0 & 0 \\ 0 & -2 & -\frac{19}{32} + 1 - z_1^2 \\ 0 & -\frac{19}{32} + 1 - z_1^2 & 1 \end{pmatrix}, \tag{6.55}$$

$$\mathbf{J}_a = \begin{pmatrix} 0 & 0 & 0 \\ 0 & 0 & -\frac{51}{32} + 1 - z_1^2 \\ 0 & \frac{51}{32} - 1 + z_1^2 & 0 \end{pmatrix}. \tag{6.56}$$

A Gram-Schmidt procedure yields $\boldsymbol{\alpha}_{n1} = (0, 1, 0)^T$ and $\boldsymbol{\alpha}_{n2} = (0, 0, 1)^T$, thus

$$\mathbf{Q}_n = \begin{pmatrix} 0 & 0 \\ 1 & 0 \\ 0 & 1 \end{pmatrix}. \tag{6.57}$$

The reduced Jacobian for the stretching in the normal plane is, from Eq. (6.20),

$$\mathbf{J}_{ns} = \begin{pmatrix} -2 & -\frac{19}{32} + 1 - z_1^2 \\ -\frac{19}{32} + 1 - z_1^2 & 1 \end{pmatrix}. \tag{6.58}$$

The eigenvalues of $\mathbf{J}_{ns}$ give the extremal normal stretching rates $\sigma_{n,i}$:

$$\sigma_{n,i} = -\frac{1}{2} \pm \frac{\sqrt{2473 - 832z_1^2 + 1024z_1^4}}{32}. \tag{6.59}$$

Evaluating, we find $\sigma_{n,1} \sim 1$ and $\sigma_{n,2} \sim -2$ for $z_1 \in [-1, 1]$. The presence of a positive normal stretching rate opens the possibility of divergence of a nearby trajectory from the CIM.

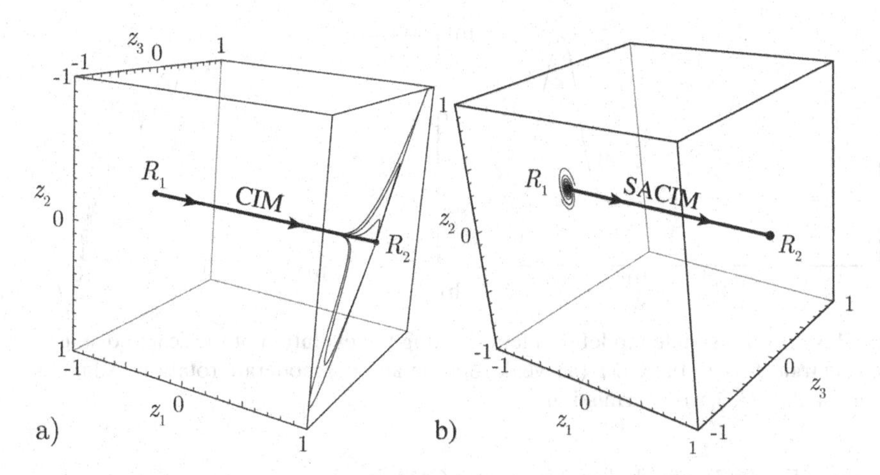


Figure 6.7. Results for a simple model problem showing the heteroclinic connections and behavior of nearby trajectories for two cases: (a) weak rotation, which induces a CIM, and (b) moderate rotation, which induces a SACIM. Adapted from Powers et al., 2015, with permission.

The reduced Jacobian for the rotation in the normal plane is, from Eq. (6.20),

$$\mathbf{J}_{na} = \begin{pmatrix} 0 & -\frac{51}{32} + 1 - z_1^2 \\ \frac{51}{32} - 1 + z_1^2 & 0 \end{pmatrix}. \tag{6.60}$$

Its magnitude, $||\mathbf{J}_{na}|| = \omega$, on the CIM ranges from $19/32$ to $51/32$.

We verify that this CIM does not lead to an ACIM by direct calculation. We present the CIM and a family of trajectories that originate in the neighborhood of $R_1$ in Fig. 6.7a. The initial conditions near $R_1$ have $z_1 = -0.99$ and $z_2, z_3$ comprising a number of equally spaced points on the circle $\sqrt{z_2^2 + z_3^2} = \epsilon = 2 \times 10^{-16}$. Clearly the trajectories remain close to the CIM at early time, but as they approach $R_2$, they suffer large deviations from the CIM. These large deviations are aligned with the direction of maximal normal stretching. Ultimately rotation combined with stability brings all trajectories to $R_2$ as $t \to \infty$. The large deviations from the CIM are consistent with the fact that the ansatz for attractiveness, Eq. (6.22), is not met here. Calculation reveals that the attractiveness parameter $\mu \in [0.563, 1.43]$. Consequently there are regions where the positive normal stretching is not overcome by sufficiently rapid rotation, allowing the trajectory to suffer a large deviation from the CIM before its ultimate return to the sink at $R_2$. This can also be considered to be an effect of what is known as nonnormality (Trefethen and Embree, 2005), a feature of dynamical systems whose Jacobians are asymmetric such as those that arise in reactive systems.

Positive normal stretching does not guarantee divergence from a CIM; it simply permits it. Because trajectories spend infinite time approaching equilibria, the time spent near a sink equilibrium in regions of negative normal stretching overwhelms the time spent in regions of positive normal stretching, which induces the ultimate return of all trajectories to the sink. However, any reduction algorithm that relies on projection onto such a CIM in regions far from equilibrium would likely induce significant error in the prediction of many state variables. Last nearby points to the CIM are either normally repelled or attracted much faster than motions on the CIM because the magnitude of each extremal $\sigma_{n,i}$ is much greater than $\sigma_t$. In this case, we have $\eta \sim 5$. We easily verify the lack of attractiveness by direct calculation of

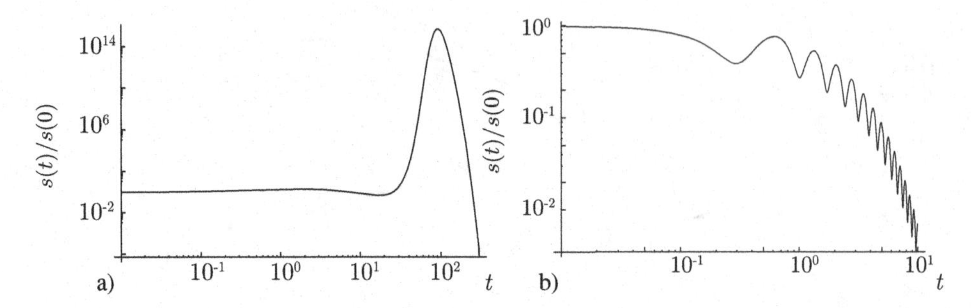


Figure 6.8. Results for a simple model problem showing the evolution of the scaled distance from a nearby trajectory to the CIM: (a) weak rotation and (b) moderate rotation. Adapted from Powers et al. (2015), with permission.

$s(t)/s(0)$ from Eq. (6.7), which, for this simple CIM, is

$$\frac{s(t)}{s(0)} = \frac{\sqrt{z_2^2(t) + z_3^2(t)}}{\epsilon}. \tag{6.61}$$

Results are shown in Fig. 6.8a. It is seen that the relative deviation becomes large for $t \sim 100$, thus verifying that this CIM is not attractive.

If we make one minor change so as to increase the rotation rate, we find the CIM becomes a SACIM, even in the presence of regions of positive normal stretching. Let us replace Eq. (6.50) with

$$\frac{dz_3}{dt} = 10z_2 + z_3. \tag{6.62}$$

The equilibria remain the same with $R_1$: $(-1, 0, 0)^T$ and $R_2$: $(1, 0, 0)^T$. The dynamics near each equilibrium have a slight change of character. The eigenvalues of $\mathbf{J}$ near $R_1$ are $\boldsymbol{\lambda} = \{-1/2 \pm i\sqrt{314}/4, 1/10\}$. Thus, $R_1$ has one linearly unstable mode and two stable oscillatory modes. The eigenvalues of $\mathbf{J}$ near $R_2$ are $\boldsymbol{\lambda} = \{-1/2 \pm i\sqrt{314}/4, -1/10\}$. The presence of oscillatory modes near $R_2$ renders this model problem less like reversible combustion problems in closed systems. Nevertheless, $R_2$ is stable, with two of its modes containing an oscillatory character. Therefore, a CIM again exists as a heteroclinic connection between $R_1$ and $R_2$; it is confined to the $z_1$ axis, with $z_1 \in [-1, 1]$. The relative volumetric stretching rate is unchanged and strictly negative on the CIM, $\overline{\ln \upsilon}^{\cdot} = -1 - z_1/10$. The tangential stretching rate on the CIM is unchanged and is $\sigma_t = -z_1/10$. The Gram-Schmidt orthogonalization is identical as well. The change induces enhanced rotation with $\omega$ ranging from $163/32$ to $195/32$. Evaluation of $\mathbf{J}_{ns}$ leads to normal stretching rates that are weakly varying functions of $z_1$ with extrema $\sigma_{n,1} \sim 4$ and $\sigma_{n,2} \sim -5$.

Direct calculation reveals that this modest enhancement in rotation rate is sufficient to render the CIM to be an ACIM. The ACIM is seen to be a SACIM, verifying our second ansatz, Eq. (6.23), as $\eta$ has a minimum value of $36.84$, well above unity. We present the SACIM and a family of trajectories that originate in the near neighborhood of $R_1$ in Fig. 6.7b. The initial conditions near $R_1$ have $z_1 = -0.99$ and $z_2, z_3$ comprised of a number of equally spaced points on the circle $\sqrt{z_2^2 + z_3^2} = \epsilon = 1 \times 10^{-1}$. These were chosen further from $R_1$ so as to illustrate their fast attraction to the SACIM. In spite of positive normal stretching, the rotation is sufficiently fast to keep nearby trajectories close to the SACIM. Calculation reveals that the range of the attractiveness parameter on the SACIM is $\mu \in [1.10, 1.65]$. While modest, it is sufficiently large to keep the trajectories from diverging. This is verified by direct calculation of $s(t)/s(0)$ with results shown in Fig. 6.8b. The oscillations are attributed to the trajectory rotating in and out of regions of positive and negative $\sigma_{n,i}$.

## 6.3 Reduction of Combustion Systems

We present two examples here of systems based on actual combustion dynamics. The first has $N' = 2$, in which case rotation will be irrelevant. The has $N' = 3$, rendering rotation to be of potential relevance.

### 6.3.1 Zel'dovich Mechanism

The Zel'dovich reaction mechanism of nitric oxide formation is adopted as used in Sec. 1.2.2. This mechanism consists of $N = 5$ species, $L = 2$ elements, and $J = 2$ reversible reactions. The mechanism and kinetic data are repeated in Table 6.1.

The dependent variables are the specific moles $z_i$, where $i = \{1, 2, 3, 4, 5\}$ corresponds to the species $\{NO, N, N_2, O, O_2\}$, respectively. The system is taken to be isothermal and isochoric. The mixture temperature and volume are assigned as $T = 6000$ K and $V = 1$ cm$^3$, respectively.

One can form five inhomogeneous ordinary differential equations to describe the time evolution of the five species, and one must specify an initial value for each species. At this stage, we take each of the five species to have a presence of $10^{-6}$ mol, though we will relax this later. As the molecular masses $M_i$ are known to be

$$M_i = \{30, 14, 28, 16, 32\} \text{ g/mol}, \tag{6.63}$$

the constant mixture mass is $m = \sum_{i=1}^{5} M_i n_i = 1.2 \times 10^{-4}$ g, and the constant mass density is $\rho = m/V = 1.2 \times 10^{-4}$ g/cm$^3$. The initial values are

$$z_i(0) = \frac{n_i(0)}{m} = \frac{1}{120} \frac{\text{mol}}{\text{g}} = 8.333 \times 10^{-3} \frac{\text{mol}}{\text{g}}, \qquad i = 1, \ldots, 5. \tag{6.64}$$

Linear combinations of three of the ordinary differential equations can be formulated into three homogeneous ordinary differential equations, which can be integrated to form three algebraic constraints, using the initial conditions to evaluate the integration constants. Two of these constraints are due to the conservation of the elements N and O. The third is a consequence of having only bimolecular reactions, rendering the total number of moles to be time-independent. The three

Table 6.1. *Zel'dovich Mechanism of Nitric Oxide Formation, Sec. 1.2.2*

| $j$ | Reaction | $a_j \left(\frac{\text{cm}^3}{\text{mol s K}^{\beta_j}}\right)$ | $\beta_j$ | $T_{a,j}$ (K) |
|---|---|---|---|---|
| 1 | $N + NO \rightleftharpoons N_2 + O$ | $2.107 \times 10^{13}$ | 0.00 | 0 |
| 2 | $N + O_2 \rightleftharpoons NO + O$ | $5.8394 \times 10^{9}$ | 1.01 | 3120 |

constraints are

$$z_1 + z_3 + 2z_4 = 4/120 = 0.0333333 \text{ mol/g}, \tag{6.65}$$

$$z_1 + z_2 + 2z_5 = 4/120 = 0.0333333 \text{ mol/g}, \tag{6.66}$$

$$z_1 + z_2 + z_3 + z_4 + z_5 = 5/120 = 0.0416667 \text{ mol/g}. \tag{6.67}$$

We use these constraints to eliminate the dependency of $\{N_2, O, O_2\}$ in the evolution equations for $\{NO, N\}$. Thus, we have $N' = 2$, and the system's dynamics can be fully described in the $\mathbb{R}^2$ reactive composition space, though the number of species is 5. Because $N' = 2, \omega = 0$, and rotation will not play a role in the system's dynamics.

The two ordinary differential equations that describe the system's evolution are

$$\begin{aligned}\frac{dz_1}{dt} &= \left(1.374236558534416 \times 10^9\right) z_2{}^2 - \left(3.835175300191297 \times 10^9\right) z_1 z_2 \\ &\quad + \left(2.218063027523232 \times 10^7\right) z_2 - \left(9.43526213966668 \times 10^5\right) z_1 \\ &\quad + 6.02760305839954 \times 10^3,\end{aligned} \tag{6.68}$$

$$\begin{aligned}\frac{dz_2}{dt} &= \left(-1.352537187524177 \times 10^9\right) z_2{}^2 - \left(1.199925328798465 \times 10^9\right) z_1 z_2 \\ &\quad - \left(2.326559882574424 \times 10^7\right) z_2 + \left(5.81870030462696 \times 10^5\right) z_1 \\ &\quad + 6.02760305839954 \times 10^3.\end{aligned} \tag{6.69}$$

Equations (6.68) and (6.69) are simply transformed versions of Eqs. (1.235) and (1.236). We include many significant digits so that results can be carefully verified. While we can certainly apply the initial conditions of Eq. (6.64), we relax these and actually choose a broader range of initial conditions for $z_1$ and $z_2$, all the while maintaining the constraints of Eqs. (6.65)–(6.67).

Similar to what was seen in Sec. 1.2.2, Eqs. (6.65)–(6.69) have three real finite equilibria:

$$R_1 \equiv (\mathbf{z}^e) = \left(-1.33721 \times 10^{-2}, -2.55004 \times 10^{-4}\right) \text{ mol/g}, \tag{6.70}$$

$$R_2 \equiv (\mathbf{z}^e) = \left(-4.31123 \times 10^{-4}, -1.70692 \times 10^{-2}\right) \text{ mol/g}, \tag{6.71}$$

$$R_3 \equiv (\mathbf{z}^e) = \left(6.11325 \times 10^{-3}, 3.08995 \times 10^{-4}\right) \text{ mol/g}. \tag{6.72}$$

Multiplication of each root by $\rho = 1.2 \times 10^{-4}$ g/cm$^3$ shows the roots are equal to those given in Eqs. (1.242)–(1.244). Here, $R_1$ and $R_2$ are nonphysical equilibria, while $R_3$ is a physical root that corresponds to the reactive system's unique physical equilibrium. Linear analysis within the neighborhood of each finite critical point reveals that $R_3$ is a sink, $R_1$ is a saddle, and $R_2$ is a spiral source. The eigenvalue spectrum associated with each finite critical point is

$$R_1 : (\boldsymbol{\lambda}) = \left(-1.193 \times 10^7, 5.434 \times 10^6\right) \text{ s}^{-1}, \tag{6.73}$$

$$R_2 : (\boldsymbol{\lambda}) = \left(4.397 \times 10^7 + 8.000 \times 10^6 i, 4.397 \times 10^7 - 8.000 \times 10^6 i\right) \text{ s}^{-1}, \tag{6.74}$$

$$R_3 : (\boldsymbol{\lambda}) = \left(-3.143 \times 10^7, -2.132 \times 10^6\right) \text{ s}^{-1}. \tag{6.75}$$

These eigenvalues are identical to those obtained for the same system in Sec. 1.2.2; see those for the physical root in Eqs. (1.260). The system's temporal stiffness near $R_3$ is $\kappa = 14.75$, as in Eq. (1.262).

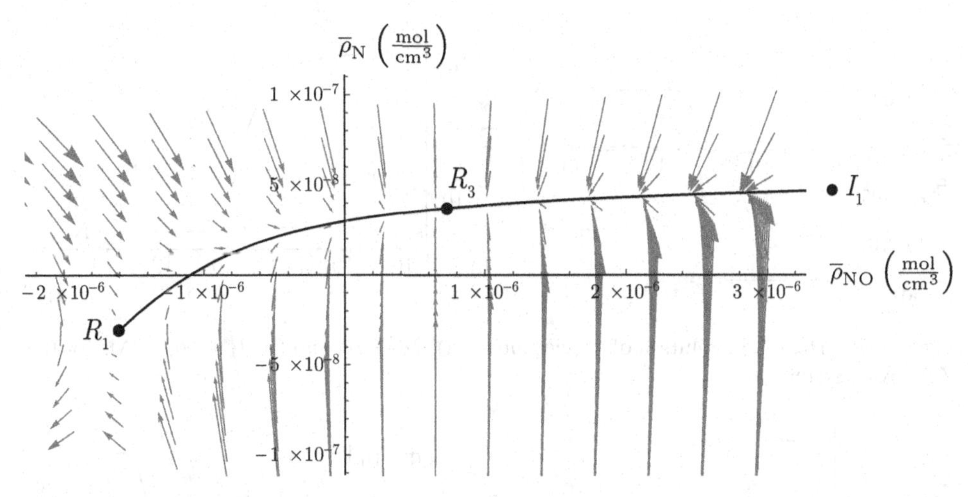


Figure 6.9. The two CIMs for the Zel'dovich system illustrated as thick lines. The solid dots are finite critical points, and the circles are critical points located at infinity; $R_3$ represents the system's physical equilibrium state, $R_2$ represents starting point of the first CIM, and $I_1$ represents the starting point of the second CIM.

Equilibria located at infinity are identified using the projective space technique described by Perko (2001). The Zel'dovich mechanism in the projective space is realized by the following transformation: $Z_1 = 1/z_1,\ Z_2 = z_2/z_1$. This yields the system

$$-\frac{dZ_1}{dt} = 6027.60Z_1^2 + 2.21806 \times 10^7 Z_1 Z_2 - 943526 Z_1 + 1.37424 \times 10^9 Z_2^2 - 3.83518 \times 10^9 Z_2, \tag{6.76}$$

$$Z_1 \frac{dZ_2}{dt} - Z_2 \frac{dZ_1}{dt} = 6027.60Z_1^2 - 2.32656 \times 10^7 Z_1 Z_2 + 581870. Z_1 - 1.35254 \times 10^9 Z_2^2 - 1.19993 \times 10^9 Z_2. \tag{6.77}$$

This transformed system has equilibria that can be identified by algebraic analysis as well. Pertinent for our analysis is

$$I_1 \equiv (\mathbf{Z}^e) = (0, 0). \tag{6.78}$$

Three other roots are found, each of which transforms back to the finite roots already identified. The root $I_1$ does not transform to finite value of $z_1$ or $z_2$. Then, two CIMs, that is, heteroclinic orbits, are generated via integrating the dynamical system starting from near $R_2$ and $I_1$ in the directions pointing toward the reactive system's physical equilibrium, $R_3$. The two CIMs presented in Fig. 6.9 are connected with $R_3$ along its slowest mode. To allow direct comparison with Fig. 1.12, we have transformed back to $\overline{\rho}_{NO}, \overline{\rho}_N$ space. The attractiveness of these two CIMs is evident in Fig. 6.9; they are attracting all nearby trajectories. The species evolution along the two CIMs is presented in Fig. 6.10, where in Fig. 6.10b the species evolution is illustrated in the original composition space by mapping back the obtained results in the projective space via employing the following transformation: $z_1 = 1/Z_1,\ z_2 = Z_2/Z_1$. All species have $z_i > 0$ within $\partial\mathbb{S}$. Moreover, both approach the same equilibrium $R_3$, though that is difficult to discern in Fig. 6.10 due to the scaling.

To investigate the attractiveness and slowness of the two constructed CIMs, stretching-based diagnostics are employed locally along the CIMs. The local relative

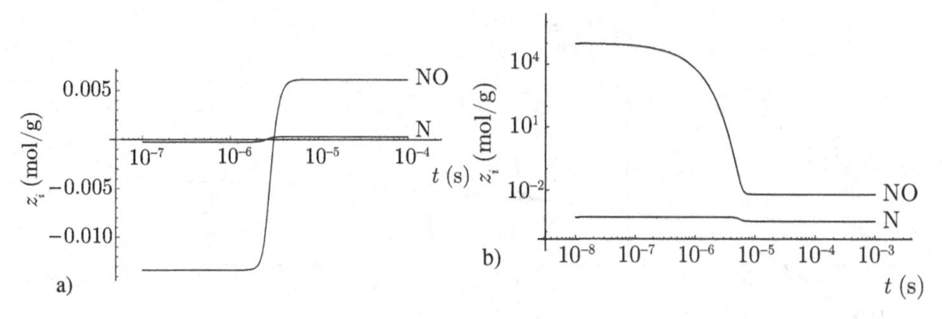


Figure 6.10. The time evolution of species along (a) $R_1 \to R_3$ and (b) $I_1 \to R_3$ CIMs for the Zel'dovich system.

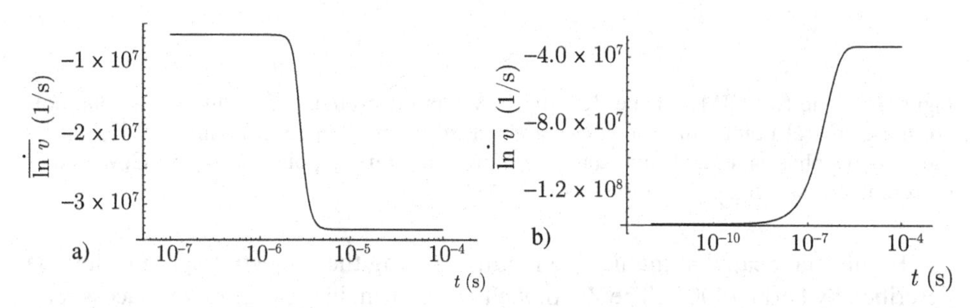


Figure 6.11. The relative volumetric deformation rate along (a) $R_1 \to R_3$ and (b) $I_1 \to R_3$ CIMs for the Zel'dovich system.

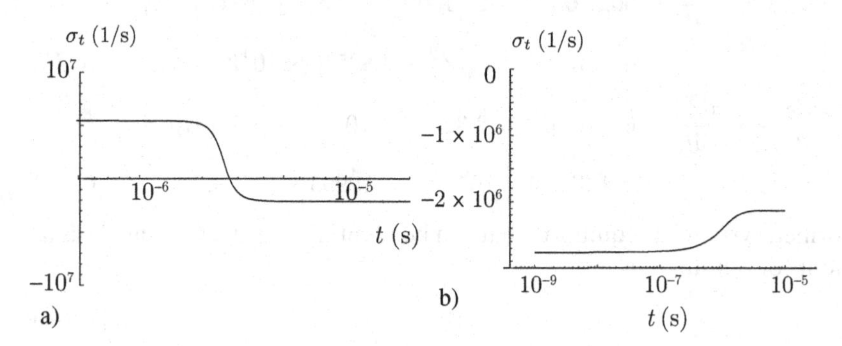


Figure 6.12. The tangential stretching rate along (a) $R_1 \to R_3$ and (b) $I_1 \to R_3$ CIMs for the Zel'dovich system.

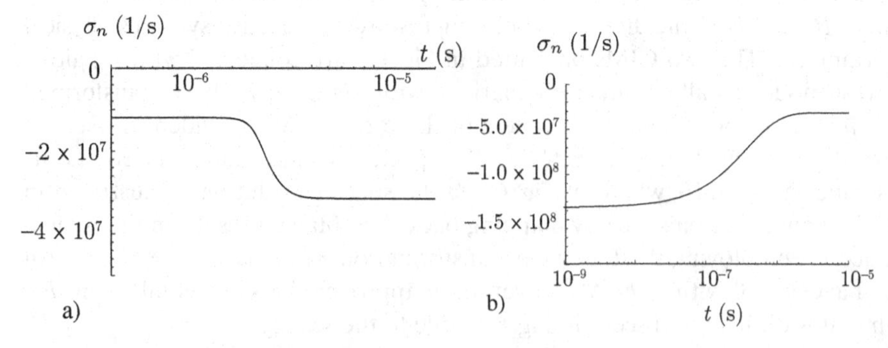


Figure 6.13. The normal stretching rate along (a) $R_1 \to R_3$ and (b) $I_1 \to R_3$ CIMs for the Zel'dovich system.

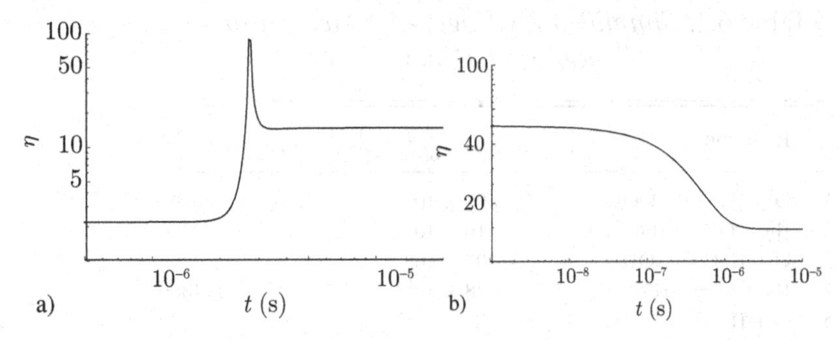


Figure 6.14. The ratio of the normal and tangential stretching rates along (a) $R_1 \to R_3$ and (b) $I_1 \to R_3$ CIMs for the Zel'dovich system.

volumetric deformation rate of the systems along both CIMs is presented in Fig. 6.11. The relative volumetric deformation rates are negative. The tangential stretching rates along the two CIMs are presented in Fig. 6.12. Along $R_1$ to $R_3$, $\sigma_t$ is initially positive, but then becomes negative as it approaches the physical equilibrium. Along $I_1$ to $R_3, \sigma_t$ is always negative. Figure 6.13 gives the single normal stretching rate and shows that both CIMs, $R_2 \to R_3$ and $I_1 \to R_3$, are attractive because $\sigma_n$ along the CIMs is always negative. Thus, the two constructed CIMs are ACIMs. For this $N' = 2$ system, rotation is irrelevant. Figure 6.14 confirms our second ansatz and shows that the dynamics on the two ACIMs are slow, because $\eta$ along the two ACIMs is greater than unity. This indicates that the motion on the two ACIMs is slower than the motion onto the ACIMs. So, these two ACIMs are SACIMs. One must recognize that we have only studied a single case and that it is possible that other conditions exist for which the CIMs are not SACIMs. For this system with $N' = 2, \eta$ is also the stiffness ratio $\kappa$. For long time, $\eta \to 14.75$, as was found earlier in Eq. (1.262).

### 6.3.2 $H_2$-Air Combustion

We next consider a more complicated system with $N' = 3$ for which rotation could be relevant. We examine the combustion problem given first by Ren et al. (2006) and was later studied by Al-Khateeb et al. (2009) and Powers et al. (2015), each of whom provide additional details. The reaction mechanism contains $N = 6$ species, $L = 3$ elements, and $J = 6$ reversible reactions; see Table 6.2. The dependent variables are the specific moles $z_i = n_i/m$, $i = \{1, 2, 3, 4, 5, 6\}$ and correspond to the species $\{H_2, O, H_2O, H, OH, N_2\}$, respectively. As is standard in reaction dynamics, M represents a generic third body. The system under consideration is isothermal and isobaric. The mixture temperature and pressure are $T = 3000$ K and $P = 1$ atm, respectively.

One can form $N = 6$ differential equations to describe the time evolution of the six species, and one must specify an initial value for each. Because of element conservation, three of the differential equations have first integrals, yielding three algebraic constraints and three ordinary differential equations. Because Reactions 4–6 do not conserve molecules, there is no constraint associated with the number of molecules. The three differential equations that describe the system's temporal evolution are of the form of Eq. (6.4) with $N' = N - L = 3$. Because they are lengthy, they are not given here. Thus, the dynamics are fully described by evolution

Table 6.2. *Simplified Hydrogen-Air Mechanism of Ren et al. (2006)*

| $j$ | Reaction | $a_j\ \left(\frac{\text{cm}^3}{\text{mol s K}^{\beta_j}}\right)$ | $\beta_j$ | $\overline{\mathcal{E}}_j\ \left(\frac{\text{cal}}{\text{mol}}\right)$ |
|---|---|---|---|---|
| 1 | $O + H_2 \rightleftharpoons H + OH$ | $5.08 \times 10^4$ | 2.7 | 6290 |
| 2 | $H_2 + OH \rightleftharpoons H_2O + H$ | $2.16 \times 10^8$ | 1.5 | 3430 |
| 3 | $O + H_2O \rightleftharpoons 2OH$ | $2.97 \times 10^6$ | 2.0 | 13400 |
| 4 | $H_2 + M \rightleftharpoons 2H + M$ | $4.58 \times 10^{19}$ | −1.4 | 104380 |
| 5 | $O + H + M \rightleftharpoons OH + M$ | $4.71 \times 10^{18}$ | −1.0 | 0 |
| 6 | $H + OH + M \rightleftharpoons H_2O + M$ | $3.80 \times 10^{22}$ | −2.0 | 0 |

*Note:* The Nonunity Third Body Collision Efficiency Coefficients are $\alpha_{j,H_2} = 2.5, \alpha_{j,H_2O} = 12$, and $j = 4, 5, 6$.

equations for $\{H_2, O, H_2O\}$, and the rest of the species, $\{H, OH, N_2\}$, are given by the $L = 3$ linear constraints:

$$2z_1 + 2z_3 + z_4 + z_5 = 1.234 \times 10^{-2} \text{ mol/g}, \tag{6.79}$$

$$z_2 + z_3 + z_5 = 4.11 \times 10^{-3} \text{ mol/g}, \tag{6.80}$$

$$2z_6 = 6.581 \times 10^{-2} \text{ mol/g}. \tag{6.81}$$

These are easily solved to give $z_4$ and $z_5$ as functions of $z_1$, $z_2$, and $z_3$. Because $N_2$ is inert, $z_6$ remains constant. As for the Zel'dovich system, we study a variety of initial conditions for $z_1$, $z_2$, and $z_3$, while maintaining the constraints of Eqs. (6.79)–(6.81).

As discussed by Al-Khateeb et al. (2009), the system has fifteen equilibrium points located within the finite domain; eight of them are complex, and seven are real. The real ones are

$$R_1 \equiv (\mathbf{z}^e) = \left(-1.67204 \times 10^{-1}, 3.03617 \times 10^{-3}, 3.53209 \times 10^{-3}\right) \text{ mol/g},$$

$$R_2 \equiv (\mathbf{z}^e) = \left(6.44204 \times 10^{-2}, 1.20566 \times 10^{-2}, -7.12337 \times 10^{-3}\right) \text{ mol/g},$$

$$R_3 \equiv (\mathbf{z}^e) = \left(-6.47244 \times 10^{-3}, -2.00868 \times 10^{-2}, -2.19220 \times 10^{-3}\right) \text{ mol/g},$$

$$R_4 \equiv (\mathbf{z}^e) = \left(1.97888 \times 10^{-3}, 5.03888 \times 10^{-3}, 9.41881 \times 10^{-3}\right) \text{ mol/g},$$

$$R_5 \equiv (\mathbf{z}^e) = \left(-1.21290 \times 10^{-3}, -4.44837 \times 10^{-3}, 5.03482 \times 10^{-3}\right) \text{ mol/g},$$

$$R_6 \equiv (\mathbf{z}^e) = \left(2.72293 \times 10^{-3}, 3.34454 \times 10^{-4}, 4.71857 \times 10^{-3}\right) \text{ mol/g},$$

$$R_7 \equiv (\mathbf{z}^e) = \left(2.02552 \times 10^{-3}, 3.10118 \times 10^{-4}, 3.06770 \times 10^{-3}\right) \text{ mol/g}.$$

Note that $R_4$ and $R_6$ are in $\mathbb{S}'$ because other species specific mol numbers obtained from the linear constraints are negative. Thus, $R_7$ is the unique physical equilibrium in $\mathbb{S}$, as indicated in Fig. 6.15, where the dashed simplex outlines the convex polytope forming $\partial\mathbb{S}$. Also, the system has two higher-dimensional equilibria located at infinity. One is one-dimensional and the other two-dimensional.

Linear analysis in the neighborhood of each real, finite critical point reveals that $R_3$ and $R_7$ are sinks, and $R_1$, $R_2$, $R_4$, $R_5$, and $R_6$ are saddles. The eigenvalue spectrum

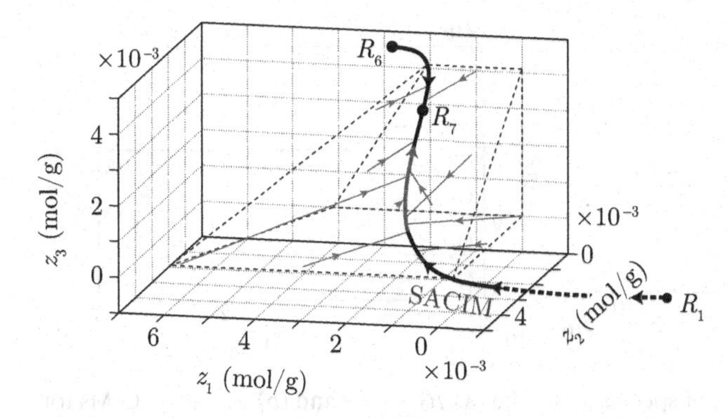


Figure 6.15. Two SACIMs, $R_1 \to R_7$ and $R_6 \to R_7$, for the simple hydrogen-air reactive system illustrated as thick lines. The solid dots are finite critical points; $R_7$ represents the system's physical equilibrium, $R_1$ represents starting point of the first SACIM, and $R_6$ represents the starting point of the second SACIM. The dashed simplex outlines the convex polytope defining $\partial\mathbb{S}$, and the thin lines illustrate several trajectories. The variables $z_1$, $z_2$, and $z_3$ are associated with $H_2$, O, and $H_2O$, respectively. Adapted from Powers et al. (2015), with permission.

associated with each finite critical point is

$$
\begin{aligned}
R_1 &: (\boldsymbol{\lambda}) = \left(2.92 \times 10^3, -6.67 \times 10^6 \pm i1.00 \times 10^8\right) \text{ s}^{-1},\\
R_2 &: (\boldsymbol{\lambda}) = \left(1.84 \times 10^{14}, -1.27 \times 10^{12}, -1.70 \times 10^{14}\right) \text{ s}^{-1},\\
R_3 &: (\boldsymbol{\lambda}) = \left(-1.03 \times 10^5, -2.97 \times 10^7 \pm i2.64 \times 10^7\right) \text{ s}^{-1},\\
R_4 &: (\boldsymbol{\lambda}) = \left(1.62 \times 10^7, 8.94 \times 10^6, -4.65 \times 10^4\right) \text{ s}^{-1},\\
R_5 &: (\boldsymbol{\lambda}) = \left(3.22 \times 10^4, -2.13 \times 10^6 \pm i6.71 \times 10^6\right) \text{ s}^{-1},\\
R_6 &: (\boldsymbol{\lambda}) = \left(1.57 \times 10^4, -6.28 \times 10^6 \pm i4.37 \times 10^6\right) \text{ s}^{-1},\\
R_7 &: (\boldsymbol{\lambda}) = \left(-5.59 \times 10^3, -9.08 \times 10^6, -1.77 \times 10^7\right) \text{ s}^{-1}.
\end{aligned}
$$

Using our earlier definition, the system's temporal stiffness is 3166. Out of the seven real finite zero-dimensional equilibria, there is one sink, $R_7$, located in $\mathbb{S}$, one sink, $R_3$, located in $\mathbb{S}'$, and four saddles with one unstable mode: $R_1$, $R_2$, $R_5$, and $R_6$, all in $\mathbb{S}'$. The root $R_4$ has two unstable modes.

Following the procedure presented by Al-Khateeb et al. (2009), these four saddles are candidate points for the one-dimensional CIM, and they are ordered based on the magnitude of their positive eigenvalue; the first candidate point is the one with the least positive eigenvalue among all candidate points. So, the first candidate point is $R_1$, the second one is $R_6$, the third one is $R_5$, and the last one is $R_2$. Then, two CIMs are generated from the first two candidates via integrating the dynamical system starting from $R_1$ and $R_6$ in the direction of the eigenvectors associated with the candidate points' positive eigenvalues pointing toward the reactive system's physical equilibrium, $R_7$. These two CIMs, presented in Fig. 6.15, connect with $R_7$ along its slowest mode. The attractiveness of these two CIMs is shown in Fig. 6.15; they are attractive to other trajectories within the physically accessible domain, $\mathbb{S}$. The species evolution along the two CIMs is presented in Fig. 6.16. The other two candidate points, $R_5$ and $R_2$, generate another two CIMs that connect with the unphysical sink $R_3$ as depicted in Fig. 6.17.

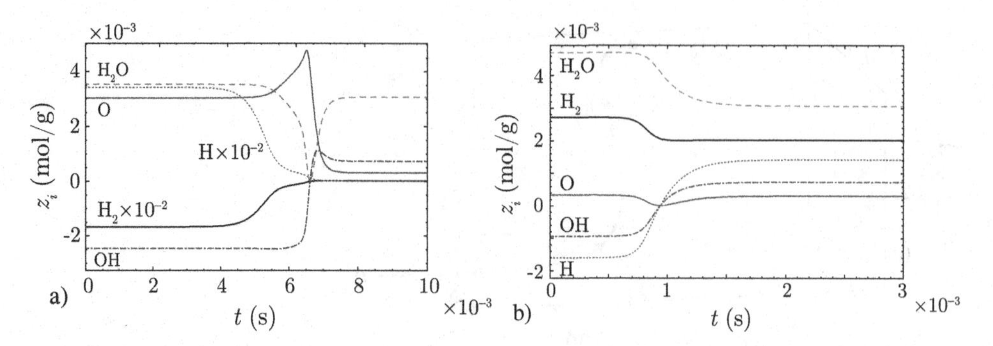


Figure 6.16. The time evolution of species along the (a) $R_1 \rightarrow R_7$ and (b) $R_6 \rightarrow R_7$ CIMs for the simple hydrogen-air system. Adapted from Powers et al. (2015), with permission.

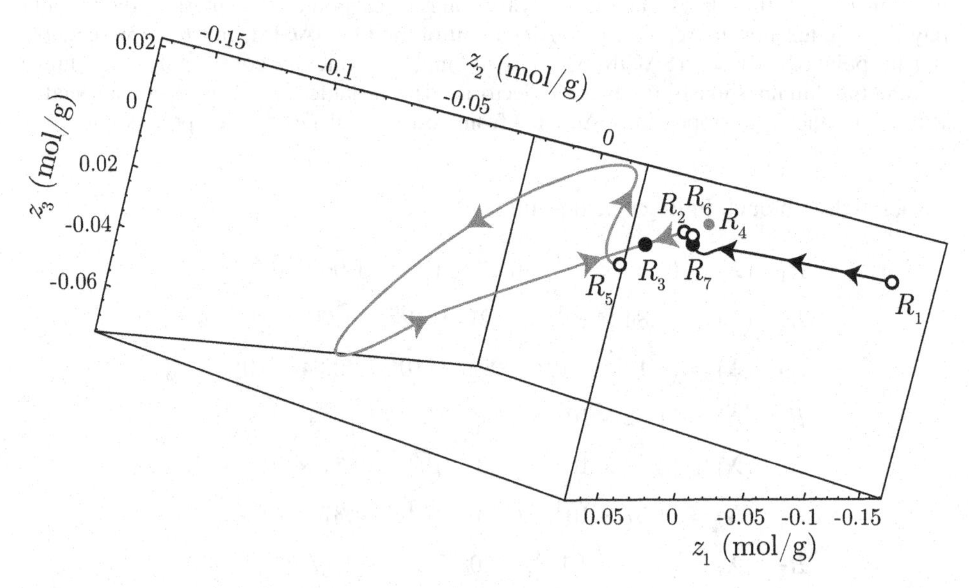


Figure 6.17. Phase space for simple hydrogen-air dynamics. The filled black circles are the dynamical system's sinks, the open black circles are the candidate points, the gray dot is a saddle with two unstable modes, the black line represents the two CIMs that connect $R_1$ and $R_6$ with the physical sink $R_7$, and the gray line illustrates the two CIMs that connect $R_5$ and $R_2$ with the unphysical sink $R_3$. Adapted from Powers et al. (2015), with permission.

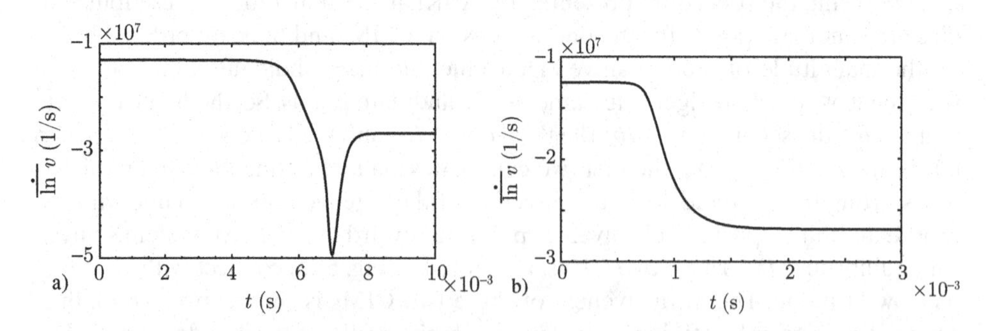


Figure 6.18. The relative volumetric deformation rate along the (a) $R_1 \rightarrow R_7$ and (b) $R_6 \rightarrow R_7$ CIMs for the simple hydrogen-air system. Adapted from Powers et al. (2015), with permission.

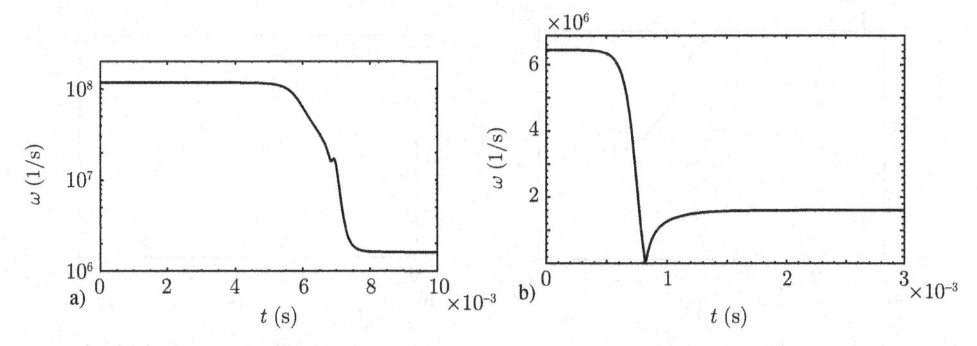


Figure 6.19. The local rotation rates along the (a) $R_1 \to R_7$ and (b) $R_6 \to R_7$ CIMs for the simple hydrogen-air system. Adapted from Powers et al. (2015), with permission.

To quantify the attractiveness and slowness of the physically relevant CIMs, $R_1 \to R_7$ and $R_6 \to R_7$, stretching-based diagnostics are employed locally along the CIMs. The local relative volumetric deformation rate is presented in Fig. 6.18. The relative volumetric deformation rates are negative. The local rotation and tangential stretching rates along the two CIMs are presented in Figs. 6.19 and 6.20, respectively. The tangential stretching rates undergo sign changes, indicated by a change in color in Fig. 6.20.

Figure 6.21 shows that the $R_1 \to R_7$ CIM may not be attractive because one of the extremal normal stretching rates $\sigma_{n,i}$ along the CIM is positive. However, Fig. 6.22 indicates that the repulsion is overcome by the local rotation rate along the $R_1 \to R_7$ CIM, because $\mu$ along the $R_1 \to R_7$ CIM is greater than unity in the section of the CIM where repulsion exists; the rotation is sufficiently fast to prevent any trajectories from diverging from the CIM. Indeed at late time $\mu < 1$; however, simultaneously all $\sigma_{n,i} < 0$, rendering the rotation irrelevant at late time. Thus, the two CIMs are ACIMs. This is confirmed by plotting several trajectories originating near both CIMs, where the results are presented in Fig. 6.23. It is shown that both CIMs are attractive. The rotational effect has a major influence along the $R_1 \to R_7$ CIM, suppressing the effect of positive normal stretching. For the $R_6 \to R_7$ CIM, the rotation is weak; however, the normal stretching is strictly negative, thus rendering the CIM attractive.

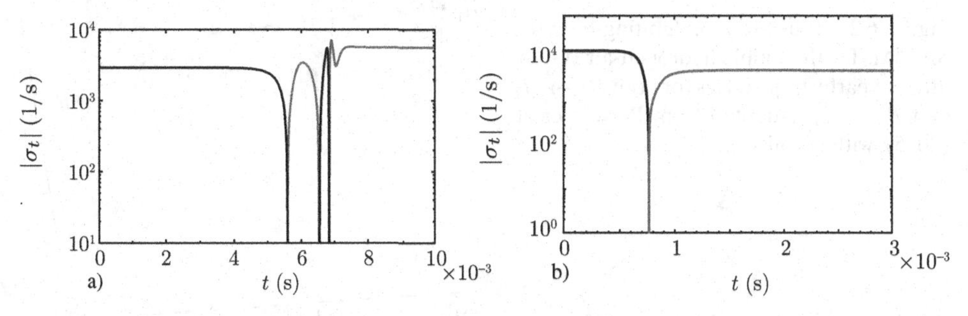


Figure 6.20. The tangential stretching rates along the (a) $R_1 \to R_7$ and (b) $R_6 \to R_7$ CIMs for the simple hydrogen-air system. The gray shade indicates negative value and the black shade indicates positive value; the spike indicates switching signs. Adapted from Powers et al. (2015), with permission.

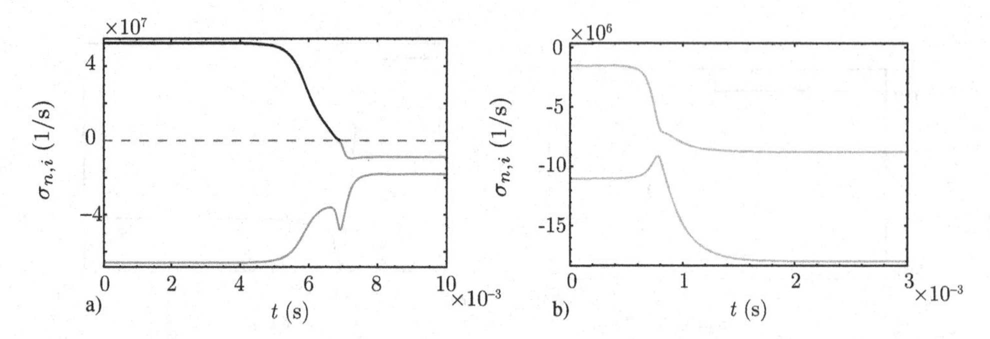


Figure 6.21. The normal stretching rates along the (a) $R_1 \rightarrow R_7$ and (b) $R_6 \rightarrow R_7$ CIMs for the simple hydrogen-air system. The gray shade indicates negative value, and the black indicates positive value. Adapted from Powers et al. (2015), with permission.

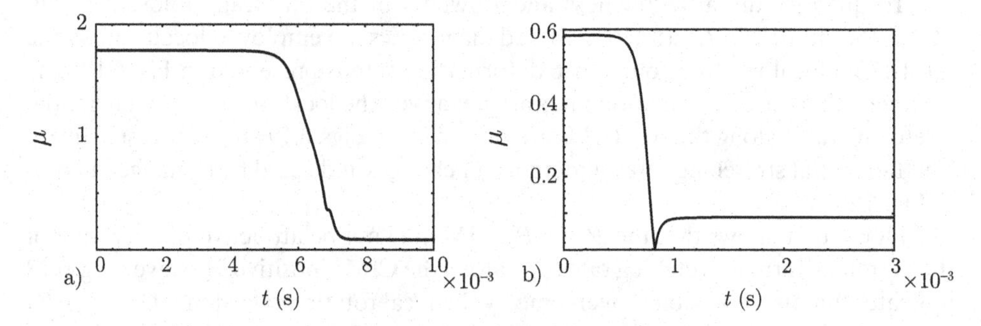


Figure 6.22. The time evolution of the ratio of rotation to normal maximum stretching rates $\mu(t)$ along the (a) $R_1 \rightarrow R_7$ and (b) $R_6 \rightarrow R_7$ CIMs for the simple hydrogen-air system. Adapted from Powers et al. (2015), with permission.

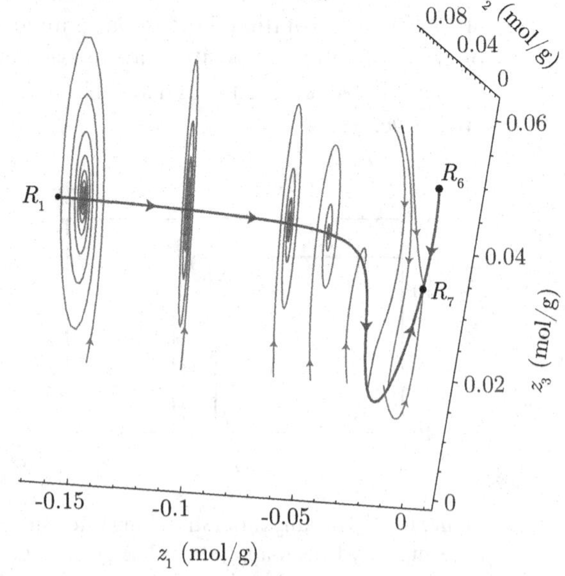


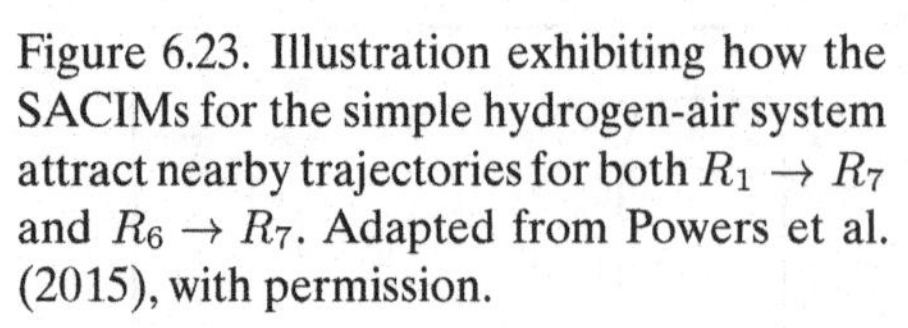
Figure 6.23. Illustration exhibiting how the SACIMs for the simple hydrogen-air system attract nearby trajectories for both $R_1 \rightarrow R_7$ and $R_6 \rightarrow R_7$. Adapted from Powers et al. (2015), with permission.

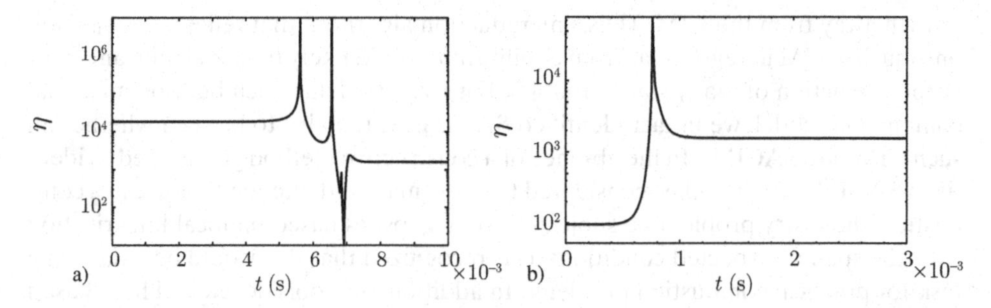


Figure 6.24. The time evolution of the ratio of minimum normal stretching rate to the tangential stretching rate, $\eta(t)$, along the (a) $R_1 \to R_7$ and (b) $R_6 \to R_7$ CIMs for the simple hydrogen-air system. The spikes indicate switching signs. Adapted from Powers et al. (2015), with permission.

Figure 6.24 indicates that the dynamics on the constructed two ACIMs are slow, because $\eta$ along the two ACIMs is greater than unity. This suggests that the motion on the two ACIMs is slower than the motion onto the two ACIMs; that is, these two ACIMs are SACIMs. Finally, in Fig. 6.25, we show the physically relevant SACIMs as they evolve in both the physical $\mathbb{S}$ and nonphysical $\mathbb{S}'$. Within $\mathbb{S}$ where $\eta \sim 10^3$, the one-dimensional manifold captures the slow dynamics of the system.

Diagnostic techniques based on local linearization near a manifold are seen to have value. These diagnostic tools can be applied to any candidate manifold, including a CIM, and answer whether the manifold is both slow and attractive in the neighborhood of the CIM. A related question that is as important in practical applications – where is the slow manifold? – remains unanswered. The CIM identified by heteroclinic connection of saddle and sink equilibria is clearly a viable candidate, but we have shown by example with a simplified three-dimensional model problem that even for a CIM on which a local volume element is shrinking, the rotation of that same element may not be sufficiently rapid to overcome localized

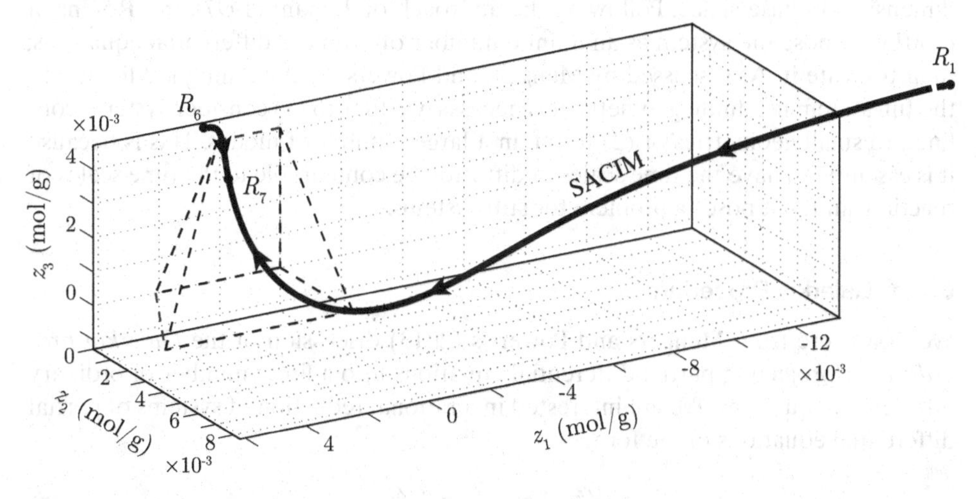


Figure 6.25. The SACIMs for the simple hydrogen-oxygen system are colored based on the relative slowness $\eta$. The solid dots are finite critical points; $R_7$ represents the system's physical equilibrium state, $R_1$ and $R_6$ represent the starting points of the SACIMs, and the dashed simplex represents the physical domain. Adapted from Powers et al. (2015), with permission.

growth away from the CIM. Thus, any reduction algorithm that relies on projection onto such a CIM in regions far from equilibrium would likely induce significant error in the prediction of many state variables. For two problems, each built on an actual combustion model, we in fact identified SACIMs. It remains to be seen whether all such CIMs are SACIMs. In the absence of a constructive method guaranteed to identify a SACIM, it is should be considered that any manifold intended for use in a combustion chemistry problem be subjected to diagnostics based on local linearization over the span of expected conditions. It is recognized that this would be a daunting task for practical combustion problems. In addition, the tools developed here based on local analysis near the CIM cannot speak to the global stability of the CIM, which remains an even more challenging problem.

## 6.4 Diffusion Effects

Until now, we have restricted all discussion to spatially homogeneous systems. To gain an appreciation for the significant complications introduced by spatial inhomogeneity, let us study a model reaction-diffusion system in the context of model reduction and dynamical systems. The problem we choose will be linear and so enable straightforward analysis. Real combustion problems are nonlinear. Those effects in the context of model reduction are not considered here. Examining any of the many studies of reduction of reaction-diffusion systems (e.g., Mengers and Powers, 2013), it can be concluded that mathematically rigorous reduction of nonlinear systems is extremely challenging. Beginning in the next chapter, we will commence discussion of more realistic spatially inhomogeneous combustion systems. Because we find here that reduction with optimal fidelity with the underlying detailed model is difficult, our future analysis will focus on full simulations of unreduced models.

To summarize this section, we will first add simple diffusion effects. Mathematically, this will convert the systems of ordinary differential equations into partial differential equations. This can be thought of as converting a system of equations that evolves in a finite-dimensional phase space into one that evolves in an infinite-dimensional phase space. Following the approach of Temam (1997) and Robinson (2001), we pose the system as an infinite number of ordinary differential equations, then truncate it. As discussed by Mengers and Powers, to maintain good fidelity to the full system in a finite geometry, it is necessary either to (1) consider systems confined to small geometries or (2) to retain a large number of modes. This is because it is essential to have the time scales of diffusion be comparable to the time scales of reaction, and that poses a problem for stiff systems.

### 6.4.1 Galerkin Procedure

We draw here from Mengers and Powers's (2013) discussion of the *Galerkin procedure* for projecting partial differential equations into a finite number of ordinary differential equations. We are interested in a rational reduction of systems of partial differential equations of the form

$$\frac{\partial \mathbf{z}}{\partial t} = \mathbf{f}(\mathbf{z}) + \mathcal{D}\frac{\partial^2 \mathbf{z}}{\partial x^2}, \tag{6.82}$$

where $\mathbf{z} \in \mathbb{R}^{N'}$ are dependent variables, $\mathbf{f}$ are nonlinear functions that represent rates of change due to reaction, $\mathcal{D}$ is a constant mass diffusivity, the temporal domain is

$t \in [0, \infty)$, and there is only one spatial Cartesian dimension, $x \in [0, L]$. Equation (6.82) can be cast as

$$\frac{\partial \mathbf{z}}{\partial t} = \mathbf{f}(\mathbf{z}) - \mathcal{L}(\mathbf{z}), \tag{6.83}$$

where $\mathcal{L} = -\mathcal{D}\ \partial^2/\partial x^2$ is a self-adjoint, positive semidefinite, linear spatial differential operator that models diffusion. We choose to study homogeneous Neumann[7] boundary conditions

$$\left.\frac{\partial \mathbf{z}}{\partial x}\right|_{x=0} = \left.\frac{\partial \mathbf{z}}{\partial x}\right|_{x=L} = \mathbf{0}. \tag{6.84}$$

This choice will enable a direct comparison of reduction methods developed for spatially homogeneous systems to those that include diffusion. More general boundary conditions could be studied at the expense of introducing thin layers into the solution.

Following Finlayson (1972), we approximate a solution to Eqs. (6.83) and (6.84) by using separation of variables coupled with the method of weighted residuals. The dependent variables, $\mathbf{z}$, are approximated by a series of the product of time-dependent amplitudes, $\zeta_m(t)$, and a set of spatial basis functions, $\phi_m(x)$:

$$\mathbf{z}(x,t) \approx \tilde{\mathbf{z}}_M(x,t) = \sum_{m=0}^{M} \zeta_m(t)\phi_m(x), \tag{6.85}$$

where $M+1$ is the number of terms, and for each $m = 0, \ldots, M$, $\zeta_m \in \mathbb{R}^{N'}$.

We can choose the basis functions to be the eigenfunctions of the diffusion operator,

$$\mathcal{L}(\phi_m) = \mu_m \phi_m, \qquad m = 0, \ldots, M, \tag{6.86}$$

where the eigenvalues, $\mu_m$, are guaranteed to be nonnegative and real, and the eigenfunctions orthonormal,

$$\langle \phi_m(x), \phi_n(x) \rangle = \delta_{mn}, \tag{6.87}$$

in a Lebesgue[8] space where the inner product is defined as

$$\langle \phi_m(x), \phi_n(x) \rangle = \int_0^L \phi_m(x)\phi_n(x)\, dx. \tag{6.88}$$

The eigenfunctions are required to match the boundary conditions of Eq. (6.84):

$$\left.\frac{d\phi_m}{dx}\right|_{x=0} = \left.\frac{d\phi_m}{dx}\right|_{x=L} = 0. \tag{6.89}$$

We also choose $\phi_0$ to be a spatially homogeneous constant whose eigenvalue is zero, $\mu_0 = 0$. We then order the subsequent eigenfunctions by increasing numerical value of the eigenvalues (i.e., $\mu_0 \le \mu_1 \le \mu_2 \ldots$).

When the approximation from Eq. (6.85) is substituted into the form from Eq. (6.83), the result does not satisfy the equation exactly, but will have a nonzero

[7] Carl Gottfried Neumann (1832–1925), German mathematician.
[8] Henri Lebesgue (1875–1941), French mathematician.

residual $\mathbf{r}(x,t)$:

$$\mathbf{r}(x,t) = \sum_{m=0}^{M} \frac{d\zeta_m}{dt}\phi_m(x) - \mathbf{f}\left(\sum_{\hat{m}=0}^{M} \zeta_{\hat{m}}(t)\phi_{\hat{m}}(x)\right) + \sum_{m=0}^{M} \mu_m \zeta_m(t)\phi_m(x), \qquad (6.90)$$

when $M$ is finite. This residual is not the error, $\mathbf{e}(x,t) = \mathbf{z}(x,t) - \sum_{m=0}^{M} \zeta_m(t)\phi_m(x)$; however, if the residual is zero, the error will be zero as well. To formulate evolution equations for the amplitudes, we take a series of $M$ spatially weighted averages of the residual and require each be zero,

$$\langle \mathbf{r}(x,t), \psi_m(x)\rangle = 0, \quad m = 0,\ldots,M, \quad t \in [0,\infty), \qquad (6.91)$$

where $\psi_m(x)$ is a set of $M$ spatial weighting functions. There are many viable choices for weighting functions. If we make the common choice of the basis functions as the weighting functions, $\psi_m(x) = \phi_m(x)$, our method is a Galerkin method. Substituting Eq. (6.90) (with dummy indices changed from $m$ to $n$) into Eq. (6.91) and distributing the inner product linear operator to each term in the residual yields

$$\left\langle \sum_{n=0}^{M} \frac{d\zeta_n}{dt}\phi_n(x), \phi_m(x)\right\rangle - \left\langle \mathbf{f}\left(\sum_{\hat{m}=0}^{M} \zeta_{\hat{m}}(t)\phi_{\hat{m}}(x)\right), \phi_m(x)\right\rangle + \left\langle \sum_{n=0}^{M} \mu_n \zeta_n(t)\phi_n(x), \phi_m(x)\right\rangle = 0. \qquad (6.92)$$

Further simplification removes spatially independent terms from the inner products and arranges the terms in the order they were in Eq. (6.83),

$$\sum_{n=0}^{M} \frac{d\zeta_n}{dt}\langle \phi_n(x), \phi_m(x)\rangle = \left\langle \mathbf{f}\left(\sum_{\hat{m}=0}^{M} \zeta_{\hat{m}}(t)\phi_{\hat{m}}(x)\right), \phi_m(x)\right\rangle - \sum_{n=0}^{M} \mu_n \zeta_n(t)\langle \phi_n(x), \phi_m(x)\rangle. \qquad (6.93)$$

Because of the orthonormality of our basis functions, Eq. (6.93) can be reformulated to express the evolution of the amplitudes as

$$\frac{d\zeta_m}{dt} = \left\langle \mathbf{f}\left(\sum_{\hat{m}=0}^{M} \zeta_{\hat{m}}(t)\phi_{\hat{m}}(x)\right), \phi_m(x)\right\rangle - \mu_m \zeta_m(t), \qquad m = 0,\ldots,M, \quad (6.94)$$

which yields a system of $N'(M+1)$ ordinary differential equations. We define the reactions' contribution to the amplitude evolution as

$$\dot{\mathbf{\Omega}}_m(\zeta_{\hat{m}}) = \left\langle \mathbf{f}\left(\sum_{\hat{m}=0}^{M} \zeta_{\hat{m}}(t)\phi_{\hat{m}}(x)\right), \phi_m(x)\right\rangle. \qquad (6.95)$$

To obtain the exact solution of Eq. (6.83), the residual must be driven to zero, which requires the limit of $M \to \infty$. In this sense, the partial differential equation of Eq. (6.83) can be considered to be an infinite set of ordinary differential equations. While the exact solution to an infinite-dimensional system is intractable, approximations with finite $M$ project the trajectories of solutions to the infinite-dimensional system onto a finite-dimensional *approximate inertial manifold* (AIM); see Robinson (2001). The dynamics on this finite-dimensional approximation are governed by a finite system of ordinary differential equations:

$$\frac{d\zeta_m}{dt} = \dot{\mathbf{\Omega}}_m(\zeta_{\hat{m}}) - \mu_m\zeta_m, \qquad m, \hat{m} = 0, \ldots, M. \tag{6.96}$$

*Equation (6.96) is itself a model reduction, taking a system from an infinite-dimensional phase space to a finite-dimensional phase space whose dimension is chosen by the user. At this stage, no appeal to traditional reduced kinetics as been made. Assuming that Eq. (6.96) accurately represents the dynamics of the system, one may ask if further reduction can be made. In the context of this chapter, one can ask if the phase space in which Eq. (6.96) is solved contains an embedded SACIM. This reduction has the advantage of making no appeal to uncoupling reaction and diffusion, as these mechanisms are often fully coupled in portions of phase space where interesting dynamics evolve.*

Robinson (p. 387) shows that under certain conditions, inertial manifolds exponentially attract all of the trajectories of solutions to partial differential equations. Because the diffusion operator we are modeling is dissipative, and because we truncate the high-frequency (and therefore fastest decaying) amplitudes in Eqs. (6.82) and (6.83), we assume the AIM will also exponentially attract all trajectories of solutions to our partial differential equations. Our approximate solutions are obtained by integrating this finite-dimensional system of ordinary differential equations to obtain the amplitude evolution, and then employing Eq. (6.85) to reconstruct an approximation of $\mathbf{z}(x,t)$. These assumptions provide realistic approximations that compare favorably to well-resolved simulations. The evolution of amplitudes in Eq. (6.96) is forced by $\mathbf{F}_m(\zeta_{\hat{m}}) = \dot{\mathbf{\Omega}}_m(\zeta_{\hat{m}}) - \mu_m\zeta_m(t)$, for $m, \hat{m} = 0, \ldots, M$. The projection of $\mathbf{f}(\mathbf{z})$ onto the AIM, $\dot{\mathbf{\Omega}}_m(\zeta_{\hat{m}})$, represents the rate of change of each amplitude due to reaction, and the projection of $\mathcal{L}(\mathbf{z})$ onto the AIM, $\mu_m\zeta_m(t)$, represents the contribution from diffusion.

In the case where the Galerkin projection is truncated at $M = 0$, the spatially homogeneous case is recovered, and Eq. (6.96) reduces to

$$\frac{d\zeta_0}{dt} = \dot{\mathbf{\Omega}}_0(\zeta_0). \tag{6.97}$$

Equation (6.97) can then be easily related to the spatially homogeneous system in the original variables. Because many manifold methods focus on the spatially homogeneous system, we consider this case for comparison. Because spatially homogeneous dynamics are present in a subspace of all truncations of this Galerkin projection, we can use the results from the higher-order truncations to identify deviations from the slow dynamics of the spatially homogeneous approximation.

This general Galerkin reduction has a long history in nonlinear dynamics. It is the method Lorenz (1963) used to reduce the Navier-Stokes equations to a set of three nonlinear ordinary differential equations whose solution displayed chaotic dynamics. For nonlinear combustion systems, the procedure can be algebraically complicated. Such examples are provided by Mengers and Powers (2013). We apply the procedure to a linear example next, which will still be of sufficient complexity to illustrate the challenges of reduction with spatial inhomogeneity.

### 6.4.2 Linear Example

Consider the system with $N' = 2$ of

$$\frac{\partial z_1}{\partial t} = -az_1 + \mathcal{D}\frac{\partial^2 z_1}{\partial x^2}, \qquad \frac{\partial z_2}{\partial t} = \gamma a(z_1 - z_2) + \mathcal{D}\frac{\partial^2 z_2}{\partial x^2}. \tag{6.98}$$

For purposes of this discussion, let us consider $z_1$, $z_2$, and $\gamma$ to be dimensionless, $t$ to have units of s, $x$ units of cm, $a$ to have units of s$^{-1}$, and $\mathcal{D}$ to have units of cm$^2$/s. We require $a > 0$, $\mathcal{D} > 0$, and $\gamma > 1$. We take homogeneous Neumann boundary conditions:

$$\left.\frac{\partial z_1}{\partial x}\right|_{x=0} = \left.\frac{\partial z_1}{\partial x}\right|_{x=L} = \left.\frac{\partial z_2}{\partial x}\right|_{x=0} = \left.\frac{\partial z_2}{\partial x}\right|_{x=L} = 0. \tag{6.99}$$

This insures no diffusive mass flux of either species exists at the system boundaries.

If we were to neglect diffusion, the so-called chemical Jacobian is given by

$$\mathbf{J} = \begin{pmatrix} -a & 0 \\ \gamma a & -\gamma a \end{pmatrix}. \tag{6.100}$$

The chemical eigenvalues are thus $\lambda_1 = -a$ and $\lambda_2 = -a\gamma$. The chemical stiffness is $\gamma$. And the SACIM based on chemistry alone can be shown to be $z_2 = \gamma z_1/(\gamma - 1)$. Critically though, the chemical stiffness is less important than the overall system stiffness, now affected by diffusion. And a chemistry-based SACIM does not account for the effects of diffusion.

Let us apply the Galerkin procedure. We first choose a set of orthonormal basis functions that satisfy the boundary conditions. Take

$$\phi_m(x) = \begin{cases} \sqrt{\dfrac{1}{L}}, & m = 0, \\ \sqrt{\dfrac{2}{L}}\cos\dfrac{m\pi x}{L}, & m = 1, \ldots, M. \end{cases} \tag{6.101}$$

We next take

$$z_1(x,t) = \sqrt{\frac{1}{L}}\zeta_{1,0}(t) + \sqrt{\frac{2}{L}}\sum_{m=1}^{M}\zeta_{1,m}(t)\cos\frac{m\pi x}{L}, \tag{6.102}$$

$$z_2(x,t) = \sqrt{\frac{1}{L}}\zeta_{2,0}(t) + \sqrt{\frac{2}{L}}\sum_{m=1}^{M}\zeta_{2,m}(t)\cos\frac{m\pi x}{L}. \tag{6.103}$$

Taking $M = 1$, we apply the Galerkin procedure and get a system of four ordinary differential equations:

$$\frac{d}{dt}\begin{pmatrix} \zeta_{1,0} \\ \zeta_{2,0} \\ \zeta_{1,1} \\ \zeta_{2,1} \end{pmatrix} = \underbrace{\begin{pmatrix} -a & 0 & 0 & 0 \\ \gamma a & -\gamma a & 0 & 0 \\ 0 & 0 & -a - \frac{\mathcal{D}\pi^2}{L^2} & 0 \\ 0 & 0 & \gamma a & -\gamma a - \frac{\mathcal{D}\pi^2}{L^2} \end{pmatrix}}_{\mathbf{J}_{AIM}}\begin{pmatrix} \zeta_{1,0} \\ \zeta_{2,0} \\ \zeta_{1,1} \\ \zeta_{2,1} \end{pmatrix}. \tag{6.104}$$

The appropriate Jacobian for the system dynamics is not $\mathbf{J}$, but instead $\mathbf{J}_{AIM}$, with

$$\mathbf{J}_{AIM} = \begin{pmatrix} -a & 0 & 0 & 0 \\ \gamma a & -\gamma a & 0 & 0 \\ 0 & 0 & -a - \frac{\mathcal{D}\pi^2}{L^2} & 0 \\ 0 & 0 & \gamma a & -\gamma a - \frac{\mathcal{D}\pi^2}{L^2} \end{pmatrix}. \tag{6.105}$$

Because here $\mathbf{J}_{AIM}$ is upper triangular, its eigenvalues are on the diagonal. They are $-a$, $-\gamma a$, $-a - \mathcal{D}\pi^2/L^2$, and $-\gamma a - \mathcal{D}\pi^2/L^2$. The time scales of the system are the reciprocals of the magnitude of the eigenvalues and are $1/a, 1/(\gamma a), 1/(a + \mathcal{D}\pi^2/L^2)$, and $1/(\gamma a + \mathcal{D}\pi^2/L^2)$. Relative to the spatially homogeneous limit, diffusion has rendered the system to be more stiff. For $M = 1$, the stiffness ratio is

$$\kappa = \gamma + \frac{\mathcal{D}\pi^2}{L^2 a}. \tag{6.106}$$

This linear system can be solved, and evolution will occur on the time scales described. For diffusion to represent a perturbation of reaction for an $M = 1$ approximation, we must have a sufficiently small geometry where

$$L < \pi\sqrt{\frac{\mathcal{D}}{\gamma a}}. \tag{6.107}$$

If the geometry is larger, we will need to retain more modes to have an accurate solution. For general $M$, the eigenvalues of the system can be found to be

$$\lambda_{i,m} = \lambda_{i,0} - \frac{m^2\pi^2\mathcal{D}}{L^2}, \qquad i = 1, \ldots, N', \quad m = 0, \ldots, M. \tag{6.108}$$

Here, $\lambda_{i,0}$ are the chemical eigenvalues. The reciprocals of these eigenvalues give the various time scales:

$$\tau_{i,m} = \left(\lambda_{i,0} - \frac{m^2\pi^2\mathcal{D}}{L^2}\right)^{-1}, \qquad i = 1, \ldots, N', \quad m = 0, \ldots, M. \tag{6.109}$$

We make some important points regarding time scales of the system of reaction-diffusion equations that applies to more complicated systems as well:

- Formally *for the system*, there are no purely chemical time scales or diffusion time scales, there are just system time scales.
- Some system time scales are dominated by reaction effects.
- Some system time scales are dominated by diffusion effects.
- Some system time scales are affected roughly the same by both reaction and diffusion.
- While purely chemical time scales available from spatially homogeneous theory are often small, there are always diffusion-dominated system time scales that are smaller; therefore, the common assertion that reaction is faster than diffusion is too simplistic.

Increasing the number of modes increases the stiffness. The eigenvalue with the smallest magnitude is $a$. That with the largest is $\gamma a + M^2\pi^2\mathcal{D}/L^2$. So the stiffness ratio is

$$\kappa = \gamma + \frac{M^2\mathcal{D}\pi^2}{L^2 a}. \tag{6.110}$$

As the number of modes $M$ increases, chemistry plays an ever-diminishing role in determining the stiffness. We can roughly estimate the number of modes necessary for accurate capture of diffusion by balancing chemistry-induced stiffness with the diffusion-induced stiffness:

$$\gamma = \frac{M^2\mathcal{D}\pi^2}{L^2 a}. \tag{6.111}$$

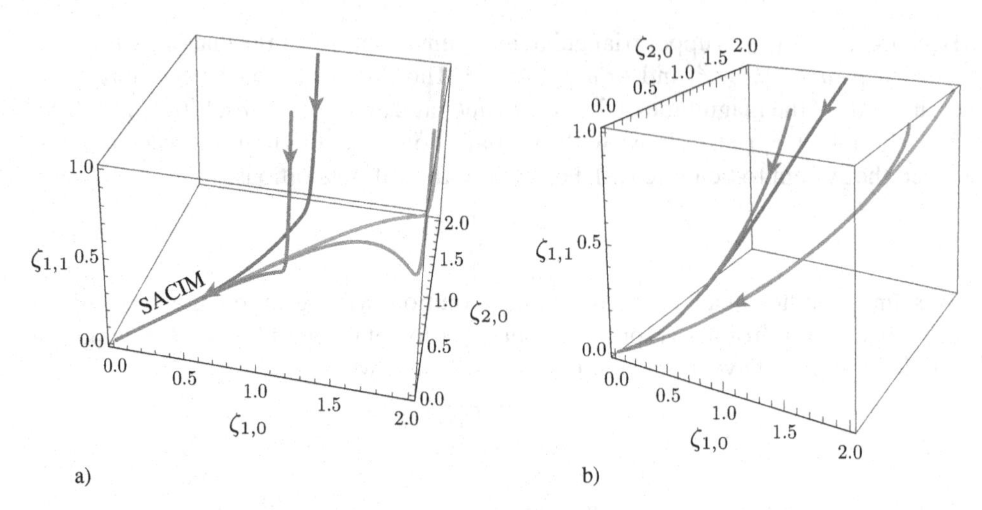


Figure 6.26. Plots of (a) fast diffusion-based relaxation to a plane where chemistry dominates and (b) relaxation to equilibrium with no relaxation to a chemical plane.

Solving for $M$, we get

$$M = \frac{L}{\pi}\sqrt{\frac{a\gamma}{\mathcal{D}}}. \tag{6.112}$$

More modes are required if the geometry is large, the reaction rate coefficient is large, the chemical stiffness is large, or the diffusion coefficient is small.

For the $M = 1$ system, when Eq. (6.107) is satisfied, trajectories in the phase space of dimension four will relax quickly to the plane where chemistry alone dominates, and the system may then move along the chemistry-defined SACIM. Consider an example for which $M = 1$, with $\gamma = 5$, $\mathcal{D} = 10\ \text{cm}^2/\text{s}$, $a = 1\ \text{s}^{-1}$, $L = 1$ cm. Then $\pi\sqrt{\mathcal{D}/\gamma/a} = 4.44\ \text{cm} > L$, so we should expect rapid attraction to the plane where chemistry dominates, and then to the SACIM. This is seen in Fig. 6.26a. Here, we show a projection of the four-dimensional trajectory in a three-dimensional phase volume. Here, fast diffusion-dominated effects rapidly pull the trajectories into the chemistry-dominated plane, and then there is relaxation to the chemical SACIM, and finally motion on the SACIM to equilibrium. Different behavior is realized if we change $\mathcal{D}$ to $1/10\ \text{cm}^2/\text{s}$, holding all other parameters equal. In this case $\pi\sqrt{\mathcal{D}/\gamma/a} = 0.44\ \text{cm} < L$. For these parameters, there is no initial relaxation to the chemistry-dominated plane, nor to the chemistry-based SACIM. There is of course relaxation to the final equilibrium point at the origin.

Omitting details given by Mengers and Powers, 2013, more complicated nonlinear examples show that ACIMs may be identified by heteroclinic trajectories connecting equilibria, but they do not appear to be SACIMs. This is likely attributable to the diffusion modes not possessing so-called *spectral gaps*. Consequently motion toward the ACIM occurs at a similar rate to motion along the ACIM, and identification of the ACIM offers no real advantage in filtering fast–time scale dynamics.

In conclusion, we are pessimistic about mathematically rigorous reduction as a useful tool for elucidating combustion dynamics. Much effort is required even for simple systems, and it is difficult to draw general conclusions. We thus turn in the next chapter, and the remainder of the book, to unreduced systems with advection, reaction, and diffusion.

## EXERCISES

**6.1.** Consider the Davis-Skodje system of Sec. 6.2.1. For $\gamma = 5$, $(z_{1o}, z_{2o}) = (1, 1)$ determine and plot as functions of $t$ (a) the exact solution $z_1(t)$, $z_2(t)$; (b) the approximate solution using the SACIM; (c) the approximate solution using the steady state assumption; and (d) the approximate solution using the ILDM assumption.

**6.2.** Extend the system of Sec. 6.2.2 so that

$$\frac{dz_1}{dt} = \frac{1}{20}(1 - z_1^2),$$
$$\frac{dz_2}{dt} = -2z_2 - \frac{35}{16}z_3 + 2(1 - z_1^2)z_3,$$
$$\frac{dz_3}{dt} = \beta z_2 + z_3,$$

where $\beta$ is an adjustable parameter. Find the critical value of $\beta$ for which the CIM is neither attracting or repelling.

**6.3.** Repeat the analysis of Sec. 6.3.1, except take $n_{O_2}(t = 0) = 2 \times 10^{-6}$ mol.

**6.4.** Repeat the analysis of Sec. 6.4.2, except take $M = 2$. Also prepare a high-accuracy solution with $M = 10$. Compute an error by subtracting the high-accuracy $M = 10$ solution from the $M = 1$ and $M = 2$ solutions and prepare plots that show the error reduction as $M$ increases.

## References

A. Adrover, F. Creta, M. Giona, and M. Valorani, 2007, Stretching-based diagnostics and reduction of chemical kinetic models with diffusion, *Journal of Computational Physics*, 225(2): 1442–1471.

A. N. Al-Khateeb, J. M. Powers, S. Paolucci, A. J. Sommese, J. A. Diller, J. D. Hauenstein, and J. D. Mengers, 2009, One-dimensional slow invariant manifolds for spatially homogeneous reactive systems, *Journal of Chemical Physics*, 131(2): 024118.

R. Aris, 1962, *Vectors, Tensors, and the Basic Equations of Fluid Mechanics*, Dover, New York.

M. J. Davis and R. T. Skodje, 1999, Geometric investigation of low-dimensional manifolds in systems approaching equilibrium, *Journal of Chemical Physics*, 111(3): 859–874.

B. A. Finlayson, 1972, *The Method of Weighted Residuals and Variational Principles with Application in Fluid Mechanics, Heat and Mass Transfer*, Academic Press, New York.

A. N. Gorban and I. V. Karlin, 2003, Method of invariant manifold for chemical kinetics, *Chemical Engineering Science*, 58(21): 4751–4768.

J. F. Griffiths, 1996, Reduced kinetic models and their application to practical combustion systems, *Progress in Energy and Combustion Science*, 21(1): 25–107.

S. H. Lam and D. A. Goussis, 1989, Understanding complex chemical kinetics with computational singular perturbation, *Symposium (International) on Combustion*, 22(1): 931–941.

E. N. Lorenz, 1963, Deterministic nonperiodic flow, *Journal of the Atmospheric Sciences*, 20(2): 130–141.

T. F. Lu and C. K. Law, 2009, Toward accommodating realistic fuel chemistry in large-scale computations, *Progress in Energy and Combustion Science*, 35(2): 192–215.

U. Maas and S. B. Pope, 1992, Simplifying chemical kinetics: Intrinsic low-dimensional manifolds in composition space, *Combustion and Flame*, 88 (3–4): 239–264.

J. D. Mengers, 2012, Slow invariant manifolds for reaction-diffusion systems, PhD dissertation, University of Notre Dame.

J. D. Mengers and J. M. Powers, 2013, One-dimensional slow invariant manifolds for fully coupled reaction and micro-scale diffusion, *SIAM Journal on Applied Dynamical Systems*, 12(2): 560–595.

L. Perko, 2001, *Differential Equations and Dynamical Systems*, 3rd ed., Springer, New York.

J. M. Powers, S. Paolucci, J. D. Mengers, and A. N. Al-Khateeb, 2015, Slow attractive canonical invariant manifolds for reactive systems, *Journal of Mathematical Chemistry*, 53(2): 737–766.

Z. Ren, S. B. Pope, A. Vladimirsky, and J. M. Guckenheimer, 2006, The invariant constrained equilibrium edge preimage curve method for the dimension reduction of chemical kinetics, *Journal of Chemical Physics*, 124(11): 114111.

J. C. Robinson, 2001, *Infinite-Dimensional Dynamical Systems*, Cambridge University Press, Cambridge, UK.

M. R. Roussel and S. J. Fraser, Invariant manifold methods for metabolic model reduction, *Chaos*, 11(1): 196–206.

S. Singh, J. M. Powers, and S. Paolucci, 2002, On slow manifolds of chemically reactive systems, *Journal of Chemical Physics*, 117(4): 1482–1496.

G. Strang, 2005, *Linear Algebra and Its Applications*, 4th ed., Cengage Learning, Stamford, CT.

R. Temam, 1997, *Infinite-Dimensional Dynamical Systems in Mechanics and Physics*, Springer, Berlin.

L. N. Trefethen and M. Embree, 2005, *Spectra and Pseudospectra: The Behavior of Nonnormal Matrices and Operators*, Princeton University Press, Princeton, NJ.

# PART II

## Advective-Reactive-Diffusive Systems

# 7 Reactive Navier-Stokes Equations

*Aerothermochemistry deals with flow phenomena of compressible fluids in which chemical reactions take place.*[*]

— Theodore von Kármán (1881–1963), "Fundamental Equations in Aerothermochemistry," in *Selected Combustion Problems, Part 2*

Up to this point we have focused on systems that are spatially homogeneous. Our brief discussion at the end of Chapter 6 introduced some of the significant challenges of spatial inhomogeneity. For the remainder of the book, we study the effects of spatial inhomogeneity on chemical dynamics. We thus now turn to important systems for which the state variables are functions of both time and space. In this chapter, we present a standard continuum model for a mixture of gases that are allowed to react, advect, and diffuse. That model is the compressible reactive Navier[1]-Stokes[2] equations for an ideal mixture of $N$ ideal gases. We do not give detailed derivations. We are guided in part by the derivations given by Merk (1959) and Aris (1962). Additional background is in other standard sources (Woods, 1975; Buckmaster and Ludford, 1982; de Groot and Mazur, 1984; Williams, 1985; Rosner, 2000; Kee et al., 2003; Kuo, 2005; Poinsot and Veynante, 2005; Bird et al., 2006; Law, 2006; Warnatz et al., 2006). The model is highly detailed, and solution by numerical means is often the only possibility for the full system. Because of this complexity, in following chapters, we study this model in simpler limits, focusing on interplay between advection, reaction, and diffusion in cases where analytic methods can be employed.

## 7.1 Evolution Axioms

### 7.1.1 Conservative Form

The conservation of mass, linear momenta, and energy, and the entropy inequality for the mixture are expressed in conservative form as

$$\frac{\partial \rho}{\partial t} + \nabla \cdot (\rho \mathbf{u}) = 0, \qquad (7.1)$$

[*]T. von Kármán, 1956, Fundamental equations in aerothermochemistry, *Selected Combustion Problems, Part 2*, Butterworths, London, 1956, pp. 167–184.

[1]Claude-Louis Navier (1785–1836), French engineer.

[2]George Gabriel Stokes (1819–1903), Anglo-Irish mathematician.

$$\frac{\partial}{\partial t}(\rho\mathbf{u}) + \nabla \cdot (\rho\mathbf{u}\mathbf{u} + P\mathbf{I} - \boldsymbol{\tau}) = \mathbf{0}, \quad (7.2)$$

$$\frac{\partial}{\partial t}\left(\rho\left(e + \frac{1}{2}\mathbf{u}\cdot\mathbf{u}\right)\right) + \nabla\cdot\left(\rho\mathbf{u}\left(e+\frac{1}{2}\mathbf{u}\cdot\mathbf{u}\right) + \mathbf{j}^q + (P\mathbf{I} - \boldsymbol{\tau})\cdot\mathbf{u}\right) = 0, \quad (7.3)$$

$$\frac{\partial}{\partial t}(\rho s) + \nabla\cdot\left(\rho s\mathbf{u} + \frac{\mathbf{j}^q}{T}\right) \geq 0. \quad (7.4)$$

New variables here are the velocity vector $\mathbf{u}$, the viscous shear tensor $\boldsymbol{\tau}$, and the diffusive heat flux vector $\mathbf{j}^q$. These equations are precisely the same one would use for a single fluid. We have neglected body forces and radiation. We have also not explicitly written an equation for angular momenta; this in fact is an independent principle that yields the result that the stress tensor must be symmetric. As our choice for stress tensor will be symmetric, we will have absorbed this principle intrinsically into the model. In this section we reserve Gibbs's boldfaced notation for vectors and tensors in ordinary three-dimensional physical space, such as velocity. Terms in chemical space will be denoted here with index notation.

The form is known as *conservative* because those quantities that are conserved, mass, linear momenta, and energy, appear within the time derivative. Moreover, it is straightforward to apply the divergence theorem to speak to conservation within a finite volume. For example, we can integrate Eq. (7.1) over a fixed volume $V$ bounded by a surface $A$ to obtain

$$\int_V \frac{\partial\rho}{\partial t}\,dV + \int_V \nabla\cdot(\rho\mathbf{u})\,dV = \underbrace{\int_V 0\,dV}_{0}. \quad (7.5)$$

We can use the divergence theorem to convert a volume integral into a surface integral:

$$\frac{d}{dt}\underbrace{\int_V \rho\,dV}_{m} + \int_A \rho\mathbf{u}\cdot\mathbf{n}\,dA = 0. \quad (7.6)$$

We next recognize (1) for the fixed volume $V$ that Leibniz's[3] rule allows us to bring the $\partial/\partial t$ operator outside the integrand as $d/dt$ and (2) that the volume integral of density is the mass $m$ within the volume, so that

$$\frac{dm}{dt} = -\int_A \rho\mathbf{u}\cdot\mathbf{n}\,dA. \quad (7.7)$$

This form states that the mass $m$ within $V$ evolves solely due to the net flux of mass through the bounding surface area $A$, where $\mathbf{n}$ is the unit outer normal to $A$. Similar interpretations could be made for linear momenta, energy, and entropy.

The evolution of molecular species is dictated by an evolution axiom, an extension of Eq. (5.21) to account for reaction, and now, additionally, advection and diffusion:

$$\frac{\partial}{\partial t}(\rho Y_i) + \nabla\cdot(\rho Y_i\mathbf{u} + \mathbf{j}_i^m) = M_i\dot{\omega}_i, \qquad i = 1,\ldots,N-1. \quad (7.8)$$

Here, the diffusive mass flux vector of species $i$ is $\mathbf{j}_i^m$. Together, Eqs. (7.1) and $N-1$ of Eq. (7.8) form $N$ equations for the evolution of the $N$ species. We insist that the

[3] Gottfried Wilhelm Leibniz (1646–1716), German mathematician and philosopher.

species diffusive mass flux be constrained by

$$\sum_{i=1}^{N} \mathbf{j}_i^m = \mathbf{0}. \tag{7.9}$$

Recalling our earlier definitions of $M_i$ and $\dot{\omega}_i$, Eqs. (5.1) and (5.18), respectively, Eq. (7.8) can be rewritten as

$$\frac{\partial}{\partial t}(\rho Y_i) + \nabla \cdot (\rho Y_i \mathbf{u} + \mathbf{j}_i^m) = \left(\sum_{l=1}^{L} \mathcal{M}_l \phi_{li}\right) \sum_{j=1}^{J} \nu_{ij} r_j. \tag{7.10}$$

Let us sum Eq. (7.10) over all species to get

$$\sum_{i=1}^{N} \frac{\partial}{\partial t}(\rho Y_i) + \sum_{i=1}^{N} \nabla \cdot (\rho Y_i \mathbf{u} + \mathbf{j}_i^m) = \sum_{i=1}^{N} \left(\sum_{l=1}^{L} \mathcal{M}_l \phi_{li}\right) \sum_{j=1}^{J} \nu_{ij} r_j, \tag{7.11}$$

$$\frac{\partial}{\partial t}\left(\rho \underbrace{\sum_{i=1}^{N} Y_i}_{1}\right) + \nabla \cdot \left(\rho \mathbf{u} \underbrace{\sum_{i=1}^{N} Y_i}_{1} + \underbrace{\sum_{i=1}^{N} \mathbf{j}_i^m}_{0}\right) = \sum_{i=1}^{N}\sum_{l=1}^{L}\sum_{j=1}^{J} \mathcal{M}_l \phi_{li} \nu_{ij} r_j, \tag{7.12}$$

$$= \sum_{j=1}^{J} r_j \sum_{l=1}^{L} \mathcal{M}_l \underbrace{\sum_{i=1}^{N} \phi_{li} \nu_{ij}}_{0}, \tag{7.13}$$

$$\frac{\partial \rho}{\partial t} + \nabla \cdot (\rho \mathbf{u}) = 0. \tag{7.14}$$

This is the extension of the spatially homogeneous Eq. (4.300). So, the summation over all species gives a redundancy with Eq. (7.1).

We can get a similar relation for the elements. Let us multiply Eq. (7.8) by our element-species matrix $\phi_{li}$ to get

$$\phi_{li} \frac{\partial}{\partial t}(\rho Y_i) + \phi_{li} \nabla \cdot (\rho Y_i \mathbf{u} + \mathbf{j}_i^m) = \phi_{li} M_i \dot{\omega}_i, \tag{7.15}$$

$$\frac{\partial}{\partial t}\left(\frac{\rho \phi_{li} Y_i}{M_i}\right) + \nabla \cdot \left(\frac{\rho \phi_{li} Y_i}{M_i} \mathbf{u} + \phi_{li} \frac{\mathbf{j}_i^m}{M_i}\right) = \phi_{li} \sum_{j=1}^{J} \nu_{ij} r_j. \tag{7.16}$$

Now sum over all species to get

$$\sum_{i=1}^{N} \frac{\partial}{\partial t}\left(\frac{\rho \phi_{li} Y_i}{M_i}\right) + \sum_{i=1}^{N} \nabla \cdot \left(\frac{\rho \phi_{li} Y_i}{M_i} \mathbf{u} + \phi_{li} \frac{\mathbf{j}_i^m}{M_i}\right) = \sum_{i=1}^{N} \phi_{li} \sum_{j=1}^{J} \nu_{ij} r_j, \tag{7.17}$$

$$\frac{\partial}{\partial t}\left(\rho \sum_{i=1}^{N} \frac{\phi_{li} Y_i}{M_i}\right) + \nabla \cdot \left(\rho \sum_{i=1}^{N} \frac{\phi_{li} Y_i}{M_i} \mathbf{u} + \sum_{i=1}^{N} \phi_{li} \frac{\mathbf{j}_i^m}{M_i}\right) = \sum_{j=1}^{J} r_j \underbrace{\sum_{i=1}^{N} \phi_{li} \nu_{ij}}_{0}, \tag{7.18}$$

$$\frac{\partial}{\partial t}\left(\rho \sum_{i=1}^{N} \frac{\phi_{li} Y_i}{M_i}\right) + \nabla \cdot \left(\rho \sum_{i=1}^{N} \frac{\phi_{li} Y_i}{M_i} \mathbf{u} + \sum_{i=1}^{N} \phi_{li} \frac{\mathbf{j}_i^m}{M_i}\right) = 0, \tag{7.19}$$

$$\mathcal{M}_l \frac{\partial}{\partial t}\left(\rho \sum_{i=1}^{N} \frac{\phi_{li} Y_i}{M_i}\right) + \mathcal{M}_l \nabla \cdot \left(\rho \sum_{i=1}^{N} \frac{\phi_{li} Y_i}{M_i}\mathbf{u} + \sum_{i=1}^{N} \phi_{li} \frac{\mathbf{j}_i^m}{M_i}\right) = 0, \quad l = 1, \ldots, L, \tag{7.20}$$

$$\frac{\partial}{\partial t}\left(\rho \mathcal{M}_l \sum_{i=1}^{N} \frac{\phi_{li} Y_i}{M_i}\right) + \nabla \cdot \left(\rho \mathcal{M}_l \sum_{i=1}^{N} \frac{\phi_{li} Y_i}{M_i}\mathbf{u} + \mathcal{M}_l \sum_{i=1}^{N} \phi_{li} \frac{\mathbf{j}_i^m}{M_i}\right) = 0, \quad l = 1, \ldots, L. \tag{7.21}$$

Let us now define the *element mass fraction* $Y_l^e, l = 1, \ldots, L$, as

$$Y_l^e \equiv \mathcal{M}_l \sum_{i=1}^{N} \frac{\phi_{li} Y_i}{M_i}. \tag{7.22}$$

Note that this can be expressed as

$$Y_l^e = \underbrace{\left(\frac{\text{mass element } l}{\text{moles element } l}\right)}_{\mathcal{M}_l} \times \sum_{i=1}^{N} \underbrace{\left(\frac{\text{moles element } l}{\text{moles species } i}\right)}_{\phi_{li}} \underbrace{\left(\frac{\text{mass species } i}{\text{total mass}}\right)}_{Y_i} \underbrace{\left(\frac{\text{moles species } i}{\text{mass species } i}\right)}_{1/M_i}, \tag{7.23}$$

$$= \frac{\text{mass element } l}{\text{total mass}}. \tag{7.24}$$

Similarly, we take the diffusive element mass flux to be

$$\mathbf{j}_l^e \equiv \mathcal{M}_l \sum_{i=1}^{N} \phi_{li} \frac{\mathbf{j}_i^m}{M_i}, \qquad l = 1, \ldots, L. \tag{7.25}$$

Substitute Eqs. (7.22) and (7.25) into Eq. (7.21) to get

$$\frac{\partial}{\partial t}(\rho Y_l^e) + \nabla \cdot (\rho Y_l^e \mathbf{u} + \mathbf{j}_l^e) = 0. \tag{7.26}$$

We also insist that

$$\sum_{l=1}^{L} \mathbf{j}_l^e = \mathbf{0}, \tag{7.27}$$

so as to keep the total mass of each element constant. This is easily seen to be guaranteed. Sum Eq. (7.25) over all elements to get

$$\sum_{l=1}^{L} \mathbf{j}_l^e = \sum_{l=1}^{L} \mathcal{M}_l \sum_{i=1}^{N} \phi_{li} \frac{\mathbf{j}_i^m}{M_i} = \sum_{i=1}^{N} \sum_{l=1}^{L} \mathcal{M}_l \phi_{li} \frac{\mathbf{j}_i^m}{M_i}, \tag{7.28}$$

$$= \sum_{i=1}^{N} \frac{\mathbf{j}_i^m}{M_i} \underbrace{\sum_{l=1}^{L} \mathcal{M}_l \phi_{li}}_{M_i} = \sum_{i=1}^{N} \mathbf{j}_i^m = \mathbf{0}. \tag{7.29}$$

In summary, we have $L-1$ conservation equations for the elements, one global mass conservation equation, and $N-L$ species evolution equations, in general. These add to form $N$ equations for the overall species evolution.

### 7.1.2 Nonconservative Form

It is often convenient to have an alternative nonconservative form of the governing equations. Let us define the *material derivative*, also known as the derivative following a material particle, the total derivative, or the substantial derivative, as

$$\frac{d}{dt} = \frac{\partial}{\partial t} + \mathbf{u}\cdot\nabla. \tag{7.30}$$

We then use this definition to aid in the recasting of each of the conservation principles.

#### Mass

Using the product rule to expand the mass equation, Eq. (7.1), we get

$$\frac{\partial \rho}{\partial t} + \nabla\cdot(\rho\mathbf{u}) = 0, \tag{7.31}$$

$$\underbrace{\frac{\partial \rho}{\partial t} + \mathbf{u}\cdot\nabla\rho}_{d\rho/dt} + \rho\nabla\cdot\mathbf{u} = 0, \tag{7.32}$$

$$\frac{d\rho}{dt} + \rho\nabla\cdot\mathbf{u} = 0. \tag{7.33}$$

#### Linear Momenta

We again use the product rule to expand the linear momenta equation, Eq. (7.2):

$$\rho\frac{\partial \mathbf{u}}{\partial t} + \mathbf{u}\frac{\partial \rho}{\partial t} + \mathbf{u}\nabla\cdot(\rho\mathbf{u}) + \rho\mathbf{u}\cdot\nabla\mathbf{u} + \nabla P - \nabla\cdot\boldsymbol{\tau} = \mathbf{0}, \tag{7.34}$$

$$\underbrace{\rho\left(\frac{\partial \mathbf{u}}{\partial t} + \mathbf{u}\cdot\nabla\mathbf{u}\right)}_{d\mathbf{u}/dt} + \mathbf{u}\underbrace{\left(\frac{\partial \rho}{\partial t} + \nabla\cdot(\rho\mathbf{u})\right)}_{0} + \nabla P - \nabla\cdot\boldsymbol{\tau} = \mathbf{0}, \tag{7.35}$$

$$\rho\frac{d\mathbf{u}}{dt} + \nabla P - \nabla\cdot\boldsymbol{\tau} = \mathbf{0}. \tag{7.36}$$

#### Energy

Now, use the product rule to expand the energy equation, Eq. (7.3):

$$\rho\frac{\partial}{\partial t}\left(e + \frac{1}{2}\mathbf{u}\cdot\mathbf{u}\right) + \rho\mathbf{u}\cdot\nabla\left(e + \frac{1}{2}\mathbf{u}\cdot\mathbf{u}\right) + \left(e + \frac{1}{2}\mathbf{u}\cdot\mathbf{u}\right)\underbrace{\left(\frac{\partial \rho}{\partial t} + \nabla\cdot(\rho\mathbf{u})\right)}_{0}$$
$$+ \nabla\cdot\mathbf{j}^q + \nabla\cdot(P\mathbf{u}) - \nabla\cdot(\boldsymbol{\tau}\cdot\mathbf{u}) = 0, \tag{7.37}$$

$$\rho\frac{\partial}{\partial t}\left(e + \frac{1}{2}\mathbf{u}\cdot\mathbf{u}\right) + \rho\mathbf{u}\cdot\nabla\left(e + \frac{1}{2}\mathbf{u}\cdot\mathbf{u}\right) + \nabla\cdot\mathbf{j}^q + \nabla\cdot(P\mathbf{u})$$
$$- \nabla\cdot(\boldsymbol{\tau}\cdot\mathbf{u}) = 0. \tag{7.38}$$

We have once again used the mass equation, Eq. (7.1), to simplify. Let us expand more using the product rule:

$$\underbrace{\rho\left(\frac{\partial e}{\partial t}+\mathbf{u}\cdot\nabla e\right)}_{de/dt}+\rho\mathbf{u}\cdot\left(\frac{\partial \mathbf{u}}{\partial t}+\mathbf{u}\cdot\nabla\mathbf{u}\right)+\nabla\cdot\mathbf{j}^q+\nabla\cdot(P\mathbf{u})$$

$$-\nabla\cdot(\boldsymbol{\tau}\cdot\mathbf{u})=0, \tag{7.39}$$

$$\rho\frac{de}{dt}+\mathbf{u}\cdot\underbrace{\left(\rho\left(\frac{\partial \mathbf{u}}{\partial t}+\mathbf{u}\cdot\nabla\mathbf{u}\right)+\nabla P-\nabla\cdot\boldsymbol{\tau}\right)}_{0}+\nabla\cdot\mathbf{j}^q+P\nabla\cdot\mathbf{u}$$

$$-\boldsymbol{\tau}:\nabla\mathbf{u}=0, \tag{7.40}$$

$$\rho\frac{de}{dt}+\nabla\cdot\mathbf{j}^q+P\nabla\cdot\mathbf{u}-\boldsymbol{\tau}:\nabla\mathbf{u}=0. \tag{7.41}$$

We have used the linear momenta equation, Eq. (7.36), to simplify. From the mass equation, Eq. (7.33), we have $\nabla\cdot\mathbf{u}=-(1/\rho)d\rho/dt$, so the energy equation, Eq. (7.41), can also be written as

$$\rho\frac{de}{dt}+\nabla\cdot\mathbf{j}^q-\frac{P}{\rho}\frac{d\rho}{dt}-\boldsymbol{\tau}:\nabla\mathbf{u}=0. \tag{7.42}$$

This energy equation can be formulated in terms of enthalpy. Use the definition, Eq. (3.73), $h=e+P/\rho$ to get an expression for $dh/dt$:

$$dh=de-\frac{P}{\rho^2}\,d\rho+\frac{1}{\rho}\,dP, \tag{7.43}$$

$$\frac{dh}{dt}=\frac{de}{dt}-\frac{P}{\rho^2}\frac{d\rho}{dt}+\frac{1}{\rho}\frac{dP}{dt}, \tag{7.44}$$

$$\rho\frac{de}{dt}-\frac{P}{\rho}\frac{d\rho}{dt}=\rho\frac{dh}{dt}-\frac{dP}{dt}. \tag{7.45}$$

So, the energy equation, Eq. (7.42), in terms of enthalpy is

$$\rho\frac{dh}{dt}+\nabla\cdot\mathbf{j}^q-\frac{dP}{dt}-\boldsymbol{\tau}:\nabla\mathbf{u}=0. \tag{7.46}$$

**Second Law**

Let us expand Eq. (7.4) to write the second law in nonconservative form:

$$\frac{\partial}{\partial t}(\rho s)+\nabla\cdot\left(\rho s\mathbf{u}+\frac{\mathbf{j}^q}{T}\right)\geq 0, \tag{7.47}$$

$$\underbrace{\rho\left(\frac{\partial s}{\partial t}+\mathbf{u}\cdot\nabla s\right)}_{ds/dt}+s\underbrace{\left(\frac{\partial \rho}{\partial t}+\nabla\cdot(\rho\mathbf{u})\right)}_{0}+\nabla\cdot\left(\frac{\mathbf{j}^q}{T}\right)\geq 0, \tag{7.48}$$

$$\rho\frac{ds}{dt}+\nabla\cdot\left(\frac{\mathbf{j}^q}{T}\right)\geq 0. \tag{7.49}$$

**Species**

Let us expand Eq. (7.8) to write species evolution in nonconservative form:

$$\frac{\partial}{\partial t}(\rho Y_i) + \nabla \cdot (\rho Y_i \mathbf{u} + \mathbf{j}_i^m) = M_i \dot{\omega}_i, \quad i = 1, \ldots, N-1, \tag{7.50}$$

$$\rho \underbrace{\left(\frac{\partial Y_i}{\partial t} + \mathbf{u} \cdot \nabla Y_i\right)}_{dY_i/dt} + Y_i \underbrace{\left(\frac{\partial \rho}{\partial t} + \nabla \cdot (\rho \mathbf{u})\right)}_{0} + \nabla \cdot \mathbf{j}_i^m = M_i \dot{\omega}_i, \quad i = 1, \ldots, N-1, \tag{7.51}$$

$$\rho \frac{dY_i}{dt} + \nabla \cdot \mathbf{j}_i^m = M_i \dot{\omega}_i, \quad i = 1, \ldots, N-1. \tag{7.52}$$

**Elements**

Let us expand Eq. (7.26) to write element conservation in nonconservative form:

$$\frac{\partial}{\partial t}(\rho Y_l^e) + \nabla \cdot (\rho Y_l^e \mathbf{u} + \mathbf{j}_l^e) = 0, \qquad l = 1, \ldots, L, \tag{7.53}$$

$$\rho \underbrace{\left(\frac{\partial Y_l^e}{\partial t} + \mathbf{u} \cdot \nabla Y_l^e\right)}_{dY_l^e/dt} + Y_l^e \underbrace{\left(\frac{\partial \rho}{\partial t} + \nabla \cdot (\rho \mathbf{u})\right)}_{0} + \nabla \cdot \mathbf{j}_l^e = 0, \qquad l = 1, \ldots, L, \tag{7.54}$$

$$\rho \frac{dY_l^e}{dt} + \nabla \cdot \mathbf{j}_l^e = 0, \qquad l = 1, \ldots, L. \tag{7.55}$$

## 7.2 Mixture Rules

We adopt the following rules from Chapter 2 for the ideal mixture:

$$P = \sum_{i=1}^{N} P_i, \quad 1 = \sum_{i=1}^{N} Y_i, \quad \rho = \sum_{i=1}^{N} \rho_i = \sum_{i=1}^{N} \frac{\overline{\rho}_i}{M_i}, \quad e = \sum_{i=1}^{N} Y_i e_i, \tag{7.56}$$

$$h = \sum_{i=1}^{N} Y_i h_i, \quad s = \sum_{i=1}^{N} Y_i s_i, \quad c_v = \sum_{i=1}^{N} Y_i c_{vi}, \quad c_P = \sum_{i=1}^{N} Y_i c_{Pi}, \tag{7.57}$$

$$V = V_i, \quad T = T_i. \tag{7.58}$$

## 7.3 Constitutive Models

The evolution axioms do not form a complete set of equations. Let us supplement these by a set of constitutive model equations appropriate for a mixture of calorically perfect ideal gases that react according the to the law of mass action with an Arrhenius kinetic reaction rate. We have seen many of these models before, and repeat them here for completeness.

For the thermal equation of state, we take the ideal gas law for the partial pressures:

$$P_i = \overline{R} T \overline{\rho}_i = \overline{R} T \frac{\rho Y_i}{M_i} = \overline{R} T \frac{\rho_i}{M_i} = R_i T \rho_i. \tag{7.59}$$

So, the mixture pressure is

$$P = \overline{R}T \sum_{i=1}^{N} \overline{\rho}_i = \overline{R}T \sum_{i=1}^{N} \frac{\rho Y_i}{M_i} = \overline{R}T \sum_{i=1}^{N} \frac{\rho_i}{M_i}. \tag{7.60}$$

For the ideal gas, the enthalpy and internal energy of each component is a function of $T$ at most. We have for the enthalpy of a component

$$h_i = h^0_{T_0,i} + \int_{T_0}^{T} c_{Pi}(\hat{T})\, d\hat{T}. \tag{7.61}$$

Recall from Eq. (4.82) that $\overline{h}^0_{T_0,i} = \overline{h}^0_{f,i}$. So, the mixture enthalpy is

$$h = \sum_{i=1}^{N} Y_i \left( h^0_{T_0,i} + \int_{T_0}^{T} c_{Pi}(\hat{T})\, d\hat{T} \right). \tag{7.62}$$

We then use the definition of enthalpy to recover the internal energy of component $i$:

$$e_i = h_i - \frac{P_i}{\rho_i} = h_i - R_i T. \tag{7.63}$$

So, the mixture internal energy is

$$e = \sum_{i=1}^{N} Y_i \left( h^0_{T_0,i} - R_i T + \int_{T_0}^{T} c_{Pi}(\hat{T})\, d\hat{T} \right), \tag{7.64}$$

$$= \sum_{i=1}^{N} Y_i \left( h^0_{T_0,i} - R_i (T - T_0) - R_i T_0 + \int_{T_0}^{T} c_{Pi}(\hat{T})\, d\hat{T} \right), \tag{7.65}$$

$$= \sum_{i=1}^{N} Y_i \left( h^0_{T_0,i} - R_i T_0 + \int_{T_0}^{T} (c_{Pi}(\hat{T}) - R_i)\, d\hat{T} \right), \tag{7.66}$$

$$= \sum_{i=1}^{N} Y_i \left( h^0_{T_0,i} - R_i T_0 + \int_{T_0}^{T} c_{vi}(\hat{T})\, d\hat{T} \right), \tag{7.67}$$

$$= \sum_{i=1}^{N} Y_i \left( e^0_{T_0,i} + \int_{T_0}^{T} c_{vi}(\hat{T})\, d\hat{T} \right). \tag{7.68}$$

The mixture entropy is

$$s = \sum_{i=1}^{N} Y_i s^0_{T_0,i} + \int_{T_0}^{T} \frac{c_P(\hat{T})}{\hat{T}}\, d\hat{T} - \sum_{i=1}^{N} Y_i R_i \ln \left( \frac{P_i}{P_0} \right). \tag{7.69}$$

Among others, Kee et al. (2003, Chapter 12) give a full development of constitutive models for viscosity, thermal conductivity, and multicomponent mass diffusion. We do not give full details here. The viscous shear stress for an isotropic Newtonian fluid that satisfies Stokes's assumption is

$$\boldsymbol{\tau} = 2\hat{\mu} \left( \frac{\nabla \mathbf{u} + (\nabla \mathbf{u})^T}{2} - \frac{1}{3} (\nabla \cdot \mathbf{u}) \mathbf{I} \right). \tag{7.70}$$

Here, $\hat{\mu}$ is the mixture viscosity coefficient that is determined from a suitable mixture rule averaging over each component. Kee et al. report a commonly used Wilke rule as

$$\hat{\mu} = \sum_{i=1}^{N} \frac{y_i \hat{\mu}_i}{\sum_{m=1}^{N} \Phi_{im} y_m}, \tag{7.71}$$

where

$$\Phi_{im} = \frac{1}{\sqrt{8}} \left(1 + \frac{M_i}{M_m}\right)^{-1/2} \left(1 + \left(\frac{\hat{\mu}_i}{\hat{\mu}_m}\right)^{1/2} \left(\frac{M_m}{M_i}\right)^{1/4}\right)^2, \tag{7.72}$$

and $\hat{\mu}_i$ is the temperature-dependent dynamic viscosity of species $i$, where the temperature-dependency is typically taken from a curve fit of data. Recall $y_i$ is the mol fraction.

The energy flux vector $\mathbf{j}^q$ is written as

$$\mathbf{j}^q = -\mathsf{k}\nabla T + \sum_{i=1}^{N} \mathbf{j}_i^m h_i - \overline{R}T \sum_{i=1}^{N} \frac{D_i^T}{M_i} \left(\frac{\nabla y_i}{y_i} + \left(1 - \frac{M_i}{M}\right)\frac{\nabla P}{P}\right). \tag{7.73}$$

Here, $\mathsf{k}$ is a suitably mixture-averaged thermal conductivity. Kee et al. give the formula

$$\mathsf{k} = \frac{1}{2}\left(\sum_{i=1}^{N} y_i \mathsf{k}_i + \frac{1}{\sum_{m=1}^{N} \frac{y_m}{\mathsf{k}_m}}\right). \tag{7.74}$$

Here, $\mathsf{k}_i$ is the temperature-dependent thermal conductivity of species $i$. The temperature-dependency is usually built around curve fits based on experimental data. The parameter $D_i^T$ is the so-called thermal diffusion coefficient.

We take the mass diffusion vector to be

$$\mathbf{j}_i^m = \rho \sum_{k=1,\, k\neq i}^{N} \frac{M_i D_{ik} Y_k}{M} \left(\frac{\nabla y_k}{y_k} - \left(1 - \frac{M_k}{M}\right)\frac{\nabla P}{P}\right) - D_i^T \frac{\nabla T}{T}. \tag{7.75}$$

The multicomponent and thermal diffusion coefficients, $D_{ij}$ and $D_i^T$, require calculation for each mixture. As we do not focus on this phenomenon, and because full exposition of the details is extensive, we omit them; one can consult Kee et al. (2003) or the literature (e.g., Dixon-Lewis, 1968).

We adopt, as before, the species molar production rate $i$, Eq. (5.18):

$$\dot{\omega}_i = \sum_{j=1}^{J} \nu_{ij} r_j. \tag{7.76}$$

Here, $r_j$ is given by the law of mass action, Eq. (5.14):

$$r_j = k_j \prod_{k=1}^{N} \overline{\rho}_k^{\nu'_{kj}} \left(1 - \frac{1}{K_{c,j}} \prod_{k=1}^{N} \overline{\rho}_k^{\nu_{kj}}\right), \tag{7.77}$$

The reaction rate coefficient $k_j$ is given by the Arrhenius kinetics rule, Eq. (5.17):

$$k_j = a_j T^{\beta_j} \exp\left(\frac{-\overline{\mathcal{E}}_j}{\overline{R}T}\right), \tag{7.78}$$

and the equilibrium constant $K_{c,j}$ for the ideal gas mixture is given by Eq. (5.7):

$$K_{c,j} = \left(\frac{P_0}{\overline{R}T}\right)^{\sum_{i=1}^{N}\nu_{ij}} \exp\left(-\frac{\Delta G_j^0}{\overline{R}T}\right). \tag{7.79}$$

### 7.4 Temperature Evolution

Because temperature $T$ has an important role in many discussions of combustion, let us formulate our energy conservation principle as a temperature evolution equation by employing a variety of constitutive laws.

Let us begin with Eq. (7.46) coupled with our constitutive law for $h$, Eq. (7.62):

$$\rho\frac{d}{dt}\underbrace{\left(\sum_{i=1}^{N} Y_i \underbrace{\left(h_{T_0,i}^0 + \int_{T_0}^{T} c_{Pi}(\hat{T})\,d\hat{T}\right)}_{h_i}\right)}_{h} + \nabla\cdot\mathbf{j}^q - \frac{dP}{dt} - \boldsymbol{\tau}:\nabla\mathbf{u} = 0, \tag{7.80}$$

$$\rho\frac{d}{dt}\sum_{i=1}^{N} Y_i h_i + \nabla\cdot\mathbf{j}^q - \frac{dP}{dt} - \boldsymbol{\tau}:\nabla\mathbf{u} = 0, \tag{7.81}$$

$$\sum_{i=1}^{N}\left(\rho Y_i\frac{dh_i}{dt} + \rho h_i\frac{dY_i}{dt}\right) + \nabla\cdot\mathbf{j}^q - \frac{dP}{dt} - \boldsymbol{\tau}:\nabla\mathbf{u} = 0, \tag{7.82}$$

$$\sum_{i=1}^{N}\left(\rho Y_i c_{Pi}\frac{dT}{dt} + \rho h_i\frac{dY_i}{dt}\right) + \nabla\cdot\mathbf{j}^q - \frac{dP}{dt} - \boldsymbol{\tau}:\nabla\mathbf{u} = 0, \tag{7.83}$$

$$\rho\frac{dT}{dt}\underbrace{\sum_{i=1}^{N} Y_i c_{Pi}}_{c_P} + \sum_{i=1}^{N} h_i \underbrace{\rho\frac{dY_i}{dt}}_{M_i\dot{\omega}_i - \nabla\cdot\mathbf{j}_i^m} + \nabla\cdot\mathbf{j}^q - \frac{dP}{dt} - \boldsymbol{\tau}:\nabla\mathbf{u} = 0, \tag{7.84}$$

$$\rho c_P\frac{dT}{dt} + \sum_{i=1}^{N} h_i\left(M_i\dot{\omega}_i - \nabla\cdot\mathbf{j}_i^m\right) + \nabla\cdot\mathbf{j}^q - \frac{dP}{dt} - \boldsymbol{\tau}:\nabla\mathbf{u} = 0. \tag{7.85}$$

Now, let us define the "thermal diffusion" flux $\mathbf{j}^T$ as

$$\mathbf{j}^T \equiv -\overline{R}T\sum_{i=1}^{N}\frac{D_i^T}{M_i}\left(\frac{\nabla y_i}{y_i} + \left(1 - \frac{M_i}{M}\right)\frac{\nabla P}{P}\right). \tag{7.86}$$

With this, our total energy diffusion flux vector $\mathbf{j}^q$, Eq. (7.73), becomes

$$\mathbf{j}^q = -\mathsf{k}\nabla T + \sum_{i=1}^{N}\mathbf{j}_i^m h_i + \mathbf{j}^T. \tag{7.87}$$

Now, substitute Eq. (7.87) into Eq. (7.85) and rearrange to get

$$\rho c_P\frac{dT}{dt} + \sum_{i=1}^{N} h_i\left(M_i\dot{\omega}_i - \nabla\cdot\mathbf{j}_i^m\right) + \nabla\cdot\left(-\mathsf{k}\nabla T + \sum_{i=1}^{N}\mathbf{j}_i^m h_i + \mathbf{j}^T\right) = \frac{dP}{dt} + \boldsymbol{\tau}:\nabla\mathbf{u}, \tag{7.88}$$

$$\rho c_P \frac{dT}{dt} + \sum_{i=1}^{N} (h_i M_i \dot{\omega}_i - h_i \nabla \cdot \mathbf{j}_i^m + \nabla \cdot (\mathbf{j}_i^m h_i)) = \nabla \cdot (\mathsf{k} \nabla T) - \nabla \cdot \mathbf{j}^T + \frac{dP}{dt} + \boldsymbol{\tau} : \nabla \mathbf{u}, \tag{7.89}$$

$$\rho c_P \frac{dT}{dt} + \sum_{i=1}^{N} (h_i M_i \dot{\omega}_i + \mathbf{j}_i^m \cdot \nabla h_i) = \nabla \cdot (\mathsf{k} \nabla T) - \nabla \cdot \mathbf{j}^T + \frac{dP}{dt} + \boldsymbol{\tau} : \nabla \mathbf{u}, \tag{7.90}$$

$$\rho c_P \frac{dT}{dt} + \sum_{i=1}^{N} (h_i M_i \dot{\omega}_i + c_{Pi} \mathbf{j}_i^m \cdot \nabla T) = \nabla \cdot (\mathsf{k} \nabla T) - \nabla \cdot \mathbf{j}^T + \frac{dP}{dt} + \boldsymbol{\tau} : \nabla \mathbf{u}. \tag{7.91}$$

So, the equation for the evolution of a material fluid particle is

$$\rho c_P \frac{dT}{dt} = -\sum_{i=1}^{N} (h_i M_i \dot{\omega}_i + c_{Pi} \mathbf{j}_i^m \cdot \nabla T) + \nabla \cdot (\mathsf{k} \nabla T) - \nabla \cdot \mathbf{j}^T + \frac{dP}{dt} + \boldsymbol{\tau} : \nabla \mathbf{u}. \tag{7.92}$$

Let us impose some more details. First, we recall that $\overline{h}_i = h_i M_i$, so

$$\rho c_P \frac{dT}{dt} = -\sum_{i=1}^{N} \overline{h}_i \dot{\omega}_i - \sum_{i=1}^{N} c_{Pi} \mathbf{j}_i^m \cdot \nabla T + \nabla \cdot (\mathsf{k} \nabla T) - \nabla \cdot \mathbf{j}^T + \frac{dP}{dt} + \boldsymbol{\tau} : \nabla \mathbf{u}. \tag{7.93}$$

Impose now Eq. (7.76) to expand $\dot{\omega}_i$:

$$\rho c_P \frac{dT}{dt} = -\sum_{i=1}^{N} \overline{h}_i \sum_{j=1}^{J} \nu_{ij} r_j - \sum_{i=1}^{N} c_{Pi} \mathbf{j}_i^m \cdot \nabla T + \nabla \cdot (\mathsf{k} \nabla T) - \nabla \cdot \mathbf{j}^T + \frac{dP}{dt} + \boldsymbol{\tau} : \nabla \mathbf{u}, \tag{7.94}$$

$$\rho c_P \frac{dT}{dt} = -\sum_{j=1}^{J} r_j \sum_{i=1}^{N} \overline{h}_i \nu_{ij} - \sum_{i=1}^{N} c_{Pi} \mathbf{j}_i^m \cdot \nabla T + \nabla \cdot (\mathsf{k} \nabla T) - \nabla \cdot \mathbf{j}^T + \frac{dP}{dt} + \boldsymbol{\tau} : \nabla \mathbf{u}. \tag{7.95}$$

Now, recall Eq. (5.30):

$$\Delta H_j \equiv \sum_{i=1}^{N} \overline{h}_i \nu_{ij}. \tag{7.96}$$

With this, Eq. (7.95) becomes, after small rearrangement,

$$\rho c_P \frac{dT}{dt} = -\sum_{j=1}^{J} r_j \Delta H_j + \frac{dP}{dt} \underbrace{- \sum_{i=1}^{N} c_{Pi} \mathbf{j}_i^m \cdot \nabla T + \nabla \cdot (\mathsf{k} \nabla T) - \nabla \cdot \mathbf{j}^T + \boldsymbol{\tau} : \nabla \mathbf{u}}_{\text{diffusion effects}}. \tag{7.97}$$

By comparing Eq. (7.97) with Eq. (4.340) or (5.29), it is easy to see the effects of variable pressure and diffusion on how temperature evolves. Interestingly, mass, momentum, and energy diffusion all influence temperature evolution. The nondiffusive terms are combinations of advection, reaction, and spatially homogeneous effects.

## 7.5 Shvab-Zel'dovich Formulation

Under some restrictive assumptions, we can simplify the energy equation considerably and recover what is known as a *Shvab-Zel'dovich formulation* (Williams, 1985, p. 10; Lam and Bellan, 2003). Let us assume the following:

- The low Mach[4] number limit is applicable (see Sec. 10.1.1), which can be shown to imply that pressure changes and work due to viscous dissipation are negligible at leading order, $dP/dt \sim 0, \boldsymbol{\tau} : \nabla \mathbf{u} \sim 0$.
- The incompressible limit applies, $d\rho/dt = 0$.
- Thermal diffusion is negligible, $D_i^T \sim 0$.
- All species have identical molecular masses, so that $M_i = M$.
- The multicomponent diffusion coefficients of all species are equal, $D_{ij} = \mathcal{D}$.
- All species possess identical specific heats, $c_{Pi} = c_P$, which is itself a constant.
- Energy diffuses at the same rate as mass so that $\mathsf{k}/c_P = \rho\mathcal{D}$.

With the low Mach number limit, the energy equation, Eq. (7.46), reduces to

$$\rho \frac{dh}{dt} + \nabla \cdot \mathbf{j}^q = 0. \tag{7.98}$$

With $D_i^T = 0$, the diffusive energy flux vector, Eq. (7.73), reduces to

$$\mathbf{j}^q = -\mathsf{k}\nabla T + \sum_{i=1}^{N} \mathbf{j}_i^m h_i. \tag{7.99}$$

Substituting Eq. (7.99) into Eq. (7.98), we get

$$\rho \frac{dh}{dt} + \nabla \cdot \left( -\mathsf{k}\nabla T + \sum_{i=1}^{N} \mathbf{j}_i^m h_i \right) = 0. \tag{7.100}$$

With all component specific heats equal and constant, $c_{Pi} = c_P$, Eq. (7.62) for the mixture enthalpy reduces to

$$h = \sum_{i=1}^{N} Y_i \left( h_{T_0,i}^0 + c_P(T - T_0) \right) = c_P(T - T_0) + \sum_{i=1}^{N} Y_i h_{T_0,i}^0. \tag{7.101}$$

Similarly, for a component, Eq. (7.61) reduces to

$$h_i = h_{T_0,i}^0 + c_P(T - T_0). \tag{7.102}$$

With Eqs. (7.101) and (7.102), Eq. (7.100) transforms to

$$\rho c_P \frac{dT}{dt} + \rho \frac{d}{dt} \left( \sum_{i=1}^{N} Y_i h_{T_0,i}^0 \right) - \nabla \cdot \left( \mathsf{k}\nabla T - \sum_{i=1}^{N} \mathbf{j}_i^m \left( h_{T_0,i}^0 + c_P(T - T_0) \right) \right) = 0, \tag{7.103}$$

[4]Ernst Mach (1838–1916), Austrian physicist.

$$\rho c_P \frac{dT}{dt} + \rho \frac{d}{dt}\left(\sum_{i=1}^{N} Y_i h^0_{T_0,i}\right) - \nabla \cdot \left( \mathsf{k}\nabla T - \sum_{i=1}^{N} \mathbf{j}_i^m h^0_{T_0,i} - c_P(T - T_0)\underbrace{\sum_{i=1}^{N} \mathbf{j}_i^m}_{0} \right) = 0, \tag{7.104}$$

$$\rho c_P \frac{dT}{dt} + \rho \frac{d}{dt}\left(\sum_{i=1}^{N} Y_i h^0_{T_0,i}\right) - \nabla \cdot \left( \mathsf{k}\nabla T - \sum_{i=1}^{N} \mathbf{j}_i^m h^0_{T_0,i} \right) = 0, \tag{7.105}$$

$$\rho c_P \frac{d}{dt}\left(T + \frac{\sum_{i=1}^{N} Y_i h^0_{T_0,i}}{c_P}\right) - \nabla \cdot \left( \mathsf{k}\nabla T - \sum_{i=1}^{N} \mathbf{j}_i^m h^0_{T_0,i} \right) = 0. \tag{7.106}$$

With $M = M_i$, we recover mass fractions to be identical to mol fractions, $Y_k = y_k$. Using this along with our assumptions of negligible thermal diffusion, $D_i^T = 0$, and equal multicomponent diffusion coefficients, $D_{ij} = \mathcal{D}$, the mass diffusion flux vector, Eq. (7.75), reduces to

$$\mathbf{j}_i^m = \rho \sum_{k=1,k\neq i}^{N} \mathcal{D}\nabla Y_k = \rho\mathcal{D} \sum_{k=1,k\neq i}^{N} \nabla Y_k = \rho\mathcal{D}\nabla \sum_{k=1,k\neq i}^{N} Y_k, \tag{7.107}$$

$$= \rho\mathcal{D}\nabla \left( -Y_i + \underbrace{\sum_{k=1}^{N} Y_k}_{1} \right) = \rho\mathcal{D}\nabla(-Y_i + 1) = -\rho\mathcal{D}\nabla Y_i. \tag{7.108}$$

Now, substitute Eq. (7.108) into Eq. (7.106) and use our equidiffusion rate assumption, $\rho\mathcal{D} = \mathsf{k}/c_P$, to get

$$\rho c_P \frac{d}{dt}\left(T + \frac{\sum_{i=1}^{N} Y_i h^0_{T_0,i}}{c_P}\right) - \nabla \cdot \left( \mathsf{k}\nabla T + \rho\mathcal{D} \sum_{i=1}^{N} \nabla Y_i h^0_{T_0,i} \right) = 0, \tag{7.109}$$

$$\rho c_P \frac{d}{dt}\left(T + \frac{\sum_{i=1}^{N} Y_i h^0_{T_0,i}}{c_P}\right) - \nabla \cdot \left( \mathsf{k}\nabla T + \frac{\mathsf{k}}{c_P} \sum_{i=1}^{N} \nabla Y_i h^0_{T_0,i} \right) = 0, \tag{7.110}$$

$$\frac{d}{dt}\left(T + \frac{\sum_{i=1}^{N} Y_i h^0_{T_0,i}}{c_P}\right) - \frac{\mathsf{k}}{\rho c_P}\nabla \cdot \left( \nabla\left(T + \frac{\sum_{i=1}^{N} Y_i h^0_{T_0,i}}{c_P}\right)\right) = 0. \tag{7.111}$$

Now, iff

- there is a spatially uniform distribution of the quantity $T + (1/c_P)\sum_{i=1}^{N} Y_i h^0_{T_0,i}$ at $t = 0$, and

- there is no flux of $T + (1/c_P)\sum_{i=1}^{N} Y_i h^0_{T_0,i}$ at the boundary of the spatial domain for all time,

then Eq. (7.111) can be satisfied for all time by the algebraic relation .

$$T + \frac{\sum_{i=1}^{N} Y_i h^0_{T_0,i}}{c_P} = T(0) + \frac{\sum_{i=1}^{N} Y_i(0) h^0_{T_0,i}}{c_P}, \tag{7.112}$$

where $T(0)$ and $Y_i(0)$ are constants at the spatially uniform initial state. We can slightly rearrange to write

$$\underbrace{c_P(T - T_0) + \sum_{i=1}^{N} Y_i h^0_{T_0,i}}_{h(T,Y_i)} = \underbrace{c_P(T(0) - T_0) + \sum_{i=1}^{N} Y_i(0) h^0_{T_0,i}}_{h(T(0),Y_i(0))}, \tag{7.113}$$

$$h(T, Y_i) = h(T(0), Y_i(0)). \tag{7.114}$$

This simply says the enthalpy function is a constant, which can be evaluated at the initial state.

## EXERCISES

**7.1.** Consider the effects of a body force on the governing equations. First add the body force per volume $\rho \mathbf{g}$ to Eq. (7.2) and the work done by that body force per volume, $\rho \mathbf{u} \cdot \mathbf{g}$, to Eq. (7.3). Then assume the body force is conservative so that it can be expressed as the gradient of a potential, $\varphi$, with $\mathbf{g} = -\nabla \varphi$. Write the linear momenta and energy equations and second law of thermodynamics in conservative and nonconservative forms, accounting for the conservative body force.

**7.2.** Write the temperature evolution equation, Eq. (7.97), in one-dimensional spherical coordinates for which all variables are functions of radius $r$ and time $t$.

**7.3.** Show for an inert fluid that use of Eq. (3.1), $de = T\,ds - P\,dv$, to eliminate $e$ in favor of $s$ allows the first law of thermodynamics to be written as

$$\rho \frac{ds}{dt} = -\nabla \cdot \left(\frac{\mathbf{j}^q}{T}\right) - \frac{1}{T^2}\mathbf{j}^q \cdot \nabla T + \frac{1}{T}\boldsymbol{\tau} : \nabla \mathbf{u}^T$$

and that the second law of thermodynamics induces the Clausius-Duhem inequality

$$-\frac{1}{T^2}\mathbf{j}^q \cdot \nabla T + \frac{1}{T}\boldsymbol{\tau} : \nabla \mathbf{u}^T \geq 0.$$

Consider the challenging problem of extending this result to a reacting fluid.

**7.4.** Consider the example from heat transfer that motivated the comment of Müller and Weiss (2005), from Sec. 5.5.5. For a steady one-dimensional incompressible material at rest with no mass diffusion but with temperature-dependent thermal conductivity, Eq. (7.73) simplifies considerably to

$$j^q = -\mathsf{k}(T)\frac{dT}{dx}.$$

The first law of thermodynamics, Eq. (7.3), also simplifies considerably to

$$\frac{dj^q}{dx} = 0.$$

Show that when combined, these two equations imply that the first law of thermodynamics for such a material induces

$$\frac{d^2T}{dx^2} + \frac{d \ln \mathsf{k}}{dT}\left(\frac{dT}{dx}\right)^2 = 0.$$

The analysis of the previous problem yields a second law restriction of

$$-\frac{1}{T^2} j^q \frac{dT}{dx} \geq 0.$$

For our $j^q$, this gives

$$\frac{\mathsf{k}(T)}{T^2}\left(\frac{dT}{dx}\right)^2 \geq 0.$$

This obviously holds for $\mathsf{k} \geq 0$. Examine then the irreversible entropy generation rate in a domain $x \in [0, L]$:

$$\dot{\sigma} = A \int_0^L \frac{\mathsf{k}(T)}{T^2}\left(\frac{dT}{dx}\right)^2 dx,$$

where $A$ is the constant cross-sectional area. Review the calculus of variations, and show that the Euler-Lagrange equation induced by extremizing $\dot{\sigma}$ is

$$\frac{d^2T}{dx^2} + \frac{d \ln\left(\frac{\sqrt{\mathsf{k}}}{T}\right)}{dT}\left(\frac{dT}{dx}\right)^2 = 0.$$

Comment on the consistency of this with the first law of thermodynamics and the validity or invalidity of either Müller and Weiss's arguments or those of Prigogine (1967).

## References


R. Aris, 1962, *Vectors, Tensors, and the Basic Equations of Fluid Mechanics*, Dover, New York.

R. B. Bird, W. E. Stewart, and E. N. Lightfoot, 2006, *Transport Phenomena*, 2nd ed., John Wiley, New York.

J. D. Buckmaster and G. S. S. Ludford, 1982, *Theory of Laminar Flames*, Cambridge University Press, Cambridge, UK, Chapter 1.

S. R. de Groot and P. Mazur, 1984, *Non-Equilibrium Thermodynamics*, Dover, New York, Chapters 1–3.

G. Dixon-Lewis, 1968, Flame structure and flame reaction kinetics. II. Transport phenomena in multicomponent systems, *Proceedings of the Royal Society of London. Series A. Mathematical and Physical Sciences*, 307(1488): 111–135.

R. J. Kee, M. E. Coltrin, and P. Glarborg, 2003, *Chemically Reacting Flow: Theory and Practice*, John Wiley, Hoboken, NJ, Chapters 2–3.

K. K. Kuo, 2005, *Principles of Combustion*, 2nd ed., John Wiley, New York, Chapter 3.

S. H. Lam and J. Bellan, 2003, On de-coupling of Shvab-Zel'dovich variables in the presence of diffusion, *Combustion and Flame*, 132(4): 691–696.

C. K. Law, 2006, *Combustion Physics*, Cambridge University Press, Cambridge, UK, Chapter 2.

H. J. Merk, 1959, The macroscopic equations for simultaneous heat and mass transfer in isotropic continuous and closed systems, *Applied Scientific Research*, 8(1): 73–99.

I. Müller and W. Weiss, 2005, *Entropy and Energy*, Springer, New York, p. 182.
T. Poinsot and D. Veynante, 2005, *Theoretical and Numerical Combustion*, 2nd ed., Edwards, Flourtown, PA, Chapter 1.
I. Prigogine, 1967, *Introduction to Thermodynamics of Irreversible Processes*, Interscience, New York.
D. E. Rosner, 2000, *Transport Processes in Chemically Reacting Flow Systems*, Dover, New York, Chapters 2,3.
J. Warnatz, U. Maas, and R. W. Dibble, 2006, *Combustion*, 4th ed., Springer, New York, Chapter 12.
F. A. Williams, 1985, *Combustion Theory*, 2nd ed., Benjamin-Cummings, Menlo Park, CA, Chapter 1.
L. C. Woods, 1975, *The Thermodynamics of Fluid Systems*, Oxford University Press, Oxford, UK.

# 8 Simple Linear Combustion

*Consider a linear cow . . .*
— Anonymous

The reactive Navier-Stokes equations have a formidable complexity that impedes causal mechanistic understanding of numerical solution features. To acquire a better understanding of the fundamentals of the interplay of the mechanisms of advection, reaction, and diffusion in combustion, a highly simplified linear model problem is employed in this chapter. The discussion is largely an extension of that given by Al-Khateeb et al. (2013).[1] We use this model to identify the relevant time and length scales of combustion and associate those time and length scales with physical parameters. Our key result is that the various length scales needed to be captured $\ell_i$ are related to the time scales of reaction by $\ell_i = \sqrt{\mathcal{D}\tau_i}$, where $\mathcal{D}$ is a generalized diffusivity. Fast reactions thus induce fine length scales. We discuss this in the context of how this should influence choices of time and space discretization in numerical methods. In general, one can conclude that to achieve a so-called Direct Numerical Simulation (DNS) one must have time and space discretizations sufficiently small to capture the physics that has been modeled. The complicated interplay of advection, reaction, and diffusion can pose challenges for DNS of many multiscale combustion problems (Hundsdorfer and Verwer, 2003; Manley et al., 2008). It is this interplay that dictates one of nature's most complicated and important phenomena, turbulent combustion. The complexity of that subject is such that it is beyond the scope of this book. Many dedicated studies exist, such as the monographs of Peters (2000), Fox (2003), and Kuo and Acharya (2012) or the compendium of Echekki and Mastorakos (2011).

## 8.1 Single Reaction

Consider the following linear advection-reaction-diffusion problem:

$$\frac{\partial}{\partial t}Y(x,t) + u\frac{\partial}{\partial x}Y(x,t) = \mathcal{D}\frac{\partial^2}{\partial x^2}Y(x,t) - a(Y(x,t) - Y_{eq}), \tag{8.1}$$

$$Y(x,0) = Y_0, \quad Y(0,t) = Y_0, \quad \frac{\partial Y}{\partial x}(\infty, t) \to 0, \tag{8.2}$$

[1] A. N. Al-Khateeb, J. M. Powers, and S. Paolucci, 2013, Analysis of the spatio-temporal scales of laminar premixed flames near equilibrium, *Combustion Theory and Modelling*, 17(1): 76–108, reprinted by permission of the publisher, Taylor & Francis Ltd., http://www.tandfonline.com/.

where the independent variables are time $t > 0$ and distance $x \in (0, \infty)$. Here, $Y(x,t) > 0$ is a scalar that can be loosely considered to be a mass fraction, $u > 0$ is a constant advective wave speed, $\mathcal{D} > 0$ is a constant diffusion coefficient, $a > 0$ is the chemical consumption rate constant, $Y_0 > 0$ is a constant, as is $Y_{eq} > 0$. We note that $Y(x,t) = Y_{eq}$ is a solution iff $Y_0 = Y_{eq}$. For $Y_0 \neq Y_{eq}$, we may expect a boundary layer in which $Y$ adjusts from its value at $x = 0$ to $Y_{eq}$, the equilibrium value.

### 8.1.1 Spatially Homogeneous Solution

The spatially homogeneous version of Eqs. (8.1) and (8.2) is

$$\frac{dY(t)}{dt} = -a(Y(t) - Y_{eq}), \qquad Y|_{t=0} = Y_0, \tag{8.3}$$

which has solution

$$Y(t) = Y_{eq} + (Y_0 - Y_{eq})e^{-at}. \tag{8.4}$$

The time scale $\tau$ over which $Y$ evolves is

$$\tau = 1/a. \tag{8.5}$$

This time scale serves as an upper bound for the required time step to capture the dynamics in a numerical simulation. Because there is only one dependent variable in this problem, the temporal spectrum contains only one time scale. Consequently, this formulation of the system is not temporally stiff.

**EXAMPLE 8.1**
For a spatially homogeneous solution, plot the solution $Y(t)$ to Eq. (8.3) if $a = 10^8\ \text{s}^{-1}$, $Y_0 = 0.1$, and $Y_{eq} = 0.001$.

For these parameters, the solution from Eq. (8.4) is

$$Y(t) = 0.001 + 0.099e^{-(10^8\ \text{s}^{-1})t}. \tag{8.6}$$

The time scale of relaxation is given by Eq. (8.5) and is

$$\tau = 1/a = 1/(10^8\ \text{s}^{-1}) = 10^{-8}\ \text{s}. \tag{8.7}$$

A plot of $Y(t)$ is given in Fig. 8.1. It is seen that for early time, $t \ll \tau, Y$ is near $Y_0$. Significant relaxation of $Y$ occurs when $t \approx \tau$. For $t \gg \tau$, we see $Y \to Y_{eq}$. The plot is presented on a log-log scale that better highlights the dynamics. In particular, when examined over orders of magnitude, the reaction event is seen in perspective as a sharp change from one state to another. Reaction dynamics are typically characterized by a near-constant, "frozen" state, seemingly in equilibrium. This pseudo-equilibrium is punctuated by a reaction event, during which the system relaxes to a final true equilibrium. The notion of "punctuated equilibrium" is also well known in modern evolutionary biology (Eldredge and Gould, 1972), usually for far longer time scale events, and has an analog with our chemical reaction dynamics.

### 8.1.2 Steady Solution

A simple means to determine the relevant length scales, and consequently, an upper bound for the required spatial grid resolution, is to obtain the steady structure $Y(x)$,

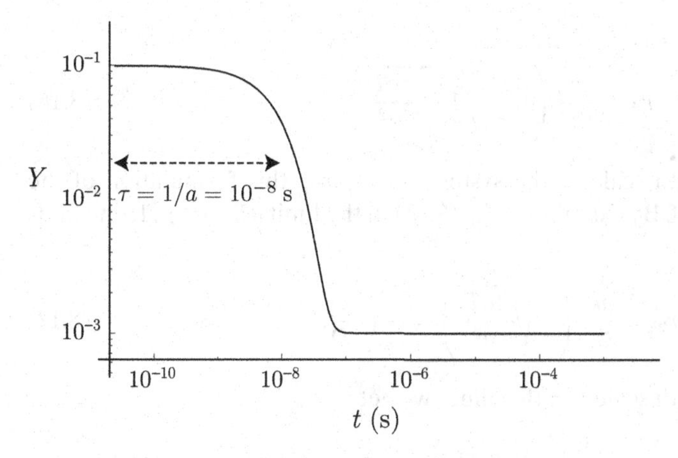


Figure 8.1. Mass fraction versus time for spatially homogeneous problem with simple one-step linear kinetics.

which is governed by the time-independent version of Eqs. (8.1):

$$u\frac{dY(x)}{dx} = \mathcal{D}\frac{d^2Y(x)}{dx^2} - a(Y(x) - Y_{eq}), \quad Y|_{x=0} = Y_0, \quad \left.\frac{dY}{dx}\right|_{x\to\infty} \to 0. \tag{8.8}$$

Assuming solutions of the form $Y(x) = Y_{eq} + Ce^{rx}$, we rewrite Eq. (8.8) as

$$uCre^{rx} = \mathcal{D}Cr^2e^{rx} - aCe^{rx} \tag{8.9}$$

and, through simplification, are led to a characteristic polynomial of

$$ur = \mathcal{D}r^2 - a, \tag{8.10}$$

which has roots

$$r = \frac{u}{2\mathcal{D}}\left(1 \pm \sqrt{1 + \frac{4a\mathcal{D}}{u^2}}\right). \tag{8.11}$$

Taking $r_1$ to denote the "plus" root, for which $r_1 > 0$, and $r_2$ to denote the "minus" root, for which $r_2 < 0$, the two solutions can be linearly combined to take the form

$$Y(x) = Y_{eq} + C_1e^{r_1x} + C_2e^{r_2x}, \tag{8.12}$$

where $C_1$ and $C_2$ are constants. Taking the spatial derivative of Eq. (8.12), we get

$$\frac{dY}{dx} = C_1r_1e^{r_1x} + C_2r_2e^{r_2x}. \tag{8.13}$$

In the limit of large positive $x$, the boundary condition at infinity in Eq. (8.8) requires the derivative to vanish giving

$$\lim_{x\to\infty}\frac{dY}{dx} = 0 = \lim_{x\to\infty}\left(C_1r_1e^{r_1x} + C_2r_2e^{r_2x}\right). \tag{8.14}$$

Because $r_1 > 0$, we must insist that $C_1 = 0$. Then, enforcing that $Y(0) = Y_0$, we find that the solution of Eq. (8.8) is

$$Y(x) = Y_{eq} + (Y_0 - Y_{eq})e^{r_2x}, \tag{8.15}$$

where

$$r_2 = \frac{u}{2\mathcal{D}}\left(1 - \sqrt{1 + \frac{4a\mathcal{D}}{u^2}}\right). \tag{8.16}$$

Hence, there is one length scale in the system, $\ell \equiv 1/|r_2|$; this formulation of the system is not spatially stiff. By examining Eq. (8.16) in the limit $a\mathcal{D}/u^2 \gg 1$, one finds that

$$r_2 \approx \frac{u}{2\mathcal{D}}\left(-\sqrt{\frac{4a\mathcal{D}}{u^2}}\right) = -\sqrt{\frac{a}{\mathcal{D}}}. \tag{8.17}$$

Thus, solving for the length scale $\ell$ in this limit, we get

$$\ell = \frac{1}{|r_2|} \approx \sqrt{\frac{\mathcal{D}}{a}} = \sqrt{\mathcal{D}\tau}, \tag{8.18}$$

where $\tau = 1/a$ is the time scale from spatially homogeneous reaction, Eq. (8.5). So, this length scale $\ell$ reflects the inherent physics of coupled advection-reaction-diffusion. In the limit of $a\mathcal{D}/u^2 \ll 1$, one finds $r_2 \to 0$, $\ell \to \infty$, and $Y(x) \to Y_0$, a constant.

**EXAMPLE 8.2**

For a steady solution, plot $Y(x)$ if $a = 10^8\ \text{s}^{-1}, u = 10^2\ \text{cm/s}, \mathcal{D} = 10^1\ \text{cm}^2/\text{s}, Y_0 = 10^{-1}$, and $Y_{eq} = 10^{-3}$.

For this system, we have from Eq. (8.16) that

$$r_2 = \frac{u}{2\mathcal{D}}\left(1 - \sqrt{1 + \frac{4a\mathcal{D}}{u^2}}\right), \tag{8.19}$$

$$= \frac{10^2\ \frac{\text{cm}}{\text{s}}}{2\left(10^1\ \frac{\text{cm}^2}{\text{s}}\right)}\left(1 - \sqrt{1 + \frac{4\left(10^8\ \text{s}^{-1}\right)\left(10^1\ \frac{\text{cm}^2}{\text{s}}\right)}{\left(10^2\ \frac{\text{cm}}{\text{s}}\right)^2}}\right) = -3.2 \times 10^3\ \text{cm}^{-1}. \tag{8.20}$$

Because $a\mathcal{D}/u^2 = 10^5 \gg 1$, $r_2$ is well estimated by Eq. (8.17):

$$r_2 \approx -\sqrt{\frac{a}{\mathcal{D}}} = -\sqrt{\frac{10^8\ \text{s}^{-1}}{10^1\ \frac{\text{cm}^2}{\text{s}}}} = -3.2 \times 10^3\ \text{cm}^{-1}. \tag{8.21}$$

Then, from Eq. (8.15), the solution is

$$Y(x) = Y_{eq} + (Y_0 - Y_{eq})e^{r_2 x} = 0.001 + 0.099e^{-(3.2\times 10^3\ \text{cm}^{-1})x}. \tag{8.22}$$

The length scale of reaction is estimated by Eq. (8.18):

$$\ell = \frac{1}{|r_2|} \approx \sqrt{\frac{\mathcal{D}}{a}} = \sqrt{\mathcal{D}\tau} = \sqrt{\left(10^1\ \frac{\text{cm}^2}{\text{s}}\right)\left(10^{-8}\ \text{s}\right)} = 3.2 \times 10^{-4}\ \text{cm}. \tag{8.23}$$

A plot of $Y(x)$ is given in Fig. 8.2.

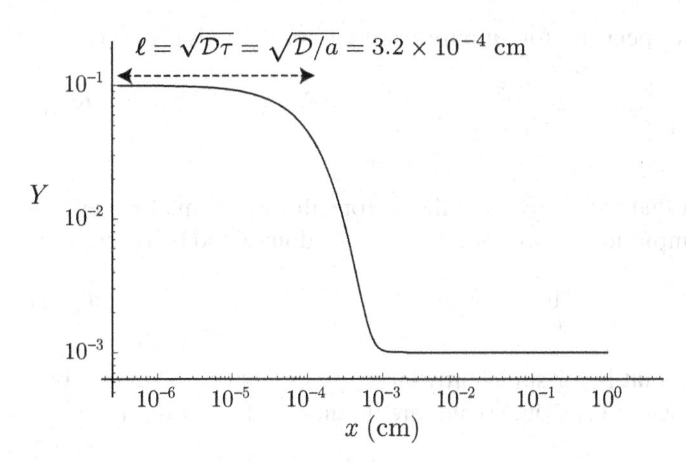


Figure 8.2. Mass fraction versus distance for steady advection-reaction-diffusion problem with simple one-step linear kinetics.

### 8.1.3 Spatiotemporal Solution

Now, for Eq. (8.1), it is possible to find a simple analytic expression for the continuous spectrum of time scales $\tau$ as a function of a particular linearly independent Fourier mode's wavenumber $\hat{k}$. A Fourier mode with wavenumber $\hat{k}$ has wavelength $\lambda = 2\pi/\hat{k}$. Assume a solution of the form

$$Y(x,t) = Y_{eq} + B(t)e^{i\hat{k}x}, \tag{8.24}$$

where $B(t)$ is the time-dependent amplitude of the chosen mode. Recall that $e^{i\hat{k}x} = \cos\hat{k}x + i\sin\hat{k}x$. We thus see spatial oscillations are built into our assumed functional form for $Y$. The fact that our chosen form also contains an imaginary part is inconsequential. It does simplify some of the notation, and one can always confine attention to the real part of the solution.

For this problem that considers a single Fourier mode, it does not make sense to impose the initial condition of Eq. (8.2). Substituting Eq. (8.24) into Eq. (8.1) gives

$$\frac{dB}{dt}e^{i\hat{k}x} + i\hat{k}uBe^{i\hat{k}x} = -\mathcal{D}\hat{k}^2Be^{i\hat{k}x} - aBe^{i\hat{k}x}, \tag{8.25}$$

$$\frac{dB}{dt} + i\hat{k}uB = -\mathcal{D}\hat{k}^2B - aB. \tag{8.26}$$

This takes the form

$$\frac{dB(t)}{dt} = -\beta B(t), \qquad B(0) = B_0, \tag{8.27}$$

where

$$\beta = a\left(1 + \frac{\mathcal{D}\hat{k}^2}{a} + \frac{i\hat{k}u}{a}\right), \tag{8.28}$$

and we have imposed $B_0$ as an initial value. This has solution

$$B(t) = B_0e^{-\beta t}. \tag{8.29}$$

The complete solution is easily shown to be

$$Y(x,t) = Y_{eq} + B_0e^{i\hat{k}(x-ut)-\mathcal{D}\hat{k}^2t-at}. \tag{8.30}$$

The continuous time scale spectrum for amplitude growth or decay is given by

$$\tau = \frac{1}{|\mathrm{Re}\,(\beta)|} = \frac{1}{a\left(1 + \frac{\mathcal{D}\hat{k}^2}{a}\right)}, \qquad 0 < \hat{k} \in \mathbb{R}. \tag{8.31}$$

From Eq. (8.31), it is clear that for $\mathcal{D}\hat{k}^2/a \ll 1$; that is, for sufficiently small wavenumbers, the time scales of amplitude growth or decay will be dominated by reaction

$$\lim_{\hat{k}\to 0} \tau = 1/a. \tag{8.32}$$

However, for $\mathcal{D}\hat{k}^2/a \gg 1$, that is, for sufficiently large wavenumbers or small wavelengths, the amplitude growth/decay time scales are dominated by diffusion:

$$\lim_{\hat{k}\to\infty} \tau = \frac{1}{\mathcal{D}\hat{k}^2} = \frac{1}{\mathcal{D}}\left(\frac{\lambda}{2\pi}\right)^2. \tag{8.33}$$

From Eq. (8.31), we see that a balance between reaction and diffusion exists for $\hat{k} = \sqrt{a/\mathcal{D}}$. In terms of wavelength, and recalling Eq. (8.18), we see the balance at

$$\lambda/(2\pi) = 1/\hat{k} = \sqrt{\mathcal{D}/a} = \sqrt{\mathcal{D}\tau} = \ell, \tag{8.34}$$

where $\ell = 1/\hat{k}$ is proportional to the wavelength.

The oscillatory behavior is of lesser importance. The continuous time scale spectrum for oscillatory mode, $\tau_O$ is given by

$$\tau_O = 1/|\mathrm{Im}(\beta)| = 1/(\hat{k}u). \tag{8.35}$$

As $\hat{k} \to 0, \tau_O \to \infty$. While $\tau_O \to 0$ as $\hat{k} \to \infty$, it approaches at a rate $\sim 1/\hat{k}$, in contrast to the more demanding time scale of diffusion that approaches zero at a faster rate $\sim 1/\hat{k}^2$. Thus, it is clear that advection does not play a role in determining the limiting values of the time scale spectrum; reaction and diffusion are the major players. Last, it is easy to show in the absence of diffusion that the length scale where reaction effects balance advection effects is found at

$$\ell = u/a = u\tau, \tag{8.36}$$

where $\tau = 1/a$ is the time scale from spatially homogeneous chemistry.

**EXAMPLE 8.3**

Examine the behavior of the time scales as a function of the length scales for the linear advective-reactive-diffusive system characterized by $a = 10^8$ 1/s, $\mathcal{D} = 10^1$ cm$^2$/s, $u = 10^2$ cm/s.

These values are loosely motivated by values for gas phase kinetics of physical systems. For these values, we find the estimate from Eq. (8.18) for the length scale where reaction balances diffusion as

$$\ell = \sqrt{\mathcal{D}\tau} = \sqrt{\frac{\mathcal{D}}{a}} = \sqrt{\left(10^1\ \frac{\mathrm{cm}^2}{\mathrm{s}}\right)(10^{-8}\ \mathrm{s})} = 3.16228 \times 10^{-4}\ \mathrm{cm}. \tag{8.37}$$

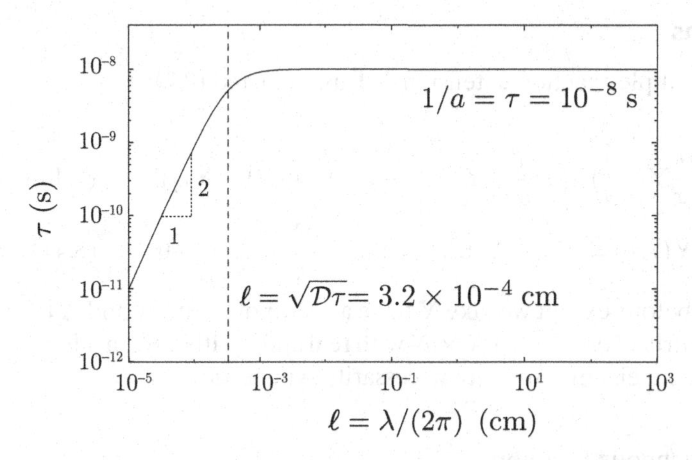


Figure 8.3. Time scale spectrum versus length scale for the simple advection-reaction-diffusion model. Adapted from Al-Khateeb et al. (2013), reprinted by permission of the publisher, Taylor & Francis Ltd., http://www.tandfonline.com/.

A plot of $\tau$ versus $\ell = \lambda/(2\pi)$ from Eq. (8.31),

$$\tau = \frac{1}{a\left(1 + \frac{\mathcal{D}\hat{k}^2}{a}\right)} = \frac{1}{a + \frac{\mathcal{D}}{\ell^2}}, \tag{8.38}$$

is given in Fig. 8.3. For long wavelengths, the time scales are determined by reaction; for fine wavelengths, the time scale's falloff is dictated by diffusion, and our simple formula for the critical $\ell = \sqrt{\mathcal{D}\tau}$, illustrated as a dashed line, predicts the transition well. For small $\ell$, it is seen that a one-decade decrease in $\ell$ induces a two-decade decrease in $\tau$, consistent with the prediction of Eq. (8.33): $\lim_{\hat{k}\to\infty} (\ln \tau) \sim 2\ln(\ell) - \ln(\mathcal{D})$ . Last, over the same range of $\ell$, the oscillatory time scales induced by advection are orders of magnitude less demanding and are thus not included in the plot.

The results of this simple analysis can be summarized as follows:

- *Long wavelength spatial disturbances have time dynamics that are dominated by chemistry; each spatial point behaves as an isolated spatially homogeneous reactor.*
- *Short wavelength spatial disturbances have time dynamics that are dominated by diffusion.*
- *Intermediate wavelength spatial disturbances have time dynamics determined by fully coupled combination diffusion and chemistry. The critical intermediate length scale where this balance exists is given by $\ell = \sqrt{\mathcal{D}\tau}$.*
- *A DNS of a combustion process with advection-reaction-diffusion requires*

$$\Delta t < \tau, \qquad \Delta x < \sqrt{\mathcal{D}\tau}. \tag{8.39}$$

*Less restrictive choices will not capture time dynamics and spatial structures inherent in the continuum model. Advection usually plays a secondary role in determining time dynamics.*

This argument is by no means new and is effectively the same as that given by Landau and Lifshitz (1959) in their chapter on combustion.

## 8.2 Multiple Reactions

Let us next consider a multiple-reaction extension to Eqs. (8.1) and (8.2):

$$\frac{\partial}{\partial t}\mathbf{Y}(x,t) + u\frac{\partial}{\partial x}\mathbf{Y}(x,t) = \mathcal{D}\frac{\partial^2}{\partial x^2}\mathbf{Y}(x,t) - \mathbf{A}\cdot(\mathbf{Y}(x,t) - \mathbf{Y}_{eq}), \tag{8.40}$$

$$\mathbf{Y}(x,0) = \mathbf{Y}_0, \quad \mathbf{Y}(0,t) = \mathbf{Y}_0, \quad \frac{\partial \mathbf{Y}}{\partial x}(\infty,t) \to \mathbf{0}. \tag{8.41}$$

Here, all variables are as before, except we take $\mathbf{Y}$ to be a vector of length $N$ and $\mathbf{A}$ to be a constant full rank matrix of dimension $N \times N$ with real and positive eigenvalues, with $N$ linearly independent eigenvectors not necessarily symmetric.

### 8.2.1 Spatially Homogeneous Solution

The spatially homogeneous version of Eqs. (8.40) and (8.41) is

$$\frac{d\mathbf{Y}}{dt} = -\mathbf{A}\cdot(\mathbf{Y} - \mathbf{Y}_{eq}), \qquad \mathbf{Y}(0) = \mathbf{Y}_0. \tag{8.42}$$

Because of the way $\mathbf{A}$ has been defined, it can be decomposed as

$$\mathbf{A} = \mathbf{S}\cdot\boldsymbol{\sigma}\cdot\mathbf{S}^{-1}, \tag{8.43}$$

where S is an $N \times N$ matrix whose columns are populated by the $N$ linearly independent eigenvectors of $\mathbf{A}$, and $\boldsymbol{\sigma}$ is the diagonal matrix with the $N$ positive eigenvalues, $\sigma_1, \ldots, \sigma_N$, of $\mathbf{A}$ on its diagonal. Substitute Eq. (8.43) into Eq. (8.40), take advantage of the fact that $d\mathbf{Y}_{eq}/dt = \mathbf{0}$, and operate to find

$$\frac{d}{dt}(\mathbf{Y} - \mathbf{Y}_{eq}) = -\underbrace{\mathbf{S}\cdot\boldsymbol{\sigma}\cdot\mathbf{S}^{-1}}_{\mathbf{A}}\cdot(\mathbf{Y} - \mathbf{Y}_{eq}), \tag{8.44}$$

$$\mathbf{S}^{-1}\cdot\frac{d}{dt}(\mathbf{Y} - \mathbf{Y}_{eq}) = -\mathbf{S}^{-1}\cdot\mathbf{S}\cdot\boldsymbol{\sigma}\cdot\mathbf{S}^{-1}\cdot(\mathbf{Y} - \mathbf{Y}_{eq}), \tag{8.45}$$

$$\frac{d}{dt}\left(\mathbf{S}^{-1}\cdot(\mathbf{Y} - \mathbf{Y}_{eq})\right) = -\boldsymbol{\sigma}\cdot\mathbf{S}^{-1}\cdot(\mathbf{Y} - \mathbf{Y}_{eq}). \tag{8.46}$$

Take now

$$\mathbf{Z} = \mathbf{S}^{-1}\cdot(\mathbf{Y} - \mathbf{Y}_{eq}), \tag{8.47}$$

so that

$$\frac{d\mathbf{Z}}{dt} = -\boldsymbol{\sigma}\cdot\mathbf{Z}. \tag{8.48}$$

Our initial condition becomes

$$\mathbf{Z}(0) = \mathbf{S}^{-1}\cdot(\mathbf{Y}_0 - \mathbf{Y}_{eq}) = \mathbf{Z}_0. \tag{8.49}$$

The solution is

$$\mathbf{Z}(t) = e^{-\boldsymbol{\sigma}t}\cdot\mathbf{Z}_0, \tag{8.50}$$

$$\mathbf{S}^{-1}\cdot(\mathbf{Y}(t) - \mathbf{Y}_{eq}) = e^{\boldsymbol{\sigma}t}\cdot\mathbf{S}^{-1}\cdot(\mathbf{Y}_0 - \mathbf{Y}_{eq}), \tag{8.51}$$

$$\mathbf{Y}(t) = \mathbf{Y}_{eq} + \mathbf{S}\cdot e^{\boldsymbol{\sigma}t}\cdot\mathbf{S}^{-1}\cdot(\mathbf{Y}_0 - \mathbf{Y}_{eq}). \tag{8.52}$$

Expanded, one can say

$$\begin{pmatrix} Y_1(t) \\ \vdots \\ Y_N(t) \end{pmatrix} = \begin{pmatrix} Y_{1e} \\ \vdots \\ Y_{Ne} \end{pmatrix} + \begin{pmatrix} \vdots & \vdots & \vdots \\ \mathbf{s}_1 & \vdots & \mathbf{s}_N \\ \vdots & \vdots & \vdots \end{pmatrix} \begin{pmatrix} e^{-\sigma_1 t} & 0 & 0 \\ 0 & \ddots & 0 \\ 0 & 0 & e^{-\sigma_N t} \end{pmatrix}$$

$$\cdot \begin{pmatrix} \dots & \mathbf{s}_1^{-1} & \dots \\ \dots & \dots & \dots \\ \dots & \mathbf{s}_N^{-1} & \dots \end{pmatrix} \begin{pmatrix} Y_{1o} - Y_{1eq} \\ \vdots \\ Y_{No} - Y_{Neq} \end{pmatrix}. \quad (8.53)$$

Here, $\mathbf{s}_i, i = 1, \dots, N$, are eigenvectors of $\mathbf{A}$. There are $N$ time scales $\tau_i = 1/\sigma_i, i = 1, \dots, N$, on which the solution evolves. Each dependent variable $Y_i(t), i = 1, \dots, N$, can evolve on each of the time scales.

**EXAMPLE 8.4**

For a case where $N = 2$, examine the solution to Eqs. (8.42) if

$$\mathbf{A} = \begin{pmatrix} 1000000\ \mathrm{s}^{-1} & -99000000\ \mathrm{s}^{-1} \\ -99000000\ \mathrm{s}^{-1} & 99010000\ \mathrm{s}^{-1} \end{pmatrix}, \quad \mathbf{Y}_0 = \begin{pmatrix} 10^{-2} \\ 10^{-1} \end{pmatrix}, \quad \mathbf{Y}_{eq} = \begin{pmatrix} 10^{-5} \\ 10^{-6} \end{pmatrix}. \quad (8.54)$$

Thus, solve

$$\frac{dY_1}{dt} = -(1000000\ \mathrm{s}^{-1})(Y_1 - 10^{-5}) + (99000000\ \mathrm{s}^{-1})(Y_2 - 10^{-6}), \quad Y_1(0) = 10^{-2}, \quad (8.55)$$

$$\frac{dY_2}{dt} = (99000000\ \mathrm{s}^{-1})(Y_1 - 10^{-5}) - (99010000\ \mathrm{s}^{-1})(Y_2 - 10^{-6}), \quad Y_2(0) = 10^{-1}. \quad (8.56)$$

Straightforward calculation reveals the eigenvalues of $\mathbf{A}$ to be

$$\sigma_1 = 10^8\ \mathrm{s}^{-1}, \qquad \sigma_2 = 10^4\ \mathrm{s}^{-1}. \quad (8.57)$$

Thus the time scales of reaction $\tau_i = 1/\sigma_i$ are

$$\tau_1 = 10^{-8}\ \mathrm{s}, \qquad \tau_2 = 10^{-4}\ \mathrm{s}. \quad (8.58)$$

Clearly the ratio of time scales is large with a stiffness ratio of $10^4$; thus, this is obviously a multiscale problem. It is not easy to infer either the time scales or the stiffness ratio from simple examination of the numerical values of $\mathbf{A}$. Instead, one must perform the eigenvalue calculation.

It is easily shown that a diagonal decomposition of $\mathbf{A}$ is given by

$$\mathbf{A} = \underbrace{\begin{pmatrix} -1 & 1 \\ 1 & \frac{1}{100} \end{pmatrix}}_{\mathbf{S}} \underbrace{\begin{pmatrix} 10^8 & 0 \\ 0 & 10^4 \end{pmatrix}}_{\boldsymbol{\sigma}} \underbrace{\begin{pmatrix} -\frac{1}{101} & \frac{100}{101} \\ \frac{100}{101} & \frac{100}{101} \end{pmatrix}}_{\mathbf{S}^{-1}}. \quad (8.59)$$

Detailed calculation as given in the preceding section shows that the exact solution is given by

$$Y_1(t) = -\frac{9891 e^{-(10^8\ \mathrm{s}^{-1})t}}{100000} + \frac{1089 e^{-(10^4\ \mathrm{s}^{-1})t}}{10000} + 10^{-5}, \quad (8.60)$$

$$Y_2(t) = \frac{9891 e^{-(10^8\ \mathrm{s}^{-1})t}}{100000} + \frac{1089 e^{-(10^4\ \mathrm{s}^{-1})t}}{1000000} + 10^{-6}. \quad (8.61)$$

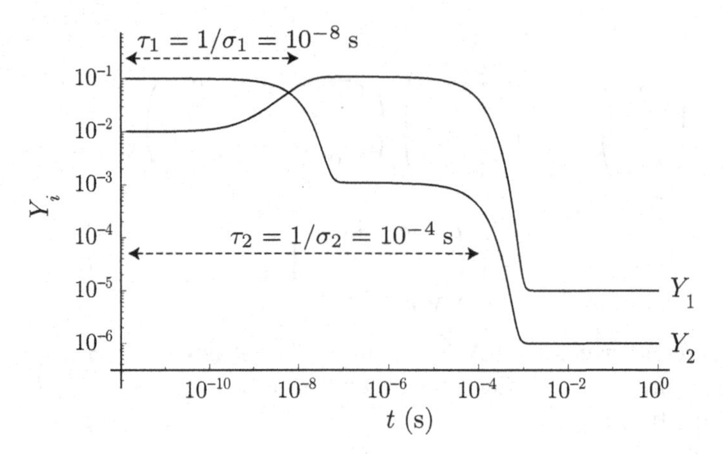


Figure 8.4. Mass fraction versus time for spatially homogeneous problem with simple two-step linear kinetics.

A plot of $Y_1(t)$ and $Y_2(t)$ is given in Fig. 8.4. Clearly, for $t < \tau_1 = 10^{-8}$ s, both $Y_1(t)$ and $Y_2(t)$ are frozen at the initial values. When $t \approx \tau_1 = 10^{-8}$ s, the first reaction mode begins to have an effect. Both $Y_1$ and $Y_2$ then maintain intermediate pseudo-equilibrium values for $t \in [\tau_1, \tau_2]$. When $t \approx \tau_2 = 10^{-4}$ s, both $Y_1$ and $Y_2$ rapidly approach their true equilibrium values.

### 8.2.2 Steady Solution

The time-independent version of Eqs. (8.40) and (8.41) is

$$u\frac{d\mathbf{Y}}{dx} = \mathcal{D}\frac{d^2\mathbf{Y}}{dx^2} - \mathbf{A}\cdot(\mathbf{Y} - \mathbf{Y}_{eq}), \quad \mathbf{Y}(0) = \mathbf{Y}_0, \quad \lim_{x\to\infty}\frac{d\mathbf{Y}}{dx} \to \mathbf{0}. \tag{8.62}$$

Let us again employ Eq. (8.43) and the fact that $d\mathbf{Y}_{eq}/dx = \mathbf{0}$ to recast Eq. (8.62) as

$$u\frac{d}{dx}(\mathbf{Y} - \mathbf{Y}_{eq}) = \mathcal{D}\frac{d^2}{dx^2}(\mathbf{Y} - \mathbf{Y}_{eq}) - \mathbf{S}\cdot\boldsymbol{\sigma}\cdot\mathbf{S}^{-1}\cdot(\mathbf{Y} - \mathbf{Y}_{eq}). \tag{8.63}$$

Next operate on both sides of Eq. (8.63) with the constant matrix $\mathbf{S}^{-1}$ and then use Eq. (8.47) to get

$$u\mathbf{S}^{-1}\cdot\frac{d}{dx}(\mathbf{Y} - \mathbf{Y}_{eq}) = \mathcal{D}\mathbf{S}^{-1}\cdot\frac{d^2}{dx^2}(\mathbf{Y} - \mathbf{Y}_{eq}) - \mathbf{S}^{-1}\cdot\mathbf{S}\cdot\boldsymbol{\sigma}\cdot\mathbf{S}^{-1}\cdot(\mathbf{Y} - \mathbf{Y}_{eq}), \tag{8.64}$$

$$u\frac{d}{dx}\left(\mathbf{S}^{-1}\cdot(\mathbf{Y} - \mathbf{Y}_{eq})\right) = \mathcal{D}\frac{d^2}{dx^2}\left(\mathbf{S}^{-1}\cdot(\mathbf{Y} - \mathbf{Y}_{eq})\right) - \boldsymbol{\sigma}\cdot\mathbf{S}^{-1}\cdot(\mathbf{Y} - \mathbf{Y}_{eq}), \tag{8.65}$$

$$u\frac{d\mathbf{Z}}{dx} = \mathcal{D}\frac{d^2\mathbf{Z}}{dx^2} - \boldsymbol{\sigma}\cdot\mathbf{Z}. \tag{8.66}$$

Similar to Eq. (8.49), the boundary conditions become

$$\mathbf{Z}(0) = \mathbf{Z}_0, \quad \lim_{x\to\infty}\frac{d\mathbf{Z}}{dx} \to \mathbf{0}. \tag{8.67}$$

Importantly, these equations are now uncoupled. For example, the $i$th equation and boundary conditions become

$$u\frac{dZ_i(x)}{dx} = \mathcal{D}\frac{d^2Z_i(x)}{dx^2} - \sigma_i Z_i(x), \qquad Z_i|_{x=0} = Z_{i0}, \quad \left.\frac{dZ_i}{dx}\right|_{x\to\infty} \to 0. \quad (8.68)$$

The solution can then be directly inferred from Eqs. (8.15) and (8.16) to be

$$Z_i(x) = Z_{i,o}e^{r_{2,i}x}, \qquad i = 1,\ldots,N, \quad (8.69)$$

where

$$r_{2,i} = \frac{u}{2\mathcal{D}}\left(1 - \sqrt{1 + \frac{4\sigma_i\mathcal{D}}{u^2}}\right), \qquad i = 1,\ldots,N. \quad (8.70)$$

Analogously, in the limit where $\sigma_i\mathcal{D}/u^2 \gg 1$, we can infer

$$\ell_i = \sqrt{\mathcal{D}\tau_i}, \qquad i = 1,\ldots,N, \quad (8.71)$$

with the reaction time scale $\tau_i$ taken as

$$\tau_i = 1/\sigma_i, \qquad i = 1,\ldots,N. \quad (8.72)$$

Then, knowing $\mathbf{Z}(x)$, one can use Eq. (8.47) to form

$$\mathbf{Y}(x) = \mathbf{Y}_{eq} + \mathbf{S}\cdot\mathbf{Z}(x). \quad (8.73)$$

Thus, any $Y_i(x)$ can be expected to relax over all $N$ values of length scales $\ell_i$.

**EXAMPLE 8.5**

For a case where $N = 2$, $\mathcal{D} = 10^1$ cm$^2$/s, $u = 10^2$ cm/s, examine the solution to Eqs. (8.62) if

$$\mathbf{A} = \begin{pmatrix} 1000000\,\mathrm{s}^{-1} & -99000000\,\mathrm{s}^{-1} \\ -99000000\,\mathrm{s}^{-1} & 99010000\,\mathrm{s}^{-1}\end{pmatrix}, \quad \mathbf{Y}_0 = \begin{pmatrix}10^{-2}\\10^{-1}\end{pmatrix}, \quad \mathbf{Y}_{eq} = \begin{pmatrix}10^{-5}\\10^{-6}\end{pmatrix}. \quad (8.74)$$

Thus, solve

$$\left(10^2\,\frac{\mathrm{cm}}{\mathrm{s}}\right)\frac{dY_1}{dx} = \left(10^1\,\frac{\mathrm{cm}^2}{\mathrm{s}}\right)\frac{d^2Y_1}{dx^2} - (1000000\,\mathrm{s}^{-1})(Y_1 - 10^{-5}) + (99000000\,\mathrm{s}^{-1})(Y_2 - 10^{-6}), \quad (8.75)$$

$$\left(10^2\,\frac{\mathrm{cm}}{\mathrm{s}}\right)\frac{dY_2}{dx} = \left(10^1\,\frac{\mathrm{cm}^2}{\mathrm{s}}\right)\frac{d^2Y_2}{dx^2} + (99000000\,\mathrm{s}^{-1})(Y_1 - 10^{-5}) - (99010000\,\mathrm{s}^{-1})(Y_2 - 10^{-6}), \quad (8.76)$$

$$Y_1(0) = 10^{-2}, \quad Y_2(0) = 10^{-1}, \quad \lim_{x\to\infty}\frac{dY_1}{dx} = 0, \quad \lim_{x\to\infty}\frac{dY_2}{dx} = 0. \quad (8.77)$$

Employing the transformation from $\mathbf{Y}$ to $\mathbf{Z}$ along with $\mathbf{S}^{-1}$ as given in Eq. (8.59), our system can be rewritten as

$$\left(10^2\,\frac{\mathrm{cm}}{\mathrm{s}}\right)\frac{dZ_1}{dx} = \left(10^1\,\frac{\mathrm{cm}^2}{\mathrm{s}}\right)\frac{d^2Z_1}{dx^2} - (10^8\,\mathrm{s}^{-1})Z_1, \quad (8.78)$$

$$\left(10^2\,\frac{\mathrm{cm}}{\mathrm{s}}\right)\frac{dZ_2}{dx} = \left(10^1\,\frac{\mathrm{cm}^2}{\mathrm{s}}\right)\frac{d^2Z_2}{dx^2} - (10^4\,\mathrm{s}^{-1})Z_2, \quad (8.79)$$

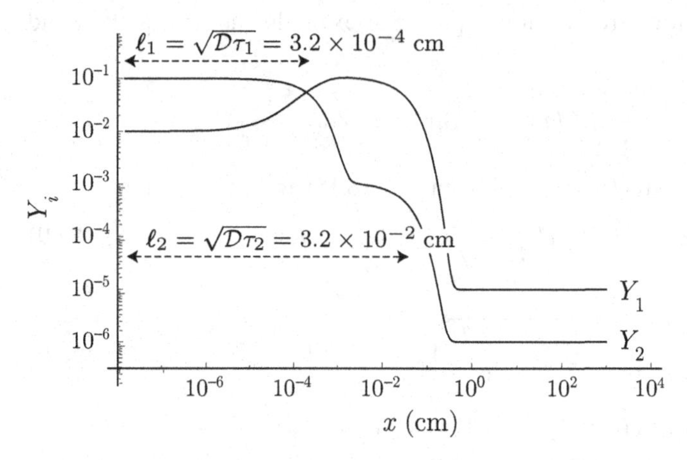


Figure 8.5. Mass fraction versus distance for advection-reaction-diffusion problem with simple two-step linear kinetics.

$$Z_1(0) = \frac{9891}{100000}, \quad Z_2(0) = \frac{1089}{10000}, \quad \lim_{x\to\infty} \frac{dZ_1}{dx} = 0, \quad \lim_{x\to\infty} \frac{dZ_2}{dx} = 0. \tag{8.80}$$

These have solution

$$Z_1(x) = \frac{9891 e^{\left(\left(5-5\sqrt{400001}\right)\text{cm}^{-1}\right)x}}{100000} = 0.0981 e^{-(3.2\times10^3\ \text{cm}^{-1})x}, \tag{8.81}$$

$$Z_2(x) = \frac{1089 e^{\left(\left(5-5\sqrt{41}\right)\text{cm}^{-1}\right)x}}{10000} = 0.1089 e^{-(2.7\times10^1\ \text{cm}^{-1})x}. \tag{8.82}$$

The relevant length scales are

$$\ell_1 = \frac{1}{\left(5-5\sqrt{400001}\right)\text{cm}^{-1}} = 3.2\times10^{-4}\ \text{cm}, \tag{8.83}$$

$$\ell_2 = \frac{1}{\left(5-5\sqrt{41}\right)\text{cm}^{-1}} = 3.7\times10^{-2}\ \text{cm}. \tag{8.84}$$

Especially for $\ell_1$, these are both well estimated by the simple formulae of Eq. (8.71):

$$\ell_1 \approx \sqrt{\mathcal{D}\tau_1} = \sqrt{\left(10^1\ \frac{\text{cm}^2}{\text{s}}\right)(10^{-8}\ \text{s})} = 3.2\times10^{-4}\ \text{cm}, \tag{8.85}$$

$$\ell_2 \approx \sqrt{\mathcal{D}\tau_2} = \sqrt{\left(10^1\ \frac{\text{cm}^2}{\text{s}}\right)(10^{-4}\ \text{s})} = 3.2\times10^{-2}\ \text{cm}. \tag{8.86}$$

For the slower Reaction 2, advection plays a larger role, rendering the diffusion-based estimate to have a small but noticeable error.

Forming $\mathbf{Y}$ via $\mathbf{Y} = \mathbf{Y}_{eq} + \mathbf{S}\cdot\mathbf{Z}$, we find the steady solution to be

$$Y_1(x) = 10^{-5} - 0.09891 e^{-(3.2\times10^3\ \text{cm}^{-1})x} + 0.1089 e^{-(2.7\times10^1\ \text{cm}^{-1})x}, \tag{8.87}$$

$$Y_2(x) = 10^{-6} + 0.09891 e^{-(3.2\times10^3\ \text{cm}^{-1})x} + 0.001089 e^{-(2.7\times10^1\ \text{cm}^{-1})x}. \tag{8.88}$$

Both variables evolve over two distinct length scales as they relax to their distinct equilibria. A plot of $Y_1(t)$ and $Y_2(t)$ is given in Fig. 8.5. Similar to the time-dependent version of this system, a frozen state near $x = 0$ first undergoes a reaction to a pseudo-equilibrium state near $x = \ell_1$. Near $x = \ell_2$, the system relaxes to its true equilibrium.

### 8.2.3 Spatiotemporal Solution

Let us next study solutions with dependency on both time and distance. We extend the analysis and nomenclature of Sec. 8.1.3 so as to take

$$\mathbf{Y}(x,t) = \mathbf{Y}_{eq} + \mathbf{B}(t)e^{i\hat{k}x}, \tag{8.89}$$

so that Eq. (8.40) becomes

$$\frac{d\mathbf{B}}{dt}e^{i\hat{k}x} + i\hat{k}u\mathbf{B}e^{i\hat{k}x} = -\mathcal{D}\hat{k}^2\mathbf{B}e^{i\hat{k}x} - \mathbf{A}\cdot\mathbf{B}e^{i\hat{k}x}, \tag{8.90}$$

$$\frac{d\mathbf{B}}{dt} + i\hat{k}u\mathbf{B} = -\mathcal{D}\hat{k}^2\mathbf{B} - \mathbf{A}\cdot\mathbf{B}, \tag{8.91}$$

$$\frac{d\mathbf{B}}{dt} = -\left(\left(i\hat{k}u + \mathcal{D}\hat{k}^2\right)\mathbf{I} + \mathbf{A}\right)\cdot\mathbf{B}. \tag{8.92}$$

Here, $\mathbf{I}$ is the identity matrix. Now it is the real part of the eigenvalues of the matrix $-\left(\left(i\hat{k}u + \mathcal{D}\hat{k}^2\right)\mathbf{I} + \mathbf{A}\right)$ that dictates whether the amplitudes grow or decay. With the operator "eig" operating on a matrix to yield its eigenvalues, it is a well-known result from linear algebra that eig $(\alpha\mathbf{I} + \mathbf{A}) = \alpha + \text{eig}\,\mathbf{A}$. Now, $\mathbf{A}$ is dictated by chemical kinetics alone, and is known to have $N$ real and positive eigenvalues, $\sigma_i$, $i = 1,\ldots,N$. Our eigenvalues $\beta_i$ are thus seen to be

$$\beta_i = i\hat{k}u + \mathcal{D}\hat{k}^2 + \sigma_i, \qquad i = 1,\ldots,N. \tag{8.93}$$

It is only the real part of $\beta_i$ that dictates growth or decay of a mode. Because

$$\text{Re}\,(\beta_i) = \mathcal{D}\hat{k}^2 + \sigma_i > 0, \quad \forall i = 1,\ldots,N, \tag{8.94}$$

we see that all modes are decaying, and that diffusion induces them to decay more rapidly. The time scales of decay $\tau_i$ are again given by the reciprocals of the eigenvalues and are seen to be

$$\tau_i = \frac{1}{\text{Re}\,(\beta_i)} = \frac{1}{\sigma_i\left(1 + \dfrac{\mathcal{D}\hat{k}^2}{\sigma_i}\right)} = \frac{1}{\sigma_i + \dfrac{\mathcal{D}}{\ell^2}}, \quad i = 1,\ldots,N, \tag{8.95}$$

using $\ell = 1/\hat{k}$ from Eq. (8.34).

From Eq. (8.95), it is clear that for $\mathcal{D}\hat{k}^2/\sigma_i \ll 1$; that is, for sufficiently small wavenumbers or long wavelengths, the time scales of amplitude growth or decay will be dominated by reaction:

$$\lim_{\hat{k}\to 0} \tau_i = 1/\sigma_i. \tag{8.96}$$

However, for $\mathcal{D}\hat{k}^2/\sigma_i \gg 1$, that is, for sufficiently large wavenumbers or small wavelengths, the amplitude growth/decay time scales are dominated by diffusion:

$$\lim_{\hat{k}\to\infty} \tau_i = \frac{1}{\mathcal{D}\hat{k}^2} = \frac{1}{\mathcal{D}}\left(\frac{\lambda}{2\pi}\right)^2. \tag{8.97}$$

From Eq. (8.95), we see that a balance between reaction and diffusion exists for $\hat{k} = \hat{k}_i = \sqrt{\sigma_i/\mathcal{D}}$. In terms of wavelength, and recalling Eq. (8.72), we see the balance at

$$\lambda/(2\pi) = 1/\hat{k}_i = \sqrt{\mathcal{D}/\sigma_i} = \sqrt{\mathcal{D}\tau_i} = \ell_i. \tag{8.98}$$

Here, $\ell_i$ is the $\ell$ for which the balance exists.

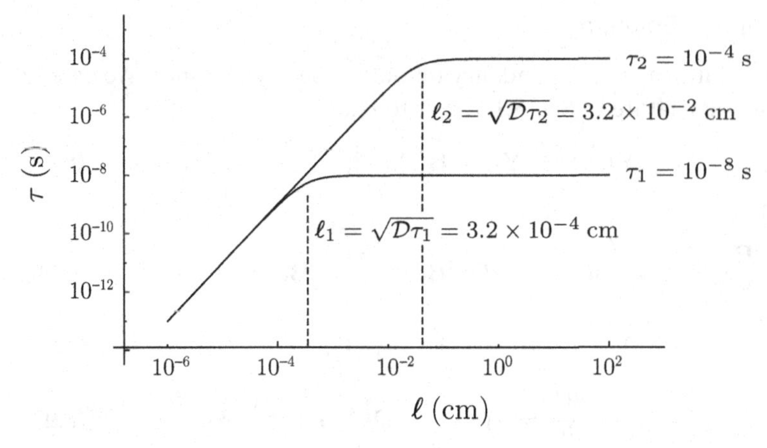


Figure 8.6. Time scale spectrum versus length scale for the simple advection-reaction-diffusion model with two-step linear kinetics.

**EXAMPLE 8.6**

For a case where $N = 2, \mathcal{D} = 10^1$ cm$^2$/s, $u = 10^2$ cm/s, examine time scales as a function of the length scales when considering solutions to Eq. (8.40) if

$$\mathbf{A} = \begin{pmatrix} 1000000\ \text{s}^{-1} & -99000000\ \text{s}^{-1} \\ -99000000\ \text{s}^{-1} & 99010000\ \text{s}^{-1} \end{pmatrix}. \tag{8.99}$$

We have examined this matrix earlier and know from Eqs. (8.57) and (8.58) that the eigenvalues and spatially homogeneous reaction time scales are

$$\sigma_1 = 10^8\ \text{s}^{-1}, \quad \tau_1 = 10^{-8}\ \text{s}, \qquad \sigma_2 = 10^4\ \text{s}^{-1}, \quad \tau_2 = 10^{-4}\ \text{s}. \tag{8.100}$$

From Eq. (8.95), we get expressions for the effects of diffusion on the two time scales:

$$\tau_1 = \frac{1}{\sigma_1 + \frac{\mathcal{D}}{\ell^2}} = \frac{1}{(10^8\ \text{s}^{-1}) + \frac{10^1\ \frac{\text{cm}^2}{\text{s}}}{\ell^2}}, \tag{8.101}$$

$$\tau_2 = \frac{1}{\sigma_2 + \frac{\mathcal{D}}{\ell^2}} = \frac{1}{(10^4\ \text{s}^{-1}) + \frac{10^1\ \frac{\text{cm}^2}{\text{s}}}{\ell^2}}. \tag{8.102}$$

The length scales where reaction and diffusion balance are given by Eq. (8.98):

$$\ell_1 = \sqrt{\mathcal{D}\tau_1} = \sqrt{\left(10^1\ \frac{\text{cm}^2}{\text{s}}\right)(10^{-8}\ \text{s})} = 3.2 \times 10^{-4}\ \text{cm}, \tag{8.103}$$

$$\ell_2 = \sqrt{\mathcal{D}\tau_2} = \sqrt{\left(10^1\ \frac{\text{cm}^2}{\text{s}}\right)(10^{-4}\ \text{s})} = 3.2 \times 10^{-2}\ \text{cm}. \tag{8.104}$$

This behavior is displayed in Fig. 8.6. We see that for large $\ell$, the time scales are dictated by those given by a spatially homogeneous theory. As $\ell$ is reduced, diffusion first plays a role in modulating the time scale of the slow reaction. As $\ell$ is further reduced, diffusion also modulates the time scale of the fast reaction. It is the fast reaction that dictates the time scale that needs to be considered to capture the advection-reaction-diffusion dynamics.

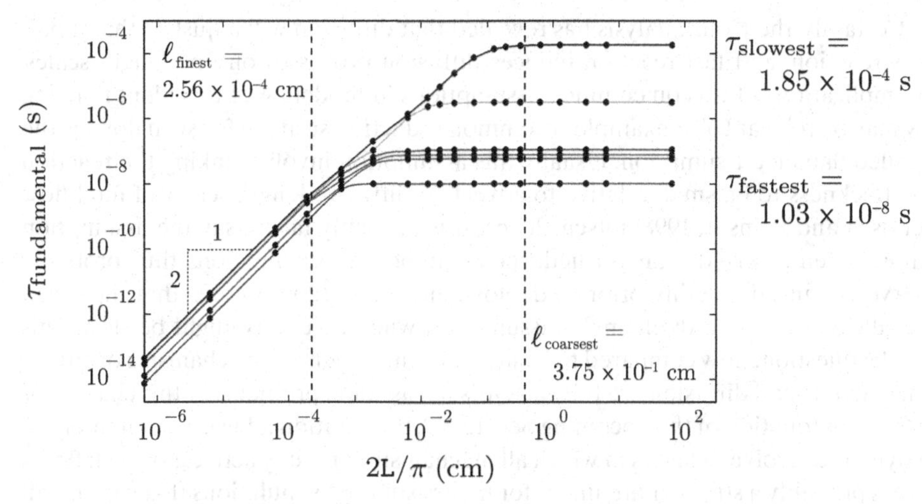


Figure 8.7. Time scales associated with the fundamental modes for a nine-species, nineteen-step stoichiometric, $P_0 = 1$ atm, $T_0 = 800$ K, hydrogen-air advection-reaction-diffusion system versus the length $2L/\pi$. Adapted from Al-Khateeb et al. (2013), reprinted by permission of the publisher, Taylor & Francis Ltd., http://www.tandfonline.com/.

## 8.3 $H_2$-Air Near Equilibrium

We show analogous results for a detailed kinetics advection-reaction-diffusion combustion system near equilibrium in Fig. 8.7. The analysis necessary to generate these results is complicated, is based on the approach of Kirkby and Schmitz (1966), and is given by Al-Khateeb et al. (2013). To summarize, the time-dependent one-dimensional governing equations were considered. These equations were discretized in space so as to yield a large set of ordinary differential equations in time. Next a spatially homogeneous chemically equilibrated solution was subjected to a small spatiotemporal perturbation. The large system of ordinary differential equations in time was thus given a set of initial values. This system was linearized about the spatially homogeneous, chemically equilibrated state, and the resulting eigenvalues and eigenfunctions were determined. The time scales are the reciprocals of the real parts of the eigenvalues. The results of Fig. 8.7 were obtained for an $N = 9$-species, $L = 3$-element, $J = 19$-reaction model of hydrogen-air kinetics augmented by both advection and multicomponent mass and energy diffusion.

For long length scales, the time scales are effectively identical to those obtained by analysis of a spatially homogeneous system. Thus we see clearly $N - L = 6$ independent time scales. The fastest is $\tau_{\text{fastest}} = 1.03 \times 10^{-8}$ s and the slowest is $\tau_{\text{slowest}} = 1.85 \times 10^{-4}$ s. As the length scale is decreased, one finds that diffusive effects first become important in describing dynamics associated with the slowest reaction model. As the length scale continues to decrease, diffusion becomes more and more important for each progressively faster reaction mode, until the fastest reaction mode is reached. With the extreme diffusion coefficients selected from the multicomponent diffusion matrix, we find $\mathcal{D}_{\text{min}} = 6.376\ \text{cm}^2/\text{s}$ and $\mathcal{D}_{\text{max}} = 760.73\ \text{cm}^2/\text{s}$, the coarsest and finest transition length scales, $\ell_{\text{coarsest}}$ and $\ell_{\text{finest}}$ are found to be

$$\ell_{\text{finest}} = \sqrt{\mathcal{D}_{\text{min}}\tau_{\text{fastest}}} = 2.56 \times 10^{-4}\ \text{cm}, \tag{8.105}$$

$$\ell_{\text{coarsest}} = \sqrt{\mathcal{D}_{\text{max}}\tau_{\text{slowest}}} = 3.75 \times 10^{-1}\ \text{cm}. \tag{8.106}$$

Certainly, then, our analysis has revealed that diffusion will adjust so that it balances reaction, and fast reaction induces diffusion processes on fine length scales. The implications of this on common assumptions in modeling of turbulent reacting flow may be relevant. For example, a common reduction strategy for such flows is the so-called flamelet assumption. Usually this assumption involves taking the reaction zone thickness to be small relative to advective-diffusive length scales of fluid flow (Vervisch and Poinsot, 1998; Pitsch, 2006). It is probably fair to say the assumption is more often asserted than verified; such common approximations thus probably deserve enhanced scrutiny prior to deployment. Riley (2006) notes that there can be challenges in using the flamelet assumption when reactions might be slow. This begs the question, slow compared to what? Certainly when the mechanisms of advection, reaction, and diffusion are fully active, such as some problems of this chapter or unsteady detonation of the upcoming Sec. 12.3.2, the solution adjusts so that interesting dynamics evolve in layers in which all mechanisms are fully active. For such flows, there is probably a stronger argument for high resolution simulations that capture all of the length and time scales present.

## EXERCISES

**8.1.** Solve Eq. (8.1) for $u = 0$ with the following initial and boundary conditions:

$$Y(x,0) = Y_0\left(1 + \frac{x}{L}\left(1 - \frac{x}{L}\right)\right), \qquad Y(0,t) = Y(L,t) = Y_0.$$

Choose a set of numerical parameters and generate plots that illustrate solution features.

**8.2.** Consider the example problem in Sec. 8.1.2. Find a numerical approximation to the solution via a discretization of your choice, for example, finite difference or finite element. You will need to choose a large but finite length $L$ to model the semi-infinite domain as well as a spatial discretization size $\Delta x$. Compare the numerical approximate solution with the exact solution. Plot the maximum error magnitude as a function of $\Delta x$ and show the error is converging toward zero at a rate consistent with the order of accuracy of the chosen numerical method.

**8.3.** Repeat the example of Sec. 8.2.3 with $u = 10^0$ cm/s and $u = 10^5$ cm/s.

## References

A. N. Al-Khateeb, J. M. Powers, and S. Paolucci, 2013, Analysis of the spatio-temporal scales of laminar premixed flames near equilibrium, *Combustion Theory and Modelling*, 17(1): 76–108.

T. Echekki and E. Mastorakos, eds., 2011, *Turbulent Combustion Modeling*, Springer, Heidelberg.

N. Eldredge and S. J. Gould, 1972, Punctuated equilibria: An alternative to phyletic gradualism, in *Models in Paleobiology*, edited by T. J. M. Schopf, Freeman-Cooper, San Francisco, pp. 82–115.

R. O. Fox, 2003, *Computational Models for Turbulent Reacting Flow*, Cambridge University Press, Cambridge, UK.

W. Hundsdorfer and J. G. Verwer, 2003, *Numerical Solution of Time-Dependent Advection-Diffusion-Reaction Equations*, Springer, Berlin.

L. L. Kirkby and R. A. Schmitz, 1966, An analytical study of stability of a laminar diffusion flame, *Combustion and Flame*, 10(3): 205–220.

K. K. Kuo and R. Acharya, 2012, *Fundamentals of Turbulent and Multiphase Combustion*, John Wiley, Hoboken, NJ.

L. D. Landau and E. M. Lifshitz, 1959, *Fluid Mechanics*, Pergamon Press, London, p. 475.

D. K. Manley, A. McIlroy, and C. A. Taatjes, 2008, Research needs for future internal combustion engines, *Physics Today*, 61(11): 47–52.

N. Peters, 2000, *Turbulent Combustion*, Cambridge University Press, Cambridge, UK.

H. Pitsch, 2006, Large-eddy simulation of turbulent combustion, *Annual Reviews of Fluid Mechanics*, 38: 453–482.

J. J. Riley, 2006, Review of large-eddy simulation of non-premixed turbulent combustion, *Journal of Fluids Engineering*, 128(2): 209–215.

L. Vervisch and T. Poinsot, 1998, Direct numerical simulation of non-premixed turbulent flames, *Annual Review of Fluid Mechanics*, 30: 655–691.

# 9 Idealized Solid Combustion

*If, on the other hand, one considers that heat transfer . . . occurs purely by conduction, a certain temperature distribution . . . will be obtained; the highest temperature will be at the center of the vessel where inflammation ought, therefore, to start.*

– David Albertovich Frank-Kamenetskii (1910–1970), *Diffusion and Heat Transfer in Chemical Kinetics*

Here, we add effects of diffusion to the simple thermal explosion theory of Sec. 1.3.1 that balanced unsteady evolution against reaction. Starting from a fully unsteady formulation, we focus first on cases that are steady, resulting in a balance between reaction and diffusion. We then reintroduce unsteady effects to see how they are influenced by reaction and diffusion. Such a theory is known after its founder as Frank-Kamenetskii[1] theory. In particular, we look for transitions from a low-temperature reaction to a high-temperature reaction. Additional background is given by Buckmaster and Ludford (1983), Kanury (1975), Bebernes and Eberly (1989), Kapila (1983), Glassman et al. (2014), and Griffiths and Barnard (1995).

We will consider a model of the combustion of an idealized solid. The model is not appropriate for quantitative predictions of actual combustible solids such as coal or wood. However, the incompressibility and zero advection assumptions we will take are more characteristic of solid combustion than gaseous combustion. Moreover, diffusion in gases is almost always accompanied by advection. Thus, the idealized model we consider here is better thought of as for a solid than a gas.

## 9.1 Simple Planar Model

Let us consider a slab of an idealized solid premixed fuel/oxidizer mixture. The material is modeled to be of infinite extent in the $y$ and $z$ directions and has length $2L$ in the $x$ direction. The temperature at each end, $x = \pm L$, is held fixed at $T = T_0$. The slab is initially unreacted. Exothermic conversion from reactants to products will generate an elevated temperature within the slab $T > T_0$, for $x \in (-L, L)$. If the thermal energy generated diffuses rapidly enough, the temperature within the slab will be $T \sim T_0$, and the reaction rate will be low. If the energy generated by local reaction diffuses slowly, it will accumulate in the interior, accelerate the local reaction rate, and induce rapid energy release and high temperature.

[1] David Albertovich Frank-Kamenetskii (1910–1970), Soviet physicist.

Let us assume, extending the earlier discussions in Secs. 1.3.1 and 4.7 that

- The material is an immobile, incompressible solid with constant specific heat and thermal conductivity; thus,
  - $\rho, c_P = c_v$, and k are all constant.
  - There is no advective transport of mass, linear momenta, or energy: $\mathbf{u} = \mathbf{0}$.
- The material has variation only with $x$ and $t$.
- The reaction can be modeled as $A \to B$, where $A$ and $B$ have identical molecular masses.
- The reaction is irreversible and exothermic.
- Initially only $A$ is present.

Let us take, as we did in thermal explosion theory, Sec. 1.3.1,

$$Y_A = 1 - \lambda, \qquad Y_B = \lambda. \tag{9.1}$$

We interpret $\lambda$ as a *reaction progress variable* that has $\lambda \in [0, 1]$. For $\lambda = 0$, the material is all $A$; when $\lambda = 1$, the material is all $B$.

### 9.1.1 Model Equations

Our simple model for reaction is

$$\frac{\partial \lambda}{\partial t} = a e^{-\overline{\mathcal{E}}/\overline{\mathcal{R}}/T}(1 - \lambda), \tag{9.2}$$

$$\rho \frac{\partial e}{\partial t} = -\frac{\partial \mathsf{q}}{\partial x}, \tag{9.3}$$

$$\mathsf{q} = -\mathsf{k}\frac{\partial T}{\partial x}, \tag{9.4}$$

$$e = c_v T - \lambda q. \tag{9.5}$$

Equation (9.2) is our reaction kinetics law. It is the equivalent of the earlier derived Eq. (1.291) in the irreversible limit, $K_c \to \infty$. Equation (9.3) is our energy conservation expression. It is a limiting form of Eq. (7.41), with q playing the role of the heat flux $\mathbf{j}^q$. Equation (9.4) is the constitutive law for heat flux; it is the equivalent of Eq. (7.73) when mass diffusion is neglected and amounts to Fourier's[2] law. Equation (9.5) is our caloric equation of state; it is Eq. (1.321) with $q = \overline{e}^0_{T_0,A} - \overline{e}^0_{T_0,B}$ and $Y_B = \lambda$. Equations (9.2)–(9.5) are completed by initial and boundary conditions, which are

$$T(-L, t) = T(L, t) = T_0, \qquad T(x, 0) = T_0, \qquad \lambda(x, 0) = 0. \tag{9.6}$$

### 9.1.2 Simple Planar Derivation

Let us perform a simple "control volume" derivation of the energy equation, Eq. (9.3). Consider a small volume of dimension $A$ by $\Delta x$. At the left boundary $x_1$, we have heat flux q, in which we notate as $\mathsf{q}|_{x_1}$. At the right boundary, $x_1 + \Delta x$, we have heat flux out of $\mathsf{q}|_{x_1+\Delta x}$. Recall the units of heat flux are J/m$^2$/s. A sketch is shown in Fig 9.1. The first law of thermodynamics is

$$\text{change in extensive energy} = \text{heat in} - \underbrace{\text{work out}}_{0}. \tag{9.7}$$

[2]Jean Baptiste Joseph Fourier (1768–1830), French mathematician.

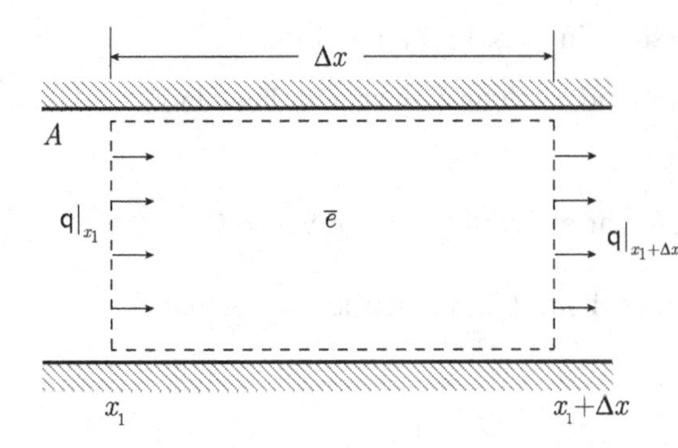


Figure 9.1. Sketch of one-dimensional energy diffusion.

There is no work for our system because there is no volume change. But there is heat flux over system boundaries. In a combination of symbols and words, we can say

$$\underbrace{\text{extensive energy @ } t + \Delta t - \text{extensive energy @ } t}_{\text{unsteady}}$$
$$= \underbrace{\text{energy flux in} - \text{energy flux out}}_{\underbrace{\text{advection}}_{0} \text{ and diffusion}}. \tag{9.8}$$

Energy fluxes in general are due to advection and diffusion; however, we have assumed there is no advection. As such, there is no kinetic energy to consider, and we neglect potential energy as well. Mathematically, we can say

$$E|_{t+\Delta t} - E|_t = -\left(E_{flux\ out} - E_{flux\ in}\right), \tag{9.9}$$

$$\underbrace{\rho A \Delta x}_{\text{kg}} \underbrace{\left(e|_{t+\Delta t} - e|_t\right)}_{\text{J/kg}} = -\underbrace{\left(\mathsf{q}|_{x_1+\Delta x} - \mathsf{q}|_{x_1}\right)}_{\text{J/m}^2\text{/s}} \underbrace{A \Delta t}_{(\text{m}^2\ \text{s})}, \tag{9.10}$$

$$\rho \frac{e|_{t+\Delta t} - e|_t}{\Delta t} = -\left(\frac{\mathsf{q}|_{x_1+\Delta x} - \mathsf{q}|_{x_1}}{\Delta x}\right). \tag{9.11}$$

Now, let $\Delta x \to 0$ and $\Delta t \to 0$ and get

$$\rho \frac{\partial e}{\partial t} = -\frac{\partial \mathsf{q}}{\partial x}. \tag{9.12}$$

Standard constitutive theory gives Fourier's law to specify the heat flux, Eq. (9.4). So, Eq. (9.4), along with the thermal state equation, Eq. (9.5), when substituted into Eq. (9.12), yield

$$\rho \frac{\partial}{\partial t}\left(c_v T - \lambda q\right) = \mathsf{k} \frac{\partial^2 T}{\partial x^2}, \tag{9.13}$$

$$\rho c_v \frac{\partial T}{\partial t} - \rho q \frac{\partial \lambda}{\partial t} = \mathsf{k} \frac{\partial^2 T}{\partial x^2}, \tag{9.14}$$

$$\rho c_v \frac{\partial T}{\partial t} - \rho q a e^{-\overline{\mathcal{E}}/\overline{R}/T}(1-\lambda) = \mathsf{k} \frac{\partial^2 T}{\partial x^2}, \tag{9.15}$$

$$\rho c_v \frac{\partial T}{\partial t} = \mathsf{k} \frac{\partial^2 T}{\partial x^2} + \rho q a e^{-\overline{\mathcal{E}}/\overline{R}/T}(1-\lambda). \tag{9.16}$$

So, our complete system is two equations in two unknowns with appropriate initial and boundary conditions:

$$\rho c_v \frac{\partial T}{\partial t} = \mathsf{k}\frac{\partial^2 T}{\partial x^2} + \rho q a e^{-\overline{\mathcal{E}}/\overline{R}/T}(1-\lambda), \tag{9.17}$$

$$\frac{\partial \lambda}{\partial t} = a e^{-\overline{\mathcal{E}}/\overline{R}/T}(1-\lambda), \tag{9.18}$$

$$T(-L,t) = T(L,t) = T_0, \quad T(x,0) = T_0, \quad \lambda(x,0) = 0. \tag{9.19}$$

### 9.1.3 Ad Hoc Approximation

Let us consider an ad hoc approximation to yield a system much like Eqs. (9.17)–(9.19) but that has the advantage of being one equation and one unknown.

#### Planar Formulation

*If* there were no diffusion, Eq. (9.12) would yield $\partial e/\partial t = 0$ and would lead us to conclude that $e(x,t) = e(x)$. And because we have nothing now to introduce a spatial inhomogeneity, there is no reason to take $e(x)$ to be anything other than a constant $e_0$. That would lead us to $e(x,t) = e_0$, so

$$e_0 = c_v T - \lambda q. \tag{9.20}$$

Now, at $t = 0$, we have $\lambda = 0$ and $T = T_0$, so $e_0 = c_v T_0$; thus,

$$c_v T_0 = c_v T - \lambda q, \qquad \lambda = \frac{c_v(T - T_0)}{q}. \tag{9.21}$$

We also get the final temperature at $\lambda = 1$ to be

$$T(\lambda = 1) = T_0 + \frac{q}{c_v}. \tag{9.22}$$

So, at complete reaction, the temperature has risen as a consequence of chemical energy release. We shall adopt for now Eq. (9.21) as our model for $\lambda$ in place of Eq. (9.18). Had we admitted species diffusion, we could more rigorously have arrived at a similar result, but it would be more difficult to justify treating the material as a solid.

Let us use Eq. (9.21) to eliminate $\lambda$ in Eq. (9.17) to get a single equation for $T$:

$$\rho c_v \frac{\partial T}{\partial t} = \mathsf{k}\frac{\partial^2 T}{\partial x^2} + \underbrace{\rho q a e^{-\overline{\mathcal{E}}/\overline{R}/T}\left(1 - \frac{c_v(T-T_0)}{q}\right)}_{\text{reaction source term}}. \tag{9.23}$$

The first two terms in Eq. (9.23) are nothing more than the classical heat equation. The nonclassical term is an algebraic source term due to chemical reaction. When $T = T_0$, the reaction source term is $\rho q a \exp(-\overline{\mathcal{E}}/\overline{R}/T) > 0$. This induces is a tendency for the temperature to increase. When $\lambda = 1$, so $T = T_0 + q/c_v$, the reaction source term is zero, so there is no local heat release. This effectively accounts for reactant depletion.

Scaling the equation by $\rho c_v$ and employing the formula for thermal diffusivity

$$\alpha = \frac{\mathsf{k}}{\rho c_P} \sim \frac{\mathsf{k}}{\rho c_v}, \tag{9.24}$$

for the solid, we get

$$\frac{\partial T}{\partial t} = \alpha \frac{\partial^2 T}{\partial x^2} + a\left(\frac{q}{c_v} - (T - T_0)\right) e^{-\overline{\mathcal{E}}/\overline{R}/T}, \tag{9.25}$$

$$T(-L,t) = T(L,t) = T_0, \qquad T(x,0) = T_0. \tag{9.26}$$

From here on, we shall consider Eqs. (9.25) and (9.26) and cylindrical and spherical variants to be the full problem. We shall consider solutions to it in various limits.

**More General Coordinate Systems**

Equation (9.26) can be extended to general coordinate systems via

$$\frac{\partial T}{\partial t} = \alpha \nabla^2 T + a\left(\frac{q}{c_v} - (T - T_0)\right) e^{-\overline{\mathcal{E}}/\overline{R}/T}. \tag{9.27}$$

Appropriate initial and boundary conditions for the particular coordinate system would be necessary. For one-dimensional solutions in planar, cylindrical, and spherical coordinates, one can summarize the formulations as

$$\frac{\partial T}{\partial t} = \alpha \frac{1}{x^m}\frac{\partial}{\partial x}\left(x^m \frac{\partial T}{\partial x}\right) + a\left(\frac{q}{c_v} - (T - T_0)\right) e^{-\overline{\mathcal{E}}/\overline{R}/T}. \tag{9.28}$$

Here, we have $m = 0$ for a planar coordinate system, $m = 1$ for cylindrical, and $m = 2$ for spherical. For cylindrical and spherical systems, the Dirichlet[3] boundary condition at $x = -L$ would be replaced with a boundedness condition on $T$ at $x = 0$.

## 9.2 Nondimensionalization

Let us nondimensionalize Eq. (9.27). The scaling we will choose is not unique. We can define nondimensional variables

$$T_* = \frac{c_v(T - T_0)}{q}, \qquad x_* = \frac{x}{L}, \qquad t_* = at. \tag{9.29}$$

With these choices, Eq. (9.27) transforms to

$$\begin{aligned}\frac{qa}{c_v}\frac{\partial T_*}{\partial t_*} &= \frac{q}{c_v}\alpha\frac{1}{L^2}x_*^{-m}\frac{\partial}{\partial x_*}\left(x_*^m \frac{\partial T_*}{\partial x_*}\right)\\ &\quad + a\left(\frac{q}{c_v} - \frac{q}{c_v}T_*\right)\exp\left(-\frac{\overline{\mathcal{E}}}{\overline{R}}\frac{1}{T_0 + (q/c_v)T_*}\right),\end{aligned} \tag{9.30}$$

$$\begin{aligned}\frac{\partial T_*}{\partial t_*} &= \frac{\alpha}{aL^2}x_*^{-m}\frac{\partial}{\partial x_*}\left(x_*^m \frac{\partial T_*}{\partial x_*}\right)\\ &\quad + (1 - T_*)\exp\left(-\frac{\overline{\mathcal{E}}}{\overline{R}T_0}\frac{1}{\left(1 + \left(\frac{q}{c_v T_0}\right)T_*\right)}\right).\end{aligned} \tag{9.31}$$

Now, both sides are dimensionless. Let us define the dimensionless parameters

$$\mathfrak{D} = \frac{aL^2}{\alpha}, \qquad \Theta = \frac{\overline{\mathcal{E}}}{\overline{R}T_0}, \qquad Q = \frac{q}{c_v T_0}. \tag{9.32}$$

[3] Johann Peter Gustav Lejeune Dirichlet (1805–1859), German mathematician.

Here, $\mathfrak{D}$ is our definition of a so-called Damköhler[4] number. We recall a diffusion time scale $\tau_d$ to be, similar to Eq. (8.33),

$$\tau_d = L^2/\alpha, \tag{9.33}$$

and take a first estimate (which will be shown to be crude) of the reaction time scale, $\tau_r$, to be

$$\tau_r = 1/a. \tag{9.34}$$

We then see that our Damköhler number is the ratio of the thermal diffusion time to the crude reaction time (ignoring activation energy effects!):

$$\mathfrak{D} = \frac{L^2/\alpha}{1/a} = \frac{\tau_d}{\tau_r} = \frac{\text{thermal diffusion time}}{\text{reaction time}}. \tag{9.35}$$

The combustion literature has a variety of definitions for Damköhler number. Law (2006) takes one similar to our definition but also admits one based on flow velocity. Strehlow (1984) lists five Damköhler numbers, four of them introduced by Damköhler himself. Fast reaction, relative to any of a variety of physical time scales, typically implies large Damköhler number. We can think of $\Theta$ and $Q$ as ratios as well:

$$\Theta = \frac{\text{activation energy}}{\text{ambient energy}}, \qquad Q = \frac{\text{exothermic heat release}}{\text{ambient energy}}. \tag{9.36}$$

The initial and boundary conditions scale to

$$T_*(-1, t_*) = T_*(1, t_*) = 0, \qquad T_*(x_*, 0) = 0, \quad \text{if } m = 0, \tag{9.37}$$

$$T_*(1, t_*) = 0, \quad T_*(0, t_*) < \infty, \quad T_*(x_*, 0) = 0, \quad \text{if } m = 1, 2. \tag{9.38}$$

### 9.2.1 Final Form

Let us now drop the $*$ notation and understand that all variables are dimensionless. So, our Eqs. (9.31), (9.37), and (9.38) become

$$\frac{\partial T}{\partial t} = \frac{1}{\mathfrak{D}} x^{-m} \frac{\partial}{\partial x}\left(x^m \frac{\partial T}{\partial x}\right) + (1 - T)\exp\left(\frac{-\Theta}{1 + QT}\right),$$

$$T(-1, t) = T(1, t) = 0, \quad T(x, 0) = 0, \quad \text{if } m = 0,$$

$$T(1, t) = 0, \quad T(0, t) < \infty, \quad T(x, 0) = 0, \quad \text{if } m = 1, 2. \tag{9.39}$$

The initial and boundary conditions are homogeneous. The only inhomogeneity lives in the exothermic reaction source term. It is nonlinear due to the $\exp(-1/T)$ term. Also, it will prove to be the case that a symmetry boundary condition at $x = 0$ suffices, though our original formulation is more rigorous. Such equivalent boundary conditions are

$$T(1, t) = 0, \qquad \frac{\partial T}{\partial x}(0, t) = 0, \qquad T(x, 0) = 0, \qquad m = 1, 2, 3. \tag{9.40}$$

### 9.2.2 Integral Form

As an aside, let us consider the evolution of energy within the domain. To do so, we integrate a differential volume element through the entire volume. We recall

[4] Gerhard Damköhler (1908–1944), German chemist.

$dV \sim x^m\,dx$, for $m = 0, 1, 2$, (planar, cylindrical, spherical):

$$x^m \frac{\partial T}{\partial t}\,dx = \frac{1}{\mathfrak{D}} x^m x^{-m} \frac{\partial}{\partial x}\left(x^m \frac{\partial T}{\partial x}\right)\,dx$$
$$+ x^m (1-T) \exp\left(\frac{-\Theta}{1+QT}\right)\,dx, \tag{9.41}$$

$$\int_0^1 x^m \frac{\partial T}{\partial t}\,dx = \int_0^1 \frac{1}{\mathfrak{D}} \frac{\partial}{\partial x}\left(x^m \frac{\partial T}{\partial x}\right)\,dx$$
$$+ \int_0^1 x^m (1-T) \exp\left(\frac{-\Theta}{1+QT}\right)\,dx, \tag{9.42}$$

$$\underbrace{\frac{d}{dt}\int_0^1 x^m T\,dx}_{\text{thermal energy change}} = \underbrace{\frac{1}{\mathfrak{D}} \left.\frac{\partial T}{\partial x}\right|_{x=1}}_{\text{boundary heat flux}}$$
$$+ \underbrace{\int_0^1 x^m (1-T) \exp\left(\frac{-\Theta}{1+QT}\right)\,dx}_{\text{internal conversion}}. \tag{9.43}$$

The total thermal energy in our domain changes due (1) diffusive energy flux at the isothermal boundary and (2) internal conversion of chemical energy to thermal energy.

### 9.2.3 Infinite Damköhler Limit

For $\mathfrak{D} \to \infty$, diffusion becomes unimportant in Eq. (9.39), and we recover a balance between unsteady effects and reaction:

$$\frac{dT}{dt} = (1-T) \exp\left(\frac{-\Theta}{1+QT}\right), \qquad T(0) = 0. \tag{9.44}$$

This is the problem we have already considered in thermal explosion theory, Sec. 1.3.1. We recall that thermal explosion theory predicts significant acceleration of reaction when

$$t \to \frac{e^{\Theta}}{Q\Theta}. \tag{9.45}$$

There is nothing more to add to this that has not already been discussed earlier.

## 9.3 Steady Solutions

Let us seek solutions to the planar ($m = 0$) version of Eqs. (9.39) that are steady, so that $\partial/\partial t = 0$, and a balance between reaction and energy diffusion is attained. In that limit, Eqs. (9.39) reduce to the following two-point boundary value problem:

$$0 = \frac{1}{\mathfrak{D}} \frac{d^2 T}{dx^2} + (1-T) \exp\left(\frac{-\Theta}{1+QT}\right), \tag{9.46}$$

$$0 = T(-1) = T(1). \tag{9.47}$$

This problem is difficult to solve analytically because of the strong nonlinearity in the reaction source term. We next seek an approximate means of solution that will lead us to insights into how reaction balances with diffusion.

### 9.3.1 High-Activation-Energy Asymptotics

Motivated by our earlier success in getting approximate solutions to a similar problem in spatially homogeneous thermal explosion theory, let us take a similar approach here.

Let us seek low-temperature solutions where the nonlinearity may be weak. So, let us define a small parameter $\epsilon$ with $0 \leq \epsilon \ll 1$. We do not yet specify a physical interpretation of $\epsilon$. And now let us assume a power series expansion of $T$ of the form

$$T(x) = T_0(x) + \epsilon T_1(x) + \epsilon^2 T_2(x) + \cdots \tag{9.48}$$

Because we focus on low-temperature solutions, we will insist that $T_0(x) = 0$, giving

$$T = \epsilon T_1(x) + \epsilon^2 T_2(x) + \cdots \tag{9.49}$$

Here, we assume $T_1(x) \sim \mathcal{O}(1), T_2(x) \sim \mathcal{O}(1), \ldots$. We focus attention on getting an approximate solution for $T_1(x)$.

With the assumption of Eq. (9.49), Eq. (9.46) expands to

$$\epsilon \frac{d^2 T_1}{dx^2} + \cdots = -\mathfrak{D}(1 - \epsilon T_1 - \cdots) \exp(-\Theta(1 - \epsilon Q T_1 + \cdots)). \tag{9.50}$$

Ignoring higher-order terms and simplifying, we get

$$\epsilon \frac{d^2 T_1}{dx^2} = -\mathfrak{D}(1 - \epsilon T_1) \exp(-\Theta) \exp(\epsilon \Theta Q T_1). \tag{9.51}$$

Once again, take the high-activation-energy limit and insist that $\epsilon$ be defined such that

$$\epsilon \equiv \frac{1}{Q\Theta}. \tag{9.52}$$

This definition is similar to the one adopted in Eq. (1.351). This gives us

$$\frac{d^2 T_1}{dx^2} = -\frac{\mathfrak{D}}{\epsilon}(1 - \epsilon T_1) \exp(-\Theta) \exp(T_1). \tag{9.53}$$

We have gained an analytic advantage once again by moving the temperature into the numerator of the argument of the exponential.

Now, let us neglect $\epsilon T_1$ as small relative to 1 and define a new parameter $\delta$ such that

$$\delta \equiv \frac{\mathfrak{D}}{\epsilon} \exp(-\Theta) = \mathfrak{D} Q \Theta \exp(-\Theta). \tag{9.54}$$

Our governing equation system then reduces to

$$\frac{d^2 T_1}{dx^2} = -\delta e^{T_1}, \qquad T_1(-1) = T_1(1) = 0. \tag{9.55}$$

It is not clear how to solve this. One might assume $T_1$ should have symmetry about $x = 0$. If so, we might also presume that the gradient of $T_1$ is zero at $x = 0$:

$$\left. \frac{dT_1}{dx} \right|_{x=0} = 0, \tag{9.56}$$

which induces $T_1(0)$ to take on an extreme value, say, $T_1(0) = T_1^m$, where $m$ could denote maximum or minimum. Certainly $T_1^m$ is unknown at this point.

Let us explore the appropriate phase space solution. To this end, we define $\mathsf{q}$ as

$$\mathsf{q} \equiv -\frac{dT_1}{dx}. \tag{9.57}$$

We included the minus sign so that $\mathsf{q}$ has a physical interpretation of heat flux. With this assumption, Eq. (9.55) can be written as two autonomous ordinary differential equations in two unknowns:

$$\frac{d\mathsf{q}}{dx} = \delta e^{T_1}, \qquad \mathsf{q}(0) = 0, \tag{9.58}$$

$$\frac{dT_1}{dx} = -\mathsf{q}, \qquad T_1(0) = T_1^m. \tag{9.59}$$

This is slightly unsatisfying because we do not know $T_1^m$. But we presume it exists and is a constant. Let us see if the analysis can reveal it by pressing forward.

Let us scale Eq. (9.59) by Eq. (9.58) to get

$$\frac{dT_1}{d\mathsf{q}} = -\frac{\mathsf{q}}{\delta e^{T_1}}, \qquad T_1|_{\mathsf{q}=0} = T_1^m. \tag{9.60}$$

Now, Eq. (9.60) can be solved by separating variables. Doing so and solving, we get

$$-\delta e^{T_1}\, dT_1 = \mathsf{q}\, d\mathsf{q}, \tag{9.61}$$

$$-\delta e^{T_1} = \frac{\mathsf{q}^2}{2} + C. \tag{9.62}$$

Applying the initial condition, we get

$$-\delta e^{T_1^m} = C, \tag{9.63}$$

$$\frac{\mathsf{q}^2}{2} = \delta(e^{T_1^m} - e^{T_1}), \tag{9.64}$$

$$\mathsf{q} = \sqrt{2\delta(e^{T_1^m} - e^{T_1})}. \tag{9.65}$$

The plus square root is taken here. This will correspond to $x \in [0, 1]$. The negative square root will correspond to $x \in [-1, 0]$.

We can now substitute Eq. (9.65) into Eq. (9.59) to get

$$\frac{dT_1}{dx} = -\sqrt{2\delta(e^{T_1^m} - e^{T_1})}. \tag{9.66}$$

Once again, separate variables to get

$$\frac{dT_1}{\sqrt{2\delta(e^{T_1^m} - e^{T_1})}} = -dx. \tag{9.67}$$

Computer algebra reveals that this can be integrated to form

$$-\frac{\sqrt{2}\tanh^{-1}\left(\sqrt{1 - \frac{e^{T_1}}{e^{T_1^m}}}\right)}{e^{T_1^m/2}\sqrt{\delta}} = -x + C. \tag{9.68}$$

Now, when $x = 0$, we have $T_1 = T_1^m$. The inverse hyperbolic tangent has an argument of zero there, and it evaluates to zero. Thus, we must have $C = 0$, so

$$\frac{\sqrt{2}\tanh^{-1}\left(\sqrt{1-\frac{e^{T_1}}{e^{T_1^m}}}\right)}{e^{T_1^m/2}\sqrt{\delta}} = x, \tag{9.69}$$

$$\tanh^{-1}\left(\sqrt{1-\frac{e^{T_1}}{e^{T_1^m}}}\right) = e^{T_1^m/2}\sqrt{\frac{\delta}{2}}x. \tag{9.70}$$

Now, recall that

$$\tanh^{-1}\sqrt{1-\beta^2} = \operatorname{sech}^{-1}\beta,$$

so Eq. (9.70) becomes

$$\operatorname{sech}^{-1}\left(e^{\frac{T_1-T_1^m}{2}}\right) = e^{T_1^m/2}\sqrt{\frac{\delta}{2}}x, \tag{9.71}$$

$$e^{\frac{T_1-T_1^m}{2}} = \operatorname{sech}\left(e^{T_1^m/2}\sqrt{\frac{\delta}{2}}x\right), \tag{9.72}$$

$$e^{\frac{T_1}{2}} = e^{T_1^m/2}\operatorname{sech}\left(e^{T_1^m/2}\sqrt{\frac{\delta}{2}}x\right), \tag{9.73}$$

$$T_1 = 2\ln\left(e^{T_1^m/2}\operatorname{sech}\left(e^{T_1^m/2}\sqrt{\frac{\delta}{2}}x\right)\right). \tag{9.74}$$

So, we have an exact solution for $T_1(x)$. This is an achievement, but we are not sure it satisfies the boundary condition at $x = 1$, nor do we know the value of $T_1^m$. We must choose $T_1^m$ such that $T_1(1) = 0$, which we have not yet enforced. So

$$0 = 2\ln\left(e^{T_1^m/2}\operatorname{sech}\left(e^{T_1^m/2}\sqrt{\frac{\delta}{2}}\right)\right). \tag{9.75}$$

Only when the argument of a logarithm is unity does it map to zero. So

$$e^{T_1^m/2}\operatorname{sech}\left(e^{T_1^m/2}\sqrt{\frac{\delta}{2}}\right) = 1, \tag{9.76}$$

$$e^{T_1^m/2}\sqrt{\frac{\delta}{2}} = \operatorname{sech}^{-1}\left(e^{-T_1^m/2}\right), \tag{9.77}$$

$$e^{T_1^m/2}\sqrt{\frac{\delta}{2}} = \cosh^{-1}\left(e^{T_1^m/2}\right), \tag{9.78}$$

$$\sqrt{\frac{\delta}{2}} = e^{-T_1^m/2}\cosh^{-1}\left(e^{T_1^m/2}\right). \tag{9.79}$$

Equation (9.79) gives a direct relationship between $\delta$ and $T_1^m$. Given $T_1^m$, we get $\delta$ explicitly. Thus, we can easily generate a plot; see Fig 9.2a. The inverse cannot be achieved analytically but can be done numerically via iterative techniques. We notice a *critical value* of $\delta$, $\delta_c = 0.878458$. When $\delta = \delta_c = 0.878458$, we find $T_1^m = 1.118684$. For $\delta < \delta_c$, there are *two* admissible values of $T_1^m$. That is to say the solution is

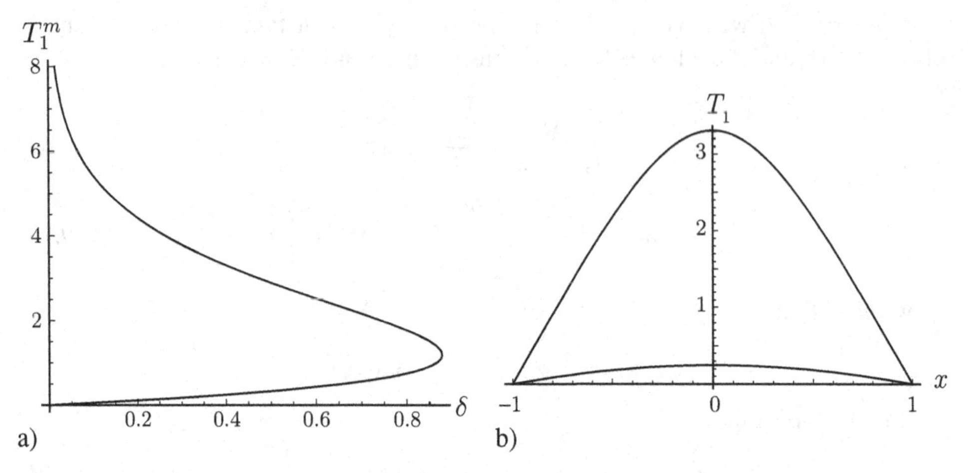


Figure 9.2. Plots of (a) maximum temperature perturbation $T_1^m$ versus reaction rate constant $\delta$ for steady reaction-diffusion in the high-activation-energy limit and (b) $T_1(x)$ for $\delta = 0.4$ showing the two admissible steady solutions.

*nonunique* for $\delta < \delta_c$. Moreover, both solutions are physical. For $\delta = \delta_c$, there is a single solution. For $\delta > \delta_c$, there are no steady solutions.

Note also that because we have from Eq. (9.49) that $T \sim \epsilon T_1$ and from Eq. (9.52) that $\epsilon = 1/\Theta/Q$, we can say

$$T^m \sim \frac{T_1^m}{\Theta Q}. \tag{9.80}$$

Presumably reintroduction of neglected processes would aid in determining which solution is realized in nature. We shall see in Sec. 9.4 that transient stability analysis aids in selecting the correct solution when there are choices. We shall also see in Sec. 9.3.3 that reintroduction of reactant depletion into the model induces additional physical solutions, including those for $\delta > \delta_c$.

For $\delta = 0.4 < \delta_c$ we get two solutions. Both are plotted in Fig 9.2b. The high-temperature solution has $T_1^m = 3.3079$, and the low-temperature solution has $T_1^m = 0.24543$. At this point we are not sure if either solution is temporally stable to small perturbations. We shall demonstrate in Sec. 9.4.1 that the low-temperature solution is stable and the higher-temperature solution is unstable. Certainly for $\delta > \delta_c$, there are no low-temperature solutions. We will see in Sec. 9.4.1 that upon reintroduction of missing physics, there are stable high-temperature solutions. *To prevent high-temperature solutions, we need* $\delta < \delta_c$.

Now, recall from Eq. (9.54) that

$$\delta = \mathfrak{D} Q \Theta e^{-\Theta}. \tag{9.81}$$

Now, to prevent the high-temperature solution, we demand

$$\delta = \mathfrak{D} Q \Theta e^{-\Theta} < 0.878458. \tag{9.82}$$

Let us bring back our dimensional parameters to examine this criterion:

$$\mathfrak{D} Q \Theta e^{-\Theta} < 0.878458, \tag{9.83}$$

$$\underbrace{\frac{L^2 a}{\alpha}}_{\mathfrak{D}} \underbrace{\frac{q}{c_v T_0}}_{Q} \underbrace{\frac{\overline{\mathcal{E}}}{\overline{R} T_0}}_{\Theta} \exp\left(\frac{-\overline{\mathcal{E}}}{\overline{R} T_0}\right) < 0.878458. \tag{9.84}$$

Thus, the factors that tend to prevent thermal explosion are

- High thermal diffusivity; this removes the thermal energy rapidly.
- Small length scales; the energy thus has less distance to diffuse.
- Slow reaction rate kinetics.
- High activation energy.

### 9.3.2 Method of Weighted Residuals

The method of the previous section was challenging. Let us approach the problem of calculating the temperature distribution with a powerful alternate approximate method: the method of weighted residuals (Finlayson, 1972), using so-called Dirac[5] delta functions as weighting functions. We first briefly review the Dirac delta function, then move on to solving the Frank-Kamenetskii problem with the method of weighted residuals.

The Dirac delta function, $\delta_D$, is defined by

$$\int_{\alpha}^{\beta} f(x)\delta_D(x-a)\,dx = \begin{cases} 0 & \text{if } a \notin [\alpha,\beta] \\ f(a) & \text{if } a \in [\alpha,\beta] \end{cases}. \tag{9.85}$$

From this, it follows by considering the special case in which $f(x) = 1$ that

$$\delta_D(x-a) = 0, \text{ if } x \neq a, \tag{9.86}$$

$$\int_{-\infty}^{\infty} \delta_D(x-a)\,dx = 1. \tag{9.87}$$

The Dirac delta, $\delta_D$, has no relation to the parameter of this problem, $\delta$.

Let us first return to Eq. (9.55), rearranged as

$$\frac{d^2T_1}{dx^2} + \delta e^{T_1} = 0, \qquad T_1(-1) = T_1(1) = 0. \tag{9.88}$$

We shall approximate $T_1(x)$ by

$$T_1(x) \approx T_a(x) = \sum_{i=1}^{N} c_i\phi_i(x). \tag{9.89}$$

We have not yet specified the so-called *trial functions* $\phi_i(x)$ nor the constants $c_i$.

#### One-Term Collocation Solution

Let us choose $\phi_i(x)$ to be linearly independent functions that satisfy the boundary conditions. Moreover, let us consider the simplest of approximations for which $N = 1$. A simple function that can satisfy the boundary conditions is a polynomial. The polynomial needs to be at least quadratic to be nontrivial. So, let us take

$$\phi_1(x) = 1 - x^2. \tag{9.90}$$

This gives $\phi_1(-1) = \phi_1(1) = 0$. So, our one-term approximate solution takes the form

$$T_a(x) = c_1(1 - x^2). \tag{9.91}$$

[5]Paul Adrien Maurice Dirac (1902–1984), English physicist; 1933 Nobel laureate in Physics.

We still do not yet have a value for $c_1$. Let us choose $c_1$ so as to minimize a residual error of our approximation. The residual error of our approximation $r(x)$ will be

$$r(x) = \frac{d^2T_a}{dx^2} + \delta e^{T_a(x)}, \tag{9.92}$$

$$= \frac{d^2}{dx^2}\left(c_1(1-x^2)\right) + \delta \exp(c_1(1-x^2)), \tag{9.93}$$

$$= -2c_1 + +\delta \exp(c_1(1-x^2)). \tag{9.94}$$

If we could choose $c_1$ in such a way that $r(x)$ were exactly 0 for $x \in [-1, 1]$, the problem would be solved. But one cannot find such a $c_1$. So, let us choose $c_1$ to drive a weighted domain-averaged residual to zero. That is, let us demand that

$$\int_{-1}^{1} \psi_1(x) r(x)\, dx = 0. \tag{9.95}$$

We have introduced here a weighting function $\psi_1(x)$. Many choices exist for the weighting function. If we choose $\psi_1(x) = \phi_1(x)$, our method is a Galerkin method. Let us choose instead another common weighting, $\psi_1(x) = \delta_D(x)$. This method is known as a *collocation* method. We have chosen a single collocation point at $x = 0$. With this choice, Eq. (9.95) becomes

$$\int_{-1}^{1} \delta_D(x)\left(-2c_1 + \delta \exp(c_1(1-x^2))\right) dx = 0. \tag{9.96}$$

The evaluation of this integral is particularly simple due to the choice of the Dirac weighting. We simply evaluate the integrand at the collocation point $x = 0$ and get

$$-2c_1 + \delta \exp(c_1) = 0, \tag{9.97}$$

$$\delta = 2c_1 e^{-c_1}. \tag{9.98}$$

Once again, $\delta$ is a physical parameter with no relation the Dirac delta function $\delta_D$. With our approximation $T_a(x) = c_1(1-x^2)$, the maximum value of $T_a$ is $T_a^m = c_1$ at $x = 0$. So, we can say

$$\delta = 2T_a^m e^{-T_a^m}. \tag{9.99}$$

We can plot $T_a^m$ as a function of $\delta$; see Fig 9.3a. The predictions of Fig. 9.3a are remarkably similar to those of Fig. 9.2a. For example, when $\delta = 0.4$, numerical solution of Eq. (9.99) yields two roots:

$$T_a^m = 0.259171, \qquad T_a^m = 2.54264. \tag{9.100}$$

Thus, we get explicit approximations for the high- and low-temperature solutions for a one-term collocation approximation:

$$T_a(x) = 0.259171(1-x^2), \tag{9.101}$$

$$T_a(x) = 2.54264(1-x^2). \tag{9.102}$$

Plots of the one-term collocation approximation $T_a(x)$ for high- and low-temperature solutions are given in Fig 9.3b. We can easily find $\delta_c$ for this approximation. Differentiating Eq. (9.99) gives

$$\frac{d\delta}{dT_a^m} = 2e^{-T_a^m}(1 - T_a^m). \tag{9.103}$$

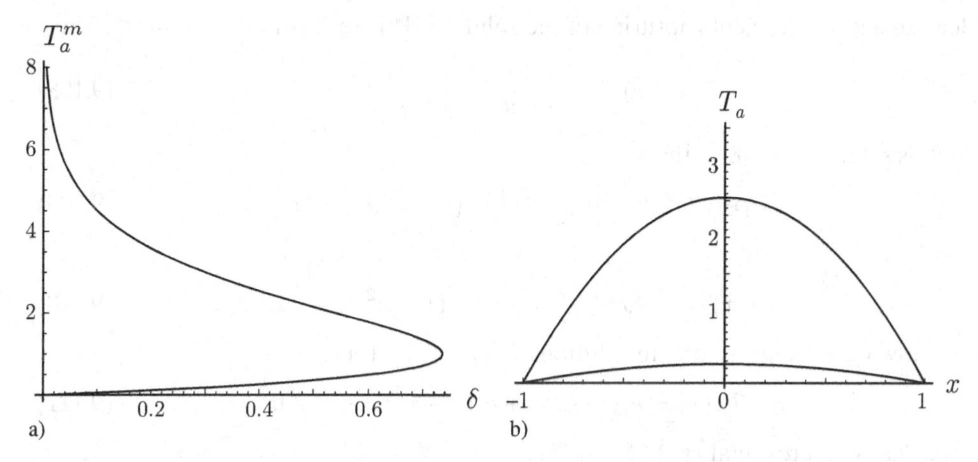


Figure 9.3. Plot of (a) maximum temperature perturbation $T_a^m$ versus reaction rate constant $\delta$ for steady reaction-diffusion using a one-term collocation method and (b) high- and low-temperature distributions $T_a(x)$ using a one-term collocation method.

A critical point exists when $d\delta/dT_a^m = 0$. We find this exists when $T_a^m = 1$. So, $\delta_c = 2(1)e^{-1} = 0.735769$.

### Two-Term Collocation Solution

We can improve accuracy by including more basis functions. Let us take $N = 2$. We have a wide variety of acceptable choices for basis function that (1) satisfy the boundary conditions and (2) are linearly independent. Let us focus on polynomials. We can select the first as before with $\phi_1(x) = 1 - x^2$ as the lowest-order nontrivial polynomial that satisfies both boundary conditions. We could multiply this by an arbitrary constant, but that would not be of any particular use. Leaving out details, if we were to select the second basis function as a cubic polynomial, we would find the coefficient on the cubic basis function to be $c_2 = 0$. That is a consequence of the oddness of the cubic basis function and the evenness of the solution we are simulating. So, it turns out that the lowest-order nontrivial basis function is a quartic polynomial, taken to be of the form

$$\phi_2(x) = a_0 + a_1 x + a_2 x^2 + a_3 x^3 + a_4 x^4. \tag{9.104}$$

We can insist that $\phi_2(-1) = 0$ and $\phi_2(1) = 0$. This gives

$$a_0 + a_1 + a_2 + a_3 + a_4 = 0, \qquad a_0 - a_1 + a_2 - a_3 + a_4 = 0. \tag{9.105}$$

We solve these for $a_0$ and $a_1$ to get $a_0 = -a_2 - a_4, a_1 = -a_3$. Our approximation is

$$\phi_2(x) = (-a_2 - a_4) - a_3 x + a_2 x^2 + a_3 x^3 + a_4 x^4. \tag{9.106}$$

Motivated by the fact that a cubic approximation did not contribute to the solution, let us select $a_3 = 0$ so as to get

$$\phi_2(x) = (-a_2 - a_4) + a_2 x^2 + a_4 x^4. \tag{9.107}$$

Let us now make the choice that $\int_{-1}^{1} \phi_1(x)\phi_2(x)\, dx = 0$. This guarantees an orthogonal basis, although these basis functions are not eigenfunctions of any relevant self-adjoint linear operator for this problem. Often orthogonality of basis functions can lead to a more efficient capturing of the solution. This results in $a_4 = -7a_2/8$. Thus,

$$\phi_2(x) = a_2 \left( -\frac{1}{8} + x^2 - \frac{7}{8} x^4 \right). \tag{9.108}$$

Let us selcct $a_2 = -8$ so that

$$\phi_2(x) = 1 - 8x^2 + 7x^4 = (1 - x^2)(1 - 7x^2) \tag{9.109}$$

and

$$\phi_1(x) = 1 - x^2, \qquad \phi_2(x) = (1 - x^2)(1 - 7x^2). \tag{9.110}$$

So, now we seek approximate solutions $T_A(x)$ of the form

$$T_A(x) = c_1(1 - x^2) + c_2(1 - x^2)(1 - 7x^2). \tag{9.111}$$

This leads to a residual $r(x)$ of

$$r(x) = \frac{d^2 T_A}{dx^2} + \delta e^{T_A(x)}, \tag{9.112}$$

$$= \frac{d^2}{dx^2} \left( c_1(1 - x^2) + c_2(1 - x^2)(1 - 7x^2) \right) + \delta \exp(c_1(1 - x^2) + c_2(1 - x^2)(1 - 7x^2)), \tag{9.113}$$

$$= -2c_1 + 56c_2 x^2 - 2c_2(1 - 7x^2) - 14c_2(1 - x^2) + \delta \exp(c_1(1 - x^2) + c_2(1 - x^2)(1 - 7x^2)). \tag{9.114}$$

Now, we drive two weighted residuals to zero:

$$\int_{-1}^{1} \psi_1(x) r(x)\, dx = 0, \qquad \int_{-1}^{1} \psi_2(x) r(x)\, dx = 0. \tag{9.115}$$

Let us once again choose the weighting functions $\psi_i(x)$ to be Dirac delta functions so that we have a two-term collocation method. Let us choose unevenly distributed collocation points so as to generate independent equations taking $x = 0$ and $x = 1/2$. Symmetric choices would lead to a linearly dependent set of equations. Other unevenly distributed choices would work as well. So, we get

$$\int_{-1}^{1} \delta_D(x) r(x)\, dx = 0, \qquad \int_{-1}^{1} \delta_D(x - 1/2) r(x)\, dx = 0, \tag{9.116}$$

or

$$r(0) = 0, \qquad r(1/2) = 0. \tag{9.117}$$

Expanding, these equations are

$$-2c_1 - 16c_2 + \delta \exp(c_1 + c_2) = 0, \tag{9.118}$$

$$-2c_1 + 5c_2 + \delta \exp\left( \frac{3}{4} c_1 - \frac{9}{16} c_2 \right) = 0. \tag{9.119}$$

For $\delta = 0.4$, we find a high-temperature solution via numerical methods,

$$c_1 = 3.07054, \qquad c_2 = 0.683344, \tag{9.120}$$

and a low-temperature solution as well:

$$c_1 = 0.243622, \qquad c_2 = 0.00149141. \tag{9.121}$$

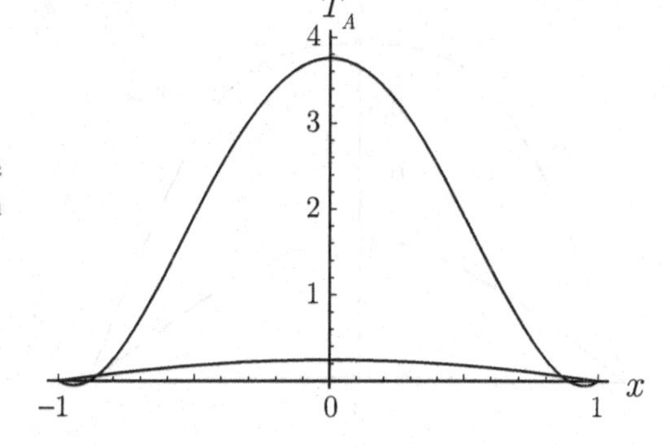


Figure 9.4. Plots of high- and low-temperature distributions $T_A(x)$ using a two-term collocation method.

So, the high-temperature distribution is

$$T_A(x) = 3.07054(1 - x^2) + 0.683344(1 - x^2)(1 - 7x^2). \qquad (9.122)$$

The peak temperature of the high-temperature distribution is $T_A^m = 3.75388$. This is an improvement over the one-term approximation of $T_a^m = 2.54264$ and closer to the high-temperature solution found via exact methods of $T_I^m = 3.3079$.

The low-temperature distribution is

$$T_A(x) = 0.243622(1 - x^2) + 0.00149141(1 - x^2)(1 - 7x^2). \qquad (9.123)$$

The peak temperature of the low-temperature distribution is $T_A^m = 0.245113$. This in an improvement over the one-term approximation of $T_a^m = 0.259171$ and compares favorably to the low-temperature solution found via exact methods of $T_I^m = 0.24543$.

Plots of the two-term collocation approximation $T_A(x)$ for high- and low-temperature solutions are given in Fig 9.4. While the low-temperature solution is accurate, the high-temperature solution exhibits a small negative portion near the boundary.

### 9.3.3 Steady Solution with Reactant Depletion

Let us return to the version of steady state reaction with reactant depletion without resorting to the high-activation-energy limit, Eq. (9.47):

$$0 = \frac{1}{\mathfrak{D}} \frac{d^2T}{dx^2} + (1 - T)\exp\left(\frac{-\Theta}{1 + QT}\right), \qquad 0 = T(-1) = T(1). \qquad (9.124)$$

Rearrange to get

$$\frac{d^2T}{dx^2} = -\mathfrak{D}\,(1 - T)\exp\left(\frac{-\Theta}{1 + QT}\right), \qquad (9.125)$$

$$= -\mathfrak{D}\frac{Q\Theta\exp(-\Theta)}{Q\Theta\exp(-\Theta)}(1 - T)\exp\left(\frac{-\Theta}{1 + QT}\right). \qquad (9.126)$$

Now, simply adapting our earlier definition of $\delta = \mathfrak{D}Q\Theta\exp(-\Theta)$, which does not imply that we have taken any high-activation-energy limits, we get

$$\frac{d^2T}{dx^2} = -\delta\frac{\exp(\Theta)}{Q\Theta}(1 - T)\exp\left(\frac{-\Theta}{1 + QT}\right), \qquad T(-1) = T(1) = 0. \quad (9.127)$$

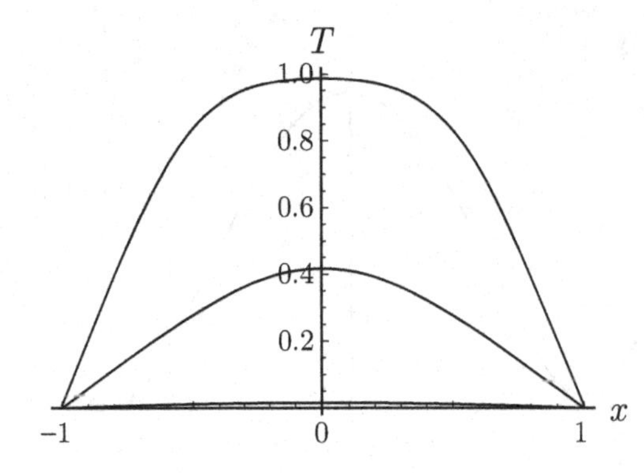


Figure 9.5. Plots of high, low, and intermediate-temperature distributions $T(x)$ for $\delta = 0.4$, $Q = 1$, $\Theta = 15$.

Equations (9.127) can be solved by a numerical trial-and-error method where we demand that $dT/dx(x = 0) = 0$ and guess $T(0)$. We keep guessing $T(0)$ until we have satisfied the boundary conditions.

When we do this with $\delta = 0.4, \Theta = 15, Q = 1$ (so $\mathfrak{D} = \delta e^{\Theta}/Q/\Theta = 87173.8$ and $\epsilon = 1/\Theta/Q = 1/15$), we find *three* steady solutions. One is at low temperature with $T^m = 0.016$. This is essentially equal that predicted by our high activation energy limit estimate of $T^m \sim \epsilon T_1^m = 0.24542/15 = 0.016$. We find a second intermediate-temperature solution with $T^m = 0.417$. This is a little higher than that predicted in the high-activation-energy limit of $T^m \sim \epsilon T_1^m = 3.3079/15 = 0.221$. And we find a high-temperature solution with $T^m = 0.987$. There is no counterpart to this solution from the high-activation-energy limit analysis. Plots of $T(x)$ for high-, low-, and intermediate-temperature solutions are given in Fig 9.5.

We can use a one-term collocation approximation to estimate the relationship between $\delta$ and $T^m$. Let us estimate that

$$T_a(x) = c_1(1 - x^2). \tag{9.128}$$

With that choice, we get a residual of

$$r(x) = -2c_1 + \frac{\delta}{Q\Theta} \exp\left(\Theta - \frac{\Theta}{1 + c_1 Q(1 - x^2)}\right)(1 - c_1(1 - x^2)). \tag{9.129}$$

We choose a one-term collocation method with $\psi_1(x) = \delta_D(x)$. Then, setting the weighted residual to zero, $\int_{-1}^{1} \psi_1(x) r(x)\, dx = 0$, gives

$$r(0) = -2c_1 + \frac{\delta}{Q\Theta} \exp\left(\Theta - \frac{\Theta}{1 + c_1 Q}\right)(1 - c_1) = 0. \tag{9.130}$$

We solve for $\delta$ and get

$$\delta = \frac{2c_1}{1 - c_1} \frac{Q\Theta}{e^{\Theta}} \exp\left(\frac{\Theta}{1 + c_1 Q}\right). \tag{9.131}$$

The maximum temperature of the approximation is given by $T_a^m = c_1$ and occurs at $x = 0$. A plot of $T_a^m$ versus $\delta$ is given in Fig 9.6. For $\delta < \delta_{c1} \sim 0.2$, one low-temperature solution exists. For $\delta_{c1} < \delta < \delta_{c2} \sim 0.84$, three solutions exist. For $\delta > \delta_{c2}$, one high-temperature solution exists.

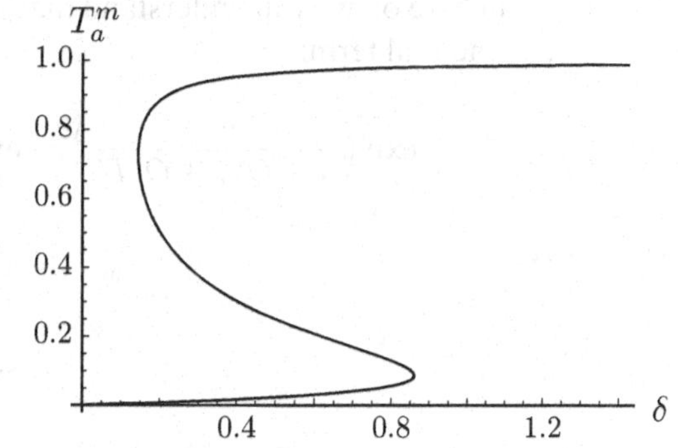


Figure 9.6. Plot of $T_a^m$ versus $\delta$, with $Q = 1, \Theta = 15$ from a one-term collocation approximate solution.

## 9.4 Unsteady Solutions

Let us now study the effects of time dependency. Let us consider the planar, $m = 0$, version of Eqs. (9.39):

$$\frac{\partial T}{\partial t} = \frac{1}{\mathfrak{D}}\frac{\partial^2 T}{\partial x^2} + (1 - T)\exp\left(\frac{-\Theta}{1 + QT}\right),$$
$$T(-1, t) = T(1, t) = 0, \qquad T(x, 0) = 0. \tag{9.132}$$

### 9.4.1 Linear Stability

We first consider small deviations from the steady solutions found earlier and see if those deviations grow or decay with time. This will allow us to make a definitive statement about the linear stability of those steady solutions.

#### Formulation

First, recall that we have independently determined three exact numerical steady solutions to the time-independent version of Eq. (9.132). Let us call any of these $T_e(x)$. By construction, $T_e(x)$ satisfies the boundary conditions on $T$.

Let us subject a steady solution to a small perturbation and consider that to be our initial condition for an unsteady calculation. Take, then,

$$T(x, 0) = T_e(x) + \epsilon A(x), \qquad A(-1) = A(1) = 0, \tag{9.133}$$
$$A(x) = \mathcal{O}(1), \qquad 0 < \epsilon \ll 1. \tag{9.134}$$

Here, $A(x)$ is some function that satisfies the same boundary conditions as $T(x, t)$, and $\epsilon$ is a small parameter, unrelated to any previous version.

Now, let us assume that

$$T(x, t) = T_e(x) + \epsilon T'(x, t), \quad \text{with} \quad T'(x, 0) = A(x). \tag{9.135}$$

Here, $T'$ is an $\mathcal{O}(1)$ quantity; the "prime" notation does not imply any differentiation. We then substitute our Eq. (9.135) into Eq. (9.132) to get

$$\frac{\partial}{\partial t}\left(T_e(x) + \epsilon T'(x, t)\right) = \frac{1}{\mathfrak{D}}\frac{\partial^2}{\partial x^2}\left(T_e(x) + \epsilon T'(x, t)\right)$$
$$+ \left(1 - T_e(x) - \epsilon T'(x, t)\right)$$
$$\times \exp\left(\frac{-\Theta}{1 + QT_e(x) + Q\epsilon T'(x, t)}\right). \tag{9.136}$$

From here on we will understand that $T_e$ is $T_e(x)$ and $T'$ is $T'(x,t)$. Now, consider the exponential term:

$$\exp\left(\frac{-\Theta}{1+QT_e+Q\epsilon T'}\right) = \exp\left(\frac{-\Theta}{1+QT_e}\frac{1}{1+\frac{Q}{1+QT_e}\epsilon T'}\right), \tag{9.137}$$

$$\sim \exp\left(\frac{-\Theta}{1+QT_e}\left(1-\frac{Q}{1+QT_e}\epsilon T'\right)\right), \tag{9.138}$$

$$\sim \exp\left(\frac{-\Theta}{1+QT_e}\right) \times \exp\left(\frac{\epsilon\Theta Q}{(1+QT_e)^2}T'\right), \tag{9.139}$$

$$\sim \exp\left(\frac{-\Theta}{1+QT_e}\right) \times \left(1+\frac{\epsilon\Theta Q}{(1+QT_e)^2}T'\right). \tag{9.140}$$

Significantly, we have brought $T'$ out of the denominator of an exponential and recast it so that it appears as a purely linear term in the limits we study. So, our Eq. (9.136) can be rewritten as

$$\begin{aligned}
\frac{\partial}{\partial t}(T_e+\epsilon T') &= \frac{1}{\mathfrak{D}}\frac{\partial^2}{\partial x^2}(T_e+\epsilon T') \\
&\quad + (1-T_e-\epsilon T')\exp\left(\frac{-\Theta}{1+QT_e}\right)\left(1+\frac{\epsilon\Theta Q}{(1+QT_e)^2}T'\right) \\
&= \frac{1}{\mathfrak{D}}\frac{\partial^2}{\partial x^2}(T_e+\epsilon T') \\
&\quad + \exp\left(\frac{-\Theta}{1+QT_e}\right)(1-T_e-\epsilon T')\left(1+\frac{\epsilon\Theta Q}{(1+QT_e)^2}T'\right) \\
&= \frac{1}{\mathfrak{D}}\frac{\partial^2}{\partial x^2}(T_e+\epsilon T') + \exp\left(\frac{-\Theta}{1+QT_e}\right) \\
&\quad \times \left((1-T_e)+\epsilon T'\left(-1+\frac{(1-T_e)\Theta Q}{(1+QT_e)^2}\right)+\mathcal{O}(\epsilon^2)\right) \\
&= \frac{\epsilon}{\mathfrak{D}}\frac{\partial^2 T'}{\partial x^2} + \underbrace{\frac{1}{\mathfrak{D}}\frac{\partial^2 T_e}{\partial x^2} + \exp\left(\frac{-\Theta}{1+QT_e}\right)(1-T_e)}_{0} \\
&\quad + \exp\left(\frac{-\Theta}{1+QT_e}\right)\left(\epsilon T'\left(-1+\frac{(1-T_e)\Theta Q}{(1+QT_e)^2}\right)+\mathcal{O}(\epsilon^2)\right).
\end{aligned} \tag{9.141}$$

Now, we recognize the bracketed term as zero because $T_e(x)$ is constructed to satisfy the steady state equation. We also recognize that $\partial T_e(x)/\partial t = 0$. So, neglecting $\mathcal{O}(\epsilon^2)$ terms and canceling $\epsilon$, our equation reduces to

$$\frac{\partial T'}{\partial t} = \frac{1}{\mathfrak{D}}\frac{\partial^2 T'}{\partial x^2} + \exp\left(\frac{-\Theta}{1+QT_e}\right)\left(-1+\frac{(1-T_e)\Theta Q}{(1+QT_e)^2}\right)T'. \tag{9.142}$$

Equation (9.142) is a *linear* partial differential equation for $T'(x,t)$. It is of the form

$$\frac{\partial T'}{\partial t} = \frac{1}{\mathfrak{D}}\frac{\partial^2 T'}{\partial x^2} + B(x)T', \tag{9.143}$$

with

$$B(x) \equiv \exp\left(\frac{-\Theta}{1+QT_e(x)}\right)\left(-1+\frac{(1-T_e(x))\Theta Q}{(1+QT_e(x))^2}\right). \tag{9.144}$$

**Separation of Variables**

Let us use the technique of separation of variables to solve Eq. (9.143). First assume that $T'$ can be separated into the product of two functions, one of $x$ and the other of $t$:

$$T'(x,t) = H(x)K(t). \tag{9.145}$$

So, Eq. (9.143) becomes

$$H(x)\frac{dK(t)}{dt} = \frac{1}{\mathfrak{D}}K(t)\frac{d^2H(x)}{dx^2} + B(x)H(x)K(t), \tag{9.146}$$

$$\frac{1}{K(t)}\frac{dK(t)}{dt} = \frac{1}{\mathfrak{D}}\frac{1}{H(x)}\frac{d^2H(x)}{dx^2} + B(x) = -\lambda. \tag{9.147}$$

Because the left side is a function of $t$ and the right side is a function of $x$, the only way they to achieve equality is if both are the same constant. We will call that constant $-\lambda$. Now, Eq. (9.147) really contains two equations, the first of which is

$$\frac{dK(t)}{dt} + \lambda K(t) = 0. \tag{9.148}$$

This has solution

$$K(t) = C\exp(-\lambda t), \tag{9.149}$$

where $C$ is some arbitrary constant. Clearly, if $\lambda > 0$, this solution is stable, with time constant of relaxation $\tau = 1/\lambda$.

The second differential equation contained within Eq. (9.147) is

$$\frac{1}{\mathfrak{D}}\frac{d^2H(x)}{dx^2} + B(x)H(x) = -\lambda H(x), \tag{9.150}$$

$$\left(-\frac{1}{\mathfrak{D}}\frac{d^2}{dx^2} - B(x)\right)H(x) = \lambda H(x). \tag{9.151}$$

This is of the classical eigenvalue form for a linear operator $\mathcal{L}$; that is, $\mathcal{L}(H(x)) = \lambda H(x)$, where

$$\mathcal{L} = -\frac{1}{\mathfrak{D}}\frac{d^2}{dx^2} - B. \tag{9.152}$$

We also must have

$$H(-1) = H(1) = 0 \tag{9.153}$$

to satisfy the spatially homogeneous boundary conditions on $T'(x,t)$. If $B \equiv 0$, the operator $\mathcal{L}$ is positive definite, all the eigenvalues are positive, and the system is stable. The fact that $B \neq 0$ gives rise to the possibility that $\mathcal{L}$ is not positive definite, and the system may be unstable.

This eigenvalue problem is difficult to solve because of the complicated nature of $B(x)$. Let us see how the solution would proceed in the limiting case of $B$ as a constant. We will generalize later. If $B$ is a constant, we have

$$\frac{d^2H}{dx^2} + (B + \lambda)\mathfrak{D}H = 0, \qquad H(-1) = H(1) = 0. \tag{9.154}$$

The following mapping simplifies the problem somewhat:

$$y = \frac{x+1}{2}. \tag{9.155}$$

This takes our domain of $x \in [-1, 1]$ to $y \in [0, 1]$. By the chain rule,

$$\frac{dH}{dx} = \frac{dH}{dy}\frac{dy}{dx} = \frac{1}{2}\frac{dH}{dy}.$$

So,

$$\frac{d^2H}{dx^2} = \frac{1}{4}\frac{d^2H}{dy^2}.$$

Our eigenvalue problem thus transforms to

$$\frac{d^2H}{dy^2} + 4\mathfrak{D}(B + \lambda)H = 0, \qquad H(0) = H(1) = 0. \tag{9.156}$$

This has solution

$$H(y) = C_1 \cos\left(\left(\sqrt{4\mathfrak{D}(B+\lambda)}\,\right)y\right) + C_2 \sin\left(\left(\sqrt{4\mathfrak{D}(B+\lambda)}\right)y\right). \tag{9.157}$$

At $y = 0$ we have, then,

$$H(0) = 0 = C_1(1) + C_2(0), \tag{9.158}$$

so $C_1 = 0$. Thus,

$$H(y) = C_2 \sin\left(\left(\sqrt{4\mathfrak{D}(B+\lambda)}\right)y\right). \tag{9.159}$$

At $y = 1$, we have the other boundary condition:

$$H(1) = 0 = C_2 \sin\left(\left(\sqrt{4\mathfrak{D}(B+\lambda)}\right)\right). \tag{9.160}$$

Because $C_2 \neq 0$ to avoid a trivial solution, we must require that

$$\sin\left(\left(\sqrt{4\mathfrak{D}(B+\lambda)}\right)\right) = 0. \tag{9.161}$$

For this to occur, the argument of the sin function must be an integer multiple of $\pi$:

$$\sqrt{4\mathfrak{D}(B+\lambda)} = n\pi, \qquad n = 1, 2, 3, \ldots \tag{9.162}$$

Thus,

$$\lambda = \frac{n^2\pi^2}{4\mathfrak{D}} - B. \tag{9.163}$$

We need $\lambda > 0$ for stability. For large $n$ and $\mathfrak{D} > 0$, we have stability. Depending on the value of $B$, low $n$, which corresponds to low-frequency modes, could be unstable. Looked at another way, when combustion is not present $B = 0$, and we are guaranteed stability for the purely diffusive system. The presence of combustion, with nonzero $B$, may or may not induce instability, depending on its magnitude and the wavenumber of the disturbance.

**Numerical Eigenvalue Solution**

Let us return to the full problem where $B = B(x)$. Let us solve the eigenvalue problem via the method of finite differences. Let us take our domain $x \in [-1, 1]$ and discretize into $N$ points with

$$\Delta x = \frac{2}{N-1}, \qquad x_i = (i-1)\Delta x - 1. \tag{9.164}$$

Note that when $i = 1$, $x_i = -1$, and when $i = N$, $x_i = 1$. Let us define $B(x_i) = B_i$ and $H(x_i) = H_i$. We can rewrite Eq. (9.151) as

$$-\frac{1}{\mathfrak{D}}\frac{d^2H(x)}{dx^2} - (B(x) + \lambda)H(x) = 0, \qquad H(-1) = H(1) = 0. \tag{9.165}$$

Now, let us apply an appropriate equation at each node. At $i = 1$, we must satisfy the boundary condition so

$$H_1 = 0. \tag{9.166}$$

At $i = 2$, we discretize Eq. (9.165) with a second-order central difference to obtain

$$-\frac{1}{\mathfrak{D}}\frac{H_1 - 2H_2 + H_3}{\Delta x^2} - (B_2 + \lambda)H_2 = 0. \tag{9.167}$$

We could have chosen a higher- or lower-order differencing scheme. We get a similar equation at a general interior node $i$:

$$-\frac{1}{\mathfrak{D}}\frac{H_{i-1} - 2H_i + H_{i+1}}{\Delta x^2} - (B_i + \lambda)H_i = 0. \tag{9.168}$$

At the $i = N - 1$ node, we have

$$-\frac{1}{\mathfrak{D}}\frac{H_{N-2} - 2H_{N-1} + H_N}{\Delta x^2} - (B_{N-1} + \lambda)H_{N-1} = 0. \tag{9.169}$$

At the $i = N$ node, we have the boundary condition

$$H_N = 0. \tag{9.170}$$

These represent a linear tridiagonal system of equations of the form

$$\underbrace{\begin{pmatrix} \left(\frac{2}{\mathfrak{D}\Delta x^2} - B_2\right) & -\frac{1}{\mathfrak{D}\Delta x^2} & 0 & 0 & \dots & 0 \\ -\frac{1}{\mathfrak{D}\Delta x^2} & \left(\frac{2}{\mathfrak{D}\Delta x^2} - B_3\right) & -\frac{1}{\mathfrak{D}\Delta x^2} & 0 & \dots & 0 \\ 0 & -\frac{1}{\mathfrak{D}\Delta x^2} & \dots & \dots & \dots & \dots \\ \dots & \dots & \dots & \dots & \dots & \dots \\ 0 & 0 & \dots & \dots & \dots & \dots \end{pmatrix}}_{\mathbf{L}} \underbrace{\begin{pmatrix} H_2 \\ H_3 \\ \vdots \\ \vdots \end{pmatrix}}_{\mathbf{h}} = \lambda \begin{pmatrix} H_2 \\ H_3 \\ \vdots \\ \vdots \end{pmatrix}. \tag{9.171}$$

This is of the classical linear algebraic eigenvalue form $\mathbf{L} \cdot \mathbf{h} = \lambda \mathbf{h}$. All one need do after discretization is find the eigenvalues of the matrix $\mathbf{L}$. These will be good approximations to the eigenvalues of the differential operator $\mathcal{L}$. The eigenvectors of $\mathbf{L}$ will be good approximations of the eigenfunctions of $\mathcal{L}$. To get a better approximation, one need only reduce $\Delta x$.

Because the matrix $\mathbf{L}$ is symmetric, the eigenvalues are guaranteed real, and the eigenvectors are guaranteed orthogonal. This is actually a consequence of the original problem being in Sturm-Liouville form, which is guaranteed to be self-adjoint with real eigenvalues and orthogonal eigenfunctions.

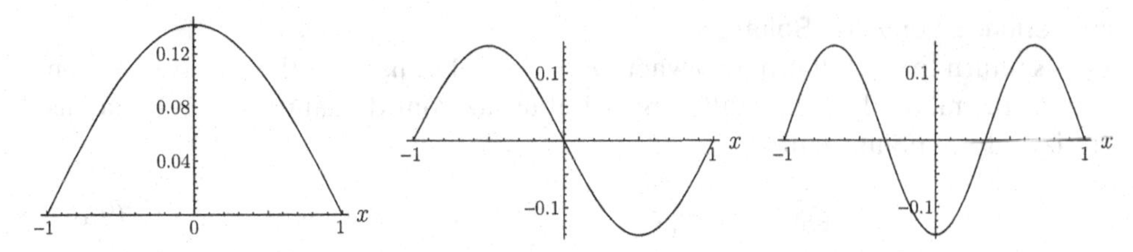


Figure 9.7. Plots of first, second, and third harmonic modes of eigenfunctions versus $x$, with $\delta = 0.4, Q = 1$, and $\Theta = 15$, for low-temperature steady solution $T_e(x)$.

**LOW-TEMPERATURE TRANSIENTS.** For our case of $\delta = 0.4$, $Q = 1$, $\Theta = 15$ (so $\mathfrak{D} = 87173.8$), we can calculate the stability of the low-temperature solution. Choosing $N = 101$ points to discretize the domain, we find a set of eigenvalues. They are all positive, so the solution is stable. The first few are

$$\lambda = 0.0000232705, 0.000108289, 0.000249682, 0.000447414, \ldots. \tag{9.172}$$

The first few eigenvalues can be approximated by inert theory with $B(x) = 0$, see Eq. (9.163):

$$\lambda \sim \frac{n^2\pi^2}{4\mathfrak{D}} = 0.0000283044, 0.000113218, 0.00025474, 0.00045287, \ldots. \tag{9.173}$$

The first eigenvalue is associated with the longest time scale $\tau = 1/0.0000232705 = 42972.9$ and a low-frequency mode, whose shape is given by the associated eigenvector, plotted in Fig. 9.7. This represents the fundamental mode, also known as the *first harmonic mode*. Shown also in Fig. 9.7 are the second and third harmonic modes.

**INTERMEDIATE-TEMPERATURE TRANSIENTS.** For the intermediate-temperature solution with $T^m = 0.417$, we find the first few eigenvalues to be

$$\lambda = -0.0000383311, 0.0000668221, 0.000209943, \ldots \tag{9.174}$$

Except for the first, all the eigenvalues are positive. The first eigenvalue of $\lambda = -0.0000383311$ is associated with an *unstable* fundamental mode. All the higher-order harmonic modes are stable. We plot the first three modes in Fig. 9.8.

**HIGH-TEMPERATURE TRANSIENTS.** For the high-temperature solution with $T^m = 0.987$, we find the first few eigenvalues to be

$$\lambda = 0.000146419, 0.00014954, 0.000517724, \ldots \tag{9.175}$$

All the eigenvalues are positive, so all modes are stable. We plot the first three modes in Fig. 9.9.

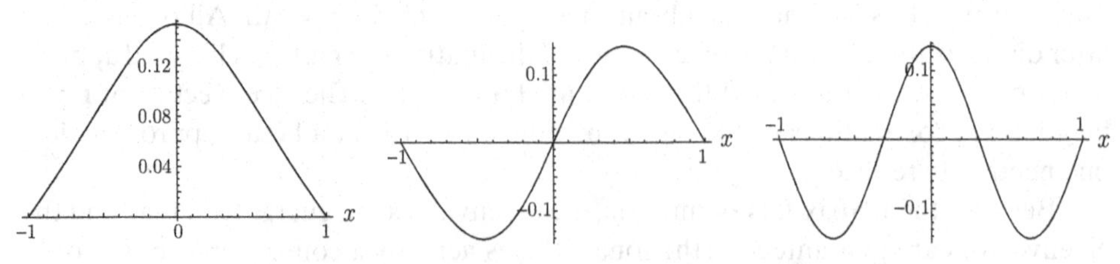


Figure 9.8. Plot of first, second, and third harmonic modes of eigenfunctions versus $x$, with $\delta = 0.4, Q = 1, \Theta = 15$, for intermediate-temperature steady solution $T_e(x)$.

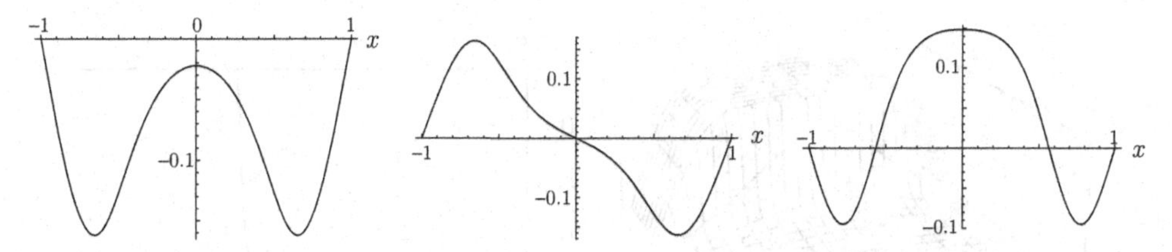


Figure 9.9. Plot of first, second, and third harmonic modes of eigenfunctions versus $x$, with $\delta = 0.4, Q = 1, \Theta = 15$, for high-temperature steady solution $T_e(x)$.

### 9.4.2 Full Transient Solution

We can get a full transient solution to Eqs. (9.132) with numerical methods. We omit details that can be found in standard texts (e.g., Iserles, 2008).

#### Low-Temperature Solution

For our case of $\delta = 0.4, Q = 1, \Theta = 15$ (so $\mathfrak{D} = 87173.8$), we show a plot of the full transient solution in Fig. 9.10. Also seen in Fig. 9.10 is that the centerline temperature $T(0,t)$ relaxes to the long time value predicted by the low-temperature steady solution:

$$\lim_{t\to\infty} T(0,t) = 0.016. \tag{9.176}$$

#### High-Temperature Solution

We next select a value of $\delta = 1.2 > \delta_c$. This should induce transition to a high-temperature solution. We maintain $\Theta = 15, Q = 1$. We get $\mathfrak{D} = \delta e^{\Theta}/\Theta/Q = 261521$. The full transient solution is shown in Fig. 9.11. Also shown in Fig. 9.11 is the centerline temperature $T(0,t)$. We see it relaxes to the long time value predicted by the high-temperature steady solution:

$$\lim_{t\to\infty} T(0,t) = 0.9999185. \tag{9.177}$$

It is seen that there is a rapid acceleration of the reaction for $t \sim 10^6$. This compares with the prediction of Eq. (1.371), the induction time from the infinite Damköhler number, $\mathfrak{D} \to \infty$, thermal explosion theory to occur when

$$t \to \frac{e^{\Theta}}{Q\Theta} = \frac{e^{15}}{(1)(15)} = 2.17934 \times 10^5. \tag{9.178}$$

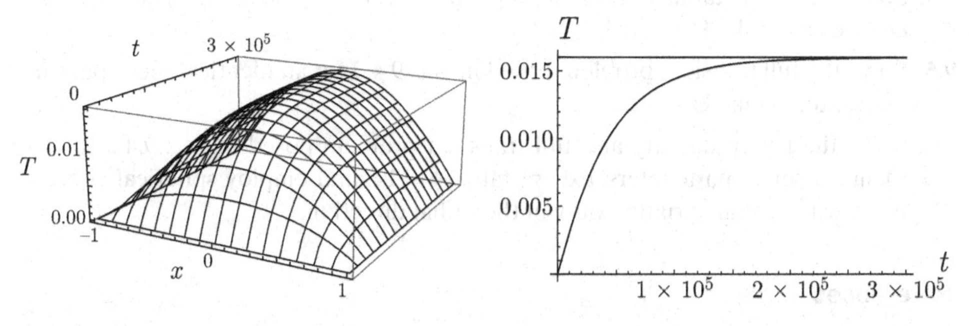


Figure 9.10. Plot of $T(x,t)$ and plot of $T(0,t)$ along with the long time exact low-temperature centerline solution, $T_e(0)$, with $\delta = 0.4, Q = 1, \Theta = 15$.

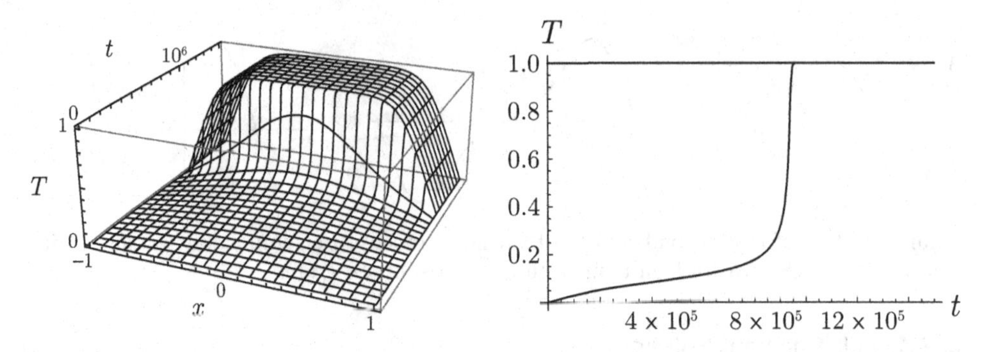


Figure 9.11. Plot of $T(x,t)$ and plot of $T(0,t)$ along with the long time exact low-temperature centerline solution, $T_e(0)$, with $\delta = 1.2, Q = 1, \Theta = 15$.

The estimate underpredicts the value by a factor of five. This is likely due to (1) cooling of the domain due to the low-temperature boundaries at $x = \pm 1$ and (2) effects of finite activation energy.

In summary, this detailed nonlinear example shows clear consequences of the competition between reaction and diffusion as they simultaneously influence the system dynamics. The simplifications employed enabled a wide variety of independent analytical and numerical tools to be employed. That each of these methods yielded consistent predictions is useful to achieve confidence that the numerical approximations to the nonlinear dynamics are in fact accurate.

## EXERCISES

**9.1.** Pose the problem of Sec. 9.3.1 in polar coordinates and consider variation in the radial direction only. Find $\delta_c$ for this configuration. Approximate $\delta_c$ by use of a one-term collocation method similar to that described in Sec. 9.3.2.

**9.2.** Pose the problem of Sec. 9.3.1 in spherical coordinates and consider variation in the radial direction only. Find $\delta_c$ for this configuration. Approximate $\delta_c$ by use of a two-term collocation method similar to that described in Sec. 9.3.2.

**9.3.** Solve the problem posed in Sec. 9.3.3 with one- and two-term collocation methods for an identical set of parameters, except take $\Theta = 14$.

**9.4.** Solve the linear stability problem posed in Sec. 9.4.1 for an identical set of parameters, except take $\Theta = 14$.

**9.5.** Solve the full transient problem posed in Sec. 9.4.2 for an identical set of parameters, except take $\Theta = 14$.

**9.6.** Solve the linear stability and full transient problem posed in Sec. 9.4.2 for an identical set of parameters, except take $\Theta = 14$, and employ spherical coordinates with spatial variation only in the radial direction.

## References

J. Bebernes and D. Eberly, 1989, *Mathematical Problems from Combustion Theory*, Springer, Berlin.

J. D. Buckmaster and G. S. S. Ludford, 1983, *Lectures on Mathematical Combustion*, SIAM, Philadelphia, Chapter 1.
B. A. Finlayson, 1972, *The Method of Weighted Residuals and Variational Principles*, Academic Press, New York.
D. A. Frank-Kamenetskii, 1969, *Diffusion and Heat Transfer in Chemical Kinetics*, 2nd ed., Plenum Press, New York.
I. Glassman, R. A. Yetter, and N. G. Glumac, 2014, *Combustion*, 5th ed., Academic Press, New York, Chapter 7.
J. F. Griffiths and J. A. Barnard, 1995, *Flame and Combustion*, 3rd ed., Chapman and Hall, Glasgow.
A. Iserles, 2008, *A First Course in the Numerical Analysis of Differential Equations*, Cambridge University Press, Cambridge, UK.
A. M. Kanury, 1975, *Introduction to Combustion Phenomena*, Gordon and Breach, Amsterdam.
A. K. Kapila, 1983, *Asymptotic Treatment of Chemically Reacting Systems*, Pitman, Boston.
C. K. Law, 2006, *Combustion Physics*, Cambridge University Press, New York, p. 189.
R. A. Strehlow, 1984, *Combustion Fundamentals*, McGraw-Hill, New York, p. 120.

# 10 Premixed Laminar Flame

*There is no more open door by which you can enter into the study of natural philosophy than by considering the physical phenomena of a candle.*[*]

— Michael Faraday (1791–1867), *The Chemical History of a Candle*

Here, we will consider premixed one-dimensional steady laminar flames. These flames are even simpler than candle flames, which are complicated by buoyancy, multidimensionality, lack of premixing, and pyrolysis of solid to gaseous fuel. Background is available in many sources (e.g., Bone and Townend, 1927; Hirschfelder et al., 1954; Lewis and von Elbe, 1961; Chigier, 1981; Buckmaster and Ludford, 1982; Strehlow, 1984; Williams, 1985; Kee et al., 2003; Poinsot and Veynante, 2005; Law, 2006; Warnatz et al., 2006). We will not discuss unsteady flames and acoustics, for which a rich literature exists (e.g., Matalon, 2007; Lieuwen, 2012).

This topic is broad and widely studied; in a similar spirit as that of Layzer (1954a, 1954b), we restrict attention to the simplest cases that illustrate the nonlinear dynamics of coupled advection, reaction, and diffusion. The methods of this section are impractical for realistic premixed laminar flames. Moreover, there are many classes of laminar flames that we do not discuss. Instead, our emphasis is to give an example of how to exploit dynamical system theory to better understand a simple flame. We consider a simple reversible kinetics model

$$A \leftrightharpoons B, \tag{10.1}$$

where $A$ and $B$ have identical molecular masses, $M_A = M_B = M$, and are both calorically perfect ideal gases with the same specific heats, $c_{PA} = c_{PB} = c_P$. Because we are modeling the system as premixed, we consider species $A$ to be composed of molecules that have their own fuel and oxidizer. This is not common in hydrocarbon kinetics, but is more so in the realm of explosives. More common in gas phase kinetics are situations in which cold fuel and cold oxidizer react together to form hot products. In such a situation, one can also consider nonpremixed flames in which streams of fuel first must mix with streams of oxidizer before significant reaction can commence. Such flames will not be considered.

The introduction of advection introduces an unusual mathematical difficulty in properly modeling reacting flow on the doubly infinite spatial domains we consider. We envision our flame as that in which a cold fluid particle at $x \to -\infty$ moves in the

[*]M. Faraday, 1861, *The Chemical History of a Candle*, Griffin and Bohn, London, 1861, p. 1.

direction of increasing $x$, undergoes vigorous reaction near $x = 0$, and then relaxes to a hot equilibrium state as $x \to \infty$. If small finite reaction rates are permitted at the cold boundary at $-\infty$, the fluid particle will have infinite time to react. We need to find a way to formally drive the reaction rate to zero at the cold boundary. This will be achieved in a way that is unappealing, but nevertheless both common and useful. We shall introduce an ignition temperature $T_{ig}$ in our reaction kinetics law to suppress all reaction for $T < T_{ig}$. This will render the cold boundary to be a true mathematical equilibrium point of the model. This is not a traditional chemical equilibrium, but it is an equilibrium in the formal mathematical sense. We close the chapter by comparing to an analogous flame in a hydrogen-air system modeled by equations of similar complexity to those given in Chapter 7.

## 10.1 Governing Equations

### 10.1.1 Evolution Equations

Consider conservation and evolution equations, independent of any specific material.

#### Conservative Form

Let us commence with the one-dimensional, one-step kinetics versions of the conservative form Eqs. (7.1)–(7.3) and (7.8):

$$\frac{\partial \rho}{\partial t} + \frac{\partial}{\partial x}(\rho u) = 0, \tag{10.2}$$

$$\frac{\partial}{\partial t}(\rho u) + \frac{\partial}{\partial x}\left(\rho u^2 + P - \tau\right) = 0, \tag{10.3}$$

$$\frac{\partial}{\partial t}\left(\rho\left(e + \frac{1}{2}u^2\right)\right) + \frac{\partial}{\partial x}\left(\rho u\left(e + \frac{1}{2}u^2\right) + j^q + (P - \tau)u\right) = 0, \tag{10.4}$$

$$\frac{\partial}{\partial t}(\rho Y_B) + \frac{\partial}{\partial x}\left(\rho u Y_B + j_B^m\right) = M\dot{\omega}_B. \tag{10.5}$$

#### Nonconservative Form

Using standard reductions similar to those made to achieve Eqs. (7.33), (7.36), (7.41), and (7.52), the nonconservative form of the governing equations, Eqs. (10.2)–(10.5), is

$$\frac{\partial \rho}{\partial t} + u\frac{\partial \rho}{\partial x} + \rho\frac{\partial u}{\partial x} = 0, \tag{10.6}$$

$$\rho\frac{\partial u}{\partial t} + \rho u\frac{\partial u}{\partial x} + \frac{\partial P}{\partial x} - \frac{\partial \tau}{\partial x} = 0, \tag{10.7}$$

$$\rho\frac{\partial e}{\partial t} + \rho u\frac{\partial e}{\partial x} + \frac{\partial j^q}{\partial x} + P\frac{\partial u}{\partial x} - \tau\frac{\partial u}{\partial x} = 0, \tag{10.8}$$

$$\rho\frac{\partial Y_B}{\partial t} + \rho u\frac{\partial Y_B}{\partial x} + \frac{\partial j_B^m}{\partial x} = M\dot{\omega}_B. \tag{10.9}$$

Equations (10.6)–(10.9) can be written more compactly using the material derivative, $d/dt = \partial/\partial t + u\partial/\partial x$:

$$\frac{d\rho}{dt} + \rho\frac{\partial u}{\partial x} = 0, \tag{10.10}$$

$$\rho\frac{du}{dt} + \frac{\partial P}{\partial x} - \frac{\partial \tau}{\partial x} = 0, \tag{10.11}$$

$$\rho\frac{de}{dt} + \frac{\partial j^q}{\partial x} + P\frac{\partial u}{\partial x} - \tau\frac{\partial u}{\partial x} = 0, \tag{10.12}$$

$$\rho\frac{dY_B}{dt} + \frac{\partial j_B^m}{\partial x} = M\dot{\omega}_B. \tag{10.13}$$

### Formulation Using Enthalpy

It will be more useful to formulate the equations using enthalpy. Use the definition of enthalpy, Eq. (3.73), $h = e + Pv = e + P/\rho$, to get an expression for $dh$:

$$dh = de - \frac{P}{\rho^2}d\rho + \frac{1}{\rho}dP, \tag{10.14}$$

$$\frac{dh}{dt} = \frac{de}{dt} - \frac{P}{\rho^2}\frac{d\rho}{dt} + \frac{1}{\rho}\frac{dP}{dt}, \tag{10.15}$$

$$\rho\frac{de}{dt} - \frac{P}{\rho}\frac{d\rho}{dt} = \rho\frac{dh}{dt} - \frac{dP}{dt}. \tag{10.16}$$

Now, using the mass equation, Eq. (10.10), to eliminate $P\partial u/\partial x$ in the energy equation, Eq. (10.12), the energy equation can be rewritten as

$$\rho\frac{de}{dt} + \frac{\partial j^q}{\partial x} - \frac{P}{\rho}\frac{d\rho}{dt} - \tau\frac{\partial u}{\partial x} = 0. \tag{10.17}$$

Next, use Eq. (10.16) to simplify Eq. (10.17):

$$\rho\frac{dh}{dt} + \frac{\partial j^q}{\partial x} - \frac{dP}{dt} - \tau\frac{\partial u}{\partial x} = 0. \tag{10.18}$$

### Low Mach Number Limit

In the limit of low Mach number, one can do a formal asymptotic expansion with the reciprocal of the Mach number squared as a perturbation parameter (Paolucci, 1982; Kee et al., 2003, pp. 120–123). All variables take the form $\psi = \psi_0 + \mathsf{M}^2\psi_1 + \cdots$, where $\psi$ is a general variable and M is the Mach number, not to be confused with the molecular mass $M$. In this limit, the linear momentum equation can be shown to reduce at leading order to $\partial P_0/\partial x = 0$, giving rise to $P_0 = P_0(t)$.

We shall ultimately be concerned only with time-independent flows where $P = P_0$. We adopt the constant pressure assumption now. Also in the low Mach number limit, it can be shown that viscous work is negligible, so $\tau\partial u/\partial x \sim 0$. Our evolution equations, Eqs. (10.10), (10.18), and (10.13), then in the low Mach number, constant pressure limit are

$$\frac{\partial \rho}{\partial t} + u\frac{\partial \rho}{\partial x} + \rho\frac{\partial u}{\partial x} = 0, \tag{10.19}$$

$$\rho\frac{\partial h}{\partial t} + \rho u\frac{\partial h}{\partial x} + \frac{\partial j^q}{\partial x} = 0, \tag{10.20}$$

$$\rho\frac{\partial Y_B}{\partial t} + \rho u\frac{\partial Y_B}{\partial x} + \frac{\partial j_B^m}{\partial x} = M\dot{\omega}_B. \tag{10.21}$$

We have effectively removed the linear momentum equation. It would reappear in a nontrivial way were we to formulate the equations at the next order. Equations (10.19)–(10.21) take on the conservative form

$$\frac{\partial \rho}{\partial t} + \frac{\partial}{\partial x}(\rho u) = 0, \tag{10.22}$$

$$\frac{\partial}{\partial t}(\rho h) + \frac{\partial}{\partial x}\left(\rho u h + j^q\right) = 0, \tag{10.23}$$

$$\frac{\partial}{\partial t}(\rho Y_B) + \frac{\partial}{\partial x}\left(\rho u Y_B + j_B^m\right) = M\dot{\omega}_B. \tag{10.24}$$

### 10.1.2 Constitutive Models

Equations (10.19)–(10.21) are supplemented by simple constitutive models:

$$P_0 = \rho R T, \tag{10.25}$$

$$h = c_P(T - T_0) - Y_B q, \tag{10.26}$$

$$j_B^m = -\rho \mathcal{D} \frac{\partial Y_B}{\partial x}, \tag{10.27}$$

$$j^q = -\mathsf{k}\frac{\partial T}{\partial x} + \rho \mathcal{D} q \frac{\partial Y_B}{\partial x}, \tag{10.28}$$

$$\dot{\omega}_B = a T^{\beta} e^{-\mathcal{E}/(RT)} \underbrace{\frac{\rho}{M}(1 - Y_B)}_{\overline{\rho}_A} \left(1 - \frac{1}{K_c}\underbrace{\left(\frac{Y_B}{1 - Y_B}\right)}_{\overline{\rho}_B/\overline{\rho}_A}\right) H(T - T_{ig}), \tag{10.29}$$

$$K_c = e^{q/(RT)}. \tag{10.30}$$

Thus, we have nine equations for the nine unknowns, $\rho, u, h, T, j^q, Y_B, j_B^m, \dot{\omega}_B, K_c$. Many of these are obvious. Some are not. First, we note the new factor $H(T - T_{ig})$ in our kinetics law, Eq. (10.29). Here, $H$ is a Heaviside[1] unit step function:

$$H(T - T_{ig}) = \begin{cases} 0, & T < T_{ig}, \\ 1, & T \geq T_{ig}. \end{cases} \tag{10.31}$$

Next, let us see how to get Eq. (10.27) from the more general Eq. (7.75). We first take the thermal diffusion coefficient to be zero, $D_i^T = 0$. Because $M_A = M_B = M$, $y_k = Y_k$; that is, mol fractions are the same as mass fractions. So, Eq. (7.75) simplifies considerably to

$$j_i^m = \rho \sum_{k=1, k\neq i}^{N} D_{ik} \frac{\partial Y_k}{\partial x}. \tag{10.32}$$

Next, we take $D_{ik} = \mathcal{D}$ and write for each of the two diffusive mass fluxes that

$$j_A^m = \rho \mathcal{D} \frac{\partial Y_B}{\partial x}, \qquad j_B^m = \rho \mathcal{D} \frac{\partial Y_A}{\partial x}. \tag{10.33}$$

[1] Oliver Heaviside (1850–1925), English engineer.

Because $Y_A + Y_B = 1$, we have also

$$j_A^m = -\rho \mathcal{D} \frac{\partial Y_A}{\partial x}, \qquad j_B^m = -\rho \mathcal{D} \frac{\partial Y_B}{\partial x}. \tag{10.34}$$

Note that $j_A^m + j_B^m = 0$, as required by Eq. (7.9).

Under the same assumptions, Eq. (7.73) reduces to

$$j^q = -\mathsf{k}\frac{\partial T}{\partial x} + j_A^m h_A + j_B^m h_B, \tag{10.35}$$

$$= -\mathsf{k}\frac{\partial T}{\partial x} + j_A^m (h_{A,T_0} + c_P(T - T_0)) + j_B^m (h_{B,T_0} + c_P(T - T_0)), \tag{10.36}$$

$$= -\mathsf{k}\frac{\partial T}{\partial x} + j_A^m h_{A,T_0} + j_B^m h_{B,T_0} + \underbrace{(j_A^m + j_B^m)}_{0} c_P(T - T_0), \tag{10.37}$$

$$= -\mathsf{k}\frac{\partial T}{\partial x} - j_B^m \underbrace{(h_{A,T_0} - h_{B,T_0})}_{q} = -\mathsf{k}\frac{\partial T}{\partial x} - j_B^m q, \tag{10.38}$$

$$= -\mathsf{k}\frac{\partial T}{\partial x} + \rho \mathcal{D} q \frac{\partial Y_B}{\partial x}. \tag{10.39}$$

Here, we have defined a heat release per unit mass, $q$, as

$$q = h_{A,T_0} - h_{B,T_0}. \tag{10.40}$$

For the equilibrium constant $K_c$, we specialize Eq. (4.208), recalling $\nu_A + \nu_B = 1 - 1 = 0$ for our simple reaction kinetics, and get

$$K_c = \exp\left(\frac{-\Delta G^0}{\overline{R}T}\right) = \exp\left(\frac{g_A^0 - g_B^0}{RT}\right), \tag{10.41}$$

$$= \exp\left(\frac{h_{A,T_0}^0 - h_{B,T_0}^0}{RT}\right) \exp\left(\frac{s_{B,T_0}^0 - s_{A,T_0}^0}{R}\right). \tag{10.42}$$

Here, because of constant specific heats for $A$ and $B$, many terms have canceled. Let us take now $s_{B,T_0}^0 = s_{A,T_0}^0$ and our definition of the heat release, Eq. (10.40), to get

$$K_c = \exp\left(\frac{q}{RT}\right). \tag{10.43}$$

### 10.1.3 Alternate Forms

Further analysis of the evolution equations combined with the constitutive equations can yield forms that give physical insight.

#### Species Equation

If we combine our species evolution equation, Eq. (10.21), with the constitutive law, Eq. (10.27), we get

$$\rho \frac{\partial Y_B}{\partial t} + \rho u \frac{\partial Y_B}{\partial x} - \frac{\partial}{\partial x}\left(\rho \mathcal{D} \frac{\partial Y_B}{\partial x}\right) = M \dot{\omega}_B. \tag{10.44}$$

Note that for flows with no advection ($u = 0$), constant density ($\rho =$ constant), and no reaction ($\dot{\omega}_B = 0$), this reduces to the classical heat equation from mathematical physics $\partial Y_B / \partial t = \mathcal{D}(\partial^2 Y_B / \partial x^2)$.

**Energy Equation**

Here, we specialize the analysis of Sec. 7.4 to get an evolution equation for $T$. Consider the energy equation, Eq. (10.20), using Eqs. (10.26) and (10.28) to eliminate $h$ and $j^q$:

$$\rho \frac{\partial}{\partial t} \underbrace{(c_P(T - T_0) - Y_B q)}_{h}$$

$$+ \rho u \frac{\partial}{\partial x} \underbrace{(c_P(T - T_0) - Y_B q)}_{h} + \frac{\partial}{\partial x} \underbrace{\left( -\mathsf{k} \frac{\partial T}{\partial x} + \rho \mathcal{D} q \frac{\partial Y_B}{\partial x} \right)}_{j^q} = 0, \quad (10.45)$$

$$\rho \frac{\partial}{\partial t} \left( T - Y_B \frac{q}{c_P} \right)$$

$$+ \rho u \frac{\partial}{\partial x} \left( T - Y_B \frac{q}{c_P} \right) - \frac{\partial}{\partial x} \left( \frac{\mathsf{k}}{c_P} \frac{\partial T}{\partial x} - \rho \mathcal{D} \frac{\partial Y_B}{\partial x} \frac{q}{c_P} \right) = 0, \quad (10.46)$$

$$\rho \frac{\partial T}{\partial t} + \rho u \frac{\partial T}{\partial x} - \frac{\mathsf{k}}{c_P} \frac{\partial^2 T}{\partial x^2}$$

$$- \frac{q}{c_P} \underbrace{\left( \rho \frac{\partial Y_B}{\partial t} + \rho u \frac{\partial Y_B}{\partial x} - \frac{\partial}{\partial x} \left( \rho \mathcal{D} \frac{\partial Y_B}{\partial x} \right) \right)}_{M \dot{\omega}_B} = 0. \quad (10.47)$$

We notice that the terms involving $Y_B$ simplify via Eq. (10.44) to yield an equation for temperature evolution

$$\rho \frac{\partial T}{\partial t} + \rho u \frac{\partial T}{\partial x} - \frac{\mathsf{k}}{c_P} \frac{\partial^2 T}{\partial x^2} - \frac{q}{c_P} M \dot{\omega}_B = 0, \quad (10.48)$$

$$\frac{\partial T}{\partial t} + u \frac{\partial T}{\partial x} = \underbrace{\frac{\mathsf{k}}{\rho c_P}}_{\alpha} \frac{\partial^2 T}{\partial x^2} + \frac{q}{\rho c_P} M \dot{\omega}_B. \quad (10.49)$$

We note (see Eq. (9.24)), that thermal diffusivity $\alpha = \mathsf{k}/\rho c_P$. This convenient definition is slightly nontraditional as it involves the weakly varying property $\rho$. So, the reduced energy equation is

$$\frac{\partial T}{\partial t} + u \frac{\partial T}{\partial x} = \alpha \frac{\partial^2 T}{\partial x^2} + \frac{q}{\rho c_P} M \dot{\omega}_B. \quad (10.50)$$

Equation (10.50) is closely related to the more general Eq. (7.97). For exothermic reaction, $q > 0$ accompanied with production of product $B$, $\dot{\omega}_B > 0$, induces a temperature rise of a material particle. The temperature change is modulated by energy diffusion. In the inert zero advection limit, we recover the heat equation $\partial T/\partial t = \alpha(\partial^2 T/\partial x^2)$.

**Shvab-Zel'dovich Form**

Let us now adopt the assumption that mass and energy diffuse at the same rate. This is not difficult to believe as both are molecular collision-based phenomena in gas flames. Such an assumption will allow us to write the energy equation in the Shvab-Zel'dovich form such as studied in a more general sense in Sec. 7.5. The dimensionless ratio of energy diffusivity $\alpha$ to mass diffusivity $\mathcal{D}$ is known as the Lewis[2] number, $Le$:

$$Le = \alpha/\mathcal{D}. \tag{10.51}$$

If we insist that mass and energy diffuse at the same rate, we have $Le = 1$, which gives

$$\mathcal{D} = \alpha = \frac{\mathsf{k}}{\rho c_P}. \tag{10.52}$$

Thus $\mathcal{D}$ also has a weak variation with $\rho$ if we take $Le = 1$. With this assumption, the energy equation, Eq. (10.46), takes the form

$$\rho\frac{\partial}{\partial t}\left(T - Y_B\frac{q}{c_P}\right) + \rho u\frac{\partial}{\partial x}\left(T - Y_B\frac{q}{c_P}\right) - \frac{\partial}{\partial x}\left(\frac{\mathsf{k}}{c_P}\frac{\partial T}{\partial x} - \frac{\mathsf{k}}{c_P}\frac{\partial Y_B}{\partial x}\frac{q}{c_P}\right) = 0. \tag{10.53}$$

Using the material derivative and rearranging, this becomes

$$\frac{d}{dt}\left(T - Y_B\frac{q}{c_P}\right) - \frac{\mathsf{k}}{\rho c_P}\frac{\partial}{\partial x}\left(\frac{\partial}{\partial x}\left(T - Y_B\frac{q}{c_P}\right)\right) = 0. \tag{10.54}$$

Equation (10.54) holds that for a material fluid particle, the quantity $T - Y_B q/c_P$ changes only in response to local spatial gradients. Now, if we consider an initial value problem in which $T$ and $Y_B$ are initially spatially uniform and there are no gradients of either $T$ or $Y_B$ at $x \to \pm\infty$, there will no tendency for *any* material particle to have its value of $T - Y_B q/c_P$ change. Let us assume that at $t = 0$, we have $T = T_0$ and no product, so $Y_B = 0$. Then the following relation holds for all space and time:

$$T - Y_B\frac{q}{c_P} = T_0. \tag{10.55}$$

Solving for $Y_B$, we get

$$Y_B = \frac{c_P(T - T_0)}{q}. \tag{10.56}$$

For this chapter, we are mainly interested in steady waves. We can imagine that our steady waves are the long time limit of a situation just described that was initially spatially uniform. Compare Eq. (10.56) to our related ad hoc assumption for reactive solids, Eq. (9.21). They are essentially equivalent, especially when one recalls that $\lambda$ plays the same role as $Y_B$ and for the reactive solid $c_P \sim c_v$.

We can use Eq. (10.56) to eliminate $Y_B$ in the species equation, Eq. (10.44) to get a single equation for temperature evolution. First adopt the equal diffusion assumption, Eq. (10.52), in Eq. (10.44) and also use the material derivative:

$$\rho\frac{dY_B}{dt} - \frac{\partial}{\partial x}\left(\frac{\mathsf{k}}{c_P}\frac{\partial Y_B}{\partial x}\right) = M\dot{\omega}_B. \tag{10.57}$$

Next use Eq. (10.56) to eliminate $Y_B$:

$$\rho\frac{d}{dt}\left(\frac{c_P(T - T_0)}{q}\right) - \frac{\partial}{\partial x}\left(\frac{\mathsf{k}}{c_P}\frac{\partial}{\partial x}\left(\frac{c_P(T - T_0)}{q}\right)\right) = M\dot{\omega}_B. \tag{10.58}$$

Simplifying, we get

$$\rho c_P\frac{dT}{dt} - \mathsf{k}\frac{\partial^2 T}{\partial x^2} = qM\dot{\omega}_B. \tag{10.59}$$

[2] Warren Kendall Lewis (1882–1975), American chemical engineer.

Now, let us use Eq. (10.29) to eliminate $\dot{\omega}_B$:

$$\rho c_P \frac{dT}{dt} - \mathsf{k}\frac{\partial^2 T}{\partial x^2} = \rho q a T^{\beta} e^{-\mathcal{E}/(RT)}(1 - Y_B)\left(1 - \frac{Y_B/K_c}{1 - Y_B}\right) H(T - T_{ig}). \quad (10.60)$$

Now, use Eq. (10.43) to eliminate $K_c$ and Eq. (10.56) to eliminate $Y_B$ so to get

$$\rho c_P \frac{dT}{dt} - \mathsf{k}\frac{\partial^2 T}{\partial x^2} = \rho q a T^{\beta} e^{-\mathcal{E}/(RT)}\left(1 - \frac{c_P(T - T_0)}{q}\right)$$
$$\times \left(1 - e^{-q/(RT)} \frac{\frac{c_P(T-T_0)}{q}}{1 - \frac{c_P(T-T_0)}{q}}\right) H(T - T_{ig}). \quad (10.61)$$

Equation (10.61) is remarkably similar to our earlier Eq. (9.23) for temperature evolution in a heat conducting reactive solid. The differences are as follows. We have (1) used $c_P$ instead of $c_v$, (2) included advection, (3) accounted for finite $\beta$, (4) accounted for reversible reaction, and (5) imposed an ignition temperature.

### 10.1.4 Equilibrium Conditions

We can gain further insights by examining when Eq. (10.61) is in equilibrium in the spatially homogeneous limit. There is one potential equilibria, the state of chemical

Table 10.1. *Numerical Values of Parameters for Simple Premixed Laminar Flame Calculation*

| Parameter | Value | Units |
|---|---|---|
| Fundamental Dimensional | | |
| $T_0$ | 300 | K |
| $c_P$ | 1000 | J/kg/K |
| $R$ | 287 | J/kg/K |
| $q$ | $1.5 \times 10^6$ | J/kg |
| $\mathcal{E}$ | $1.722 \times 10^5$ | J/kg |
| $\alpha_0$ | $10^{-5}$ | $m^2/s$ |
| $a$ | $10^5$ | 1/s |
| $u_0$ | 1.4142 | m/s |
| $P_0$ | $10^5$ | Pa |
| $T_{ig}$ | 600 | K |
| Secondary Dimensional | | |
| $\rho_0$ | 1.1614 | $kg/m^3$ |
| k | 0.0116114 | J/m/K/s |
| Fundamental Dimensionless | | |
| $\gamma$ | 1.4025 | |
| $\mathcal{Q}$ | 5 | |
| $\Theta$ | 2 | |
| $\mathfrak{D}$ | 2 | |
| $T_{IG}$ | 0.2 | |

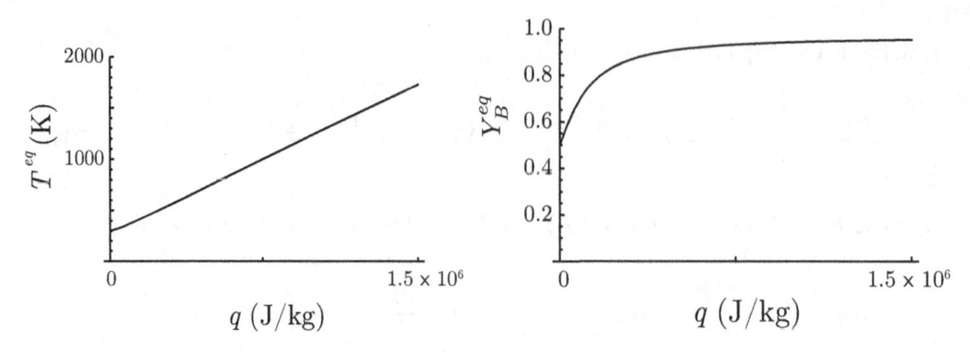


Figure 10.1. Variation of chemical equilibrium temperature and product mass fraction with heat release.

equilibrium where $K_c = Y_B/(1 - Y_B) = Y_B/Y_A$. This occurs when

$$1 - e^{-q/(RT)} \frac{\frac{c_P(T - T_0)}{q}}{1 - \frac{c_P(T - T_0)}{q}} = 0. \tag{10.62}$$

This is a transcendental equation for $T$. Let us examine how the equilibrium $T$ varies for some sample parameters. For this and later calculations, we give a set of parameters in Table 10.1. The chemical equilibrium flame temperature and product mass fraction are plotted as functions of $q$ in Fig. 10.1. For $q = 0$, the equilibrium temperature is the ambient temperature. As $q$ increases, the equilibrium temperature increases monotonically. When $q = 0$, the equilibrium mass fraction is $Y_B = 0.5$; that is, there is no tendency for either products or reactants. As $q$ increases, the tendency for increased product mass fraction increases, and $Y_B \to 1$.

There is also the state of complete reaction when $Y_B = 1$ that corresponds to

$$1 - \frac{c_P(T - T_0)}{q} = 0, \qquad T = T_0 + \frac{q}{c_P}. \tag{10.63}$$

Numerically, at our maximum value of $q = 1.5 \times 10^6$ J/kg, we get complete reaction when $T = 1800$ K. This state is not an equilibrium state unless $K_c \to \infty$. As $x \to \infty$, we expect a state of chemical equilibrium and a corresponding high temperature. At some intermediate point, the temperature will have its value at ignition, $T_{ig}$. We will ultimately transform our domain so the ignition temperature is realized at $x = 0$, which we can associate with a burner surface. We say that the flame is thus anchored to the burner. Without such a prescribed anchor, our equations are such that any translation in $x$ will yield an invariant solution, so the origin of $x$ is arbitrary.

## 10.2 Steady Burner-Stabilized Flames

Let us consider an important problem in laminar flame theory: that of a burner-stabilized flame (Margolis, 1978; Smooke, 1982). We shall consider a doubly infinite domain, $x \in (-\infty, \infty)$. We will assume that as $x \to -\infty$, we have a fresh unburned stream of reactant $A$, $Y_A = 1$, $Y_B = 0$, with known velocity $u_0$, density $\rho_0$ and temperature $T_0$. The values of $\rho_0$ and $T_0$ will be consistent with a state equation so that the pressure has a value of $P_0$.

### 10.2.1 Formulation

Let us consider our governing equations in the steady wave frame where there is no variation with $t$: $\partial/\partial t = 0$. This, coupled with our other reductions, yields the system

$$\frac{d}{dx}(\rho u) = 0, \tag{10.64}$$

$$\begin{aligned}\rho u c_P \frac{dT}{dx} - \mathsf{k}\frac{d^2T}{dx^2} &= \rho q a T^{\beta} e^{-\mathcal{E}/(RT)} \left(1 - \frac{c_P(T-T_0)}{q}\right) \\ &\times \left(1 - e^{-q/(RT)} \frac{\frac{c_P(T-T_0)}{q}}{1 - \frac{c_P(T-T_0)}{q}}\right) \\ &\times H(T - T_{ig}),\end{aligned} \tag{10.65}$$

$$P_0 = \rho R T. \tag{10.66}$$

Now, we have three equations for the three remaining unknowns, $\rho, u, T$.

We can integrate the mass equation and apply the conditions at $x \to -\infty$ to get

$$\rho u = \rho_0 u_0. \tag{10.67}$$

We then rewrite the remaining differential equation as

$$\begin{aligned}\rho_0 u_0 c_P \frac{dT}{dx} - \mathsf{k}\frac{d^2T}{dx^2} &= \frac{P_0}{R} q a T^{\beta-1} e^{-\mathcal{E}/(RT)} \left(1 - \frac{c_P(T-T_0)}{q}\right) \\ &\times \left(1 - e^{-q/(RT)} \frac{\frac{c_P(T-T_0)}{q}}{1 - \frac{c_P(T-T_0)}{q}}\right) \\ &\times H(T - T_{ig}).\end{aligned} \tag{10.68}$$

Our boundary conditions for this second-order differential equation are

$$\frac{dT}{dx} \to 0, \quad x \to \pm\infty. \tag{10.69}$$

We will ultimately need to integrate Eq. (10.68) numerically, and this poses some difficulty because the boundary conditions are applied at $\pm\infty$. We can overcome this in the following way. Our equations are invariant under a translation in $x$. We will choose some large value of $x$ to commence numerical integration. We shall initially assume this is near the chemical equilibrium point. We shall integrate backward in $x$. We shall note the value of $x$ where $T = T_{ig}$. Beyond that point, the reaction rate is zero, and we can obtain an exact solution for $T$. We will then spatially translate all our results so that the ignition temperature is realized at $x = 0$.

Let us scale temperature so that

$$T = T_0\left(1 + \frac{q}{c_P T_0} T_*\right). \tag{10.70}$$

Inverting, we see that

$$T_* = \frac{c_P(T - T_0)}{q}. \tag{10.71}$$

We get a dimensionless ignition temperature $T_{IG}$ of

$$T_{IG} = \frac{c_P(T_{ig} - T_0)}{q}. \tag{10.72}$$

Let us also define a dimensionless heat release $Q$ as

$$Q = \frac{q}{c_P T_0}. \tag{10.73}$$

Thus

$$\frac{T}{T_0} = 1 + QT_*. \tag{10.74}$$

Note that $T_* = 1$ corresponds to our complete reaction point where $Y_B = 1$. Let us also restrict ourselves to the case where $\beta = 0$. With these scalings, and with

$$\Theta = \frac{\mathcal{E}}{RT_0}, \tag{10.75}$$

our system becomes

$$\begin{aligned} \rho_0 u_0 c_P T_0 Q \frac{dT_*}{dx} - \mathsf{k} T_0 Q \frac{d^2 T_*}{dx^2} &= \frac{P_0}{R} qa \exp\left(\frac{-\Theta}{1 + QT_*}\right)\left(\frac{1 - T_*}{T_0(1 + QT_*)}\right) \\ &\times \left(1 - \exp\left(-\frac{\frac{\gamma}{\gamma-1}Q}{1 + QT_*}\right)\frac{T_*}{1 - T_*}\right) \\ &\times H(T_* - T_{IG}), \end{aligned} \tag{10.76}$$

$$\frac{dT_*}{dx} \to 0, \quad x \to \pm\infty. \tag{10.77}$$

Let us now scale both sides by $\rho_0 u_0 c_P T_0$ to get

$$\begin{aligned} Q\frac{dT_*}{dx} - \frac{\mathsf{k}}{\rho_0 c_P} \frac{1}{u_0} Q \frac{d^2 T_*}{dx^2} &= \underbrace{\frac{P_0}{\rho_0 R T_0}}_{1} \underbrace{\frac{q}{c_P T_0}}_{Q} \frac{a}{u_0} \exp\left(\frac{-\Theta}{1 + QT_*}\right)\left(\frac{1 - T_*}{1 + QT_*}\right) \\ &\times \left(1 - \exp\left(-\frac{\frac{\gamma}{\gamma-1}Q}{1 + QT_*}\right)\frac{T_*}{1 - T_*}\right) \\ &\times H(T_* - T_{IG}). \end{aligned} \tag{10.78}$$

Realizing that $P_0 = \rho_0 R T_0$, the ambient thermal diffusivity, $\alpha_0 = \mathsf{k}/(\rho_0 c_P)$, and canceling $Q$, we get

$$\begin{aligned} \frac{dT_*}{dx} - \alpha_0 \frac{1}{u_0} \frac{d^2 T_*}{dx^2} &= \frac{a}{u_0} \exp\left(\frac{-\Theta}{1 + QT_*}\right)\left(\frac{1 - T_*}{1 + QT_*}\right) \\ &\times \left(1 - \exp\left(-\frac{\frac{\gamma}{\gamma-1}Q}{1 + QT_*}\right)\frac{T_*}{1 - T_*}\right) \\ &\times H(T_* - T_{IG}). \end{aligned} \tag{10.79}$$

Now, let us scale $x$ by a characteristic length, $L = u_0/a$. This length scale is dictated by a balance between reaction and advection. When the reaction is fast, $a$ is large, and the length scale is reduced. When the incoming velocity is fast, $u_0$ is large, and the length scale is increased,

$$x_* = x/L = ax/u_0. \tag{10.80}$$

With this choice of length scale, Eq. (10.79) becomes

$$\frac{dT_*}{dx_*} - \alpha_0 \frac{a}{u_0^2} \frac{d^2T_*}{dx_*^2} = \exp\left(\frac{-\Theta}{1+QT_*}\right)\left(\frac{1-T_*}{1+QT_*}\right) \times \left(1 - \exp\left(-\frac{\frac{\gamma}{\gamma-1}Q}{1+QT_*}\right)\frac{T_*}{1-T_*}\right) \times H(T_* - T_{IG}). \tag{10.81}$$

Now, similar to Eq. (9.32), we take the Damköhler number $\mathfrak{D}$ to be

$$\mathfrak{D} = \frac{aL^2}{\alpha_0} = \frac{u_0^2}{a\alpha_0}. \tag{10.82}$$

In contrast to Eq. (9.32) the Damköhler number here includes the effects of advection. With this choice, Eq. (10.81) becomes

$$\frac{dT_*}{dx_*} - \frac{1}{\mathfrak{D}} \frac{d^2T_*}{dx_*^2} = \exp\left(\frac{-\Theta}{1+QT_*}\right)\left(\frac{1-T_*}{1+QT_*}\right) \times \left(1 - \exp\left(-\frac{\frac{\gamma}{\gamma-1}Q}{1+QT_*}\right)\frac{T_*}{1-T_*}\right) \times H(T_* - T_{IG}). \tag{10.83}$$

Last, let us dispose with the $*$ notation and understand that all variables are dimensionless. Our differential equation and boundary conditions become

$$\frac{dT}{dx} - \frac{1}{\mathfrak{D}} \frac{d^2T}{dx^2} = \exp\left(\frac{-\Theta}{1+QT}\right)\left(\frac{1-T}{1+QT}\right)\left(1 - \exp\left(-\frac{\frac{\gamma}{\gamma-1}Q}{1+QT}\right)\frac{T}{1-T}\right) \times H(T - T_{IG}), \tag{10.84}$$

$$\left.\frac{dT}{dx}\right|_{x\to\pm\infty} \to 0. \tag{10.85}$$

Equation (10.84) is similar to the Frank-Kamenetskii problem embodied in Eq. (9.46). The major differences are reflected by Eq. (10.84)'s accounting for (1) advection, (2) variable density, (3) reversible reaction, and (4) a doubly infinite spatial domain.

### 10.2.2 Solution Procedure

There are some unusual challenges in the numerical solution of the equations for laminar flames. Formally we are solving a two-point boundary value problem on a doubly infinite domain. The literature is not always coherent on solutions methods or the interpretations of results. The heart of the challenge is the cold boundary difficulty. We have patched this problem via the use of an ignition temperature built into a Heaviside function; see Eq. (10.29). We shall see that this patch has advantages and disadvantages. Here, we will not dwell on nuances, but will present a result that is mathematically sound and offer interpretations. Our result will use standard techniques of nonlinear analysis from dynamical systems theory. We will pose the problem as a coupled system of first-order nonlinear differential equations, find their equilibria, use local linear analysis to ascertain the stability of the fixed point, and use numerical integration to calculate the nonlinear laminar flame structure. We will also see the unusual features introduced at the equilibrium at the cold boundary.

### Model Linear System

Before commencing with Eq. (10.84), let us consider a related linear model system:

$$3\frac{dT}{dx} - \frac{d^2T}{dx^2} = 4(1-T)H(T - T_{IG}), \qquad \left.\frac{dT}{dx}\right|_{x\to\pm\infty} \to 0. \tag{10.86}$$

We require $0 < T_{IG} \ll 1$. The forcing is removed when either $T = 1$ or $T < T_{IG}$. Defining $\mathsf{q} \equiv -dT/dx$, we rewrite our model equation as a nearly linear system of first-order equations:

$$\frac{d\mathsf{q}}{\mathsf{dx}} = 3\mathsf{q} + 4(1-T)H(T - T_{IG}), \qquad \mathsf{q}(\infty) = 0, \tag{10.87}$$

$$\frac{dT}{dx} = -\mathsf{q}, \qquad \mathsf{q}(-\infty) = 0. \tag{10.88}$$

Formally, the system has a nonlinear source term in the Heaviside function; this will not be a difficulty, and the system is otherwise linear. The system has an equilibrium point at $\mathsf{q} = 0$ and $T = 1$. The system is also in equilibrium when $\mathsf{q} = 0$ and $T < T_{IG}$, but we will not focus on this now. Taking $\hat{T} = T - 1$, the system near the equilibrium point is

$$\frac{d}{dx}\begin{pmatrix} \mathsf{q} \\ \hat{T} \end{pmatrix} = \begin{pmatrix} 3 & -4 \\ -1 & 0 \end{pmatrix}\begin{pmatrix} \mathsf{q} \\ \hat{T} \end{pmatrix}. \tag{10.89}$$

The local Jacobian matrix has two eigenvalues $\lambda = 4$ and $\lambda = -1$. Thus, the equilibrium is a saddle.

Considering the solution away from the equilibrium point, but still for $T > T_{IG}$, and returning from $\hat{T}$ to $T$, the solution takes the form

$$\begin{pmatrix} \mathsf{q} \\ T \end{pmatrix} = \begin{pmatrix} 0 \\ 1 \end{pmatrix} + \frac{1}{5}e^{-x}\begin{pmatrix} C_1 + 4C_2 \\ C_1 + 4C_2 \end{pmatrix} + \frac{1}{5}e^{4x}\begin{pmatrix} 4C_1 - 4C_2 \\ -C_1 + C_2 \end{pmatrix}. \tag{10.90}$$

We imagine as $x \to \infty$ that $T$ approaches unity from below. We wish this equilibrium to be achieved. Thus, we must suppress the unstable $\lambda = 4$ mode; this is achieved by requiring $C_1 = C_2$. This yields

$$\begin{pmatrix} \mathsf{q} \\ T \end{pmatrix} = \begin{pmatrix} 0 \\ 1 \end{pmatrix} + \frac{1}{5}e^{-x}\begin{pmatrix} 5C_1 \\ 5C_1 \end{pmatrix} = \begin{pmatrix} 0 \\ 1 \end{pmatrix} + C_1 e^{-x}\begin{pmatrix} 1 \\ 1 \end{pmatrix}. \tag{10.91}$$

We can force $T(x_{IG}) = T_{IG}$ by taking $C_1 = -(1 - T_{IG})e^{x_{IG}}$. The solution for $x > x_{IG}$ is

$$T = 1 - (1 - T_{IG})e^{-(x - x_{IG})}, \qquad \mathsf{q} = -(1 - T_{IG})e^{-(x - x_{IG})}. \tag{10.92}$$

At the interface $x = x_{IG}$, we have $T(x_{IG}) = T_{IG}$ and $\mathsf{q}(x_{IG}) = -(1 - T_{IG})$.

For $x < x_{IG}$, our system reduces to

$$3\frac{dT}{dx} - \frac{d^2T}{dx^2} = 0, \qquad T(x_{IG}) = T_{IG}, \qquad \left.\frac{dT}{dx}\right|_{x\to-\infty} = 0. \tag{10.93}$$

Solutions in this region take the form

$$T = T_{IG}e^{3(x-x_{IG})} + C\left(1 - e^{3(x-x_{IG})}\right), \tag{10.94}$$

$$\mathsf{q} = -3T_{IG}e^{3(x-x_{IG})} + 3Ce^{3(x-x_{IG})}. \tag{10.95}$$

All of them have the property that $\mathsf{q} \to 0$ as $x \to -\infty$. One is faced with the question of how to choose the constant $C$. Let us choose it to match the energy flux predicted at $x = x_{IG}$. At the interface $x = x_{IG}$, we get $T = T_{IG}$ and $\mathsf{q} = -3T_{IG} + 3C$. Matching values of $\mathsf{q}$ at the interface, we get

$$-(1 - T_{IG}) = -3T_{IG} + 3C. \tag{10.96}$$

Solving for $C$, we get

$$C = \frac{4T_{IG} - 1}{3}. \tag{10.97}$$

So, we find for $x < x_{IG}$ that

$$T = T_{IG}e^{3(x-x_{IG})} + \left(\frac{4T_{IG} - 1}{3}\right)\left(1 - e^{3(x-x_{IG})}\right), \tag{10.98}$$

$$\mathsf{q} = -3T_{IG}e^{3(x-x_{IG})} + (4T_{IG} - 1)\, e^{3(x-x_{IG})}. \tag{10.99}$$

### System of First-Order Equations

Let us apply this technique to the full nonlinear Eq. (10.84). First, again define the nondimensional Fourier heat flux $\mathsf{q}$ as

$$\mathsf{q} = -\frac{dT}{dx}. \tag{10.100}$$

Note this is a mathematical convenience. It is not the full diffusive energy flux as it makes no account for mass diffusion effects. However, it is physically intuitive. With this definition, we can rewrite Eq. (10.100) along with Eq. (10.84) as

$$\begin{aligned}\frac{d\mathsf{q}}{dx} &= \mathfrak{D}\mathsf{q} + \mathfrak{D}\exp\left(\frac{-\Theta}{1+QT}\right)\left(\frac{1-T}{1+QT}\right)\\ &\times\left(1 - \exp\left(-\frac{\frac{\gamma}{\gamma-1}Q}{1+QT}\right)\frac{T}{1-T}\right)H(T - T_{IG}),\end{aligned} \tag{10.101}$$

$$\frac{dT}{dx} = -\mathsf{q}. \tag{10.102}$$

We have two boundary conditions for this problem:

$$\mathsf{q}(\pm\infty) = 0. \tag{10.103}$$

We are also going to fix our coordinate system so that $T = T_{IG}$ at $x = x_{IG} \equiv 0$. We require continuity of $T$ and $dT/dx$ at $x = x_{IG} = 0$. Our nonlinear system has the form

$$\frac{d}{dx}\begin{pmatrix}\mathsf{q}\\ T\end{pmatrix} = \begin{pmatrix}f(\mathsf{q}, T)\\ g(\mathsf{q}, T)\end{pmatrix}. \tag{10.104}$$

### Equilibrium

Our equilibrium condition is $f(\mathsf{q}, T) = 0, g(\mathsf{q}, T) = 0$. By inspection, equilibrium is

$$\mathsf{q} = 0, \tag{10.105}$$

$$1 - \exp\left(-\frac{\frac{\gamma}{\gamma-1}Q}{1+QT}\right)\frac{T}{1-T} = 0, \qquad \text{or} \qquad T < T_{IG}. \tag{10.106}$$

We will focus most attention on the isolated equilibrium point that corresponds to traditional chemical equilibrium. We have considered the dimensional version of the same equilibrium condition for $T$ in Sec. 10.1.4. The solution is found via solving a transcendental equation for $T$.

### Linear Analysis Near Equilibrium

Linear analysis near the equilibrium point can be achieved by first examining the eigenvalues of the Jacobian matrix $\mathbf{J}$:

$$\mathbf{J} = \left.\begin{pmatrix} \frac{\partial f}{\partial \mathsf{q}} & \frac{\partial f}{\partial T} \\ \frac{\partial g}{\partial \mathsf{q}} & \frac{\partial g}{\partial T} \end{pmatrix}\right|_{eq} = \begin{pmatrix} \mathfrak{D} & \left.\frac{\partial f}{\partial T}\right|_{eq} \\ -1 & 0 \end{pmatrix}. \tag{10.107}$$

This Jacobian has eigenvalues of

$$\lambda = \frac{\mathfrak{D}}{2}\left(1 \pm \sqrt{1 - \frac{4}{\mathfrak{D}^2}\left.\frac{\partial f}{\partial T}\right|_{eq}}\right). \tag{10.108}$$

For $\mathfrak{D} \gg \sqrt{\partial f/\partial T}$, the eigenvalues are approximated well by

$$\lambda_1 \sim \mathfrak{D}, \qquad \lambda_2 \sim \frac{1}{\mathfrak{D}}\left.\frac{\partial f}{\partial T}\right|_{eq}. \tag{10.109}$$

This gives rise to a stiffness ratio, valid in the limit of $\mathfrak{D} \gg \sqrt{\partial f/\partial T}$, of

$$\left|\frac{\lambda_1}{\lambda_2}\right| \sim \frac{\mathfrak{D}^2}{\left|\frac{\partial f}{\partial T}\right|}. \tag{10.110}$$

We adopt the parameters of Table 10.1. With these, we find the dimensionless chemical equilibrium temperature at

$$T^{eq} = 0.953. \tag{10.111}$$

This corresponds to a dimensional value of 1730.24 K. Our parameters induce a dimensional length scale of $u_0/a = 1.4142 \times 10^{-5}$ m. This is smaller than actual flames, which have much larger values of $\mathfrak{D}$. So as to destiffen the system, we have selected a smaller than normal value for $\mathfrak{D}$. For these values, we get a Jacobian matrix of

$$\mathbf{J} = \begin{pmatrix} 2 & -0.287 \\ -1 & 0 \end{pmatrix}. \tag{10.112}$$

We get eigenvalues near the equilibrium point of

$$\lambda = 2.134, \quad \lambda = -0.134. \tag{10.113}$$

The equilibrium is a saddle point. The negative value of $\partial f/\partial T$ at equilibrium gives rise to the saddle character of the equilibrium. The actual stiffness ratio is given by the

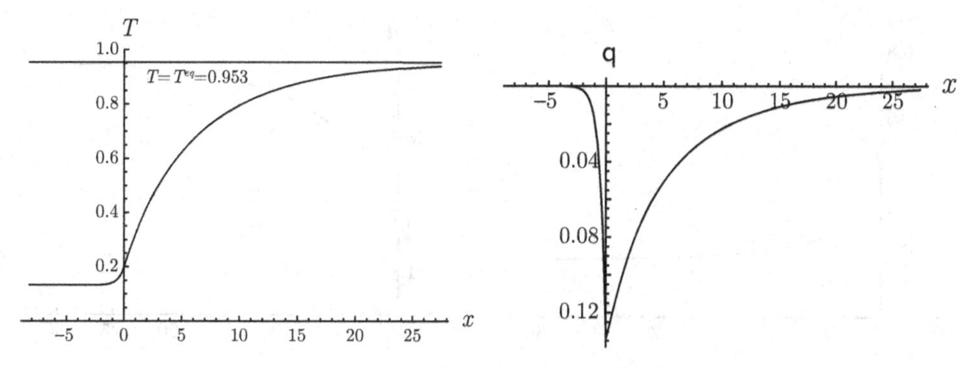


Figure 10.2. Dimensionless temperature and heat flux as a function of position for a burner-stabilized premixed laminar flame, $T_{IG} = 0.2$.

ratio of the eigenvalues' magnitudes $|2.134/(-0.134)| = 15.9$. This is well predicted by our simple formula, Eq. (10.110), which gives $|2^2/(-0.287)| = 13.9$.

In principle, we could also analyze the set of equilibrium points along the cold boundary, where $\mathsf{q} = 0$ and $T < T_{IG}$. Near such points, linearization techniques fail because of the nature of the Heaviside function. We would have to perform a more robust analysis. As an alternative, we shall simply visually examine the results of calculations and infer the stability of this set of fixed points.

### Laminar Flame Structure

**IGNITION TEMPERATURE: $T_{IG} = 0.2$.** Let us get a numerical solution by integrating backward in space from the equilibrium point. Here, we will use the parameters of Table 10.1, in particular to contrast with a later calculation, the somewhat elevated ignition temperature of $T_{IG} = 0.2$. We shall commence the solution near the isolated equilibrium point corresponding to chemical equilibrium. We could choose the initial condition to lie just off the saddle along the eigenvector associated with the negative eigenvalue. It will work just as well to choose a point on the correct side of the equilibrium. Let us approximate $x \to \infty$ by $x_B$, require $\mathsf{q}(x_B) = 0, T(x_B) = T^{eq} - \epsilon$, where $0 < \epsilon \ll 1$. That is, we perturb the temperature to be just less than its equilibrium value. We integrate from the large positive $x = x_B$ back toward $x = -\infty$. When we find $T = T_{IG}$, we record the value of $x$. We translate the plots so that the ignition point is reached at $x = 0$ and give results.

Predictions of $T(x)$ and $\mathsf{q}(x)$ are shown in in Fig. 10.2. We see the temperature has $T(0) = 0.2 = T_{IG}$. As $x \to \infty$, the temperature approaches its equilibrium value. For $x < 0$, the temperature continues to fall until it comes to a final value of $T \sim 0.132$. This value cannot be imposed by the boundary conditions, because we enforce no other condition on $T$ except to anchor it at the ignition temperature at $x = 0$. The heat flux $\mathsf{q}(x)$ is always negative. And it clearly relaxes to zero as $x \to \pm\infty$. *This indicates that energy released in combustion overcomes advection and diffuses its way back into the fresh mixture, triggering combustion of the fresh material. This is the essence of the laminar flame.* Predictions of $\rho(x) \sim 1/(1 + QT(x))$ and $u(x) \sim 1 + QT(x)$ are shown in in Fig. 10.3. We are somewhat troubled because $\rho(x)$ nowhere takes on dimensionless value of unity, despite it being scaled by $\rho_0$. Had we anchored the flame at an ignition temperature close to zero, we in fact would have seen $\rho$ approach unity at the anchor point of the flame. We chose an elevated ignition temperature so as to display the actual idiosyncrasies of the model. Even for a flame anchored, in

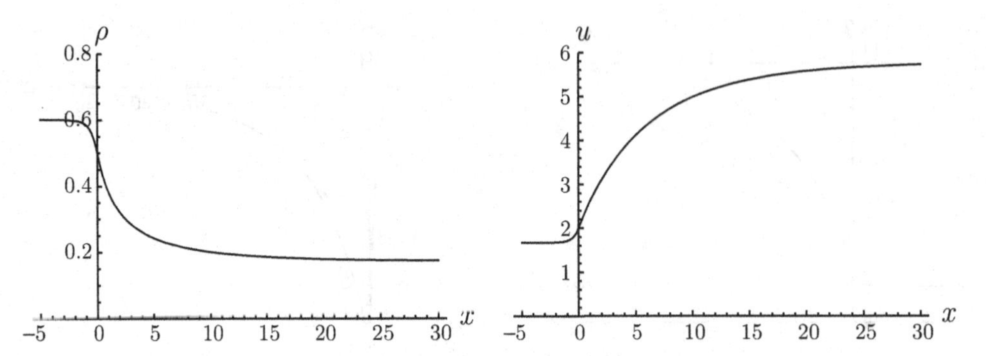


Figure 10.3. Dimensionless density $\rho$ and fluid particle velocity $u$ as functions of distance for a burner-stabilized premixed laminar flame, $T_{IG} = 0.2$.

dimensional variables, at $T_{ig} \sim T_0$, we would find a significant decay of density in the cold region of the flow for $x < x_{ig}$. Similar to $\rho$, we are somewhat troubled that the dimensionless velocity does not take a value near unity. Once again, had we set $T_{ig} \sim T_0$, we would have found the $u$ at $x = x_{ig}$ would have taken a value near unity. But it also would modulate significantly in the cold region $x < x_{ig}$. We also would have discovered that had $T_{ig} \sim T_0$ that the temperature in the cold region would drop significantly below $T_0$. That would be an unusual result. Alternatively, we could iterate on the parameter $T_{ig}$ until we found a value for which $\lim_{x \to -\infty} T \to 0$. For such a value, we would also find $u$ and $\rho$ to approach unity as $x \to -\infty$.

We can better understand the flame structure by considering the $(T, \mathsf{q})$ phase plane as shown in Fig. 10.4. Here, gray denotes equilibria. We see the isolated equilibrium point at $(T, \mathsf{q}) = (0.953, 0)$. And we see a continuous one-dimensional set of equilibria for $\{(T, \mathsf{q}) | \mathsf{q} = 0, T \in (-\infty, T_{IG}]\}$. Thin lines are trajectories in the phase

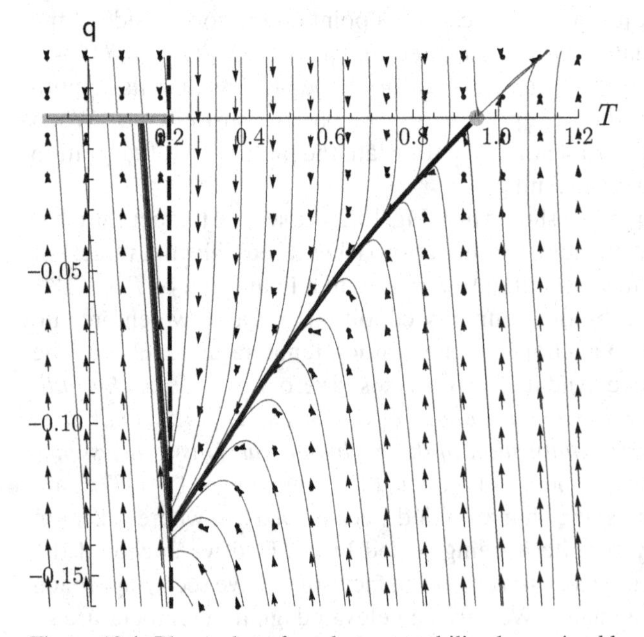


Figure 10.4. Phase plane for a burner-stabilized premixed laminar flame, $T_{IG} = 0.2$.

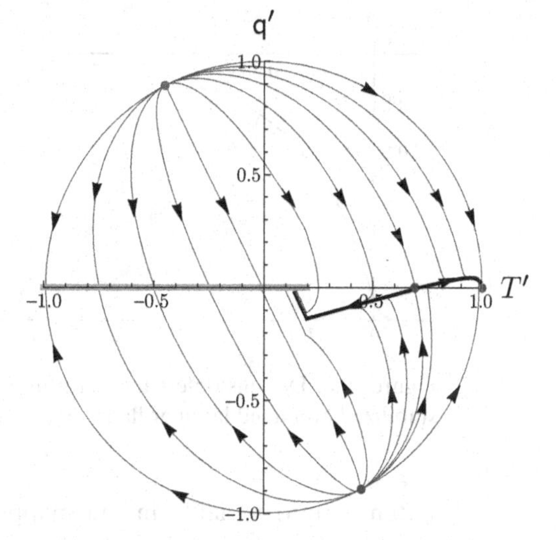


Figure 10.5. Projection of trajectories on Poincaré sphere onto the $(T', \mathsf{q}')$ phase plane for a burner-stabilized premixed laminar flame, $T_{IG} = 0.2$.

space. The arrows have been associated with movement in the $-x$ direction. This is because this is the direction in which it is easiest to construct the flame structure.

The heteroclinic orbit shown here in the thick black line is the CIM that gives the actual flame structure. It connects the saddle point, shown in gray, at $(T, \mathsf{q}) = (0.953, 0)$ to a stable point on the one-dimensional continuum of equilibria (shown in a gray line) at $(T, \mathsf{q}) = (0.132, 0)$. It originates on the eigenvector associated with the negative eigenvalue at the equilibrium point. Moreover, it appears that the CIM is an ACIM, as nearby trajectories appear to be attracted. The thick dashed line represents our chosen ignition temperature, $T_{IG} = 0.2$. Had we selected a different ignition temperature, the heteroclinic trajectory originating from the saddle would be the same for $T > T_{IG}$. However it would have turned at a different location and relaxed to a different cold equilibrium on the one-dimensional continuum of equilibria.

Had we attempted to construct our solution by integrating forward in $x$, our task would have been more difficult. We would likely have chosen the initial temperature to be just greater than $T_{IG}$. But we would have to had guessed the initial value of $\mathsf{q}$. And because of the saddle nature of the equilibrium point, a guess on either side of the correct value would cause the solution to diverge as $x$ became large. It would be possible to construct a trial and error procedure to hone the initial guess so that on one side of a critical value, the solution diverged to $\infty$, while on the other it diverged to $-\infty$. Our procedure, however, has the clear advantage, as no guessing is required.

As an aside, there is one additional heteroclinic orbit admitted mathematically; however, its physical relevance is far from clear. It seems there is another attracting trajectory for $T > 0.953$. This trajectory is associated with the same eigenvector as the physical trajectory. Along this trajectory, $T$ increases beyond unity, at which point the mass fraction becomes greater than unity, and is thus nonphysical. Mathematically this trajectory continues until it reaches an equilibrium at $(T, \mathsf{q}) \to (\infty, 0)$. These dynamics can be revealed using the mapping of the Poincaré sphere; see Fig. 10.5. Details can be found in some dynamical systems texts (e.g., Perko, 2001). In short, the mapping $(T, \mathsf{q}) \to (T', \mathsf{q}')$ is introduced via

$$T' = \frac{T}{\sqrt{1 + T^2 + \mathsf{q}^2}}, \qquad \mathsf{q}' = \frac{\mathsf{q}}{\sqrt{1 + T^2 + \mathsf{q}^2}}. \tag{10.114}$$

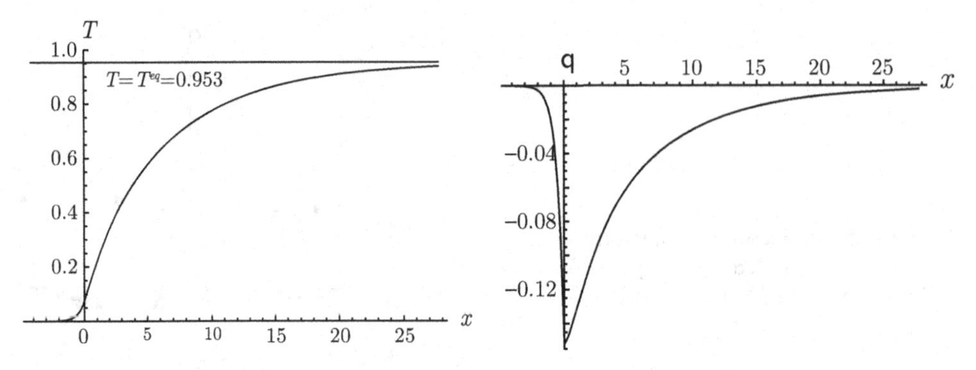


Figure 10.6. Dimensionless temperature and heat flux as a function of position for a burner-stabilized premixed laminar flame, $T_{IG} = 0.076$.

Often a third variable in the mapping is introduced $Z = 1/\sqrt{1 + T^2 + \mathsf{q}^2}$. This induces $T'^2 + \mathsf{q}'^2 + Z^2 = 1$, the equation of a sphere, known as the Poincaré[3] sphere.

As it is not clear that physical relevance can be found for this, we leave out most of the mathematical details and briefly describe the results. This mapping takes points at infinity in physical space onto the unit circle. Equilibria are marked in thick gray lines and dots. Trajectories are in thin gray lines. We notice in Fig. 10.5 that the saddle is evident around $(T', \mathsf{q}') \sim (0.7, 0)$. We also see new equilibria on the unit circle, which corresponds to points at infinity in the original space. One of these new equilibria is at $(T', \mathsf{q}') = (1, 0)$. It is a sink. The other two, one in the second quadrant, the other in the fourth, are sources. The heteroclinic trajectories that connect to the saddle from the sources in the second and fourth quadrants form the boundaries for the basins of attraction. To the left of this boundary, trajectories are attracted to the continuous set of equilibria. To the right, they are attracted to the point $(1, 0)$.

**IGNITION TEMPERATURE: $T_{IG} = 0.076$.** We can modulate the flame structure by altering the ignition temperature. In particular, it is possible to iterate on the ignition temperature in such a way that as $x \to -\infty, T \to 0, \rho \to 1$, and $u \to 1$. Thus, the cold flow region takes on its ambient value. This has some aesthetic appeal. It is, however, unsatisfying in that one would like to think that an ignition temperature, if it truly existed, would be a physical property of the system and not just a parameter to adjust to meet some other criterion. That duly noted, we present flame structures using all the parameters of Table 10.1, except we take $T_{ig} = 414$ K, so that $T_{IG} = 0.076$.

Predictions of $T(x)$ and $\mathsf{q}(x)$ are shown in in Fig. 10.6. We see the temperature has $T(0) = 0.076 = T_{IG}$. As $x \to \infty$, the temperature approaches its chemical equilibrium value. For $x < 0$, the temperature continues to fall until it comes to a final value of $T \sim 0$. Thus, in dimensional terms, the temperature has arrived at $T_0$ in the cold region. Predictions of $\rho(x) \sim 1/(1 + QT(x))$ and $u(x)$ are shown in in Fig. 10.7. We see that this special value of $T_{IG}$ has allowed the dimensionless density to take on a value of unity as $x \to -\infty$. In contrast to our earlier result, for this special value of $T_{IG}$, we have been able to allow $u = 1$ as $x \to -\infty$. We can better understand the flame structure by considering the $(T, \mathsf{q})$ phase plane as shown in Fig. 10.8. Figure 10.8 is essentially the same as Fig. 10.4, except the ignition point has been moved.

[3] Jules Henri Poincaré (1854–1912), French mathematician.

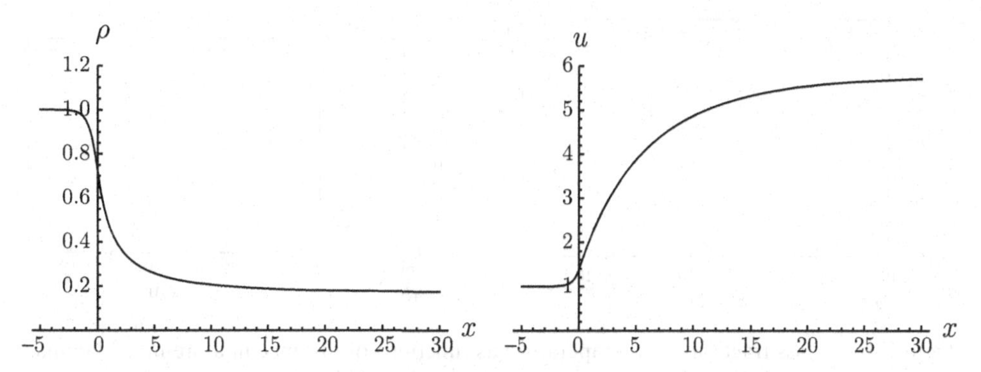


Figure 10.7. Dimensionless density $\rho$ and fluid particle velocity $u$ as functions of distance for a burner-stabilized premixed laminar flame, $T_{IG} = 0.076$.

### 10.2.3 Detailed $H_2$-Air Kinetics

We give brief results for a premixed laminar flame in a mixture of calorically imperfect ideal gases that obey mass action kinetics with Arrhenius kinetics. Multicomponent diffusion is modeled as is thermal diffusion. We consider the kinetics model of Table 1.2. The numerical solution method is standard and is described in detail in a series of reports (Kee et al., 1998, 2000a, 2000b, 2000c). The methods employed in the solution here are slightly different. In short, a large, but finite domain defined. Then, the ordinary differential equations describing the flame structure are discretized. This leads to a large system of nonlinear algebraic equations. These are solved by iterative methods, seeded with an appropriate initial guess. The temperature is pinned at $T_0$ at one end of the domain, which is a slightly different boundary condition than we employed previously. At an intermediate value of $x$, $x = x_f$, the temperature is pinned at $T = T_f$. Full details are in the cited references.

For $T_0 = 298$ K, $P_0 = 1.01325 \times 10^5$ Pa, and a stoichiometric unreacted mixture of $2H_2 + O_2 + 3.76N_2$, along with a small amount of minor species, we give plots of $Y_i(x)$ in Fig. 10.9a and $T(x)$ in Fig. 10.9b. There is, on a log-log scale, an induction zone

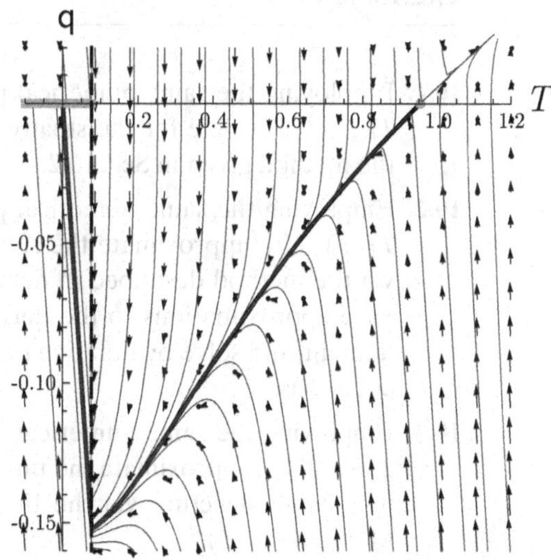


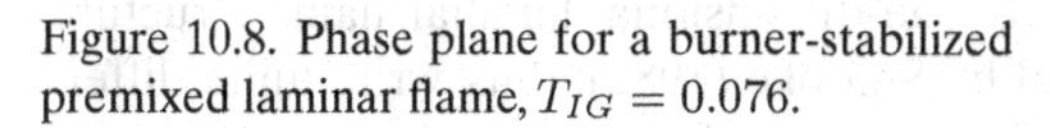
Figure 10.8. Phase plane for a burner-stabilized premixed laminar flame, $T_{IG} = 0.076$.

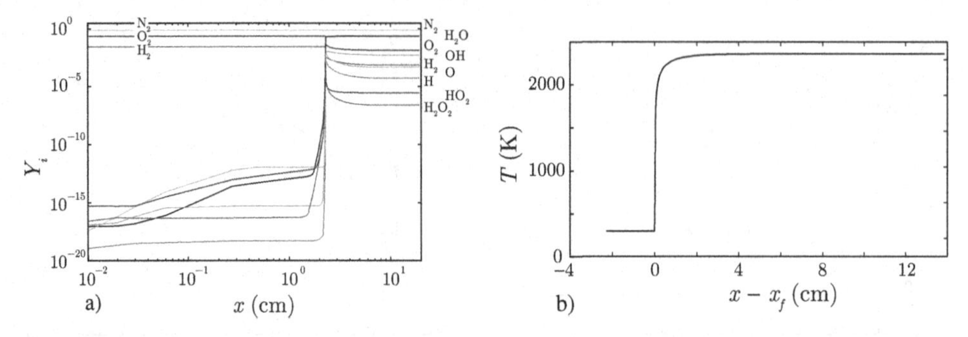


Figure 10.9. Mass fractions and temperature as functions of distance in a premixed laminar flame with detailed $H_2$-air kinetics.

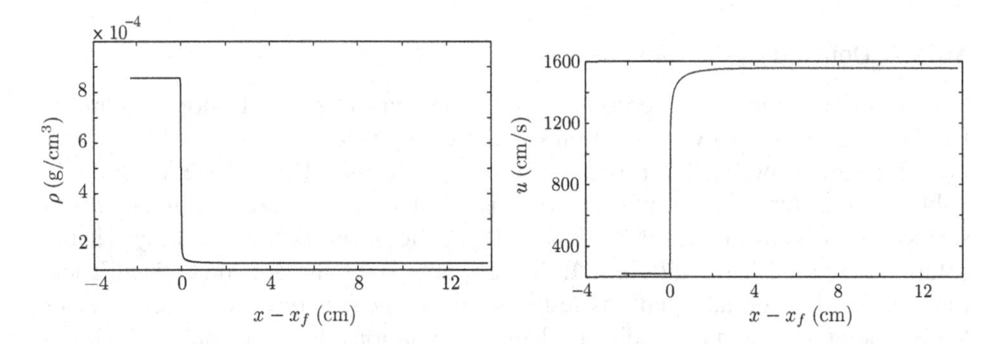


Figure 10.10. Density and fluid particle velocity as functions of distance in a premixed laminar flame with detailed $H_2$-air kinetics.

spanning a few orders of magnitude of length. Just past $x = 2$ cm, vigorous reaction commences, and all species relax to a final equilibrium. Density and fluid particle velocity as functions of distance are shown in Fig. 10.10.

## EXERCISES

**10.1.** Employing the same numerical parameters and model as in Sec. 10.2, except $T_{IG} = 0.01$, solve for the steady one-dimensional laminar flame structure via the method given in Sec. 10.2.

**10.2.** Employing the same numerical parameters and model as in Sec. 10.2, except $T_{IG} = 0.01$, approximate the steady one-dimensional laminar flame structure via the method described in detail by Smooke (1982); thus, write finite difference approximations to the derivatives, approximate the solution on a finite domain, and solve by iterative methods the resulting set of nonlinear algebraic equations.

**10.3.** Employing the same numerical parameters and model as in Sec. 10.2, except $T_{IG} = 0.01$, approximate the time relaxation to a steady one-dimensional laminar flame structure via the time-dependent method described in detail by Margolis (1978).

## References

W. A. Bone and D. T. A. Townend, 1927, *Flame and Combustion in Gases*, Longmans, London.
J. D. Buckmaster and G. S. S. Ludford, 1982, *Theory of Laminar Flames*, Cambridge University Press, Cambridge, UK.
N. Chigier, 1981, *Energy, Combustion, and Environment*, McGraw-Hill, New York.
M. Faraday, 1861, *The Chemical History of a Candle*, Griffin and Bohn, London.
J. O. Hirschfelder, C. F. Curtis, and R. B. Bird, 1954, *Molecular Theory of Gases and Liquids*, John Wiley, New York, Chapter 11.
R. J. Kee, G. Dixon-Lewis, J. Warnatz, M. E. Coltrin, and J. A. Miller, 1998, A Fortran computer code package for the evaluation of gas-phase multicomponent transport properties, Sandia National Laboratory, Report SAND86-8246, Livermore, CA.
R. J. Kee et al., 2000a, *Chemkin: A Software Package for the Analysis of Gas-Phase Chemical and Plasma Kinetics*, from the Chemkin Collection, Release 3.6, Reaction Design, San Diego, CA.
R. J. Kee et al., 2000b, *The Chemkin Thermodynamic Data Base*, part of the Chemkin Collection Release 3.6, Reaction Design, San Diego, CA.
R. J. Kee et al., 2000c, *Premix: A Fortran Program for Modeling Steady, Laminar, One Dimensional Premixed Flames*, from the Chemkin Collection, Release 3.6, Reaction Design, San Diego, CA.
R. J. Kee, M. E. Coltrin, and P. Glarborg, 2003, *Chemically Reacting Flow: Theory and Practice*, John Wiley, Hoboken, NJ.
C. K. Law, 2006, *Combustion Physics*, Cambridge University Press, New York, Chapter 7.
D. Layzer, 1954a, Theory of linear flame propagation. Part I. Existence, uniqueness, and stability of the steady state, *Journal of Chemical Physics*, 22(2): 222–229.
D. Layzer, 1954b, Theory of linear flame propagation. Part II. Structure of the steady state, *Journal of Chemical Physics*, 22(2): 229–232.
B. Lewis and G. von Elbe, 1961, *Combustion, Flames and Explosions of Gases*, 2nd ed., Academic Press, New York, Chapter 5.
T. C. Lieuwen, 2012, *Unsteady Combustor Physics*, Cambridge University Press, New York.
S. B. Margolis, 1978, Time-dependent solution of a premixed laminar flame, *Journal of Computational Physics*, 27(3): 410–427.
M. Matalon, 2007, Intrinsic flame instabilities in premixed and nonpremixed combustion, *Annual Review of Fluid Mechanics*, 39: 163–191.
S. Paolucci, 1982, On the filtering of sound from the Navier-Stokes equations, Technical Report SAND82-8257, Sandia National Laboratories, Livermore, CA.
L. Perko, 2001, *Differential Equations and Dynamical Systems*, 3rd ed., Springer, New York.
T. Poinsot and D. Veynante, 2005, *Theoretical and Numerical Combustion*, 2nd ed., Edwards, Flourtown, PA, Chapter 2.
M. D. Smooke, 1982, Solution of burner-stabilized premixed laminar flames by boundary value methods, *Journal of Computational Physics*, 48(1): 72–105.
R. A. Strehlow, 1984, *Combustion Fundamentals*, McGraw-Hill, New York, Chapter 8.
J. Warnatz, U. Maas, and R. W. Dibble, 2006, *Combustion*, 4th ed., Springer, New York, Chapter 8.
F. A. Williams, 1985, *Combustion Theory*, 2nd ed., Benjamin-Cummings, Menlo Park, CA, Chapter 5.

# 11 Oscillatory Combustion

*There are many circumstances in nature in which something is "oscillating" and in which the resonance phenomenon occurs.*

– Richard Phillips Feynman (1918–1988), *The Feynman Lectures on Physics*

In contrast to previous chapters that focused on systems that relaxed to a fixed equilibrium state, we consider here systems whose long time behavior is oscillatory. Specifically we address (1) spatially homogeneous reaction systems that oscillate in time only and (2) spatially inhomogeneous reaction-diffusion systems that may oscillate in time and space. We have seen earlier that spatially homogeneous reaction systems closed to mass exchange typically have a unique stable equilibrium. However, when the system is open to mass exchange, dynamics unavailable to closed systems may be realized; in open systems there is the possibility of multiple physical equilibria, limit cycles, pattern formation, or chaotic behavior. We further recognize that a system with spatial inhomogeneity may be thought of as an infinite set of spatially homogeneous control volumes, open to mass exchange via advection or diffusion with their neighboring volumes.

The seminal work of Turing (1952) gives the best-known example of oscillatory combustion; experimental evidence of oscillatory combustion, including chaotic dynamics, exists as well (e.g., Schmitz et al., 1977). Extensive general discussion on reaction-diffusion oscillatory dynamics and pattern formation can be found in many sources (Kuramoto, 1984; Aris, 1989, 1999; Mei, 2000; Hundsdorfer and Verwer, 2003; Cross and Greenside, 2009; Adamatzky, 2013; Volpert, 2014). Open systems may be spatially inhomogeneous, such as the flame configurations considered in Chapter 10 that were confined to steady combustion. Or they may be spatially homogeneous. We consider both. We model spatially homogeneous systems as so-called *continuously stirred tank reactors* (CSTR). In a CSTR, there is a feed flow rate of mass of fresh reactants and an equal mass flow rate of removal of products. Thus, the system mass remains constant. The continual stirring insures spatial homogeneity. The spatially inhomogeneous system we consider is easy to pose but difficult to solve.

We evaluate two reaction mechanisms as CSTRs: (1) the Gray-Scott mechanism (Gray and Scott, 1983, 1984, 1985) and (2) a hydrogen-air mechanism similar to that studied by Kalamatianos and Vlachos (1995). Both of these systems can display limit cycle behavior when modeled as spatially homogeneous. We also use the Gray-Scott system to display pattern formation in one- and two-dimensional unsteady reaction-diffusion problems. In principle, the hydrogen-air system could display pattern formation, but severe stiffness renders its simulation difficult. Most of the text and figures in this chapter have been adapted and condensed from the dissertation of Mengers (2012). It can be consulted for additional details.

## 11.1 Gray-Scott Mechanism

We examine the behavior of a Gray-Scott reaction-diffusion system that has been widely studied in the literature (e.g., Pearson, 1993; Hsu et al., 1996; Doelman et al., 1997; Marchant, 2002; Hundsdorfer and Verwer, 2003). The Gray-Scott reaction mechanism consists of $J = 2$ irreversible reactions,

$$U + 2V \rightarrow 3V, \qquad V \rightarrow P, \tag{11.1}$$

where $U$, $V$, and $P$ are the $N = 3$ species. The literature often presents a dimensionless model with little supporting discussion. We provide some details here as to how such a model can arise in the context of the nomenclature of combustion dynamics. First, to conserve mass, we require all molecular masses to be equal: $M = M_U = M_V = M_P$. The reaction rates are given by the law of mass action with irreversible kinetics:

$$r_1 = k_1 \overline{\rho}_U \overline{\rho}_V^2 = k_1 \left(\frac{\rho Y_U}{M}\right)\left(\frac{\rho Y_V}{M}\right)^2, \qquad r_2 = k_2 \overline{\rho}_V = k_2 \left(\frac{\rho Y_V}{M}\right). \tag{11.2}$$

Here, $\rho Y_i / M$ is the molar species concentration $\overline{\rho}_i$ of species $i$, and $k_j$ is the reaction rate coefficient of reaction $j$, assumed constant for what amounts to isothermal combustion. Taking $U, V$, and $P$ as species $1, 2$, and $3$, respectively, the matrix of net stoichiometric coefficients is

$$\nu_{ij} = \begin{pmatrix} -1 & 0 \\ 1 & -1 \\ 0 & 1 \end{pmatrix}. \tag{11.3}$$

Then, using Eq. (5.18), we can identify the species molar production rates as

$$\dot{\omega}_i = \sum_{j=1}^{2} \nu_{ij} r_j = \begin{pmatrix} -1 & 0 \\ 1 & -1 \\ 0 & 1 \end{pmatrix} \begin{pmatrix} r_1 \\ r_2 \end{pmatrix} = \begin{pmatrix} -r_1 \\ r_1 - r_2 \\ r_2 \end{pmatrix}. \tag{11.4}$$

We employ Fick's law of diffusion in two spatial dimensions with unequal diffusion coefficients. At this stage, the system is modeled as spatially inhomogeneous, and the evolution of species mass fraction depends on reaction, feed flow, and diffusion. The model equations are a modified version of Eq. (7.52), $\rho dY_i/dt + \nabla \cdot \mathbf{j}_i^m = M_i \dot{\omega}_i$. For our analysis, we (1) neglect advection so that the material derivative $d/dt$ becomes a partial derivative $\partial/\partial t$, (2) add an algebraic term modeling the addition and removal of mass from the system, and (3) recognize that our diffusive flux vector is a mass flux, and omit the superscript $m$. We get

$$\rho \frac{\partial Y_U}{\partial t} = -M r_1 + \frac{\dot{m}}{V}\left(Y_U^f - Y_U\right) - \nabla \cdot \mathbf{j}_U, \tag{11.5}$$

$$\rho \frac{\partial Y_V}{\partial t} = M(r_1 - r_2) + \frac{\dot{m}}{V}\left(Y_V^f - Y_V\right) - \nabla \cdot \mathbf{j}_V, \tag{11.6}$$

$$\rho \frac{\partial Y_P}{\partial t} = M r_2 + \frac{\dot{m}}{V}\left(Y_P^f - Y_P\right) - \nabla \cdot \mathbf{j}_P, \tag{11.7}$$

where $\dot{m}$ is the mass feed flow rate and $V$ is the volume of the reactor, not to be confused with species $V$. The term $Y_i^f$ is the specified mass fraction of each species of the feed flow. Summing Eqs. (11.5)–(11.7) we get

$$\rho\frac{\partial}{\partial t}\left(\underbrace{Y_U+Y_V+Y_P}_{1}\right)=\frac{\dot{m}}{V}\left(\underbrace{Y_U^f+Y_V^f+Y_P^f}_{1}-(\underbrace{Y_U+Y_V+Y_P}_{1})\right)$$
$$-\nabla\cdot\left(\underbrace{\mathbf{j}_U+\mathbf{j}_V+\mathbf{j}_P}_{\mathbf{0}}\right). \tag{11.8}$$

To maintain a global mass balance, we require the obvious Eq. (2.133) giving $Y_U+Y_V+Y_P=1$ and $Y_U^f+Y_V^f+Y_P^f=1$ as well as Eq. (7.9) giving $\mathbf{j}_U+\mathbf{j}_V+\mathbf{j}_P=\mathbf{0}$.

The diffusive flux for each species is given by

$$\mathbf{j}_i=-\rho\mathcal{D}_i\nabla Y_i,\quad \text{for } i=U,V,\qquad \mathbf{j}_P=-\mathbf{j}_U-\mathbf{j}_V, \tag{11.9}$$

where $\mathcal{D}_i$ are the constant mass diffusivities of species $i$, for $i=U,V$. We require Eq. (11.9) to satisfy Eq. (7.9). One cannot have an equivalent relation of the form $\mathbf{j}_P=-\rho\mathcal{D}_P\nabla Y_P$ and simultaneously satisfy the mixture constraint of Eq. (7.9).

We require periodic boundary conditions for each species on a domain length $\ell$ in both spatial dimensions:

$$Y_i(0,y)=Y_i(\ell,y),\quad \text{and}\quad Y_i(x,0)=Y_i(x,\ell),\quad \text{for } i=U,V,P, \tag{11.10}$$

$$\left.\frac{\partial Y_i}{\partial x}\right|_{x=0}=\left.\frac{\partial Y_i}{\partial x}\right|_{x=\ell}\quad \text{and}\quad \left.\frac{\partial Y_i}{\partial y}\right|_{y=0}=\left.\frac{\partial Y_i}{\partial y}\right|_{y=\ell},\quad \text{for } i=U,V,P. \tag{11.11}$$

The algebraic constraint on mass fraction can replace the evolution equation of $Y_P$, Eq. (11.7), resulting in differential equations for two variables: $Y_U$ and $Y_V$.

We now scale the evolution equations so that they are dimensionless. Three useful time scales of this problem are the residence time of a fluid particle in the reactor, $\tau_{\mathcal{R}}=\rho V/\dot{m}$, the time scale of the first reaction, $\tau_{k1}=M^2/k_1/\rho^2$, and the time scale of the second reaction, $\tau_{k2}=1/k_2$. There is also a characteristic length that is yet to be determined; we therefore use $\ell/\mathsf{L}$ as a length scale to transform our equations to a domain with dimensionless length $\mathsf{L}$. Indeed, while it would seem natural to select $\mathsf{L}\equiv 1$, we admit nonunity values so as to cleanly compare with standard results such as given by Pearson (1993).

Diffusion time scales associated with the first harmonic within the domain are $\tau_{\mathcal{D}U}=\ell^2/\mathcal{D}_U/\mathsf{L}^2$ and $\tau_{\mathcal{D}V}=\ell^2/\mathcal{D}_V/\mathsf{L}^2$. Using these time scales, we can transform Eqs. (11.5)–(11.7) into dimensionless equations in terms of the parameters: dimensionless feed rate, $\mathsf{F}=\tau_{k1}/\tau_{\mathcal{R}}$; ratio of reaction time scales, $\mathsf{k}=\tau_{k1}/\tau_{k2}$; and dimensionless diffusion coefficients, $\mathsf{D}_U=\tau_{k1}/\tau_{\mathcal{D}U}$ and $\mathsf{D}_V=\tau_{k1}/\tau_{\mathcal{D}V}$. These dimensionless diffusion coefficients are reciprocal Damköhler numbers (see Eq. (9.35)). We can also use the time and length scales to transform the coordinates to be dimensionless: $\mathsf{t}=t/\tau_{k1}$, $\mathsf{x}=\mathsf{L}x/\ell$, and $\mathsf{y}=\mathsf{L}y/\ell$. Assuming the inflow conditions of $Y_U^f=1, Y_V^f=0$, and $Y_P^f=0$, Eqs. (11.5) and (11.6) become

$$\frac{\partial Y_U}{\partial \mathsf{t}}=-Y_UY_V^2+\mathsf{F}(1-Y_U)+\mathsf{D}_U\left(\frac{\partial^2 Y_U}{\partial \mathsf{x}^2}+\frac{\partial^2 Y_U}{\partial \mathsf{y}^2}\right), \tag{11.12}$$

$$\frac{\partial Y_V}{\partial \mathsf{t}}=Y_UY_V^2-(\mathsf{F}+\mathsf{k})\,Y_V+\mathsf{D}_V\left(\frac{\partial^2 Y_V}{\partial \mathsf{x}^2}+\frac{\partial^2 Y_V}{\partial \mathsf{y}^2}\right), \tag{11.13}$$

with boundary conditions

$$Y_i(\mathsf{x},\mathsf{y}) = Y_i(\mathsf{x}+\mathsf{L},\mathsf{y}) \quad \text{and} \quad Y_i(\mathsf{x},\mathsf{y}) = Y_i(\mathsf{x},\mathsf{y}+\mathsf{L}), \quad \text{for } i = U, V, \tag{11.14}$$

where $\mathsf{x} \in [0, \mathsf{L})$ and $\mathsf{y} \in [0, \mathsf{L})$. This has a similar form to that which has been studied by others (Doelman et al., 1997; Marchant, 2002; Hundsdorfer and Verwer, 2003); it is identical to that studied by Pearson (1993).

### 11.1.1 Spatially Homogeneous

We first study Eqs. (11.12) and (11.13) in the spatially homogeneous limit,

$$\frac{dY_U}{dt} = -Y_U Y_V^2 + \mathsf{F}(1 - Y_U), \quad \frac{dY_V}{dt} = Y_U Y_V^2 - (\mathsf{F}+\mathsf{k})\, Y_V. \tag{11.15}$$

Varying the dimensionless parameters $\mathsf{F}$ and $\mathsf{k}$ results in dynamics of varying character. When $\mathsf{F} = 0$, the system is closed, and the Gray-Scott system does not behave quite like traditional combustion systems in that it does not have a unique equilibrium. For $\mathsf{F} = 0$, there is a continua of equilibria along the line $Y_V = 0$. We limit our study to $\mathsf{F} > 0$ and $\mathsf{k} > 0$ for positive feed flow and reaction rates. By inspection, we find an equilibrium to Eqs. (11.15) at $Y_U = 1$ and $Y_V = 0$, which we label $R_1$. By evaluating the Jacobian,

$$\mathbf{J} = \begin{pmatrix} -Y_V^2 - \mathsf{F} & -2Y_U Y_V \\ Y_V^2 & 2Y_U Y_V - \mathsf{F} - \mathsf{k} \end{pmatrix}, \tag{11.16}$$

we find eigenvalues in the neighborhood of $R_1$ to be $\boldsymbol{\lambda} = \{-\mathsf{F}, -\mathsf{F}-\mathsf{k}\}$, which are negative because $\mathsf{F} > 0$ and $\mathsf{k} > 0$. Therefore, $R_1$ is a sink.

We find two additional finite equilibria for Eqs. (11.15): $R_2$ at

$$Y_U = \frac{\mathsf{F} + \sqrt{\mathsf{F}(\mathsf{F} - 4(\mathsf{F}+\mathsf{k})^2)}}{2\mathsf{F}}, \quad Y_V = \frac{\mathsf{F} - \sqrt{\mathsf{F}(\mathsf{F} - 4(\mathsf{F}+\mathsf{k})^2)}}{2(\mathsf{F}+\mathsf{k})}, \tag{11.17}$$

and $R_3$ at

$$Y_U = \frac{\mathsf{F} - \sqrt{\mathsf{F}(\mathsf{F} - 4(\mathsf{F}+\mathsf{k})^2)}}{2\mathsf{F}}, \quad Y_V = \frac{\mathsf{F} + \sqrt{\mathsf{F}(\mathsf{F} - 4(\mathsf{F}+\mathsf{k})^2)}}{2(\mathsf{F}+\mathsf{k})}. \tag{11.18}$$

If $\mathsf{F}(\mathsf{F} - 4(\mathsf{F}+\mathsf{k})^2) \geq 0$, then $R_2$ and $R_3$ are real; otherwise, they are complex. This indicates a saddle-node bifurcation (Guckenheimer and Holmes, 1983), which occurs where $\mathsf{F} - 4(\mathsf{F}+\mathsf{k})^2 = 0$ (because we have demanded $\mathsf{F} > 0$). The range of parameters for which $R_2$ and $R_3$ are real is shown in gray in Fig. 11.1.

The character of the equilibria $R_2$ and $R_3$ can change depending on the values of the parameters $\mathsf{F}$ and $\mathsf{k}$. Therefore, we evaluate the eigenvalues of the Jacobian in Eq. (11.16) in terms of the dependent variables and parameters:

$$\lambda = \frac{1}{2}\left(-2\mathsf{F} - \mathsf{k} + 2Y_U Y_V - Y_V^2 \pm \sqrt{\mathsf{k}^2 + Y_V^2(-2Y_U + Y_V)^2 - 2\mathsf{k}Y_V(2Y_U + Y_V)}\right). \tag{11.19}$$

In the neighborhood of the equilibrium $R_2$, we find that both eigenvalues are real (one is negative, and one is positive) for the entire range of parameter space where $R_2$ is real. This means that $R_2$ has the character of a saddle with one positive eigenvalue. In the neighborhood of the equilibrium $R_3$, we find that the real parts of both eigenvalues have the same sign throughout the parameter space where $R_3$ is real. We

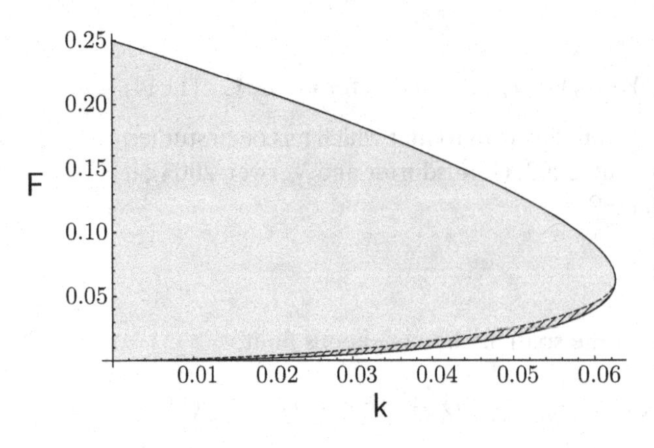


Figure 11.1. Behavior of equilibria for the spatially homogeneous Gray-Scott system. Adapted from Mengers (2012), reprinted with permission.

find a Hopf[1] bifurcation (Guckenheimer and Holmes, 1983) where the real parts of $R_3$'s complex conjugate pair of eigenvalues change sign; this occurs at

$$\mathsf{F} = \hat{\mathsf{F}} = \frac{1}{2}\left(\sqrt{\mathsf{k}} - 2\mathsf{k} - \sqrt{\mathsf{k} - 4\mathsf{k}^{3/2}}\right). \tag{11.20}$$

That is shown as a dashed line in Fig. 11.1. Where $\mathsf{F} < \hat{\mathsf{F}}$ (shown as a hatched area in Fig. 11.1), $R_3$ is a source; in the remainder of the parameter space where $R_3$ is real, it is a sink. While $R_3$ is a real spiral source, there is no guarantee of a stable limit cycle. Pearson (1993) finds this to occur only for $\mathsf{k} < 0.035$ for values of $\mathsf{F}$ that are marginally less than $\hat{\mathsf{F}}$, while for $\mathsf{k} > 0.035$, there are unstable limit cycles for values of $\mathsf{F}$ that are marginally greater than $\hat{\mathsf{F}}$. In the language of dynamical systems theory, this Hopf bifurcation can be either supercritical for $\mathsf{k} < 0.035$ or subcritical for $\mathsf{k} > 0.035$.

We examine the dynamics of the spatially homogeneous Gray-Scott mechanism for three sets of parameters: (1) $\mathsf{F} = 2.5 \times 10^{-2}$ and $\mathsf{k} = 5.5 \times 10^{-2}$, (2) $\mathsf{F} = 9.16 \times 10^{-3}$ and $\mathsf{k} = 3.1 \times 10^{-2}$, and (3) $\mathsf{F} = 4.1 \times 10^{-3}$ and $\mathsf{k} = 2.0 \times 10^{-2}$. Pattern formation in case 1 is described by Pearson (1993). We use the Poincaré sphere projection,

$$\eta_1 = \frac{Y_U}{\sqrt{1 + Y_U^2 + Y_V^2}}, \quad \eta_2 = \frac{Y_V}{\sqrt{1 + Y_U^2 + Y_V^2}}, \tag{11.21}$$

to display the dynamics at infinity. We identify six equilibria at infinity that are independent of $\mathsf{F}$ and $\mathsf{k}$. These equilibria's coordinates in the Poincaré sphere mapping and corresponding mass fractions are shown in Table 11.1. The equilibria $I_3$ and $I_6$ are both found along the line $Y_V = -Y_U$ in mass fraction space as $Y_U \to +\infty$ and $Y_U \to -\infty$, respectively. Evaluating the character of the infinite equilibria, we find that $I_1$ and $I_4$ are saddle-node nonhyperbolic equilibria, $I_2$ and $I_5$ are saddles with one positive eigenvalue, and $I_3$ and $I_6$ are sources; $I_1, I_2, I_4$, and $I_5$ are candidates for SACIM seeds, as discussed in Chapter 6.

We show the dynamics for these three sets of parameters in Fig. 11.2. The left column shows the Poincaré sphere mapping of the phase space, and the right column shows the mass fraction phase space. The physical domain is shaded in gray, heteroclinic orbits and limit cycles are thick, and other trajectories are thin gray. Although

[1]Eberhard Frederich Ferdinand Hopf (1902–1983), Austrian-German-American mathematician.

Table 11.1. *Gray-Scott Infinite Equilibria in the Poincaré Sphere Mapping*

| | $\eta_1$ | $\eta_2$ | $Y_U$ | $Y_V$ |
|---|---|---|---|---|
| $I_1$ | 1 | 0 | $+\infty$ | 0 |
| $I_2$ | 0 | 1 | 0 | $+\infty$ |
| $I_3$ | $-\sqrt{2}/2$ | $\sqrt{2}/2$ | $-\infty$ | $+\infty$ |
| $I_4$ | $-1$ | 0 | $-\infty$ | 0 |
| $I_5$ | 0 | $-1$ | 0 | $-\infty$ |
| $I_6$ | $\sqrt{2}/2$ | $-\sqrt{2}/2$ | $+\infty$ | $-\infty$ |

our main objective is understanding the dynamics, not reducing the system, the language of reduced kinetics from Chapter 6 is useful here. For case 1, the only physical equilibrium is $R_1$. We see several CIMs connecting to $R_1$ from $I_1, I_2, I_4, I_5$, and $I_6$.

Figure 11.2b shows three physical equilibria, $R_1, R_2$, and $R_3$. We identify another CIM originating at $R_2$, and we find it has two relevant branches; these branches originate along the unstable eigenvector of $R_2$ in either direction. Along one CIM branch from $R_2$, whose initial rate of change has $Y_U$ increasing and $Y_V$ decreasing, the species follow a direct path through phase space to $R_1$. Along the other branch from $R_2$, the species evolve further away from $R_1$, and approach the CIM branch between $I_2$ and $R_1$ before decaying to the physical equilibrium, $R_1$.

For the parameters of case 2, there are also two limit cycles: an inner stable limit cycle encompassing the spiral source at $R_3$, and an outer unstable limit cycle encompassing the stable limit cycle. The unstable limit cycle can be identified by integrating backward in time, because it stable in reverse time. The outer unstable limit cycle defines the boundary of the basin of attraction of the inner limit cycle. A sketch is provided in Fig. 11.3, where the growth and decay away from and toward the limit cycles is exaggerated for clarity.

When we evaluate the system with the parameters from case 2 and initial conditions $Y_U = 0.325$ and $Y_V = 0.290$, we predict the long time limit of the mass fractions will display oscillations, because the initial conditions are within the basin of attraction of the stable limit cycle. If we use a reduction method and project onto any of the branches of the CIM, notated as $\mathcal{P}(Y_i)$, the long time dynamics will not be accurately predicted. This is shown in Fig. 11.4, with a phase space plot on the left and a time evolution plot on the right. In the phase space, the initial conditions are indicated with a cross, the time integrated solution is shown as a gray line, the projection onto the CIM branch that connects $I_2$ to $R_1$ is shown as a black dashed line, the CIM is shown as a black line, and the basin of attraction of the stable limit cycle is shown as an area hatched with gray lines. Figure 11.4 shows the error incurred from projecting the dynamics across the boundary of the basin of attraction. In the time evolution plot, we see the governing equations predict a solution that decays to a stable limit cycle; however, the solution projected onto the CIM, while initially relatively accurate, does not capture the limit cycle behavior, erroneously predicting a decay to the stable equilibrium, $R_1$. Any initial conditions within the basin of attraction of the limit cycle projected to a CIM that is found using the heteroclinic orbit construction technique will result in an inaccurate prediction; however, a projection onto the attractive limit cycle will accurately predict the long time dynamics of the system.

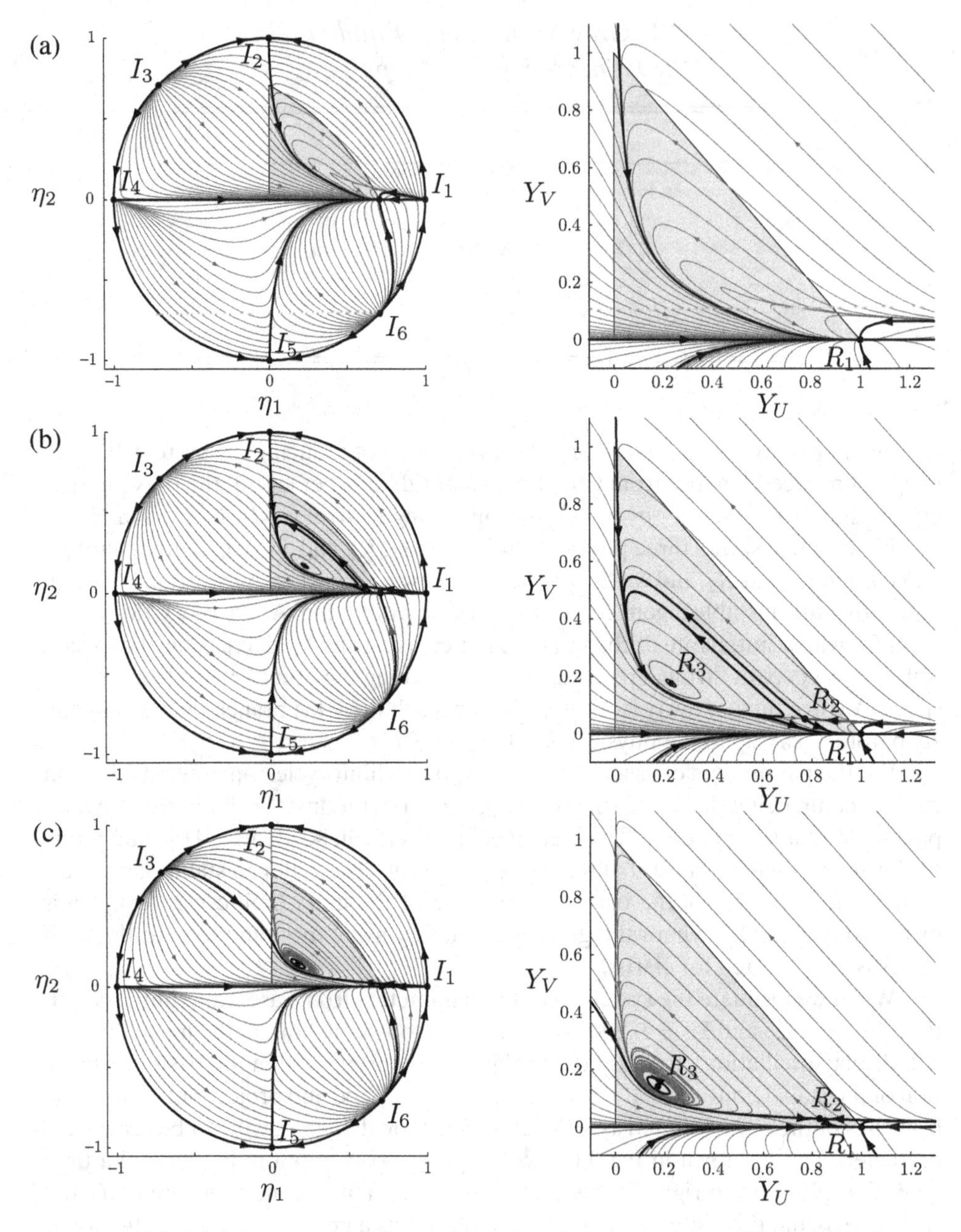


Figure 11.2. Poincaré sphere projection and mass fraction phase spaces of the spatially homogeneous Gray-Scott system. Adapted from Mengers (2012), reprinted with permission.

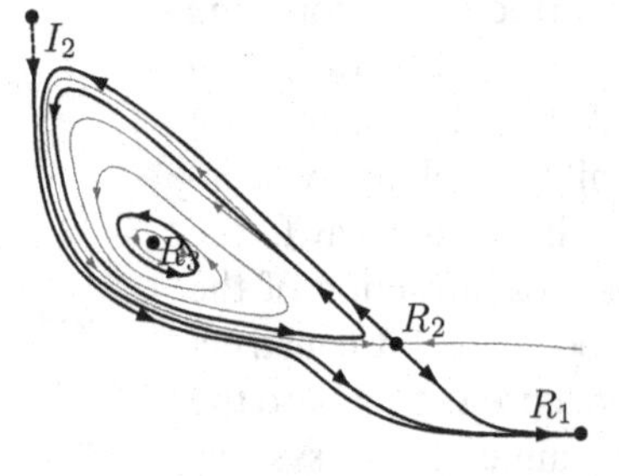


Figure 11.3. Sketch of the stable and unstable limit cycles in the spatially homogeneous Gray-Scott system with $\mathsf{F} = 9.16 \times 10^{-3}$ and $\mathsf{k} = 3.1 \times 10^{-2}$. Adapted from Mengers (2012), reprinted with permission.

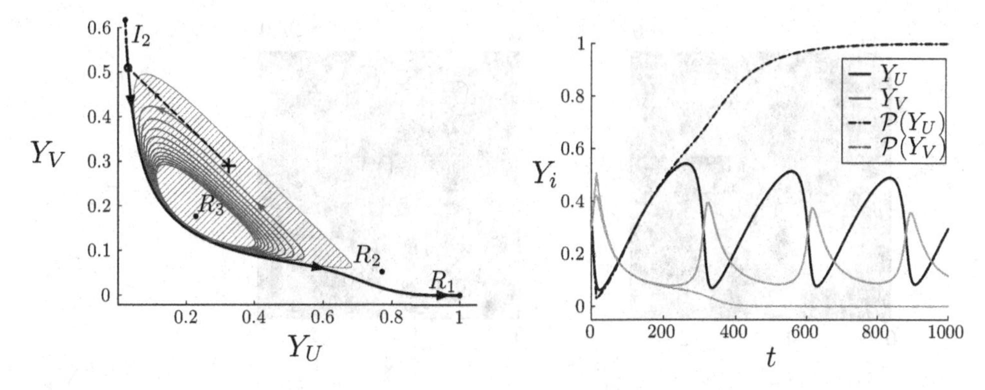


Figure 11.4. Naïve projection onto a CIM in the Gray-Scott system with $\mathsf{F} = 9.16 \times 10^{-3}$ and $\mathsf{k} = 3.1 \times 10^{-2}$. Adapted from Mengers (2012), reprinted with permission.

Results for case 3, Fig. 11.2c, are similar to those of case 2, except there is only one stable limit cycle.

### 11.1.2 Spatial Variations and Pattern Formation

We return to the full reaction-diffusion system from Eqs. (11.12)–(11.14), and following Pearson (1993), we solve the system on a domain $\mathsf{x} \in [0, \mathsf{L})$ and $\mathsf{y} \in [0, \mathsf{L})$ with periodic boundary conditions, where the dimensionless length and diffusion coefficients are $\mathsf{L} = 2.5$, $\mathsf{D}_U = 2 \times 10^{-5}$, and $\mathsf{D}_V = 10^{-5}$. The governing equations are evaluated on a discrete spatial grid of 256×256 uniformly spaced points; the diffusion operator is modeled using a second-order central finite difference operator, and the time evolution is evaluated using the algorithm embodied in `LSODE` (Hindmarsh, 1983). We choose to examine the parameters from case 1 in Sec. 11.1.1, $\mathsf{F} = 2.5 \times 10^{-2}$ and $\mathsf{k} = 5.5 \times 10^{-2}$. For this set of parameters, our results show pattern formation that is similar to the pattern labeled as "$\gamma$" by Pearson. The mass fraction evolution for this system is shown in Fig. 11.5 at six dimensionless times. The initial conditions are $Y_U = 1$ and $Y_V = 0$ for the entire domain, except a small region, $0.3125 \leq \mathsf{x} \leq 0.9375$ and $0.3125 \leq \mathsf{y} \leq 0.9375$, which has the concentration $Y_U = 0.5$ and $Y_V = 0.25$. To break the symmetry of the system, the initial conditions at each grid point are perturbed by a random amount, $\epsilon$, which is uniformly distributed between the values of $-0.01$ and $0.01$. As the system evolves, we see predominately striped pattern formation that propagates outward throughout the domain. These patterns encompass the entire domain by $\mathsf{t} = 5000$. The solution at $\mathsf{t} = 5000$ still shows symmetries from the initial condition; as the solution evolves further (by $\mathsf{t} = 2 \times 10^5$), the perturbations in the initial conditions have broken most of those symmetries, and the solution appears more isotropic. Mengers (2012) provides additional analysis based on image processing of the two-dimensional structures that loosely correlates the average wavelength of the spatial oscillations with the time scales of reaction motivated by the relation of Eq. (8.18): $\ell = \sqrt{\mathcal{D}\tau}$.

## 11.2 $H_2$-Air Mechanism

We next to turn oscillations predicted by a system of physically realistic chemical kinetics in a spatially homogeneous CSTR.

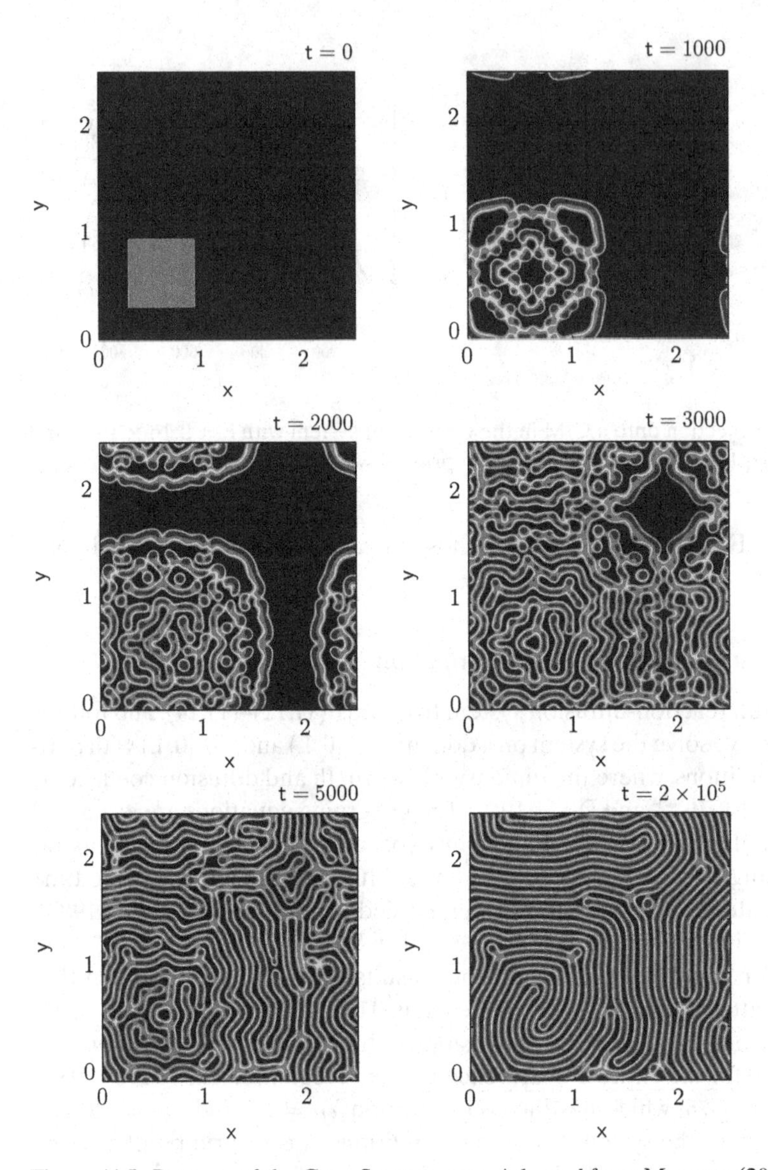


Figure 11.5. Patterns of the Gray-Scott system. Adapted from Mengers (2012), reprinted with permission.

We consider the $N = 9$ species, $J = 20$ reaction, $L = 3$ element hydrogen-air mechanism extracted from Miller and Bowman (1989), which is shown in Table 11.2. We examine conditions a similar to a lean isothermal isobaric system that Kalamatianos and Vlachos (1995) show to have a limit cycle. While we also evaluate this mechanism as isothermal, $T = 993$ K, we consider it isochoric, $\rho = 9.25 \times 10^{-4}$ g/cm$^3$, to simplify our method of solving for equilibria using the algebraic root-finding algorithm in BERTINI (Bates et al., 2013). To exhibit a similar limit cycle in the modified system, we consider a feed flow rate corresponding to a residence time of $\tau_{\mathcal{R}} = V\rho/\dot{m} = 1.1 \times 10^{-3}$ s; this feed flow rate is decreased slightly from the one considered in Kalamatianos and Vlachos that had a residence time of $\tau_{\mathcal{R}} = 1 \times 10^{-3}$ s. The inflow is a hydrogen-air mixture with an equivalence ratio of $\Phi = 0.5$: for every

Table 11.2. *$H_2$-Air Reaction Mechanism*

| $j$ | Reaction | $a_j$ $\left[\dfrac{\left(\dfrac{\text{mol}}{\text{cm}^3}\right)^{\left(1-\nu'_{M,j}-\sum_{i=1}^{N}\nu'_{i,j}\right)}}{\text{s K}^{\beta_j}}\right]$ | $\beta_j$ | $\bar{\mathcal{E}}_j$ $\left[\dfrac{\text{cal}}{\text{mol}}\right]$ |
|---|---|---|---|---|
| 1 | $H_2 + O_2 \rightleftharpoons 2OH$ | $0.170 \times 10^{14}$ | 0.000 | 47780 |
| 2 | $H_2 + OH \rightleftharpoons H_2O + H$ | $0.117 \times 10^{10}$ | 1.300 | 3626 |
| 3 | $OH + O \rightleftharpoons O_2 + H$ | $0.400 \times 10^{15}$ | −0.500 | 0 |
| 4 | $H_2 + O \rightleftharpoons OH + H$ | $0.506 \times 10^{5}$ | 2.670 | 6290 |
| 5 | $O_2 + H + M \rightleftharpoons HO_2 + M^a$ | $0.361 \times 10^{18}$ | −0.720 | 0 |
| 6 | $OH + HO_2 \rightleftharpoons O_2 + H_2O$ | $0.750 \times 10^{13}$ | 0.000 | 0 |
| 7 | $H + HO_2 \rightleftharpoons 2OH$ | $0.140 \times 10^{15}$ | 0.000 | 1073 |
| 8 | $O + HO_2 \rightleftharpoons O_2 + OH$ | $0.140 \times 10^{14}$ | 0.000 | 1073 |
| 9 | $2OH \rightleftharpoons H_2O + O$ | $0.600 \times 10^{9}$ | 1.300 | 0 |
| 10 | $2H + M \rightleftharpoons H_2 + M^b$ | $0.100 \times 10^{19}$ | −1.000 | 0 |
| 11 | $2H + H_2 \rightleftharpoons 2H_2$ | $0.920 \times 10^{17}$ | −0.600 | 0 |
| 12 | $2H + H_2O \rightleftharpoons H_2 + H_2O$ | $0.600 \times 10^{20}$ | −1.250 | 0 |
| 13 | $OH + H + M \rightleftharpoons H_2O + M^c$ | $0.160 \times 10^{23}$ | −2.000 | 0 |
| 14 | $H + O + M \rightleftharpoons OH + M^c$ | $0.620 \times 10^{17}$ | −0.600 | 0 |
| 15 | $2O + M \rightleftharpoons O_2 + M$ | $0.189 \times 10^{14}$ | 0.000 | −1788 |
| 16 | $H + HO_2 \rightleftharpoons H_2 + O_2$ | $0.125 \times 10^{14}$ | 0.000 | 0 |
| 17 | $2HO_2 \rightleftharpoons O_2 + H_2O_2$ | $0.200 \times 10^{13}$ | 0.000 | 0 |
| 18 | $H_2O_2 + M \rightleftharpoons 2OH + M$ | $0.130 \times 10^{18}$ | 0.000 | 45500 |
| 19 | $H + H_2O_2 \rightleftharpoons H_2 + HO_2$ | $0.160 \times 10^{13}$ | 0.000 | 3800 |
| 20 | $OH + H_2O_2 \rightleftharpoons H_2O + HO_2$ | $0.100 \times 10^{14}$ | 0.000 | 1800 |

All third body molecule collision coefficients are unity, except where noted.
[a]Reaction 5: $M_{H_2} = 2.86, M_{H_2O} = 18.6, M_{N_2} = 1.26$.
[b]Reaction 10: $M_{H_2} = 0.0, M_{H_2O} = 0.0$.
[c]Reactions 13 and 14: $M_{H_2O} = 5.0$.

1 mol of $O_2$, there will be $2\Phi = 1$ mol of $H_2$ and 3.67 mol[2] of $N_2$. For these parameters, employing the molecular masses of $O_2$, $H_2$, and $N_2$, the inflow in specific moles has

$$z_1 = \frac{1}{(1)(31.9988) + (1)(2.01588) + (3.67)(28.0134)} = 7.3087 \times 10^{-3} \text{ mol/g}, \quad (11.22)$$

$$z_2 = \frac{1}{(1)(31.9988) + (1)(2.01588) + (3.67)(28.0134)} = 7.3087 \times 10^{-3} \text{ mol/g}, \quad (11.23)$$

$$z_4 = \frac{3.67}{(1)(31.9988) + (1)(2.01588) + (3.67)(28.0134)} = 2.6823 \times 10^{-2} \text{ mol/g}. \quad (11.24)$$

The inflow has a pressure of $P = 3.1648 \times 10^6$ dyne/cm$^2$; at completion of combustion in the isochoric chamber, the pressure is $P = 2.6066 \times 10^6$ dyne/cm$^2$. We use a

[2]Chosen arbitrarily to be slightly different than the common standard for air of 3.76 mol.

standard data base (Kee et al., 2000) to evaluate the species' thermodynamic properties. By demanding that the initial conditions' concentrations have the same elemental composition as the inflow, this system has $L = 3$ algebraic constraints: conservation of each element, H, O, and N; therefore, the system has $R = N - L = 6$ dimensions. We model species $i = \{1, 2, 3, 5, 6, 7\}$ and use the algebraic constraints to solve for the concentrations of species $i = \{4, 8, 9\}$. Following the reduction technique first described in Sec. 1.2.2, we identify the algebraic constraints as

$$z_4 = 2.6823 \times 10^{-2} \text{ mol/g}, \tag{11.25}$$

$$z_8 = -2z_1 + 2z_2 + z_3 + z_6 - z_7, \tag{11.26}$$

$$z_9 = z_1 - 2z_2 - \frac{3}{2}z_3 - \frac{1}{2}z_5 - z_6 + \frac{1}{2}z_7 + 7.3086 \times 10^{-3} \text{ mol/g}. \tag{11.27}$$

We model hydrogen-air reaction and seek to construct the branches of the spatially homogeneous CIM. Using the BERTINI software package (Bates et al., 2013), with the input file described by Mengers (2012), we identify 97 real finite equilibria of this system. Of these equilibria, three have positive concentrations for all nine species, making them physical, and 13 have one positive eigenvalue, making them candidates for the origins of CIMs; the physical equilibria and CIM origin candidates are listed by Mengers (2012).

We evaluate the character of the physical equilibria: $R_4$ has all negative real eigenvalues and is a physical sink; $R_{69}$ has one positive real eigenvalue, making it both a physical equilibrium and a CIM origin candidate; and $R_1$ has four negative real eigenvalues and one complex conjugate pair of eigenvalues with positive real part, making it a saddle. The long time dynamics of systems with initial conditions in the neighborhood of $R_1$ exhibit limit cycle behavior. The eigenvalues of the physical equilibria are given in detail by Mengers (2012).

We integrate the system with initial conditions perturbed along $R_{69}$'s unstable eigenvector in either direction. In one direction, the trajectory approaches $R_4$ along its slowest eigenvector; this heteroclinic orbit is a branch of the CIM. In the other direction, the trajectory collapses onto the stable limit cycle. These trajectories are shown in a projection of phase space in Fig. 11.6, where the small branch of the CIM is in black and the trajectory that collapses onto the stable limit cycle is in gray. The time evolution is shown in Fig. 11.7, where on the left is the evolution along the branch of the CIM, and on the right is the stable limit cycle. We deduce that $R_{69}$ lies on the boundary of the basin of attraction between the stable limit cycle and the steady state equilibrium, $R_4$.

In conclusion, nonlinear oscillatory combustion both in space and time is a paradigm problem that is likely necessary to understand before more complicated combustion instabilities and even turbulence can be fathomed. Because otherwise small effects are often amplified by nonlinearity that affects limit cycle behavior, one should be cautious in applying filtering strategies as they may dramatically alter the predicted patterns.

## EXERCISES

**11.1.** Study the dynamics of the spatially homogeneous Gray-Scott system for $\mathsf{k} = 0.03$ and a variety of values of $\mathsf{F}$. Describe how the dynamics change as $\mathsf{F}$

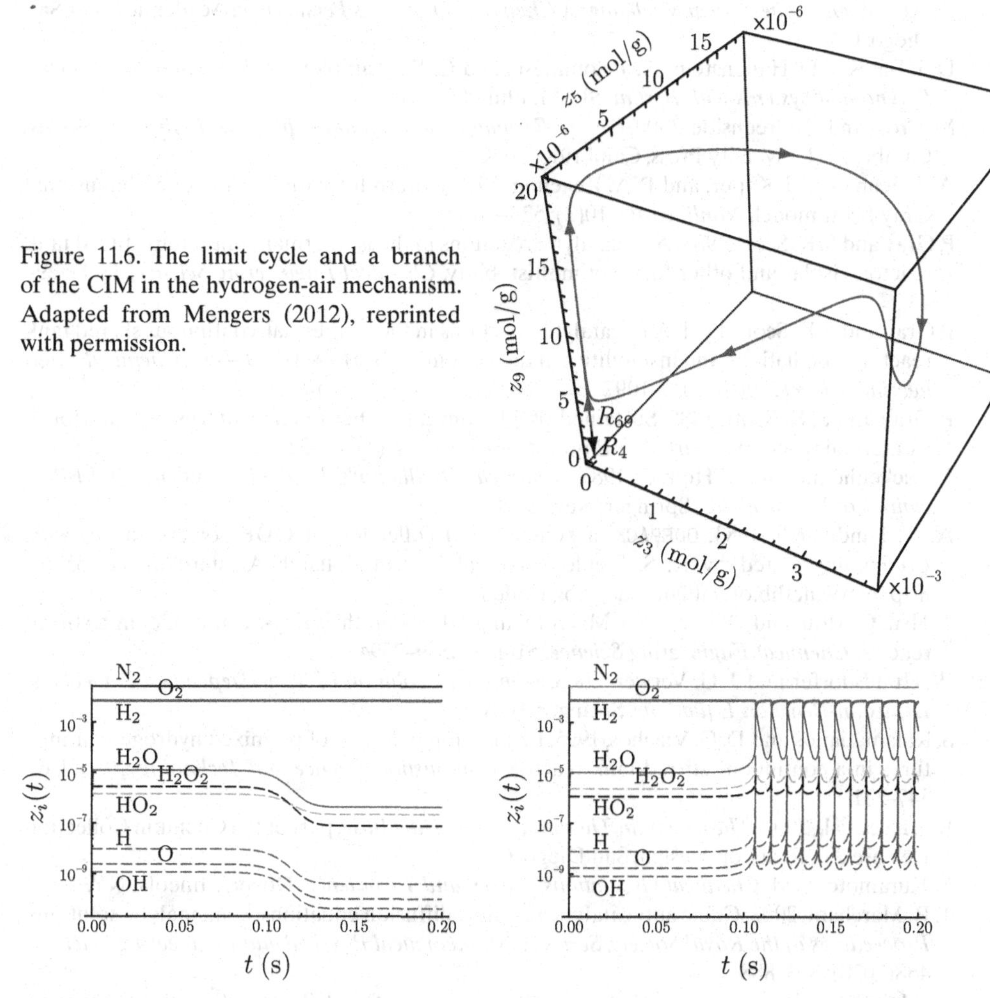


Figure 11.6. The limit cycle and a branch of the CIM in the hydrogen-air mechanism. Adapted from Mengers (2012), reprinted with permission.

Figure 11.7. Evolution of hydrogen-air mechanism along either direction of the $R_{69}$ unstable eigenvector. Adapted from Mengers (2012), reprinted with permission.

changes. Select initial conditions that effectively display the varieties of behaviors that may exist.

**11.2.** Reproduce the results of Sec. 11.1.2, and then show how the patterns may or may not change depending on the resolution of the spatial grid. Find the minimum grid size necessary to display the essential dynamics.

**11.3.** Pose and solve numerically a problem in which a CSTR is operated with a calorically perfect ideal gas that obeys a one-step irreversible Arrhenius kinetics law. When the mass flow rate is zero, use numerical solution to reproduce the results of Sec. 1.3.1. Show how varying the mass flow rate affects the reaction dynamics.

## References

A. Adamatzky, 2013, *Reaction-Diffusion Automata: Phenomenology, Localisations, Computation*, Springer, Berlin.

R. Aris, 1989, *Elementary Chemical Reactor Analysis*, Butterworths, Boston.

R. Aris, 1999, *Mathematical Modeling: A Chemical Engineer's Perspective*, Academic Press, San Diego, CA.

D. J. Bates, J. D. Hauenstein, A. J. Sommese, and C. W. Wampler, 2013, *Numerically Solving Polynomial Systems with Bertini*, SIAM, Philadelphia.

M. Cross and H. Greenside, 2009, *Pattern Formation and Dynamics in Nonequilibrium Systems*, Cambridge University Press, Cambridge, UK.

A. Doelman, T. J. Kaper, and P. A. Zegeling, 1997, Pattern formation in the one-dimensional Gray-Scott model, *Nonlinearity*, 10(2): 523–563.

P. Gray and S. K. Scott, 1983, Autocatalytic reactions in the isothermal, continuous stirred tank reactor—isolas and other forms of multistability, *Chemical Engineering Science*, 38(1): 29–43.

P. Gray and S. K. Scott, 1984, Autocatalytic reactions in the isothermal, continuous stirred tank reactor—oscillations and instabilities in the system $A + 2B \to 3B; B \to C$, *Chemical Engineering Science*, 39(6): 1087–1097.

P. Gray and S. K. Scott, 1985, Sustained oscillations and other exotic patterns of behavior in isothermal reactions, *Journal of Physical Chemistry*, 89(1): 22–32.

J. Guckenheimer and P. Holmes, 1983, *Nonlinear Oscillation's, Dynamical Systems, and Bifurcations of Vector Fields*, Springer, New York.

A. C. Hindmarsh, 1983, `ODEPACK`, a systematized collection of ODE solvers, in *Scientific Computing*, edited by R. S. Stepleman et al., North-Holland, Amsterdam, pp. 55–64. http://www.netlib.org/alliant/ode/prog/lsode.f.

T. Hsu, C. Mou, and D. Lee, 1996, Macromixing effects on the Gray-Scott model in a stirred reactor, *Chemical Engineering Science*, 51(11): 2589–2594.

W. Hundsdorfer and J. G. Verwer, 2003, *Numerical Solution of Time-Dependent Advection-Diffusion-Reaction Equations*, Springer, Berlin.

S. Kalamatianos and D. G. Vlachos, 1995, Bifurcation behavior of premixed hydrogen/air mixtures in a continuous stirred tank reactor, *Combustion Science and Technology*, 109(1–6): 347–371.

R. J. Kee et al., 2000, *The Chemkin Thermodynamic Data Base*, part of the Chemkin Collection Release 3.6, Reaction Design, San Diego, CA.

Y. Kuramoto, 1984, *Chemical Oscillations, Waves, and Turbulence*, Dover, Mineola, NY.

T. R. Marchant, 2002, Cubic autocatalytic reaction-diffusion equations: Semianalytic solutions, *Proceedings of the Royal Society, Series A, Mathematical Physical and Engineering Sciences*, 458(2020): 873–888.

Z. Mei, 2000, *Numerical Bifurcation Analysis for Reaction-Diffusion Equations*, Springer, Berlin.

J. D. Mengers, 2012, Slow invariant manifolds for reaction-diffusion systems, PhD dissertation, University of Notre Dame.

J. A. Miller and C. T. Bowman, 1989, Mechanism and modeling of nitrogen chemistry in combustion, *Progress in Energy and Combustion Science*, 15(4): 287–338.

J. E. Pearson, 1993, Complex patterns in a simple system, *Science*, 261(5118): 189–192.

R. A. Schmitz, K. R. Graziani, and J. L. Hudson, 1977, Experimental evidence of chaotic states in the Belousov-Zhabotinskii reaction, *Journal of Chemical Physics*, 67(7): 3040–3044.

A. M. Turing, 1952, The chemical basis of morphogenesis, *Philosophical Transactions of the Royal Society of London, Series B, Biological Sciences*, 237(641): 37–72.

V. Volpert, 2014, *Elliptic Partial Differential Equations, Volume 2: Reaction-Diffusion Equations*, Springer, Basel.

# 12 Detonation

*Plato says, my friend, that society cannot be saved until either the Professors of Greek take to making gunpowder, or else the makers of gunpowder become Professors of Greek.*
– Undershaft to Cusins in George Bernard Shaw's (1856–1950) *Major Barbara*

Let us consider fundamental aspects of detonation theory. Detonation is defined as a shock-induced combustion process. Most features of detonations are well modeled by considering only the mechanisms of advection and reaction. Diffusion plays a more subtle role. Its inclusion is not critical to the main physics; however, among other things, it is potentially important in determining stability boundaries, as well as in correctly capturing multidimensional instabilities and patterns. The definitive text is that of Fickett and Davis (1979). The present chapter is strongly influenced by this monograph. Useful exposition is also given in many standard sources (Bone and Townend, 1927; Hirschfelder et al., 1954; Zel'dovich and Kompaneets, 1960; Lewis and von Elbe, 1961; Toong, 1982; Strehlow, 1984; Williams, 1985; Zel'dovich et al., 1985; Korobeinikov, 1989; Courant and Friedrichs, 1999; Dremin, 1999; Zel'dovich and Raizer, 2002; Kuo, 2005; Law, 2006; Keating, 2007; Terao, 2007; Lee, 2008; Liberman, 2008; Oppenheim, 2008; Shepherd, 2009; Gelfand et al., 2012; Glassman et al., 2014). We do not consider the rich field of condensed phase detonations (e.g., Butler and Krier, 1985; Chéret, 1993; Bdzil and Stewart, 2007).

We begin with a presentation of the reactive Euler equations for a simple one-step irreversible kinetics model. The Euler model neglects diffusion but includes full compressibility. This dramatically affects the analysis. Mathematically, the governing partial differential equations are hyperbolic. Those considered in previous chapters were parabolic or elliptic. This affects the type of initial and boundary conditions necessary for a well-posed problem (Zauderer, 1989). Hyperbolicity renders finite speed signaling to be important and also admits shock waves into the solution. We then examine solutions of the steady reactive Euler equations, often turning to dynamical systems theory to aid in interpreting the results.

We close the book by turning to a series of unsteady problems to highlight the coupling of nonlinear dynamics with thermodynamics for detonating systems. We study inviscid and viscous models with both simple and detailed kinetics. Each of these can be shown to undergo a bifurcation process and transition to chaos as physical parameters are varied. We finish by focusing on how diffusion perturbs the nonlinear reactive-advective dynamics. Often, its effects are small; however, near bifurcation points, the perturbative effect of diffusion can be such that it can more significantly affect the behavior. Furthermore, it will be seen that to capture these multiscale phenomena, fine scale discretization is essential. For this chapter, we mainly draw on a series of papers (Powers and Gonthier, 1992; Grismer and Powers, 1996; Singh et al., 2001; Powers and Paolucci, 2005; Powers, 2006; Henrick et al., 2006; Romick et al., 2012, 2015) and the dissertation of Romick (2015).

## 12.1 Reactive Euler Equations

### 12.1.1 One-Step Irreversible Kinetics

Let us focus on one-dimensional solutions in which all diffusion processes are neglected. We shall also here only be concerned with simple one-step irreversible kinetics in which

$$A \rightarrow B. \tag{12.1}$$

Much of this preliminary one-step kinetics model development is repeated from Sec. 4.7; however, because the detonation literature invokes some conflicting notation relative to more standard combustion literature, we repeat for clarity some of the details, taking care to note where they diverge.

We adopt the assumption that $A$ and $B$ are materials with identical properties; thus, $M_A = M_B = M, c_{PA} = c_{PB} = c_P$, but $A$ is endowed with chemical energy that is released as it forms $B$. So here, we have $N = 2$ species and $J = 1$ reaction. The number of elements will not be important. Extension to multistep kinetics is straightforward but involves details that obscure many of the key points. At this stage, we will leave the state equations relatively general, but we will soon take them to be ideal and calorically perfect. Because we have only a single reaction, our reaction rate vector $r_j$, is a scalar that we will call $r$. Our matrix of net stoichiometric coefficients $\nu_{ij}$ is of dimension $2 \times 1$ and is

$$\nu_{ij} = \begin{pmatrix} -1 \\ 1 \end{pmatrix}. \tag{12.2}$$

Here, the first entry is associated with $A$ and the second with $B$. Thus, our species molar production rate vector, from Eq. (5.18), $\dot{\omega}_i = \sum_{j=1}^{J} \nu_{ij} r_j$, reduces to

$$\begin{pmatrix} \dot{\omega}_A \\ \dot{\omega}_B \end{pmatrix} = \begin{pmatrix} -1 \\ 1 \end{pmatrix} (r) = \begin{pmatrix} -r \\ r \end{pmatrix}. \tag{12.3}$$

Now, the right side of Eq. (7.8) takes the form $M_i \dot{\omega}_i$. Let us focus on the products and see how this form expands:

$$M_B \dot{\omega}_B = M_B r = M r = M k \overline{\rho}_A = M k \left( \frac{\rho Y_A}{M} \right) = \rho k Y_A = \rho k (1 - Y_B). \tag{12.4}$$

For a reaction that commences with all $A$, let us define the reaction progress variable $\lambda$ as $\lambda = Y_B = 1 - Y_A$. And let us define $\mathsf{r}$ for this problem such that

$$\mathsf{r} = k(1 - Y_B) = k(1 - \lambda). \tag{12.5}$$

Expanding $k$ to show Arrhenius temperature sensitivity, we can say

$$\mathsf{r} = \underbrace{a T^{\beta} \exp\left( -\frac{\overline{\mathcal{E}}}{\overline{R} T} \right)}_{k} (1 - \lambda) = a T^{\beta} \exp\left( -\frac{\mathcal{E}}{R T} \right) (1 - \lambda). \tag{12.6}$$

Here, we have defined $\mathcal{E} = \overline{\mathcal{E}}/M$. The units of $\mathsf{r}$ for this problem will be 1/s, whereas the units for $r$ must be mol/cm$^3$/s, and

$$r = \rho \mathsf{r}/M. \tag{12.7}$$

The use of two different forms of the reaction rate, $r$ and $\mathsf{r}$ is unfortunate. One is nearly universal in the physical chemistry literature, $r$; the other is nearly universal in the one-step detonation chemistry community, $\mathsf{r}$. The two should be distinguished.

### 12.1.2 Sound Speed and Thermicity

To analyze all advective processes in a detonation, we need an expression for the frozen sound speed, $c$. Let us specialize the definition, Eq. (3.248), for our one-step chemistry:

$$c = v\left(\left(P + \left.\frac{\partial e}{\partial v}\right|_{P,\lambda}\right) \Big/ \left.\frac{\partial e}{\partial P}\right|_{v,\lambda}\right)^{1/2}. \tag{12.8}$$

This gives, using $v = 1/\rho$,

$$\rho^2 c^2 = \left(P + \left.\frac{\partial e}{\partial v}\right|_{P,\lambda}\right) \Big/ \left.\frac{\partial e}{\partial P}\right|_{v,\lambda}. \tag{12.9}$$

Let us, as is common in the detonation literature, define the *thermicity* $\sigma$ via the relation

$$\rho c^2 \sigma = -\left.\frac{\partial e}{\partial \lambda}\right|_{P,v} \Big/ \left.\frac{\partial e}{\partial P}\right|_{v,\lambda}. \tag{12.10}$$

As should be obvious, the notation $\sigma$ used in this chapter for thermicity has no connection to the same notation used in earlier chapters for irreversible entropy production or for eigenvalues. We now specialize our general mathematical relation, Eq. (3.56), to get

$$\left.\frac{\partial P}{\partial e}\right|_{v,\lambda} \left.\frac{\partial e}{\partial \lambda}\right|_{P,v} \left.\frac{\partial \lambda}{\partial P}\right|_{v,e} = -1, \tag{12.11}$$

so that one gets

$$-\left.\frac{\partial e}{\partial \lambda}\right|_{P,v} \Big/ \left.\frac{\partial e}{\partial P}\right|_{v,\lambda} = \left.\frac{\partial P}{\partial \lambda}\right|_{v,e}. \tag{12.12}$$

Thus, using Eq. (12.12), we can rewrite Eq. (12.10) as

$$\sigma = \frac{1}{\rho c^2} \left.\frac{\partial P}{\partial \lambda}\right|_{v,e}. \tag{12.13}$$

For our one-step reaction, we see that as the reaction moves forward, that is, as $\lambda$ increases, $\sigma$ is a measure of how much the pressure increases, because $\rho > 0, c^2 > 0$.

### 12.1.3 Parameters for $H_2$-Air

Let us postulate some parameters that yield predictions that loosely match results of the detailed kinetics calculation of Powers and Paolucci (2005) for $H_2$-air detonations. Rough estimates that allow one-step kinetics models with calorically perfect

Table 12.1. *Numerical Values of Parameters That Roughly Model $H_2$-Air Detonation*

| Parameter | Value | Units |
|---|---|---|
| $\gamma$ | 1.4 | |
| $M$ | 20.91 | kg/kmol |
| $R$ | 397.58 | J/kg/K |
| $P_0$ | $1.01325 \times 10^5$ | Pa |
| $T_0$ | 298 | K |
| $\rho_0$ | 0.85521 | kg/m$^3$ |
| $v_0$ | 1.1693 | m$^3$/kg |
| $q$ | $1.89566 \times 10^6$ | J/kg |
| $\mathcal{E}$ | $8.29352 \times 10^6$ | J/kg |
| $a$ | $5 \times 10^9$ | 1/s |
| $\beta$ | 0 | |

ideal gas assumptions to approximate the results of detailed kinetics models with calorically imperfect ideal gas mixtures are given in Table 12.1.

### 12.1.4 Conservative Form

Let us first consider a three-dimensional conservative form of the governing equations in the inviscid limit:

$$\frac{\partial \rho}{\partial t} + \nabla \cdot (\rho \mathbf{u}) = 0, \tag{12.14}$$

$$\frac{\partial}{\partial t}(\rho \mathbf{u}) + \nabla \cdot (\rho \mathbf{u}\mathbf{u} + P\mathbf{I}) = \mathbf{0}, \tag{12.15}$$

$$\frac{\partial}{\partial t}\left(\rho\left(e + \frac{1}{2}\mathbf{u}\cdot\mathbf{u}\right)\right) + \nabla \cdot \left(\rho \mathbf{u}\left(e + \frac{1}{2}\mathbf{u}\cdot\mathbf{u} + \frac{P}{\rho}\right)\right) = 0, \tag{12.16}$$

$$\frac{\partial}{\partial t}(\rho \lambda) + \nabla \cdot (\rho \mathbf{u} \lambda) = \rho \mathsf{r}, \tag{12.17}$$

$$e = e(P, \rho, \lambda), \tag{12.18}$$

$$\mathsf{r} = \mathsf{r}(P, \rho, \lambda). \tag{12.19}$$

For mass conservation, Eq. (12.14) is identical to the earlier Eq. (7.1). For linear momenta conservation, Eq. (12.15) is Eq. (7.2) with the viscous stress, $\boldsymbol{\tau} = \mathbf{0}$. For energy conservation, Eq. (12.16) is Eq. (7.3) with viscous stress $\boldsymbol{\tau} = \mathbf{0}$, and diffusive energy transport $\mathbf{j}^q = \mathbf{0}$. For species evolution, Eq. (12.19) is Eq. (7.8) with diffusive mass flux $\mathbf{j}_i^m = 0$. Equations (12.14)–(12.19) form eight equations in the eight unknowns, $\rho, \mathbf{u}, P, e, \lambda, \mathsf{r}$. Note that $\mathbf{u}$ has three unknowns, and Eq. (12.15) gives three equations. We assume the functional forms of $e$ and $\mathsf{r}$ are given.

The conservative form of the equations is the most useful and one of the more fundamental forms. It is the form that arises from the even more fundamental integral form, which admits discontinuities. We shall take advantage of this in later forming shock jump equations. The disadvantage of the conservative form is that it is unwieldy and masks simpler causal relations.

### 12.1.5 Nonconservative Form

We can reveal some of the physics described by Eqs. (12.14)–(12.16) by writing them in a nonconservative form. Here, we specialize the analysis given for the more general reactive Navier-Stokes equations in Sec. 7.1.2.

#### Mass

We can use the product rule to expand Eq. (12.14) as

$$\underbrace{\frac{\partial \rho}{\partial t} + \mathbf{u} \cdot \nabla \rho}_{d\rho/dt} + \rho \nabla \cdot \mathbf{u} = 0. \tag{12.20}$$

We recall the material derivative from Eq. (7.30), $d/dt \equiv \partial/\partial t + \mathbf{u} \cdot \nabla$. Using it, we can rewrite the mass equation as

$$\frac{d\rho}{dt} + \rho \nabla \cdot \mathbf{u} = 0. \tag{12.21}$$

Often, Newton's notation for derivatives, the dot, is used to denote the material derivative. We can also recall the definition of the divergence operator, div, to be used in place of $\nabla \cdot$, so that the mass equation can be written compactly as

$$\dot{\rho} + \rho \,\text{div}\, \mathbf{u} = 0. \tag{12.22}$$

The density of a material particle changes in response to the divergence of the velocity field. That can be correlated with the rate of volume expansion of the material region.

#### Linear Momenta

We can use the product rule to expand the linear momenta equations, Eq. (12.15), as

$$\rho \frac{\partial \mathbf{u}}{\partial t} + \mathbf{u} \frac{\partial \rho}{\partial t} + \mathbf{u}(\nabla \cdot (\rho \mathbf{u})) + \rho \mathbf{u} \cdot \nabla \mathbf{u} + \nabla P = \mathbf{0}, \tag{12.23}$$

$$\rho \frac{\partial \mathbf{u}}{\partial t} + \underbrace{\mathbf{u} \left( \frac{\partial \rho}{\partial t} + \nabla \cdot (\rho \mathbf{u}) \right)}_{0} + \rho \mathbf{u} \cdot \nabla \mathbf{u} + \nabla P = \mathbf{0}, \tag{12.24}$$

$$\underbrace{\rho \left( \frac{\partial \mathbf{u}}{\partial t} + \mathbf{u} \cdot \nabla \mathbf{u} \right)}_{d\mathbf{u}/dt} + \nabla P = \mathbf{0}, \tag{12.25}$$

$$\rho \frac{d\mathbf{u}}{dt} + \nabla P = \mathbf{0}. \tag{12.26}$$

Or in terms of our simplified notation with specific volume $v = 1/\rho$, we have

$$\dot{\mathbf{u}} + v \,\text{grad}\, P = \mathbf{0}. \tag{12.27}$$

The fluid particle accelerates in response to a pressure gradient.

### Energy

We can apply the product rule to the energy equation, Eq. (12.16), to get

$$\rho\frac{\partial}{\partial t}\left(e+\frac{1}{2}\mathbf{u}\cdot\mathbf{u}\right)+\left(e+\frac{1}{2}\mathbf{u}\cdot\mathbf{u}\right)\frac{\partial\rho}{\partial t}+\left(e+\frac{1}{2}\mathbf{u}\cdot\mathbf{u}\right)\nabla\cdot(\rho\mathbf{u})$$
$$+\rho\mathbf{u}\cdot\nabla\left(e+\frac{1}{2}\mathbf{u}\cdot\mathbf{u}\right)+\nabla\cdot(P\mathbf{u})=0, \tag{12.28}$$

$$\rho\frac{\partial}{\partial t}\left(e+\frac{1}{2}\mathbf{u}\cdot\mathbf{u}\right)+\left(e+\frac{1}{2}\mathbf{u}\cdot\mathbf{u}\right)\underbrace{\left(\frac{\partial\rho}{\partial t}+\nabla\cdot(\rho\mathbf{u})\right)}_{0}$$
$$+\rho\mathbf{u}\cdot\nabla\left(e+\frac{1}{2}\mathbf{u}\cdot\mathbf{u}\right)+\nabla\cdot(P\mathbf{u})=0, \tag{12.29}$$

$$\rho\frac{\partial}{\partial t}\left(e+\frac{1}{2}\mathbf{u}\cdot\mathbf{u}\right)+\rho\mathbf{u}\cdot\nabla\left(e+\frac{1}{2}\mathbf{u}\cdot\mathbf{u}\right)+\nabla\cdot(P\mathbf{u})=0. \tag{12.30}$$

This is simpler, but there is more that can be done by taking advantage of the linear momenta equations. Let us continue working with the product rule once more along with some simple rearrangements:

$$\rho\frac{\partial e}{\partial t}+\rho\mathbf{u}\cdot\nabla e+\rho\frac{\partial}{\partial t}\left(\frac{1}{2}\mathbf{u}\cdot\mathbf{u}\right)+\rho\mathbf{u}\cdot\nabla\left(\frac{1}{2}\mathbf{u}\cdot\mathbf{u}\right)+\nabla\cdot(P\mathbf{u})=0, \tag{12.31}$$

$$\rho\frac{\partial e}{\partial t}+\rho\mathbf{u}\cdot\nabla e+\rho\mathbf{u}\cdot\frac{\partial\mathbf{u}}{\partial t}+\rho\mathbf{u}\cdot(\mathbf{u}\cdot\nabla)\mathbf{u}+\mathbf{u}\cdot\nabla P+P\nabla\cdot\mathbf{u}=0, \tag{12.32}$$

$$\underbrace{\rho\frac{\partial e}{\partial t}+\rho\mathbf{u}\cdot\nabla e}_{\rho\, de/dt}+\mathbf{u}\cdot\underbrace{\left(\rho\frac{\partial\mathbf{u}}{\partial t}+\rho(\mathbf{u}\cdot\nabla)\mathbf{u}+\nabla P\right)}_{0}+P\nabla\cdot\mathbf{u}=0, \tag{12.33}$$

$$\rho\frac{de}{dt}+P\nabla\cdot\mathbf{u}=0. \tag{12.34}$$

Now, from Eq. (12.21), we have $\nabla\cdot\mathbf{u}=-(1/\rho)d\rho/dt$. Eliminating the divergence of the velocity field from Eq. (12.34) and scaling by $\rho$, we get

$$\frac{de}{dt}-\frac{P}{\rho^2}\frac{d\rho}{dt}=0. \tag{12.35}$$

With $\rho=1/v$, so $d\rho/dt=-(1/v^2)dv/dt=-\rho^2 dv/dt$, our energy equation becomes

$$\frac{de}{dt}+P\frac{dv}{dt}=0. \tag{12.36}$$

Or in terms of our dot notation, we get

$$\dot{e}+P\dot{v}=0. \tag{12.37}$$

The internal energy of a fluid particle changes in response to the work done by the pressure force. In terms of differentials removing time, this is equivalent to saying

$$de=-P\,dv. \tag{12.38}$$

This is the Gibbs equation, Eq. (3.1), in the limit of $ds=0$ for a nonreactive material. One should not infer though that the reactive Euler equations are isentropic. There are two sources of irreversible entropy production: (1) chemical reaction and

(2) shock waves. That shock waves generate irreversible entropy production is not evident in the partial differential equation form of the Euler equations. It is only evident upon returning to a more primitive integral form that will be considered in Sec. 12.1.8.

**Reaction**

Using the product rule on Eq. (12.19), we get

$$\rho \frac{\partial \lambda}{\partial t} + \lambda \frac{\partial \rho}{\partial t} + \rho \mathbf{u} \nabla \lambda + \lambda \nabla \cdot (\rho \mathbf{u}) = \rho \mathsf{r}, \tag{12.39}$$

$$\underbrace{\rho \left( \frac{\partial \lambda}{\partial t} + \mathbf{u} \cdot \nabla \lambda \right)}_{d\lambda/dt} + \lambda \underbrace{\left( \frac{\partial \rho}{\partial t} + \nabla \cdot (\rho \mathbf{u}) \right)}_{0} = \rho \mathsf{r}, \tag{12.40}$$

$$\frac{d\lambda}{dt} = \mathsf{r}, \tag{12.41}$$

$$\dot{\lambda} = \mathsf{r}. \tag{12.42}$$

The mass fraction of product species changes according to the forward reaction rate.

**Summary**

In summary, our nonconservative equations are

$$\dot{\rho} + \rho \,\mathrm{div}\, \mathbf{u} = 0, \tag{12.43}$$

$$\dot{\mathbf{u}} + v \,\mathrm{grad}\, P = 0, \tag{12.44}$$

$$\dot{e} + P\dot{v} = 0, \tag{12.45}$$

$$\dot{\lambda} = \mathsf{r}, \tag{12.46}$$

$$e = e(P, \rho, \lambda), \tag{12.47}$$

$$\mathsf{r} = \mathsf{r}(P, \rho, \lambda), \tag{12.48}$$

$$\rho = 1/v. \tag{12.49}$$

### 12.1.6 One-Dimensional Form

Here, we consider the equations to be restricted to one-dimensional planar geometries.

**Conservative Form**

In the one-dimensional planar limit, Eqs. (12.14)–(12.19) reduce to

$$\frac{\partial \rho}{\partial t} + \frac{\partial}{\partial x}(\rho u) = 0, \tag{12.50}$$

$$\frac{\partial}{\partial t}(\rho u) + \frac{\partial}{\partial x}\left(\rho u^2 + P\right) = 0, \tag{12.51}$$

$$\frac{\partial}{\partial t}\left(\rho\left(e + \frac{1}{2}u^2\right)\right) + \frac{\partial}{\partial x}\left(\rho u \left(e + \frac{1}{2}u^2 + \frac{P}{\rho}\right)\right) = 0, \tag{12.52}$$

$$\frac{\partial}{\partial t}(\rho \lambda) + \frac{\partial}{\partial x}(\rho u \lambda) = \rho \mathsf{r}, \tag{12.53}$$

$$e = e(P, \rho, \lambda), \tag{12.54}$$

$$\mathsf{r} = \mathsf{r}(P, \rho, \lambda). \tag{12.55}$$

### Nonconservative Form

The nonconservative Eqs. (12.22)–(12.49) reduce to

$$\dot{\rho} + \rho \frac{\partial u}{\partial x} = 0, \tag{12.56}$$

$$\dot{u} + v \frac{\partial P}{\partial x} = 0, \tag{12.57}$$

$$\dot{e} + P\dot{v} = 0, \tag{12.58}$$

$$\dot{\lambda} = \mathsf{r}, \tag{12.59}$$

$$e = e(P, \rho, \lambda), \tag{12.60}$$

$$\mathsf{r} = \mathsf{r}(P, \rho, \lambda), \tag{12.61}$$

$$\rho = 1/v. \tag{12.62}$$

Equations (12.56)–(12.62) can be expanded using the definition of the material derivative:

$$\left(\frac{\partial \rho}{\partial t} + u\frac{\partial \rho}{\partial x}\right) + \rho \frac{\partial u}{\partial x} = 0, \tag{12.63}$$

$$\left(\frac{\partial u}{\partial t} + u\frac{\partial u}{\partial x}\right) + v \frac{\partial P}{\partial x} = 0, \tag{12.64}$$

$$\left(\frac{\partial e}{\partial t} + u\frac{\partial e}{\partial x}\right) + P\left(\frac{\partial v}{\partial t} + u\frac{\partial v}{\partial x}\right) = 0, \tag{12.65}$$

$$\left(\frac{\partial \lambda}{\partial t} + u\frac{\partial \lambda}{\partial x}\right) = \mathsf{r}, \tag{12.66}$$

$$e = e(P, \rho, \lambda), \tag{12.67}$$

$$\mathsf{r} = \mathsf{r}(P, \rho, \lambda), \tag{12.68}$$

$$\rho = 1/v. \tag{12.69}$$

### Reduction of Energy Equation

Let us use standard results from calculus of many variables to expand the caloric equation of state, Eq. (12.67):

$$de = \left.\frac{\partial e}{\partial P}\right|_{v,\lambda} dP + \left.\frac{\partial e}{\partial v}\right|_{P,\lambda} dv + \left.\frac{\partial e}{\partial \lambda}\right|_{P,v} d\lambda. \tag{12.70}$$

Taking the time derivative gives

$$\dot{e} = \left.\frac{\partial e}{\partial P}\right|_{v,\lambda} \dot{P} + \left.\frac{\partial e}{\partial v}\right|_{P,\lambda} \dot{v} + \left.\frac{\partial e}{\partial \lambda}\right|_{P,v} \dot{\lambda}. \tag{12.71}$$

Note this is simply a time derivative of the caloric state equation; it says nothing about energy conservation. Next let us use Eq. (12.71) to eliminate $\dot{e}$ in the first law of thermodynamics, Eq. (12.58), to get

$$\underbrace{\left.\frac{\partial e}{\partial P}\right|_{v,\lambda}\dot{P}+\left.\frac{\partial e}{\partial v}\right|_{P,\lambda}\dot{v}+\left.\frac{\partial e}{\partial \lambda}\right|_{P,v}\dot{\lambda}}_{\dot{e}}+P\dot{v}=0, \tag{12.72}$$

$$\dot{P}+\underbrace{\left(\frac{P+\left.\frac{\partial e}{\partial v}\right|_{P,\lambda}}{\left.\frac{\partial e}{\partial P}\right|_{v,\lambda}}\right)}_{\rho^2c^2}\dot{v}+\underbrace{\frac{\left.\frac{\partial e}{\partial \lambda}\right|_{P,v}}{\left.\frac{\partial e}{\partial P}\right|_{v,\lambda}}}_{-\rho c^2\sigma}\dot{\lambda}=0, \tag{12.73}$$

$$\dot{P}+\rho^2c^2\dot{v}-\rho c^2\sigma\dot{\lambda}=0, \tag{12.74}$$

$$\dot{P}=-\rho^2c^2\dot{v}+\rho c^2\sigma\dot{\lambda}, \tag{12.75}$$

$$\dot{P}=\rho c^2\left(\sigma \mathsf{r}-\rho\dot{v}\right). \tag{12.76}$$

Now, because $\dot{v}=-(1/\rho^2)\dot{\rho}$, we get

$$\dot{P}=c^2\dot{\rho}+\rho c^2\sigma\mathsf{r}. \tag{12.77}$$

Note that if either $\mathsf{r}=0$ or $\sigma=0$, the pressure changes will be restricted to those from classical isentropic thermo-acoustics: $\dot{P}=c^2\dot{\rho}$. If $\sigma\mathsf{r}>0$, reaction induces positive pressure changes. If $\sigma\mathsf{r}<0$, reaction induces negative pressure changes. Moreover, Eq. (12.77) is not restricted to calorically perfect ideal gases. It is valid for general state equations.

### 12.1.7 Characteristic Form

Let us obtain a standard form known as *characteristic form* for the one-dimensional unsteady equations. We follow the procedure described by Whitham (1974). For this form, let us use Eq. (12.77) and generalized forms for sound speed $c$ and thermicity $\sigma$ to recast our governing equations, Eqs. (12.63)–(12.69), as

$$\frac{\partial \rho}{\partial t}+u\frac{\partial \rho}{\partial x}+\rho\frac{\partial u}{\partial x}=0, \tag{12.78}$$

$$\frac{\partial u}{\partial t}+u\frac{\partial u}{\partial x}+\frac{1}{\rho}\frac{\partial P}{\partial x}=0, \tag{12.79}$$

$$\frac{\partial P}{\partial t}+u\frac{\partial P}{\partial x}-c^2\left(\frac{\partial \rho}{\partial t}+u\frac{\partial \rho}{\partial x}\right)=\rho c^2\sigma\mathsf{r}, \tag{12.80}$$

$$\frac{\partial \lambda}{\partial t}+u\frac{\partial \lambda}{\partial x}=\mathsf{r}, \tag{12.81}$$

$$c^2=c^2(P,\rho), \tag{12.82}$$

$$\mathsf{r}=\mathsf{r}(P,\rho,\lambda), \tag{12.83}$$

$$\sigma=\sigma(P,\rho,\lambda). \tag{12.84}$$

Let us write the differential equations in matrix form:

$$\begin{pmatrix} 1 & 0 & 0 & 0\\ 0 & 1 & 0 & 0\\ -c^2 & 0 & 1 & 0\\ 0 & 0 & 0 & 1\end{pmatrix}\begin{pmatrix}\frac{\partial \rho}{\partial t}\\ \frac{\partial u}{\partial t}\\ \frac{\partial P}{\partial t}\\ \frac{\partial \lambda}{\partial t}\end{pmatrix}+\begin{pmatrix} u & \rho & 0 & 0\\ 0 & u & \frac{1}{\rho} & 0\\ -c^2u & 0 & u & 0\\ 0 & 0 & 0 & u\end{pmatrix}\begin{pmatrix}\frac{\partial \rho}{\partial x}\\ \frac{\partial u}{\partial x}\\ \frac{\partial P}{\partial x}\\ \frac{\partial \lambda}{\partial x}\end{pmatrix}=\begin{pmatrix}0\\ 0\\ \rho c^2\sigma\mathsf{r}\\ \mathsf{r}\end{pmatrix}. \tag{12.85}$$

These take the general form

$$A_{ij}\frac{\partial w_j}{\partial t} + B_{ij}\frac{\partial w_j}{\partial x} = C_i. \tag{12.86}$$

Let us attempt to cast this the left hand side of this system in the form $\partial w_j/\partial t + \mu\partial w_j/\partial x$. Here, $\mu$ is a scalar that has the units of velocity. To do so, we shall seek vectors $\ell_i$ such that

$$\ell_i A_{ij}\frac{\partial w_j}{\partial t} + \ell_i B_{ij}\frac{\partial w_j}{\partial x} = \ell_i C_i = m_j\left(\frac{\partial w_j}{\partial t} + \mu\frac{\partial w_j}{\partial x}\right). \tag{12.87}$$

For $\ell_i$ to have the desired properties, we will insist that

$$\ell_i A_{ij} = m_j, \tag{12.88}$$

$$\ell_i B_{ij} = \mu m_j. \tag{12.89}$$

Using Eq. (12.88) to eliminate $m_j$ in Eq. (12.89), we get

$$0 = \ell_i(\mu A_{ij} - B_{ij}). \tag{12.90}$$

Equation (12.90) has the trivial solution $\ell_i = 0$. For a nontrivial solution, standard linear algebra tells us that we must enforce the condition that the determinant of the coefficient matrix be zero:

$$|\mu A_{ij} - B_{ij}| = 0. \tag{12.91}$$

Specializing Eq. (12.91) for Eq. (12.85), we find

$$\begin{vmatrix} \mu - u & -\rho & 0 & 0 \\ 0 & \mu - u & -\frac{1}{\rho} & 0 \\ -c^2(\mu - u) & 0 & \mu - u & 0 \\ 0 & 0 & 0 & \mu - u \end{vmatrix} = 0. \tag{12.92}$$

It is easily verified by co-factor expansion of the determinant that there are four roots to this equation; two are repeated:

$$\mu = u, \quad \mu = u, \quad \mu = u + c, \quad \mu = u - c. \tag{12.93}$$

Let us find the left eigenvector $\ell_i$ associated with the eigenvalues $\mu = u \pm c$. So, Eq. (12.90) reduces to

$$(\ell_1 \quad \ell_2 \quad \ell_3 \quad \ell_4)\begin{pmatrix} (u \pm c) - u & -\rho & 0 & 0 \\ 0 & (u \pm c) - u & -\frac{1}{\rho} & 0 \\ -c^2((u \pm c) - u) & 0 & (u \pm c) - u & 0 \\ 0 & 0 & 0 & (u \pm c) - u \end{pmatrix} = (0 \quad 0 \quad 0 \quad 0). \tag{12.94}$$

Simplifying,

$$(\ell_1 \quad \ell_2 \quad \ell_3 \quad \ell_4)\begin{pmatrix} \pm c & -\rho & 0 & 0 \\ 0 & \pm c & -\frac{1}{\rho} & 0 \\ \mp c^3 & 0 & \pm c & 0 \\ 0 & 0 & 0 & \pm c \end{pmatrix} = (0 \quad 0 \quad 0 \quad 0). \tag{12.95}$$

We thus find four equations, with linear dependencies:

$$\pm c\ell_1 \mp c^3\ell_3 = 0, \tag{12.96}$$

$$-\rho\ell_1 \pm c\ell_2 = 0, \tag{12.97}$$

$$-\frac{1}{\rho}\ell_2 \pm c\ell_3 = 0, \tag{12.98}$$

$$\pm c\ell_4 = 0. \tag{12.99}$$

Now, $c \neq 0$, so Eq. (12.99) insists that

$$\ell_4 = 0. \tag{12.100}$$

We expect a linear dependency. That implies that we are free to set at least one of the remaining $\ell_i$ to an arbitrary value. Let us see if we can get a solution with $\ell_3 = 1$. With that Eqs. (12.96)–(12.98) reduce to

$$\pm c\ell_1 \mp c^3 = 0, \tag{12.101}$$

$$-\rho\ell_1 \pm c\ell_2 = 0, \tag{12.102}$$

$$-\frac{1}{\rho}\ell_2 \pm c = 0. \tag{12.103}$$

Solving Eq. (12.101) gives

$$\ell_1 = c^2. \tag{12.104}$$

Then, Eq. (12.102) becomes $-\rho c^2 \pm c\ell_2 = 0$. Solving gives

$$\ell_2 = \pm\rho c. \tag{12.105}$$

This is redundant with solving Eq. (12.103), which also gives $\ell_2 = \pm\rho c$. So, we have for $\mu = u \pm c$ that

$$\ell_i = \begin{pmatrix} c^2 & \pm\rho c & 1 & 0 \end{pmatrix}. \tag{12.106}$$

So, Eq. (12.85) becomes, after multiplication by $\ell_i$ from Eq. (12.106),

$$\begin{aligned}
&\begin{pmatrix} c^2 & \pm\rho c & 1 & 0 \end{pmatrix}\begin{pmatrix} 1 & 0 & 0 & 0 \\ 0 & 1 & 0 & 0 \\ -c^2 & 0 & 1 & 0 \\ 0 & 0 & 0 & 1 \end{pmatrix}\begin{pmatrix} \frac{\partial \rho}{\partial t} \\ \frac{\partial u}{\partial t} \\ \frac{\partial P}{\partial t} \\ \frac{\partial \lambda}{\partial t} \end{pmatrix} \\
&+\begin{pmatrix} c^2 & \pm\rho c & 1 & 0 \end{pmatrix}\begin{pmatrix} u & \rho & 0 & 0 \\ 0 & u & \frac{1}{\rho} & 0 \\ -c^2 u & 0 & u & 0 \\ 0 & 0 & 0 & u \end{pmatrix}\begin{pmatrix} \frac{\partial \rho}{\partial x} \\ \frac{\partial u}{\partial x} \\ \frac{\partial P}{\partial x} \\ \frac{\partial \lambda}{\partial x} \end{pmatrix} \\
&= \begin{pmatrix} c^2 & \pm\rho c & 1 & 0 \end{pmatrix}\begin{pmatrix} 0 \\ 0 \\ \rho c^2 \sigma \mathsf{r} \\ \mathsf{r} \end{pmatrix}.
\end{aligned} \tag{12.107}$$

Performing the vector-matrix multiplication operations, we get

$$\left(0 \quad \pm\rho c \quad 1 \quad 0\right)\begin{pmatrix} \frac{\partial \rho}{\partial t} \\ \frac{\partial u}{\partial t} \\ \frac{\partial P}{\partial t} \\ \frac{\partial \lambda}{\partial t} \end{pmatrix} + \left(0 \quad \pm\rho c(u \pm c) \quad u \pm c \quad 0\right)\begin{pmatrix} \frac{\partial \rho}{\partial x} \\ \frac{\partial u}{\partial x} \\ \frac{\partial P}{\partial x} \\ \frac{\partial \lambda}{\partial x} \end{pmatrix}$$

$$= \left(c^2 \quad \pm\rho c \quad 1 \quad 0\right)\begin{pmatrix} 0 \\ 0 \\ \rho c^2 \sigma \mathsf{r} \\ \mathsf{r} \end{pmatrix}. \tag{12.108}$$

Simplifying,

$$\pm\,\rho c\left(\frac{\partial u}{\partial t} + (u \pm c)\frac{\partial u}{\partial x}\right) + \left(\frac{\partial P}{\partial t} + (u \pm c)\frac{\partial P}{\partial x}\right) = \rho c^2 \sigma \mathsf{r}. \tag{12.109}$$

Now, let us confine our attention to lines in $x - t$ space on which

$$\frac{dx}{dt} = u \pm c. \tag{12.110}$$

On such lines, Eq. (12.109) can be written as

$$\pm\,\rho c\left(\frac{\partial u}{\partial t} + \frac{dx}{dt}\frac{\partial u}{\partial x}\right) + \left(\frac{\partial P}{\partial t} + \frac{dx}{dt}\frac{\partial P}{\partial x}\right) = \rho c^2 \sigma \mathsf{r}. \tag{12.111}$$

Consider now a variable, say, $u$, that is really $u(x, t)$. From calculus, we have

$$du = \frac{\partial u}{\partial t}\,dt + \frac{\partial u}{\partial x}\,dx, \qquad \frac{du}{dt} = \frac{\partial u}{\partial t} + \frac{dx}{dt}\frac{\partial u}{\partial x}. \tag{12.112}$$

If we insist $dx/dt = u \pm c$, let us call the derivative $du/dt_{\pm}$, so that Eq. (12.111) becomes

$$\pm\,\rho c\frac{du}{dt_{\pm}} + \frac{dP}{dt_{\pm}} = \rho c^2 \sigma \mathsf{r}. \tag{12.113}$$

In the inert limit, after additional analysis, Eq. (12.113) reduces to the form $d\psi/dt_{\pm} = 0$, which shows that $\psi$ is maintained as a constant on lines where $dx/dt = u \pm c$. Thus, one can say that a signal is propagated in $(x, t)$ space at speed $u \pm c$. So, we see from the characteristic analysis how the thermodynamic property $c$ has the added significance of influencing the speed at which signals propagate.

Let us now find $\ell_i$ for $\mu = u$. For this root, we find

$$\left(\ell_1 \quad \ell_2 \quad \ell_3 \quad \ell_4\right)\begin{pmatrix} 0 & -\rho & 0 & 0 \\ 0 & 0 & -\frac{1}{\rho} & 0 \\ 0 & 0 & 0 & 0 \\ 0 & 0 & 0 & 0 \end{pmatrix} = \left(0 \quad 0 \quad 0 \quad 0\right). \tag{12.114}$$

It can be seen by inspection that two independent solutions $\ell_i$ satisfy Eq. (12.114):

$$\ell_i = \left(0 \quad 0 \quad 1 \quad 0\right), \qquad \ell_i = \left(0 \quad 0 \quad 0 \quad 1\right). \tag{12.115}$$

These two eigenvectors induce the characteristic form that was already obvious from the initial form of the energy and species equations:

$$\frac{\partial P}{\partial t} + u\frac{\partial P}{\partial x} - c^2\left(\frac{\partial \rho}{\partial t} + u\frac{\partial \rho}{\partial x}\right) = \rho c^2 \sigma \mathsf{r}, \qquad \frac{\partial \lambda}{\partial t} + u\frac{\partial \lambda}{\partial x} = \mathsf{r}. \tag{12.116}$$

On lines where $dx/dt = u$, that is to say, on material particle pathlines, these reduce to $\dot{P} - c^2\dot{\rho} = \rho c^2 \sigma \mathsf{r}$ and $\dot{\lambda} = \mathsf{r}$. Because all of the eigenvalues $\mu$ are real, and because we were able to find a set of four linearly independent left eigenvectors $\ell_i$ so as to transform our four partial differential equations into characteristic form, we can say that our system is *strictly hyperbolic* (Zauderer, 1989). Thus, it

- Is well posed for initial value problems given that initial data is provided on a noncharacteristic curve.
- Admits discontinuous solutions described by a set of Rankine-Hugoniot jump conditions that arise from a more primitive form of the governing equations.

In summary, we can write our equations in characteristic form as

$$\frac{dP}{dt_+} + \rho c\frac{du}{dt_+} = \rho c^2 \sigma \mathsf{r}, \qquad \text{on} \qquad \frac{dx}{dt} = u + c, \tag{12.117}$$

$$\frac{dP}{dt_-} - \rho c\frac{du}{dt_-} = \rho c^2 \sigma \mathsf{r}, \qquad \text{on} \qquad \frac{dx}{dt} = u - c, \tag{12.118}$$

$$\frac{dP}{dt} - c^2\frac{d\rho}{dt} = \rho c^2 \sigma \mathsf{r}, \qquad \text{on} \qquad \frac{dx}{dt} = u, \tag{12.119}$$

$$\frac{d\lambda}{dt} = \mathsf{r}, \qquad \text{on} \qquad \frac{dx}{dt} = u. \tag{12.120}$$

### 12.1.8 Rankine-Hugoniot Jump Conditions

As described by LeVeque (1992), the proper way to arrive at what are known as Rankine[1]-Hugoniot[2] jump equations describing discontinuities is to use a more primitive form of the conservation laws, expressed in terms of integrals of conservative form quantities balanced by fluxes and source terms of those quantities. If $\mathbf{q}$ is a set of conservative form variables, and $\mathbf{f}(\mathbf{q})$ is the flux of $\mathbf{q}$ (e.g., for mass conservation, $\rho$ is a conserved variable and $\rho u$ is the flux), and $\mathbf{s}(\mathbf{q})$ is the internal source term, then the primitive form of the conservation law can be written as

$$\frac{d}{dt}\int_{x_1}^{x_2} \mathbf{q}(x,t)\,dx = \mathbf{f}(\mathbf{q}(x_1,t)) - \mathbf{f}(\mathbf{q}(x_2,t)) + \int_{x_1}^{x_2} \mathbf{s}(\mathbf{q}(x,t))\,dx. \tag{12.121}$$

Here, we have considered flow into and out of a one-dimensional box for $x \in [x_1, x_2]$. For our reactive Euler equations, we have

$$\mathbf{q} = \begin{pmatrix} \rho \\ \rho u \\ \rho\left(e + \frac{1}{2}u^2\right) \\ \rho\lambda \end{pmatrix}, \quad \mathbf{f} = \begin{pmatrix} \rho u \\ \rho u^2 + P \\ \rho u\left(e + \frac{1}{2}u^2 + \frac{P}{\rho}\right) \\ \rho u\lambda \end{pmatrix}, \quad \mathbf{s} = \begin{pmatrix} 0 \\ 0 \\ 0 \\ \rho \mathsf{r} \end{pmatrix}. \tag{12.122}$$

[1] William John Macquorn Rankine (1820–1872), Scottish engineer.
[2] Pierre Henri Hugoniot (1851–1887), French mathematician.

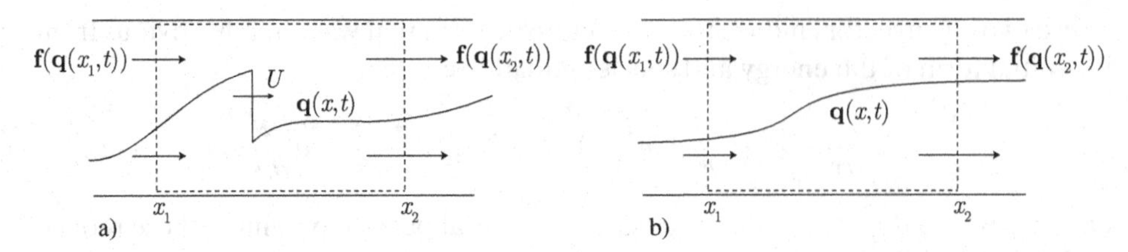


Figure 12.1. Schematic of general flux **f** into and out of finite volume in which general variable **q** evolves.

If we assume there is a discontinuity in the region $x \in [x_1, x_2]$ propagating at speed $U$, see Fig. 12.1a, we can find the Cauchy[3] principal value of the integral by splitting it into

$$\frac{d}{dt}\int_{x_1}^{x_1+Ut^-} \mathbf{q}(x,t)\,dx + \frac{d}{dt}\int_{x_1+Ut^+}^{x_2} \mathbf{q}(x,t)\,dx$$
$$= \mathbf{f}(\mathbf{q}(x_1,t)) - \mathbf{f}(\mathbf{q}(x_2,t)) + \int_{x_1}^{x_2} \mathbf{s}(\mathbf{q}(x,t))\,dx. \tag{12.123}$$

Here, $x_1 + Ut^-$ lies just before the discontinuity and $x_1 + Ut^+$ lies just past the discontinuity. Using Leibniz's rule, we get

$$\mathbf{q}(x_1 + Ut^-, t)U - 0 + \int_{x_1}^{x_1+Ut^-} \frac{\partial \mathbf{q}}{\partial t}\,dx + 0 - \mathbf{q}(x_1 + Ut^+, t)U$$
$$+ \int_{x_1+Ut^+}^{x_2} \frac{\partial \mathbf{q}}{\partial t}\,dx = \mathbf{f}(\mathbf{q}(x_1,t)) - \mathbf{f}(\mathbf{q}(x_2,t)) + \int_{x_1}^{x_2} \mathbf{s}(\mathbf{q}(x,t))\,dx. \tag{12.124}$$

Now, if we assume that $x_2 - x_1 \to 0$, and that on either side of the discontinuity the volume of integration is sufficiently small so that the time and space variation of **q** is negligibly small, we get

$$U\left(\mathbf{q}(x_1) - \mathbf{q}(x_2)\right) = \mathbf{f}(\mathbf{q}(x_1)) - \mathbf{f}(\mathbf{q}(x_2)). \tag{12.125}$$

The contribution of the source term **s** is negligible as $x_2 - x_1 \to 0$. Defining next the notation for a jump as

$$[\![\mathbf{q}(x)]\!] \equiv \mathbf{q}(x_2) - \mathbf{q}(x_1), \tag{12.126}$$

the jump conditions are rewritten as

$$U\,[\![\mathbf{q}(x)]\!] = [\![\mathbf{f}(\mathbf{q}(x))]\!]. \tag{12.127}$$

If $U = 0$, as is the case when we transform to the wave frame, we recover

$$[\![\mathbf{f}(\mathbf{q}(x))]\!] = \mathbf{0}. \tag{12.128}$$

That is, the fluxes on either side of the discontinuity are equal. We also get a more general result for $U \neq 0$:

$$U = \frac{\mathbf{f}(\mathbf{q}(x_2)) - \mathbf{f}(\mathbf{q}(x_1))}{\mathbf{q}(x_2) - \mathbf{q}(x_1)} = \frac{[\![\mathbf{f}(\mathbf{q}(x))]\!]}{[\![\mathbf{q}(x)]\!]}. \tag{12.129}$$

[3] Augustin-Louis Cauchy (1789–1857), French mechanician.

The general Rankine-Hugoniot equation then for the one-dimensional reactive Euler equations across a nonstationary jump is given by

$$U \begin{pmatrix} \rho_2 - \rho_1 \\ \rho_2 u_2 - \rho_1 u_1 \\ \rho_2\left(e_2 + \frac{1}{2}u_2^2\right) - \rho_1\left(e_1 + \frac{1}{2}u_1^2\right) \\ \rho_2\lambda_2 - \rho_1\lambda_1 \end{pmatrix}$$

$$= \begin{pmatrix} \rho_2 u_2 - \rho_1 u_1 \\ \rho_2 u_2^2 + P_2 - \rho_1 u_1^2 - P_1 \\ \rho_2 u_2\left(e_2 + \frac{1}{2}u_2^2 + \frac{P_2}{\rho_2}\right) - \rho_1 u_1\left(e_1 + \frac{1}{2}u_1^2 + \frac{P_1}{\rho_1}\right) \\ \rho_2 u_2 \lambda_2 - \rho_2 u_1 \lambda_1 \end{pmatrix}. \qquad (12.130)$$

If there is no discontinuity (see Fig. 12.1b), Eq. (12.121) reduces to the reactive Euler equations. We can rewrite Eq. (12.121) as

$$\left(\frac{d}{dt}\int_{x_1}^{x_2} \mathbf{q}(x,t)\,dx\right) + \mathbf{f}(\mathbf{q}(x_2,t)) - \mathbf{f}(\mathbf{q}(x_1,t)) = \int_{x_1}^{x_2} \mathbf{s}(\mathbf{q}(x,t))\,dx. \qquad (12.131)$$

Now, if we assume *continuity* of all fluxes and variables, we can use Taylor series expansion and Leibniz's rule to say

$$\left(\int_{x_1}^{x_2} \frac{\partial}{\partial t}\mathbf{q}(x,t)\,dx\right) + \left(\mathbf{f}(\mathbf{q}(x_1,t)) + \frac{\partial \mathbf{f}}{\partial x}(x_2 - x_1) + \cdots\right) - \mathbf{f}(\mathbf{q}(x_1,t))$$

$$= \int_{x_1}^{x_2} \mathbf{s}(\mathbf{q}(x,t))\,dx. \qquad (12.132)$$

Now let $x_2 \to x_1$ so that

$$\left(\int_{x_1}^{x_2} \frac{\partial}{\partial t}\mathbf{q}(x,t)\,dx\right) + \left(\frac{\partial \mathbf{f}}{\partial x}(x_2 - x_1)\right) = \int_{x_1}^{x_2} \mathbf{s}(\mathbf{q}(x,t))\,dx, \qquad (12.133)$$

$$\left(\int_{x_1}^{x_2} \frac{\partial}{\partial t}\mathbf{q}(x,t)\,dx\right) + \int_{x_1}^{x_2} \frac{\partial \mathbf{f}}{\partial x}\,dx = \int_{x_1}^{x_2} \mathbf{s}(\mathbf{q}(x,t))\,dx. \qquad (12.134)$$

Combining all terms under a single integral, we get

$$\int_{x_1}^{x_2} \left(\frac{\partial \mathbf{q}}{\partial t} + \frac{\partial \mathbf{f}}{\partial x} - \mathbf{s}\right) dx = \mathbf{0}. \qquad (12.135)$$

Now, this integral must be zero for an arbitrary $x_1$ and $x_2$, so the integrand itself must be zero, and we get our partial differential equation,

$$\frac{\partial \mathbf{q}}{\partial t} + \frac{\partial \mathbf{f}}{\partial x} - \mathbf{s} = \mathbf{0}, \qquad (12.136)$$

which applies away from jumps.

### 12.1.9 Galilean Transformation

We know that Newtonian mechanics have been constructed so as to be invariant under a Galilean[4] transformation that takes one from a fixed laboratory frame to a

[4] After Galileo Galilei (1564–1642), Italian polymath.

constant velocity frame with respect to the fixed frame. The Galilean transformation is such that our original coordinate system in the laboratory frame, $(x, t)$, transforms to a steady traveling wave frame, $(\hat{x}, \hat{t})$, via

$$\hat{x} = x - Dt, \qquad \hat{t} = t. \tag{12.137}$$

We now seek to see how our derivatives transform under this coordinate change. We first take differentials of Eqs. (12.137) and get

$$d\hat{x} = \frac{\partial \hat{x}}{\partial x}\, dx + \frac{\partial \hat{x}}{\partial t}\, dt = dx - D\, dt, \tag{12.138}$$

$$d\hat{t} = \frac{\partial \hat{t}}{\partial x}\, dx + \frac{\partial \hat{t}}{\partial t}\, dt = dt. \tag{12.139}$$

Scaling $d\hat{x}$ by $d\hat{t}$ gives us, then,

$$\frac{d\hat{x}}{d\hat{t}} = \frac{dx}{dt} - D. \tag{12.140}$$

Taking as usual the particle velocity in the fixed frame to be $u = dx/dt$ and defining the particle velocity in the laboratory frame to be $\hat{u} = d\hat{x}/d\hat{t}$, we see that

$$\hat{u} = u - D. \tag{12.141}$$

Now, a dependent variable $\psi$ has a representation in the original space of $\psi(x, t)$ and in the transformed space as $\psi(\hat{x}, \hat{t})$. And they must both map to the same value of $\psi$ at the same point. And there is a differential of $\psi$ in both spaces that must be equal:

$$d\psi = \left.\frac{\partial \psi}{\partial x}\right|_t dx + \left.\frac{\partial \psi}{\partial t}\right|_x dt = \left.\frac{\partial \psi}{\partial \hat{x}}\right|_{\hat{t}} d\hat{x} + \left.\frac{\partial \psi}{\partial \hat{t}}\right|_{\hat{x}} d\hat{t}, \tag{12.142}$$

$$= \left.\frac{\partial \psi}{\partial \hat{x}}\right|_{\hat{t}} (dx - Ddt) + \left.\frac{\partial \psi}{\partial \hat{t}}\right|_{\hat{x}} dt, \tag{12.143}$$

$$= \left.\frac{\partial \psi}{\partial \hat{x}}\right|_{\hat{t}} dx + \left( \left.\frac{\partial \psi}{\partial \hat{t}}\right|_{\hat{x}} - D \left.\frac{\partial \psi}{\partial \hat{x}}\right|_{\hat{t}} \right) dt. \tag{12.144}$$

Now, consider Eq. (12.144) for constant $x$, thus $dx = 0$, and divide by $dt$:

$$\left.\frac{\partial \psi}{\partial t}\right|_x = \left.\frac{\partial \psi}{\partial \hat{t}}\right|_{\hat{x}} - D \left.\frac{\partial \psi}{\partial \hat{x}}\right|_{\hat{t}}. \tag{12.145}$$

More generally, the partial with respect to $t$ becomes

$$\left.\frac{\partial}{\partial t}\right|_x = \left.\frac{\partial}{\partial \hat{t}}\right|_{\hat{x}} - D \left.\frac{\partial}{\partial \hat{x}}\right|_{\hat{t}}. \tag{12.146}$$

Now, consider Eq. (12.144) for constant $t$, thus $dt = 0$, and divide by $dx$:

$$\left.\frac{\partial \psi}{\partial x}\right|_t = \left.\frac{\partial \psi}{\partial \hat{x}}\right|_{\hat{t}}. \tag{12.147}$$

More generally,

$$\left.\frac{\partial}{\partial x}\right|_t = \left.\frac{\partial}{\partial \hat{x}}\right|_{\hat{t}}. \tag{12.148}$$

**Transformed Equations**

We can write our nonconservative form, Eqs. (12.56)–(12.59), in the transformed frame. First, let us consider how a material derivative transforms:

$$\left.\frac{\partial}{\partial t}\right|_x + u\left.\frac{\partial}{\partial x}\right|_t = \left(\left.\frac{\partial}{\partial \hat{t}}\right|_{\hat{x}} - D\left.\frac{\partial}{\partial \hat{x}}\right|_{\hat{t}}\right) + (\hat{u}+D)\left.\frac{\partial}{\partial \hat{x}}\right|_{\hat{t}}, \tag{12.149}$$

$$= \left.\frac{\partial}{\partial \hat{t}}\right|_{\hat{x}} + \hat{u}\left.\frac{\partial}{\partial \hat{x}}\right|_{\hat{t}}. \tag{12.150}$$

With this, we can then say

$$\frac{\partial \rho}{\partial \hat{t}} + \hat{u}\frac{\partial \rho}{\partial \hat{x}} + \rho\frac{\partial \hat{u}}{\partial \hat{x}} = 0, \tag{12.151}$$

$$\frac{\partial \hat{u}}{\partial \hat{t}} + \hat{u}\frac{\partial \hat{u}}{\partial \hat{x}} + \frac{1}{\rho}\frac{\partial P}{\partial \hat{x}} = 0, \tag{12.152}$$

$$\frac{\partial e}{\partial \hat{t}} + \hat{u}\frac{\partial e}{\partial \hat{x}} - P\left(\frac{\partial v}{\partial \hat{t}} + \hat{u}\frac{\partial v}{\partial \hat{x}}\right) = 0, \tag{12.153}$$

$$\frac{\partial \lambda}{\partial \hat{t}} + \hat{u}\frac{\partial \lambda}{\partial \hat{x}} = \mathrm{r}. \tag{12.154}$$

Moreover, we can write these equations in conservative form. Leaving out the details, which amounts to reversing our earlier steps that led to the nonconservative form, we get our transformed model to have an identical form as it had in the original frame:

$$\frac{\partial \rho}{\partial \hat{t}} + \frac{\partial}{\partial \hat{x}}(\rho\hat{u}) = 0, \tag{12.155}$$

$$\frac{\partial}{\partial \hat{t}}(\rho\hat{u}) + \frac{\partial}{\partial \hat{x}}(\rho\hat{u}^2 + P) = 0, \tag{12.156}$$

$$\frac{\partial}{\partial \hat{t}}\left(\rho\left(e + \frac{1}{2}\hat{u}^2\right)\right) + \frac{\partial}{\partial \hat{x}}\left(\rho\hat{u}\left(e + \frac{1}{2}\hat{u}^2 + \frac{P}{\rho}\right)\right) = 0, \tag{12.157}$$

$$\frac{\partial}{\partial \hat{t}}(\rho\lambda) + \frac{\partial}{\partial \hat{x}}(\rho\hat{u}\lambda) = \rho\mathrm{r}. \tag{12.158}$$

## 12.2 One-Dimensional, Steady Solutions

In nature one observes disturbances that propagate either steadily or unsteadily. We will defer considerations of those that are unsteady until Sec. 12.3. There we will see that some steady detonations have an unstable response in time to perturbation. Here, though, let us confine attention to a steadily propagating disturbance in a one-dimensional compressible reactive inviscid flow field. In the laboratory frame, the disturbance is taken to propagate with constant velocity $D$. Let us analyze such a disturbance.

### 12.2.1 Steady Shock Jumps

In the steady frame that travels with the wave at speed $D$, our jump conditions, Eq. (12.130), have $U = 0$ and reduce to

$$\begin{pmatrix} 0 \\ 0 \\ 0 \\ 0 \end{pmatrix} = \begin{pmatrix} \rho_2\hat{u}_2 - \rho_1\hat{u}_1 \\ \rho_2\hat{u}_2^2 + P_2 - \rho_1\hat{u}_1^2 - P_1 \\ \rho_2\hat{u}_2\left(e_2 + \frac{1}{2}\hat{u}_2^2 + \frac{P_2}{\rho_2}\right) - \rho_1\hat{u}_1\left(e_1 + \frac{1}{2}\hat{u}_1^2 + \frac{P_1}{\rho_1}\right) \\ \rho_2\hat{u}_2\lambda_2 - \rho_1\hat{u}_1\lambda_1 \end{pmatrix}. \tag{12.159}$$

The mass jump equation can be used to quickly simplify the energy and species jump equations to get the revised set:

$$\begin{aligned} \rho_2\hat{u}_2 &= \rho_1\hat{u}_1, \\ \rho_2\hat{u}_2^2 + P_2 &= \rho_1\hat{u}_1^2 + P_1, \\ e_2 + \frac{1}{2}\hat{u}_2^2 + \frac{P_2}{\rho_2} &= e_1 + \frac{1}{2}\hat{u}_1^2 + \frac{P_1}{\rho_1}, \\ \lambda_2 &= \lambda_1. \end{aligned} \tag{12.160}$$

### 12.2.2 Ordinary Differential Equations of Motion

#### Conservative Form

Let us now assert that in the steady laboratory frame that no variable has dependence on $\hat{t}$, so $\partial/\partial\hat{t} = 0$, and $\partial/\partial\hat{x} = d/d\hat{x}$. With this assumption, our partial differential equations of motion, Eqs. (12.155)–(12.158), become ordinary differential equations:

$$\frac{d}{d\hat{x}}\left(\rho\hat{u}\right) = 0, \tag{12.161}$$

$$\frac{d}{d\hat{x}}\left(\rho\hat{u}^2 + P\right) = 0, \tag{12.162}$$

$$\frac{d}{d\hat{x}}\left(\rho\hat{u}\left(e + \frac{1}{2}\hat{u}^2 + \frac{P}{\rho}\right)\right) = 0, \tag{12.163}$$

$$\frac{d}{d\hat{x}}(\rho\hat{u}\lambda) = \rho \mathsf{r}, \tag{12.164}$$

$$e = e(P, \rho, \lambda), \tag{12.165}$$

$$\mathsf{r} = \mathsf{r}(P, \rho, \lambda). \tag{12.166}$$

We take the following conditions for the undisturbed fluid just before the shock:

$$\rho|_{\hat{x}=0_-} = \rho_0, \ \hat{u}|_{\hat{x}=0_-} = -D, \ P|_{\hat{x}=0_-} = P_0, \ e|_{\hat{x}=0_-} = e_0, \ \lambda|_{\hat{x}=0_-} = 0. \tag{12.167}$$

In the laboratory frame, this corresponds to a material at rest because $u = \hat{u} + D = (-D) + D = 0$.

#### Unreduced Nonconservative Form

We can gain insights into how Eqs. (12.161)–(12.164) behave by writing them in a nonconservative form. Let us take advantage of the reductions we used to acquire the characteristic form, Eqs. (12.78)–(12.81), to write those equations after transformation to the steady Galilean transformed frame as

$$\rho \frac{d\hat{u}}{d\hat{x}} + \hat{u}\frac{d\rho}{d\hat{x}} = 0, \tag{12.168}$$

$$\hat{u}\frac{d\hat{u}}{d\hat{x}} + \frac{1}{\rho}\frac{dP}{d\hat{x}} = 0, \tag{12.169}$$

$$\hat{u}\frac{dP}{d\hat{x}} - c^2\hat{u}\frac{d\rho}{d\hat{x}} = \rho c^2 \sigma \mathsf{r}, \tag{12.170}$$

$$\hat{u}\frac{d\lambda}{d\hat{x}} = \mathsf{r}. \tag{12.171}$$

Let us attempt next to get explicit representations for the first derivatives of each equation. Shapiro (1953) gives a related analysis for what he calls "generalized one-dimensional continuous flow," for inert flows. In matrix form, we can say our system is

$$\begin{pmatrix} \hat{u} & \rho & 0 & 0 \\ 0 & \hat{u} & \frac{1}{\rho} & 0 \\ -c^2\hat{u} & 0 & \hat{u} & 0 \\ 0 & 0 & 0 & \hat{u} \end{pmatrix} \begin{pmatrix} \frac{d\rho}{d\hat{x}} \\ \frac{d\hat{u}}{d\hat{x}} \\ \frac{dP}{d\hat{x}} \\ \frac{d\lambda}{d\hat{x}} \end{pmatrix} = \begin{pmatrix} 0 \\ 0 \\ \rho c^2 \sigma \mathsf{r} \\ \mathsf{r} \end{pmatrix}. \tag{12.172}$$

We can use Cramer's[5] rule to invert the coefficient matrix to solve for the evolution of each state variable. This first requires the determinant of the coefficient matrix $\Delta$:

$$\Delta = \begin{vmatrix} \hat{u} & \rho & 0 & 0 \\ 0 & \hat{u} & \frac{1}{\rho} & 0 \\ -c^2\hat{u} & 0 & \hat{u} & 0 \\ 0 & 0 & 0 & \hat{u} \end{vmatrix} = \hat{u}^2(\hat{u}^2 - c^2). \tag{12.173}$$

As $\Delta$ appears in the denominator after application of Cramer's rule, we see immediately that *when the particle velocity $\hat{u}$ becomes locally sonic ($\hat{u} = c$), our system of differential equations is potentially singular.*

Now, by Cramer's rule, $d\rho/d\hat{x}$ is found via

$$\frac{d\rho}{d\hat{x}} = \frac{\begin{vmatrix} 0 & \rho & 0 & 0 \\ 0 & \hat{u} & \frac{1}{\rho} & 0 \\ \rho c^2 \sigma \mathsf{r} & 0 & \hat{u} & 0 \\ \mathsf{r} & 0 & 0 & \hat{u} \end{vmatrix}}{\Delta} = \frac{\rho c^2 \sigma \mathsf{r}}{\hat{u}(\hat{u}^2 - c^2)}. \tag{12.174}$$

Similarly, solving for $d\hat{u}/d\hat{x}$, we get

$$\frac{d\hat{u}}{d\hat{x}} = \frac{\begin{vmatrix} \hat{u} & 0 & 0 & 0 \\ 0 & 0 & \frac{1}{\rho} & 0 \\ -c^2\hat{u} & \rho c^2 \sigma \mathsf{r} & \hat{u} & 0 \\ 0 & \mathsf{r} & 0 & \hat{u} \end{vmatrix}}{\Delta} = -\frac{c^2 \sigma \mathsf{r}}{\hat{u}^2 - c^2}. \tag{12.175}$$

[5] Gabriel Cramer (1704–1752), Swiss mathematician.

For $dP/d\hat{x}$, we get

$$\frac{dP}{d\hat{x}} = \frac{\begin{vmatrix} \hat{u} & \rho & 0 & 0 \\ 0 & \hat{u} & 0 & 0 \\ -c^2\hat{u} & 0 & \rho c^2 \sigma \mathsf{r} & 0 \\ 0 & 0 & \mathsf{r} & \hat{u} \end{vmatrix}}{\Delta} = \frac{\hat{u}\rho c^2 \sigma \mathsf{r}}{\hat{u}^2 - c^2}. \tag{12.176}$$

Last, we have by inspection

$$\frac{d\lambda}{d\hat{x}} = \frac{\mathsf{r}}{\hat{u}}. \tag{12.177}$$

We can employ the local Mach number in the steady wave frame,

$$\hat{\mathsf{M}}^2 \equiv \frac{\hat{u}^2}{c^2}, \tag{12.178}$$

to write our system of ordinary differential equations as

$$\frac{d\rho}{d\hat{x}} = -\frac{\rho\sigma\mathsf{r}}{\hat{u}(1-\hat{\mathsf{M}}^2)}, \tag{12.179}$$

$$\frac{d\hat{u}}{d\hat{x}} = \frac{\sigma\mathsf{r}}{1-\hat{\mathsf{M}}^2}, \tag{12.180}$$

$$\frac{dP}{d\hat{x}} = -\frac{\rho\hat{u}\sigma\mathsf{r}}{1-\hat{\mathsf{M}}^2}, \tag{12.181}$$

$$\frac{d\lambda}{d\hat{x}} = \frac{\mathsf{r}}{\hat{u}}. \tag{12.182}$$

When the flow is locally sonic, $\hat{\mathsf{M}} = 1$, our equations may be singular. Recall an analogy from compressible flow with area change (Shapiro, 1953), which is really an application of l'Hôpital's[6] rule. For the flow to be locally sonic, we must insist that simultaneously the numerator must be zero; thus, we might demand that at a sonic point, $\sigma\mathsf{r} = 0$. So, an end state with $\mathsf{r} = 0$ may in fact be a sonic point. For multistep reactions, each with their own thermicity and reaction progress, we require the generalization $\boldsymbol{\sigma} \cdot \mathbf{r} = 0$ at a local sonic point. Here, $\boldsymbol{\sigma}$ is the vector of thermicities, and $\mathbf{r}$ is the vector of reaction rates. This condition will be important in Sec. 12.2.7 when we consider two-step detonations.

### Reduced Nonconservative Form

Let us use the mass equation, Eq. (12.161), to simplify the reaction equation, Eq. (12.164), then integrate our differential equations for mass momentum and energy conservation, apply the initial conditions, and thus reduce our system of four differential and two algebraic equations to one differential and five algebraic equations:

$$\rho\hat{u} = -\rho_0 D, \tag{12.183}$$

$$P + \rho\hat{u}^2 = P_0 + \rho_0 D^2, \tag{12.184}$$

$$\rho\hat{u}\left(e + \frac{1}{2}\hat{u}^2 + \frac{P}{\rho}\right) = -\rho_0 D\left(e_0 + \frac{1}{2}D^2 + \frac{P_0}{\rho_0}\right), \tag{12.185}$$

[6]Guillaume François Antoine, Marquis de l'Hôpital (1661–1704), French mathematician.

$$\frac{d\lambda}{d\hat{x}} = \frac{1}{\hat{u}}\mathsf{r}, \tag{12.186}$$

$$e = e(P, \rho, \lambda), \tag{12.187}$$

$$\mathsf{r} = \mathsf{r}(P, \rho, \lambda). \tag{12.188}$$

With some effort, we can unravel these equations to form one ordinary differential equation in one unknown. But let us delay that analysis until after we have fully examined the consequences of the algebraic constraints.

### 12.2.3 Rankine-Hugoniot Analysis

Let us first analyze our steady mass, momentum, and energy equations, Eqs. (12.183)–(12.185). Our analysis here will be valid both within ($\lambda \in [0, 1]$) and at the end of the reaction zone ($\lambda = 1$).

#### Rayleigh Line

Let us get what is known as the *Rayleigh*[7] *line* by considering only the mass and linear momentum equations, Eqs. (12.183) and (12.184). Let us first rewrite Eq. (12.184) as

$$P + \frac{\rho^2\hat{u}^2}{\rho} = P_0 + \frac{\rho_0^2 D^2}{\rho_0}. \tag{12.189}$$

Then, the mass equation, Eq. (12.14), allows us to rewrite the momentum equation, Eq. (12.189), as

$$P + \frac{\rho_0^2 D^2}{\rho} = P_0 + \frac{\rho_0^2 D^2}{\rho_0}. \tag{12.190}$$

Rearranging to solve for $P$, we find

$$P = P_0 - \rho_0^2 D^2 \left(\frac{1}{\rho} - \frac{1}{\rho_0}\right). \tag{12.191}$$

In terms of $v = 1/\rho$, and a slight rearrangement, Eq. (12.191) can be restated as

$$\frac{P}{P_0} = 1 - \frac{D^2}{P_0 v_0}\left(\frac{v}{v_0} - 1\right). \tag{12.192}$$

Note:

- This is a line in $(P, 1/\rho)$ space, the Rayleigh line.
- The slope of the Rayleigh line is strictly negative.
- The magnitude of the slope of the Rayleigh line is proportional the square of the wave speed; high wave speeds induce steep slopes.
- The Rayleigh line passes through the ambient state $(P_0, 1/\rho_0)$.
- The Rayleigh line admits negative pressures, volumes, and densities, all of which are unphysical.
- Small volume leads to high pressure.
- These conclusions are a consequence of mass and momentum conservation *alone*. No consideration of energy has been made.

[7] John William Strutt, third Baron Rayleigh (1842–1919), English physicist; 1904 Nobel laureate in Physics.

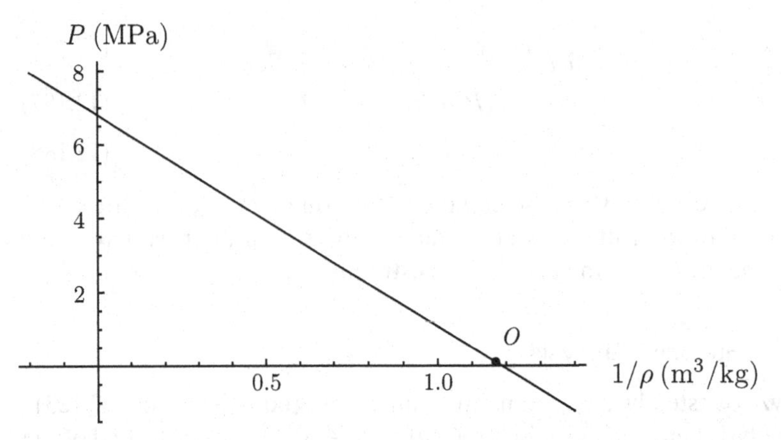


Figure 12.2. Plot of Rayleigh line for parameters of Table 12.1 and $D = 2800$ m/s. Its slope is $-\rho_0^2 D^2 < 0$.

- The Rayleigh line equation is valid at all stages of the reaction: the inert state, a shocked stated, an intermediate reacted state, and a completely reacted state. It is always the same line.

A plot of a Rayleigh line for $D = 2800$ m/s and the parameters of Table 12.1 is shown in Fig. 12.2. Here, the point labeled "O" is the ambient state $(1/\rho_0, P_0)$.

### Hugoniot Curve

Let us next focus on the energy equation, Eq. (12.185). We shall use the mass and momentum equations (12.183) and (12.184) to cast the energy equation in a form that is independent of both velocities and wave speeds. From Eq. (12.183), we can easily see that Eq. (12.185) first reduces to

$$e + \frac{1}{2}\hat{u}^2 + \frac{P}{\rho} = e_0 + \frac{1}{2}D^2 + \frac{P_0}{\rho_0}. \tag{12.193}$$

Now, the mass equation (12.183) also tells us that

$$\hat{u} = -\frac{\rho_0}{\rho}D, \tag{12.194}$$

so Eq. (12.193) can be recast as

$$e + \frac{1}{2}\left(\frac{\rho_0}{\rho}\right)^2 D^2 + \frac{P}{\rho} = e_0 + \frac{1}{2}D^2 + \frac{P_0}{\rho_0}. \tag{12.195}$$

Rearranging gives

$$e - e_0 + \frac{1}{2}D^2\left(\frac{(\rho_0 - \rho)(\rho_0 + \rho)}{\rho^2}\right) + \frac{P}{\rho} - \frac{P_0}{\rho_0} = 0. \tag{12.196}$$

Now, the Rayleigh line, Eq. (12.191), can be used to solve for $D^2$,

$$D^2 = \frac{P_0 - P}{\rho_0^2}\left(\frac{1}{\rho} - \frac{1}{\rho_0}\right)^{-1} = \frac{P_0 - P}{\rho_0^2}\left(\frac{\rho\rho_0}{\rho_0 - \rho}\right). \tag{12.197}$$

Now, use Eq. (12.197) to eliminate $D^2$ in Eq. (12.196):

$$e - e_0 + \frac{1}{2}\underbrace{\left(\frac{P_0 - P}{\rho_0^2}\left(\frac{\rho\rho_0}{\rho_0 - \rho}\right)\right)}_{D^2}\left(\frac{(\rho_0 - \rho)(\rho_0 + \rho)}{\rho^2}\right) + \frac{P}{\rho} - \frac{P_0}{\rho_0} = 0. \quad (12.198)$$

After lengthy algebra, this reduces to the simple

$$\underbrace{e - e_0}_{\text{change in energy}} = \underbrace{-\frac{P + P_0}{2}}_{\text{-average pressure}} \times \underbrace{\left(\frac{1}{\rho} - \frac{1}{\rho_0}\right)}_{\text{change in volume}}. \quad (12.199)$$

This is the Hugoniot equation for a general material. It applies for solid, liquid, or gas. Note the following:

- This form of the Hugoniot does not depend on the state equation.
- The Hugoniot has no dependency on particle velocity or wave speed.
- The Hugoniot is valid for all $e(\lambda)$; that is, it is valid for inert, partially reacted, and totally reacted material. Different degrees of reaction $\lambda$ will induce shifts in the curve.

We can write the Hugoniot as the discrete jump

$$[\![e]\!] = -P_{ave}\,[\![v]\!], \quad (12.200)$$

which has the same form as the differential change of Eq. (12.38), $de = -P\,dv$.

Now, let us specify an equation of state. For convenience and to more easily illustrate the features of the Rankine-Hugoniot analysis, let us focus on the simplest physically based state equation, the calorically perfect ideal gas. With that we adopt the perfect caloric state equation studied earlier, Eq. (9.5):

$$e = c_v T - \lambda q. \quad (12.201)$$

We also adopt an ideal gas assumption

$$P = \rho R T. \quad (12.202)$$

Solving for $T$, we get $T = P/\rho/R$. Thus, $c_v T = (c_v/R)(P/\rho)$. Recalling that $R = c_P - c_v$, we then get $c_v T = (c_v/(c_P - c_v))(P/\rho)$. And recalling that for the calorically perfect ideal gas $\gamma = c_P/c_v$, we get $c_v T = 1/(\gamma - 1)(P/\rho)$. Thus, our caloric equation of state, Eq. (12.201), for our simple model becomes

$$e(P, \rho, \lambda) = \frac{1}{\gamma - 1}\frac{P}{\rho} - \lambda q. \quad (12.203)$$

In terms of $v$, Eq. (12.203) is

$$e(P, v, \lambda) = \frac{1}{\gamma - 1} P v - \lambda q. \quad (12.204)$$

As an aside, for this equation of state, the sound speed can be deduced from Eq. (12.9) as

$$\rho^2 c^2 = \frac{P + \left.\frac{\partial e}{\partial v}\right|_{P,\lambda}}{\left.\frac{\partial e}{\partial P}\right|_{v,\lambda}} = \frac{P + \frac{1}{\gamma-1}P}{\frac{1}{\gamma-1}v} = \frac{\gamma P}{v}. \quad (12.205)$$

$$c^2 = \gamma \frac{P}{v}\frac{1}{\rho^2} = \gamma P v = \gamma \frac{P}{\rho}. \quad (12.206)$$

For the thermicity, we note that

$$P = \rho RT = \frac{R}{v}T = \frac{R}{v}\frac{e + \lambda q}{c_v}, \tag{12.207}$$

$$\left.\frac{\partial P}{\partial \lambda}\right|_{v,e} = \frac{R}{c_v}\frac{q}{v} = (\gamma - 1)\rho q, \tag{12.208}$$

$$\sigma = \frac{1}{\rho c^2}\left.\frac{\partial P}{\partial \lambda}\right|_{v,e} = \frac{1}{\rho c^2}(\gamma - 1)\rho q = (\gamma - 1)\frac{q}{c^2} = \frac{\gamma - 1}{\gamma}\frac{\rho q}{P}. \tag{12.209}$$

Now, at the initial state, we have $\lambda = 0$, and so Eq. (12.203) reduces to

$$e_0 = \frac{1}{\gamma - 1}\frac{P_0}{\rho_0}. \tag{12.210}$$

We now use our caloric state relations, Eqs. (12.203) and (12.210), to specialize our Hugoniot relation, Eq. (12.199), to

$$\frac{1}{\gamma - 1}\left(\frac{P}{\rho} - \frac{P_0}{\rho_0}\right) - \lambda q = -\frac{P + P_0}{2}\left(\frac{1}{\rho} - \frac{1}{\rho_0}\right). \tag{12.211}$$

Employing $v = 1/\rho$, we get a more compact form:

$$\frac{1}{\gamma - 1}\left(Pv - P_0 v_0\right) - \lambda q = -\frac{1}{2}(P + P_0)\left(v - v_0\right). \tag{12.212}$$

Lengthy algebra yields the exact result

$$P = \frac{2\lambda q + P_0\left(\frac{\gamma+1}{\gamma-1}v_0 - v\right)}{\left(\frac{\gamma+1}{\gamma-1}v - v_0\right)}. \tag{12.213}$$

Note that as

$$v \to \frac{\gamma - 1}{\gamma + 1}v_0, \qquad P \to \infty. \tag{12.214}$$

For $\gamma = 7/5$, this gives $v \to v_0/6$ and induces infinite pressure. In fact, an ideal gas cannot be compressed by a single shock beyond this limit, known as the *strong shock limit*. As

$$v \to \infty, \qquad -\frac{\gamma - 1}{\gamma + 1}P_0 < 0. \tag{12.215}$$

So, large volumes induce nonphysical pressures. Even more lengthy algebra yields the compact form for the Hugoniot:

$$\left(\frac{P}{P_0} + \frac{\gamma - 1}{\gamma + 1}\right)\left(\frac{v}{v_0} - \frac{\gamma - 1}{\gamma + 1}\right) = \frac{\gamma - 1}{\gamma + 1}\frac{2\lambda q}{P_0 v_0} + \frac{4\gamma}{(\gamma + 1)^2}. \tag{12.216}$$

Equation (12.216) represents a hyperbola in the $(P, v) = (P, 1/\rho)$ plane. As $\lambda$ proceeds from 0 to 1, the Hugoniot moves. A plot of a series of Hugoniot curves for values of $\lambda = 0, 1/2, 1$, along with two Rayleigh lines for $D = 2800$ m/s and $D = 1991.1$ m/s, is shown in Fig. 12.3. The $D = 1991.1$ m/s Rayleigh line happens to be exactly tangent to the $\lambda = 1$ Hugoniot curve. We will see this has special significance.

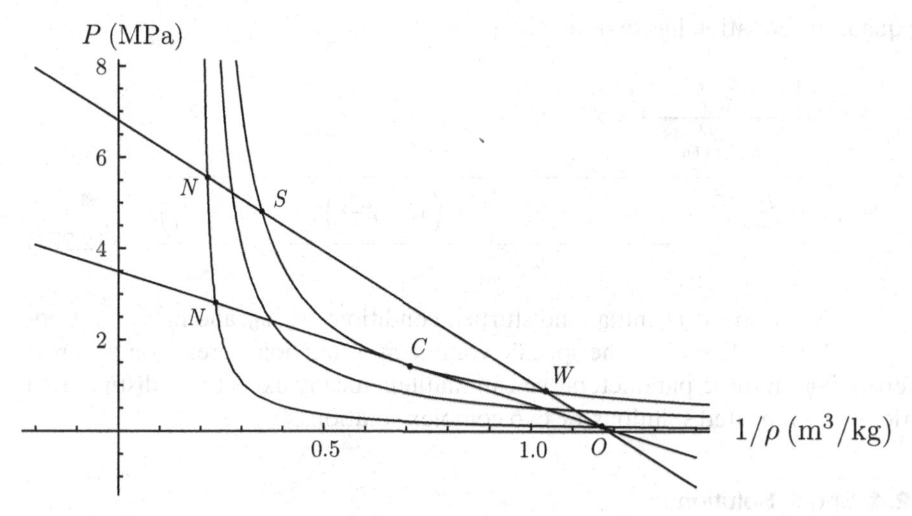


Figure 12.3. Plot of $\lambda = 0, 1/2, 1$ Hugoniot curves and two Rayleigh lines for $D = 2800$ m/s, $D = 1991.1$ m/s and parameters of Table 12.1.

Let us define, for convenience, the parameter $\hat{\mu}^2$ as

$$\hat{\mu}^2 \equiv \frac{\gamma - 1}{\gamma + 1}. \tag{12.217}$$

With this definition, we have

$$1 - \hat{\mu}^4 = 1 - \left(\frac{\gamma - 1}{\gamma + 1}\right)^2 = \frac{4\gamma}{(\gamma + 1)^2}. \tag{12.218}$$

With this definition, Eq. (12.216), the Hugoniot becomes

$$\left(\frac{P}{P_0} + \hat{\mu}^2\right)\left(\frac{v}{v_0} - \hat{\mu}^2\right) = 2\hat{\mu}^2 \frac{\lambda q}{P_0 v_0} + 1 - \hat{\mu}^4. \tag{12.219}$$

Now, let us seek to intersect the Rayleigh line with the Hugoniot curve and find the points of intersection. To do so, let us use the Rayleigh line, Eq. (12.192), to eliminate pressure in the Hugoniot, Eq. (12.219):

$$\left(1 - \frac{D^2}{P_0 v_0}\left(\frac{v}{v_0} - 1\right) + \hat{\mu}^2\right)\left(\frac{v}{v_0} - \hat{\mu}^2\right) = 2\hat{\mu}^2 \frac{\lambda q}{P_0 v_0} + 1 - \hat{\mu}^4. \tag{12.220}$$

We now structure Eq. (12.220) so that it can be solved for $v$:

$$\left(1 + \hat{\mu}^2 - \frac{D^2}{P_0 v_0}\frac{v}{v_0} + \frac{D^2}{P_0 v_0}\right)\left(\frac{v}{v_0} - \hat{\mu}^2\right) = 2\hat{\mu}^2 \frac{\lambda q}{P_0 v_0} + 1 - \hat{\mu}^4. \tag{12.221}$$

Now, let us regroup to form

$$\left(\frac{v}{v_0}\right)^2 \left(-\frac{D^2}{P_0 v_0}\right) + \left(\frac{v}{v_0}\right)\left((1 + \hat{\mu}^2)\left(1 + \frac{D^2}{P_0 v_0}\right)\right)$$
$$- \hat{\mu}^2\left(1 + \frac{D^2}{P_0 v_0}\right) - 2\hat{\mu}^2 \frac{\lambda q}{P_0 v_0} - 1 = 0. \tag{12.222}$$

This quadratic equation has two roots:

$$\left(\frac{v}{v_0}\right) = \frac{\left(1+\frac{D^2}{P_0v_0}\right)(1+\hat{\mu}^2)}{2\frac{D^2}{P_0v_0}} \pm \frac{\sqrt{\left(1+\frac{D^2}{P_0v_0}\right)^2(1+\hat{\mu}^2)^2 - 4\frac{D^2}{P_0v_0}\left(1+\left(1+\frac{D^2}{P_0v_0}\right)\hat{\mu}^2 - 2\hat{\mu}^2\frac{\lambda q}{P_0v_0}\right)}}{2\frac{D^2}{P_0v_0}}. \tag{12.223}$$

For a given wave speed $D$, initial undisturbed conditions, $P_0, v_0$, and material properties, $\hat{\mu}, q$, Eq. (12.223) gives the specific volume as a function of reaction progress $\lambda$. Depending on these parameters, we can mathematically expect two distinct real solutions, two repeated solutions, or two complex solutions.

### 12.2.4 Shock Solutions

We know from Eq. (12.160) that $\lambda$ does not change through a shock. So, if $\lambda = 0$ before the shock, it has the same value after. So, we can get the shock state by enforcing the Rankine-Hugoniot jump conditions with $\lambda = 0$:

$$\left(\frac{v}{v_0}\right) = \frac{\left(1+\frac{D^2}{P_0v_0}\right)(1+\hat{\mu}^2)}{2\frac{D^2}{P_0v_0}} \pm \frac{\sqrt{\left(1+\frac{D^2}{P_0v_0}\right)^2(1+\hat{\mu}^2)^2 - 4\frac{D^2}{P_0v_0}\left(1+\left(1+\frac{D^2}{P_0v_0}\right)\hat{\mu}^2\right)}}{2\frac{D^2}{P_0v_0}}. \tag{12.224}$$

With considerable effort, or alternatively, by direct calculation via computational algebra, the two roots of Eq. (12.224) can be shown to reduce to

$$\frac{v}{v_0} = 1, \qquad \frac{v}{v_0} = \hat{\mu}^2 + \frac{P_0v_0}{D^2}(1+\hat{\mu}^2). \tag{12.225}$$

The first is the ambient solution; the second is the shock solution. The shock solution can also be expressed as

$$\frac{v}{v_0} = \frac{\gamma-1}{\gamma+1} + \frac{2\gamma}{\gamma+1}\frac{P_0v_0}{D^2}. \tag{12.226}$$

In the limit as $P_0v_0/D^2 \to 0$, the so-called strong shock limit, we find

$$\frac{v}{v_0} \to \frac{\gamma-1}{\gamma+1}. \tag{12.227}$$

The reciprocal gives the density ratio in the strong shock limit:

$$\frac{\rho}{\rho_0} \to \frac{\gamma+1}{\gamma-1}. \tag{12.228}$$

With the solutions for $v/v_0$, we can employ the Rayleigh line, Eq. (12.192), to get the pressure. Again, we find two solutions, inert and shock:

$$\frac{P}{P_0} = 1, \qquad \frac{P}{P_0} = \frac{D^2}{P_0v_0}(1-\hat{\mu}^2) - \hat{\mu}^2. \tag{12.229}$$

The shock solution is rewritten as

$$\frac{P}{P_0} = \frac{2}{\gamma+1}\frac{D^2}{P_0 v_0} - \frac{\gamma-1}{\gamma+1}. \tag{12.230}$$

In the strong shock limit $P_0 v_0/D^2 \to 0$, the shock pressure reduces to

$$\frac{P}{P_0} \to \frac{2}{\gamma+1}\frac{D^2}{P_0 v_0}. \tag{12.231}$$

Note that this can also be rewritten as

$$\frac{P}{P_0} \to \frac{2\gamma}{\gamma+1}\frac{D^2}{\gamma P_0 v_0} = \frac{2\gamma}{\gamma+1}\frac{D^2}{c_0^2}. \tag{12.232}$$

Last, the particle velocity can be obtained via the mass equation, Eq. (12.194):

$$\frac{\hat{u}}{D} = -\frac{\rho_0}{\rho} = -\frac{v}{v_0} = -\frac{\gamma-1}{\gamma+1} - \frac{2\gamma}{\gamma+1}\frac{P_0 v_0}{D^2}. \tag{12.233}$$

In the strong shock limit, $P_0 v_0/D^2 \to 0$, we get

$$\frac{\hat{u}}{D} \to -\frac{\gamma-1}{\gamma+1}. \tag{12.234}$$

In the laboratory frame, one gets

$$\frac{u-D}{D} \to -\frac{\gamma-1}{\gamma+1}, \qquad \frac{u}{D} \to \frac{2}{\gamma+1}. \tag{12.235}$$

### 12.2.5 Equilibrium Solutions

Our remaining differential equation, Eq. (12.186) is in equilibrium when r = 0, which, for one-step irreversible kinetics, Eq. (12.5), occurs when $\lambda = 1$. So, for equilibrium end states, we enforce $\lambda = 1$ in Eq. (12.223) and get

$$\left(\frac{v}{v_0}\right) = \frac{\left(1+\frac{D^2}{P_0 v_0}\right)(1+\hat{\mu}^2)}{2\frac{D^2}{P_0 v_0}} \pm \frac{\sqrt{\left(1+\frac{D^2}{P_0 v_0}\right)^2 (1+\hat{\mu}^2)^2 - 4\frac{D^2}{P_0 v_0}\left(1+\left(1+\frac{D^2}{P_0 v_0}\right)\hat{\mu}^2 - 2\hat{\mu}^2\frac{q}{P_0 v_0}\right)}}{2\frac{D^2}{P_0 v_0}}. \tag{12.236}$$

This has only one free parameter, $D$. There are two solutions for $v/v_0$ at complete reaction. They can be distinct and real, repeated and real, or complex, depending on the value of $D$. We are most interested in $D$ for which the solutions are real; these will be physically realizable.

#### Chapman-Jouguet Solutions

Let us first consider solutions for which the two roots of Eq. (12.236) are repeated. This is known as a Chapman[8]-Jouguet[9] (CJ) solution for their early independent work (Chapman, 1899; Jouguet, 1905). Lesser known is the related earlier work of Mikhel'son (1890). For a CJ solution, the Rayleigh line is tangent to the Hugoniot at

[8] David Leonard Chapman (1869–1958), English chemist.
[9] Jacques Charles Émile Jouguet (1871–1943), French engineer.

$\lambda = 1$ if the reaction is driven by one-step irreversible kinetics. We can find values of $D$ for which the solutions are CJ by requiring the discriminant under the square root operator in Eq. (12.236) to be zero. We label such solutions with a $CJ$ subscript and say

$$\left(1+\frac{D_{CJ}^2}{P_0 v_0}\right)^2 (1+\hat{\mu}^2)^2 - 4\frac{D_{CJ}^2}{P_0 v_0}\left(1+\left(1+\frac{D_{CJ}^2}{P_0 v_0}\right)\hat{\mu}^2 - 2\hat{\mu}^2\frac{q}{P_0 v_0}\right) = 0. \tag{12.237}$$

Equation (12.237) is quartic in $D_{CJ}$ and quadratic in $D_{CJ}^2$. It has solutions

$$\frac{D_{CJ}^2}{P_0 v_0} = \frac{1 + 4\hat{\mu}^2\frac{q}{P_0 v_0} - \hat{\mu}^4 \pm 2\sqrt{\frac{2q}{P_0 v_0}\hat{\mu}^2(1+2\hat{\mu}^2-\hat{\mu}^4)}}{(\hat{\mu}^2-1)^2}. \tag{12.238}$$

The "+" root corresponds to a large value of $D_{CJ}$. This is known as the *detonation* branch. For the parameter values of Table 12.1, we find by substitution that $D_{CJ} = 1991.1$ m/s for our $H_2$-air mixture. It corresponds to a pressure increase and a volume decrease. The "−" root corresponds to a small value of $D_{CJ}$. It corresponds to a pressure decrease and a volume increase. It is known as the *deflagration* branch. For our $H_2$-air mixture, we find $D_{CJ} = 83.306$ m/s on the deflagration branch.

Here, we are most concerned with the detonation branch. The deflagration branch may be of interest, but neglected mechanisms, such as diffusion, may be of more importance for this branch. In fact laminar flames in hydrogen move much more slowly than that predicted by the CJ deflagration speed. Also, for $q \to 0$, we have $D_{CJ}^2/(P_0 v_0) = (1+\hat{\mu}^2)/(1-\hat{\mu}^2) = \gamma$. Thus, for $q \to 0$, we have $D_{CJ}^2 \to \gamma P_0 v_0$, and the wave speed is the ambient sound speed.

Taylor series expansion of the detonation branch in the strong shock limit, $P_0 v_0/D_{CJ}^2 \to 0$ shows that

$$D_{CJ}^2 \to 2q(\gamma^2-1), \tag{12.239}$$

$$v_{CJ} \to \frac{\gamma}{\gamma+1} v_0, \tag{12.240}$$

$$P_{CJ} \to 2(\gamma-1)\frac{q}{v_0}, \tag{12.241}$$

$$\hat{u}_{CJ} \to -\frac{\gamma}{\gamma+1}\sqrt{2q(\gamma^2-1)}, \tag{12.242}$$

$$u_{CJ} \to \frac{\sqrt{2q(\gamma^2-1)}}{\gamma+1}. \tag{12.243}$$

Importantly, the Mach number in the wave frame at the CJ state is

$$\hat{\mathsf{M}}_{CJ}^2 = \frac{\hat{u}_{CJ}^2}{c_{CJ}^2} = \frac{\hat{u}_{CJ}^2}{\gamma P_{CJ} v_{CJ}} = \frac{\frac{\gamma^2}{(\gamma+1)^2}2q(\gamma^2-1)}{\gamma\frac{2q}{v_0}(\gamma-1)v_0\frac{\gamma}{\gamma+1}}, \tag{12.244}$$

$$= \frac{\frac{\gamma^2}{(\gamma+1)^2}(\gamma^2-1)}{\gamma(\gamma-1)\frac{\gamma}{\gamma+1}} = \frac{\frac{\gamma^2}{(\gamma+1)^2}(\gamma+1)(\gamma-1)}{\gamma(\gamma-1)\frac{\gamma}{\gamma+1}} = 1. \tag{12.245}$$

In the strong shock limit, the local Mach number in the wave frame is sonic at the end of the reaction zone. This holds away from the strong shock limit as well.

### Weak and Strong Solutions

For $D < D_{CJ}$, there are no real solutions on the detonation branch. For $D > D_{CJ}$, there are two real solutions. These are known as the *weak* and *strong* solution. These solutions represent the intersection of the Rayleigh line with the complete reaction Hugoniot at two points. The higher pressure solution is known as the strong solution. The lower pressure solution is known as the weak solution. Equilibrium analysis cannot determine which of the solutions, strong or weak, is preferred if $D > D_{CJ}$. A consideration of the dynamics in the reaction zone is required to draw conclusions; see Sec. 12.2.6.

We can understand much about detonations, weak, strong, and CJ, by considering how they behave as the final velocity in the laboratory frame is changed. We can think of the final velocity in the laboratory frame as that of a piston that is pushing the detonation. While we could analyze this on the basis of the theory we have already developed, the algebra is complicated. Let us instead return to a more primitive form. Consider the Rankine-Hugoniot jump equations, Eqs. (12.183)–(12.185), with caloric state equation, Eq. (12.203). Take the final state, denoted by the subscript $f$, being $\lambda = 1$ and the initial state being $\lambda = 0$, Eq. (12.183) being used to simplify Eq. (12.185), and the laboratory frame velocity $u$ used in place of $\hat{u}$:

$$\rho_f(u_f - D) = -\rho_0 D, \tag{12.246}$$

$$P_f + \rho_f(u_f - D)^2 = P_0 + \rho_0 D^2, \tag{12.247}$$

$$\underbrace{\frac{1}{\gamma-1}\frac{P_f}{\rho_f}}_{e_f} - q + \frac{1}{2}(u_f - D)^2 + \frac{P_f}{\rho_f} = \underbrace{\frac{1}{\gamma-1}\frac{P_0}{\rho_0}}_{e_0} + \frac{1}{2}D^2 + \frac{P_0}{\rho_0}. \tag{12.248}$$

Let us consider the unknowns to be $P_f, \rho_f$, and $D$. Analysis of these three equations yields two sets of solutions. The relevant physical branch has a solution for $D$ of

$$D = u_f\left(\frac{\gamma+1}{4} + \frac{\gamma-1}{2}\frac{q}{u_f^2} + \sqrt{\gamma\frac{P_0}{\rho_0 u_f^2} + \left(\frac{\gamma+1}{4} + \frac{\gamma-1}{2}\frac{q}{u_f^2}\right)^2}\right). \tag{12.249}$$

We give a plot of $D$ as a function of the supporting piston velocity $u_f$ in Fig. 12.4. We notice on Fig. 12.4 that there is a clear minimum $D$. This value of $D$ is the CJ value of $D_{CJ} = 1991.1$ m/s. It corresponds to a piston velocity of $u_f = 794.9$ m/s.

A piston driving at $u_f = 794.9$ m/s will just drive the wave at the CJ speed. At this piston speed, all of the energy to drive the wave is supplied by the combustion process itself. As $u_f$ increases beyond 794.9 m/s, the wave speed $D$ increases. For such piston velocities, the piston itself is supplying energy to drive the wave. For $u_f <$ 794.9 m/s, our formula predicts an increase in $D$, but that is not what is observed in experiment. Instead, a wave propagating at $D_{CJ}$ is observed. In the inert, $q = 0$, small piston velocity, $u_f \to 0$, limit, Eq. (12.249) reduces to $D = \sqrt{\gamma P_0/\rho_0}$. That is, the wave speed is the ambient sound speed.

### Summary of Solution Properties

Here is a summary of the properties of solutions for various values of $D$ that can be obtained by equilibrium end state analysis:

- $D < D_{CJ}$: No Rayleigh line intersects a complete reaction Hugoniot on the detonation branch. There is no real steady equilibrium detonation solution.

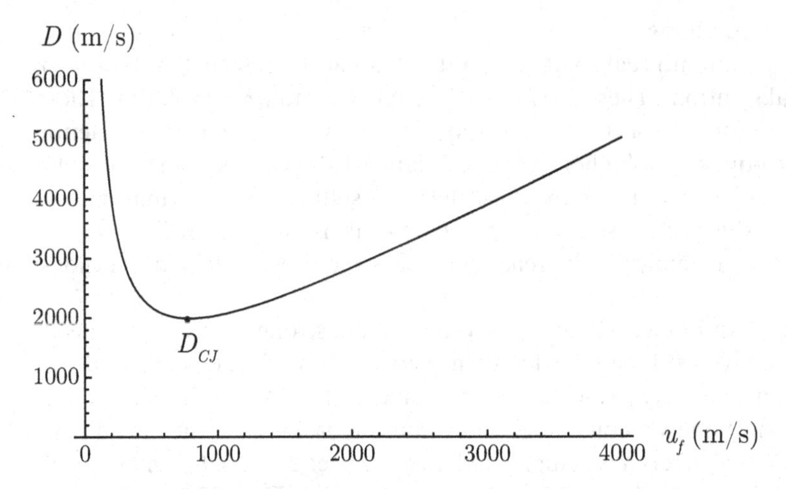


Figure 12.4. Plot of detonation wave speed $D$ versus piston velocity $u_f$; parameters from Table 12.1 loosely model $H_2$-air detonation.

- $D = D_{CJ}$: There are a two repeated solutions at a single point, which we will call $C$, the Chapman-Jouguet point. At $C$, the Rayleigh line is tangent to the complete reaction Hugoniot. The solution has
  - Sonic character, $\hat{u}_{CJ}/c_{CJ} = 1$.
  - A unique speed of propagation of a wave without piston support if the reaction is one-step irreversible.
  - All the energy from the reaction being *just sufficient* to drive the wave.
  - The feature that downstream acoustic disturbances cannot overtake the reaction zone.
- $D > D_{CJ}$: Two solutions are admitted at the equilibrium end state, the strong solution at point $S$ and the weak solution at point $W$.
  - The strong solution $S$ has
    - Subsonic character, $\hat{u}/c < 1$.
    - Necessity for piston support to drive the wave forward.
    - Some energy to drive the wave coming from the reaction; some coming from the piston.
    - The feature that if the piston support is withdrawn, acoustic disturbances will overtake and weaken the wave.
  - The weak solution $W$ has
    - Supersonic character, $\hat{u}/c > 1$.
    - A character often thought to be nonphysical, at least for one-step irreversible kinetics because of no initiation mechanism, though exceptions exist.

### 12.2.6 ZND Solutions: One-Step Irreversible Kinetics

We next consider the structure of the reaction zone. Structure was first considered contemporaneously and independently by Zel'dovich (1940), von Neumann[10] (1942), and Döring[11] (1943). Equation (12.223) gives $v(\lambda)$ for either the strong or weak

[10] John von Neumann (1903–1957), Hungarian-American mathematician, physicist, and chemical engineer.

[11] Werner Döring (1911–2006), German physicist.

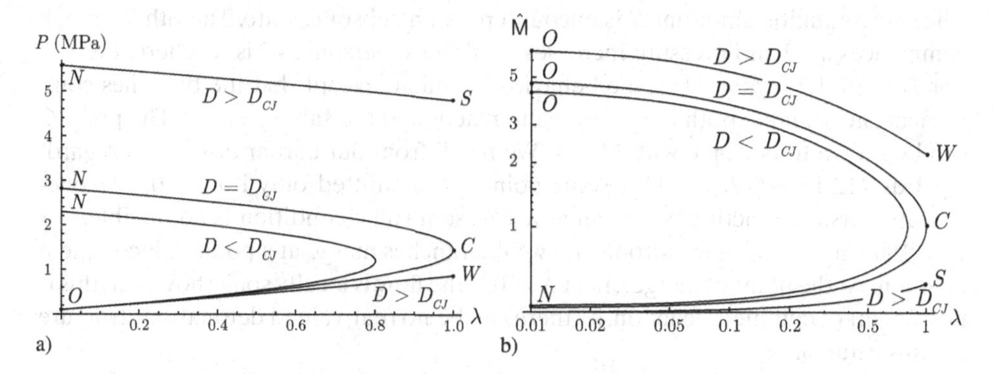


Figure 12.5. (a) $P$ versus $\lambda$ and (b) $\hat{\mathsf{M}}$ versus $\lambda$ from Rankine-Hugoniot analysis for one-step irreversible reaction for $D = 2800$ m/s $> D_{CJ}$, $D = 1991.1$ m/s $= D_{CJ}$, $D = 1800$ m/s $< D_{CJ}$ for $H_2$/air-based parameters of Table 12.1.

branches of the solution. Knowing $v(\lambda)$ and thus $\rho(\lambda)$, because $v = 1/\rho$, we can use the integrated mass equation, Eq. (12.183), to write an explicit equation for $\hat{u}(\lambda)$. Our Rankine-Hugoniot analysis also gives $T(\lambda)$. These can be employed in the reaction kinetics equation, Eqs. (12.186) and (12.6), to form a single ordinary differential equation for the evolution of $\lambda$ of the form

$$\frac{d\lambda}{d\hat{x}} = \frac{a(T(\lambda))^{\beta} \exp\left(-\frac{\mathcal{E}}{RT(\lambda)}\right)(1-\lambda)}{\hat{u}(\lambda)}, \qquad \lambda(0) = 0. \tag{12.250}$$

We consider the initial unshocked state to have the label $O$. We label the point after the shock $N$ for the Neumann point, named after von Neumann. Recall $\lambda = 0$ both at $O$ and at $N$. But the state variables, for example, $P, \rho, \hat{u}$, change from $O$ to $N$.

Before we actually solve the differential equations, we can learn much by considering how $P$ varies with $\lambda$ in the reaction zone by using Rankine-Hugoniot analysis. Consider the Rankine-Hugoniot equations, Eqs. (12.183)–(12.185), with caloric state equation, Eq. (12.203), Eq. (12.183) being used to simplify Eq. (12.185):

$$\rho\hat{u} = -\rho_0 D, \tag{12.251}$$

$$P + \rho\hat{u}^2 = P_0 + \rho_0 D^2, \tag{12.252}$$

$$\underbrace{\frac{1}{\gamma-1}\frac{P}{\rho} - \lambda q}_{e} + \frac{1}{2}\hat{u}^2 + \frac{P}{\rho} = \underbrace{\frac{1}{\gamma-1}\frac{P_0}{\rho_0}}_{e_0} + \frac{1}{2}D^2 + \frac{P_0}{\rho_0}. \tag{12.253}$$

Detailed algebraic analysis reveals the solution for $P(\lambda)$ to be

$$P(\lambda) = \frac{1}{\gamma+1}\left(P_0 + \rho_0 D^2\left(1 \pm \sqrt{\left(1 - \frac{\gamma P_0}{\rho_0 D^2}\right)^2 - \frac{2(\gamma^2-1)\lambda q}{D^2}}\right)\right). \tag{12.254}$$

In Fig. 12.5a, we plot $P$ versus $\lambda$ for three different values of $D$: $D = 2800$ m/s $> D_{CJ}$, $D = 1991.1$ m/s $= D_{CJ}$, $D = 1800$ m/s $< D_{CJ}$. We can also via detailed analysis get an algebraic expression for $\hat{\mathsf{M}}$ as a function of $\lambda$. We omit that here, but do present a plot for our system (Fig. 12.5b). The plot of $\hat{\mathsf{M}}$ versus $\lambda$ is log-log so that the sonic condition may be clearly exhibited.

For $D = 2800$ m/s $> D_{CJ}$, there are two branches, the strong and the weak. The strong branch commences at $N$ where $\lambda = 0$ and proceeds to decrease to $\lambda = 1$

where the equilibrium point $S$ is encountered at a subsonic state. The other branch commences at $O$ and pressure increases until the supersonic $W$ is reached at $\lambda = 1$. For $D = 1991.1$ m/s $= D_{CJ}$, the behavior is similar, except that the branches commencing at $N$ and $O$ both reach complete reaction at the same point $C$. The point $C$ can be shown to be sonic with $\hat{\mathsf{M}} = 1$. We recall from our earlier discussion regarding Eqs. (12.179)–(12.182) that sonic points are admitted only if $\sigma \mathsf{r} = 0$. For one-step irreversible reaction, $\mathsf{r} = 0$ when $\lambda = 1$, so a sonic condition is admissible. For $D = 1800$ m/s $< D_{CJ}$, the strong and weak branches merge at a point of incomplete reaction. At the point of merger, near $\lambda = 0.8$, the flow is locally sonic; however, this is not a point of complete reaction, so there can be no real-valued detonation structure for this value of $D$.

Finally, we write an alternate system of differential-algebraic equations that can be integrated for the detonation structure:

$$\rho \hat{u} = -\rho_0 D, \tag{12.255}$$

$$P + \rho \hat{u}^2 = P_0 + \rho_0 D^2, \tag{12.256}$$

$$e + \frac{1}{2}\hat{u}^2 + \frac{P}{\rho} = e_0 + \frac{1}{2}D^2 + \frac{P_0}{\rho_0}, \tag{12.257}$$

$$e = \frac{1}{\gamma - 1}\frac{P}{\rho} - \lambda q, \tag{12.258}$$

$$P = \rho R T, \tag{12.259}$$

$$\frac{d\lambda}{d\hat{x}} = \frac{1-\lambda}{\hat{u}} a \exp\left(-\frac{\mathcal{E}}{RT}\right). \tag{12.260}$$

We need the condition $\lambda(0) = 0$. These form six equations for the six unknowns $\rho$, $\hat{u}$, $P$, $e$, $T$, and $\lambda$.

### CJ ZND Structures

We now fix $D = D_{CJ} = 1991.1$ m/s and integrate Eq. (12.250) from the shocked state $N$ to a complete reaction point, $C$, the Chapman-Jouguet detonation state. We could also integrate from $O$ to $C$, but this is not observed in nature. After obtaining $\lambda(\hat{x})$, we can use our Rankine-Hugoniot analysis results to plot all variables as functions of $\hat{x}$.

In Fig. 12.6, we plot the density and pressure versus $\hat{x}$. The density first jumps discontinuously from $O$ to its shocked value at $N$. From there it slowly drops through the reaction zone until it relaxes near $\hat{x} \sim -0.01$ m to its equilibrium value at complete reaction. Thus, the reaction zone has a thickness of roughly 1 cm. Similar behavior is seen for the pressure. The wave frame–based fluid particle velocity is also shown in Fig. 12.6. Because the unshocked fluid is at rest in the laboratory frame with $u = 0$ m/s, the fluid in the wave frame has velocity $\hat{u} = 0 - 1991.1$ m/s $= -1991.1$ m/s. It is shocked to a lower velocity and then relaxes to its equilibrium value. The structure of the velocity profiles is easier to understand in the laboratory frame, as also shown in Fig. 12.6. Here, we see the unshocked fluid velocity of $u = 0$ m/s. The fluid is shocked to a high velocity, which then decreases to a value at the end of the reaction zone. The final velocity can be associated with that of a supporting piston, $u_f = 794.9$ m/s.

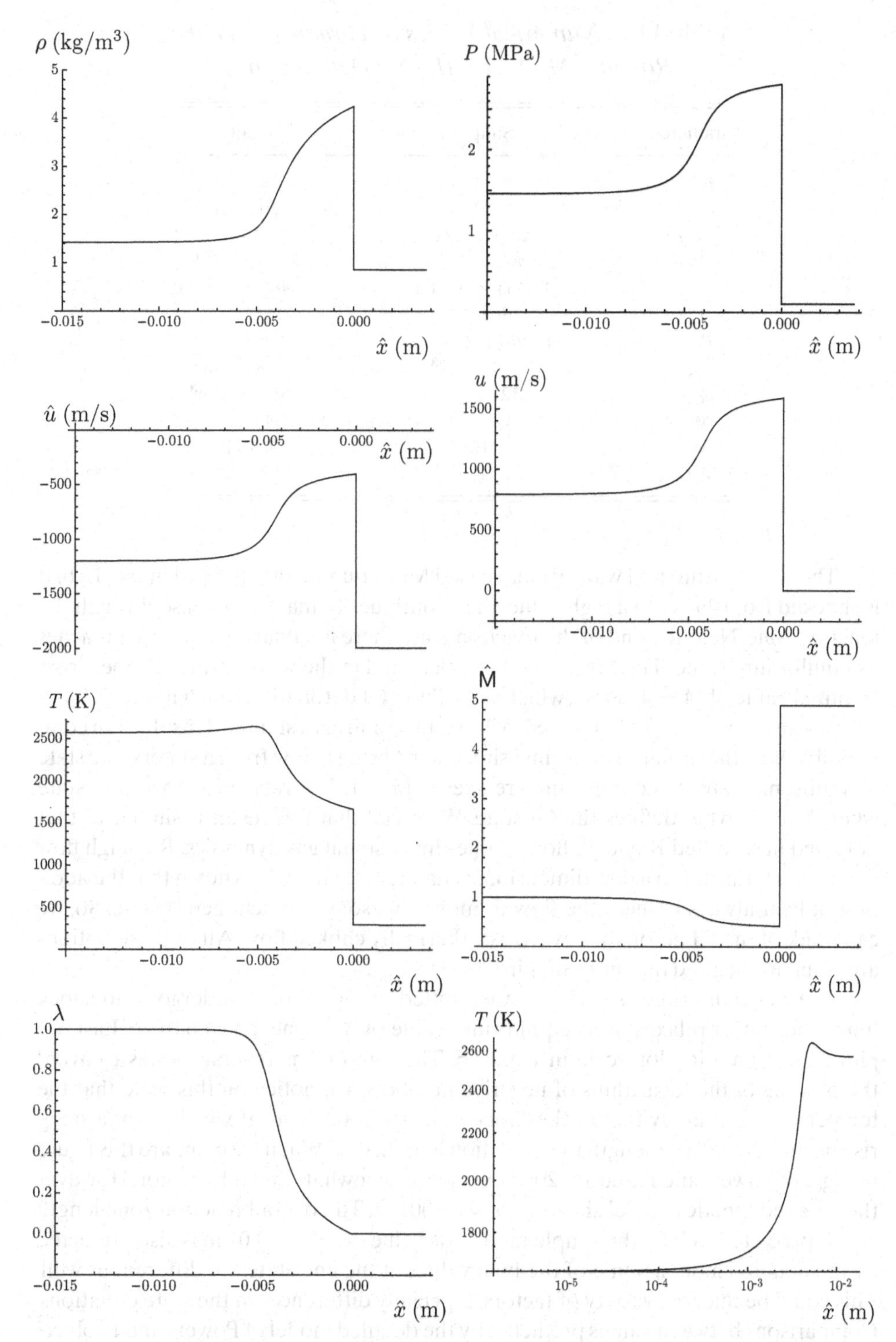


Figure 12.6. ZND structure of $\rho$, $P$, wave frame–based fluid particle velocity, $\hat{u}$ laboratory frame–based particle velocity $u, T$, wave frame–based Mach number $\hat{\mathsf{M}}$, $\lambda$, and $T$ on a log-log scale for $D = D_{CJ} = 1991.1$ m/s for one-step irreversible reaction for $H_2$/air-based parameters of Table 12.1.

Table 12.2. *Numerical Values of Parameters That Roughly Model CJ $H_2$-Air Detonation*

| Parameter | Simple | Detailed |
|---|---|---|
| $\ell_{rxn}$ | $10^{-2}$ m | $10^{-2}$ m |
| $\ell_{ind}$ | $10^{-3}$ m | $10^{-4}$ m |
| $D_{CJ}$ | 1991.1 m/s | 1979.7 m/s |
| $P_s$ | $2.80849 \times 10^6$ Pa | $2.8323 \times 10^6$ Pa |
| $P_{CJ}$ | $1.4553 \times 10^6$ Pa | $1.6483 \times 10^6$ Pa |
| $T_s$ | 1664.4 K | 1542.7 K |
| $T_{CJ}$ | 2570.86 K | 2982.1 K |
| $\rho_s$ | 4.244 kg/m$^3$ | 4.618 kg/m$^3$ |
| $\rho_{CJ}$ | 1.424 kg/m$^3$ | 1.5882 kg/m$^3$ |
| $\hat{M}_0$ | 4.88887 | 4.8594 |
| $\hat{M}_s$ | 0.41687 | 0.40779 |
| $\hat{M}_{CJ}$ | 1 | 0.93823 |

The temperature and wave frame–based Mach number are plotted in Fig. 12.6. It is shocked from 298 K to a high value, then continues to mainly increase through the reaction zone. Near the end of the reaction zone, there is a final decrease as it reaches its equilibrium value. The Mach number calculated in the wave frame, $\hat{M}$, goes from an initial value of $\hat{M} = 4.88887$, which we call the CJ detonation Mach number, $M_{CJ}$, to a postshock value of $\hat{M} = 0.41687$. This result confirms a standard result from compressible flow that a standing normal shock must bring a flow from a supersonic state to a subsonic state. At equilibrium it relaxes to $\hat{M} = 1$. This relaxation to a sonic state when $\lambda = 1$ is what defines the CJ state. We recall that this result is similar to that obtained in so-called Rayleigh flow of one-dimensional gas dynamics. Rayleigh flow admits heat transfer to a one-dimensional channel, and it is well known that the addition of heat always induces the flow to move to a sonic (or "choked") state. So, we can think of the CJ detonation wave as a thermally choked flow. All of these notions are standard in gas dynamics (Shapiro, 1953).

The reaction progress variable $\lambda$ is plotted in Fig. 12.6. It undergoes no shock jump and simply relaxes to its equilibrium value of $\lambda = 1$ near $\hat{x} = 0.01$ m. Last, we plot $T(-\hat{x})$ on a log-log scale in Fig. 12.6. The sign of $\hat{x}$ is reversed so as to avoid the plotting of the logarithms of negative numbers. We notice on this scale that the temperature is roughly that of the shock until $-\hat{x} = 0.001$ m, at which point a steep rise begins. We call this length the induction length, $\ell_{ind}$. When we compare this figure to Fig. 2 of Powers and Paolucci (2005), we see a somewhat similar behavior. However the detailed kinetics model shows $\ell_{ind} \sim 0.0001$ m. The overall reaction zone length $\ell_{rxn}$ is predicted well by the simple model. Its value of $\ell_{rxn} \sim 0.01$ m is also predicted by the detailed model. Some of the final values at the end state are different as well. This could be due to a variety of factors, especially differences in the state equations. Comparisons between values predicted by the detailed model of Powers and Paolucci (2005), against those of the simple model here are given in Table 12.2.

### Strong ZND Structures

We increase the detonation velocity $D$ to $D > D_{CJ}$ and obtain strong detonation structures. These have a path from $O$ to $N$ to the equilibrium point $S$. For very high

$D$, there is no induction zone, as the shock is sufficiently strong to overcome the activation energy barrier. These detonations require piston support to propagate, as the energy supplied by heat release alone is insufficient to maintain their steady speed. Similar to our plots of the CJ structures, we give plots of the strong, $D = 2800$ m/s structures of $\rho(\hat{x})$, $P(\hat{x})$, $\hat{u}(\hat{x})$, $u(\hat{x})$, $T(\hat{x})$, $\hat{\mathsf{M}}(\hat{x})$, $\lambda(\hat{x})$ in Fig. 12.7. The behavior of the plots is qualitatively similar to that for CJ detonations. We see, however, that the reaction zone has become significantly thinner, $\ell_{rxn} \sim 0.0001$ m. This is because the higher temperatures associated with the stronger shock induce faster reactions, thus thinning the reaction zone. Comparison with Fig. 12.3 reveals that the shocked and final values of pressure agree with those of the Rankine-Hugoniot jump analysis. The final value of $\hat{\mathsf{M}}$ is subsonic. This allows information to propagate from the supporting piston all the way to the shock front.

#### Weak ZND Structures

For the simple one-step irreversible kinetics model, there is no path from $O$ through the shocked state $N$ to the weak solution $W$. There is a direct path from $O$ to $W$; however, it is physically unrealistic, because the reaction zone is extraordinarily large.

#### Piston Problem

We can understand the physics of the one-step kinetics problem better in the context of a piston problem, where the supporting piston connects to the final laboratory frame fluid velocity $u_p = u(\hat{x} \to -\infty)$. Let us consider pistons with high velocity and then lower them and examine the changes of structure.

- $u_p > u_{p,CJ}$. This high velocity piston will drive a strong shock into the fluid at a speed $D > D_{CJ}$. The solution will proceed from $O$ to $N$ to $S$ and be subsonic throughout. Therefore, changes at the piston face will be able to be communicated all the way to the shock front. The energy to drive the wave comes from a combination of energy released during combustion and energy supplied by the piston support.
- $u_p = u_{p,CJ}$. At a critical value of piston velocity, $u_{p,CJ}$, the solution will go from $O$ to $N$ to $C$, where it is locally sonic.
- $u_p < u_{p,CJ}$. For such flows, the detonation wave is self-supporting. There is no means to communicate with the supporting piston. The detonation wave proceeds at $D = D_{CJ}$. *We note that $D_{CJ}$ is not a function of the specific kinetic mechanism. So, for a one-step irreversible kinetics model*, the conclusion the $D_{CJ}$ is the unique speed of propagation of an unsupported wave is verified. As shown in detail by Fickett and Davis, nearly any complication added to the model, for example, reversibility, multistep kinetics, multidimensionality, diffusion, and so on, can potentially alter this conclusion.

### 12.2.7 Detonation Structure: Two-Step Irreversible Kinetics

Consider a small change to the one-step model of the previous sections: we will now examine a two-step irreversible kinetics model. The first reaction will be exothermic and the second endothermic. Both reactions will be driven to completion, and when they are complete, the global heat release will be identical to that of the one-step reaction. All other parameters will remain the same from the one-step model. This model is discussed in detail by Fickett and Davis, and in a two-dimensional

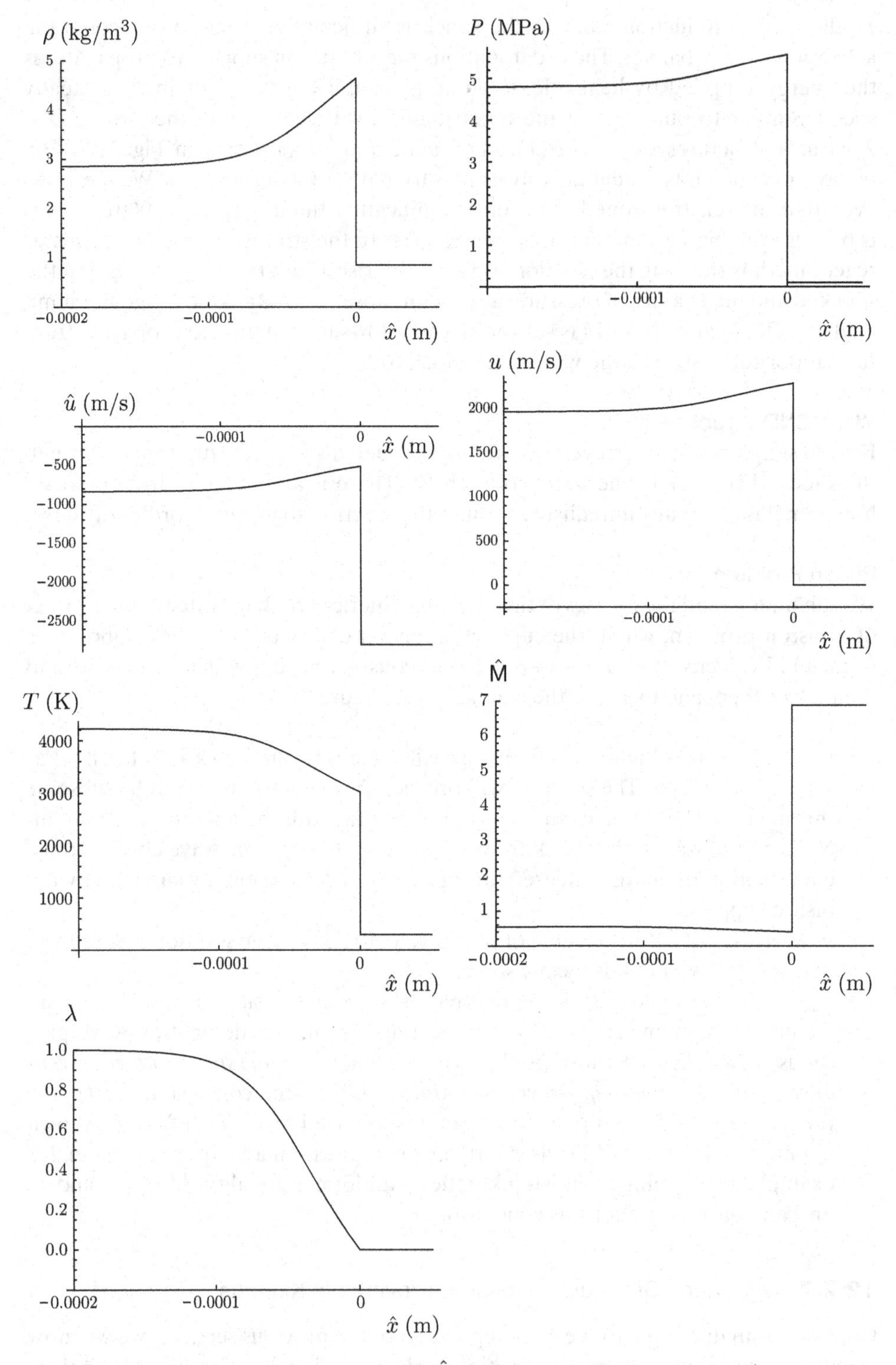


Figure 12.7. ZND structure of $\rho$ $P$, $\hat{u}$, $u$, $T$, $\hat{\mathsf{M}}$, and $\lambda$ for strong $D = 2800$ m/s $> D_{CJ}$ for one-step irreversible reaction for $H_2$/air-based parameters of Table 12.1.

extension by Powers and Gonthier (1992). We will see that this simple modification has profound effects on what is a preferred detonation structure. In particular, we will see that *for such a two-step model*

- The CJ structure is no longer the preferred state of an unsupported detonation.
- The steady speed of the unsupported detonation wave is unique and greater than the CJ speed.
- There is a path from the unshocked state $O$ to the shocked state $N$ through a sonic incomplete reaction pathological point $P$ to the weak equilibrium end state $W$.
- There is a strong analog to steady compressible one-dimensional inert flow with area change, that is, rocket nozzle flow (Shapiro, 1953).

Let us pose the two-step irreversible kinetics model of

$$A \rightarrow B, \qquad B \rightarrow C. \tag{12.261}$$

Let us insist that $A$, $B$, and $C$ each have the same molecular mass, $M_A = M_B = M_C = M$, and the same constant specific heats, $c_{PA} = c_{PB} = c_{PC} = c_P$. Let us also insist that both reactions have the same kinetic parameters, $\mathcal{E}_1 = \mathcal{E}_2 = \mathcal{E}, a_1 = a_2 = a$, $\beta_1 = \beta_2 = 0$. Therefore, the reaction rate coefficients are such that

$$k_1 = k_2 = k. \tag{12.262}$$

Let us assume at the initial state that we have all $A$ and no $B$ or $C$: $\overline{\rho}_A(0) = \hat{\overline{\rho}}_A$, $\overline{\rho}_B(0) = 0, \overline{\rho}_C(0) = 0$. Because $J = 2$, we have a reaction rate vector $r_j$ of length 2:

$$r_j = \begin{pmatrix} r_1 \\ r_2 \end{pmatrix}. \tag{12.263}$$

Our matrix of net stoichiometric coefficients $\nu_{ij}$ has dimension $3 \times 2$ because $N = 3$ and $J = 2$:

$$\nu_{ij} = \begin{pmatrix} -1 & 0 \\ 1 & -1 \\ 0 & 1 \end{pmatrix}. \tag{12.264}$$

For species molar production rates, from $\dot{\omega}_i = \sum_{j=1}^{J} \nu_{ij} r_j$, we have

$$\frac{d}{dt}\begin{pmatrix} \overline{\rho}_A/\rho \\ \overline{\rho}_B/\rho \\ \overline{\rho}_C/\rho \end{pmatrix} = \frac{1}{\rho}\begin{pmatrix} \dot{\omega}_A \\ \dot{\omega}_B \\ \dot{\omega}_C \end{pmatrix} = \frac{1}{\rho}\begin{pmatrix} -1 & 0 \\ 1 & -1 \\ 0 & 1 \end{pmatrix}\begin{pmatrix} r_1 \\ r_2 \end{pmatrix} = \frac{1}{\rho}\begin{pmatrix} -r_1 \\ r_1 - r_2 \\ r_2 \end{pmatrix}. \tag{12.265}$$

We recall here that $d/dt$ denotes the material derivative following a fluid particle, $d/dt = \partial/\partial t + \hat{u}\partial/\partial\hat{x}$. For our steady waves, we will have $d/dt = \hat{u}d/d\hat{x}$. We next recall that $\overline{\rho}_i/\rho = Y_i/M_i$, which for us is $Y_i/M$, because the molecular masses are constant. So

$$\frac{d}{dt}\begin{pmatrix} Y_A \\ Y_B \\ Y_C \end{pmatrix} = \frac{M}{\rho}\begin{pmatrix} \dot{\omega}_A \\ \dot{\omega}_B \\ \dot{\omega}_C \end{pmatrix} = \frac{M}{\rho}\begin{pmatrix} -1 & 0 \\ 1 & -1 \\ 0 & 1 \end{pmatrix}\begin{pmatrix} r_1 \\ r_2 \end{pmatrix} = \frac{M}{\rho}\begin{pmatrix} -r_1 \\ r_1 - r_2 \\ r_2 \end{pmatrix}. \tag{12.266}$$

Elementary row operations gives us the row echelon form

$$\frac{d}{dt}\begin{pmatrix} Y_A \\ Y_A + Y_B \\ Y_A + Y_B + Y_C \end{pmatrix} = \frac{M}{\rho}\begin{pmatrix} -1 & 0 \\ 0 & -1 \\ 0 & 0 \end{pmatrix}\begin{pmatrix} r_1 \\ r_2 \end{pmatrix}. \tag{12.267}$$

We can integrate the homogeneous third equation and apply the initial condition to get

$$Y_A + Y_B + Y_C = 1. \tag{12.268}$$

This can be thought of as an unusual matrix equation:

$$\begin{pmatrix} 1 & 1 & 1 \end{pmatrix} \begin{pmatrix} Y_A \\ Y_B \\ Y_C \end{pmatrix} = \begin{pmatrix} 1 \end{pmatrix}. \tag{12.269}$$

We can perform an analogous exercise to finding the form of Eq. (1.55): $\overline{\rho} = \hat{\overline{\rho}} + \mathbf{D} \cdot \overline{\xi}$:

$$\begin{pmatrix} Y_A \\ Y_B \\ Y_C \end{pmatrix} = \begin{pmatrix} 1 \\ 0 \\ 0 \end{pmatrix} + \underbrace{\begin{pmatrix} -1 & 0 \\ 1 & -1 \\ 0 & 1 \end{pmatrix}}_{\mathbf{F}} \begin{pmatrix} \lambda_1 \\ \lambda_2 \end{pmatrix}. \tag{12.270}$$

The column vectors of $\mathbf{F}$ are linearly independent and lie in the right null space of the coefficient matrix $(1, 1, 1)$. The choices for $\mathbf{F}$ are not unique but are convenient. We can think of the independent variables $\lambda_1, \lambda_2$ as reaction progress variables. Thus, for Reaction 1, we have $\lambda_1$, and for Reaction 2, we have $\lambda_2$. Both $\lambda_1(0) = 0$ and $\lambda_2(0) = 0$. The mass fraction of each species can be related to the reaction progress via

$$Y_A = 1 - \lambda_1, \tag{12.271}$$

$$Y_B = \lambda_1 - \lambda_2, \tag{12.272}$$

$$Y_C = \lambda_2. \tag{12.273}$$

When the reaction is complete, we have $\lambda_1 \to 1$, $\lambda_2 \to 1$, and $Y_A \to 0$, $Y_B \to 0$, $Y_C \to 1$. Now, our reaction law is

$$\frac{d}{dt}\begin{pmatrix} Y_A \\ Y_B \\ Y_C \end{pmatrix} = \frac{M}{\rho}\begin{pmatrix} -r_1 \\ r_1 - r_2 \\ r_2 \end{pmatrix} = \frac{M}{\rho}\begin{pmatrix} -k\overline{\rho}_A \\ k\overline{\rho}_A - k\overline{\rho}_B \\ k\overline{\rho}_B \end{pmatrix} = k\begin{pmatrix} -Y_A \\ Y_A - Y_B \\ Y_B \end{pmatrix}. \tag{12.274}$$

Eliminating $Y_A, Y_B$, and $Y_C$ in favor of $\lambda_1$ and $\lambda_2$, we get

$$\frac{d}{dt}\begin{pmatrix} 1 - \lambda_1 \\ \lambda_1 - \lambda_2 \\ \lambda_2 \end{pmatrix} = k\begin{pmatrix} -(1 - \lambda_1) \\ (1 - \lambda_1) - (\lambda_1 - \lambda_2) \\ \lambda_1 - \lambda_2 \end{pmatrix}. \tag{12.275}$$

This reduces to

$$\frac{d}{dt}\begin{pmatrix} \lambda_1 \\ \lambda_1 - \lambda_2 \\ \lambda_2 \end{pmatrix} = k\begin{pmatrix} 1 - \lambda_1 \\ 1 - 2\lambda_1 + \lambda_2 \\ \lambda_1 - \lambda_2 \end{pmatrix}. \tag{12.276}$$

The second of these equations is the difference of the first and the third, so it is redundant, and we need only consider

$$\frac{d}{dt}\begin{pmatrix} \lambda_1 \\ \lambda_2 \end{pmatrix} = k\begin{pmatrix} 1 - \lambda_1 \\ \lambda_1 - \lambda_2 \end{pmatrix}. \tag{12.277}$$

In the steady wave frame, this is written as

$$\hat{u}\frac{d\lambda_1}{d\hat{x}} = (1 - \lambda_1)k, \qquad \hat{u}\frac{d\lambda_2}{d\hat{x}} = (\lambda_1 - \lambda_2)k. \tag{12.278}$$

Because the rates $k_1 = k_2 = k$ have been taken to be identical, we can actually get $\lambda_2(\lambda_1)$. Dividing our two kinetic equations gives

$$\frac{d\lambda_2}{d\lambda_1} = \frac{\lambda_1 - \lambda_2}{1 - \lambda_1}. \tag{12.279}$$

Because $\lambda_1(0) = 0$ and $\lambda_2(0) = 0$, we can say that $\lambda_2(\lambda_1 = 0) = 0$. We rearrange this differential equation to get

$$\frac{d\lambda_2}{d\lambda_1} + \frac{1}{1 - \lambda_1}\lambda_2 = \frac{\lambda_1}{1 - \lambda_1}. \tag{12.280}$$

This equation is first order and linear. It has an integrating factor of

$$\exp\left(\int \frac{d\lambda_1}{1 - \lambda_1}\right) = \exp\left(-\ln(1 - \lambda_1)\right) = \frac{1}{1 - \lambda_1}.$$

Multiplying both sides by the integrating factor, we get

$$\frac{1}{1 - \lambda_1}\frac{d\lambda_2}{d\lambda_1} + \left(\frac{1}{1 - \lambda_1}\right)^2 \lambda_2 = \frac{\lambda_1}{(1 - \lambda_1)^2}. \tag{12.281}$$

Using the product rule, we then get

$$\frac{d}{d\lambda_1}\left(\frac{\lambda_2}{1 - \lambda_1}\right) = \frac{\lambda_1}{(1 - \lambda_1)^2}, \tag{12.282}$$

$$\frac{\lambda_2}{1 - \lambda_1} = \int \frac{\lambda_1}{(1 - \lambda_1)^2}\, d\lambda_1. \tag{12.283}$$

Taking $u = \lambda_1$ and $dv = d\lambda_1/(1 - \lambda_1)^2$ and integrating the right side by parts, we get

$$\frac{\lambda_2}{1 - \lambda_1} = \frac{\lambda_1}{1 - \lambda_1} - \int \frac{d\lambda_1}{1 - \lambda_1}, \tag{12.284}$$

$$= \frac{\lambda_1}{1 - \lambda_1} + \ln(1 - \lambda_1) + C, \tag{12.285}$$

$$\lambda_2 = \lambda_1 + (1 - \lambda_1)\ln(1 - \lambda_1) + C(1 - \lambda_1). \tag{12.286}$$

Now, because $\lambda_2(\lambda_1 = 0) = 0$, we get $C = 0$, so

$$\lambda_2(\hat{x}) = \lambda_1(\hat{x}) + (1 - \lambda_1(\hat{x}))\ln(1 - \lambda_1(\hat{x})). \tag{12.287}$$

Leaving out details of the derivation, our state equation becomes

$$e(T, \lambda_1, \lambda_2) = c_v(T - T_0) - \lambda_1 q_1 - \lambda_2 q_2. \tag{12.288}$$

We find it convenient to define $Q(\lambda_1, \lambda_2)$ as

$$Q(\lambda_1, \lambda_2) \equiv \lambda_1 q_1 + \lambda_2 q_2. \tag{12.289}$$

So, the equation of state can be written as

$$e(T, \lambda_1, \lambda_2) = c_v(T - T_0) - Q(\lambda_1, \lambda_2). \tag{12.290}$$

The frozen sound speed remains

$$c^2 = \gamma P v = \gamma \frac{P}{\rho}. \tag{12.291}$$

Table 12.3. *Numerical Values of Parameters for Two-Step Irreversible Kinetics*

| Parameter | Value | Units |
|---|---|---|
| $\gamma$ | 1.4 | |
| $M$ | 20.91 | kg/kmol |
| $R$ | 397.58 | J/kg/K |
| $P_0$ | $1.01325 \times 10^5$ | Pa |
| $T_0$ | 298 | K |
| $\rho_0$ | 0.85521 | kg/m$^3$ |
| $v_0$ | 1.1693 | m$^3$/kg |
| $q_1$ | $7.58265 \times 10^6$ | J/kg |
| $q_2$ | $-5.68698 \times 10^6$ | J/kg |
| $\mathcal{E}$ | $8.29352 \times 10^6$ | J/kg |
| $a$ | $5 \times 10^9$ | 1/s |
| $\beta$ | 0 | |

There are now two thermicities:

$$\sigma_1 = \frac{1}{\rho c^2} \frac{\partial P}{\partial \lambda_1}\bigg|_{v,e,\lambda_2} = \frac{\gamma}{\gamma - 1} \frac{\rho q_1}{P}, \tag{12.292}$$

$$\sigma_2 = \frac{1}{\rho c^2} \frac{\partial P}{\partial \lambda_2}\bigg|_{v,e,\lambda_1} = \frac{\gamma}{\gamma - 1} \frac{\rho q_2}{P}. \tag{12.293}$$

Parameters for our two-step model are identical to those of our one-step model, except for the heat releases. The parameters are listed in Table 12.3. At complete reaction, we have $Q(\lambda_1, \lambda_2) = Q(1, 1) = q_1 + q_2 = 1.89566 \times 10^6$ J/kg. Thus, the overall heat release at complete reaction $\lambda_1 = \lambda_2 = 1$ is identical to our earlier one-step model.

Let us do some new Rankine-Hugoniot analysis. We can write a set of mass, momentum, energy, and state equations as

$$\rho \hat{u} = -\rho_0 D, \tag{12.294}$$

$$P + \rho \hat{u}^2 = P_0 + \rho_0 D^2, \tag{12.295}$$

$$e + \frac{1}{2}\hat{u}^2 + \frac{P}{\rho} = e_0 + \frac{1}{2}D^2 + \frac{P_0}{\rho_0}, \tag{12.296}$$

$$e = \frac{1}{\gamma - 1}\frac{P}{\rho} - \lambda_1 q_1 - \lambda_2 q_2, \tag{12.297}$$

$$\lambda_2 = \lambda_1 + (1 - \lambda_1)\ln(1 - \lambda_1). \tag{12.298}$$

Let us consider $D$ and $\lambda_1$ to be unspecified but known parameters for this analysis. These equations are five equations for the five unknowns, $\rho, \hat{u}, P, e$, and $\lambda_2$. They can be solved for $\rho(D, \lambda_1), \hat{u}(D, \lambda_1), P(D, \lambda_1), e(D, \lambda_1)$, and $\lambda_2(\lambda_1)$.

The solution is lengthy, but the plot is revealing. For three different values of $D$, pressure as a function of $\lambda_1$ is shown in Fig. 12.8. There are three important classes of $D$, each shown in Fig. 12.8, depending on how $D$ compares to a critical value, $\tilde{D}$.

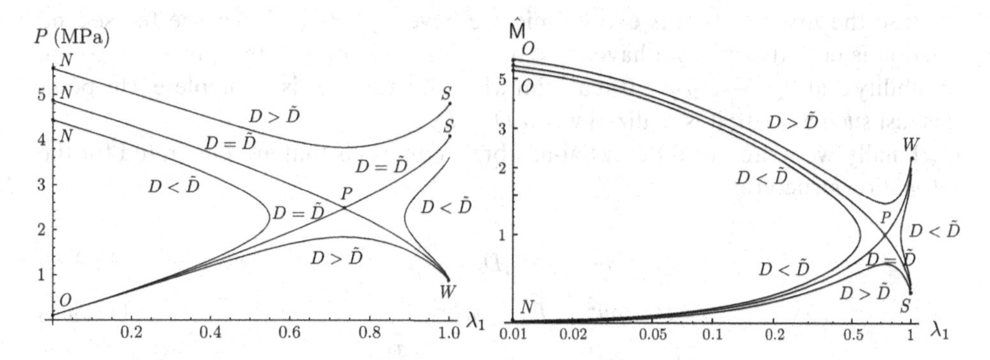


Figure 12.8. Pressure $P$ and Mach number $\hat{M}$ versus $\lambda_1$ for two-step kinetics problem for three different values of $D$: $D = 2800$ m/s $> \tilde{D}$, $D = 2616.5$ m/s $= \tilde{D}$, $D = 2500$ m/s $< \tilde{D}$; parameters are from Table 12.3

- $D > \tilde{D}$. There are two potential paths here. The important physical branch starts at point $O$, and is immediately shocked to state $N$, the Neumann point. From $N$ the pressure first decreases as $\lambda_1$ increases. Near $\lambda_1 = 0.75$, the pressure reaches a local minimum, and then increases to the complete reaction point at $S$. This is a strong solution. There is a second branch that commences at $O$ and is unshocked. On this branch the pressure increases to a maximum, then decreases to the end state at $W$. While this branch is admissible mathematically, its length scales are unphysically long, and this branch is discarded.
- $D = \tilde{D}$. Let us only consider branches that are shocked from $O$ to $N$. The unshocked branches are again nonphysical. On this branch, the pressure decreases from $N$ to the pathological point $P$. At $P$, the flow is locally sonic, with $\hat{M} = 1$. Here, the pressure can take two distinct paths. The one chosen will depend on the velocity of the supporting piston at the end state. On one path the pressure increases to its final value at the strong point $S$. On the other the pressure decreases to its final value at the weak point $W$.
- $D < \tilde{D}$. For such values of $D$, there is no physical structure for the entire reaction zone $0 < \lambda_1 < 1$. This branch is discarded.

Mach number in the wave frame, $\hat{M}$ as a function of $\lambda_1$ is also shown in in Fig. 12.8. The ambient point $O$ is always supersonic, and the Neumann point $N$ is always subsonic. For flows originating at $N$, if $D > \tilde{D}$, the flow remains subsonic throughout until its termination at $S$. For $D = \tilde{D}$, the flow can undergo a subsonic to supersonic transition at the pathological point $P$. The weak point $W$ is a supersonic end state.

The important Fig. 12.8 bears remarkable similarity to curves of $P(x)$ in compressible inert flow in a converging-diverging nozzle (Shapiro, 1953). We recall that for such flows, a subsonic to supersonic transition is only realized at an area minimum. This can be explained because the equation for evolution of pressure for such flows takes the form $dP/dx \sim (dA/dx)/(1 - M^2)$. So, if the flow is locally sonic, it must encounter a critical point in area, $dA/dx = 0$ to avoid infinite pressure gradients. This is what is realized in actual nozzles. For us, the analogous equation is the two-step version of Eq. (12.181), which can be shown to be

$$\frac{dP}{d\hat{x}} = -\frac{\rho\hat{u}(\sigma_1 r_1 + \sigma_2 r_2)}{1 - \hat{M}^2}. \tag{12.299}$$

Because the first reaction is exothermic, we have $\sigma_1 > 0$, and because the second reaction is endothermic, we have $\sigma_2 < 0$. With $r_1 > 0$, $r_2 > 0$, this gives rise to the possibility that $\sigma_1 r_1 + \sigma_2 r_2 = 0$ at a point where the reaction is incomplete. The point $P$ is just such a point; it is realized when $D = \tilde{D}$.

Finally, we write our differential-algebraic equations that are integrated for the detonation structure:

$$\rho \hat{u} = -\rho_0 D, \tag{12.300}$$

$$P + \rho \hat{u}^2 = P_0 + \rho_0 D^2, \tag{12.301}$$

$$e + \frac{1}{2}\hat{u}^2 + \frac{P}{\rho} = e_0 + \frac{1}{2}D^2 + \frac{P_0}{\rho_0}, \tag{12.302}$$

$$e = \frac{1}{\gamma - 1}\frac{P}{\rho} - \lambda_1 q_1 - \lambda_2 q_2, \tag{12.303}$$

$$P = \rho RT, \tag{12.304}$$

$$\lambda_2 = \lambda_1 + (1 - \lambda_1)\ln(1 - \lambda_1), \tag{12.305}$$

$$\frac{d\lambda_1}{d\hat{x}} = \frac{1 - \lambda_1}{\hat{u}} a \exp\left(-\frac{\mathcal{E}}{RT}\right). \tag{12.306}$$

We need the condition $\lambda_1(0) = 0$. These form seven equations for the seven unknowns $\rho$, $\hat{u}$, $P$, $e$, $T$, $\lambda_1$, and $\lambda_2$. We also realize that the algebraic solutions are multivalued and must take special care to be on the proper branch. This becomes particularly important for solutions that pass through $P$.

**Strong Structures**

Here, we consider strong structures for two cases: $D > \tilde{D}$ and $D = \tilde{D}$. All of these will proceed from $O$ to $N$ through a pressure minimum and finish at $S$.

***$D > \tilde{D}$.*** Structures for a strong detonation with $D = 2800$ m/s $> \tilde{D}$ are given in Fig. 12.9. The structure of all of these can be compared directly to those of the one-step kinetics model at the same $D = 2800$ m/s (Fig. 12.7). The shock values are identical. The reaction zone thicknesses are similar as well at $\ell_{rxn} \sim 0.0001$ m. The structures themselves have some differences; most notably, the two-step model structures display interior critical points before complete reaction. We take special note of the pressure plot of Fig. 12.9, which can be compared with Fig. 12.8. We see in both figures the shock from $O$ to $N$, followed by a drop of pressure to a minimum, followed by a final relaxation to an equilibrium value at $S$. The two curves have the opposite sense of direction as $\lambda_1$ commences at 0 and goes to 1, while $\hat{x}$ commences at 0 and goes to $-0.0002$ m. We can also compare the $\hat{M}(\hat{x})$ plot of Fig. 12.9 with that of $\hat{M}(\lambda_1)$ of Fig. 12.8. In both the supersonic $O$ is shocked to a subsonic $N$. The Mach number rises slightly then falls in the reaction zone to its equilibrium value at $S$. It never returns to a supersonic state.

***$D = \tilde{D}$.*** For $D = \tilde{D} = 2616.5$ m/s, we can find a strong structure with a path from $O$ to $N$ to $S$. Pressure $P$ and Mach number $\hat{M}$ are plotted in Fig. 12.10. At an interior point in the structure, a cusp in the $P$ and $\hat{M}$ profile is seen. At this point, the flow is locally sonic with $\hat{M} = 1$.

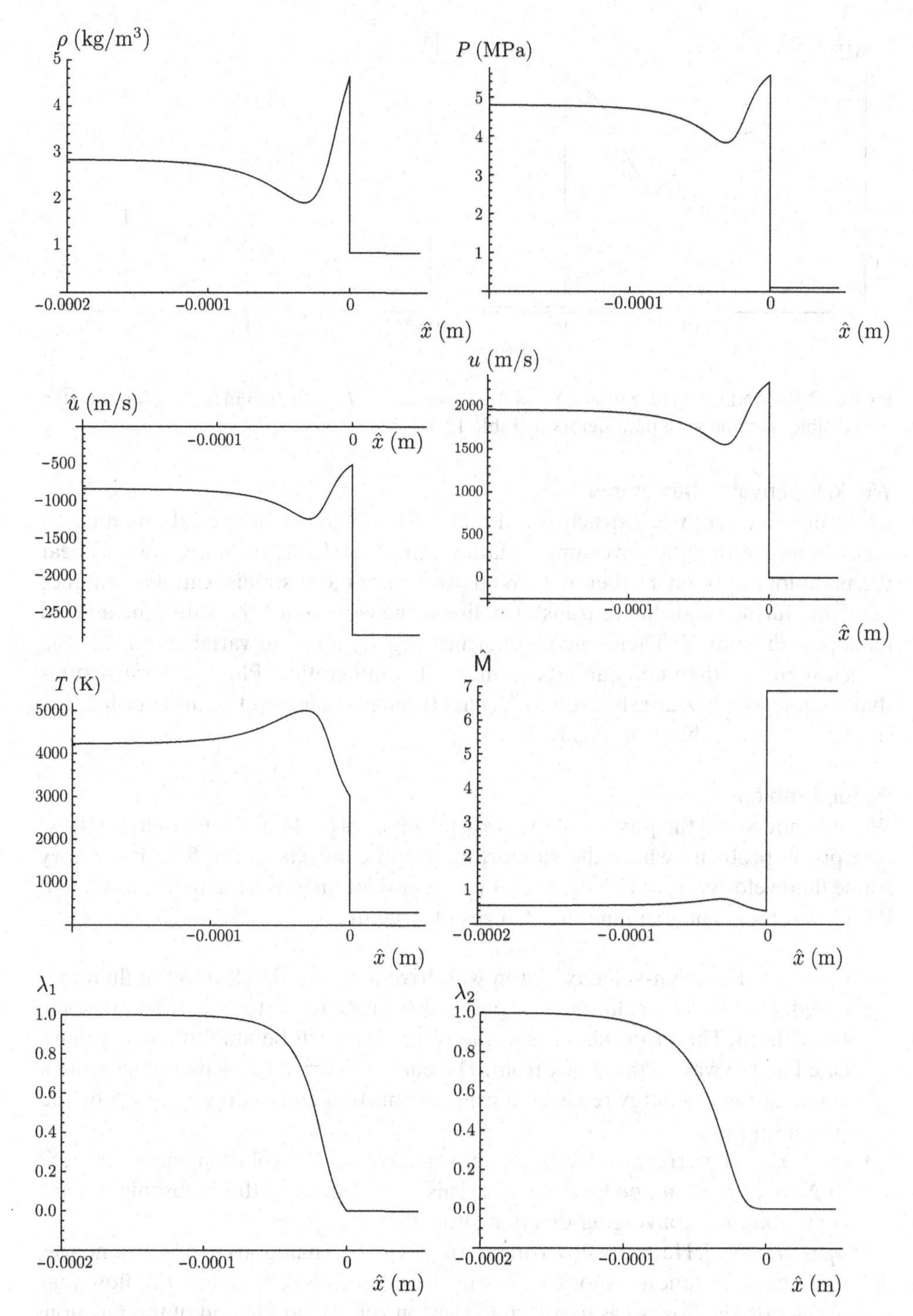


Figure 12.9. ZND structure of $\rho$, $P$, $\hat{u}$, $u$, $T$, $\hat{\mathsf{M}}$, $\lambda_1$, and $\lambda_2$ for strong $D = 2800$ m/s $> \tilde{D}$ for two-step irreversible reaction with parameters of Table 12.3.

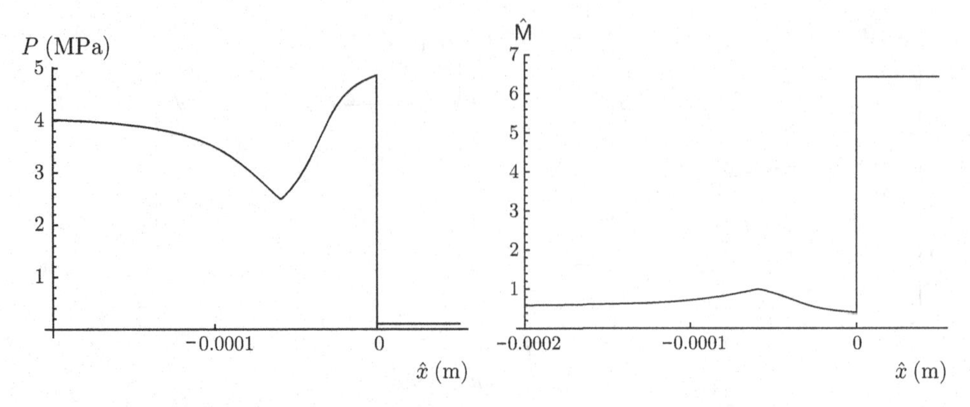


Figure 12.10. ZND structure of $P(\hat{x})$ and $\hat{M}(\hat{x})$ for strong $D = 2616.5$ m/s $= \tilde{D}$ for two-step irreversible reaction with parameters of Table 12.3.

### Weak, Eigenvalue Structures

Let us now consider weak structures with $D = \tilde{D} = 2616.5$ m/s. Special care must be taken in integrating the governing equations. In general, one must integrate to near the pathological point $P$, then halt. While there are more sophisticated techniques involving further coordinate transformations, one can record the values near $P$ on its approach from $N$. Then, one can perturb slightly all state variables so that the $\hat{M}$ is just greater than unity and recommence the integration. Plots of the structures that commence at $O$, are shocked to $N$, pass through sonic point $P$, and finish at the supersonic $W$ are shown in Fig. 12.11.

### Piston Problem

We can understand the physics of the two-step kinetics problem better in the context of a piston problem, where the supporting piston connects to the final laboratory frame fluid velocity $u_p = u(\hat{x} \to -\infty)$. Let us consider pistons with high velocity and then lower them and examine the changes of structure.

- $u_p > \tilde{u}_{ps}$. This high-velocity piston will drive a strong shock into the fluid at a speed $D > \tilde{D}$. The solution will proceed from $O$ to $N$ to $S$ and be subsonic throughout. Therefore, changes at the piston face will be able to be communicated all the way to the shock front. The energy to drive the wave comes from a combination of energy released during combustion and energy supplied by the piston support.
- $u_p = \tilde{u}_{ps}$. At a critical value of piston velocity, $\tilde{u}_{ps}$, the solution will go from $O$ to $N$ to $P$ to $S$, and be locally sonic. This is analogous to the "subsonic design" condition for a converging-diverging nozzle.
- $u_p \in [\tilde{u}_{ps}, \tilde{u}_{pw}]$. Here, the flow can be complicated. Analogous to flow in a nozzle, there can be standing shock waves in the supersonic portion of the flow that decelerate the flow so as to match the piston velocity at the end of the reaction. Such flows will proceed from $O$ to $N$ through $P$, and then are shocked back onto the subsonic branch to terminate at $S$.
- $u_p = \tilde{u}_{pw}$. This state is analogous to the "supersonic design" condition of flow in a converging-diverging nozzle. The fluid proceeds from $O$ to $N$ through $P$ and terminates at $W$. All the energy to propagate the wave comes from reaction.
- $u_p < \tilde{u}_{pw}$. For such flows, the detonation wave is self-supporting. There is no means to communicate with the supporting piston. The detonation proceeds at

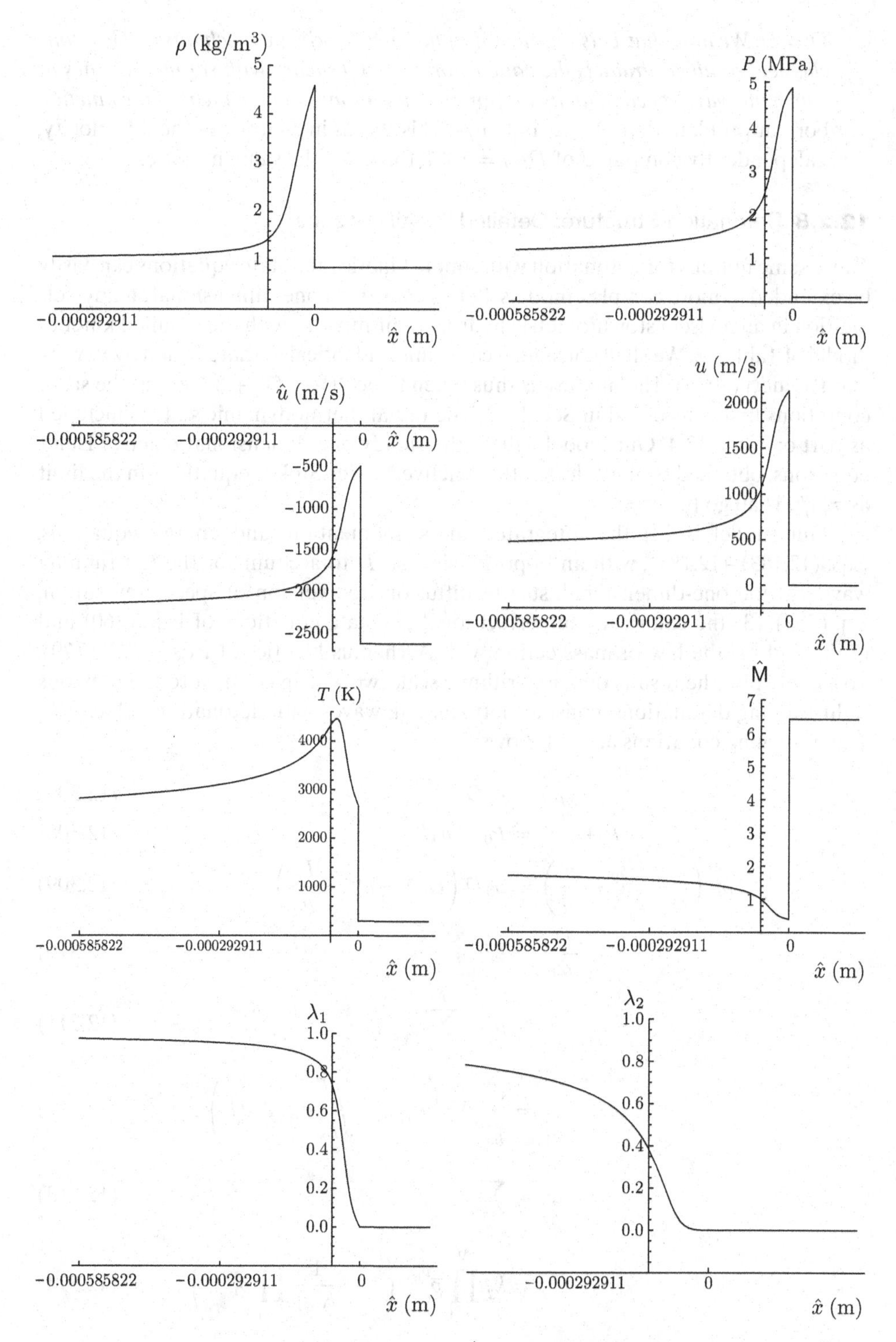


Figure 12.11. ZND structure of $\rho$, $P(\hat{x})$, $\hat{u}$, $u$, $T$, $\hat{\mathsf{M}}$, $\lambda_1$, $\lambda_2$ for weak $D = 2616.5$ m/s $= \tilde{D}$ for two-step irreversible reaction with parameters of Table 12.3.

$D = \tilde{D}$. *We note that* $\tilde{D}$ *is a function of the specific kinetic mechanism. This non-classical result contradicts the conclusion from CJ theory with simpler kinetics in which the wave speed of an unsupported detonation is independent of the kinetics.* For our problem that $\tilde{D} = 2616.5$ m/s. This stands in contrast to the CJ velocity, independently computed of $D_{CJ} = 1991.1$ m/s for the same mixture.

### 12.2.8 Detonation Structure: Detailed $H_2$-Air Kinetics

These same notions for detonation with simple kinetics and state equations can easily be extended to more complex models. Let us consider a one-dimensional steady detonation in an inviscid stoichiometric hydrogen-air mixture with the detailed kinetics model of Table 1.2. We shall consider a case almost identical to that studied by Powers and Paolucci (2005). The mixture is thus taken to be $2H_2 + O_2 + 3.76N_2$ at the same conditions first considered in Sec. 1.3.2. The initial thermodynamic state is included as part of Table 12.4. Our model will be the steady one-dimensional reactive Euler equations, obtained by considering the reactive Navier-Stokes equations in the limit as $\boldsymbol{\tau}, \mathbf{j}_i^m, \mathbf{j}^q$ all go to zero.

Our model is (1) the integrated mass, momentum, and energy equations, Eqs. (12.183)–(12.185), with an opposite sign on $D$ to account for the left-running wave, (2) the one-dimensional, steady, diffusion-free version of species evolution, Eq. (7.52), (3) the calorically imperfect ideal gas state equations of Eqs. (7.60) and (7.68), and (4) the law of mass action with Arrhenius kinetics of Eqs. (7.76)–(7.79). To cleanly plot the results on a logarithmic scale, we will, in contrast to the previous right-running detonations, consider left-running waves with detonation velocity $D$. The governing equations are as follows:

$$\rho \hat{u} = \rho_0 D, \tag{12.307}$$

$$P + \rho \hat{u}^2 = P_0 + \rho_0 D^2, \tag{12.308}$$

$$\rho \hat{u}\left(e + \frac{1}{2}\hat{u}^2 + \frac{P}{\rho}\right) = \rho_0 D\left(e_0 + \frac{1}{2}D^2 + \frac{P_0}{\rho_0}\right), \tag{12.309}$$

$$\rho \hat{u}\frac{dY_i}{d\hat{x}} = M_i \dot{\omega}_i, \tag{12.310}$$

$$P = \rho \overline{R} T \sum_{i=1}^{N} \frac{Y_i}{M_i}, \tag{12.311}$$

$$e = \sum_{i=1}^{N} Y_i\left(e^0_{T_0,i} + \int_{T_0}^{T} c_{vi}(\hat{T})d\hat{T}\right), \tag{12.312}$$

$$\dot{\omega}_i = \sum_{j=1}^{J} \nu_{ij} r_j, \tag{12.313}$$

$$r_j = k_j \prod_{k=1}^{N} \overline{\rho}_k^{\nu'_{kj}}\left(1 - \frac{1}{K_{c,j}} \prod_{k=1}^{N} \overline{\rho}_k^{\nu_{kj}}.\right), \tag{12.314}$$

$$k_j = a_j T^{\beta_j} \exp\left(\frac{-\mathcal{E}_j}{\overline{R}T}\right), \tag{12.315}$$

$$K_{c,j} = \left(\frac{P_0}{\overline{R}T}\right)^{\sum_{i=1}^{N} \nu_{ij}} \exp\left(-\frac{\Delta G_j^0}{\overline{R}T}\right). \tag{12.316}$$

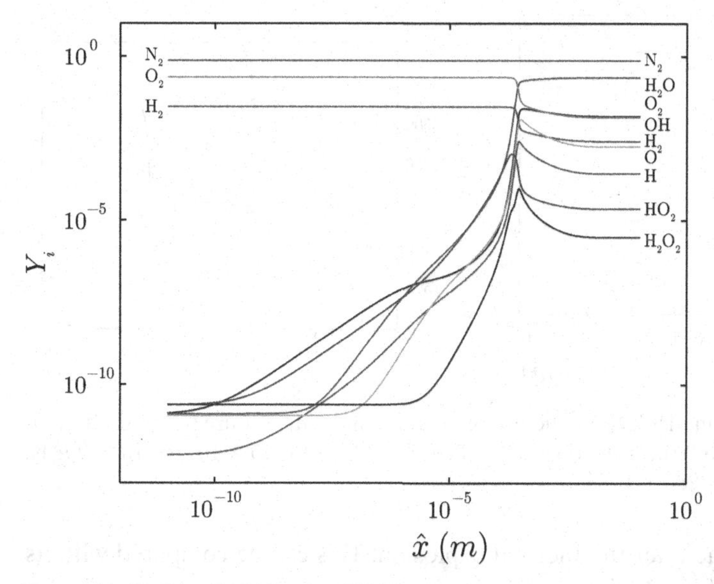


Figure 12.12. Detailed kinetics ZND structure of species mass fractions for near CJ detonation, $D_{CJ} \sim D = 1979.70$ m/s, in $2H_2 + O_2 + 3.76N_2$, $P_0 = 1.01325 \times 10^5$ Pa, $T_0 = 298$ K, $T_s = 1542.7$ K, $P_s = 2.83280 \times 10^6$ Pa.

This differential-algebraic system must be solved numerically. One can use standard methods to achieve this. Alternatively, and with significant effort, one can remove all of the algebraic constraints. Part of this requires a numerical iteration to find certain roots. After this effort, one can in principle write a set of $N - L$ differential equations for evolution of the $N - L$ independent species.

Here, we chose $D \sim D_{CJ} = 1979.70$ m/s. Because this model is not a simple one-step model, we cannot expect to find the equilibrium state to be exactly sonic. However, it is possible to slightly overdrive the wave and achieve a nearly sonic state at the equilibrium state. Here, our final Mach number was $\hat{M} = 0.9382$. Had we weakened the overdrive further, we would have encountered an interior sonic point at a nonequilibrium point, thus inducing a nonphysical sonic singularity.

Numerical solution for species mass fraction is given in Fig. 12.12. The figure is plotted on a log-log scale because of the wide range of length scales and mass fraction scales encountered. The minor species begin to change at a small length scale. At a value of $\hat{x} \sim 2.6 \times 10^{-4}$ m, a significant event occurs, known as a thermal explosion. This length is known as the induction length, $\ell_{ind} = 2.6 \times 10^{-4}$ m. We get a rough estimate of the induction time by the formula $t_{ind} \sim \ell_{ind}/\hat{u}_s = 7.9 \times 10^{-7}$ s. Here, $\hat{u}_s$ is the postshock velocity in the wave frame. Its value is $\hat{u}_s = 330.54$ m/s. All species contribute to the reaction dynamics here. This is followed by a relaxation to chemical equilibrium, achieved around 0.1 m.

The pressure and temperature profiles are given in Fig. 12.13. We artificially located the shock just away from $\hat{x} = 0$, so as to enable the log-log plot. The pressure is shocked from its atmospheric value to $2.83280 \times 10^6$ Pa (see Table 12.2). After the shock, the pressure holds nearly constant for several decades of distance. Once the thermal explosion commences, the pressure relaxes to its equilibrium value. This figure can be compared with its one-step equivalent of Fig. 12.7. The temperature is shocked to $T_s = 1542.7$ K, stays constant in the induction zone, and then increases

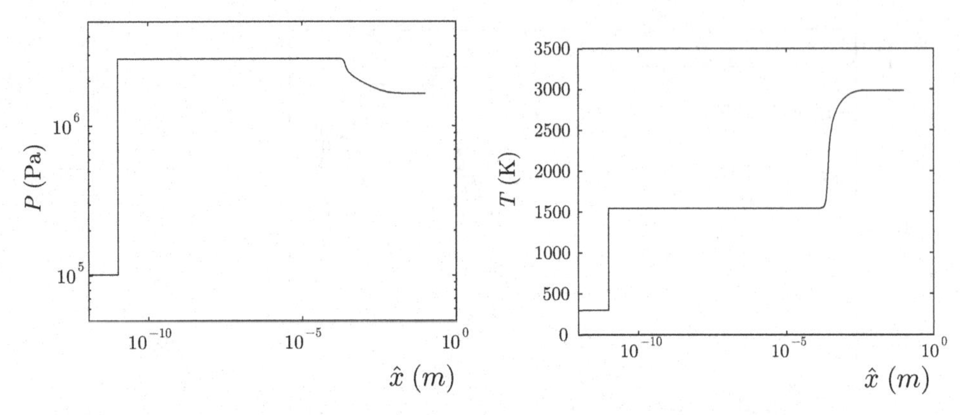


Figure 12.13. Detailed kinetics ZND structure of pressure and temperature for near CJ detonation, $D_{CJ} \sim D = 1979.70\,\text{m/s}$, in $2H_2 + O_2 + 3.76N_2$, $P_O = 1.01325 \times 10^5$ Pa, $T_O = 298$ K, $T_s = 1542.7$ K, $P_s = 2.83280 \times 10^6$ Pa.

to its equilibrium value after the thermal explosion. This can be compared with its one-step equivalent of Fig. 12.7.

It is interesting to compare these results to those obtained earlier in Sec. 1.3.2. There, an isochoric, adiabatic combustion of precisely the same stoichiometric hydrogen-air mixture with precisely the same kinetics was conducted. The adiabatic/isochoric mixture had an initial temperature and pressure identical to the post-shock pressure and temperature here. We compare the induction time of the spatially homogeneous problem, $t_{ind} = 6.6 \times 10^{-7}$ s, Eq. (1.377), to our estimate from the detonation found earlier, $t_{ind} \sim \ell_{ind}/\hat{u}_s = 7.9 \times 10^{-7}$ s. The two are remarkably similar.

We compare some other relevant values in Table 12.4. These mixtures are identical at the onset of the calculation. The detonating mixture has reached the same initial state after the shock. And a fluid particle advecting through the detonation reaction zone with undergo a thermal explosion at nearly the same time a particle that was stationary in the closed vessel will. After that the two fates are different. This is because there is no kinetic energy in the spatially homogeneous problem. Thus, all the chemical energy is transformed into thermal energy. This is reflected in the higher final temperature and pressure of the spatially homogeneous problem relative to the detonating flow. Because the final temperature is different, the two systems relax to a different chemical equilibrium, as reflected in the different mass fractions. For example, the cooler detonating flow has a higher final mass fraction of $H_2O$.

## 12.3 Nonlinear Dynamics and Transition to Chaos

We close the book with a section that exposes the links between combustion thermodynamics and nonlinear dynamics. This is done in the context of four one-dimensional detonation problems: (1) one-step kinetics with no diffusion, (2) one-step kinetics with diffusion, (3) multistep kinetics with diffusion, and (4) multistep kinetics with no diffusion. Much of this section is drawn from Romick et al. (2012, 2015) and Henrick et al. (2006), where full literature reviews can be found. The discussion will expose a remarkable link between detonation theory and chaos theory, discussed earlier by Sharpe and Falle (2000) and Ng et al. (2005).

Table 12.4. *Comparison of Relevant Predictions of a Spatially Homogeneous Model with Those of a Near CJ Detonation in the Same Mixture*

| Parameter | Spatially Homogeneous | Detonation |
|---|---|---|
| $T_0$ | 1542.7 K | 298 K |
| $T_s$ | – | 1542.7 K |
| $P_0$ | $2.83280 \times 10^6$ Pa | $1.01325 \times 10^5$ Pa |
| $P_s$ | – | $2.83280 \times 10^6$ Pa |
| $t_{ind}$ | $6.6 \times 10^{-7}$ s | $7.9 \times 10^{-7}$ s |
| $T^{eq}$ | 3382.3 K | 2982.1 K |
| $P^{eq}$ | $5.53 \times 10^6$ Pa | $1.65 \times 10^6$ Pa |
| $\rho^{eq}$ | 4.62 kg/m$^3$ | 1.59 kg/m$^3$ |
| $\hat{u}^{eq}$ | 0 m/s | 1066 m/s |
| $Y_{O_2}^{eq}$ | $1.85 \times 10^{-2}$ | $1.38 \times 10^{-2}$ |
| $Y_{H}^{eq}$ | $5.41 \times 10^{-4}$ | $2.71 \times 10^{-4}$ |
| $Y_{OH}^{eq}$ | $2.45 \times 10^{-2}$ | $1.48 \times 10^{-2}$ |
| $Y_{O}^{eq}$ | $3.88 \times 10^{-3}$ | $1.78 \times 10^{-3}$ |
| $Y_{H_2}^{eq}$ | $3.75 \times 10^{-3}$ | $2.57 \times 10^{-3}$ |
| $Y_{H_2O}^{eq}$ | $2.04 \times 10^{-1}$ | $2.22 \times 10^{-1}$ |
| $Y_{HO_2}^{eq}$ | $6.84 \times 10^{-5}$ | $2.23 \times 10^{-5}$ |
| $Y_{H_2O_2}^{eq}$ | $1.04 \times 10^{-5}$ | $3.08 \times 10^{-6}$ |
| $Y_{N_2}^{eq}$ | $7.45 \times 10^{-1}$ | $7.45 \times 10^{-1}$ |

### 12.3.1 One-Step Kinetics, With and Without Diffusion

We first consider a model that employs simple one-step irreversible chemical kinetics along with a calorically perfect ideal gas assumption.[12] We study the model with and without mass, momentum, and energy diffusion.

#### Mathematical Model

We consider a viscous analog of the equations presented in Sec. 12.1.6. The equations adopted are the one-dimensional reactive Navier-Stokes equations with one-step kinetics in a reference frame moving at constant velocity, $D$:

$$\frac{\partial \rho}{\partial t} + \frac{\partial}{\partial x}\left(\rho\left(u - D\right)\right) = 0, \qquad (12.317)$$

$$\frac{\partial}{\partial t}\left(\rho u\right) + \frac{\partial}{\partial x}\left(\rho u\left(u - D\right) + P - \tau\right) = 0, \qquad (12.318)$$

$$\frac{\partial}{\partial t}\left(\rho\left(e + \frac{u^2}{2}\right)\right) + \frac{\partial}{\partial x}\left(\rho\left(e + \frac{u^2}{2}\right)\left(u - D\right) + j_q + \left(P - \tau\right)u\right) = 0, \qquad (12.319)$$

$$\frac{\partial}{\partial t}\left(\rho\lambda\right) + \frac{\partial}{\partial x}\left(\rho\lambda\left(u - D\right) + j_m\right) = \rho r, \qquad (12.320)$$

[12] This section is adapted largely from C. M. Romick, T. D. Aslam, and J. M. Powers, 2012, The effect of diffusion on the dynamics of unsteady detonations, *Journal of Fluid Mechanics*, 699: 453–464, and A. K. Henrick, T. D. Aslam, and J. M. Powers, 2006, Simulations of pulsating one-dimensional detonations with true fifth order accuracy, *Journal of Computational Physics*, 213(1): 311–329, both reproduced with permission.

where $x$ and $t$ are the spatial and temporal coordinates, respectively; $\rho$ the mass density, $u$ the particle velocity, $P$ the pressure, $\tau$ the viscous stress, $e$ the specific internal energy, $j_q$ the diffusive heat flux, $\lambda$ the reaction progress variable, $j_m$ the diffusive mass flux, and $\mathsf{r}$ the reaction rate. Equations (12.317)–(12.319) describe the conservation of mass, linear momentum, and energy; Eq. (12.320) describes the evolution of reaction products.

The constitutive relations chosen for mass, momentum, and energy diffusion are

$$j_m = -\rho\mathcal{D}\frac{\partial \lambda}{\partial x}, \quad \tau = \frac{4}{3}\mu\frac{\partial u}{\partial x}, \quad j_q = -\mathsf{k}\frac{\partial T}{\partial x} + \rho\mathcal{D}q\frac{\partial \lambda}{\partial x}, \tag{12.321}$$

where $\mathcal{D}$ is the mass diffusion coefficient, $\mu$ the dynamic viscosity, $\mathsf{k}$ the thermal conductivity, $T$ the temperature, and $q$ the heat release of reaction. Equations (12.321) are Fick's law for binary diffusion, the Newtonian stress-strain rate relation, and an extended Fourier's law. A calorically perfect ideal gas model is adopted for an ideal mixture in which the molecular masses and specific heats of both reactant and product gases are identical:

$$P = \rho R T, \quad e = \frac{P}{(\gamma - 1)\,\rho} - q\lambda, \tag{12.322}$$

where $R$ is the gas constant and $\gamma$ the ratio of specific heats.

The irreversible one-step reaction model, $A \rightarrow B$, was chosen, where $A$ and $B$ are reactant and product, respectively. In the undisturbed state only $A$ is present; the mass fractions of $A$ and $B$ are given by $1 - \lambda$ and $\lambda$, respectively. The reaction rate, $\mathsf{r}$, is given by the law of mass action with Arrhenius kinetics:

$$\mathsf{r} = a\,(1 - \lambda)\exp\left(-\frac{\mathcal{E}}{P/\rho}\right) H\,(P - P_s)\,, \tag{12.323}$$

where $a$ is the collision frequency coefficient, $\mathcal{E}$ the activation energy, and $H(P - P_s)$ a Heaviside function that suppresses reaction when $P < P_s$ where $P_s$ is a selected pressure. Also, the ambient density and pressure are taken to be $\rho_0$ and $P_0$, respectively. Similar trends could be expected had a model with greater fidelity to realistic gas mixtures been chosen.

### Initialization and Problem Parameters

The simulations were initialized with the inviscid ZND solution in a frame traveling at the CJ speed. Each simulation is integrated in time to determine the long time behavior. By selecting the diffusion coefficient, $\mathcal{D} = 10^{-4}$ m$^2$/s, thermal conductivity, $\mathsf{k} = 10^{-1}$ W/m/K, specific heat at constant pressure $c_P = 10^3$ J/kg/K, and viscosity, $\mu = 10^{-4}$ N s/m$^2$ the Lewis, $Le = \mathsf{k}/\rho_0/c_P/\mathcal{D}$, Prandtl,[13] $Pr = \mu c_P/\mathsf{k}$, and Schmidt,[14] $Sc = \mu/\rho_0/\mathcal{D}$ numbers evaluated at the ambient density, $\rho_0 = 1$ kg/m$^3$, are unity. These parameters are within an order of magnitude of those of gases at a slightly elevated temperature. In the inviscid detonation, the activation energy controls the stability of the system; the rate constant merely introduces a length scale, the half reaction length, $L_{1/2}$ (the distance between the inviscid shock and the location at which $\lambda = 1/2$). If $L_{1/2}$ is fixed, the effect of diffusion on the system can

[13] Ludwig Prandtl (1875–1953), German engineer.
[14] Ernst Heinrich Wilhelm Schmidt (1892–1975), German engineer.

be explored. Using simple dimensional analysis of advection and diffusion parameters ($U = 1000$ m/s was chosen as a typical velocity scale) gives rise to an approximate length scale of mass diffusion, $\mathcal{D}/U = 10^{-7}$ m, and likewise for momentum and energy diffusion $\mu/\rho_0/U = 10^{-7}$ m, and $\mathsf{k}/\rho_0/c_P/U = 10^{-7}$ m. Because all the diffusion length scales are the same, let this scale be denoted as $L_\mu = 10^{-7}$ m. The parameters are taken to be $P_0 = 0.101325$ MPa, $P_s = 0.200$ MPa, $q = 5066250$ (m/s)$^2$, $\gamma = 6/5, \mathcal{E} \in [2533125, 3232400]$ (m/s)$^2$, and $c_P = (\gamma - 1)\,R/\gamma = 1000$ J/kg/K. With this heat release, $D_{CJ}$ is

$$D_{CJ} = \sqrt{\gamma\frac{P_0}{\rho_0} + \frac{q\,(\gamma^2-1)}{2}} + \sqrt{\frac{q\,(\gamma^2-1)}{2}} = 2167.56\ \frac{\text{m}}{\text{s}}. \tag{12.324}$$

The selection of $P_s$ is arbitrary, because there is minimal effect on the system over the range of 0.102 MPa to 1.010 MPa. To compare directly with previous work in the inviscid limit, the activation energies will be presented in dimensionless form, $E = \mathcal{E}/(P_0/\rho_0)$, thus, $E \in [25, 32]$. Using these parameters allows for the interaction of diffusion and reaction effects to be studied and induces a set of scales similar to those given in reactive Navier-Stokes models with detailed chemical kinetics. Unless otherwise stated, the calculations presented are for a ratio of $L_\mu/L_{1/2} = 1/10$, such that $L_{1/2} = 10^{-6}$ m, which is similar to the finest reaction length scale of $H_2$-air detonations.

The coarsest scales in $H_2$-air detonations are much larger than the chosen $L_{1/2}$; as seen in Sec. 12.2.8, an ambient mixture of $H_2$-air at atmospheric pressure has an induction zone of approximately $2 \times 10^{-4}$ m. And in the more realistic detailed kinetics systems, the main heat release occurs over the coarse length scales. It must thus be recognized that the chosen length scale on which the heat is released is much finer than expected in a realistic physical system; the main reason for this choice is to lessen the stiffness of the system so as to enable a tractable computation of a fully resolved multiscale detonation.

### Inviscid Results

In this section, solutions to the reactive Euler equations are considered. We first present high accuracy shock-fitted solutions described in detail by Henrick et al. (2006). These results are briefly compared to less accurate solutions obtained by the common method of shock capturing. Shock-capturing schemes introduce significant artificial viscous effects into the solution, sometimes masking physical dynamics.

**SHOCK FITTING.** In the shock-fitting method, one explicitly numerically solves for the shock front velocity $D$ as a function of $t$. We adopt the scaling given in detail by Henrick et al. to convert to dimensionless $\hat{D}$ and $\hat{t}$. Using a numerical resolution for which the number of points $N_{1/2}$ within the half reaction zone width is 20, we plot in Fig. 12.14 for $E = 25$, $\hat{q} = q/(P_0/\rho_0) = 50$, $\gamma = 1.2$ the behavior of $\hat{D}(\hat{t})$, for the initial conditions given by Henrick et al. Here, the dimensionless $\hat{D} = D/\sqrt{P_0/\rho_0}$. Thus $\hat{D}_{CJ} = (2167.56\ \text{m/s})/\sqrt{(101325\ \text{Pa})(1\ \text{kg/m}^3)} = 6.80947$. This solution exhibits what appear to be stable damped oscillations. The damping is not due to either physical or numerical viscosity. The numerically predicted long time value of $\hat{D}$ is close to its exact solution. Moreover, it can be shown that as numerical resolution is increased, the long time value of $\hat{D}$ asymptotically approaches the exact value.

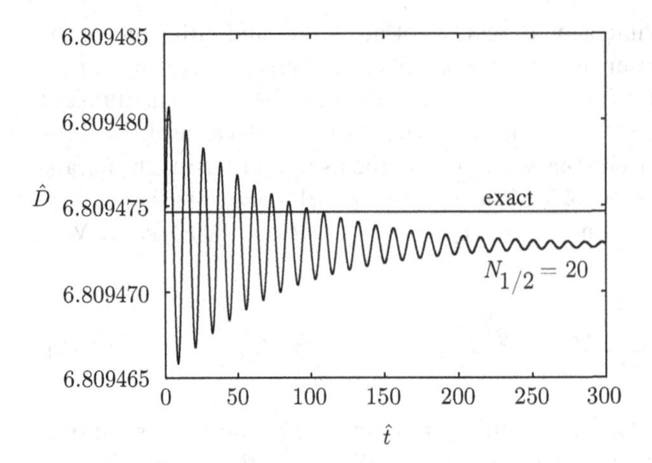


Figure 12.14. Numerically generated detonation velocity, $\hat{D}$ versus $\hat{t}$, using the high-order shock-fitting scheme, $E = 25$, $\hat{q} = 50$, $\gamma = 1.2$, with $N_{1/2} = 20$. From Henrick et al. (2006), reproduced with permission from Elsevier.

If the activation energy is slightly raised, holding all other parameters constant, one predicts different long time behavior. For the slightly higher $E = 26$, we show in Fig. 12.15 the behavior of $\hat{D}(\hat{t})$ and $d\hat{D}/d\hat{t}$ versus $\hat{D}$. The second of these is a type of phase plane. Here, we see that small oscillations in $\hat{D}$ grow and then relax into a long time limit cycle behavior. While linear stability theory has value in describing the early time growth of $\hat{D}$, the long time limit cycle is only available via numerical analysis. We call this a "period 1" limit cycle because in the long time limit, it has no folds in the $(\hat{D}, d\hat{D}/d\hat{t})$ plane. We will soon do a Fourier analysis and find many active modes; their combination is such that the limit cycle appears to have a single dominant frequency.

A slight increase of activation energy to $E = 27.25$ introduces more structure into the long time limit cycle solution as is evident from examining Fig. 12.16. Here, at long times, we see two distinct peaks in the $\hat{D}(\hat{t})$ plot; this is manifested in a fold appearing in the $(\hat{D}, d\hat{D}/d\hat{t})$ phase plane. We call this a "period 2" oscillation.

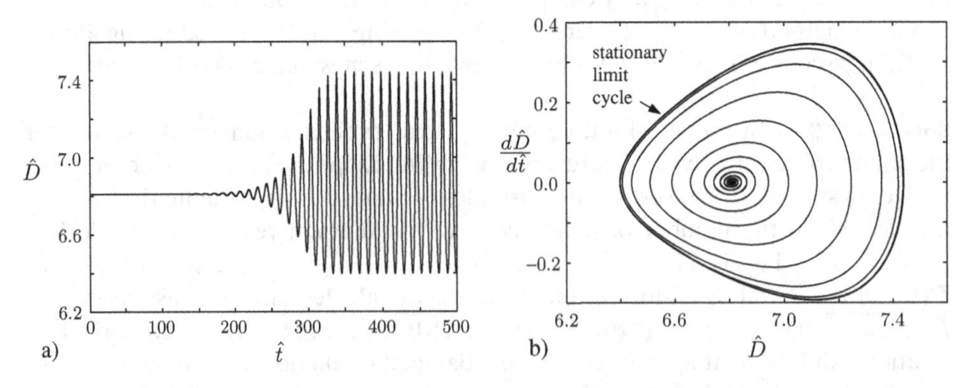


Figure 12.15. Numerically generated (a) detonation velocity, $\hat{D}$ versus $\hat{t}$ and (b) detonation acceleration $d\hat{D}/d\hat{t}$ versus $\hat{D}$, using the high-order shock-fitting scheme, $E = 26$, $\hat{q} = 50$, $\gamma = 1.2$, with $N_{1/2} = 20$. Period 1 oscillations shown. From Henrick et al. (2006), reproduced with permission from Elsevier.

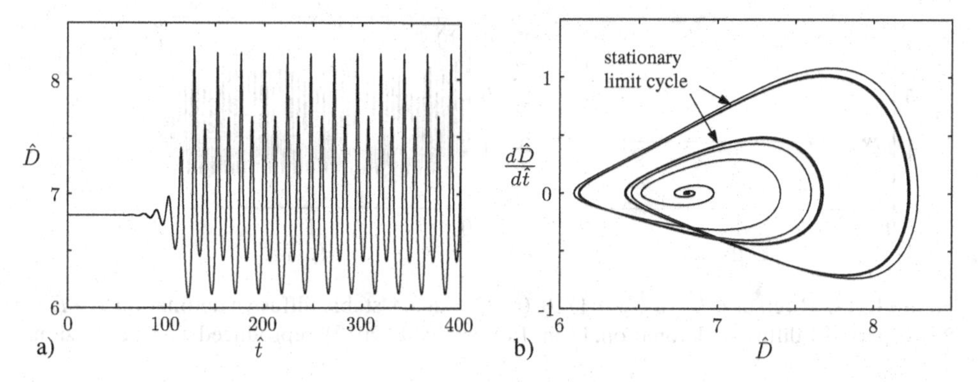


Figure 12.16. Numerically generated (a) detonation velocity, $\hat{D}$ versus $\hat{t}$ and (b) detonation acceleration $d\hat{D}/d\hat{t}$ versus $\hat{D}$, using the high-order shock-fitting scheme, $E = 27.35$, $\hat{q} = 50$, $\gamma = 1.2$, with $N_{1/2} = 20$. Period 2 oscillations shown. From Henrick et al. (2006), reproduced with permission from Elsevier.

**SHOCK CAPTURING.** In addition to using the reactive Euler equations for modeling detonations, the use of shock-capturing techniques and moving reference frames is also common. When using the Euler equations with any shock-capturing technique, a shock moving slowly relative to the numerical grid will have low-frequency numerical perturbations (Quirk, 1994). These low-frequency perturbations can be lessened by refining the grid. For an activation energy of $E = 26.64$, a simple period 1 limit cycle detonation is predicted using shock fitting; using shock capturing with the same resolution, the predicted behavior of a period 1 detonation is in agreement with that of shock fitting with a relative difference of the peak pressure of 2.1%. Increasing the resolution lessens this relative difference as shown in Fig. 12.17a. At $E = 27.82$, shock fitting predicts a period 8 limit cycle detonation, whereas shock capturing, using the higher resolution of 40 points in the half reaction length, predicts a period 4 detonation. This difference can be reconciled by increasing the resolution, demonstrated in Fig. 12.17b. It was found that $N_{1/2} > 80$ was needed in this regime. The resolution requirement to accurately predict the correct dominant frequencies in high-frequency oscillatory detonations may be even more stringent, as seen in Ng et al. This suggests that numerical diffusion is playing an important role in determining

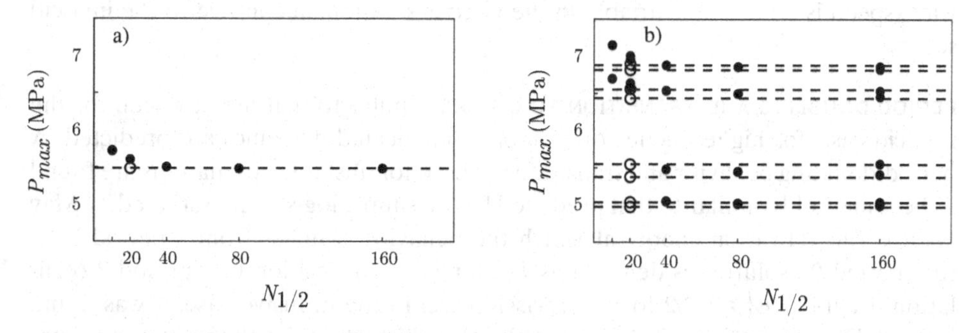


Figure 12.17. Peak inviscid detonation pressures versus $N_{1/2}$ for (a) $E = 26.64$ and (b) $E = 27.82$. Shock-capturing predictions are given by the filled circles and shock-fitting $(N_{1/2} = 20)$ prediction is represented by open circles and dashed lines. From Romick et al. (2012), reproduced with permission.

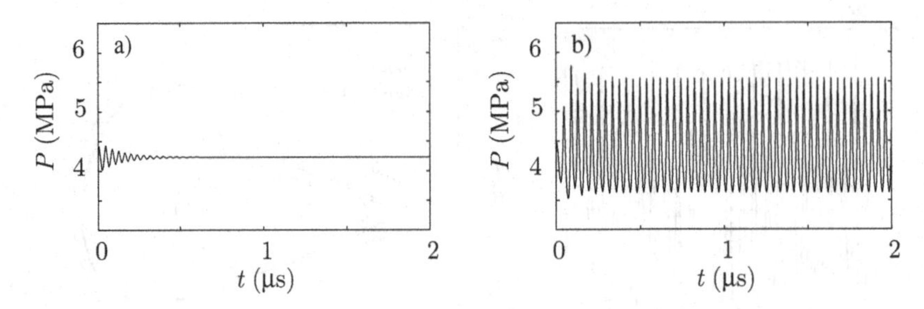


Figure 12.18. $P$ versus $t$, $L_\mu/L_{1/2} = 1/10$, (a) $E = 26.64$, stable diffusive detonation, (b) $E = 29.00$, period 1 diffusive detonation. From Romick et al. (2012), reproduced with permission.

the behavior of the system at lower resolutions for shock-capturing schemes. Relevant discussion to resolution issues is also given by Hwang et al. (2000).

### Viscous Results

The plausible yet erroneous predictions due to the inherent numerical diffusion in the model can be remedied by increasing the resolution. However, for high-frequency instabilities, the necessary resolution tends toward infinity for the inviscid model. A preferable approach is to include explicit physical diffusion and so introduce a cutoff length scale below which physical diffusion properly serves to dampen oscillations.

**STABILITY LIMIT.** In the inviscid case, linear stability analysis by Lee and Stewart (1990) revealed that for $E < 25.26$, the steady ZND wave is linearly stable and is otherwise linearly unstable. The activation energy at this stability boundary is labeled $E_0$. Henrick et al. numerically found the stability limit, for the inviscid case, at $E_0^i = 25.265 \pm 0.005$. Here, a diffusive case well above the inviscid stability limit was examined, $E = 26.64$, that Henrick et al. found to relax to a period 1 limit cycle for an inviscid simulation. In the diffusive simulation, it can be seen from Fig. 12.18a that there is no limit cycle behavior, and the detonation predicted by diffusive theory is in fact a stable steadily propagating wave. The stability boundary for the diffusive case is found at $E_0^d \approx 27.14$. A period 1 limit cycle may be realized in the diffusive case by increasing the activation energy above $E_0^d$; an example is shown in Fig. 12.18b with $E = 29.00$. In the viscous simulations, we plot the local maximum of $P$ near the wavefront versus $t$ because it is easy to identify $P$ from the viscous solution. The wave speed is not a state variable in the viscous solution, in contrast to the inviscid solution.

**PERIOD DOUBLING AND TRANSITION TO CHAOS.** Similar to that already seen for the inviscid case, for higher values of $E$, more complicated dynamics are predicted. A period-doubling behavior and transition to chaos for unstable detonations are found to be remarkably similar to that predicted by the simple logistic map studied by May (1976). The activation energy at which the behavior switches from a period $2^{n-1}$ to a period $2^n$ solution is denoted as $E_n$, for $n \geq 1$. Transition to a period 2 oscillation occurs at $E_1^i \approx 27.2$ for the inviscid case. In the diffusive case, it was found instead $E_1^d \approx 29.32$; Fig. 12.19a shows the time history of the detonation pressure for a higher $E = 29.50$, which shows in the long time limit two distinct relative maxima, $P \approx 6.117$ MPa and $P \approx 5.358$ MPa. Increasing further to $E = 29.98$, another period doubling is realized, and a period 4 oscillating detonation is achieved as seen

Table 12.5. *Numerically Determined Bifurcation Points for Inviscid and Diffusive Detonation, and Approximations to Feigenbaum's Constant*

| $n$ | Inviscid $E_n^i$ | Inviscid $\delta_n^i$ | Diffusive $E_n^d$ | Diffusive $\delta_n^d$ |
|---|---|---|---|---|
| 0 | 25.2650 | – | 27.14 | – |
| 1 | 27.1875 | 3.86 | 29.32 | 3.89 |
| 2 | 27.6850 | 4.26 | 29.88 | 4.67 |
| 3 | 27.8017 | 4.66 | 30.00 | – |
| 4 | 27.82675 | – | – | – |

in Fig. 12.19b. The bifurcation points for both models are listed in Table 12.5 along with approximations for Feigenbaum's constant, $\delta_\infty$:

$$\delta_\infty = \lim_{n\to\infty} \delta_n = \lim_{n\to\infty} \frac{E_n - E_{n-1}}{E_{n+1} - E_n}. \tag{12.325}$$

For an independent nonlinear system, Feigenbaum (1979) predicted $\delta_\infty \approx 4.669201$, and its characterization of how bifurcation points accumulate describes a wide variety of nonlinear dynamical systems, including our detonation dynamics. Diffusive and inviscid models predict $\delta_\infty$ well.

**CHAOS AND ORDER.** Figure 12.20a gives the bifurcation diagram for the inviscid limit using a shock-fitting algorithm with negligible numerical diffusion. Figure 12.20b gives the diffusive analog. It was constructed by sampling 351 points with $E \in [25, 32]$, with a spacing of $\Delta E = 0.02$. Simulations were integrated to $t = 10$ $\mu$s, and relative

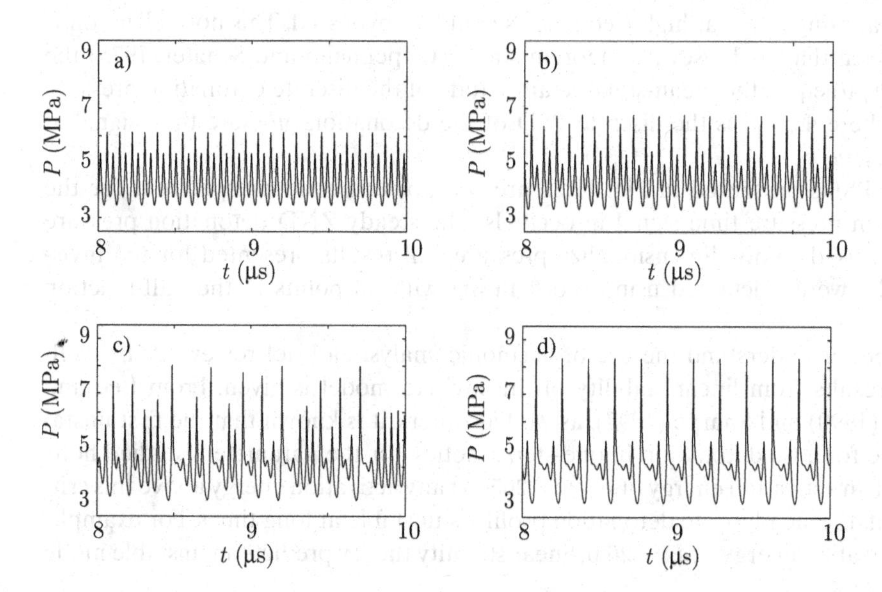


Figure 12.19. $P$ versus $t$ for diffusive detonation with $L_\mu/L_{1/2} = 1/10$: (a) $E = 29.50$, period 2, (b) $E = 29.98$, period 4, (c) $E = 30.74$, chaotic, and (d) $E = 30.86$, period 3. From Romick et al. (2012), reproduced with permission.

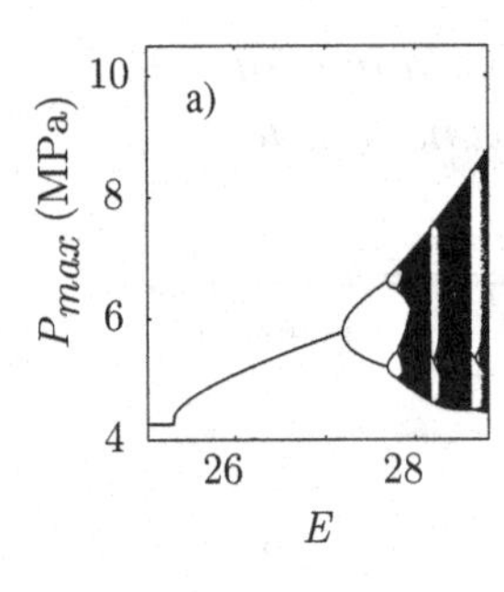


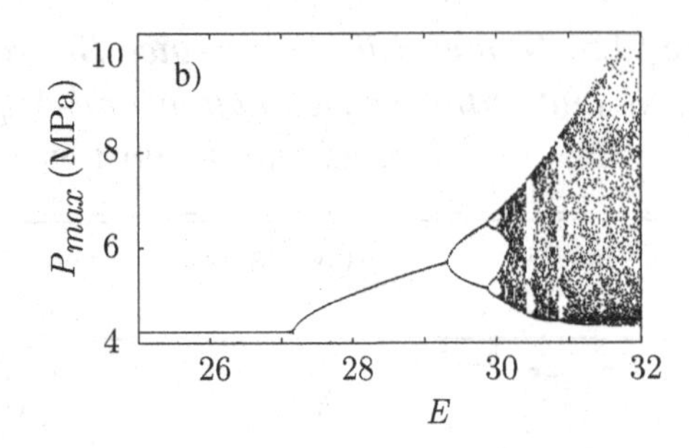


Figure 12.20. Comparison of numerically generated bifurcation diagrams: (a) inviscid detonation with shock fitting and (b) diffusive detonation with $L_\mu/L_{1/2} = 1/10$. From Romick et al. (2012), reproduced with permission.

maxima in $P$ were recorded for $t \geq 7.5$ $\mu$s. In the diffusive case, the period-doubling bifurcations occur up to $E_\infty^d \approx 30.03$. Beyond this point, there exists a region that is densely populated in relative maxima that is most likely a chaotic regime. Increasing the activation energy yet further, one comes to regions with a small number of oscillatory modes with periods of 3, 5, and 6. A chaotic detonation is shown in Fig. 12.19c; at a higher activation energy, a solution with period 3 is found and is shown in Fig. 12.19d.

**POWER SPECTRAL DENSITY.** To further clarify this issue, the power spectral density (PSD) is used.[15] The PSD of a signal describes how the variance (or power) is distributed in frequency, and it is real-valued for any real signal. It can be used to reveal possible periodicities in a complex signal. The PSD is simply defined as the Fourier transformation of the auto-correlation of a signal (Hamilton, 1994; Billinger, 2001). Moreover, it can be written as the magnitude squared of the Fourier transformation of the signal by using the Wiener-Khinchin theorem. For the work here, the discrete one-sided mean-squared amplitude PSD is used. The single-sided PSD is chosen so that the aliasing effect at high frequencies could be bypassed. This normalization is chosen such that, as Parseval's theorem states (Oppenheim and Schafer, 1975) the sum of $\Phi_d$ to equal the mean-squared amplitude of the discrete detonation pressure signal, where $\Phi_d(\nu_k)$ is the discrete PSD of the detonation pressure-time signal at frequency, $\nu_k$.

The PSDs presented in Fig. 12.21 are the power-frequency spectrum for the detonation pressure time signal in decibels. The steady ZND detonation pressure has been used to nondimensionalize pressure. All results presented for the inviscid model were calculated using shock fitting with 40 points in the half-reaction length.

To better understand the use of harmonic analysis, a brief review of the well-known results from linear stability of the inviscid model is given. From Lee and Stewart (1990) and Sharpe (1997), as well as others, it is known that the first unstable mode for a CJ detonation in one-step kinetics, for the parameters studied here, occurs at an activation energy of $E \approx 25.265$. At any activation energy above this critical point, the steady-state detonation profile is unstable at long times. For example, at an activation energy of $E = 26.0$, linear stability theory predicts an unstable mode

[15] This subsection is adapted from a portion of C. M. Romick, 2015, On the effect of diffusion on gaseous detonation, PhD dissertation, University of Notre Dame, reproduced with permission.

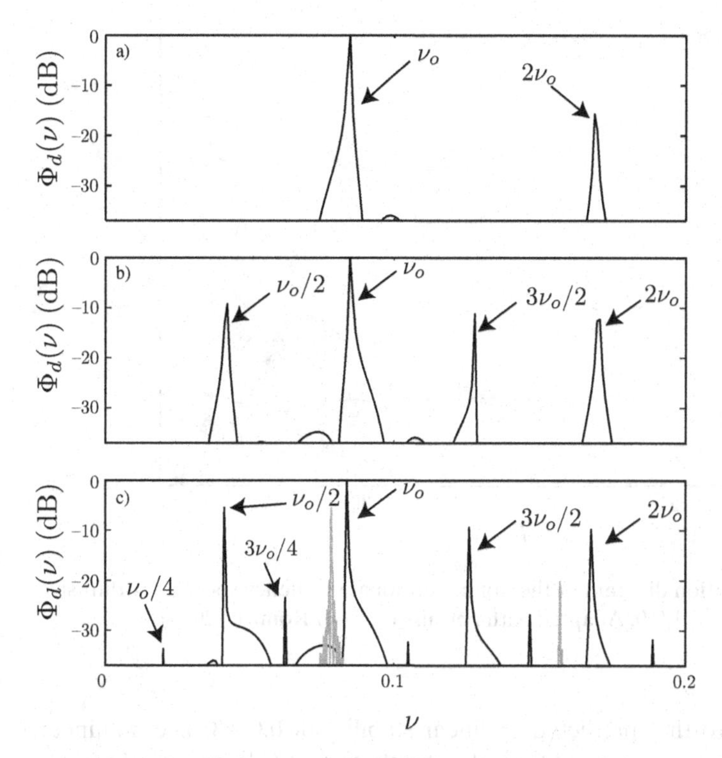


Figure 12.21. The PSD spectra in decibels of the long time behavior for (a) $E = 26.0$, (b) $E = 27.5$, and (c) $E = 27.7$. The inviscid case is indicated by the black line and if present, the viscous case is indicated by the gray line. Adapted with permission from Romick (2015).

at the nondimensional first harmonic frequency of $\nu_0 \approx 0.0879$ (see discussion in Henrick et al., 2006). The frequency has been nondimensionalized by $\sqrt{P_0/(\rho_0 L_{1/2}^2)}$. At early times, the linear stability frequency and growth rates were matched well by Henrick et al. (2006). Furthermore at this activation energy in the inviscid limit, a period 1 detonation is in fact predicted at long times (Ng et al., 2005; Henrick et al., 2006).

Now, looking at the long time behavior at this activation energy in the frequency domain, it is clear from Fig. 12.21a that nearly all of the energy is contained at a first harmonic frequency, $\nu_0 = 0.0849$. This is a relative difference of 3.41% versus the linear stability frequency. This difference is attributed to the saturation of nonlinear effects at long times. Additionally, the harmonics of the fundamental mode also contain energy of the detonation, though they show a power law decrease in energy carried.

Both Ng et al. and Henrick et al. report a subharmonic bifurcation process in the inviscid limit; additionally, Romick et al. (2012) report a similar behavior in the viscous case. This subharmonic bifurcation process is indicated by the appearance of lower frequencies developing as the activation energy is increased. As an example, for an activation energy of $E = 27.5$, a pulsating detonation with two distinct peaks in the detonation pressure-time signal is predicted. The inviscid PSD for this activation energy, which is shown in Fig. 12.21b, demonstrates the appearance of this subharmonic frequency. In fact these subharmonics are indicated by spikes at the odd multiples of $\nu_0/2$. The predicted first harmonic frequency of $\nu_0 = 0.0842$, is

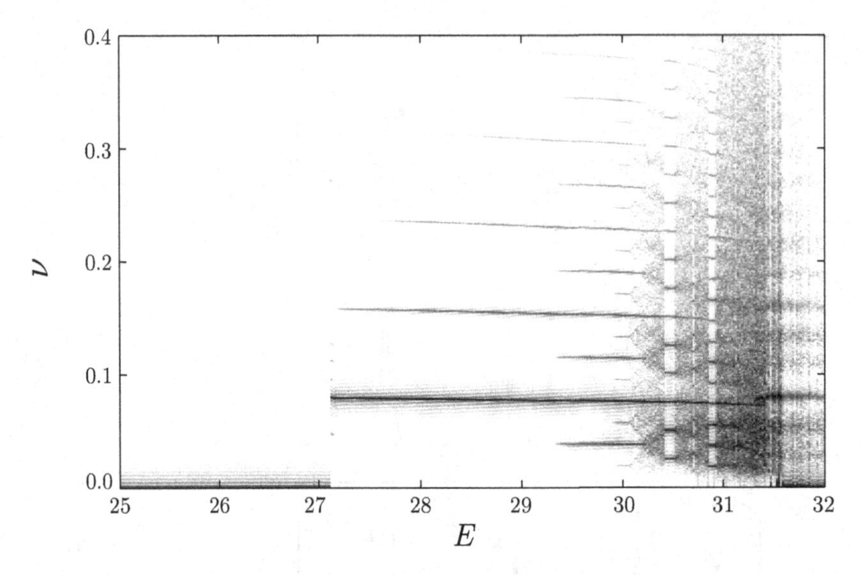


Figure 12.22. The bifurcation diagram of the nondimensional frequency spectra for diffusive detonation with $L_\mu/L_{1/2} = 1/10$. Adapted with permission from Romick (2015).

6.23% larger relative to that predicted by linear stability of 0.0793. The nonlinear effect on the frequency has increased from the strictly period 1 detonation. Furthermore, it is manifested in the multimode nature of the detonation. As a brief aside, Massa et al. (2013), who used an eigenvalue decomposition of perturbations to a multidimensional detonation wave in the one-step model, found that the least stable perturbations occurred near this first subharmonic, suggesting that even in multiple dimensions the dominant mechanism is similar to that seen in galloping case of one dimension.

At a slightly higher activation, $E = 27.7$, a period 4 detonation is predicted in the inviscid limit. This transition is indicated in the PSD spectrum by the appearance of a second subharmonic group as shown in Fig. 12.21c. This second set of subharmonics occur at the odd multiples of $\nu_0/4$. Furthermore, the first set of subharmonics now carries a more appreciable ratio of the energy.

The addition of physical viscosity alters the long time behavior at a given activation energy. In addition to the changes in the time domain, there can be significant modifications of the behavior in frequency domain. At this activation energy of $E = 27.7$, the viscous PSD does not indicate any of the subharmonics predicted in the inviscid limit. Additionally, there is a shift in the first harmonic frequency from 0.0839 in the inviscid case to 0.0787 in the viscous analog as shown in Fig. 12.21c. The addition of viscosity shifts the dominant frequency much closer to the prediction from linear stability theory of 0.0786. Moreover, there has also been a significant reduction in the amplitude of the peak in the spectrum. Note that here instead of the traditional definition of decibels $[10 \times \log_{10}(PSD/\max(PSD))]$, both spectra shown in Fig. 12.21c have been scaled by the maximum of the inviscid spectrum so the magnitude of the different spectra would not be lost.

Figure 12.22 shows how the nondimensional frequency spectra evolve versus the activation energy for the viscous case studied for $L_\mu/L_{1/2} = 1/10$. For detonation pressure time signals with deviations larger than 0.04 atm from the mean, the mean detonation pressure is subtracted for calculation of the PSDs. The strength of peaks are indicated by the shade of gray; the stronger the peak the darker the shade of gray is. The transition from a stable to an unstable detonation is indicated by the jump in the main frequency from 0 to approximately 0.0801. Within the region where the detonation pressure time signal has a single local maxima, the first harmonic frequency shifts from 0.801 to 0.0771. That is the same trend that is predicted by linear stability theory. Moreover, as the activation energy increases, so does the energy ratio of the pulsations that higher harmonics of the fundamental frequency carry; this is indicated by the appearance of lines at $2\nu_0, 3\nu_0$, and higher multiples.

As mentioned previously, the second bifurcation is indicated by the appearance of subharmonic frequencies. That occurs at $E = 29.32$ in the viscous case studied here. In fact, the subharmonic frequency appears at $\nu_0/2 = 0.0382$. Furthermore, at the further bifurcations additional subharmonic frequencies begin to appear, at $\nu_0/4$ and $\nu_0/8$. Additionally, the first harmonic frequency continues to shift to slightly lower frequencies; thus at $E = 30.00$, $\nu_0 = 0.0756$.

Within in this region of likely chaotically propagating pulsating detonation, there are pockets of order. The largest of these regions are for period 3 detonations, one pocket of order occurs near $E = 30.42$, and a second is near $E = 30.86$. These regions are indicated in the frequency spectra by the peaks at multiples of $\nu_0/3$ and $2\nu_0/3$. At the highest activation energies, there are many active modes though the strongest mode is near the zero frequency, which indicates detonation failure. This suggests that eventually the detonation becomes an uncoupled lead shock and a trailing reaction wave.

To conclude this section, investigation of the one-step kinetic model of unsteady one-dimensional detonation with mass, momentum, and energy diffusion shows that the dynamics can be nontrivially influenced in the region of instability relative to its inviscid counterpart. As in the inviscid limit, bifurcation and transition to chaos are predicted. Certainly, the addition of diffusion delays the onset of instability. The inviscid approximation indeed gives a good approximation when in the stable or weakly unstable regimes. However, for activation energies large enough to induce complicated limit cycle behavior, ordinary shock-capturing methods applied to inviscid models using an underresolved grid can fail to capture the correct long time dynamics.

We will see in the next section, for which a physically realistic hydrogen-air model is employed, that similar qualitative conclusions may be drawn. But because the relative influence of diffusion will be less, the differences in predictions of the theory with and without diffusion will not differ as dramatically.

### 12.3.2 Detailed Kinetics, With and Without Diffusion

We finish by considering a model that employs our detailed chemical kinetics model along with a calorically imperfect ideal gas model.[16] We study the model with and without mass, momentum, and energy diffusion. In contrast to the previous Sec. 12.3.1, and because we believe it is viscous solutions that are more interesting, we first present the viscous solutions and later briefly compare to inviscid results.

[16]This section is adapted largely from C. M. Romick, T. D. Aslam, and J. M. Powers, 2015, Verified and validated calculation of unsteady dynamics of viscous hydrogen-air detonations, *Journal of Fluid Mechanics*, 769: 154–181, reproduced with permission.

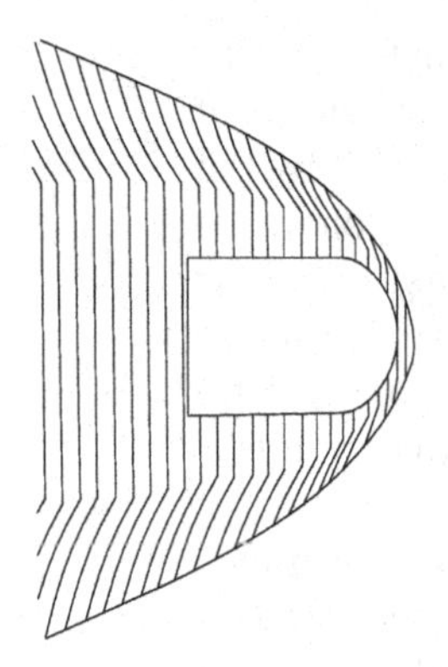

Figure 12.23. Sketch of 1.04 MHz detonation instability in hydrogen-air mixture initially at 0.421 atm, 293.15 K, observed by Lehr (1972). Sketch adapted from Lehr (1972), with permission.

The model is that presented in Chapter 7, not repeated here, for a hydrogen-air mixture. Both its complexity and mulitscale nature render it a challenging model to solve numerically. The solutions presented will be so-called direct numerical solutions, with no kinetic model reduction or other filtering applied.

### Validation and Verification

The exercises of validation, a task focusing on how well numerical predictions agree with experiment, and verification, a task focusing on if numerical approximations are consistent with the underlying continuum model are always useful. Background is given by Roache (1998) and Oberkampf and Roy (2010). As our model has been restricted to one dimension, there are limited means of validation. However, in experiments of shock-induced combustion flow around spherical projectiles in a hydrogen-air mixture initially at 0.421 atm and 293.15 K, Lehr (1972) observed longitudinal oscillations, similar to the sketch of Fig. 12.23. For an inflow condition corresponding to an overdrive $f \approx 1.10$, Lehr observed a frequency of $\nu_0 = 1.04$ MHz, where the overdrive $f = (D/D_{CJ})^2$. Starting the viscous calculation with the inviscid steady state profile with a superimposed smooth transition from the shocked state to the ambient condition over $5 \times 10^{-4}$ cm, a frequency of $\nu_0 = 0.97$ MHz is predicted. Thus, it seems that the instability observed by Lehr in multiple dimensions is captured well by a one-dimensional model. The predicted frequency here is only 6.7% different from that measured by Lehr; the discrepancy is likely due the one-dimensional assumption and uncertainty in chemical kinetic parameters.

The numerical method chosen, a wavelet-based method (Paolucci et al., 2014a, 2014b), is a self-converging method; that means that as a user-selected error–threshold parameter, $\epsilon$, is reduced, the overall error is reduced. The wavelet-based method is especially good at simultaneously capturing coarse and fine-scale solution features in a computationally efficient manner. To verify that in fact the procedure is convergent regime, several values of $\epsilon$ are examined for $\overline{u_p} = 1.500 \times 10^5$ cm/s. The long time behavior at this supporting piston velocity is a stable, steadily traveling detonation. Figure 12.24a shows the long time detonation pressure versus $\epsilon$. The detonation pressure is converging to 36.68 atm. The standard deviations of the detonation pressure are indicated by the vertical lines. As the error–threshold parameter is reduced, the standard deviation is reduced around the detonation pressure point indicated by the dots. In fact, at the two most accurate solutions, the standard deviation in the detonation pressure is difficult to identify. Additionally, the difference in the long time detonation pressure is calculated from the most accurate solution; this

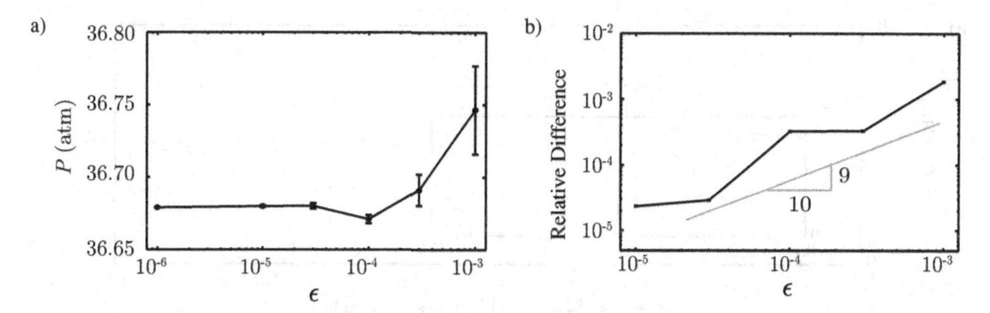


Figure 12.24. (a) Detonation pressure versus $\epsilon$ and (b) difference in detonation pressure between $\epsilon = 1 \times 10^{-6}$ and $\epsilon$ for $\overline{u_p} = 1.500 \times 10^5$ cm/s. Adapted with permission from Romick et al. (2015).

is shown in Fig. 12.24b. As the error threshold parameter is reduced the difference decreases near $O\left(\epsilon^{0.9}\right)$, as indicated on the log-log plot. Furthermore, the largest percent difference is a 0.2% giving a good indication that the numerical method is in the convergent regime.

### Behavior as Overdrive Is Varied

In this section, a study of the long time behavior of the propagating detonation is performed as the final supporting piston velocity, $\overline{u_p}$, is varied. This is done first in the time domain, and then harmonic analysis is used to examine the active frequencies of the pulsating detonations. Several comparisons between the viscous and inviscid calculations are performed in both the time and frequency domains. A series of one-dimensional, piston-driven flows of an initially stoichiometric mixture of hydrogen-air ($2H_2 + O_2 + 3.76N_2$) at ambient conditions of 293.15 K and 1 atm is considered. The detailed kinetics mechanism employed is that of Table 1.2.

**STABLE DETONATIONS.** For sufficiently high $\overline{u_p}$, a steadily traveling detonation arises and persists at long times. The detonation pressure versus time curve for a stable detonation, at $\overline{u_p} = 1.500 \times 10^5$ cm/s, is shown in Fig. 12.25. By $10 \times 10^{-6}$ s the detonation relaxes to a steadily traveling piston-supported detonation traveling to the right at $2.244 \times 10^5$ cm/s. Spatial pressure profiles after the detonation relaxes to the stable detonation are shown in Fig. 12.26a for $t = 10 \times 10^{-6}$ s, $t = 35 \times 10^{-6}$ s and $t = 60 \times 10^{-6}$ s. The later time profiles have been shifted in space using the steady wave speed. There are only minuscule differences between the front locations;

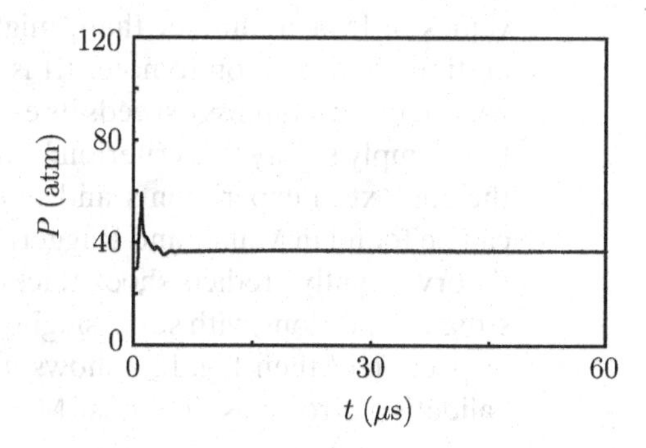


Figure 12.25. Detonation pressure versus time curve $\overline{u_p} = 1.500 \times 10^5$ cm/s for stable hydrogen-air detonation. Adapted with permission from Romick et al. (2015).

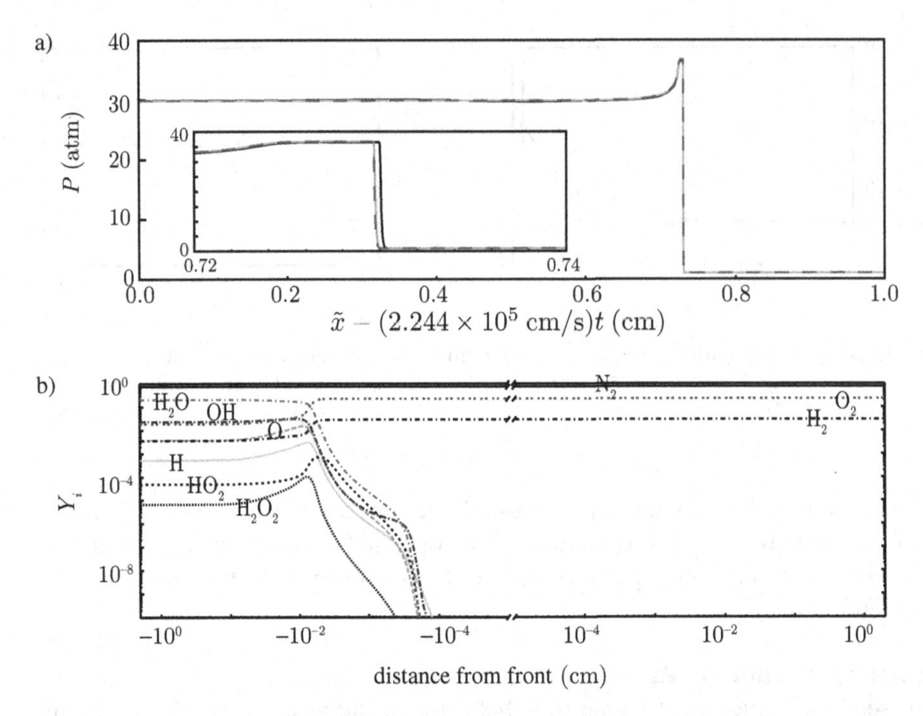


Figure 12.26. (a) Several snapshots of spatial pressure profile (solid black line, $10 \times 10^{-6}$ s; solid light gray line, $35 \times 10^{-6}$ s and dashed gray line, $60 \times 10^{-6}$ s) and (b) typical spatial profile of mass fractions at a $\overline{u_p} = 1.500 \times 10^5$ cm/s. Adapted with permission from Romick et al. (2015).

these differences are more clearly shown in the insert. However, the largest difference between the front locations is still only $2.5 \times 10^{-4}$ cm. Figure 12.26b shows the spatial mass fraction at $t = 50 \times 10^{-6}$ s that is representative of the steadily traveling detonation front. As a particle passes through the detonation, it first encounters a thin viscous shock accompanied by rapid pressure and temperature rise. Then, its pressure and temperature remain relatively constant as it traverses a short induction zone. In this zone, radicals are generated. When a sufficient number of radicals are present, the fluid particle enters a thin zone in which vigorous reaction commences. Here, pressure and temperature vary rapidly. Finally, it passes into a thick relaxation zone, where all state variables equilibrate.

It is often argued that one cannot use a continuum theory to model a shock wave in gases. However, as noted by Vincenti and Kruger (1965, p. 415), "comparisons with experiment show that the Navier-Stokes solution is accurate for larger values of [Mach number than] might be expected from purely theoretical considerations." They go on to note, "It is sometimes said that the test of a good theory is whether its usefulness exceeds its expected range of validity; the Navier-Stokes equations amply satisfy this criterion." An extensive discussion of viscous shock waves in the context of experiments, and supporting continuum and noncontinuum theories can be found in Müller and Ruggeri (1998), where it is demonstrated that continuum theory actually predicts shock thickness well for an unexpectedly large range of free stream conditions, with surprisingly good agreement achieved for $1 < \mathsf{M} < 11$. Visual inspection of their Fig. 12.2 shows the correct trends as M is varied, and a maximum validation error of $\sim 20\%$ near $\mathsf{M} = 4$.

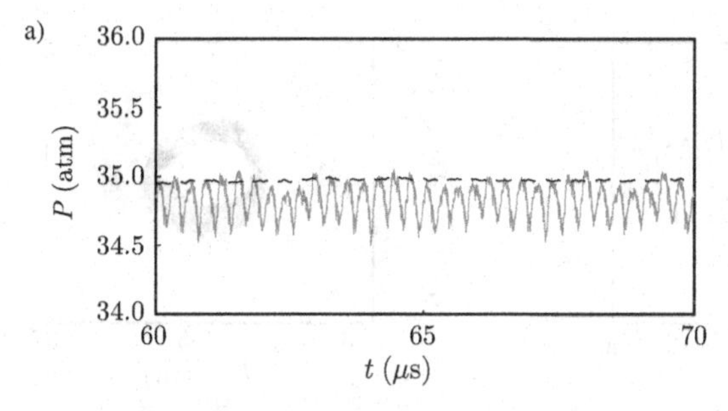


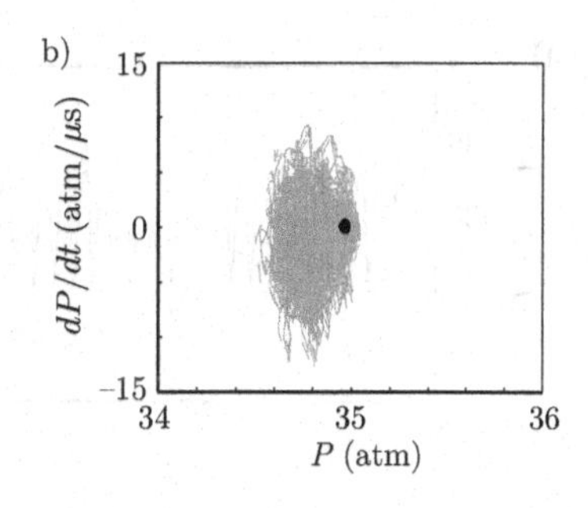


Figure 12.27. (a) Detonation pressure versus time and (b) phase space plot for both $\overline{u_p} = 1.420 \times 10^5$ cm/s (dashed black line) and $\overline{u_p} = 1.410 \times 10^5$ cm/s (solid gray line). The phase space plot for $\overline{u_p} = 1.420 \times 10^5$ cm/s has been enlarged by 10 times. Adapted with permission from Romick et al. (2015).

**HIGH-FREQUENCY PULSATING DETONATIONS.** After $\overline{u_p}$ is lowered below a critical value, the long time behavior of the propagating detonation undergoes a transition from a steadily traveling wave to a pulsating detonation. This transition occurs between $\overline{u_p} = 1.420 \times 10^5$ cm/s and $\overline{u_p} = 1.410 \times 10^5$ cm/s. Figure 12.27a shows the detonation pressure versus time curves for a supporting piston velocity just above and just below the transition point. These pulsations are caused by the slight detachment between the pressure wave and the reaction wave, which in turn elongates the induction zone. The phase space plot for both the stable and unstable case is shown in Fig. 12.27b. For the stable case, the phase space plot is a black dot located at $P = 34.95$ atm, the dot is enlarged by 10 times for ease of viewing. At $\overline{u_p} = 1.410 \times 10^5$ cm/s, it becomes clear that detonation is pulsating; however, it is difficult to discern whether it is near cyclic from the phase space plot.

In contrast to the clear periodic limit cycles predicted by Henrick et al. for the simple one-step model in the CJ limit, the pulsating detonations here do not produce nearly as smooth limit cycles. This is likely influenced by several factors. First, the piston-driven flows in this study are overdriven in nature; as such, the positively moving characteristic waves travel through different decaying N-waves in the negatively moving characteristic field emanating from the detonation front. The likelihood of these positively moving characteristic waves and decaying negatively moving characteristic N-waves being synchronized is extremely low, and thus precludes precisely periodic cycles. These positively moving characteristics in the overdriven case clearly reach the detonation shock front. In the CJ case, there is a sonic locus that remains a finite distance behind the front. As demonstrated by Kasimov and Stewart (2004), for the one-step model, this sonic locus acts as an information barrier. It only allows characteristics in front of it to propagate toward the front. Additionally, the one-step model has only a single length scale of reaction, whereas the detailed hydrogen-air mechanism has reaction length scales that span several orders of magnitude. Furthermore, the one-step model is irreversible, while the detailed kinetics has reversible reactions.

As shown in Fig. 12.28a, when $\overline{u_p}$ is lowered further below the bifurcation point, the oscillations grow in amplitude. Additionally, the frequency shifts toward lower frequencies. As the pulsations become larger in amplitude, it becomes clearer that they are nearly periodic as demonstrated by successive pulsations nearly coinciding in the phase plot of Fig. 12.28b.

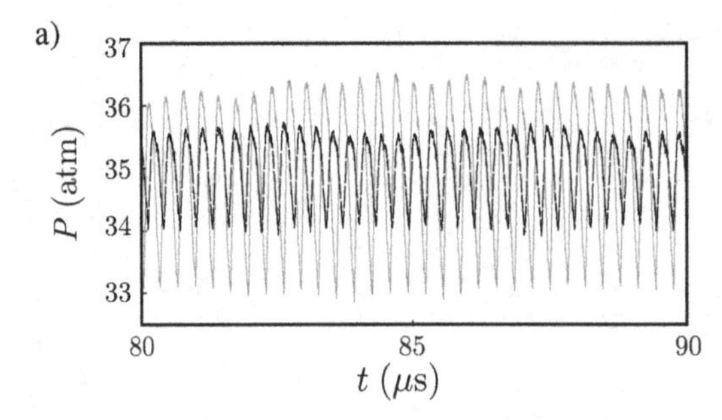


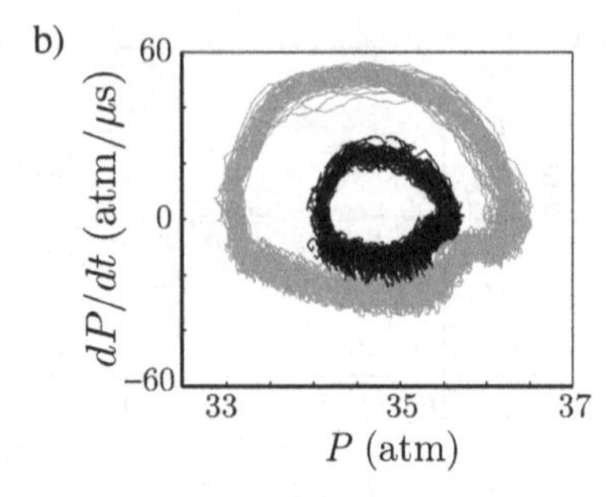


Figure 12.28. (a) Detonation pressure versus time and (b) phase space plot for both $\overline{u_p} = 1.400 \times 10^5$ cm/s (dashed black line) and $\overline{u_p} = 1.370 \times 10^5$ cm/s (solid gray line). Adapted with permission from Romick et al. (2015).

**MULTIPLE MODE PULSATING DETONATIONS.** The behavior becomes even more complex at lower supporting piston velocities with a dual mode behavior arising below a second bifurcation point. An example of this type of propagating detonation is shown in Fig. 12.29a for $\overline{u_p} = 1.310 \times 10^5$ cm/s. It is apparent that the dual mode behavior persists at long times. Although these dual mode detonations do not repeat in a clean limit cycle, it is still obvious that it is stably bounded at long times that is demonstrated by the phase plane plot shown in Fig. 12.29b.

**LOW-FREQUENCY-DOMINATED PULSATING DETONATIONS AND CHAOS.** At yet even lower supporting piston velocities, this dual mode behavior relaxes into a mode that is dominated by a single low-frequency pulsating flow. Figure 12.30a shows this relaxation to a nearly periodic limit cycle at long times. However, the phase space plot, shown in Fig. 12.30b, indicates that even at long times there is still some variation in the cycle. Once this low-frequency mode becomes the dominant mode, a behavior similar to period doubling is predicted. As shown in Fig. 12.30c, a nearly period 2 detonation is predicted at $\overline{u_p} = 1.230 \times 10^5$ cm/s. At this supporting piston velocity, the relative maxima can be grouped into two distinct groups; the first at $P_1 = 47.56 \pm 0.68$ atm and the second being $P_2 = 50.9 \pm 0.84$ atm. Figure 12.30d clearly exhibits the distinct two-lobe phase space for a period 2 detonation. This phenomenon is exhibited even more prominently at $\overline{u_p} = 1.220 \times 10^5$ cm/s, as shown in Fig. 12.30e. However, the higher relative maxima is more erratic as indicated by the wider spread in right-most lobe shown in the phase space plot of Fig. 12.30f.

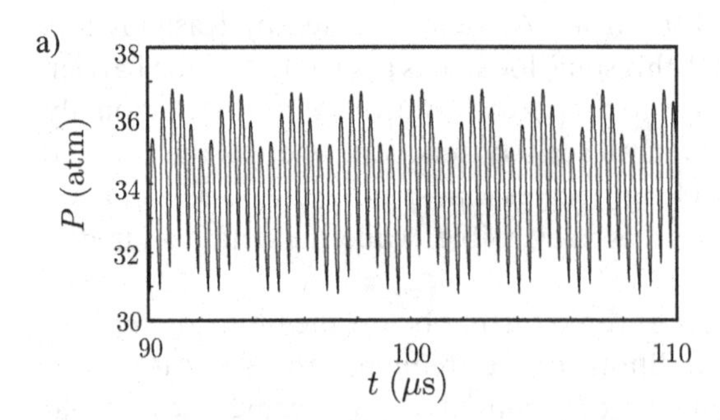


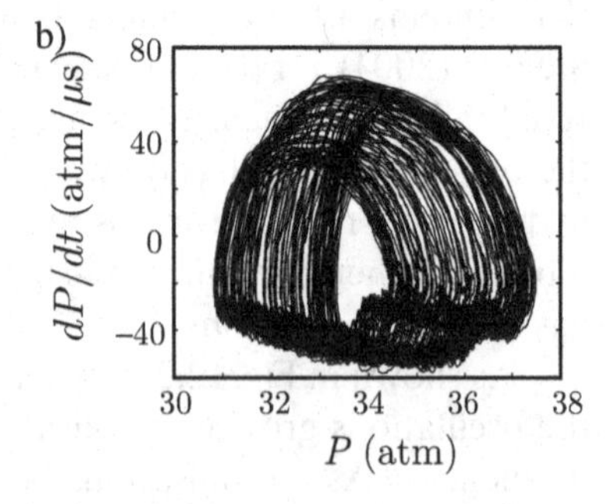


Figure 12.29. (a) Detonation pressure versus time and (b) phase space plot for a $\overline{u_p} = 1.310 \times 10^5$ cm/s. Adapted with permission from Romick et al. (2015).

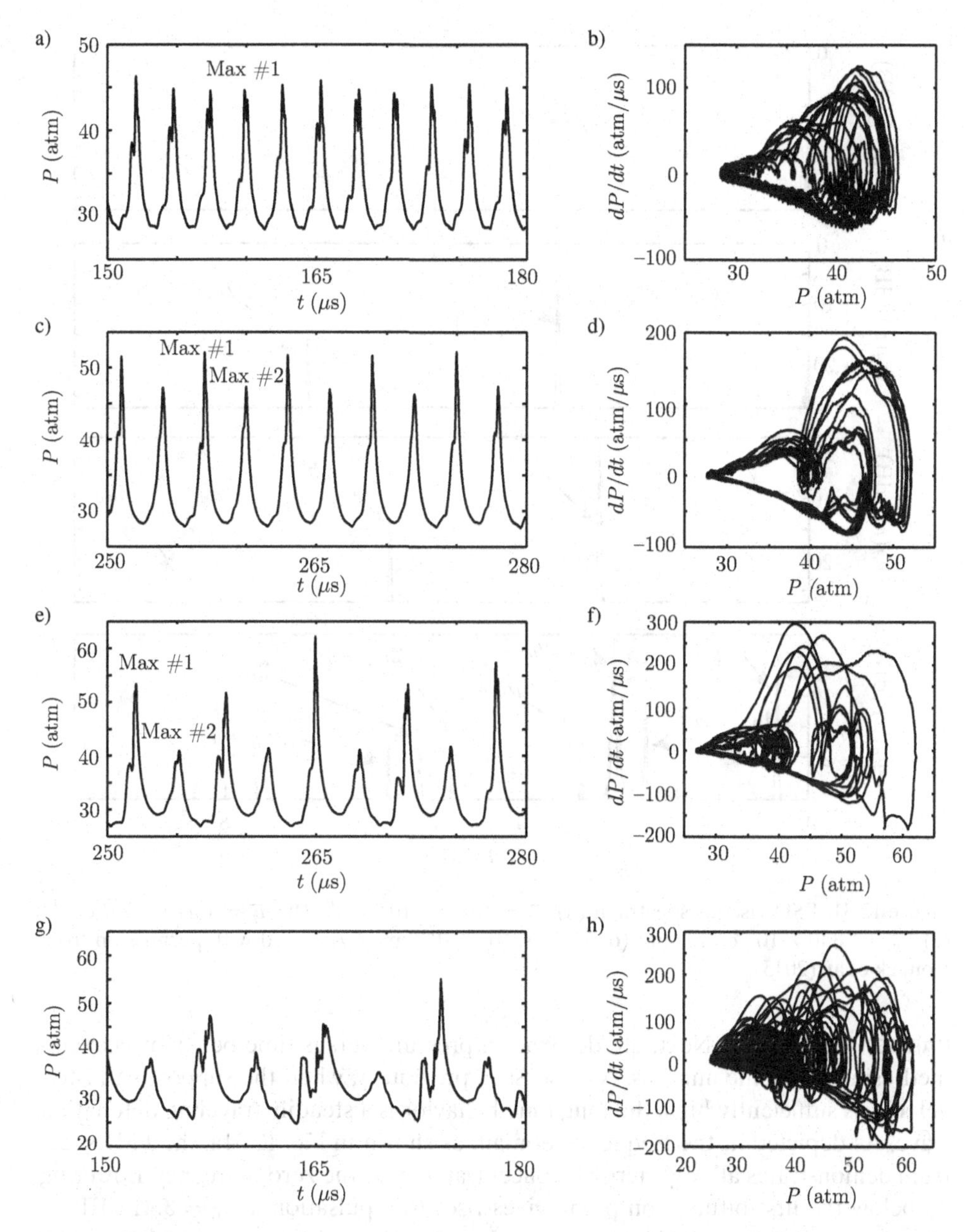


Figure 12.30. Detonation pressure versus time and phase plots for (a, b) $\overline{u_p} = 1.250 \times 10^5$ cm/s, (c, d) $\overline{u_p} = 1.230 \times 10^5$ cm/s, (e, f) $\overline{u_p} = 1.220 \times 10^5$ cm/s, and (g, h) $\overline{u_p} = 1.200 \times 10^5$ cm/s. Adapted with permission from Romick et al. (2015).

This period-doubling behavior is more clearly seen in the frequency domain, to be discussed. After this period-doubling regime, the detonation pressure versus time curve exhibits many more relative maxima; that is shown in Fig. 12.30g for $\overline{u_p} = 1.200 \times 10^5$ cm/s. This is further elucidated by examining the phase space plot, shown in Fig. 12.30h, where no consistent cycle is visible. The system likely underwent a transition to chaos. However, to definitively categorize the system as chaotic further analysis would be needed.

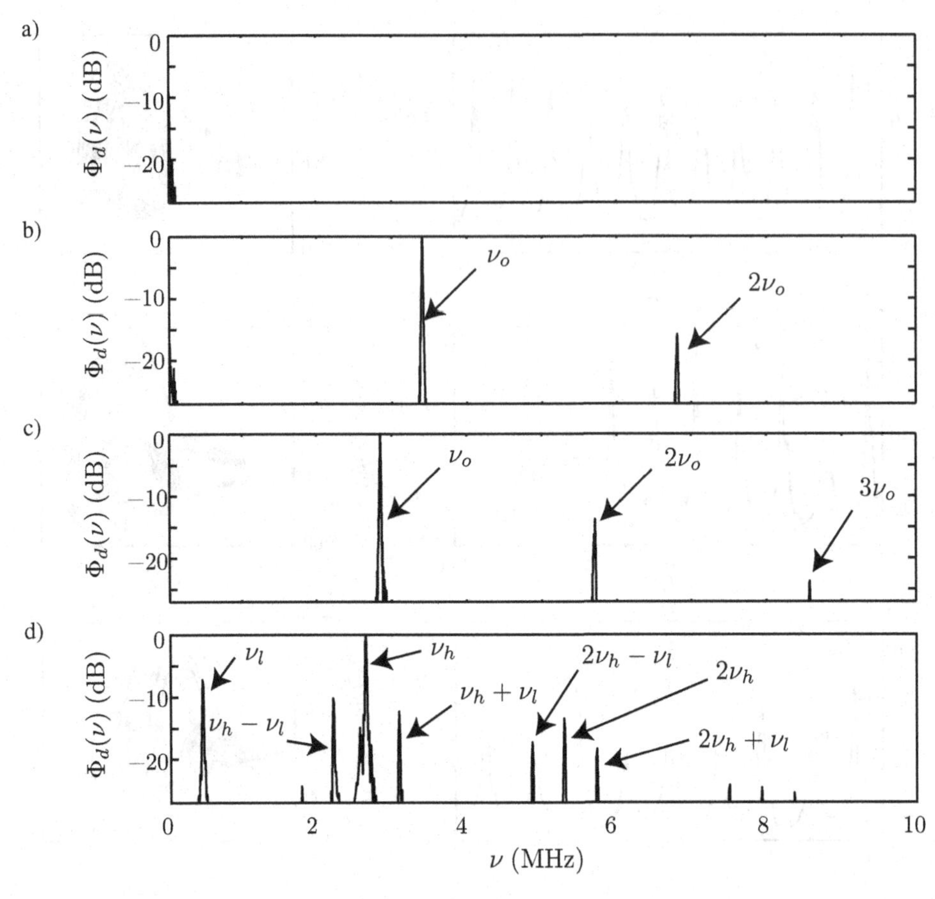


Figure 12.31. PSD viscous spectra at (a) $\overline{u_p} = 1.500 \times 10^5$ cm/s, (b) $\overline{u_p} = 1.410 \times 10^5$ cm/s, (c) $\overline{u_p} = 1.340 \times 10^5$ cm/s, and (d) $\overline{u_p} = 1.310 \times 10^5$ cm/s. Adapted with permission from Romick et al. (2015).

**HARMONIC ANALYSIS.** Next, the detonation pressure versus time behavior is examined using harmonic analysis. As discussed previously, when the supporting piston velocity is sufficiently high, the long time behavior is a steadily traveling detonation wave. As depicted in the frequency domain as shown in Fig. 12.31a, the PSD spectrum demonstrates all the energy is concentrated near the zero frequency. Lowering $\overline{u_p}$ below the first bifurcation point, gives rises to a pulsation at $\nu_0 = 3.41$ MHz at $\overline{u_p} = 1.410 \times 10^5$ cm/s. This is a higher value of $\nu_0$ than that reported earlier for the validation with Lehr's experiments for which the overdrive, ambient pressure, and ambient temperature were different. In Fig. 12.31b it is clear that the majority of the pulsation energy is carried at a single frequency. However, the second harmonic frequency also carries energy; this results in slight differences in the relative maxima in detonation pressure in cycle. As the supporting piston velocity is lowered, the frequency spectrum blue shifts. At $\overline{u_p} = 1.340 \times 10^5$ cm/s, shown in Fig. 12.31c, the first harmonic frequency is now located at $\nu_0 = 2.85$ MHz, and the harmonics have shifted as well. In fact the ratio of the amount of energy being concentrated at higher harmonics has increased. That is demonstrated by the appearance of the third harmonic in the plot. Examining a $\overline{u_p}$ further below the neutral stability boundary, it becomes clear there is a low-frequency mode that is now playing an important

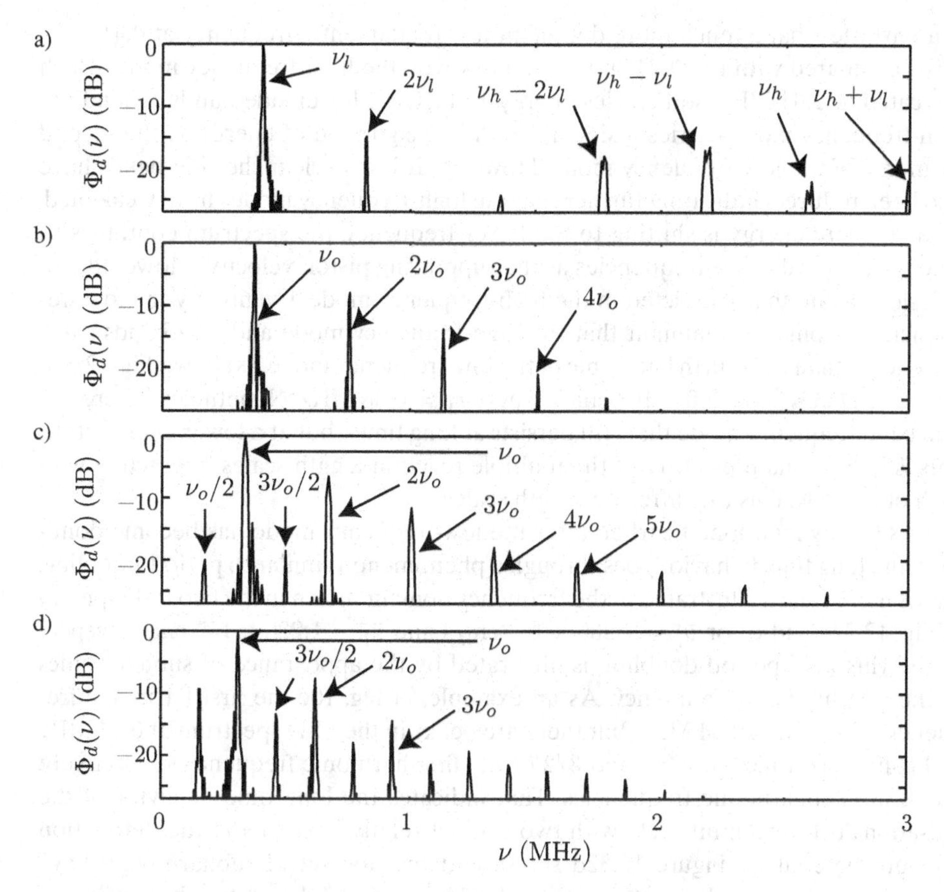


Figure 12.32. PSD viscous spectra at (a) $\overline{u_p} = 1.290 \times 10^5$ cm/s, (b) $\overline{u_p} = 1.260 \times 10^5$ cm/s, (c) $\overline{u_p} = 1.230 \times 10^5$ cm/s, and (d) $\overline{u_p} = 1.220 \times 10^5$ cm/s. Adapted with permission from Romick et al. (2015).

role in the long time behavior of the pulsations as shown for $\overline{u_p} = 1.310 \times 10^5$ cm/s in Fig. 12.31d. At this $\overline{u_p}$, the low-frequency mode occurs at $\nu_l = 0.44$ MHz and carries a significant amount of energy. However, the high-frequency mode, which occurs at $\nu_h = 2.65$ MHz, is still the dominant mode. These modes remain stationary at a smaller error–threshold parameter, once again confirming the numerical method is capturing the long time behavior. Furthermore in this regime where there are two dominant modes; the modes interact giving rise to many more modes that carry energy.

This interaction of the two dominant modes gives rise to a modulation instability. This modulation instability phenomenon occurs in many other physical systems due to the inherent nonlinearity of the physical world (Zakharov and Ostrovsky, 2009). In these pulsating detonations it manifests itself as active modes, called side bands, at multiples of the low frequency around the high frequency and its harmonics. These active modes surrounding the high-frequency mode and its harmonics form envelope waves that persist at long times.

After the appearance of the dual mode behavior, the ratio of the energy present in the pulsation carried at the high fundamental frequency continues to decrease as $\overline{u_p}$ is lowered further. Figure 12.32a shows the PSD for a $\overline{u_p} = 1.290 \times 10^5$ cm/s and demonstrates that a much more dominant low fundamental frequency at 0.41 MHz exists compared with that of Fig. 12.31d. However, the high-frequency mode, which is located at 2.64 MHz, still carries energy. In fact, the lower side bands around the high-frequency mode carries a similar order of magnitude of energy as the second harmonic of the low-frequency mode. However, it is also clear the side bands have also been reduced, indicating further that the high-frequency modes have weakened. Even as more energy is shifting to the lower frequency, the spectrum continues to blue shift toward lower frequencies as the supporting piston velocity is lowered, but at a slower rate than predicted in the high-frequency mode. Eventually the low frequencies become so dominant that the high-frequency mode and side bands carry less energy than the fourth harmonic of the low-frequency mode, as shown Fig. 12.32b by the PSD at $\overline{u_p} = 1.260 \times 10^5$ cm/s, where $\nu_0 = 0.38$ MHz. Nonetheless, there is a side band frequency mode that still persists at long times, but at a lower energy state. This is likely a manifestation of the multiple reaction length scales interacting with each other as well as the diffusion length scales.

As briefly mentioned earlier, after the low-frequency mode has become dominant, the long time behavior goes through a phenomenon similar to period-doubling. This is more clearly illustrated in the frequency domain shown in the two PSD spectra in Fig. 12.32c and d for $\overline{u_p} = 1.230 \times 10^5$ cm/s and $\overline{u_p} = 1.220 \times 10^5$ cm/s, respectively. This near period doubling is illustrated by the appearance of subharmonics of the first harmonic frequency. As an example, in Fig. 10c the first harmonic frequency is located at 0.34 MHz, but there are peaks in the PSD spectrum at 0.17 MHz and 0.49 MHz, which are 1/2 and 3/2 of the first harmonic frequency, respectively. These are subharmonic frequencies. That indicates the long time behavior of the pulsations is near a limit cycle with two distinct relative maxima in the detonation pressure time curve. Figure 12.32d shows that the first set of subharmonics have grown in amplitude indicating the strength of the period 2 detonation has grown.

**BIFURCATION DIAGRAM.** A bifurcation diagram is constructed showing the various propagation modes. It has been created with 31 supporting piston velocities spaced at $1.0 \times 10^3$ cm/s and as such is a coarse approximation of the full diagram. Figure 12.33a shows how the maximum detonation pressure evolves versus the supporting piston velocity. The peak detonation pressure has been scaled by the average detonation pressure. As the peak detonation pressure varies from cycle to cycle, the standard deviation of peaks in the stable, high-frequency-dominated modes, and low-frequency-dominated modes are indicated by vertical lines. In both the stable and high-frequency-dominated modes, the standard deviations are difficult to distinguish from the peak detonation pressure. The region in which there are two active modes is indicated by the dense number of points near $\overline{u_p} = 1.300 \times 10^5$ cm/s; likewise, the dense region near $\overline{u_p} = 1.200 \times 10^5$ cm/s is indicative of a detonation with many active modes, which is likely chaotic. This is more clearly understood by looking at the bifurcation plot in of active frequencies, shown in Fig 12.33b, in which the shade of the points indicates the magnitude with the darkest being the most dominant mode and the lightest being the weakest. In the high-frequency mode, there are three active frequencies: the first harmonic frequency, the second harmonic, and the third harmonic. The blue shift of the frequency spectrum is most clearly seen in the third harmonic. In the dual mode region, it is apparent that side banding occurs near the high-frequency mode and its harmonics; however, there are still just two dominant modes. The side banding continues in the low-frequency mode, but at

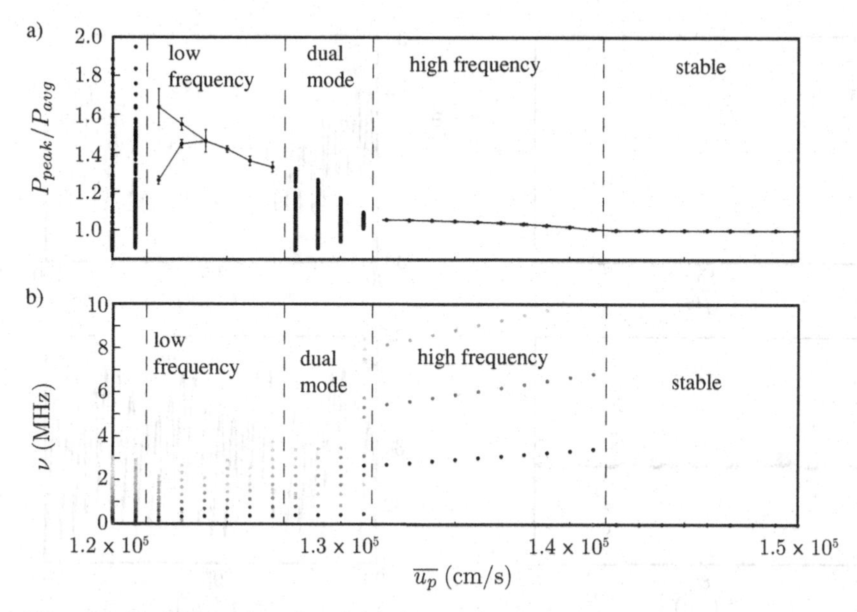


Figure 12.33. Bifurcation plot of (a) the maximum detonation pressure scaled by the average detonation pressure and (b) active frequencies versus supporting piston velocity where the darker shade of the point indicates larger magnitude. Adapted with permission from Romick et al. (2015).

weaker strengths than that of the dual mode. Additionally, subharmonics appear at $\frac{1}{2}$ and 3/2 at both $\overline{u_p} = 1.230 \times 10^5$ cm/s and $\overline{u_p} = 1.220 \times 10^5$ cm/s. At this lower supporting piston velocity, further subharmonics appear at the half intervals as well as the previously mentioned subharmonics grow in strength. In the lowest supporting piston velocities studied, many frequencies are active indicating that it is likely that the detonation is in a chaotic regime.

**COMPARISON TO THE INVISCID ANALOG.** Several supporting piston velocities are examined in the inviscid limit to elucidate the effects of physical diffusion on a detonation predicted with a detailed kinetics mechanism where instabilities are manifested as pulsations. As in the viscous case, when the supporting piston velocity is sufficiently high, a stable steadily traveling detonation is formed and persists at long times. Figure 12.34a shows both the viscous and inviscid detonation pressure versus time curves at $\overline{u_p} = 1.500 \times 10^5$ cm/s. The inviscid case relaxes to a detonation pressure of 36.68 atm. That is less than 0.1% different from the viscous analog at this piston velocity. However, at $\overline{u_p} = 1.430 \times 10^5$ cm/s the inviscid detonation begins to pulsate with an oscillation amplitude of $\sim 1$ atm whereas the viscous analog remains stable as shown in Fig. 12.34b. This pulsation amplitude is larger than that of the viscous case at $\overline{u_p} = 1.410 \times 10^5$ cm/s. Thus, the addition of diffusion to the model has added a slightly stabilizing effect, shifting the transition to a pulsating detonation by greater than 1.5%, but less than 2% with respect to the supporting piston velocity. Figure 12.34c shows the long time behavior at $\overline{u_p} = 1.400 \times 10^5$ cm/s for both the inviscid and viscous cases. The relative maxima in detonation pressure are $P = 35.66 \pm 0.10$ atm and $P = 36.62 \pm 0.005$ atm, for the viscous and inviscid cases, respectively. In addition to the reduction of the maximum detonation pressure, the amplitude of oscillations has also been reduced by 40% by the addition

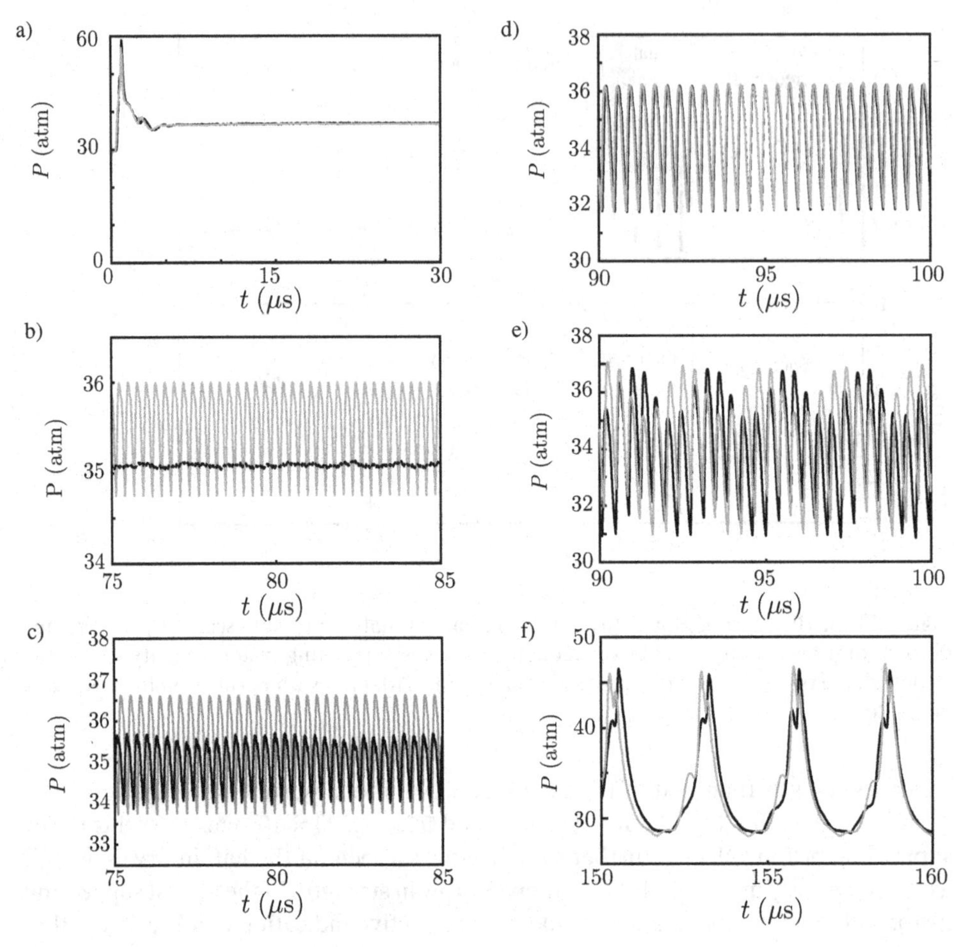


Figure 12.34. Detonation pressure versus time for both viscous (black lines) and inviscid (gray lines) cases at (a) $\overline{u_p} = 1.500 \times 10^5$ cm/s, (b) $\overline{u_p} = 1.430 \times 10^5$ cm/s, (c) $\overline{u_p} = 1.400 \times 10^5$ cm/s, (d) $\overline{u_p} = 1.320 \times 10^5$ cm/s, (e) $\overline{u_p} = 1.310 \times 10^5$ cm/s, and (f) $\overline{u_p} = 1.250 \times 10^5$ cm/s. Adapted with permission from Romick et al. (2015).

of viscosity. However, as the pulsations become stronger, the effect of viscosity is reduced as demonstrated in Fig. 12.34d for $\overline{u_p} = 1.320 \times 10^5$ cm/s. The pulsation amplitude reduction due to diffusion is weakened to less than 0.1% near the transition point to the dual mode behavior. Figure 12.34e shows the detonation pressure versus time curve for both the inviscid and viscous cases at $\overline{u_p} = 1.310 \times 10^5$ cm/s, which is in the dual mode pulsating behavior in both cases. It is difficult to identify differences in the time domain due to the interacting modes; the frequency domain will be discussed later. The average detonation pressure and the average maximum detonation for the viscous case are $P = 34.32$ atm and $P = 36.2 \pm 0.7$ atm. Likewise for the inviscid case, the average detonation pressure and the average maximum detonation are $P = 34.33$ atm and $P = 36.1 \pm 0.7$ atm. In the low-frequency-dominated mode, the effect of viscosity is nearly negligible, which is demonstrated in Fig. 12.34f for $\overline{u_p} = 1.250 \times 10^5$ cm/s. The local maxima are $P = 46.5 \pm 0.4$ atm and $P = 46.6 \pm 0.5$ atm, for the viscous and inviscid cases; respectively. This is a relative difference of 0.2%, and it is clear that the maxima overlap.

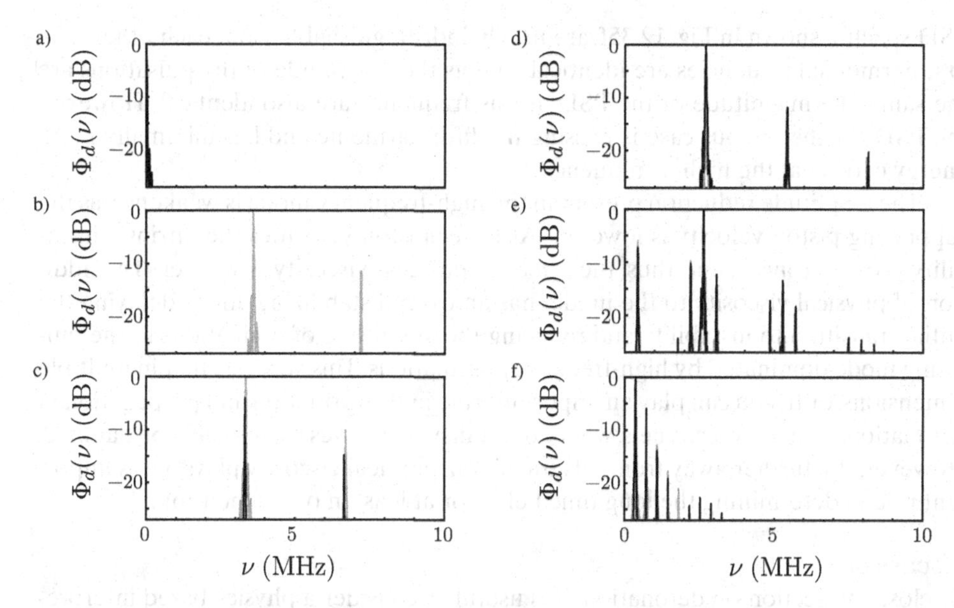


Figure 12.35. PSD spectra for viscous (black lines) and inviscid (gray lines) cases at (a) $\overline{u_p} = 1.500 \times 10^5$ cm/s, (b) $\overline{u_p} = 1.430 \times 10^5$ cm/s, (c) $\overline{u_p} = 1.400 \times 10^5$ cm/s, (d) $\overline{u_p} = 1.320 \times 10^5$ cm/s, (e) $\overline{u_p} = 1.310 \times 10^5$ cm/s, and (f) $\overline{u_p} = 1.250 \times 10^5$ cm/s. Adapted with permission from Romick et al. (2015).

The PSD spectra are calculated and compared using the average inviscid detonation pressure to scale both the inviscid and viscous detonation pressures; furthermore, the PSD is calculated in decibels using the maximum value of either case. Supporting piston velocities ranging from $1.250 \times 10^5$ cm/s to $1.500 \times 10^5$ cm/s are shown in Fig. 12.35. At $\overline{u_p} = 1.500 \times 10^5$ cm/s, both of the PSD spectra are concentrated around the zero frequency, as shown in Fig. 12.35a, indicating that the detonation is stable at long times. Figure 12.35b shows the PSD spectra at $\overline{u_p} = 1.430 \times 10^5$ cm/s; it is clear the inviscid case is pulsating at $\nu_0 = 3.60$ MHz, but the viscous PSD is still concentrated around the zero frequency indicating a stable detonation. At $\overline{u_p} = 1.400 \times 10^5$ cm/s, the first harmonic frequency in the inviscid case is $\nu_0 = 3.36$ MHz, whereas in the viscous case it is minimally shifted to a lower frequency by 1%. The shift is more apparent in the second harmonic, which is shifted by 0.06 MHz. Figure 12.35c shows that the magnitude of first harmonic frequency is larger in the inviscid case indicating the reduction in pulsation amplitude. The addition of viscosity affects more the size of the pulsation than the frequency of pulsations. Near the transition point to the dual mode behavior, at $\overline{u_p} = 1.320 \times 10^5$ cm/s, the frequency shift is reduced to 0.01 MHz. Figure 12.35d indicates that the first harmonic frequency peaks in the inviscid and viscous case are closer in magnitude than at the higher supporting piston velocities, giving another indication that the pulsation amplitude is nearly identical. In the dual mode, as shown in Fig. 12.35e the active modes are only barely distinguishable from each other; however, the strength of the low-frequency mode is stronger in the inviscid case, and the high-frequency mode is weaker. This indicates that, though small, the addition of physical viscosity to the model is still playing a role and slightly delays the transition to the dual mode behavior. Additionally, the high-frequency mode is shifted, but only by 0.3%. When a low-frequency-dominated mode ($\overline{u_p} = 1.250 \times 10^5$ cm/s) is examined, it is seen that the

PSD spectra, shown in Fig. 12.35f, are nearly indistinguishable from each other. The first harmonic frequencies are identical, and as the magnitude of the pulsations are the same, the magnitude of the PSD at this frequency are also identical. However, the PSD for the viscous case is missing the fifth harmonic and has minimally more energy carried at the higher frequencies.

The amplitude reduction present in the high-frequency mode is weakened as the supporting piston velocity is lowered. At lower piston velocities the intrinsic instability grows stronger, and thus, the effect of physical viscosity is weaker. The addition of physical viscosity to the model has an overall stabilizing effect, delaying the initial transition to instability and reducing the amplitude of oscillations in the pulsating mode dominated by high-frequency oscillations. This suggests that in multiple dimensions, diffusion can play an important role in the formation and propagation in detonations in narrow channels, where the transverse waves can possibly be damped. However, the further away from this transition, physical viscosity plays a less important role in determining the long time behavior at least in one dimension.

#### Discussion

To close the section on detonation, it is useful to consider a physics-based interpretation of these detailed unsteady detonation dynamics. The discussion presents an argument for using thermodynamics and dynamics to understand patterns predicted by the interplay of advection, reaction and diffusion. The interpretation we give is supported either directly by the current results or plausible hypotheses that could guide future studies. The discussion will be mainly cast in the framework of the current one-dimensional piston-supported detonations in a viscous hydrogen-air mixture; to quantitatively illustrate these important points for viscous limit cycle detonations, an appeal is made to the simpler model of one-step kinetics where the CJ limit cycle is well quantified.

It is well understood that the compressible reactive Navier-Stokes equations admit steady traveling wave solutions in response to a driving piston. The steady wave is driven by a combination of mechanical energy input from the driving piston and chemical energy input from the exothermic heat release in the subsonic region following the thin lead shock. For sufficiently high piston speeds such that the kinetic energy imparted by the piston is much greater than the chemical energy, the wave behaves similarly to an inert shock wave. As the piston velocity is lowered, the chemical energy makes an ever-increasing relative contribution to driving the wave. At a critical CJ piston velocity, all of the energy to drive the wave is available from the chemical energy, and the wave becomes self-propagating.

These steady waves can respond differently to small perturbations in the various regimes of supporting piston velocity. The key question is whether such perturbations grow or decay, and if they initially grow, what physical mechanisms prevent unbounded growth. In general terms, there are two physical mechanisms present that induce dissipation of structured mechanical and chemical energy into unstructured thermal energy: diffusion and the irreversible part of chemical reaction. Simultaneously, there are physical mechanisms present that induce the growth and resonance of various oscillatory structures predicted in some cases: amplification of selected modes by exothermic reaction combined with the effects of advection and diffusion. Ultimately, nonlinearity has a role. It is often the case that modes that grow linearly away from equilibrium can move into a region where nonlinear effects become important and serve to either suppress further growth or induce some variety of catastrophic growth. The action of these various physical mechanisms is a strong function of the various length and time scales in play as the driving piston velocity is varied.

For sufficiently high piston velocity, the effect of exothermic chemistry is minimal. One might imagine that a small sinusoidal disturbance near the shock front would segregate into one entropic and two acoustic modes, traveling near the local particle and acoustic speeds, respectively. Diffusion would act to reduce the amplitude of the disturbance. High frequency modes would dissipate more rapidly than low-frequency modes, but ultimately all would stabilize, and the system would relax to a steady propagating wave. Such is what is predicted for $\overline{u}_p > 1.420 \times 10^5$ cm/s.

For lower piston velocities, for example, $1.400 \times 10^5$ cm/s, it is obvious that limit cycle–like behavior is predicted. Thus, at this piston velocity, nature favors a partition of the chemical and kinetic energy of the fluid into a pattern in which some of the energy resides in the two modes displayed in Fig. 12.35c. Under these conditions, there is little difference between the viscous and inviscid predictions, so it is inferred that the physics are best understood in the context of a reactive Euler model. This piston velocity is likely favorable for the establishment of an organ–pipe type resonance influenced by a balance between reaction and advection. The relevant length is the induction zone, that is, the region between the lead shock and the point where significant chemical reaction commences. The reaction kinetics are such that the induction zone length $\ell_{ind} \simeq 10^{-2}$ cm. And the material properties are such that the postshock acoustic speed $c \simeq 10^5$ cm/s. A rough estimate of the first harmonic resonant frequency is thus $\nu \simeq c/\ell_{ind} = 10$ MHz. This is of the same order of magnitude as that predicted. This scaling argument is consistent with the results of Short (1996), who showed that perturbations within the induction zone were linearly unstable while examining one-step square wave detonations at high activation energy. Moreover for this case, Fig. 12.35c reveals that diffusion induces a small amplitude reduction in the resonant modes, as well as a small shift in the resonant frequencies. This is consistent with what is found in ordinary nonlinear mass–spring–damper systems (Strogatz, 2015). Again, similar to what is found in a nonlinear mass–spring–damper system, it is most likely that nonlinear effects serve a much stronger role in suppressing the growth of the resonant modes. For higher frequency modes, it is likely that diffusion plays the dominant role in amplitude suppression. With the finest length scale of reaction for this mixture of $\ell_{finest} \simeq 10^{-4}$ cm, and the diffusivity of the mixture $D_{mix} \simeq 10$ cm$^2$/s, one can estimate the frequency of the disturbance for which diffusion clearly dominates as $\nu \simeq D_{mix}/\ell_{finest}^2 = 1$ GHz. Useful insights on the relative importance of advection, reaction, and diffusion in hydrogen-air chemistry is given by Al-Khateeb et al. (2013) in the context of a laminar flame. There, it is shown that diffusion clearly influences the various reaction time scales on length scales given by a classical Maxwellian model, $\ell_i \simeq \sqrt{D_{mix}\tau_i}$, where $\ell_i$ is the reaction length scale associated with the chemical time scale $\tau_i$.

As the piston velocity is lowered further, nonlinearity plays a more prominent role as seen in Figs. 12.31d and 12.32a–d. As the oscillatory modes are dominated here by ever–lower frequencies as the piston velocity is lowered, it is likely that diffusion is playing even less of a role in the dominant low-frequency dynamics, with its main effect being confined to much higher frequency modes. Even then, the presence of diffusion is important in providing a physically based cutoff mechanism for high-frequency modes. Lack of such a mechanism can admit approximations that do not converge as the grid discretization scale is reduced, thus rendering the results to be potentially strongly influenced by the size of the discretization and the selected numerical method.

## 12.4 Closing Comments

In summary, this book has presented arguments for and examples of the use of deterministic methods of classical thermodynamics and nonlinear dynamical systems to better elucidate the physics of combustion. Such methods are in fact ideally suited for revealing both structure and patterns in nature and lend themselves to rendering combustion a repeatable science. That said, the multiscale challenges of realistic combustion are severe; many important problems require deterministic methods to be augmented with any of a variety of methods with more inherent uncertainty. However, as numerical modeling methods become more sophisticated and computational hardware more powerful, one should recognize that the realm where deterministic methods may be applied is broadened, sometimes widely. And combustion science should fully exploit these newly sharpened tools to explore important unsolved problems that were accessible previously only with less deterministic methods. Moreover, the value of verifiable, repeatable, theoretical approaches to combustion, and science in general, should not be understated. Having in hand verified solutions is a prerequisite for validating theories via comparison with experiment so as to distinguish theories that simply look backward and *correlate* to known results from those are useful to look forward and *predict* the unknown.

## EXERCISES

**12.1.** Show all steps necessary to reduce the conservative form of the reactive Euler equations with one-step kinetics, Eqs. (12.50)–(12.55), to the nonconservative form of Eqs. (12.56)–(12.62).

**12.2.** Using the methods of Sec. 12.1.7, write as many as possible of the reactive Navier-Stokes equations, Eqs. (12.317)–(12.323), in characteristic form. Determine whether the system is strictly hyperbolic.

**12.3.** Verify Eq. (12.174).

**12.4.** For the parameters of Table 12.1, find the postshock pressure, density, and temperature if $D = 2500$ m/s.

**12.5.** For the one-step kinetics model of Sec. 12.2.6 and the parameters of Table 12.1, find the steady spatial detonation structure of pressure, density, and temperature if $D = 2500$ m/s.

**12.6.** For the two-step kinetics model of Sec. 12.2.7 and the parameters of Table 12.3, find the steady spatial detonation structure of pressure, density, and temperature if $D = 2500$ m/s.

## References

A. N. Al-Khateeb, J. M. Powers, and S. Paolucci, 2013, Analysis of the spatio-temporal scales of laminar premixed flames near equilibrium, *Combustion Theory and Modelling*, 17(1): 560–595.

J. B. Bdzil and D. S. Stewart, 2007, The dynamics of detonation in explosive systems, *Annual Review of Fluid Mechanics*, 39: 263–292.

D. R. Billinger, 2001, *Time Series: Data Analysis and Theory*, Society for Industrial and Applied Mathematics, Philadelphia.

W. A. Bone and D. T. A. Townend, 1927, *Flame and Combustion in Gases*, Longmans, London.

P. B. Butler and H. Krier, 1985, Analysis of deflagration to detonation transition in high energy solid propellants, *Combustion and Flame*, 63(1–2): 31–48.

D. L. Chapman, 1899, On the rate of explosion in gases, *Philosophical Magazine*, 47(284): 90–104.

R. Chéret, 1993, *Detonation of Condensed Explosives*, Springer, New York.

R. Courant and K. O. Friedrichs, 1999, *Supersonic Flow and Shock Waves*, Springer, New York.

W. Döring, 1943, Über den Detonationsvorgang in Gasen, *Annalen der Physik*, 435(6–7): 421–436.

A. N. Dremin, 1999, *Toward Detonation Theory*, Springer, New York.

M. J. Feigenbaum, 1979, The universal metric properties of nonlinear transformations, *Journal of Statistical Physics*, 21(6): 669–706.

W. Fickett and W. C. Davis, 1979, *Detonation*, University of California Press, Berkeley, CA.

B. E. Gelfand, M. V. Silnikov, S. P. Medvedev, and S. V. Khomik, 2012, *Thermo-Gas Dynamics of Hydrogen Combustion and Explosion*, Springer, Berlin.

I. Glassman, R. A. Yetter, and N. G. Glumac, 2014, *Combustion*, 5th ed., Academic Press, New York, Chapter 5.

M. J. Grismer and J. M. Powers, 1996, Numerical predictions of oblique detonation stability boundaries, *Shock Waves*, 6(3): 147–156.

J. D. Hamilton, 1994, *Time Series Analysis*, Princeton University Press, Princeton, NJ.

A. K. Henrick, T. D. Aslam, and J. M. Powers, 2006, Simulations of pulsating one-dimensional detonations with true fifth order accuracy, *Journal of Computational Physics*, 213(1): 311–329.

J. O. Hirschfelder, C. F. Curtis, and R. B. Bird, 1954, *Molecular Theory of Gases and Liquids*, John Wiley, New York, Chapter 11.

P. Hwang, R. P. Fedkiw, B. Merriman, T. D. Aslam, A. R. Karagozian, and S. J. Osher, 2000, Numerical resolution of pulsating detonation waves, *Combustion Theory and Modelling*, 4(3): 217–240.

J. C. E. Jouguet, 1905, Sur la propagation des réactions chimiques dans les gaz, *Journal des Mathématiques Pures et Appliquées 6ᵉ Serie*, 1: 347–425; continued in 1906, 2: 5–86.

A. R. Kasimov and D. S. Stewart, 2004, On the dynamics of self-sustained one-dimensional detonations: A numerical study in the shock-attached frame, *Physics of Fluids*, 16(10): 3566–3578.

E. L. Keating, 2007, *Applied Combustion*, 2nd ed., CRC Press, Boca Raton, FL.

V. P. Korobeinikov, 1989, *Unsteady Interaction of Shock and Detonation Waves in Gases*, Taylor and Francis, New York.

K. K. Kuo, 2005, *Principles of Combustion*, 2nd ed., John Wiley, New York, Chapter 4.

C. K. Law, 2006, *Combustion Physics*, Cambridge University Press, Cambridge, UK, Chapter 14.

J. H. S. Lee, 2008, *The Detonation Phenomenon*, Cambridge University Press, New York.

H. I. Lee and D. S. Stewart, 1990, Calculation of linear detonation instability: one-dimensional instability of plane detonation, *Journal of Fluid Mechanics*, 216: 103–132.

R. J. LeVeque, 1992, *Numerical Methods for Conservation Laws*, Birkhäuser, Basel.

H. F. Lehr, 1972, Experiments on shock-induced combustion, *Astronautica Acta*, 17(4–5): 589–597.

B. Lewis and G. von Elbe, 1961, *Combustion, Flames and Explosions of Gases*, 2nd ed., Academic Press, New York, Chapter 8.

M. A. Liberman, 2008, *Introduction to Physics and Chemistry of Combustion*, Springer, Berlin, Chapter 8.

L. Massa, R. Kumar, and P. Ravindran, 2013, Dynamic mode decomposition analysis of detonation waves, *Physics of Fluids*, 24(6): 066101.

R. M. May, 1976, Simple mathematical models with very complicated dynamics, *Nature*, 261(5560): 459–467.

V. A. Mikhel'son, 1890, On the normal ignition velocity of explosive gaseous mixtures, PhD dissertation, Moscow University, reprinted, 1893, *Scientific Papers of the Moscow Imperial University on Mathematics and Physics*, 10: 1–93.

I. Müller and T. Ruggeri, 1998, *Rational Extended Thermodynamics*, 2nd ed., Springer, New York, pp. 277–308.

H. D. Ng, A. J. Higgins, C. B. Kiyanda, M. I. Radulescu, J. H. S. Lee, K. R. Bates, and N. Nikiforakis, 2005, Nonlinear dynamics and chaos analysis of one-dimensional pulsating detonations, *Combustion Theory and Modelling*, 9(1): 159–170.

W. L. Oberkampf and C. J. Roy, 2010, *Verification and Validation in Scientific Computing*, Cambridge University Press, Cambridge, UK.

A. K. Oppenheim, 2008, *Dynamics of Combustion Systems*, 2nd Ed., Springer, Berlin.

A. V. Oppenheim and R. W. Schafer, 1975, *Digital Signal Processing*, Prentice Hall, Englewood Cliffs, NJ.

S. Paolucci, Z. J. Zikoski, and D. Wirasaet, 2014a, WAMR: An adaptive wavelet method for the simulation of compressible reactive flow. Part I. Accuracy and efficiency of the algorithm, *Journal of Computational Physics*, 272: 814–841.

S. Paolucci, Z. J. Zikoski, and T. Grenga, 2014b, WAMR: An adaptive wavelet method for the simulation of compressible reactive flow. Part II. The parallel algorithm, *Journal of Computational Physics*, 272: 842–864.

J. M. Powers and K. A. Gonthier, 1992, Reaction zone structure for strong, weak overdriven, and weak underdriven oblique detonations, *Physics of Fluids A*, 4(9): 2082–2089.

J. M. Powers and S. Paolucci, 2005, Accurate spatial resolution estimates for reactive supersonic flow with detailed chemistry, *AIAA Journal*, 43(5): 1088–1099.

J. M. Powers, 2006, Review of multiscale modeling of detonation, *Journal of Propulsion and Power*, 22(6): 1217–1229.

J. J. Quirk, 1994, A contribution to the great Riemann solver debate, *International Journal for Numerical Methods in Fluids*, 18(6): 555–574.

P. J. Roache, 1998, *Verification and Validation in Computational Science and Engineering*, Hermosa, Albuquerque, NM.

C. M. Romick, 2015, On the effect of diffusion on gaseous detonation, PhD dissertation, University of Notre Dame.

C. M. Romick, T. D. Aslam, and J. M. Powers, 2012, The effect of diffusion on the dynamics of unsteady detonations, *Journal of Fluid Mechanics*, 699: 453–464.

C. M. Romick, T. D. Aslam, and J. M. Powers, 2015, Verified and validated calculation of unsteady dynamics of viscous hydrogen-air detonations, *Journal of Fluid Mechanics*, 769: 154–181.

A. H. Shapiro, 1953, *The Dynamics and Thermodynamics of Compressible Fluid Flow*, John Wiley, New York, Chapter 8.

G. J. Sharpe, 1997, Linear stability of idealized detonations, *Proceedings of the Royal Society of London, Series A: Mathematical, Physical and Engineering Sciences*, 453(1967): 2603–2625.

G. J. Sharpe and S. A. E. G. Falle, 2000, Numerical simulations of pulsating detonations: I. Nonlinear stability of steady detonations, *Combustion Theory and Modelling*, 4(4): 557–574.

J. E. Shepherd, 2009, Detonation in gases, *Proceedings of the Combustion Institute*, 32: 83–98.

M. Short, 1996, An asymptotic derivation of the linear stability of the square-wave detonation using the Newtonian limit, *Proceedings of the Royal Society of London, Series A: Mathematical, Physical and Engineering Sciences*, 452(1953): 2203–2224.

S. Singh, Y. Rastigejev, S. Paolucci, and J. M. Powers, 2001, Viscous detonation in $H_2/O_2/Ar$ using intrinsic low dimensional manifolds and wavelet adaptive multilevel representation, *Combustion Theory and Modelling*, 5(2): 163–184.

R. A. Strehlow, 1984, *Combustion Fundamentals*, McGraw-Hill, New York, Chapter 9.

S. H. Strogatz, 2015, *Nonlinear Dynamics and Chaos: With Applications to Physics, Biology, Chemistry, and Engineering*, Westview Press, Boulder, CO.

K. Terao, 2007, *Irreversible Phenomena: Ignitions, Combustion, and Detonation Waves*, Springer, Berlin.

T.-Y. Toong, 1982, *Combustion Dynamics: The Dynamics of Chemically Reacting Fluids*, McGraw-Hill, New York.

W. G. Vincenti and C. H. Kruger, 1965, *Introduction to Physical Gas Dynamics*, John Wiley, New York, p. 415.

J. von Neumann, 1942, Theory of detonation waves. Progress Report to the National Defense Research Committee Div. B, OSRD-549 (April 1, 1942. PB 31090), in *John von Neumann: Collected Works*, edited by A. H. Taub, Vol. 6, Pergamon, New York, 1963.

G. B. Whitham, 1974, *Linear and Nonlinear Waves*, John Wiley, New York.

F. A. Williams, 1985, *Combustion Theory*, 2nd ed., Benjamin-Cummings, Menlo Park, CA, Chapter 6.

V. E. Zakharov and L. A. Ostrovsky, 2009, Modulation instability: The beginning, *Physica D: Nonlinear Phenomena*, 238 (5): 540–548.

E. Zauderer, 1989, *Partial Differential Equations of Applied Mathematics*, 2nd ed., John Wiley, New York.

Ya. B. Zel'dovich, 1940, On the theory of the propagation of detonation in gaseous systems, *Zhumal Eksperimental'noi i Teoreticheskoi Fiziki*, 10: 542–568 (English translation: NACA TM 1261, 1960).

Ya. B. Zel'dovich and A. S. Kompaneets, 1960, *Theory of Detonation*, Academic Press, New York.

Ya. B. Zel'dovich and Yu. P. Raizer, 2002, *Physics of Shock Waves and High-Temperature Hydrodynamic Phenomena*, Dover, New York.

Ya. B. Zel'dovich, G. I. Barenblatt, V. B. Librovich, and G. M. Makhviladze, 1985, *The Mathematical Theory of Combustion and Explosions*, Plenum, New York.

# Author Index

# Subject Index